OCR
A LEVEL

Michael Raw
David Barker
Helen Harris
Andy Palmer
Peter Stiff

GEOGRAPHY

INCLUDES AS

Second Edition

HODDER
EDUCATION
AN HACHETTE UK COMPANY

This resource is endorsed by OCR for use with specification H481 A Level GCE Geography and specification H081 AS Level GCE Geography. In order to gain OCR endorsement, this resource has undergone an independent quality check. Any references to assessment and/or assessment preparation are the publisher's interpretation of the specification requirements and are not endorsed by OCR. OCR recommends that a range of teaching and learning resources are used in preparing learners for assessment. OCR has not paid for the production of this resource, nor does OCR receive any royalties from its sale. For more information about the endorsement process, please visit the OCR website, www.ocr.org.uk

Although every effort has been made to ensure that website addresses are correct at time of going to press, Hodder Education cannot be held responsible for the content of any website mentioned in this book. It is sometimes possible to find a relocated web page by typing in the address of the home page for a website in the URL window of your browser.

Hachette UK's policy is to use papers that are natural, renewable and recyclable products and made from wood grown in sustainable forests. The logging and manufacturing processes are expected to conform to the environmental regulations of the country of origin.

Orders: please contact Bookpoint Ltd, 130 Milton Park, Abingdon, Oxon OX14 4SB. Telephone: +44 (0)1235 827720. Fax: +44 (0)1235 400454. Email education@bookpoint. co.uk Lines are open from 9 a.m. to 5 p.m., Monday to Saturday, with a 24-hour message answering service. You can also order through our website: www.hoddereducation.co.uk

ISBN: 978 1 4718 5870 3

© Michael Raw, David Barker, Helen Harris, Andy Palmer, Peter Stiff

First published in 2016 by
Hodder Education,
An Hachette UK Company
Carmelite House
50 Victoria Embankment
London EC4Y 0DZ
www.hoddereducation.co.uk

Impression number   10 9 8 7 6

Year   2020 2019

Cover photo: tr3gi/Fotolia
Illustrations by Aptara, Inc. and Barking Dog Art
Typeset in India by Aptara, Inc.
Printed in Italy

A catalogue record for this title is available from the British Library.

Contents

Introduction

This textbook has been written specifically to support the OCR AS and A Level Geography specifications, introduced for first teaching in September 2016. It is the outcome of close collaboration between OCR, Hodder Education and a small team of specialist writers. All the writers have extensive experience of teaching OCR A Level Geography, have published widely, and are either current or former senior A Level examiners. They bring a unique insight into the resources that students and teachers need to support the new examination. The book has been designed as a comprehensive resource and should give confidence, particularly in those topics which are new to AS and A Level Geography. However, at the same time, we also appreciate the need to encourage students to read and research beyond the limits of the book itself.

The textbook has a number of exceptional features. As well as providing detailed and up-to-date information on the content of the AS and A Level specifications, it is lavishly illustrated with **case studies**, **charts**, **maps** and **photos**.

Other distinctive key features include:

- a summary of specialised concepts to underline their importance in the context of specific topics (pages vi–vii)
- a range of **activities** to test **knowledge**, **understanding** and **skills**, with some designed to **stretch and challenge** the most able students
- **review questions**, formed of simple question and answer tasks, which encourage students to revisit topics and consolidate understanding
- a comprehensive **glossary** providing definitions of key words and phrases
- **practice questions** at the end of each chapter, split into AS and A Level where topics are covered in each specification
- where appropriate, **fieldwork ideas** based on the authors' own teaching.

Fieldwork methods and other geographical techniques, including guidance on independent investigations, form the last two chapters. They provide a reference for essential quantitative and qualitative skills such as data presentation, data collection and data analysis, required both for the independent investigation and for answering questions on geographical techniques in the written examination.

Finally, we hope that this new textbook presents geography as a thoroughly relevant and worthwhile field of study. We also hope that it provides young people with an insight into the dynamic physical and human processes, as well as some of the most important issues, that shape our modern world.

Michael Raw (Author and Editor)

Schemes of assessment

This title covers both the AS and A Level Geography specifications, including all the optional content.

Summary of the OCR AS Geography specification and its coverage in this textbook

OCR AS Geography Paper 1: Landscape and place

82 marks; timing: 1 hour 45 minutes

Content	Chapter
Topic 1.1 Landscape systems One of these topics must be studied: ● Option A Coastal landscapes ● Option B Glaciated landscapes ● Option C Dryland landscapes	 Chapter 1 Chapter 2 Chapter 3
Topic 1.2 Changing spaces; making places	Chapter 5
Fieldwork and geographical skills	Chapter 15

OCR AS Geography Paper 2: Geographical debates

68 marks; timing: 1 hour 30 minutes

Content	Chapter
Topic 2.1 Climate change	Chapter 10
Topic 2.2 Disease dilemmas	Chapter 11
Topic 2.3 Exploring oceans	Chapter 12
Topic 2.4 Future of food	Chapter 13
Topic 2.5 Hazardous Earth	Chapter 14
Geographical skills	Chapter 15

Summary of the OCR A Level Geography specification and its coverage in this textbook

OCR A Level Geography Paper 1: Physical systems

66 marks; timing: 1 hour 30 minutes

Content	Chapter
Topic 1.1 Landscape systems One of these topics must be studied: • Option A Coastal landscapes • Option B Glaciated landscapes • Option C Dryland landscapes	 Chapter 1 Chapter 2 Chapter 3
Topic 1.2 Earth's life support systems	Chapter 4
Geographical skills	Chapter 15

OCR A Level Geography Paper 2: Human interactions

66 marks; timing: 1 hour 30 minutes

Content	Chapter
Topic 2.1 Changing spaces; making places	Chapter 5
Topic 2.2 Global connections *Global systems* One of these topics must be studied: • Option A Trade in the contemporary world • Option B Global migration *Global governance* One of these topics must be studied: • Option C Human rights • Option D Power and borders	 Chapter 6 Chapter 7 Chapter 8 Chapter 9
Geographical skills	Chapter 15

OCR A Level Geography Paper 3: Geographical debates

108 marks; timing: 2 hours 30 minutes

Content	Chapter
Two of these topics must be studied: Topic 3.1 Climate change Topic 3.2 Disease dilemmas Topic 3.3 Exploring oceans Topic 3.4 Future of food Topic 3.5 Hazardous Earth	 Chapter 10 Chapter 11 Chapter 12 Chapter 13 Chapter 14
Geographical skills	Chapter 15

OCR A Level Geography: Independent investigation

60 marks; non-examination assessment

Guidance on the independent investigation can be found in Chapter 16, pages 545–550.

Country classifications

Throughout the OCR specification the following three terms are used to classify countries. These terms are the ones used by the International Monetary Fund (IMF). The IMF regularly reappraises which group a country is placed in and adjusts its lists accordingly.

- Advanced countries (ACs) share a number of important economic development characteristics including well-developed financial markets, high degrees of financial organisation linking demand and supply of capital, goods and information, and diversified economic structures with rapidly growing service sectors. About 30 countries are in this group.
- Emerging and developing countries (EDCs) neither share all the economic development characteristics required to be an AC nor are eligible for the Poverty Reduction and Growth Trust, an IMF plan to provide financial support to developing countries. About 80 countries are in this group.
- Low-income developing countries (LIDCs) are eligible for the Poverty Reduction and Growth Trust. About 70 countries are in this group.

Specialised concepts

Adaptation – responses to environmental change which aim to modify human behaviour and economic systems permanently. For instance, declining rainfall and water shortages might encourage farmers to introduce more efficient drip irrigation, grow drought-resistant crops or convert arable to livestock enterprises. The increased frequency of heat waves in large urban areas might encourage planners to expand areas of parkland, open water and trees. Flood risks could be reduced by 'setting back' settlements on lowland coasts and preventing vulnerable new development locating on floodplains.

Causality – the relationship between cause and effect. A causal agent creates change in a dependent variable – for example, torrential rainfall (cause) leads to widespread flooding (effect). In geography most issues, such as global poverty, international migration, disease epidemics and deforestation, are complex and have multiple causes. In these situations isolating the causal factors and assessing their relative importance is often both difficult and contentious.

Equilibrium – the state of stability in a system achieved when a balance exists between inputs and outputs. For instance, in the Earth–atmosphere system, if inputs of short-wave solar radiation balance outputs of long-wave terrestrial radiation and reflection of solar radiation from clouds, snow and other surfaces, the global climate in the long-term will be stable. Equilibrium in glacial systems is achieved when there is a mass balance between inputs of snow and ice, and outputs of water through melting, evaporation and sublimation. Although fluctuations in inputs and outputs will occur from year to year, glaciers will show no long-term trend towards advance or retreat.

Feedback – an automatic internal response to change in systems. Negative feedback restores a system to balance; positive feedback amplifies change, which causes further disequilibrium. Global warming creates both negative and positive feedback responses. Rising global temperatures increase evaporation and cloud cover, which reflects more incoming solar radiation and eventually lowers temperatures. This is negative feedback. On the other hand, higher rates of evaporation increase the volume of water vapour in the atmosphere. Vapour, a potent greenhouse gas, absorbs more out-going terrestrial radiation, raising global temperatures and further increasing evaporation. This is a positive feedback effect.

Globalisation – the multiple interconnections and linkages between nations, groups of people, businesses and individuals which make up the modern world system. Time–space compression has transformed the ways that economic, social, political and cultural processes operate. Perhaps the key element of globalisation is the integration of human activities across the globe. Events, decisions and activities in one part of the world can have significant consequences for people far away. Globalisation is a 'contested concept'. People disagree about whether it is an inevitable force and if it means that the national or local are no longer important. It is clear that the processes of globalisation are not the same everywhere and that there are winners and losers among people and places from its effects.

Identity – how something is recognised. For geographers the concept of identity is closely associated with place. The identity of an individual or of groups living in a particular town, region or country is shaped by place, which in turn is shaped by identity. People have different identities because the places they live in are different, whether an inner city or suburb, urban or rural area, LIDC or AC. Often identity has an emotional connection with specific places, especially the place we call 'home'. National identities can draw people together in a positive way but at the same time can be exploited to turn against people considered to be outsiders with a different identity.

Inequality – the unequal distribution of resources, opportunities, well-being etc. within society. Some inequalities raise moral questions such as contrasts in wealth, access to health care, employment, education and so on. Geographers are particularly concerned with spatial inequalities at scales which range from intra-urban and regional to global. Analysis of spatial inequalities raises issues of social justice and poses questions concerning their causes and their economic, social, political and environmental impact.

Interdependence – at global scale, the mutual dependence of two or more countries in which there is a reciprocal relationship. Economic interdependence through trade is where a country, such as an AC, exports manufactured goods to another, such as an LIDC, and imports raw materials in return. Socio-economic interdependence through migration occurs where economic migrants provide labour in a country and on return bring newly acquired skills, ideas and values to their home country. Increasingly countries are becoming interdependent in their effects on the global environment and in their political relationships. Interdependence is a growing feature of globalisation and important in the development process.

Mitigation: action which is taken to lessen the impact of natural hazards on people, economy and society. It does not aim to prevent hazards. Mitigation responses vary from investment in hard structures such as sea walls, dams and quake-resistant buildings to reforestation of upland catchments and the development of early warning systems and emergency planning.

Representation – the ways by which meanings are given to the world. We try to understand the world around us by describing it in ways that make sense to us. This is influenced by our perceptions, which in turn are moulded by factors such as our age, gender or educational experiences. The ways we represent (literally re-present) the world is also strongly influenced by the culture we have grown up in. A teenager based in an AC urban setting will represent the world differently from a teenager from an LIDC rural environment. Representations include prose, pictures, architecture, and landscapes – in fact anything humans have had an influence on.

Resilience – the ability of countries, communities, households and environmental systems to resist, absorb and recover from the effects of shocks or stresses such as earthquakes, drought or violent conflict. Social resilience is the ability of communities to cope with disturbances as a result of social, political or environmental change. The resilience of a state depends on the fragility of its state apparatus and its capacity to respond to conflict or political upheaval. Ecological resilience is the characteristic of ecosystems to maintain themselves by resisting damage and recovering through negative feedback in the face of disturbances. Resilience building is central to current approaches to addressing disasters both natural and man-made.

Risk – the probability of a range of possible outcomes resulting from specific events such as economic shock or hazard. In economic geography this includes risks for companies involved in global supply chains such as political and economic instability, cyber-crime, or terrorism and piracy. For natural hazards, risk depends on the type and nature of the hazard, the probability of its occurrence, its magnitude, and the relative vulnerability of the people and environment that might be affected. Risk affects how we live and interact in society; it is likely to be exacerbated by climate change and land-use change and the effects of globalisation.

Sustainability – the use of resources that is environmentally and economically viable in the long term. Sustainability ensures the integrity, productivity and health of environmental and economic systems. Sustainable systems achieve a balance between supply and demand without having adverse impact on the physical and economic environment. Examples include the development of renewable energy such as solar and wind power, which are inexhaustible and do not impact the global energy budget. In contrast, reliance on fossil fuels is unsustainable. Not only are fossil fuels finite, their combustion contributes to rising levels of atmospheric carbon dioxide, global warming and climate change.

Systems – systems are groups of related objects, whether physical or human. An ecosystem, for example, comprises living organisms and the relationships both between organisms and between organisms and the environment. Physical systems are powered by inputs of solar energy, and most are open, with inputs and outputs of both energy and materials. The relationships between a system's components bind it together so that change in one component often has far-reaching impacts throughout the system. This quality of 'wholeness' is known as holisticity. Systems are found at all scales: from a garden pond, to a drainage basin, to the global atmospheric and global economic systems.

Thresholds – critical 'tipping points' in a system, which if exceeded result in massive and irreversible change. Climate scientists regard an increase in global warming beyond 2°C by 2100 as a tipping point, resulting in damaging and irreversible consequences for climate and other global environmental systems. A smaller scale tipping point is seen in livestock farming in dryland ecosystems. Overstocking of pastures beyond a certain threshold can cause severe damage to vegetation, leading to soil erosion, land degradation and a permanent reduction in energy flux and ecosystem productivity.

Acknowledgements

The publishers would like to thank the following for permission to reproduce photographs:

p.1 Michael Raw; **p.16** David Pick/Alamy Stock Photo; **p.20** Garrett Nagle; **p.21** Michael Raw; **p.23** Daniel Greenhouse/Alamy Stock Photo; **p.26** Michael Raw; **p.27** *all* Michael Raw; **p.29** acceleratorhams/iStock/Thinkstock; **p.30** sannawicks/iStock/Thinkstock; **p.31** 2004 TopFoto/Woodmansterne; **p.32** Jane Buekett; **p.33** Ian West; **p.35** Andy Palmer; **p.36** Andy Palmer; **p.51** robertharding/Alamy Stock Photo; **p.52** Alastair Schouten/Moment Open/Getty Images; **p.55** Alan Young; **p.62** *left* Hubert Stadler/Corbis Documentary/Getty Images; *right* Andy Palmer; **p.64** 2005 TopFoto/Gardner; **p.67** *t* Steven Kazlowski/RGB Ventures/SuperStock/Alamy Stock Photo; *b* mediacolor's/Alamy Stock Photo; **p.73** Michael Raw; **p.75** Danita Delimont/Alamy Stock Photo. **p.78** *both* Michael Raw; **p.80** *both* Michael Raw; **p.82** robertharding/Alamy Stock Photo; **p.84** *both* Michael Raw; **p.85** PytyCzech/iStock/ThinkStock; **p.87** muha04/iStock/ThinkStock; **p.90** David Mark/123RF; **p.93** Michael Raw; **p.94** Thomas Hallstein/Alamy Stock Photo; **p.95** Michele Falzone/Alamy Stock Photo; **p.96** *both* Michael Raw; **p.100** NASA/JPL; **p.107** *left* LowePhoto/Alamy Stock Photo; *right* Michael Raw; **p.109** Figure 7.8 from Climate Change 2007: The Physical Science Basis. Working Group I Contribution to the Fourth Assessment Report of the Intergovernmental Panel on Climate Change [Solomon, S., D. Qin, M. Manning, Z. Chen, M. Marquis, K.B. Averyt, M. Tignor and H. L. Miller (eds.)]. Cambridge University Press, Cambridge, United Kingdom and New York, NY, USA; **p.117** GeorgeBurba/iStock/ThinkStock; **p.121** *both* Michael Raw; **p.134** Michael Raw; **p.139** Peter Stiff; **p.149** Thomas Garcia/Alamy Stock Photo; **p.150** sedmak/iStock/ThinkStock; **p.151** *left* Simone80/Fotolia; *middle left* Samir Hussein/Redferns/Getty Images; *middle right* Ingram Publishing/ThinkStock; *right* Stadium Bank/Alamy Stock Photo; **p.153** Hemis/Alamy Stock Photo; **p.154** PrometheanSky/iStock/ThinkStock; **p.165** Edward Moss/Alamy Stock Photo; **p.169** BAY ISMOYO/Getty Images; **p.171** Peter Stiff; **p.177** Peter Stiff; **p.179** Ian Forsyth/Getty Images News/Getty Images; **p.189** Peter Stiff; **p.191** ZUMA Press, Inc/Alamy Stock Photo; **p.201** Jeremy Durkin/REX/Shutterstock; **p.206** Songquan Deng/iStock/Thinkstock; **p.213** Tommy Trenchard/Alamy Stock Photo; **p.217** Martin R. Berry/age fotostock/Getty Images; **p.223** *left* ZUMA Press, Inc./Alamy Stock Photo; *right* Anadolu Agency/Getty Images; **p.233** Andre Penner/AP/Press Association Images; **p.237** ams images/Alamy Stock Photo; **p.243** NATO; **p.250** Womankind Worldwide; **p.256** *t* MENAHEM KAHANA/AFP/Getty Images; *b* Eitan Simanor/Alamy Stock Photo; **p.257** dpa picture alliance/Alamy Stock Photo; **p.258** robertharding/Alamy Stock Photo; **p.261** Peter Treanor/Alamy Stock Photo; **p.271** Germano Assad/AP/Press Association Images; **p.274** Ander Gillenea/AFP/Getty Images; **p.278** David Barker; **p.281** David Barker; **p.283** Joerg Boethling/Alamy Stock Photo; **p.285** *t* frans lemmens/Alamy Stock Photo; *b* John Warburton-Lee Photography/Alamy Stock Photo; **p.288** *both* UNMISS Photo; **p.289** Fotolia; **p.297** *t* William O. Field, NSIDC, WDC/Science Photo Library; *b* Bruce F. Molnia, NSIDC, WDC/Science Photo Library; **p.313** Flip Nicklin/Minden Pictures/Getty Images; **p.324** *t* Landscapes, Seascapes, Jewellery & Action Photographer/Getty Images; *b* Global Warming Images/Alamy Stock Photo; **p.348** Adam Jones/Danita Delimont Creative/Alamy Stock Photo; **p.351** John Warburton-Lee Photography/Alamy Stock Photo; **p.360** Peerasith Chaisanit/123RF; **p.367** Stockbyte/Thinkstock; **p.369** Michael Schmeling/Alamy Stock Photo; **p.370** NASA; **p.389** US Coast Guard Photo/Alamy Stock Photo; **p.414** Mar Photographics/Alamy Stock Photo; **p.417** *t* www.worldmapper.org; *b* Planetpix/Alamy Stock Photo; **p.419** Edwin Remsberg/The Image Bank/Getty Images; **p.425** STR/AFP/Getty Images; **p.435** www.polyp.org.uk; **p.437** PRAKASH MATHEMA/AFP/Getty Images; **p.440** Pete Mcbride/Getty Images; **p.454** MARTIN BUREAU/AFP/Getty Images; **p.458** *left* Diego Azubel/epa european pressphoto agency b.v./Alamy Stock Photo; *right* WFP/Judith Schuler; **p.463** John Birdsall/Alamy Stock Photo; **p.465** Chris Leachman/Alamy Stock Photo; **p.470** Peter Stiff; **p.482** Aurora Photos/Alamy Stock Photo; **p.483** *left* Chip HIRES/Gamma-Rapho via Getty Images; *right* MPAK/Alamy Stock Photo; **p.486** AP/Press Association Images; **p.489** epa european pressphoto agency b.v./Alamy Stock Photo; **p.490** epa european pressphoto agency b.v./Alamy Stock Photo; **p.492** Maciej Dakowicz/Alamy Stock Photo; **p.500** Flirt/Alamy Stock Photo; **p.501** epa european pressphoto agency b.v./Alamy Stock Photo; **p.513** *all* Michael Raw; **p.524** Keith Erskine/Alamy Stock Photo; **p.525** Michael Raw; **p.531** *both* Michael Raw.

Maps on **pp.27,57,507**: © Crown copyright 2016 Ordnance Survey. Licence number 100036470.

pp.505,509: Experian Goad Digital Plans include mapping data licensed from Ordnance Survey with the permission of the Controller of Her Majesty's Stationery Office. © Crown Copyright and Experian Copyright. All rights reserved. Licence number PU 100017316.Further information about Goad plans can be found at Experian Ltd: goad.sales@uk.experian.com, 0845 6016011

The publishers would also like to thank the following for permission to reproduce diagrams:

p.35 *t* Carter, D., Bray, M., & Hooke, J. (2004) SCOPAC Sediment Transport Study. Department of Geography, University of Portsmouth. Report to SCOPAC, hosted on www.scopac.org.uk/sediment-transport.htm; **p.60** Future Temperatures And Event-Based Precipitation At Yucca Mountain, by Austin Long. © WM Symposia; **p.91** *t* © Summit Technologies, Inc. Retrieved from http://graphs.water-data.com/ucsnowpack; **p.111** Reprinted with permission from Global change and our common future: papers from a forum, 1989, by the National Academy of Sciences, Courtesy of the National Academies Press, Washington D.C.; **p.113** *t* © Vast Pipelines in Amazon Face Challenges Over Protecting Rights and Rivers, by Larry Rohter. Retrieved from http://www.nytimes.com/2007/01/21/world/americas/21pipeline.html; **p.116** *t* © Climate-Data.org. Retrieved from http://en.climate-data.org/location/1251; **p.129** © Philippe Rekacewicz, UNEP/GRID-Arendal; **p.190** © WTO International Trade Statistics, 2014. Retrieved from https://www.wto.org/english/res_e/statis_e/its2014_e/its2014_e.pdf; **p.192** *t* © OPEC Annual Statistical Bulletin, 2014, page 50. Retrieved from https://www.opec.org/opec_web/static_files_project/media/downloads/publications/ASB2014.pdf; **p.216** *t* © Statista; **p.217** *b* © Labour Force Survey. Retrieved from http://www.migrationobservatory.ox.ac.uk/briefings/migrants-uk-overview; **p.218** *t* © Labour Force Survey. Retrieved from http://www.migrationobservatory.ox.ac.uk/briefings/migrants-uk-overview; **p.228** *b* © UNHCR; **p.229** © International Labour Organization (ILO). Retrieved from: http://www.ilo.org/wcmsp5/groups/public/---asia/---ro-bangkok/---ilo-islamabad/documents/publication/wcms_241600.pdf; **p.234** © Migration Policy Institute. Covered under Creative Commons Attribution 3.0 licence; **p.244** *b* © The Washington Post; **p.245** *b* © World Health Organization; **p.246** © Benjamin D. Hennig; **p.247** © World Economic Forum; **p.265** *b* © Stockholm International Peace Research Institute (SIPRI); **p.266** © The Fund for Peace. **p.267** *t* © The Fund for Peace. **p.285** © Template Awesome Inc. Retrieved from http://trama-e-ordito.blogspot.in/2012/04/tuareg-lo-spirito-degli-uomini-blu.html; **p.291** *b* From: Earth's Climate: Past and Future 3rd edn. By William F. Ruddiman,© 2013 by WH Freeman and Company. Used with permission of the publisher.; **p.292** *t* © TeAra - the Encyclopaedia of New Zealand; **p.300** *b* © Contribution of Working Group I to the Fourth Assessment Report of the Intergovernmental Panel on Climate Change, IPCC, Geneva, Switzerland, p104. Used with permission from IPCC; **p.303** © Earth's Global Energy Budget, American Meteorological Society, 2009, 311-323. Used with permission; **p.330** © Guardian News and Media Limited 2016; **p.336** *t* © World Malaria Report, WHO; **p.337** *t* © Tuberculosis, estimated new cases, WHO, 2010; **p.337** *b* © International Diabetics Federation, 2009. Used with permission; **p.338** *t* © Global Atlas on cardiovascular disease prevention and control, World Health Organization, 2011; **p.355** *t* © Influenza A(H1N1) – WHO, 2009; **p.363** © Eliminating Malaria, Case-study 4, preventing reintroduction in Mauritius, World Health Organization 2012; **p.411** *t* Adapted from the IFPRI (2015); **p.419** © World Food Programme. Retrieved from http://reliefweb.int/sites/reliefweb.int/files/resources/wfp278943.pdf; **p.423** © Adapted from von Bruan and Meinzen-Dick (2009); **p.433** *c* © University of Washington; **p.435** © World Food Programme, 2010; **p.443** *t* © Macmillan Magazines Ltd; **p.446** © TES Global Ltd. Retrieved from https://openclassmariajesuscampos.wikispaces.com/Elena+Iba%C3%B1ez+Zazo+3%C2%BAD; **p.451** *tl* © Guardian News and Media Limited; **p.451** *tr* and *b* © Telegraph Media Group Limited; **p.458** © World Trade Organization. Retrieved from https://www.wto.org/english/tratop_e/dispu_e/dispu_maps_e.htm; **p.545** © Royal Geographical Society (with IBG)/D. Holmes.

Every effort has been made to trace and contact all copyright holders. The Publishers will be pleased to rectify any omissions brought to their attention at the earliest opportunity.

Part 1

Physical systems

Chapter 1

Coastal landscapes

1.1 How can coastal landscapes be viewed as systems?

> **Key idea**
> → Coastal landscapes can be viewed as systems

The development of a coastal **landscape** over time can be viewed within a systems framework. A **system** is a set of interrelated objects comprising components (stores) and processes (links) that are connected together to form a working unit or unified whole. Coastal landscape systems store and transfer energy and material on time scales that can vary from a few days to millennia (thousands of years).

The energy available to a coastal landscape system may be **kinetic**, **potential** or **thermal**. It is this energy that enables work to be carried out by the natural, **geomorphic** processes that shape the landscape.

The material found in a coastal landscape system is predominantly the sediment found on beaches, in estuaries and in the relatively shallow waters of the **nearshore zone**.

The components of open systems

Coastal landscape systems are recognised as being **open systems** (Figure 1.1). This means that energy and matter can be transferred from neighbouring systems as an **input**. It can also be transferred to neighbouring systems as an **output**. A good example of this is the input of fluvial sediment from a river, as it deposits its load at the mouth when available energy decreases.

In systems terms, a coastal landscape has:

- inputs – including kinetic energy from wind and waves, thermal energy from the heat of the Sun and potential energy from the position of material on slopes; material from marine **deposition**, **weathering** and **mass movement** from cliffs.
- outputs – including marine and wind **erosion** from beaches and rock surfaces; **evaporation**.
- throughputs – which consist of stores, including beach and nearshore sediment accumulations; and

flows (transfers), such as the movement of sediment along a beach by **longshore drift**.

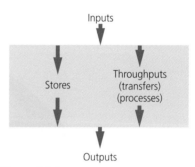

Figure 1.1 Open systems

System feedback in coastal landscapes

When a system's inputs and outputs are equal, a state of **equilibrium** exists within it. In a coastal landscape, this could happen when the rate at which sediment is being added to a beach equals the rate at which sediment is being removed from the beach; the beach will therefore remain the same size.

When this equilibrium is disturbed, the system undergoes self-regulation and changes its form in order to restore the equilibrium. This is known as **dynamic equilibrium**, as the system produces its own response to the disturbance. This is an example of **negative feedback**.

> **Activity**
>
> Construct a flow diagram to show an example of negative feedback in a coastal landscape system.

Sediment cells

A **sediment cell** is a stretch of coastline and its associated nearshore area within which the movement of coarse sediment, sand and shingle is largely self-contained. A sediment cell is generally regarded as a **closed system**, which suggests that no sediment is transferred from one cell to another. There are eleven large sediment cells around the coast of England and Wales, as shown in Figure 1.2. The boundaries of sediment cells are determined by the topography and shape of the coastline. Large physical features, such as Land's End, act as huge natural barriers that prevent the transfer of sediment to adjacent cells. In reality, however, it is unlikely that sediment cells are completely closed. With variations in wind direction and the presence of tidal currents, it is inevitable that some sediment is transferred between neighbouring cells. There are also many sub-cells of a smaller scale existing within the major cells.

Figure 1.2 Sediment cells of England and Wales

> **Activity**
>
> Describe the distribution of sediment cells shown in Figure 1.2.

> ✓ **Review questions**
>
> 1 Define the term 'millennia'.
> 2 What is the difference between kinetic and potential energy?
> 3 Outline the differing elements that make up the throughputs of a coastal landscape system.
> 4 Explain the concept of dynamic equilibrium.
> 5 What determines the boundaries of sediment cells?
> 6 Explain why sediment cells are unlikely to be completely closed systems.

> **Key idea**
> → Coastal landscape systems are influenced by a range of physical factors

The development of coastal landscapes and their operation as systems are influenced by a range of physical factors. These factors influence the way processes work, and consequently affect the shaping of the landscape. They also vary in terms of their spatial (from place to place) and temporal (over time) impacts. In any one location, or at any one time, some factors will have greater significance than others, and sometimes a factor may have very little influence at all. The factors themselves may also be interrelated, i.e. one factor may influence another.

Winds

The source of energy for coastal erosion and sediment transport is wave action. This wave energy is generated by the frictional drag of winds moving across the ocean surface. The higher the wind speed and the longer the **fetch**, the larger the waves and the more energy they possess. Onshore winds, blowing from the sea towards the land, are particularly effective at driving waves towards the coast. If winds blow at an oblique angle towards the coast, the resultant waves will also approach obliquely and generate longshore drift.

Wind is a moving force and as such is able to carry out erosion, **transportation** and deposition itself. These **aeolian** processes contribute to the shaping of many coastal landscapes.

> **Skills focus**
>
> 1 Using an appropriate atlas map, measure the maximum fetch for winds reaching the coast of Britain from the eight points of the compass.
> 2 Using the formula $H = 0.36\sqrt{F}$ (H in metres, F in kilometres), calculate the maximum wave height (H) that can be generated by each of these fetch distances (F).

Waves

A wave possesses potential energy as a result of its position above the wave trough, and kinetic energy caused by the motion of the water within the wave (Figure 1.3). It is important to realise that moving waves do not move the water forward, but rather the waves impart a circular motion to the individual water molecules. If you have ever seen a ball floating in the sea, you will have observed this phenomenon. As a moving wave passes beneath the ball, it rises and falls but does not move horizontally across the water surface.

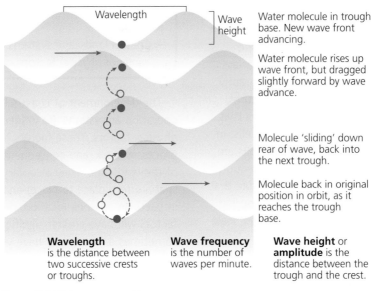

Wavelength

Wave height

Water molecule in trough base. New wave front advancing.

Water molecule rises up wave front, but dragged slightly forward by wave advance.

Molecule 'sliding' down rear of wave, back into the next trough.

Molecule back in original position in orbit, as it reaches the trough base.

Wavelength is the distance between two successive crests or troughs.

Wave frequency is the number of waves per minute.

Wave height or **amplitude** is the distance between the trough and the crest.

Figure 1.3 Water movement in a wave

The amount of energy in a wave in deep water is approximated by the formula:

$$P = H^2T$$

where P is the power in kilowatts per metre of wave front, H is wave height in metres and T is the time interval between wave crests in seconds, known as **wave period**. The relationship between wave height and wave energy is non-linear.

Table 1.1 shows that Atlantic waves are eight times higher than English Channel waves, but have 70 times more energy. The data show that wave height is a more important factor than wave period in determining wave energy.

Table 1.1 Atlantic Ocean and English Channel waves

Location	Wave height (m)	Wave period (sec)	Energy (kW per m)
Atlantic Ocean	5.0	8	200
English Channel	0.6	6	2.16

Wave anatomy

Wave anatomy is quite simple (Figure 1.4). The highest surface part of a wave is called the crest, and the lowest part is the trough. The vertical distance between the crest and the trough is the wave height. The horizontal distance between two adjacent crests or troughs is known as the wavelength. All waves have this same basic anatomy, but wave behaviour is complex and influenced by many factors, such as the shape and gradient of the sea floor and the irregularity of the coastline.

Waves formed in open oceans can travel huge distances from where they are generated. These **swell waves** generally have a long wavelength with a wave period of up to 20 seconds. In contrast, a locally generated **storm wave** typically has a short wavelength, greater height and a shorter wave period.

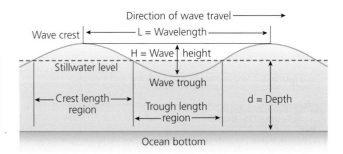

Figure 1.4 Wave anatomy (Source: P. French)

Breaking waves

When waves move into shallow water, their behaviour changes markedly. The definition of shallow water depends on the size of the wave, but it is typically at a depth of half the wavelength. At this depth the deepest circling water molecules come in contact with the seafloor. Friction between the seafloor and the water profoundly changes the speed, direction and shape of waves. Firstly, waves slow down as they drag across the bottom. The wavelength decreases, and successive waves start to bunch up. The deepest part of the wave slows down more than the top of the wave. The wave begins to steepen as the crest advances ahead of the base. Eventually, when water depth is less than 1.3 × wave height, the wave topples over and breaks against the shore. It is only at this point that there is significant forward movement of water as well as energy.

Breaking waves can be categorised as one of three types (Figure 1.5):

- Spilling – steep waves breaking onto gently sloping beaches; water spills gently forward as the wave breaks.
- Plunging – moderately steep waves breaking onto steep beaches; water plunges vertically downwards as the crest curls over.
- Surging – low-angle waves breaking onto steep beaches; the wave slides forward and may not actually break.

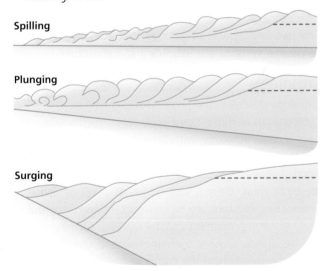

Figure 1.5 Types of breaking wave

After a wave has broken, water moves up the beach as **swash**, driven by the transfer of energy that occurs when the wave breaks. The speed of this water movement will decrease the further it travels due to friction and the uphill gradient of the beach. When it has no more available energy to move forward, the water is drawn back down the beach as **backwash**. The energy for this movement comes from gravity and always occurs perpendicular to the coastline, down the steepest slope angle.

Constructive and destructive waves

Constructive waves (Figure 1.6) tend to be quite low in height, have a long wavelength and a low frequency, typically around six to eight per minute. They usually break as spilling waves, and the strong swash travels a long way up the gently sloping beach. Due to the long wavelength, backwash returns to the sea before the

next wave breaks, and so the next swash movement is uninterrupted and thus retains its energy. A key feature of these waves is, therefore, that swash energy exceeds backwash energy.

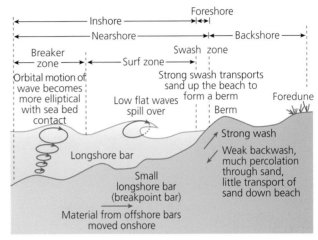

Figure 1.6 Constructive waves

In contrast, destructive waves (Figure 1.7) have greater height, shorter wavelengths and a higher frequency, often about twelve to fourteen per minute. They tend to break as plunging waves and so there is little forward transfer of energy to move water up the steeply sloping beach as swash. Friction from the steep beach slows the swash and so it does not travel far before returning down the beach as backwash. With a short wavelength, the swash of the next wave is often slowed by the frictional effects of meeting the returning backwash of the previous wave. In these waves, swash energy is less than backwash energy.

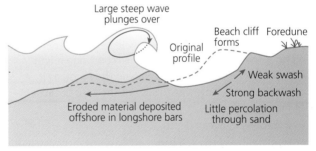

Figure 1.7 Destructive waves

There is an important but complex relationship between beach gradient and wave type. High-energy waves, often occurring during winter months, tend to remove material from the top of a beach and transport it to the offshore zone, reducing beach gradient. In contrast, low-energy waves, typical of summer months, build up the beach face, steepening the profile. Wave steepness is thought to be a critical factor in this relationship, but the angle of wave approach and sediment particle size are also important.

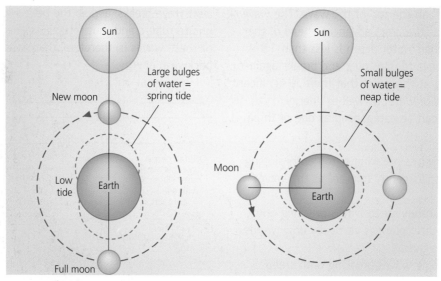

Figure 1.8 The formation of tides

Tides

Tides are the periodic rise and fall of the sea surface and are produced by the gravitational pull of the Moon and, to a lesser extent, the Sun (Figure 1.8). The Moon pulls the water towards it, creating a high tide, and there is a compensatory bulge on the opposite side of the Earth. At locations between the two bulges, there will be a low tide. As the Moon orbits the Earth, the high tides follow it. The highest tides will occur when the Moon, Sun and Earth are all aligned and so the gravitational pull is at its strongest. This happens twice each lunar month and results in spring tides with a high tidal range. Also twice a month, the Moon and the Sun are at right angles to each other and the gravitational pull is therefore at its weakest, producing neap tides with a low range.

Stretch and challenge

As tides are largely produced by the Moon, one might expect high tides to be directly under the Moon as it orbits the Earth. However, this is not the case because of factors such as:
- variations in ocean depth
- topography of the sea bed
- shapes of continental land masses.

Tidal range (Figure 1.9) can be a significant factor in the development of coastal landscapes. In enclosed seas, such as the Mediterranean, tidal ranges are low and so wave action is restricted to a narrow area of land. In places where the coast is funnelled, such as the Severn Estuary, tidal range can be as high as 14 m.

The tidal range therefore influences where wave action occurs, the weathering processes that happen on land exposed between tides and the potential scouring effect of waves along coasts with a high tidal range.

Geology

The two key aspects of geology that influence coastal landscape systems are **lithology** and **structure**.

Lithology

Lithology describes the physical and chemical composition of rocks. Some rock types, such as clay, have a weak lithology, with little resistance to erosion, weathering and mass movements. This is because the bonds between the particles that make up the rock are quite weak. Others, such as basalt, made of dense interlocking crystals, are highly resistant and are more likely to form prominent coastal features such as cliffs and headlands. Others, such as chalk and carboniferous limestone (predominantly composed of calcium carbonate), are soluble in weak acids and thus vulnerable to the chemical weathering process of carbonation.

Structure

Structure concerns the properties of individual rock types such as jointing, bedding and faulting. It also includes the permeability of rocks. In porous rocks, such as chalk, tiny air spaces (pores) separate the mineral particles. These pores can absorb and store water – a property known as primary permeability. Carboniferous limestone is also permeable, but for a different reason. Water seeps into limestone because of its many joints. This is known as secondary permeability. The joints are easily enlarged by solution.

Structure is an important influence on the planform of coasts at a regional scale. Rock outcrops that are uniform, or run parallel to the coast, tend to produce straight coastlines. These are known as **concordant** coasts. Where rocks lie at right angles to the coast they create a **discordant** planform: the more resistant rocks form headlands; the weaker rocks form bays.

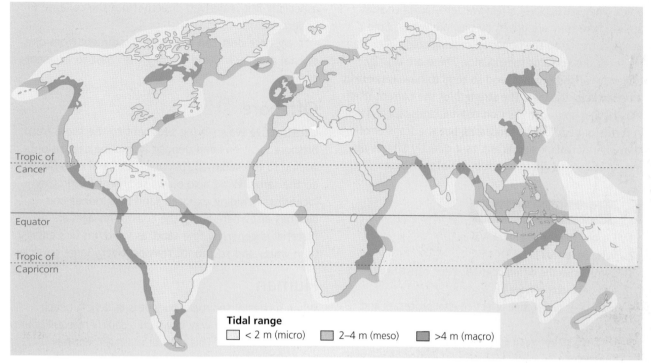

Figure 1.9 Tidal ranges

Structure also includes the angle of dip of rocks and can have a strong influence on cliff profiles. Both horizontally bedded and landward-dipping strata support cliffs with steep, vertical profiles. Where strata incline seawards cliff profiles tend to follow the angle of dip of the bedding planes.

Currents

Nearshore and offshore currents have an influence on coastal landscape systems.

Rip currents play an important role in the transport of coastal sediment. They are caused either by tidal motion or by waves breaking at right angles to the shore. A cellular circulation is generated by differing wave heights parallel to the shore. Water from the top of breaking waves with a large height travels further up the shore and then returns through the adjacent area where the lower height waves have broken. Once rip currents form, they modify the shore profile by creating **cusps** which help perpetuate the rip current, channelling flow through a narrow neck.

Ocean currents (Figure 1.10) are much larger scale phenomena, generated by the Earth's rotation and by convection, and are set in motion by the movement of winds across the water surface. Warm ocean currents transfer heat energy from low latitudes towards the

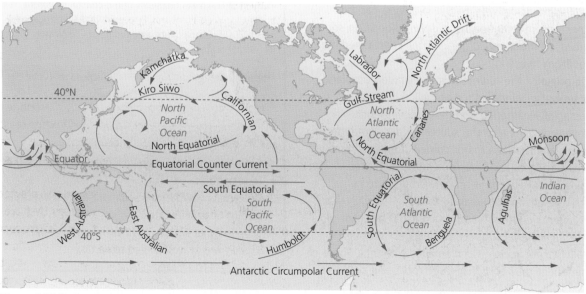

Figure 1.10 Ocean currents

poles. They particularly affect western-facing coastal areas where they are driven by onshore winds. Cold ocean currents do the opposite, moving cold water from polar regions towards the Equator. These are usually driven by offshore winds, and so tend to have less effect on coastal landscapes. The strength of the current itself may have a limited impact on coastal landscape systems in terms of geomorphic processes, but the transfer of heat energy can be significant, as it directly affects air temperature and, therefore, **sub-aerial processes**.

Review questions

1 Explain the formation of waves.
2 Why do waves break?
3 Summarise the differences between constructive and destructive waves.
4 How does tidal range affect coastal processes and landforms?
5 Distinguish between concordant and discordant coastlines.
6 How do ocean currents affect coastal landscape systems?

Key idea
→ Coastal sediment is supplied from a variety of sources

Terrestrial

Rivers are major sources of sediment input to the coastal **sediment budget** (Figure 1.11), and this is particularly true of coasts with a steep gradient, where rivers directly deposit their sediments at the coast. Sediment delivery to the shoreline can be intermittent, mostly occurring during floods. In some locations, as much as 80 per cent of coastal sediment comes from rivers.

The origin of the sediment is the erosion of inland areas by water, wind and ice as well as sub-aerial processes of weathering and mass movement.

Wave erosion is also the source of large amounts of sediment and makes a major contribution to coastal sediment budgets. Cliff erosion can be increased by rising sea levels and is amplified by storm surge events. The erosion of weak cliffs in high-energy wave environments contributes as much as 70 per cent of the overall material supplied to beaches, although typically it contributes much smaller amounts. Some of this sediment may comprise large rocks and boulders,

especially if it is derived directly from the collapse of undercut cliffs.

Longshore drift can also supply sediment from one coastal area by moving it along the coast to adjacent areas.

Offshore

Constructive waves bring sediment to the shore from offshore locations and deposit it (marine deposition), adding to the sediment budget. Tides and currents do the same. Wind also blows sediment from other locations, including exposed sand bars, dunes and beaches elsewhere along the coast. This aeolian material is generally fine sand, as wind has less energy than water and so cannot transport very large particles.

Human

When a coastal sediment budget is in deficit, beach nourishment is one way in which a sediment equilibrium can be maintained. This type of management has been adopted all over the world in order to preserve and protect the coastal environment. Sediment can be brought in by lorry and dumped on the beach before being spread out by bulldozers. Alternatively, sand and water can be pumped onshore by pipeline from offshore sources. Low bunds hold the mixture in place while the water drains away and leaves the sediment behind.

Wind, waves and longshore drift movements can also remove sediment from the coastal sediment budget. By subtracting the amount of sediment lost from the amount of sediment gained, it can be determined if the sediment budget is in surplus, deficit or equilibrium.

Skills focus

Table 1.2 The main elements of the coastal sediment budget for West Point spit, Prince Edward Island, Canada, 1980–2009 (Source: data adapted from ColdWater Recruiting report)

Inputs (average) m³/year	Outputs m³/year
Marine deposition = 9532	Total = 18,274
Cliff erosion = 3177	
Longshore drift = 5563	

1 Calculate the total average inputs per year for the time period.
2 Determine whether the sediment budget was in a surplus or deficit state.
3 Suggest what action might be necessary here, and explain why it would need to be taken.

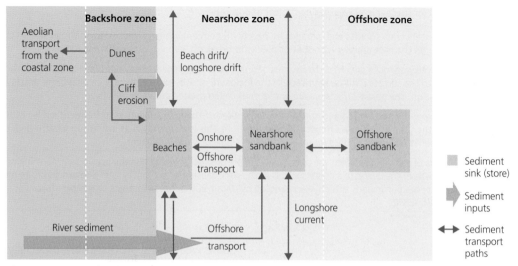

Figure 1.11 Coastal sediment budget

Fieldwork ideas

Sediment analysis can be carried out to:

- examine the sorting of beach material, either across the beach profile or across the width of the beach (linking to the process of longshore drift)
- investigate the effect of management structures, for example groynes, on the sorting of beach material
- investigate the origin of beach material
- compare sediment analysis at beaches in a range of locations and attempt to explain similarities and differences
- examine the relationship between beach sediment and other factors, for example the size and slope of the beach.

The method could involve measuring the dimensions or roundness of a sample of pebbles at a series of points along a transect, up or along the beach. The rock types of the sample pebbles could be identified and, with the aid of a geological map, their origin suggested.

✓ Review questions

1 Explain how sub-aerial processes supply coastal sediment.
2 What are 'aeolian' processes?
3 How is sediment supplied to the coastal budget from offshore?
4 Identify a human source of sediment supply.

1.2 How are coastal landforms developed?

Key idea
→ Coastal landforms develop due to a variety of interconnected climatic and geomorphic processes

Geomorphic processes

Weathering

Weathering uses energy to produce physically or chemically altered materials from surface or near surface rock. In coastal environments some types of weathering are particularly significant and influence the formation of coastal landforms.

Physical or mechanical

The breakdown of rock is largely achieved by physical weathering processes (Table 1.3) that produce smaller fragments of the same rock. No chemical alteration takes place during physical weathering. By increasing the exposed surface area of the rock, physical weathering allows further weathering to take place.

In many coastal landscapes, such as western Europe, the presence of the sea results in the moderation of temperatures and so air temperature may seldom drop below 0°C, reducing the extent of fluctuations and rendering some of the processes below ineffective.

Table 1.3 Processes of physical (or mechanical) weathering typical of coastal environments

Freeze-thaw	Water enters cracks/joints and expands by nearly 10 per cent when it freezes. In confined spaces this exerts pressure on the rock causing it to split or pieces to break off, even in very resistant rocks.
Pressure release	When overlying rocks are removed by weathering and erosion, the underlying rock expands and fractures parallel to the surface. This is significant in the exposure of sub-surface rocks such as granite and is also known as dilatation. The parallel fractures are sometimes called pseudo-bedding planes.
Thermal expansion	Rocks expand when heated and contract when cooled. If they are subjected to frequent cycles of temperature change then the outer layers may crack and flake off. This is also known as insolation weathering, although experiments have cast doubts on its effectiveness unless water is present.
Salt crystallisation	Solutions of salt can seep into the pore spaces in porous rocks. Here the salts precipitate, forming crystals. The growth of these crystals creates stress in the rock causing it to disintegrate. Sodium sulphate and sodium carbonate are particularly effective, expanding by about 300 per cent in areas of temperatures fluctuating around 26–28°C.

Chemical

The decay of rock is the result of chemical weathering (Table 1.4), which involves chemical reactions between moisture and some minerals within the rock. Chemical weathering may reduce the rock to its chemical constituents or alter its chemical and mineral composition. Chemical weathering processes produce weak residues of different material that may then be easily removed by erosion or transportation processes.

The rate of most chemical reactions increases with temperature. Van't-Hoff's Law states that a 10°C increase in temperature leads to a 2.5 times increase in the rate of chemical reaction (up to 600°C), so most chemical weathering processes occur at higher rates in tropical rather than temperate or polar regions. Thus moist tropical environments experience the fastest rates of chemical weathering and cold, dry locations the slowest. However, it is worth noting that carbonation can be more effective in low temperatures as carbon dioxide is more soluble in cold water than in warm water.

Some weathering processes are especially important when rocks are in contact with weakly acidic water. However, unless affected by pollution, sea water is typically neutral to slightly alkaline. One issue associated with climate change and increasing levels of atmospheric CO_2 is that rainfall, and therefore sea water, is becoming more acidic.

Biological

Biological weathering (Table 1.5) may consist of physical actions such as the growth of plant roots or chemical processes such as chelation by organic acids. Although this, arguably, does not fit with the precise definition of weathering, biological processes are usually classed as a separate type of weathering.

Mass movement

Mass movement occurs when the forces acting on slope material, mainly the resultant force of gravity, exceed the forces trying to keep the material on the slope, predominantly friction.

Table 1.4 Processes of chemical weathering typical of coastal environments

Oxidation	Some minerals in rocks react with oxygen, either in the air or in water. Iron is especially susceptible to this process. It becomes soluble under extremely acidic conditions and the original structure is destroyed. It often attacks the iron-rich cements that bind sand grains together in sandstone.
Carbonation	Rainwater combines with dissolved carbon dioxide from the atmosphere to produce a weak carbonic acid. This reacts with calcium carbonate in rocks such as limestone to produce calcium bicarbonate, which is soluble. This process is reversible and precipitation of calcite happens during evaporation of calcium rich water in caves to form stalactites and stalagmites.
Solution	Some salts are soluble in water. Other minerals, such as iron, are only soluble in very acidic water, with a pH of about 3. Any process by which a mineral dissolves in water is known as solution, although mineral specific processes, such as carbonation, can be identified.
Hydrolysis	This is a chemical reaction between rock minerals and water. Silicates combine with water, producing secondary minerals such as clays. Feldspar in granite reacts with hydrogen in water to produce kaolin (china clay).
Hydration	Water molecules added to rock minerals create new minerals of a larger volume. This happens when anhydrite takes up water to form gypsum. Hydration causes surface flaking in many rocks, partly because some minerals also expand by about 0.5 per cent during the chemical change because they absorb water.

Table 1.5 Processes of biological weathering typical of coastal environments

Tree roots	Tree roots grow into cracks or joints in rocks and exert outward pressure. This operates in a similar way and with similar effects to freeze-thaw. When trees topple, their roots can also exert leverage on rock and soil, bringing them to the surface and exposing them to further weathering. Burrowing animals may have a similar effect. This may be particularly significant on cliff tops and cliff faces.
Organic acids	Organic acids produced during decomposition of plant and animal litter cause soil water to become more acidic and react with some minerals in a process called chelation. Blue-green algae can have a weathering effect, producing a shiny film of iron and manganese oxides on rocks. On shore platforms, molluscs may secrete acids which produce small surface hollows in the rock.

In coastal landscape systems, the most significant mass movement processes are those acting on cliffs, which lead to the addition of material to the sediment budget by transferring rocks and **regolith** down onto the shore below. The main processes involved are:

- Rock fall: on cliffs of 40° or more, especially if the cliff face is bare, rocks may become detached from the slope by physical weathering processes. These then fall to the foot of the cliff under gravity. Wave processes usually remove this material, or it may accumulate as a relatively straight, lower angled scree slope.
- Slides: these may be linear, with movement along a straight line slip plane, such as a fault or a bedding plane between layers of rock, or rotational, with movement taking place along a curved slip plane. Rotational slides are also known as slumps. In coastal landscape systems, slides often occur due to undercutting by wave erosion at the base of the cliff which removes support for the materials above. Slumps are common in weak rocks, such as clay, which also become heavier when wet, adding to the downslope force. A layer of sand above a layer of clay may particularly encourage this, as rainwater passes through the sand but cannot penetrate the impermeable clay below, thus increasing pore pressure in the sands.

Wave processes

As explained earlier, waves are a source of energy in coastal landscape systems, and when they break onshore, the energy can be expended through geomorphic processes to shape landforms. They can also supply material to the system in the form of sediment, which is either deposited in, or transported within, the coastal system.

Erosion

Breaking waves are able to erode the coastline with a range of processes:

- Abrasion (or corrasion) is when waves armed with rock particles scour the coastline; rock rubbing against rock.
- Attrition occurs when rock particles, transported by wave action, collide with each other and with coastal rocks and progressively become worn away. They become smoother and more rounded as well as smaller, eventually producing sand.
- Hydraulic action occurs when waves break against a cliff face, and air and water trapped in cracks and crevices becomes compressed. As the wave recedes the pressure is released, the air and water suddenly expands and the crack is widened. The average pressure exerted by breaking Atlantic waves is 11,000 kg per m³.
- Pounding occurs when the mass of a breaking wave exerts pressure on the rock causing it to weaken. Forces of as much as 30 tonnes per m² can be exerted by high-energy waves.
- Solution (or corrosion) involves dissolving minerals like magnesium carbonate minerals in coastal rock. However, as the pH of sea water is invariably around 7 or 8 this process is usually of limited significance unless the water is locally polluted and acidic. Even then, only coastal rocks containing significant amounts of soluble minerals are likely to be affected by this.

Transportation
Waves, as well as tides and currents, can move material shorewards in a variety of ways:

- Solution: minerals that have been dissolved into the mass of moving water. This type of load is invisible and the minerals will remain in solution until water is evaporated and they precipitate out of solution.
- Suspension: small particles of sand, silt and clay can be carried by currents; this accounts for the brown or muddy appearance of some sea water. Larger particles can also be carried in this way, perhaps during storm events.

- Saltation: this is a series of irregular movements of material which is too heavy to be carried continuously in suspension. Turbulent flow may enable sand-sized particles to be picked up (entrained) and carried for a short distance only to drop back down again. Similarly, other particles may be dislodged by the impact, allowing water to get beneath them and cause entrainment.
- Traction: the largest particles in the load may be pushed along the sea floor by the force of the flow. Although this can be called rolling, again the movement is seldom continuous. Large boulders may undertake a partial rotation before coming to rest again.

Once deposited onshore, sediment may be moved along the shoreline by longshore drift.

Longshore drift occurs when waves approach the coast at an angle due to the direction of the dominant wind. When the waves have broken, the swash carries particles diagonally up the beach. Under the influence of gravity the backwash moves them perpendicularly back down the beach. If this movement is repeated, the net result is a movement of material along the beach. This also leads to the attrition of beach sediment so particles tend to become smaller and more rounded with increasing distance along the beach.

Deposition

Material is deposited when there is a loss of energy caused by a decrease in velocity and/or volume of water.

Deposition tends to take place in coastal landscape systems:

- where the rate of sediment accumulation exceeds the rate of removal
- when waves slow down immediately after breaking
- at the top of the swash, where for a brief moment the water is no longer moving
- during the backwash, when water percolates into the beach material
- in low-energy environments, such as those sheltered from winds and waves, e.g. estuaries.

The velocity at which sediment particles are deposited is known as the **settling velocity**. The larger and heavier particles require more energy to transport them. As flow velocity decreases, the largest particles being carried are deposited first and so on, sequentially until the finest particles are deposited (Figure 1.12).

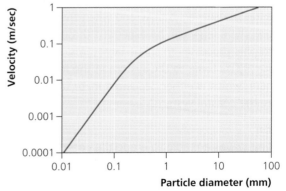

Figure 1.12 Settling velocity of sediment particles

Activities

1 Describe the relationship between particle size and settling velocity shown in Figure 1.12.
2 Suggest reasons for this relationship.

Fluvial processes

In coastal environments such as river mouths, fluvial processes often play an important part in the development of landforms. Low-energy, estuarine environments have distinctive characteristics.

Erosion

Fluvial erosion in the upper catchment is the main source of a river's sediment load. Rivers use similar erosional processes to waves, with most channel erosion occurring during high-flow, high-energy events. Sediment is also derived from weathering and mass movement processes that result in material moving into river channels from the valley sides.

Transportation

Rivers also transport sediment by traction, suspension, saltation and solution – similar processes to those of waves.

Deposition

As rivers enter the sea, there is a noticeable reduction in their velocity as the flowing water moving through the channel enters the relatively static body of sea water. Indeed, tides and currents may be moving in the opposite direction to the river flow, providing a major resistance to its forward movement. Available energy is reduced and so some, or all, of the river's sediment load is deposited. As the reduction in energy is progressive, deposition is sequential, with the largest particles being deposited first and the finest being carried further out to sea. In addition, the meeting of fresh water and salt water causes **flocculation** of clay particles. These fine, light materials clump together due to electrical charges between them in saline conditions. As a result they become heavier and sink to the sea bed.

Aeolian processes

Due to their exposure to open sea surfaces, coastal landscapes can be significantly influenced by winds, especially those blowing onshore.

Erosion

Wind is able to pick up sand particles and move them by **deflation**. At speeds of 40 km/hour, sand grains are moved by surface rolling (surface creep) and saltation. As grains of this size are relatively heavy, compared with silt and clay particles, they are seldom carried in suspension. This restricts erosion by abrasion to a height of about 1 m and has a limited effect in the erosion of rocky coastlines and cliffs. Erosive force increases exponentially with increases in wind velocity. For example, a velocity increase from 2 to 4 metres per second causes an eight-fold increase in erosive capacity.

Dry sand is much easier for wind to pick up than wet sand, as the moisture increases cohesion between particles, helping them to stick together. Attrition on land is particularly effective in wind as particles tend to be carried for much greater distances than in water, and the particles are not protected from collisions by the film of water around them.

Transportation

With the exception of solution, moving air is able to transport material using the same mechanisms as water moving in rivers and waves. Once particles have been entrained, they can be carried at velocities as low as 20 km/hour. Saltating grains are typically 0.15–0.25 mm in diameter, while those 0.26–2 mm, which are too heavy to be saltated, move by surface creep. Only the smallest grains (0.05–0.14 mm) can be carried in suspension.

Deposition

Material carried by wind will be deposited when the wind speed falls, usually as a result of surface friction. In coastal areas this will occur inland, where friction from vegetation and surface irregularities is much greater than on the open sea.

Coastal landforms

Although coastal landforms develop due to a variety of interconnected processes, each one will tend to be predominantly influenced either by erosion or deposition.

Erosional landforms

Cliffs and shore platforms

When destructive waves break repeatedly on relatively steeply sloping coastlines, undercutting can occur between the high and low tide levels where it forms a wave-cut notch. Continued undercutting weakens support for the rock **strata** above, which eventually collapses, producing a steep profile and a cliff. The regular removal of debris at the foot of the cliff by

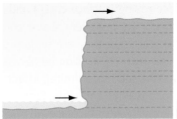

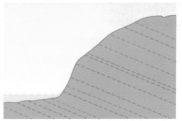

Horizontally bedded strata
Undercutting by wave action leads to rockfall; the cliffs retreat inland, parallel to the coast

Seaward-dipping strata
Undercutting by wave action removes basal support; rock layers loosened by weathering slide into the sea along the bedding planes

Landward-dipping strata
Rocks loosened by weathering and wave action are difficult to dislodge; the slope profile is gradually lowered by weathering and mass movement

Figure 1.13 The impact of geology on cliff profiles

wave action ensures that the cliff profile remains relatively steep and that the cliffs retreat inland parallel to the coast.

Cliff profiles vary depending upon their geology (Figure 1.13). Horizontally bedded and landward dipping rock strata support cliffs with a steep, near vertical profile. If the rock strata inclines seaward, the profile tends to follow the angle of the dipping strata.

As the sequence of undercutting, collapse and retreat continues, the cliff becomes higher. At its base, a gently sloping shore platform is cut into the solid rock. Although superficially appearing to be flat and even, shore platforms are often deeply dissected by abrasion due to the large amount of rock debris that is dragged across the surface by wave action.

Where rock debris is boulder-sized, it may be too large to be removed by the waves and will accumulate on the platform. Eventually, the platform will become so wide that it produces shallow water and small waves, even at high tide. Friction from the platform slows down approaching waves sufficiently for them to break on the platform rather than at the base of the cliff and so undercutting slows and eventually ceases. Research suggests that shore platforms reach a maximum width of about 500 m before this happens.

Although shore platforms are predominantly formed by erosion, weathering processes are also important in their development. Solution, freeze-thaw and salt crystallisation may all take place, depending upon the rock type and the climatic conditions of the location. Marine organisms, especially algae, can accelerate weathering when the platform is exposed at low tide. At night, algae release CO_2, as photosynthesis is not taking place. This mixes with sea water, making it more acidic, which results in higher rates of chemical weathering.

Shore platforms usually slope seawards at angles of between 0° and 3°, as wave erosion can occur anywhere between the high and low tide levels. However, as water levels are constant for longest at high and low tide, erosion is greatest at these points. This explains the formation of a ramp at the high tide level and a small cliff at the low tide level.

These features develop best if the tidal range is less than 4 m. If it is higher, then erosion is spread over a wider area of the platform; the water is at its high and low tide positions for a shorter time, and so the platform tends to be more uniform and more steeply sloping (Figure 1.14).

(a)

(b)

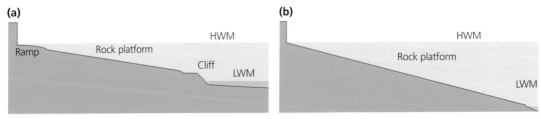

Figure 1.14 Shore platform profiles: (a) low tidal range, (b) high tidal range (HWM = high water mark; LWM = lower water mark)

Fieldwork ideas

Investigations could be carried out to:

- examine physical characteristics and features along a stretch of cliffed coastline
- identify different rock types and investigate the links between geology and physical features
- compare coastlines with different geologies
- study evidence of erosion, weathering, mass movement, human activity
- investigate and analyse strategies for protecting against erosion.

Data for cliff height could be collected using a clinometer and tape measure, followed by some simple trigonometry. The geology of the cliff could be identified and sketches drawn to identify key physical features and characteristics, as well as any management strategies being used. Photographs could also be taken and annotated later.

Bays and headlands

Bays and headlands typically form adjacent to each other, usually due to the presence of bands of rock of differing resistance to erosion. If these rock outcrops lie perpendicular to the coastline, the weaker rocks are eroded more rapidly to form bays while the more resistant rocks remain between bays as headlands. This results in the formation of a discordant coastline. The width of bays will be determined by the width of the band of weaker rock. Bay depth will depend on the differential rates of erosion between the more resistant and weaker rocks.

Rocks lying parallel to the coastline produce a concordant coastline. If the most resistant rock lies on the seaward side, it protects any weaker rocks inland from erosion. The resultant coastline is quite straight and even. However, even in this situation small bays or coves may occasionally be eroded at points of weakness, such as fault lines.

The Isle of Purbeck, Dorset, has both types of coastline (Figure 1.15). The east-facing coast is discordant, while the south-facing coast is concordant.

When waves approach an irregularly shaped coastline, **wave refraction** takes place and they develop a configuration increasingly parallel to the coastline. This is particularly true on coastlines with bays and headlands. As each wave nears the coastline, it is slowed by friction in the shallower water off the headland. At the same time, the part of the wave crest in the deeper water approaching the bay moves faster as it is not being slowed by friction. This means that the wave bends or refracts around the headland (Figure 1.16) and the **orthogonals** converge. Thus wave energy is focused on the headland and erosion is concentrated there. In the bays, the orthogonals diverge and energy is dissipated, leading to deposition. As the waves break on the sides of the headland at an angle there is a longshore movement of eroded material into the bays, adding to the build-up of beach sediment.

Geos and blowholes

Geos are narrow, steep-sided inlets. Even on coastlines with resistant geology, there may be lines of weakness such as joints and faults. These weak points are eroded more rapidly by wave action than the more resistant rock around them. Hydraulic action may be particularly important in forcing air and water into the joints and weakening the rock strata. A good example is Huntsman's

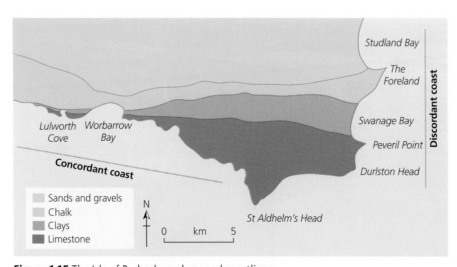

Figure 1.15 The Isle of Purbeck: geology and coastlines

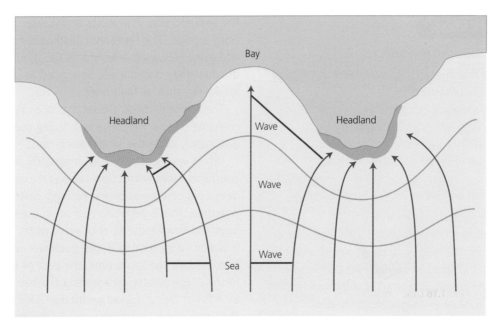

Figure 1.16 Wave refraction

Leap in Pembrokeshire, which is 35 m deep and eroded along a large joint in the carboniferous limestone.

Sometimes geos initially form as tunnel-like caves running at right angles to the cliff line, which as they become enlarged by continuing erosion may suffer from roof collapse, creating a geo. In parts of Cornwall, where they are known as zawns, they may also be associated with old tin mining shafts.

If part of the roof of a tunnel-like cave collapses along a master joint it may form a vertical shaft that reaches the cliff top. This is a blowhole. In storm conditions large waves may force spray out of the blowhole as plumes of white, aerated water. As with geos, blowholes may also be associated with mining shafts or the collapse of a cave roof, as at Trevone, Cornwall, which formed a 25 m deep blowhole (Figure 1.17).

Caves, arches, stacks and stumps

Although each of these landforms may be seen independently of each other on upland coasts, they

Figure 1.17 Blowhole, Trevone, Cornwall

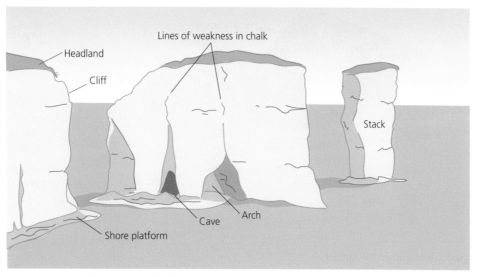

Figure 1.18 Landforms of headland erosion at Old Harry Rocks, Isle of Purbeck

also represent a sequence of erosional landforms which often develop around headlands (Figure 1.18). Due to wave refraction, energy is concentrated on the sides of headlands. Any points of weakness, such as faults or joints, are exploited by erosion processes and a small cave may develop on one side, or even both sides, of the headland. Wave attack is concentrated between high and low tide levels and it is here that caves form. If a cave enlarges to such an extent that it extends through to the other side of the headland, possibly meeting another cave, an arch is formed. Continued erosion widens the arch and weakens its support. Aided by weathering processes, the arch may collapse, leaving an isolated stack separated from the headland. Further erosion at the base of the stack may eventually cause further collapse leaving a small, flat portion of the original stack as a stump. This may only be visible at low tide.

An excellent example of these landforms is Old Harry Rocks at the seaward end of The Foreland near Swanage on the Isle of Purbeck (Figure 1.18).

Activity

Draw a sketch of Figure 1.18. Annotate it to show how these landforms have been formed.

Depositional landforms

Beaches

Beaches are the most common landform of deposition and represent the accumulation of material deposited between the lowest tides and the highest storm waves. Beach material, which consists of sand, pebbles and cobbles (Table 1.7), comes from three main sources:

- cliff erosion: typically only about 5 per cent
- offshore: combed from the sea bed, often during periods of rising sea levels; again about 5 per cent
- rivers: the remaining 90 per cent carried into the coastal system as suspended and bed load through river mouths.

Sand produces beaches with a gentle gradient; usually less than 5°, because its small particle size means that it becomes compact when wet, allowing little percolation during backwash. As little energy is lost to friction, and little volume is lost to percolation, material is carried back down the beach rather than being left at the top, resulting in a gentle gradient and the development of ridges and runnels parallel to the shore (Figure 1.19). These are occasionally breached by channels draining the water off the beach.

Shingle, a mix of pebbles and small to medium sized cobbles, produces steeper beaches because swash is stronger than backwash so there is a net movement of shingle onshore. Shingle may make up the upper part of the beach where rapid percolation due to larger air spaces means that little backwash occurs and so material is left at the top of the beach.

Table 1.7 Beach profiles and particle size

Material	Minimum diameter (mm)	Beach angle (degrees)
Cobbles	32	24
Pebbles	4	17
Coarse sand	2	7
Medium sand	0.2	5
Fine sand	0.02	3
Very fine sand	0.002	1

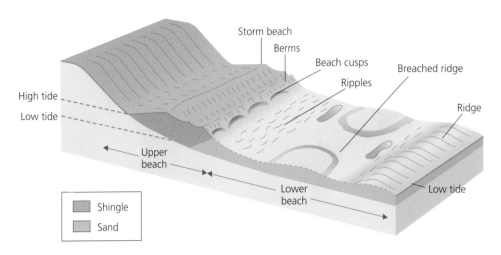

Figure 1.19 Beach profile features

Storm waves hurl pebbles and cobbles to the back of the beach, forming a storm beach or storm ridge. Berms are smaller ridges that develop at the position of the mean high tide mark, again resulting from deposition at the top of the swash (Figure 1.19). **Cusps** are small, semi-circular depressions (Figure 1.20). They are temporary features formed by a collection of waves reaching the same point and when swash and backwash have similar strength. The sides of the cusp channel incoming swash into the centre of the depression and this produces a strong backwash, which drags material down the beach from the centre of the cusp, enlarging the depression. Further down the beach, ripples may develop in the sand due to the orbital movement of water in waves.

Beaches are dynamic and their profiles change over time as wind strength and hence wave energy changes (Figure 1.21). Beaches respond to these changes by developing an equilibrium profile, with a balance between

Stretch and challenge

Cusp formation can be viewed as an example of a self-organisation model. This combines both positive and negative feedback. Investigate how this model can be applied to beach cusp formation.

Beaches also vary in their planform. Swash aligned beaches are usually straight and lack longshore drift movements as waves approach perpendicularly to the coastline and are fully refracted. These tend to be closed systems, as there is no net movement of sediment out of the coastal system. Drift aligned beaches, however, are dominated by waves approaching at an oblique angle which generates the movement of sand and shingle along the coastline by longshore drift.

erosion and deposition. High-energy, destructive waves remove sediment offshore and create flatter beach profiles. This results in shallower water, more friction and

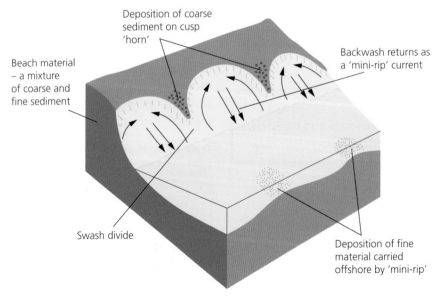

Figure 1.20 Formation of beach cusps

a reduction in wave energy. Low-energy, constructive waves transfer sediment in the opposite direction to form steeper profiles. This produces deeper water, less friction and an increase in wave energy.

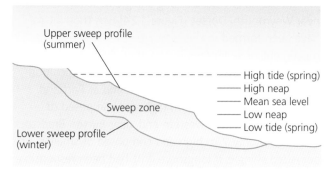

Figure 1.21 Seasonal changes in beach profile

Activities

1 Compare and contrast the summer and winter beach profiles shown in Figure 1.21.
2 Explain why they are different.

Fieldwork ideas

Beach profiles can be measured to:
- compare beaches or coastlines in different locations
- examine the effects of management on beach processes and morphology
- investigate seasonal changes in the beach profile
- examine relationships between the beach profile and other factors, for example rock type, cliff profile, sediment size or shape.

The method should involve measuring the slope angle of each section of the profile from as close to the sea as possible to the back of the beach using ranging poles, a tape measure and a clinometer.

Spits

Spits are long, narrow beaches of sand or shingle that are attached to the land at one end and extend across a bay, estuary or indentation in a coastline. They are generally formed by longshore drift occurring in one dominant direction which carries beach material to the end of the beach and then beyond into the open water. As storms build up more and larger material they make the feature more substantial and permanent. The end of the spit often becomes recurved as a result of wave refraction around the end of the spit and, possibly, the presence of a secondary wind/wave direction. Over time spits may continue to grow and a number of recurves or hooked ends may develop. If a spit forms

across an estuary, its length may be limited by the actions of the river current.

In the sheltered area behind the spit, deposition will occur as wave energy is reduced. The silt and mud deposited build up and eventually salt-tolerant vegetation may colonise, leading to the formation of a salt marsh.

A good example of a spit can be seen at Orford Ness in East Anglia. Here the coastline is east-facing (Figure 1.22) and so is largely unaffected by Britain's southwesterly prevailing winds. Instead northeasterly winds and waves are locally dominant which has resulted in longshore drift from north to south. A spit has formed across the estuary of the River Ore but the river current has prevented it reaching the land on the other side. Instead, the spit has continued to grow parallel to the coastline, diverting the river some 12 km further south. Historical maps provide evidence of this growth.

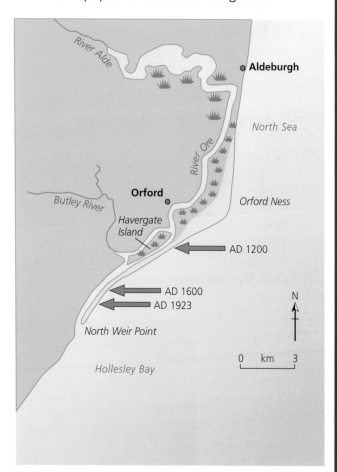

Figure 1.22 Orford Ness spit

Activities

1 Describe the direction and rate of growth of the Orford Ness spit (Figure 1.22).
2 Explain why its growth has been variable.

Onshore bars

Onshore bars can develop if a spit continues to grow across an indentation, such as a cove or bay, in the coastline until it joins onto the land at the other end (Figure 1.23). This forms a lagoon of brackish water on the landward side. The 100 m wide bar at Slapton Sands in Devon (although actually composed mainly of shingle) may have been formed in this way. However, there does not appear to be a significantly dominant direction of longshore drift on this east-facing coastline. There is no obvious pattern to the distribution of sediment sizes along the 5 km length of the bar, between Torcross in the south and Strete in the north.

The conclusion is that this feature may, at least in part, have been formed by the onshore movement of sediment, especially flint, during the post-glacial sea level rise that ended about 6000 years ago.

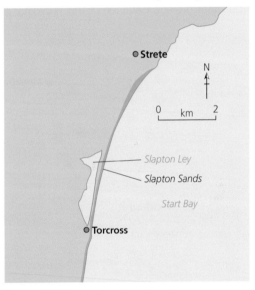

Figure 1.23 Slapton Sands onshore bar

Tombolos

Tombolos are beaches that connect the mainland to an offshore island. They are often formed from spits that have continued to grow seawards until they reach and join an island. The 30 km long shingle beach at Chesil near Weymouth, Dorset (Figure 1.24) was thought to have been formed in this way. However, the onshore movement of sediments is now thought to be the more likely cause, with it reaching its present position some 6000 years ago. At its eastern end at Portland, the ridge of shingle is 13 m high and composed of flinty pebbles. At the western end, near Burton Bradstock, the ridge is only 7 m high and composed of smaller, pea-sized shingle. If longshore drift had been responsible for the growth of Chesil it would be expected that the sediment further east would be smaller. It is more likely, therefore, that the onshore migration of shingle originally produced a uniform distribution of sediment sizes. Subsequently, strong longshore currents from the southwest have moved sediments of all sizes eastwards, while weaker longshore currents from the east have only been able to return the smaller particles westwards.

Salt marshes

Salt marshes are features of low-energy environments, such as estuaries and on the landward side of spits. The UK has 45,500 ha of salt marsh, mainly in eastern and northwest England.

Salt marshes are vegetated areas of deposited silts and clays. They are subjected twice daily to inundation and exposure as tides rise and fall. Salt-tolerant plant species such as eelgrass and spartina help trap sediment, gradually helping to increase the height of the marsh. The stems and leaves of the plants act as baffles and trap sediment swept in by

Figure 1.24 Chesil Beach

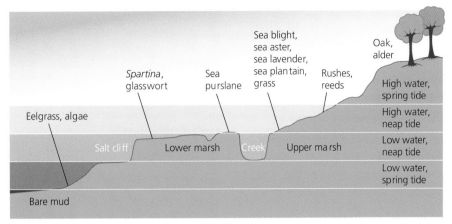

Figure 1.25 Cross-section through a typical salt marsh

tidal currents while the roots stabilise the sediment. The higher the marsh becomes, the shorter the period of daily submergence and the less saline the conditions. The low marsh, on the seaward side, is characterised by high salinity, **turbid** water and long periods of submergence. Few plant species can survive such conditions and so species diversity is poor. Further inland, conditions are not as harsh, with salinity, turbidity and submergence periods all lower. This allows a greater variety of species to survive, including sea aster, reeds and rushes (Figure 1.25).

Salt marshes have a shallow gradient which slopes seawards. A low cliff sometimes separates the salt marsh from the unvegetated mudflats on the seaward side. Even though the higher parts of the salt marsh are inundated less often, deposition rates are still quite high as at high water mark, low-energy, slack water may be present for 2–3 hours. The greater density of vegetation cover helps to trap and stabilise sediment. Extensive networks of

small, steep-sided channels, or creeks, drain the marsh at low tide and provide routes for water to enter the salt marsh as the tide rises. Between the creeks, shallow depressions are often found. These trap water when the tide falls, and these areas of salt water, called **saltpans**, are often devoid of any vegetation (Figure 1.26).

The development of salt marshes depends on the rate of accumulation of sediment, with rates of 10 cm per year quite common. Deposition of fine sediment occurs as rivers lose energy when they slow upon entering the sea. This is a key factor, but so too is flocculation. Tiny clay particles carry an electrical charge and repel each other in fresh water. In salt water, the particles are attracted to each other, combining together to form **flocs**, which being larger and heavier are unable to be carried in the river flow and so settle out of suspension, even at relatively high velocities.

Figure 1.26 Salt marsh at Flookburgh, Cumbria

Deltas

Deltas are large areas of sediment found at the mouths of many rivers. Deltaic sediments are deposited by rivers and by tidal currents. They form when rivers and tidal currents deposit sediment at a faster rate than waves and tides can remove it. Deltas typically form where:

- rivers entering the sea are carrying large sediment loads
- a broad continental shelf margin exists at the river mouth to provide a platform for sediment accumulation
- low-energy environments exist in the coastal area
- tidal ranges are low.

The structure of deltas usually consists of three distinctive components (Figure 1.27):

- The upper delta plain – furthest inland, beyond the reach of tides and composed entirely of river deposits
- The lower delta plain – in the inter-tidal zone, regularly submerged and composed of both river and marine deposits
- The submerged delta plain – lies below mean low water mark, is composed mainly of marine sediments and represents the seaward growth of the delta

Deltas are criss-crossed by a branching network of **distributaries**. Overloaded with sediment, deposition in the channel forms bars which causes the channel to split into two. This produces two channels with reduced energy levels, and so more deposition and further dividing occurs. Although these channels may be lined by **levées** on their banks, in times of flood these natural embankments are breached and deposition of lobes of sediment will then take place in the low-lying areas between the levees, called **crevasse splays**.

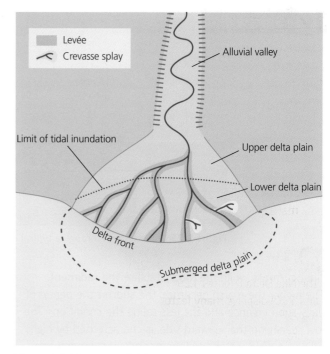

Figure 1.27 Structure of deltas

There are many different types of delta, but three of the most common types (Figure 1.28) are:

- Cuspate – a pointed extension to the coastline occurs when sediment accumulates but this is shaped by regular, gentle currents from opposite directions.
- Arcuate – sufficient sediment supply is available for the delta to grow seawards, but wave action is strong enough to smooth and trim its leading edge.
- Bird's foot – distributaries build out from the coast in a branching pattern, with river sediment supply exceeding the rates of removal by waves and currents.

(b) Arcuate delta

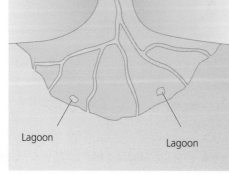

(a) Cuspate delta

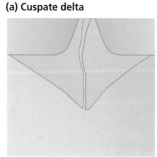

(c) Bird's foot delta

Figure 1.28 Types of delta

1 Define the terms weathering, mass movement, erosion, transportation and deposition.
2 Explain the role of erosion in the formation of shore platforms.
3 How does weathering contribute to the formation of stacks?
4 Explain how deposition forms salt marshes.
5 How do flows of energy and materials influence aeolian processes?

Key idea
→ Coastal landforms are interrelated and together make up characteristic landscapes

🔍 **Case study:** A low-energy coastal environment: the Nile Delta, Egypt

The Nile Delta (Figures 1.29 and 1.30) displays many features typical of a low-energy coastal environment, with rates of fluvial deposition having exceeded rates of marine erosion for over 3000 years. However, more recently the situation has started to change due to the influence of both physical and human factors.

An article containing further information about the features on the Nile Delta can be found at www.isesco.org.ma/ISESCO_ Technology_Vision/NUM04/doc/ Rakiby,%20Yousif%20....pdf

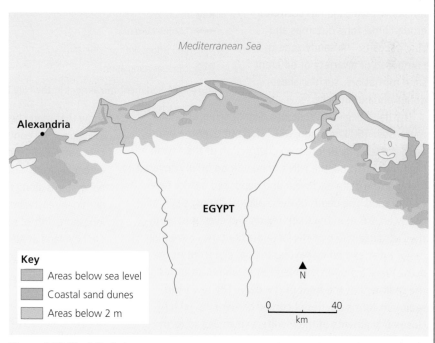

Key
■ Areas below sea level
■ Coastal sand dunes
■ Areas below 2 m

Figure 1.29 The Nile Delta

The River Nile

The Nile is, with a length of 6650 km, one of the longest rivers in the world. Its catchment area is huge, at more than 3 million km². The mean annual rainfall in the catchment is about 600 mm, yet its average discharge is less than 3000 m³/s, among the lowest of the world's great rivers. A large proportion of the Nile flow originates in Ethiopia and its summer monsoon rains. The rest comes from central Africa, as far south as Rwanda.

Despite the relatively low discharge, the river carries a huge sediment load. The suspended sediment load distribution is 30 per cent clay (<0.002 mm), 40 per cent silt (0.002–0.02 mm) and 30 per cent fine sand (0.02–0.2 mm). The annual average sediment yield is 4.26 t/ha/year and the total is 91.3 million tonnes for the whole Blue Nile Basin in Ethiopia.

Figure 1.30 The Nile Delta, Dahab Lagoon

The Nile Delta

The coastal plain occupies the northern part of the Nile Delta and runs parallel to the Mediterranean coastline. In places, three distinct subunits of the coastal plain may be recognised: the foreshore plain, the frontal plain and the sandy zone. The foreshore plain is characterised by elongated ridges, running almost parallel to the present shoreline, alternating with lagoons, salt marshes and alluvial deposits in the depressions between them. The frontal plain is located south of the foreshore plain, and has scattered eroded limestone outcrops and clay deposits. The sandy zone is composed of a variety of different sand formations, such as sheets, dunes and hummocks.

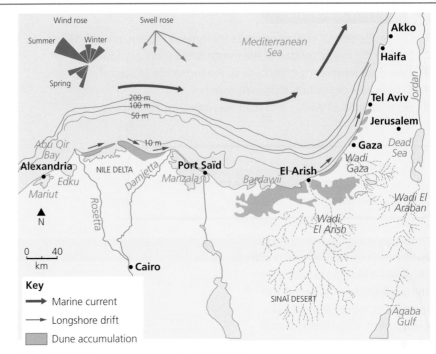

Figure 1.31 Sediment movement in the Nile Delta

The delta begins to split into distributaries at Cairo, more than 160 km inland, which fan out over the entire area of the delta. Before the construction of the Aswan High Dam in 1964, the Nile's annual flood briefly covered much of the delta each year and deposited a thick layer of silty mud. The depth of these alluvial deposits is 4 m at Aswan, 9.6 m in the Cairo region and even greater in the Delta itself. Two of the large distributaries have built major lobes extending beyond the general front of the delta. Wave action in the Mediterranean redistributes the sediment at the front of the delta. The reworked sediment forms a series of curved barrier bars, which close off segments of the Mediterranean Sea to form lagoons. The lagoons in turn form a sub-environment, and soon become filled with fine sediment.

Coastal landforms

The prevalence of northwesterly winds over the Mediterranean Sea for most of the year enhances the constant eastward movement of water and sediment. The estimated surface current velocity ranges from 9.26 to 13.5 cm/s during the summer, declines to 4.46 cm/s during the autumn, increasing sharply to 23.14 cm/s in winter due to strong winds, and declining again during calm spring weather to 8.4 cm/s. The prevailing northwesterly winds lead to waves coming from the west, northwest and north for 55–60 per cent of the time and from the northeast for only 8 per cent of the time.

The resultant nearshore features along the Nile Delta coast include underwater sand bars, typical of tideless seas. Aerial photographic analysis and field observations show that many beaches west of Abu Qir headland (shown in Figure 1.31) contain long crescentic bar systems. In contrast, parallel longshore bars exist along the delta extending from east of Abu Qir to Port Said. The parallel bar systems along the Nile Delta are generated by the dominant eastward longshore current and the associated longshore drift of sediment. The crescentic bars west of Abu Qir headland at Alexandria are associated with rip currents and negligible longshore drift.

Changes to the sediment budget

Although this delta coastline is, in general, a low-energy environment, since the building of the Aswan High Dam an imbalance has been created between two of the major forces affecting the delta: erosion and accretion. There has been a rapid reduction in the amount of sediment accreted, from 120 million tonnes/year to only trace amounts today. This has caused significant changes along the shoreline of the northwest Nile Delta with accelerated erosion and rates of coastal retreat as high as 148 m/year. Rising sea levels in the Mediterranean of 1.2 mm/year have also contributed to higher erosion rates as deeper water produces larger waves with higher amounts of energy and these reach further inland.

Activities

1 Describe the shape and structure of the Nile Delta.
2 Explain how and why the delta has changed over time.

Case study: A high-energy coastal environment: Saltburn to Flamborough Head, Yorkshire

The coastal environment between Saltburn and Flamborough Head is a rocky, upland area (Figure 1.32). This 60 km long coastal environment displays many coastal landforms and its characteristics reflect the influence of the high wave energy it receives.

Geology

The environment is strongly influenced by its geology. The adjacent North York Moors rise up to 400 m above sea level and comprise mainly sandstones, shales and limestones formed during the Jurassic period as well as some carboniferous rocks. Flamborough Head, at the southern end of this stretch of coastline, is a large chalk headland. Its spectacular cliffs are topped with till, a superficial deposit left behind by glaciers during the Devensian glacial period. Differences in rock resistance are responsible for the varied coastal scenery, notably the high cliffs and the bay and headland sequence.

Energy

The dominant waves affecting this coastline are from the north and northeast, with a fetch of over 1500 km. The most exposed parts of the coast are those that are north-facing, such as the area nearest to Saltburn, and so these receive the highest inputs of wave energy. Rates of erosion vary, partly due to these differences in wave energy inputs, but also due to variations in the resistance of the different geologies. Areas of relatively weak shale and clay experience erosion rates of 0.8 m per year on average, while the more resistant sandstones

and limestones only erode at rates of less than 0.1 m per year.

Monitoring of wave height using floating buoys in Whitby Bay during 2010–11 revealed that wave height often exceeded 4 m, even during summer months.

The high-energy inputs are also responsible for significant longshore drift from north to south along the coastline. In places, this sediment movement is interrupted by headlands, and sand and shingle then accumulate to form beaches in the bays, such as in Filey Bay.

Sediment sources

The coastline between Saltburn and Flamborough is sub-cell 1d of the major sediment cell 1, which extends south from St Abbs in southern Scotland to Flamborough. Some of the sediment in sub-cell 1d has come from the nearshore area, driven onshore as sea levels rose at the end of the last glacial period. Sediment is also supplied by cliff erosion, including sandstone and chalk from the resistant rock outcrops and the boulder clay deposits which yield significant amounts of gravel. The only large river, the Esk, enters the North Sea at Whitby. This supplies limited amounts of sediment due to the construction of weirs and reinforced banks along its course (Figure 1.33).

Beach surveys have found that there had been a net increase in beach sediment of 9245 m³ between 2008 and 2011 at Saltburn. Zones of both beach erosion and accretion were observed within Filey Bay, which reflect

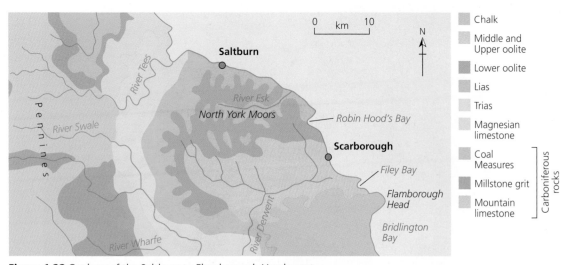

Figure 1.32 Geology of the Saltburn to Flamborough Head coast

the influence of winter storm systems with erosion at the back of the beach being particularly significant in the winter of 2010–11.

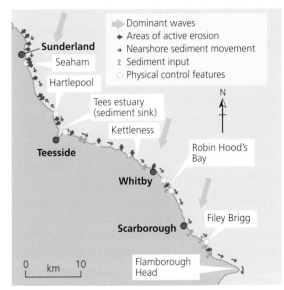

Figure 1.33 Erosion and sediment movement of the coast of northeast England

Cliffs

The sedimentary rocks of this coastline are horizontally bedded, and as a result the cliff profiles tend to have a vertical face. Most cliffs are overlain by a layer of weak glacial till, which has a much lower angle. Cliffs at Flamborough (Figure 1.34) are made of chalk, which is physically very strong with tightly bonded mineral particles. The vertical cliffs are typically 20–30 m high, with the overlying till lowered by mass movement processes to an angle of about 40°. Further north, between Robin Hood's Bay and Saltburn, the cliffs are much higher, but often with a stepped profile, reflecting the more varied geology. The steeper slope segments are formed in the more resistant sandstones and limestones, with gentler slopes corresponding to the weaker clays and shales, again lowered by mass movement processes.

Shore platforms

High-energy waves and active erosion mean that cliffs are retreating along this coastline, leaving behind rocky shore platforms. A good example can be seen at Robin Hood's Bay (Figure 1.35), eroded into Lower Lias shales. The platform slopes at a typical angle of 1°, although ramped sections are as steep as 15°. The platform has a maximum width of about 500 m, but extends much further into the off-shore zone. Based on current rates of erosion and retreat, it is quite possible that platforms such as these could have been formed within the last 6000 years, during times of predominantly stable sea level. However, some experts suggest that they are relict features, formed during earlier inter-glacial periods when sea levels were similar to those of today.

Figure 1.34 Cliffs at Flamborough Head

Figure 1.35 Cliffs and shore platform at Robin Hood's Bay

Headlands and bays

The variation in rock type along this coastline has led to the formation of a series of bays and headlands as part of a discordant planform. Robin Hood's Bay (Figure 1.35) has been eroded into relatively weak shales with more resistant bands of sandstone either side forming the headlands of Ravenscar, to the south, and Ness point, to the north. Further south, Filey Bay has developed in weak Kimmeridge Clay and is flanked by more resistant limestone and chalk. The prominent headland at Flamborough is formed of chalk, with deep bays either side formed from clay.

Landforms on headlands

As a result of wave refraction, wave energy is concentrated on resistant headlands that project into the North Sea. Weaknesses, such as large joints or faults, are then exploited by the erosive action of the waves, enlarging them to form caves and arches. These features are clearly visible in Selwick's Bay at Flamborough Head (Figure 1.36), where a master joint in the chalk has been enlarged. Green Stacks Pinnacle is an excellent example of a stack, isolated at the end of the headland following the collapse of an arch roof. Over 50 geos have formed along this coastline, with most of them aligned to the NE or NNE, facing the dominant wave direction. Blowholes have developed where vertical master joints in the chalk have been enlarged. Subsequently chalk and boulder clay have collapsed into the underlying sea caves, leaving funnel-shaped depressions on the cliff tops. On the north side of Selwick's Bay several blowholes appear to have merged and the intervening chalk has collapsed to produce a complex inlet.

Figure 1.36 Landforms at Selwick's Bay: (top) cave and arch, (bottom) stack.

Beaches

There are very few well-developed beaches along this stretch of coastline. The best examples are found in the sheltered, low-energy environments such as Scarborough and Filey Bay. Elsewhere deposits of sand and shingle accumulate slowly owing to the low input of sediment from rivers and the slow rates of erosion of the resistant rocks. High-energy waves also remove sediment before it can accumulate. Although longshore drift is considerable, the coastline lacks spits and other drift-aligned features. This is due partly to the high tidal range of around 4 m, and the lack of estuarine environments that would provide sediment sinks.

 Activities

1 Explain why the Saltburn to Flamborough Head coastline consists of a series of bays and headlands.
2 Why does this coastline have so few beaches?
3 Identify the main sediment sources in this coastal system.

1.3 How do coastal landforms evolve over time as climate changes?

Changes in the volume of water in the global ocean store are known as **eustatic** changes. These changes are influenced by variations in mean global temperatures, affecting both the amount of water in the ocean store and its density.

However, it should be appreciated that sea level change is relative as it is also affected by changes in land level. These changes, known as isostatic, are not considered here.

There are a number of physical factors that can affect changes in global temperature and the volume of water in the oceans. They include:

- variations in the Earth's orbit around the Sun, typically every 400,000 years
- variations in the amount of energy produced by the Sun, with a solar maximum every eleven years or so
- changes in the composition of the atmosphere due to major volcanic eruptions which reduce incident solar radiation
- variations in the tilt of the Earth's axis, occurring every 41,000 years.

 Activities

1 Describe the relationship between temperature and sea level shown in Figure 1.37.
2 Explain this relationship.

Key idea
→ Emergent coastal landscapes form as sea level falls

Climate change and sea level fall

A decrease in global temperature leads to more precipitation being in the form of snow. Eventually this snow turns to ice and so water is stored on the land in solid form rather than being returned to the ocean store as liquid. The result is a reduction in the volume of water in the ocean store and a worldwide fall in sea level.

As temperatures fall, water molecules contract, leading to an increased density and a reduced volume. It is estimated that a 1°C fall in mean global temperature causes sea level to fall approximately 2 m.

About 130,000 years ago, during the Tyrrhenian inter-glacial period, global mean annual temperatures were almost 3°C higher than today and sea level was about 20 m above today's position. Temperatures then fell during the onset of the Riss glacial period, reaching a minimum about 7°C lower than today about 108,000 years ago. As a result of this temperature decrease, less water was returned to the ocean store and sea levels dropped by over 100 m, making them about 83 m lower than the present day (Figure 1.37).

Emergent landforms

Landforms shaped by wave processes during times of high sea level are left exposed when sea level falls. As a result they may be found well inland, some distance from the modern coastline.

Raised beaches, marine terraces and abandoned cliffs

Raised beaches are areas of former shore platforms that are left at a higher level than the present sea level. They are often found a distance inland from the present coastline. Behind the beach along emergent coastlines it

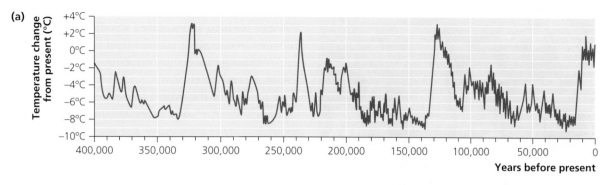

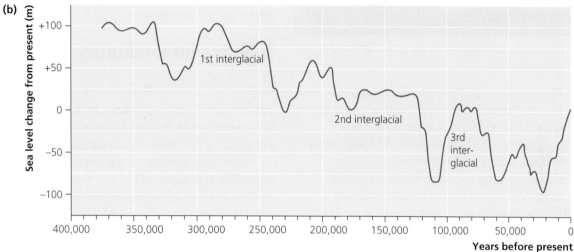

Figure 1.37 (a) Temperature change and (b) sea level during the last 400,000 years

is not uncommon to find abandoned cliffs with wave-cut notches, caves and even arches and stacks. Marine terraces are much larger scale landscape features than raised beaches, which are quite small scale and localised at the base of relic cliffs. Terraces do not necessarily have cliffs above them. Their formation, however, is essentially the same as raised beaches – marine erosion during a previous period of higher sea level.

Figure 1.38 Raised beach on the Isle of Portland, Dorset

On the southern tip of the Isle of Portland near Weymouth in Dorset there is a distinct raised beach (Figure 1.38) at a height of about 15 m above the present day sea level. This is thought to have been formed around 125,000 years ago during the Tyrrhenian inter-glacial period when sea levels were much higher than today's. The Portland limestone here was eroded by hydraulic wave action, partly through the exploitation of the bedding plane weaknesses. Erosion rates at that time are estimated to have been as much as 1 m/year. Other raised beaches at Portland are thought to date to about 210,000 years ago.

Modification of landforms

After their emergence, these landforms were no longer affected by wave processes. However, they continue to be affected by weathering and mass movement.

On top of the abandoned cliff on the Isle of Portland is a 1–1.5 m layer of frost-shattered limestone debris deposited when the area experienced periglacial conditions during the last glacial period. At the same time, the cliff face itself was gradually degraded by frost weathering processes, leading to rock fall from the cliff face. Evidence of other periglacial processes, such as **cryoturbation** (Figure 1.39), is also evident as contortions in fragmented limestone. They are the result

Figure 1.39 Cryoturbations in limestone below the raised beach, Isle of Portland, Dorset

of freezing and thawing of the permafrost in the subsoil during the late Pleistocene period, the final glacial phase.

In the post-glacial period, warmer and wetter conditions have led to the development of vegetation cover on many such exposures, often making them more difficult to recognise. With further warming of the climate predicted for the future, continued degradation is likely to occur with chemical weathering perhaps becoming more influential, especially by carbonation of limestone cliffs and platforms. Biological weathering on the raised beach may also become more significant with the colonisation of the surface by an increasing numbers of marine organisms, such as limpets and whelks.

If temperatures increase sufficiently, the associated sea level rise could lead to these emergent landforms again being found much closer to, or even at, the coastline. They would then be subjected to wave processes once more.

✓ Review questions

1 How does a cooling climate affect global sea level?
2 Explain the formation of raised beaches.
3 How might abandoned cliffs be modified in the future?

Key idea

→ Submergent coastal landscapes form as sea level rises

Climate change and sea level rise

An increase in global temperature leads to higher rates of melting of ice stored on the land in ice sheets, ice caps and valley glaciers. As a consequence there is a global increase in the volume of water in the ocean store and a consequent rise in sea level.

As temperatures rise, water molecules expand and this also leads to an increased volume. The relationship between temperature and sea level is again clear: a 1°C rise in mean global temperature results in a sea level rise of approximately 2 m.

At the end of the Würm glacial period, which happened about 25,000 years ago, temperatures were about 9°C lower than today and sea level was about 90 m lower than the present (Figure 1.37). Since then temperatures and sea level have risen to their present level. This period of significant sea level rise is known as the Flandrian Transgression.

Submergent landforms

Rias

Rias are submerged river valleys, formed as sea level rises. The lowest part of the river's course and the floodplains alongside the river may be completely drowned, but the higher land forming the tops of the valley sides and the middle and upper part of the river's course remains exposed. In cross section rias have relatively shallow water becoming increasingly deep towards the centre (Figure 1.41). The exposed valley sides are quite gently sloping. In long section they exhibit a smooth profile and water of uniform depth. In plan view they tend to be winding, reflecting the original route of the river and its valley, formed by fluvial erosion within the channel and sub-aerial processes on the valley sides.

A number of rias can be found on the south coasts of Devon and Cornwall, including those at Salcombe, Kingsbridge (Figure 1.40) and Fowey. They were formed during the post-glacial sea level rise of the Flandrian Transgression.

→ Activity

Describe the evidence seen in Figure 1.40 that suggests this landform is fluvial in origin.

Rias are typically underlain by alluvial deposits in buried channels that were eroded by the rivers that flowed down to the lower sea levels of Pleistocene glacial periods. During interglacial periods, when sea levels rose, further deposition would have occurred as the rivers had less surplus energy for erosion. Thus in the Flandrian Transgression, significant infilling of these earlier channels occurred. In some places, such as at Padstow on the north coast of Cornwall, sand was washed in from the Atlantic Ocean. At low tide the River Camel flows between broad exposed sandbanks which are only submerged at high tide.

Figure 1.40 Kingsbridge estuary, south Devon

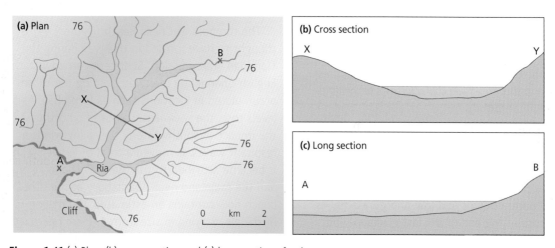

Figure 1.41 (a) Plan, (b) cross section and (c) long section of a ria

At Salcombe the ria has remained relatively deep, but it does have a shallow entrance or **threshold**, due to a sand bar having been deposited in the low-energy environment at the mouth of the Kingsbridge estuary.

Fjords

Fjords are submerged glacial valleys. They have steep, almost cliff-like, valley sides (Figure 1.42) and the water is uniformly deep, often reaching over 1000 m. The Sogne Fjord in Norway is nearly 200 km long although those in Scotland are less well developed as the ice was not as thick during the glacial period. The U-shaped cross section reflects the original shape of the glacial valley itself. They consist of a glacial rock basin with a shallower section at the end known

as the threshold. This results from lower rates of erosion at the seaward end of the valley where the ice thinned in warmer conditions. A good example is Milford Sound, New Zealand (Figure 1.43). They also tend to have much straighter planforms than rias as the glacier would have truncated any interlocking spurs present.

Due to the depth of water that occupied fjords during the Flandrian Transgression, marine erosion rates remained high and in some cases the fjords were further deepened. In others, such as those on the west coast of South Island, New Zealand, there has been some infilling with sediments. Deposited by meltwater from the glaciers of the Southern Alps, the volume of sediment has increased significantly in recent decades as glaciers have receded.

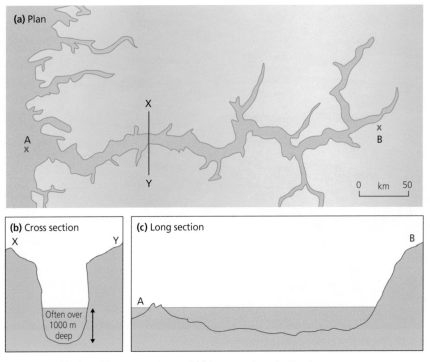

(a) Plan

X

A
x

x
B

0   km   50

(b) Cross section

X   Y

Often over
1000 m
deep

(c) Long section

B

A

Figure 1.42 (a) Plan, (b) cross section and (c) long section of a fjord

> **Activity**
>
> Using Figure 1.43, draw a sketch of Milford Sound. Label the sketch to show the main characteristics of the fjord.

Shingle beaches

When sea level falls as the volume of land-based ice grows, large areas of 'new' land emerge from the sea. Sediment accumulates on this surface, deposited by rivers, meltwater streams and low-energy waves. As sea levels rose at the end of the last glacial period, wave action pushed these sediments onshore. In some places they beached at the base of former cliff lines; elsewhere they may form tombolos and bars.

The tombolo at Chesil Beach (Figure 1.44) is thought to have formed in this way during the Flandrian Transgression. Sediment carried into the English Channel by meltwater during the Würm glacial accumulated in locations such as Lyme Bay. As sea levels rose, the sediment was carried northeast by the impact of southwesterly prevailing winds and the resultant waves. It moved a total of about 50 km until it became attached to the Isle of Portland at one end and the mainland near Abbotsbury at the other. The beach now contains an estimated 100 million tonnes of shingle, varying in size from 1–2 cm pea-sized material to 5–7 cm pebbles. It was previously thought that this tombolo was formed by the extension of a spit towards the Isle of Purbeck but the lack of recurves and

Figure 1.43 Milford Sound, New Zealand

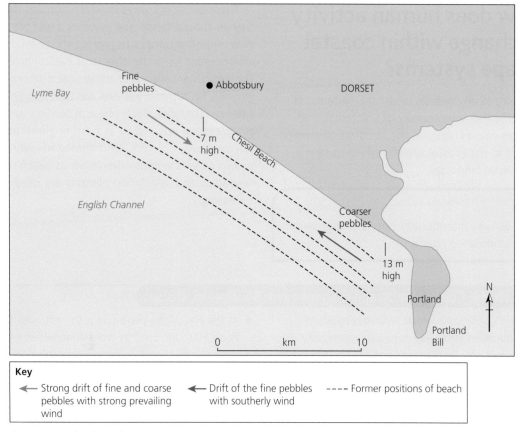

Key

← Strong drift of fine and coarse pebbles with strong prevailing wind   ← Drift of the fine pebbles with southerly wind   ---- Former positions of beach

Figure 1.44 Chesil Beach

the complex grading of pebbles suggest that it was not formed by longshore drift alone.

Modification of landforms

Both rias and fjords may be modified by the wave processes acting on their sides at the present-day sea level. The valley sides may also be affected by the operation of sub-aerial processes in today's climatic conditions or in any different climatic conditions of the future. This may eventually lead to a reduction in the steepness of the valley sides of fjords.

With sea levels predicted to rise by a further 0.6 m in the next 100 years, water depth in rias and fjords will increase. Marine erosion is also likely to increase due to stormier conditions and larger waves.

Shingle beaches, being composed of unconsolidated material, are especially vulnerable to modification. The tombolo at Chesil Beach has been significantly affected by present-day longshore drift processes and is likely to continue to be so in the future.

With further sea level rises predicted, shingle may well be moved even further to the northeast (it is currently moving at a rate of about 17 cm/year) and a breach of the tombolo is highly likely in future storm events. Recent storms have seen waves over-topping the beach.

In 2009, 1 metre-sized pieces of shelly clay were eroded from in front of the beach and washed up onto it at West Bexington in a winter storm (Figure 1.45). This sort of sediment addition will also become more common in future with higher sea levels and more storm events.

Figure 1.45 Shelly clay deposits at West Bexington after the 2009 storm. The blocks measure around 1 metre

Review questions

1 How does a warming climate affect global sea level?
2 Explain the formation of fjords.
3 How might shingle beaches be modified by future climate and sea level changes?

1.4 How does human activity cause change within coastal landscape systems?

Human activity in any coastal landscape system will cause change, either intentionally or unintentionally. Such activities cause change in the transfers of energy and sediment in the coastal system, which in turn affect the coastal landscape.

> **Key idea**
> → Human activity intentionally causes change within coastal landscape systems

There are many ways in which humans deliberately change coastal landscape systems. Most often intervention attempts to protect the coastal environment from the effects of natural processes. The implementation and construction of sea walls, groynes, rip rap and gabions are common coastal defence strategies. Such physical barriers to wave processes are referred to as **hard engineering**. Alternatively, more environmentally friendly methods, known as **soft engineering**, such as beach recharge, re-grading and vegetation planting are used increasingly.

🔍 Case study: Coastal landscape management: Sandbanks, Dorset

The Sandbanks peninsula, which separates much of Poole Harbour from Poole Bay, is heavily managed. The responsibility for its management lies collectively with Poole Harbour Commissioners, Poole Borough Council and the Environment Agency, and the strategies employed form part of the Two Bays Shoreline Management Plan, based on the sediment cell covering Poole Bay and Christchurch Bay (Figure 1.46).

The need for management

The Sandbanks peninsula is a significant part of the coastal landscape system and needs management for several reasons:

- It has a large number of high value commercial properties built on it. These include Sandbanks Hotel and Haven Hotel, both of which provide significant employment opportunities and generate spending in the local economy.
- Residential properties are in high demand and command premium prices (currently the fourth most expensive in the world per square metre). Large, detached houses command prices in excess of £10m, with many luxury apartments costing over £2m.
- The beach is a major tourist attraction. It has a Blue Flag award for water quality, and being gently sloping it is safe for family swimming.
- It provides protection and shelter from waves for Poole Harbour, which is therefore a popular and safe place for water sports, such as wind-surfing, sailing and water ski-ing. The harbour is also home to numerous yacht clubs and marinas, such as Salterns.

- At the end of the peninsula is the entrance to Poole Harbour, used by cross-channel ferries and catamarans, as well as commercial ships carrying goods such as timber. Longshore drift of beach sediment could cause the harbour entrance to become clogged and shallow.
- Climate change means that sea levels are predicted to rise here by about 0.6 m in the next 100 years. This would not only cause flooding of many properties, but could breach the peninsula at its lowest and narrowest point (only 2 m above sea level and 50 m wide) at the junction of Shore Road and Banks Road. This would effectively cut off the end of the peninsula from the mainland. It is estimated that if no management strategies are applied, £18m of damage to residential properties will occur in the next twenty years.

Management strategies and their impacts

In order to maintain a deep and wide beach, **rock groynes** have been constructed (Figure 1.47) to minimise the movements of sediment along the beach at Sandbanks by longshore drift. This not only restricts sediment from entering the harbour entrance, thereby keeping access free for shipping, but also absorbs wave energy, and reduces rates of erosion. It is estimated that without this action, erosion rates would be about 1.6 m per year.

In addition, **beach recharge** is used to conserve the beaches. Sand dredged from offshore is sprayed onto the beach, a process known as 'rainbowing' (Figure 1.48), adding to its size. This currently costs about £20/m³. However, a recent trial of dumping sediment dredged from the harbour just offshore is a much cheaper

→

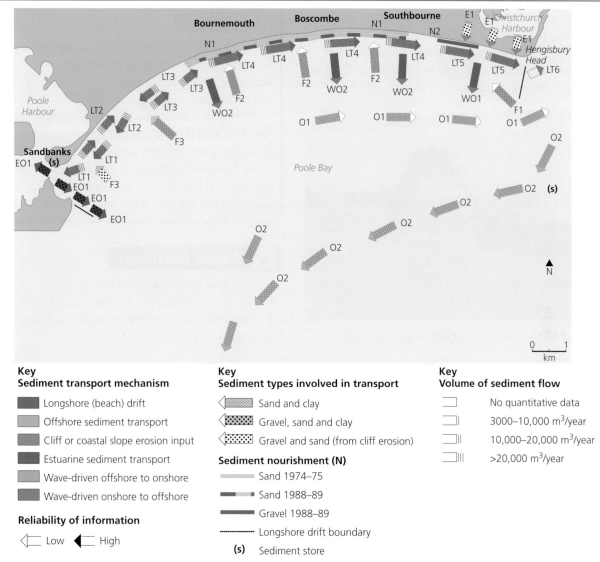

Key
Sediment transport mechanism

- Longshore (beach) drift
- Offshore sediment transport
- Cliff or coastal slope erosion input
- Estuarine sediment transport
- Wave-driven offshore to onshore
- Wave-driven onshore to offshore

Reliability of information

- Low
- High

Key
Sediment types involved in transport

- Sand and clay
- Gravel, sand and clay
- Gravel and sand (from cliff erosion)

Sediment nourishment (N)

- Sand 1974–75
- Sand 1988–89
- Gravel 1988–89
- Longshore drift boundary
- **(s)** Sediment store

Key
Volume of sediment flow

- No quantitative data
- 3000–10,000 m³/year
- 10,000–20,000 m³/year
- >20,000 m³/year

Figure 1.46 Sediment transport in Poole Bay (Adapted from SCOPAC)

Figure 1.47 Rock groynes at Sandbanks

alternative, costing only £3/m³. Natural currents will eventually transport this sand onshore where it will help to build up beaches. In total, over 3.5 million m³ of sediment have been added to Poole Bay beaches. This is an excellent example of management working with nature.

Figure 1.48 Beach recharge by rainbowing at Sandbanks

Skills focus

Using the information about beach recharge above, calculate the total difference in cost between recharging the Poole Bay beaches by rainbowing and by sediment dumping just offshore.

Although the aim of the Shoreline Management Plan is to 'hold the line' (i.e. maintain the coastline in its present position), such has been the effectiveness of the combined strategies of rock groynes and beach recharge that the width of the beach at Sandbanks is increasing and the 'line' is actually advancing slightly.

✓ Review questions

1   Why does Sandbanks need managing?
2   How is its management affecting coastal processes?
3   Assess the success of the management strategies used at Sandbanks.

Key idea
→ Economic development unintentionally causes change within coastal landscape systems

Coastal environments provide many opportunities for human activities, and the potential for economic development. Coasts not only provide attractive environments for tourism, but access to the sea for transport, trade and fishing. However, taking advantage of these opportunities can unintentionally cause change within coastal landscape systems. Sediment budgets may be affected and rates of natural processes may be altered. These changes may then require management in order to mitigate their impacts.

🔍 Case study: Sand mining along the Mangawhai–Pakiri coastline of New Zealand

Economic development

Sand is an essential mineral resource in a modern economy. It has a wide range of uses including construction, concrete making, glass manufacture and beach replenishment. A high-quality sand resource occurs in the nearshore zone at Mangawhai–Pakiri on the east coast of New Zealand's Northland Peninsula. This sand is high quality and suitable for the construction industry. Located just 50 km north of Auckland (Figure 1.49) it is convenient for New Zealand's largest and economically most dynamic metropolitan region. With a population of over 1.5 million, the Auckland region accounts for a third of New Zealand's total population and 35% of the country's GDP. Moreover, the region is growing rapidly. Apart from business, finance and high-tech industry, tourism centred on Auckland's outstanding coastal amenities is booming. 2015 saw a record 2.3 million foreign visitors.

Offshore sand mining and the sediment budget

Nearshore sand dredging on the 20 km long coastline between Mangawhai and Pakiri has operated for over 70 years. Between 1994 and 2004, 165,000 m³/year were extracted. Although mining ended at Mangawhai in 2005, it has continued at Pakiri Beach. Current rates of extraction are 75,000 m³/year until 2020. A large proportion of this sand is used for replenishing Auckland's tourist beaches.

Deposited during the Holocene (past 9000 years), sand is a non-renewable resource on the Mangawhai–Pakiri coastline. There are few sizeable rivers in the area and most sand is thought to have been derived from offshore. The coastal sediment budget is essentially a closed system. Thus outputs of sand through nearshore mining are not replaced by inputs from rivers, and waves from offshore. Indeed, extraction rates at Pakiri Beach exceed

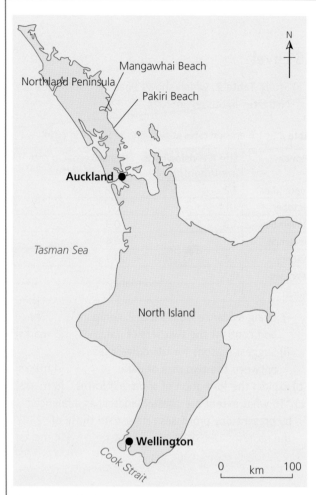

Figure 1.49 The location of Mangawhai and Pakiri Beaches

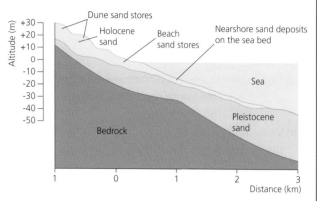

Figure 1.50 Stores of Holocene sand at Pakiri Beach

inputs by a factor of five. The effect of mining is therefore to deplete the total sand supply, stored in dunes, beaches and on the sea bed (up to 2 km offshore) (Figure 1.50). As a result, movements of sand between the major stores have diminished.

Impact on coastal landforms

Given the closed nature of the sediment cell at Mangawhai–Pakiri, current rates of sand extraction are unsustainable. Already the depletion of sand is having an impact on landforms and landscapes. Beaches starved of sediment have become wider and flatter and are less effective in absorbing waves. Higher energy waves thus erode beaches, and landforms such as dunes and spits become vulnerable. Foredune ridges are undercut by

wave action, developing steep, seaward-facing scarps. Loss of vegetation cover makes them susceptible to wind erosion. In 1978, storms caused a 28 metre breach at the base of the Mangawhai spit. This, and a second breach altered tidal currents, which led to the sedimentation of Mangawhai's harbour. Shallower water in the harbour also threatened Manawhai's waterfront community with flooding. Subsequent dredging of the harbour and groyne construction on the spit has helped restore some equilibrium.

However, studies of the Mangawhai–Pakiri coastline by the Auckland Regional Council suggest increased rates of coastal erosion are likely in future with declining natural protection from extreme storm events. Coastal retreat is already evident and is attributed partly to sand extraction, though this is complicated by climate change and rising sea level. Long-term retreat by the end of the century is estimated at 35 metres and the width of the coastal zone susceptible to erosion varies from 48 to 111 metres. Significantly, this estimate is higher than any of the Auckland region's other 123 beaches.

✔ Review questions

1 Explain why the extraction of sand at Mangawhai–Pakiri is unsustainable.
2 Describe how sand extraction has affected sand movement in the coastal system.
3 Outline the impact of the depletion of the sand store on landforms and rates of coastal retreat in Magawhai–Pakiri.

Practice questions

A Level

1 a) Explain the influence of climatic and geomorphic processes on the formation of shore platforms.

   [8 marks]

   b) Study **Table 1**, which shows beach erosion rates at a site in Sussex in 2014.

Table 1 Beach erosion rates at a site in Sussex in 2014

Month	Erosion rate (mm/month)
September	1.3
October	1.1
November	1.4
December	1.7
January	1.7
February	0.7
March	1.2

   i) Calculate the mean erosion rate for the data in Table 1. [2 marks]

   ii) Explain why beach erosion rates vary over time. [4 marks]

   c) Distinguish between offshore and terrestrial sources of coastal sediment. [3 marks]

   d) *To what extent are coastal landscapes influenced by present day processes rather than those of the past? [16 marks]

AS Level

1 a) Study **Table 2**, which shows beach erosion rates at two sites in Sussex in 2014.

Table 2 Beach erosion rates at two sites in Sussex in 2014

Month	Site A erosion rate (mm/month)	Site B erosion rate (mm/month)
September	1.2	0.2
October	1.1	0.1
November	1.3	0.3
December	1.7	0.4
January	1.7	0.3
February	0.7	0.1
March	1.1	0.1

   i) Using evidence from Table 2, compare and contrast the two sets of data. [3 marks]

   ii) Suggest reasons for the differences between the two sets of data. [4 marks]

   b) Explain the formation of shore platforms. [8 marks]

   c) *To what extent are coastal landscapes influenced by present day processes rather than those of the past? [14 marks]

Chapter 2

Glaciated landscapes

Glaciated landscapes are those parts of the Earth's surface that have been shaped, at least in part, by the action of glaciers. They include those places that are currently occupied by glaciers, in both high latitude locations, such as Antarctica and Greenland, and high altitude locations, such as the Rocky Mountains and the Himalayas. They also include those places that were glaciated in the past, such as northern Britain. Glaciated landscapes contain many distinctive landforms produced by the erosional and depositional action of glaciers.

2.1 How can glaciated landscapes be viewed as systems?

> **Key idea**
> → Glaciated landscapes can be viewed as systems

The development of a glaciated landscape over time can be viewed within a systems framework. A **system** is a set of interrelated objects comprising components (stores) and processes (links) that are connected together to form a working unit or unified whole. Glaciated landscape systems store and transfer energy and material on time scales that can vary from a few days to millennia (thousands of years).

The energy available to a glaciated landscape system may be **kinetic**, **potential** or **thermal**. It is this energy that enables work to be carried out by the natural processes that shape the landscape.

The material found in a glaciated landscape system is predominantly the sediment found on valley floors in upland areas as well as in glacial lowlands.

The components of open systems

Glaciated landscape systems are **open systems** (Figure 2.1). This means that energy and matter can be transferred from neighbouring systems as an **input**. It can also be transferred to neighbouring systems as an **output**.

In systems terms, a glaciated landscape has:

● inputs – including kinetic energy from wind and moving glaciers, thermal energy from the heat of the Sun and potential energy from the position of material on slopes; material from **deposition**, **weathering** and **mass movement** from slopes and ice from accumulated snowfall.
● outputs – including glacial and wind **erosion** from rock surfaces; **evaporation**, **sublimation** and meltwater.
● throughputs – which consist of stores, including ice, water and debris accumulations; and flows (transfers), including the movement of ice, water and debris downslope under gravity.

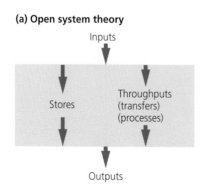

(a) Open system theory

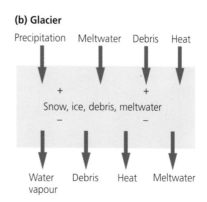

(b) Glacier

Figure 2.1 Open systems

System feedback in glaciated landscapes

When a system's inputs and outputs are equal, a state of **equilibrium** exists within it. In a glaciated landscape, this could happen when the rate at which snow and ice is being added to a glacier equals the rate at which snow and ice is being lost from the glacier by melting and sublimation; as a result the glacier will remain the same size.

When this equilibrium is disturbed, the system undergoes self-regulation and changes its form until equilibrium is restored. This is known as **dynamic equilibrium**, as the system produces its own response to the disturbance. This response is an example of **negative feedback**.

Activity

Construct a flow diagram to show an example of negative feedback in a glaciated landscape system.

Glacier mass balance

The **glacier mass balance**, or budget, is the difference between the amount of snow and ice **accumulation** and the amount of **ablation** occurring in a glacier over a one year time period.

The majority of inputs occur towards the upper reaches of the glacier and this area, where accumulation exceeds ablation, is called the accumulation zone. Most of the outputs occur at lower levels where ablation exceeds accumulation, in the ablation zone. The two zones are notionally divided by the equilibrium line where there is a balance between accumulation and ablation. This can be shown diagrammatically (Figure 2.2).

The annual budget of a glacier can be calculated by subtracting the total ablation for the year from the total accumulation. A positive figure indicates a net gain of ice through the year, increasing the volume of ice and allowing the glacier to advance or grow. The equilibrium line will, in effect, move down the valley. A negative figure indicates a net loss of ice through the year. In this situation ablation exceeds accumulation

Activities

The data below were obtained at eleven sites on the Mocho glacier in Chile during 2004–05. Figures are metres of water equivalent.

Site 1 is near the snout and site 11 is near the source of the glacier.

Site	Accumulation	Ablation
1	3.1	5.8
2	3.3	10.7
3	3.5	7.9
4	3.0	8.1
5	5.3	9.2
6	5.3	8.3
7	4.3	4.5
8	4.5	1.7
9	4.6	1.2
10	5.4	0.5
11	4.3	1.4

1 Describe the pattern of ablation along the glacier.
2 Which site do you think is closest to the equilibrium line?
3 Calculate the net value for each site.
4 Calculate the net value for the whole glacier.
5 Did this glacier advance or retreat over the year?

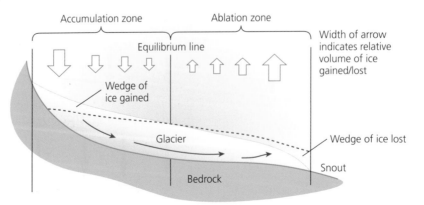

Figure 2.2 Glacier mass balance

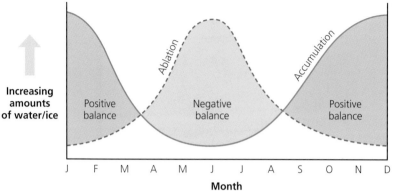

Figure 2.3 Mass balance in a northern hemisphere glacier

> **Activities**
>
> 1 Determine whether the annual mass balance of the glacier shown in Figure 2.3 is positive or negative.
> 2 How would the patterns of ablation and accumulation differ if this glacier were located in the southern hemisphere?

and the glacier will contract and retreat up-valley, along with the equilibrium line.

If the amount of accumulation equals the amount of ablation the glacier is in equilibrium and therefore remains stable in its position.

There will often be seasonal variations in the budget with accumulation exceeding ablation in the winter and vice versa in the summer (Figure 2.3). It is, therefore, possible that there will be some advance during the year even if the net budget is negative and some retreat even when it is positive. An idea that many fail to appreciate is that even when in retreat, the ice in a glacier may move forwards across the equilibrium line under gravity, so it can appear that a retreating glacier is actually advancing.

Due to changes in weather conditions from year to year, the mass balance of a glacier is not constant, but can vary quite considerably over time. Climate change can cause changes over longer-term time scales.

> **Review questions**
>
> 1 Define the term 'millennia'.
> 2 What is the difference between 'kinetic' and 'potential' energy?
> 3 Outline the differing elements that make up the throughputs of a glaciated landscape system.
> 4 Explain the concept of 'dynamic equilibrium'?
> 5 What is the mass balance of a glacier?
> 6 Why does glacier mass balance vary (a) spatially and (b) temporally?

> **Key idea**
> → Glaciated landscapes are influenced by a range of physical factors

The development of glaciated landscapes and their operation as systems are both influenced by a range of physical factors. These factors influence the way in which processes work, and consequently affect the shaping of the landscape. The factors vary in the significance of their impact spatially (from place to place) and temporally (over time). In any one location, or at any one time, some factors will have greater significance than others, and sometimes a factor may have very little influence at all. The factors themselves are often interrelated, i.e. one factor may influence another.

Climate

Wind is a moving force and as such is able to carry out erosion, **transportation** and deposition. These **aeolian** processes contribute to the shaping of glaciated landscapes, particularly acting upon fine material previously deposited by ice or meltwater.

Precipitation is a key factor in determining the mass balance of a glacier, as it provides the main input of snow, sleet and rain. In high latitude glaciated landscape systems, precipitation totals may be extremely low. For example, Vostock station in Antarctica has a mean annual precipitation total of only 4.5 mm. However, high altitude locations often have much higher totals. In Jasper National Park in the Canadian Rockies, over 600 mm per annum is typical.

The seasonal pattern of precipitation can also be very variable. In Jasper, January precipitation averages about 25 mm, while in June it is 100 mm. However, more of the January precipitation is snow, whereas in June most falls as rain. In Antarctica there is much less seasonal variation, with very little difference in precipitation between the highest and lowest months. The greater the seasonal variation in precipitation, the more varied the mass balance of the glacier will be.

Temperature is also a significant factor. If temperatures rise above 0°C, accumulated snow

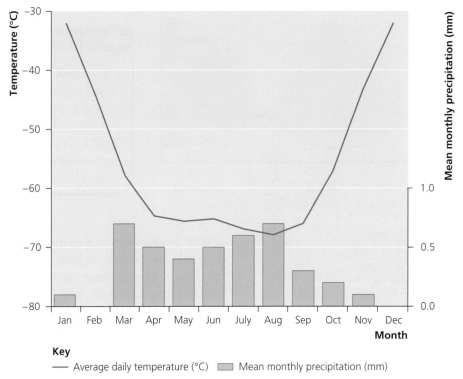

Figure 2.4 Climate graph for Vostock station, Antarctica

and ice will start to melt and become an output of the system. High altitude glaciers may experience significant periods in the summer months of above zero temperatures and melting. Whereas, in high latitude locations, temperatures may never rise above zero and so no melting occurs. This explains why ice sheets are so thick in polar regions, despite very low precipitation inputs.

> ### ▶ Activities
>
> 1 Using Figure 2.4, describe the climate of Vostock station.
> 2 Suggest how these climatic conditions are likely to influence the glacier mass balance at Vostock.

Geology

The two key aspects of geology that influence glaciated landscape systems are **lithology** and **structure**.

Lithology

Lithology describes the physical and chemical composition of rocks. Some rock types, such as clay, have a weak lithology, with little resistance to erosion, weathering and mass movements, as the bonds between the particles that make up the rock are quite weak. Others, such as basalt, made of dense interlocking crystals, are highly resistant and are more likely to form prominent glacial landforms such as arêtes and pyramidal peaks. Others, such as limestone, are predominantly composed of calcium carbonate. This is soluble in weak acids and so is vulnerable to decay by the chemical weathering process of carbonation, especially at low temperatures.

Structure

Structure concerns the properties of individual rock types such as jointing, bedding and faulting. It also includes the permeability of rocks. In porous rocks, such as chalk, tiny air spaces (pores) separate the mineral particles. These pores can absorb and store water – a property known as primary permeability. Carboniferous limestone is also permeable, but for a different reason. Water seeps into limestone because of its many joints. This is known as secondary permeability. The joints are easily enlarged by solution.

Structure also includes the angle of dip of rocks and can have a strong influence on valley side profiles. Horizontally bedded strata support steep cliffs with near vertical profiles. Where strata incline, profiles tend to follow the angle of dip of the bedding planes.

Latitude and altitude

Locations at high latitudes, most noticeably beyond the Arctic and Antarctic Circles at 66.5°N and S, tend

to have cold dry climates with little seasonal variation in precipitation. The higher the latitude, the more apparent this is (Figure 2.4). Glaciated landscapes at such latitudes tend to develop under the influence of large, relatively stable ice sheets, such as those of Greenland and Antarctica. These landscapes are quite different to those that develop under the influence of dynamic valley glaciers in lower latitude but higher altitude locations, such as the Rocky Mountains and the Himalayas. These locations tend to have higher precipitation inputs, but more variable temperatures and hence more summer melting.

Such is the decrease in temperature with altitude that glaciers are even found near the Equator in the Andes. The Pastoruri glacier in Peru lies at an altitude of 5250 m and is just 10°N of the Equator. It is a small glacier covering 8 km^2 and is about 4 km long, although it has undergone significant melting in recent years.

Relief and aspect

Although latitude and altitude are the major controls on climate, relief and aspect have an impact on microclimate and the movement of glaciers.

The steeper the relief of the landscape, the greater the resultant force of gravity and the more energy a glacier will have to move downslope.

Where air temperature is close to zero, it can have a significant influence on the melting of snow and ice and the behaviour of glacier systems. If the aspect of a slope faces away from the general direction of the Sun, temperatures are likely to remain below zero for longer, as less solar energy is received, and so less melting occurs. The mass balance of glaciers in such locations will, therefore, tend to be positive, causing them to advance. The reverse is likely to be true in areas with an aspect facing towards the Sun. These differences not only affect the mass balance, but will, as a result, influence the shaping of the landscape. Glaciers with a positive mass balance are more likely to be larger, with greater erosive power, and much more erosive than small ones and those in retreat due to a negative mass balance.

✔ Review questions

1 How do precipitation patterns influence glaciated landscape systems?
2 What is the difference between structure and lithology?
3 How might glaciated landscape systems in high latitude locations differ from those in low latitude locations?
4 How does aspect influence the microclimate of glaciated landscape systems?

Skills focus

Temperature decreases with altitude at a rate of approximately 0.6°C/100 m.

If the temperature at sea level at 10°N of the Equator is 35°C, calculate the temperature at the Pastoruri glacier.

Key idea

→ There are different types of glacier and glacier movement

The formation of glacier ice

Glaciers form when temperatures are low enough for snow that falls in one year to remain frozen throughout the year. This means that the following year, fresh snow falls on top of the previous year's snow. Fresh snow consists of flakes with an open, feathery structure and a low density of about 0.05 g/cm^3. Each new fall of snow compresses and compacts the layer beneath, causing the air to be expelled and converting low density snow into higher density ice. Snow that survives one summer is known as firn and has a density of 0.4 g/cm^3. With further compaction by subsequent years of snow fall, it becomes glacier ice with a density of between 0.83 and 0.91 g/cm^3. This process is known as diagenesis. It may take between 30–40 years and 1000 years for this to happen. True glacier ice is not encountered until a depth of about 100 m and is characterised by a bluish colour rather than the white or fresher snow which is due to the presence of air.

Valley glaciers and ice sheets

Glaciers are large, slow-moving masses of ice.

Ice sheets are the largest accumulations of ice, defined as extending for more than 50,000 km^2. There are currently only two: Antarctica and Greenland. Today these possess 96 per cent of the world's ice. During the last glacial period huge ice sheets also covered much of Europe. The Antarctic ice sheet covers 13.6 million km^2 and has a volume of about 30 million km^3. At its thickest, in eastern Antarctica, it is over 4700 m deep.

Valley glaciers are confined by valley sides. They may be outlet glaciers from ice sheets or fed by snow and ice from one or more corrie glaciers (see pages 50–51). They follow the course of existing river valleys or corridors of lower ground. They are typically between 10 and 30 km in length, although in the Karakoram Mountains of Pakistan they are as long as 60 km.

Warm-based and cold-based glaciers

Warm-based (temperate) glaciers usually have:

- high altitude locations
- steep relief
- basal temperatures at or above pressure melting point
- rapid rates of movement, typically 20–200 m per year.

Locations such as the Alps and the Rockies experience high rates of accumulation in the winter and, due to relatively high temperatures, above zero for much of the year, high rates of ablation in the summer. This makes them very active with large volumes of ice being transferred across the equilibrium line and significant seasonal advance and retreat. Not only will the rapid ice movements cause significant erosion and erosional landforms, but the ablation also produces lots of meltwater and so landforms of glacio-fluvial origin are also common.

Cold-based (polar) glaciers are characterised by:

- high latitude locations
- low relief
- basal temperatures below pressure melting point and so frozen to the bedrock
- very slow rates of movement, often only a few metres a year.

In Antarctica and Greenland, summer temperatures are below freezing and precipitation is low. This means that both accumulation and ablation are very limited and there are no great seasonal differences. The glaciers are not very dynamic and there is not only limited movement but limited landscape impact as well as little erosion, deposition and transportation take place.

The most important difference between the warm- and cold-based glaciers is their basal temperature, as shown on the temperature profiles in Figure 2.5,

Activity

Compare and contrast the temperature profiles for cold-based and warm-based glaciers shown in Figure 2.5.

and its relationship with the **pressure melting point**, as this largely determines the mechanism of movement. The pressure melting point is the temperature at which ice is on the verge of melting. At the surface this is at 0°C, but within an ice mass it will be fractionally lowered by increasing pressure. Ice at pressure melting point deforms more easily than ice below it. Most temperate glaciers are at pressure melting point at the base and sometimes within the glacier itself. Movement is facilitated by the production of meltwater. However, in polar glaciers temperatures are below pressure melting point and movement is limited.

Basal sliding and internal deformation

Glaciers move, fundamentally, due to the forces of gravity. Ice moves downslope from higher to lower altitude. In a valley glacier this involves moving from the accumulation zone, across the equilibrium line and into the ablation zone.

There are a number of factors that influence the movement of glaciers:

- Gravity – the fundamental cause of the movement of an ice mass
- Gradient – the steeper the gradient of the ground surface, the faster the ice will move if other factors are excluded
- The thickness of the ice – as this influences basal temperature and the pressure melting point

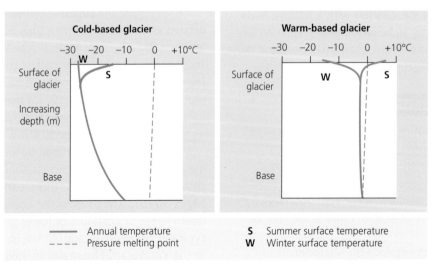

Figure 2.5 Glacier temperature profiles

- The internal temperatures of the ice – as this can allow movements of one area of ice relative to another
- The glacial budget – a positive budget (net accumulation) causes the glacier to advance

Ice moves in different ways. When it is solid and rigid it will tend to break apart as shown by crevasses. However, when under steady pressure it will deform and behave more like a plastic. There are typically two zones in a glacier (Figure 2.6) where these different movements occur:

- An upper zone where the ice is brittle and breaks
- A lower zone where under pressure the ice deforms

As a result of research on the Mer de Glace glacier in the French Alps in 1842, James Forbes concluded that the sides and base of this glacier tended to move

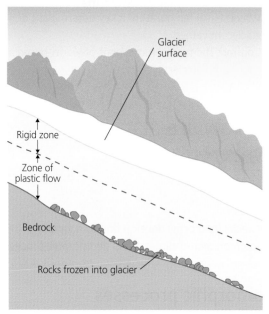

Figure 2.6 Zones of glacier movement

Activities

1 Describe the pattern surface velocity of the glacier in Figure 2.7.
2 Explain this pattern.
3 Describe the changes in velocity with depth in this glacier.
4 Why does velocity change with depth?

more slowly than the top and middle (Figure 2.7a). This was because the ice may have been frozen onto the rocks of the valley floor and sides. There may also have been obstructions that created frictional resistance and slowed down movement. It was also due to the accumulative effect of laminar flow in which each lower layer of ice not only moved itself, but carried the layers above with it (Figure 2.7b).

Basal sliding

Glaciers move differently depending on the temperature of the ice at their base.

Warm-based (temperate) glaciers mainly move by basal sliding. If the basal temperature is at or above pressure melting point a thin film of meltwater exists between the ice and the valley floor and so friction is reduced.

Basal sliding actually consists of a combination of different mechanisms:

- Slippage, where the ice slides over the valley floor as the meltwater has reduced friction between the base of the glacier (and any debris embedded in it) and the valley floor. Friction itself between the moving ice/debris and the valley floor can also lead to the creation of meltwater.

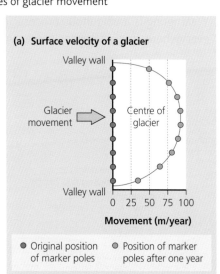

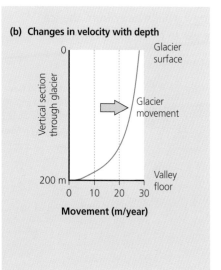

Figure 2.7 Glacier velocity

- Creep or regelation, when ice deforms under pressure due to obstructions on the valley floor. This enables it to spread around and over the obstruction, rather as a plastic, before re-freezing again when the pressure is reduced.
- Bed deformation, when the ice is carried by saturated bed sediments moving beneath it on gentle gradients. The water is under high pressure. This movement has been likened to the ice being carried on roller skates.

The Franz Josef glacier in New Zealand is warm-based and moves approximately 300 m per year. Basal sliding accounts for 45 per cent of the movement of the Salmon glacier in Canada, but can account for as much as 90 per cent in extreme cases.

Internal deformation

Cold-based (polar) glaciers are unable to move by basal sliding as the basal temperature is below pressure melting point. Instead they move mainly by internal deformation, although ice at 0°C deforms 100 times faster than ice at −20°C.

Internal deformation has two elements:

- Intergranular flow, when individual ice crystals re-orientate and move in relation to each other
- Laminar flow, when there is movement of individual layers within the glacier – often layers of annual accumulation

Both of these movements occur when the glacier is on a slope; they do not occur on level surfaces where the ice remains intact.

The Meserve glacier in Antarctica moves only 3–4 m per year at its equilibrium line and 100 per cent of this movement is by internal deformation.

When ice moves over a steep slope it is unable to deform quickly enough and so it fractures, forming crevasses. The leading ice pulls away from the ice behind it, which has yet to reach the steeper slope.

This is **extending flow**. When the gradient is reduced, **compressing flow** occurs as the ice thickens and the following ice pushes over the slower-moving leading ice. The planes of movement, called slip planes, are at different angles in each case, as seen in Figure 2.8.

Some glaciers may occasionally surge at rates of 100 m per day. This is only likely on relatively steep gradients in temperate glaciers after large inputs of snow and ice have been received. A glacier on Disko Island, Greenland, was found to have moved 10 km in four years between 1995 and 1999, with a peak movement of 30 m a day. Some surges may be triggered by tectonic activity such as earthquakes.

Review questions

1. Explain how glacier ice forms.
2. Distinguish between valley glaciers and ice sheets.
3. Why is basal temperature a key determinant of glacier movement?
4. Contrast the typical movement of warm-based and cold-based glaciers.

2.2 How are glacial landforms developed?

Key idea
→ Glacial landforms develop due to a variety of interconnected climatic and geomorphic processes

Geomorphic processes
Weathering

Weathering is a ubiquitous process, in that it happens everywhere. In glacial areas some types of weathering

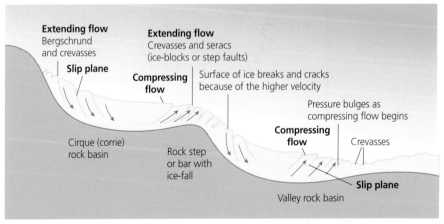

Figure 2.8 Extending and compressing flow

are particularly significant and therefore influence the formation of glacial landforms. Weathering uses heat energy to produce physically or chemically altered materials from surface or near surface rock.

Physical or mechanical

The breakdown of rock is largely achieved by physical weathering processes (Table 2.1) that produce smaller fragments of the same rock. No chemical alteration takes place during physical weathering. By increasing the exposed surface area of the rock, physical weathering allows further weathering to take place.

In many glacial landscapes, air temperature may seldom rise above 0°C, rendering some of the processes below ineffective.

Table 2.1 Processes of physical (or mechanical) weathering typical of glacial environments

Freeze-thaw	Water enters cracks/joints and expands by nearly 10 per cent when it freezes. In confined spaces this exerts pressure on the rock causing it to split or pieces to break off, even in very resistant rocks. The more frequent and regular the fluctuations of temperature around zero, the more effective this process will be.
Frost shattering	At extremely low temperatures, water trapped in rock pores freezes and expands. This creates stress which disintegrates rock to small particles.
Pressure release	When the weight of overlying ice in a glacier is lost due to melting, the underlying rock expands and fractures parallel to the surface. This is significant in the exposure of sub-surface rocks such as granite and is also known as dilatation. The parallel fractures are sometimes called pseudo-bedding planes.

Chemical

Decay of rock is the result of chemical weathering (Table 2.2), which involves chemical reaction between the elements of the weather and some minerals within the rock. It may reduce the rock to its chemical constituents or alter the chemical and mineral composition of the rock. Chemical weathering processes produce weak residues of different material that may then be easily removed by erosion or transportation processes.

The rate of most chemical reactions is faster when temperature is higher. Van't-Hoff's Law states that a 10°C increase in temperature leads to a 2.5 times increase in the rate of chemical reaction (up to 600°C), so most chemical weathering processes are most effective in warm or hot climatic regions. This is why warm, moist tropical environments experience the fastest rates of chemical weathering and cold, dry locations the slowest.

Some weathering processes are especially important when rocks are in contact with weakly acidic water. One issue associated with climate change and increasing levels of atmospheric CO_2 is that rain, and therefore ice, is becoming more acidic.

Biological

Biological weathering (Table 2.3) may consist of physical actions such as the growth of plant roots or chemical processes such as **chelation** by organic acids. Although this, arguably, does not fit with the precise definition of weathering, biological processes are usually classed as a type of weathering. Certainly the effects are very similar to some of the physical and chemical processes even if it may be difficult to directly relate them to the weather. In glacial environments, plant

Table 2.2 Processes of chemical weathering typical of glacial environments

Oxidation	Some minerals in rocks react with oxygen, either in the air or in water. Iron is especially susceptible to this process. It becomes soluble under extremely acidic conditions and the original structure is destroyed. It often attacks the iron-rich cements that bind sand grains together in sandstone.
Carbonation	Rainwater combines with dissolved carbon dioxide from the atmosphere to produce a weak carbonic acid. This reacts with calcium carbonate in rocks such as limestone to produce calcium bicarbonate, which is soluble. This process is reversible and precipitation of calcite happens during evaporation of calcium rich water in caves to form stalactites and stalagmites.
Solution	Some salts are soluble in water. Other minerals, such as iron, are only soluble in very acidic water, with a pH of about 3. Any process by which a mineral dissolves in water is known as solution, although mineral specific processes, such as carbonation, can be identified.
Hydrolysis	This is a chemical reaction between rock minerals and water. Silicates combine with water producing secondary minerals such as clays. Feldspar in granite reacts with hydrogen in water to produce kaolin (china clay).
Hydration	Water molecules added to rock minerals create new minerals of a larger volume. This happens to anhydrite forming gypsum. Hydration causes surface flaking in many rocks, partly because some minerals also expand by about 0.5 per cent during the chemical change as well because they absorb water.

Table 2.3 Processes of biological weathering typical of glacial environments

Tree roots	Tree roots grow into cracks or joints in rocks and exert outward pressure. This operates in a very similar way and with similar effects to freeze-thaw. When trees topple, their roots can also exert leverage on rock and soil, bringing them to the surface and exposing them to further weathering. Burrowing animals may have a similar effect.
Organic acids	Organic acids produced during decomposition of plant and animal litter cause soil water to become more acidic and react with some minerals in a process called chelation. Blue-green algae can have a weathering effect, producing a shiny film of iron and manganese oxides on rocks.

Stretch and challenge

It is worth noting that carbonation can be more effective in low temperatures as carbon dioxide is more readily absorbed in cold water than in warm water. This makes carbonation a significant process in glacial environments.

Stretch and challenge

Research into shear strength and shear stress. Establish what gives a slope its strength and what can exert stress on it. Investigate the relationship between shear stress and shear strength and how this relates to mass movement processes.

Tulane University provides a useful guide: www.tulane.edu/~sanelson/eens1110/massmovements.pdf

and animal activity may be severely limited by the low temperatures and so these mechanisms may be of very little significance.

Mass movement

Mass movement occurs when the forces acting on slope material (mainly the resultant force of gravity) exceed the forces trying to keep the material on the slope (predominantly friction).

In glacial landscape systems, the most significant mass movement processes are those acting on steep slopes, which lead to the addition of material to the glacier below, loading it with debris and providing the tools for abrasion. The main processes involved are:

- Rock fall: on slopes of 40° or more, especially if the surface is bare, rocks may become detached from the slope by physical weathering processes. These then fall to the foot of the slope under gravity. Transport processes may then remove this material, or it may accumulate as a relatively straight, lower angled scree slope.
- Slides: these may be linear, with movement along a straight line slip plane, such as a fault or a bedding plane between layers of rock, or rotational, with movement taking place along a curved slip plane. Rotational slides are also known as slumps. In glaciated landscape systems, slides may occur due to steepening or undercutting of valley sides by erosion at the base of the slope, adding to the downslope forces. Slumps are common in weak rocks, such as clay, which also become heavier when wet, adding to the downslope force.

Glacial processes

Moving ice in a glacier is a source of energy in glaciated landscape systems, and the energy can be expended through geomorphic processes to shape landforms. These processes can also supply material in the form of sediment, which can be deposited in, or transported within, the glacial system.

Erosion

Glacial erosion occurs as glaciers advance and this mainly occurs in upland areas. There are two main processes of erosion by glaciers:

- Plucking – this mainly happens when meltwater seeps into joints in the rocks of the valley floor/sides. This then freezes and becomes attached to the glacier. As the glacier advances it pulls pieces of rock away. A similar mechanism takes place when ice re-freezes on the down-valley side of rock obstructions. Plucking is particularly effective at the base of the glacier as the weight of the ice mass above may produce meltwater due to pressure melting. It will also be significant when the bedrock is highly jointed which allows meltwater to penetrate. Plucking is also known as quarrying.
- Abrasion – as a glacier moves across the surface, the debris embedded in its base/sides scours surface rocks, wearing them away. The process is often likened to the action of sandpapering. The coarse material will scrape, scratch and groove the rock. The finer material will tend to smooth and polish the rock. The glacial debris itself is also worn down

by this process, forming a fine **rock flour** that is responsible for the milky white appearance of glacial meltwater streams and rivers.

Rates of glacial abrasion are very variable and are influenced by a number of factors:

- Presence of basal debris – pure ice is unable to carry out abrasion of solid rock and so basal debris is an essential requirement. The rate of abrasion increases with the amount of basal debris up to a point where it produces great friction, which slows down rates of movement.
- Debris size and shape – particles embedded in ice exert a downward pressure proportional to their weight, and so larger debris is more effective in abrasion than fine material. Angular debris is also more effective as the pressure is concentrated onto a smaller area of debris–bedrock interface.
- Relative hardness of particles and bedrock – abrasion is most effective when hard, resistant rock debris at the glacier base is moved across a weak, soft bedrock. If the bedrock is more resistant than the debris, then little abrasion will be accomplished.
- Ice thickness – the greater the thickness of overlying ice, the greater the pressure exerted on the basal debris and the greater the rate of abrasion. This is, however, only true up to a point. Beyond a certain thickness the pressure becomes too great and there is too much friction between the debris and the bedrock for much movement to occur. This is not a fixed thickness, as it depends upon ice density and the nature of the debris, but it is typically 100–200 m.
- Basal water pressure – the presence of a layer of meltwater at the base of a glacier is vital if sliding and therefore abrasion is to take place. However, if the water is under pressure, perhaps because it is confined, the glacier can be buoyed up, reducing pressure and erosion.
- Sliding of basal ice – this is important as it determines whether abrasion can take place. Abrasion requires basal sliding to move the embedded debris across the rock surfaces. The greater the rate of sliding, the more potential there is to erode as more debris is passing across the rock per unit of time.
- Movement of debris to the base – abrasion does not only wear away the bedrock, it also wears away the basal debris. Debris needs to be replenished (by glacial erosion and weathering processes) if abrasion is to remain effective.

- Removal of fine debris – to maintain high rates of abrasion, rock flour (fine debris) needs to be removed so that the larger particles can abrade the bedrock. This is mainly done by meltwater.

Estimates of rates of erosion include:

- Embleton and King (1968) suggest that mean annual erosion for active valley glaciers is between 1000 and 5000 m^3.
- Boulton (1974) measured erosion on rock plates placed beneath the Breiðamerkurjökull glacier in Iceland and found that under ice 40 m thick, basalt eroded at 1 mm per year and marble at 3 mm per year. The ice had a velocity of 9.6 m per year. However, if the velocity increased to 15.4 m/year, the rate of erosion of marble increased to 3.75 mm, even though the ice was 8 m thinner. In this instance it would suggest that velocity is more important than ice thickness.
- In comparison, ice 100 m thick flowing at 250 m per year in the Glacier d'Argentière eroded a marble plate at up to 36 mm/year.

Nivation

Nivation is a glacial process that is not easily classified as erosion or weathering. This complex process is thought to include a combination of freeze-thaw action, solifluction, transport by running water and, possibly, chemical weathering. Nivation is thought to be responsible for the initial enlargement of hillside hollows and the incipient development of corries.

Transportation

Moving ice is capable of carrying huge amounts of debris. This material comes from a wide range of sources:

- Rockfall – weathered debris falls under gravity from the exposed rock above the ice down onto the edge of the glacier
- Avalanches – these often contain rock debris within the snow and ice that moves under gravity
- Debris flows – in areas of high precipitation and occasional warmer periods, melting snow or ice can combine with scree, soil and mud
- Aeolian deposits – fine material carried and deposited by wind, often blowing across outwash deposits
- Volcanic eruptions – a source of ash and dust
- Plucking – large rocks plucked from the side and base of valleys
- Abrasion – smaller material worn away from valley floors and sides

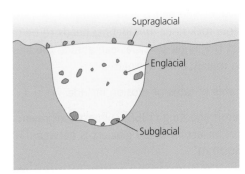

Figure 2.9 Glacial transportation

While being transported, the material may be classified according to its position in the glacier (Figure 2.9):

- Supraglacial is debris being carried on the surface of a glacier. This will most often come from weathering and rockfall.
- Englacial is debris within the ice. This may have been supraglacial material that has been covered by further snowfall, fallen into crevasses or sunk into the ice due to localised pressure melting.
- Subglacial is debris embedded in the base of the glacier which may have been derived from plucking and abrasion, or that has continued to move down through the ice as former englacial debris.

Deposition

Glaciers deposit their load when their capacity to transport material is reduced. This usually occurs as a direct result of ablation during seasonal periods of retreat or during de-glaciation. However, material can also be deposited during advance or when the glacier becomes overloaded with debris.

All material deposited during glaciation is known as **drift**. This can be subdivided into **till**, which is material deposited directly by the ice, and **outwash**, which is material deposited by meltwater. The latter is also known as **glacio-fluvial** material (see page 60).

It is estimated that glacial deposits currently cover about 8 per cent of the Earth's surface. In Europe they cover almost 30 per cent and are mainly material left by earlier ice masses that have since retreated. East Anglia has deposits that are up to 143 m thick. However, in the Gulf of Alaska they are, in places, 5000 m thick. In active glacial areas, it is possible that rates of deposition are in the order of 6 m per 100 years.

Till

There are two types of glacial till:

- Lodgement till – this is material deposited by advancing ice. Due to the downward pressure exerted by thick ice, subglacial debris may be pressed and pushed into existing valley floor material and left behind as the ice moves forward. This may

be enhanced by localised pressure melting around individual particles that are under significant weight and pressure. **Drumlins** are the main example of landforms of this type.
- Ablation till – this is material deposited by melting ice from glaciers that are stagnant or in retreat, either temporarily during a warm period or at the end of the glacial event. Most glacial depositional landforms are of this type.

Both types of till typically have three distinctive characteristics:

- Angular or sub-angular in shape – this is because it has been embedded in the ice and has not been subjected to further erosion processes, particularly by meltwater which would make it smooth and rounded, although it may have been altered in an earlier period of erosion by meltwater, before being entrained, transported and deposited by glaciers.
- Unsorted – when glaciers deposit material, all sizes are deposited en masse, together. When water deposits material, it loses energy progressively and deposits material in a size-based sequence.
- Unstratified – glacial till is dropped in mounds and ridges rather than in layers, which is typical of water-borne deposits.

👣 Fieldwork ideas

Sediment samples could be taken from areas of till and the shape and size of the particles measured. The three axes of each particle could be measured with a ruler. The shape could be described using Power's Index of Roundness, the Cailleux Index, Zingg's classification or Krumbein's Index of Sphericity. This will determine whether the sediment was deposited by water or ice, and if by ice in which direction the ice was moving.

The Field Studies Council (www.geography-fieldwork.org) details these indices.

Glacial landforms

Although glacial landforms tend to be classified according to erosional or depositional processes, they can develop due to a variety of interconnected processes.

Erosional

Corries

Corries are armchair-shaped hollows found on upland hills or mountainsides. They have a steep back wall, an over-deepened basin and often have a lip at the front, which may be solid rock or made of morainic deposits.

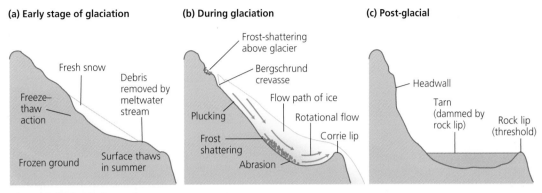

(a) Early stage of glaciation

Fresh snow

Freeze–thaw action

Debris removed by meltwater stream

Frozen ground

Surface thaws in summer

(b) During glaciation

Frost-shattering above glacier

Bergschrund crevasse

Flow path of ice

Plucking

Rotational flow

Frost shattering

Corrie lip

Abrasion

(c) Post-glacial

Headwall

Tarn (dammed by rock lip)

Rock lip (threshold)

Figure 2.10 Corrie formation

They vary in size and shape but the length to height ratio (from the lip to the top of the back wall) is usually between 2.8:1 and 3.2:1. Some are only a few hundred metres across but they can be over 15 km wide. The Walcott Corrie in Antarctica has a 3 km high back wall.

Corrie formation is the result of several interacting processes. Development starts with nivation of a small hollow on a hillside in which snow collects and accumulates year on year. Over time these hollows enlarge and contain more snow, which eventually compresses into glacier ice. At a critical depth, the ice acquires a rotational movement under its own weight. This enlarges the hollow further. Meanwhile, the rotational movement causes plucking of the back wall, making it increasingly steep. The debris derived from plucking and weathering above the hollow falls into the bergschrund or crevasse which abuts the back wall (Figure 2.10). This rock debris helps to abrade the hollow and causes it to deepen. Once the hollow has deepened, the thinner ice at the front is unable to erode so rapidly and so a higher lip is left. The lip may also consist of moraine deposited by the ice as it moves out of the corrie.

In the post-glacial landscape the corrie may become filled with water forming a small circular lake or tarn.

Skills focus

Table 2.4 The orientation of 40 corries in an area of the Cairngorms, Scotland

North	North east	East	South east	South	South west	West	North west
7	16	8	4	1	0	1	3

Using chi-squared (see Chapter 15, Geographical Skills, pages 539–40), test the hypothesis that there is a significant pattern in the orientation of corries in this area.

Arêtes and pyramidal peaks

An **arête** is a narrow, steep-sided ridge found between two corries. The ridge is often so narrow that it is described as knife-edged. Arêtes form from glacial erosion, with the steepening of slopes and the retreat of corries that are back to back or alongside each other. A good example is Striding Edge in the Lake District (Figure 2.11) which has steep slopes either side that are 200–300 m high and almost vertical in places. Striding Edge itself is so narrow that it is just wide enough for one person to walk along the footpath that runs along the crest towards the summit of Helvellyn.

Figure 2.11 Striding Edge, Lake District

Where three or more corries develop around a hill or mountain top and their back walls retreat, the remaining mass will be itself steepened to form a **pyramidal peak** (Figure 2.12). Weathering of the peak may further sharpen its shape. The Matterhorn in the Swiss Alps is an excellent example and is over 1200 m high.

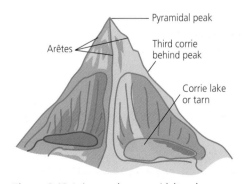

Pyramidal peak

Arêtes

Third corrie behind peak

Corrie lake or tarn

Figure 2.12 Arêtes and a pyramidal peak

Figure 2.13 Glen Avon, Cairngorms, Scotland

Troughs

Glaciers flow down pre-existing river valleys under gravity. As they move they erode the sides and floor of the valley, causing the shape to become deeper, wider and straighter. The mass of ice has far more erosive power than the river that originally cut the valley. Although they are usually described as being U-shaped, they seldom are. Rather, they are parabolic, partly due to the weathering and mass movement of the upper part of the valley sides that goes on both during the glacial period and in the subsequent periglacial period as the glacier retreats. The resultant scree slopes that accumulate at the base of the valley sides lessen the slope angle (Figure 2.13).

There are often variations in the long profile of glacial troughs. When compressing flow occurs the valley is over-deepened to form rock basins and rock steps. This process may be particularly evident where there are alternating bands of rock of different resistances on the valley floor – the weaker rocks being eroded more rapidly to form the basins.

Roche moutonnées and striations

Projections of resistant rock are sometimes found on the floor of glacial troughs. As advancing ice passes over them, there is localised pressure melting on the up-valley side. This area is smoothed and streamlined by abrasion and often has **striations** which are scratches or grooves made by debris embedded in the base of the glacier. On the down-valley side pressure is reduced and meltwater re-freezes, resulting in plucking and steepening. Roche moutonnées (Figure 2.14) can indicate the direction the ice moved through an area. They vary in size but in the Coniston area of the Lake District they are typically 1–5 m high and 5–20 m long.

Ellipsoidal basins

All of the landforms referred to so far are formed by the action of valley glaciers and their tributaries. This is known as Alpine glaciation. Significant contrasts can be seen when looking at the impact of large ice sheets on the landscape.

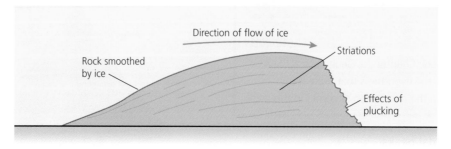

Figure 2.14 Formation of a roche moutonnée

Figure 2.15 The Laurentide ice sheet (adapted from earthguide.ucsd.edu)

Ellipsoidal basins are major erosional landforms created by ice sheets. The Laurentide ice sheet covered much of North America between about 95,000 and 12,000 years ago (Figure 2.15). At its maximum extent it spread as far south as latitude 37°N and covered an area of more than 13 million km². In some areas its thickness reached 3000 m or more.

Erosion by the Laurentide ice sheets produced a series of ellipsoidal basins in North America. The master basin holds Hudson Bay with smaller basins containing the Great Lakes. As well as the significant erosional impact of the ice, the weight of the ice sheet also led to **isostatic** lowering of the surface landscape.

Depositional

A number of different landforms are produced by sediment being directly deposited by ice: moraines, erratics, drumlins and till sheets.

Moraines

A **terminal moraine** is a ridge of till extending across a glacial trough. They are usually steeper on the up-valley side and tend to be crescent shaped, reaching further down-valley in the centre. These landforms mark the position of the maximum advance of the ice and were deposited at the glacier snout. Their crescent shape is due to the position of the snout; further advance occurs in the centre of the glacier, as there is no friction with the valley sides. The steeper up-valley side is the result of the ice behind supporting the deposits and making them less likely to collapse. The Franz Josef glacier in New Zealand has left a terminal moraine 430 m high.

A **lateral moraine** is a ridge of till running along the edge of a glacial valley. The material accumulates on top of the glacier having been weathered from the exposed valley sides. As the glacier melts or retreats, this material sinks through the ice to the ground and is deposited. A lateral moraine left by the retreating Athabasca glacier in Canada is 1.5 km long and 124 m high.

Recessional moraines are a series of ridges running transversely across glacial troughs and which are broadly parallel to each other and to the terminal moraine. They are found further up the valley than the terminal moraine (Figure 2.16). They form during a temporary still-stand in retreat. These temporary pauses are rarely prolonged; thus recessional moraines seldom exceed 100 m in height.

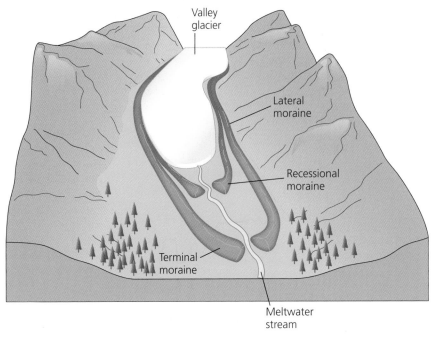

Valley glacier

Lateral moraine

Recessional moraine

Terminal moraine

Meltwater stream

Figure 2.16 Moraines

Lichenometry is a method of numerical dating that uses the size of lichen colonies on a rock surface to determine the surface's age. Lichenometry is used for rock surfaces less than about 10,000 years old.

The basic premise of lichenometry is that the diameter of the largest lichen thallus growing on a moraine, or other surface, is proportional to the length of time that the surface has been exposed to colonisation and growth. Data on lichen growth rates can enable estimates of both the age of the thallus and the period of exposure of a rock surface to be made.

Erratics

An erratic is an individual piece of rock, varying in size from a small pebble to a large boulder. What makes them distinctive is that they are composed of a different geology from that of the area in which they have been deposited. They were eroded, most likely by plucking, or added to the supraglacial debris by weathering and rockfall, in an area of one type of geology and then transported and deposited into an area of differing rock type. A good example is the erratic blocks of Silurian shale deposited on carboniferous limestone at Norber in the Yorkshire

Dales (see Figure 4.9 on page 107). These have protected the underlying rock from carbonation weathering, leaving them perched on a pedestal.

Drumlins

A drumlin is a mound of glacial debris that has been streamlined into an elongated hill. Often they are prominent landforms, sometimes more than a kilometre in length and 100 m high. In plan they are typically pear-shaped and aligned in the direction of ice flow. The higher and wider stoss, or blunt, end faces the ice flow, while the lee side is more gently tapered.

Their formation is not fully understood. They may be formed by:

- lodgement of subglacial debris as it melts out of the basal ice layers
- reshaping of previously deposited material during a subsequent re-advance
- accumulation of material around a bedrock obstruction – these are known as rock-cored drumlins
- thinning of ice as it spread out over a lowland area, reducing its ability to carry debris.

Drumlins tend to occur in large groups or swarms, forming a so-called 'basket of eggs' topography. Good examples can be seen in central Ireland, New York State and in the hills of Elslack, North Yorkshire (Figure 2.17).

Figure 2.17 Drumlin, Hills of Elslack, North Yorkshire

A tape measure could be used to record the length of the long axis of the drumlin and the width at its widest point. A common measure of their shape is the elongation ratio, which is the maximum drumlin length divided by maximum width. Typical elongation ratios are 2:1 to 7:1.

Two other components of the shape of the drumlin which are straightforward to measure are the angle of the stoss side (steep) and the angle of the lee side (less steep).

Two ranging poles, a clinometer and a tape measure could be used to record data for either side.

The ground distance between the highest point of one drumlin and the next nearest drumlin could be measured (in the field or from a map) to determine their distribution and to see if there is a pattern.

A compass could be used to find the orientation of the longest axis of each drumlin.

Till sheets

A till sheet is formed when a large mass of unstratified drift is deposited at the end of a period of ice sheet advance, which smooths the underlying surface. They may not be very conspicuous in terms of relief, but they are significant landforms because of their extent. The till itself is variable in composition, depending greatly on the nature of the rocks over which the ice has moved. If there is a high clay content, compaction by the weight of the overlying ice sheet may lead to relatively hard deposit.

In East Anglia the till is quite chalky in content, due to the rocks over which the ice passed. It is typically 30–50 m deep although it can be up to 75 m in places. It was formed by several different ice sheet advances between 480,000 and 425,000 years ago.

In Minnesota, USA, extensive till sheets were formed by deposition during the retreat of the Laurentide ice sheet.

✓ Review questions

1 Distinguish between physical and chemical weathering.
2 How is the material transported by glaciers classified?
3 What are the key characteristics of glacial till?
4 Explain how ice shapes corries.
5 How do recessional moraines and terminal moraines differ in their location in a glacial valley?
6 Describe two possible ways in which drumlins form.

Key idea
→ Glacial landforms are interrelated and together make up characteristic landscapes

The Lake District is a dramatic upland landscape in Cumbria in northwest England. It owes much of its character to glaciation over the last million years.

Geology

Three main groups of rocks are found in the Lake District: the Skiddaw Slates, the Borrowdale Volcanics and the Windermere series. The presence of these contrasting geologies has influenced the way in which the landscape has evolved over geological history.

The rocks in the Skiddaw Group are the oldest in the Lake District. They were formed as black muds and sands settling on the sea bed about 500 million years ago. They have since been raised up and folded by tectonic forces. These rocks are found mainly in the north and the mountains they form are typically smooth, with many streams occupying deep gorges.

Rocks of the Borrowdale Volcanic Group are found in the central Lake District and consist of very hard lava and ash formed in major eruptions about 450 million years ago that have withstood erosion and make up the highest mountains, such as Scafell, Helvellyn and Great Gable.

The Windermere Group are sedimentary mudstones, sandstones, siltstones and some limestone formed in the sea about 420 million years ago. These were later folded and faulted, pushed up, and eroded down to their present levels forming the gentler scenery of the southern part of the Lake District around Morecambe Bay.

The Lake District also contains two other significant, individual geologies. Huge masses of granite were intruded about 400 million years ago deep below the Lake District. Erosion has revealed outcrops in Eskdale, Ennerdale and at Shap. Approximately 320 million years ago a tropical sea covered the Lake District. The shell and skeletal remains of huge numbers of small marine animals formed the carboniferous limestone which crops out at Whitbarrow, Yewbarrow and Scout Scar in the south.

Glaciation

Glaciation of the Lake District is extremely complex. There have been many glaciations (valley glaciation to ice sheets which have submerged the landscape) in the past 400,000 years or so. All have left their mark on the landscape (particularly erosional landforms). The present landscape of the Lake District is largely the result of glaciation during the Pleistocene period, during the last million years (Figure 2.18). Over twenty glaciations occurred during the Pleistocene. However, some of the depositional landforms seen today are the result of the most recent phase of glaciation, the Loch Lomond Stadial, which took place between 12,880 and 11,500 years ago. This was a brief episode of glacial re-advance in upland Britain.

Figure 2.18 Ice sheet cover over Britain 18,000 years ago

Activity

Describe the extent of the ice sheet cover shown in Figure 2.18.

Erosional landforms

The Helvellyn Range (Figure 2.19) is an 11 km long ridge over 600 m high with numerous glacial erosional landforms. The summit of Helvellyn itself, which lies at 950 m above sea level, is often referred to as an example of a pyramidal peak. However, it is not as sharply pointed as many such landforms, having been eroded during the coldest part of the glacial period when ice completely covered this area. It also lacks more than two corries on its flanks, hence has been less sharply eroded by the retreat of their back walls. Corrie glaciers formed and re-occupied corries from earlier glaciations, moving downslope under gravity into the nearby valleys, allowing valley glaciers to form. One such corrie, now containing Red Tarn, is separated from its southerly neighbour, Nethermost Cove, by the very narrow arête, Striding Edge.

Ice from Red Tarn flowed into the valley of Glenridding, forming a valley glacier large enough to create a small glacial trough. This in turn fed into a much larger valley glacier and glacial trough, today occupied by Ullswater. From here, ice was channelled out of the central Lake District in a northeasterly direction. Although Ullswater forms a typical ribbon lake, being long and narrow, its floor is quite irregular due to the presence of resistant bands of volcanic rock. Norfolk Island is a roche moutonnée formed on an outcrop of this rock in the middle of the lake.

On the west side of the Helvellyn Range, there is a series of truncated spurs and hanging valleys, formed as a glacier moved through the trough now occupied by Thirlmere. One of the hanging valleys now contains a stream, Helvellyn Gill, which has a series of small waterfalls as it drops 500 m in just over 2 km from Low Moss into Thirlmere. Walla Crag, to its north, is a very steep truncated spur with some near vertical, bare rock outcrops.

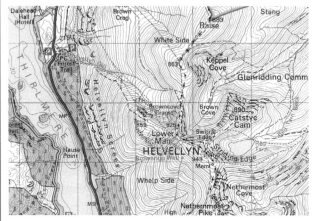

Figure 2.19 Helvellyn area

 Activity

Give the grid references for each of the erosional landforms mentioned above.

Depositional landforms

There are many drumlin fields in the Lake District, formed as ice moved radially in all directions – south to Lancashire, east into North Yorkshire and northwest to the Solway Firth. There are extensive drumlin fields just south of Kendal. The drumlins lie on carboniferous rocks and range from 50 to 125 m high; they have broad rounded tops and are frequently steep-sided. Some of the drumlins are rock-cored while others consist entirely of till. The orientation of the drumlins is NW/SW around Kendal and NNW/SSE near Furness, showing the southward movement of the ice. Fieldwork research in the area has shown that they have an average elongation ratio (long axis:short axis) of 3:1. They are rarely found above 300 m above sea level.

Many of the Lake District valleys contain moraines deposited during glacial retreat about 11,000 years ago. A good example of a lateral moraine can be found on the right bank of the Langstrath Valley, at the foot of Greenup Gill. It has retained its distinctive appearance partly due to a lack of mass movement processes on the valley side. A visible medial moraine runs down the centre of the Wythburn Valley. This was deposited during the final retreat of the valley glacier that had extended northwards down the valley from Dunmail Raise. A remarkable set of crescentic ridges of moraine are found at the end of Blea Water Tarn in Mardale.

These are recessional moraines, each one formed during a stationary period of ice retreat in the final deglaciation of the area. The Naddle valley, southeast of Keswick, has a small terminal moraine. Originally 200 to 400 m long and 10 m to 15 m high, a small lake existed behind it until about 1000 years ago. Eventually the Naddle Beck cut through it and today it is markedly eroded and degraded.

Erratics are also common in the Lake District. Some are from Scotland while others are more local in origin. A number of erratics from the Borrowdate Volcanics Group in the central part of the Lake District have been transported about 30 km southeast and deposited on carboniferous limestone at Witherslack. Although some are small, others are up to 3 m in diameter. The best known Lakeland erratics are those from Shap. They are found as far away as Cheshire, the Tees Valley and the North York Moors.

Fieldwork ideas

Well-annotated field sketches are a useful way to present information visually and to describe and explain landscape features and landforms. These could be drawn from personal field observation or using photographs. Good field sketches should include a grid-referenced location, orientation and scale. Annotations should provide detailed description and interpretation of the relevant landforms.

Geomorphic mapping allows landforms and features to be located on a base map, and information about the features to be added using geomorphic symbols. This can be used as a summary of how individual landforms interrelate to produce a distinctive landscape.

Activities

1 Explain why Helvellyn does not have a high, sharp peak.
2 Why are most Lake District corries generally oriented in a north to easterly direction?
3 Explain why morainic features are sometimes difficult to distinguish in the Lake District.
4 Why is the floor of Ullswater so uneven?
5 Explain the formation of the morainic ridges in Mardale.

Stretch and challenge

The Bowder Stone in Borrowdale is frequently given as an example of a glacial erratic. However, doubt has recently been expressed over its origin. Research this issue and consider the evidence.

The present landscape of Minnesota (Figure 2.20) resulted largely from glacial activity during the Quaternary period (2 million years ago to the present). Minnesota saw the advance and retreat of several major, successive periods of continental ice sheet activity. The gigantic Laurentide ice sheet (see Figure 2.15), centred in what is now the Hudson Bay, grew and retreated many times with climatic changes throughout the Quaternary. During colder periods, it extended southward across the upper Midwest.

Geology

Although situated in northern USA, the landscape of Minnesota is part of the Laurentian (or Canadian) Shield. Minnesota's oldest rocks lie in alternating belts in the northern half of the state and much of the Minnesota River Valley. The belts are of volcanic and sedimentary rocks; granitic rock materials lie in the areas between the belts.

Metamorphic gneiss crops out along the Minnesota River Valley dating back 3600 million years. Volcanic and sedimentary rocks began their formation 2700 million years ago, when lava escaped through rifts in what was then the sea floor. Volcanic formations lie throughout Minnesota's portion of the Laurentian Shield, some buried deep beneath glacial deposits. Volcanic debris released into the nearby seas later settled on the sea floor, forming massive layers of sedimentary rock. During this period, tectonic activity folded many of these rock formations and formed faults. Many of the volcanic rocks have metamorphosed to greenstone. At the same time, tectonic compression created a range of mountains several kilometres high in northern Minnesota.

Glaciation

Around 75,000 years ago, a series of lobes or tongues of ice extended from the main Laurentide ice sheet and spread across Minnesota (Figure 2.21). These lobes advanced and retreated a number of times, transporting and depositing till across a wide area. The different origins of these lobes resulted in tills with different characteristics and materials.

Erosional impact

The Laurentide ice sheet, over 1 km thick in places, and its lobes had a massive erosional impact on the landscape. The high mountains were worn down, such that today the highest peaks are now only a modest 500–700 m. A large, ellipsoidal basin was created by this erosion and is now studded with thousands of lakes, such as Upper and Lower Red Lakes in northern Minnesota. In the Arrowhead region of the northeast the erosional basin was particularly deep as the earlier tectonic tilting of the landscape exposed weak shale rocks which were eroded much more rapidly

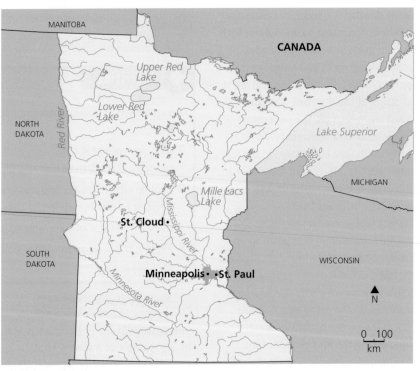

Figure 2.20 Minnesota

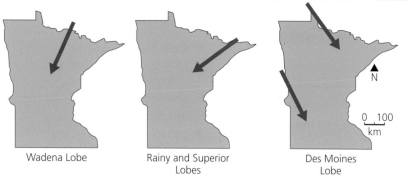

Figure 2.21 Minnesota's most recent glacial lobes (adapted from Minnesota Department of Natural Resources). Arrows show the direction of glacial lobe advance

than the resistant volcanic rocks around them. Thus the lakes of this region lie in the deeply eroded shales.

As the lobes of ice advanced they abraded striations in bare rock outcrops of gneiss and greenstone, their alignment indicating the direction of ice advance.

The far southeast of the state was not extensively covered by the ice sheet and so retains a more varied landscape of steep hills and deep valleys. Most of the rivers draining this area are tributaries of the Minnesota and Mississippi Rivers.

The erosional impact of the Laurentide ice sheet was therefore considerable and shaped the overall landscape. However, continental ice sheet erosion does not produce the spectacular landforms associated with valley glaciers and alpine glaciation.

Depositional impact

The Wadena Lobe advanced from northeast Canada and reached just south of Minneapolis. The till deposited by this lobe is characteristically red and sandy, being derived from the red sandstone and shales to the north and northeast. The Wadena Lobe first deposited the Alexandria moraine, formed the drumlin fields spanning Otter Tail, Wadena and Todd counties, and finally formed the Itasca moraine.

Ground moraine with reddish iron-rich sediments extends from St Cloud northeastward. The glaciers produced formed a set of terminal moraines which extend from northwest of St Cloud into the Twin Cities (Minneapolis and St Paul). The last advance of the Rainy and Superior Lobes left a coarse-textured till containing abundant fragments of basalts, gabbro, granite, red sandstone, slate and greenstone strewn across the northeastern half of Minnesota and as far south as the Twin Cities.

The Des Moines Lobe deposited till that is coloured tan to buff and is clay-rich and calcareous because of shale and limestone rocks at its source to the northwest. In the southwest, Prairie Coteau has a fine example of an end moraine.

Many of the till deposits in the west of the state have been found by borehole drilling to be more than 100 m thick. In the southwest, boreholes 160 m deep still had not reached bedrock.

Proglacial lakes

The edge of the giant ice sheet and its associated lobes also dammed the natural drainage of the area. This created a number of **proglacial lakes**. The largest of these was Lake Agassiz, a small part of which occupied the present Red River valley of Minnesota and North Dakota. Glaciers to the north blocked the natural northward drainage of the area. As the ice melted, a proglacial lake developed south of the ice. At its maximum this lake covered 440,000 km^2 (a similar size to the present day Black Sea). The water overflowed the watershed at Brown's Valley, Minnesota, drained through the Traverse Gap and cut the present Minnesota River valley. The amount of discharge was staggering. It helped the adjacent Mississippi River to form a very large valley in southeastern Minnesota.

The river that drained from Lake Agassiz is called the Glacial River Warren. It flowed over the top of a recessional moraine at Brown's Valley. When the lake finally drained it left behind fertile silt deposits producing the rich farmland of the Red River valley.

Activities

1 What was the main erosional impact of the Laurentide ice sheet on Minnesota?
2 Why does the southeast of Minnesota have an irregular landscape?
3 Explain why till deposits in Minnesota have different colours and materials.
4 How was Lake Agassiz formed?

2.3 How do glacial landforms evolve over time as climate changes?

> **Key idea**
> → Glacio-fluvial landforms exist as a result of climate change at the end of glacial periods

Glacio-fluvial landforms are those produced by meltwater from glaciers. They can include both erosional and depositional landforms, although only the latter are considered here. Meltwater can be released from glaciers during short, often seasonal periods of melting but mostly during deglaciation. It is at these times that most glacio-fluvial landforms visible in the contemporary landscape were formed. Those produced during seasonal melting and retreat are often severely modified or degraded during periods of subsequent re-advance.

Post-glacial climate change and its effect on geomorphic processes

Glacial periods end when global temperatures rise. They are followed by shorter inter-glacials lasting from 10,000 to 15,000 years. In the post-glacial period temperatures often increase gradually, with many fluctuations as part of a general warming trend (Figure 2.22).

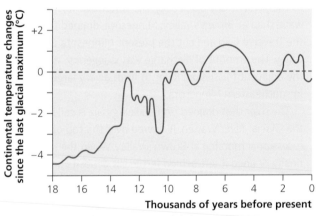

Figure 2.22 Post-glacial change in global temperature

> **Activity**
>
> Describe the pattern of change in global temperature over the past 18,000 years.

Glacio-fluvial streams and rivers deposit a distinctive sediment type known as **outwash**. In contrast to till, deposited directly by glacial ice, outwash tends to be:

- generally smaller as meltwater streams typically have less energy than glaciers and so only carry finer material
- smooth and rounded – by contact with water and by attrition
- sorted – horizontally, with the largest material found furthest up the valley and progressively finer material with distance down the valley due to the sequential nature of the deposition mechanism
- stratified – vertically, with distinctive seasonal and annual layers of sediment accumulation in many of the landforms.

> **Fieldwork ideas**
>
> Sediment samples could be taken from areas of outwash and the shape and size of the particles measured. The three axes of each particle could be measured with a ruler. The shape could be described using Power's Index of Roundness, the Cailleux Index, Zingg's classification or Krumbein's Index of Sphericity.
>
> The results could be compared with those from an area of till to see if they are significantly different, glacio-fluvial deposits typically being more rounded than till deposits.

This description is a simplification of reality, however. Meltwater rivers may have extremely high discharge in summer months enabling them to carry very large cobbles and even boulders. The Glacier d'Argentière in France, for instance, has a mean winter discharge of 0.1–1.5 cumecs, but a summer discharge of 10–11 cumecs. In Iceland, **jökulhaulps** are extreme glacial outbursts caused by geothermal or volcanic activity beneath glaciers that cause massive and sudden melting. The 2010 eruption of a volcano under the Eyjafjallajökull ice cap led to a surge of water through the Markarfljót valley with an estimated discharge of over 2500 cumecs. More than 200,000 tonnes of sediment were deposited in the valley.

> **Stretch and challenge**
>
> A distinction may be made between outwash material which is carried relatively long distances (possibly well beyond the snout and any terminal moraine), and becomes very smooth, rounded and highly sorted, and **ice-contact drift**, which is deposited under or against the ice and tends to be more sub-rounded and less well sorted.

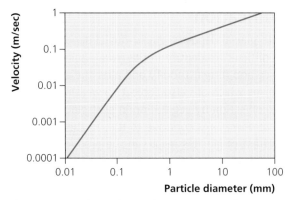

Figure 2.23 Settling velocity of sediment particles

> **Activities**
>
> 1 Describe the relationship between particle size and settling velocity shown in Figure 2.23.
> 2 Suggest reasons for this relationship.
> 3 How does this model relate to the sorting of glacio-fluvial sediments in the pro-glacial area?

Landforms of glacio-fluvial deposition

Deposition of outwash results in the formation of distinctive glacio-fluvial landforms: kames, eskers and outwash plains.

Kames

A kame is a hill or hummock composed of stratified sand and gravel laid down by glacial meltwater. There are two types of kame.

Delta kames form in different ways. Some are formed by en-glacial streams emerging at the snout of the glacier. They lose energy at the base of the glacier and deposit their load. Others are the result of supraglacial streams depositing material on entering ice-marginal lakes, losing energy as they enter the static body of water. Some also form as debris-filled crevasses collapse during ice retreat. Kames are widespread in East Lothian, Scotland.

Kame terraces are ridges of material running along the edge of the valley floor. Supraglacial streams on the edge of the glacier pick up and carry lateral moraine which is later deposited on the valley floor as the glacier retreats (Figure 2.24). The streams form due to the melting of ice warmed in contact with the valley sides as a result of friction and the heat-retaining properties of the valley-side slopes. Although they may look similar to lateral moraines, unlike moraines they are composed of glacio-fluvial deposits that are more rounded and sorted. In the Kingsdale valley of the

Yorkshire Dales, a kame terrace extends for about 2 km along the north side of the valley. It is approximately 2 m high for most of its length.

Eskers

An **esker** is long, sinuous ridge composed of stratified sand and gravel laid down by glacial meltwater. Material is deposited in sub-glacial tunnels as the supply of meltwater decreases at the end of the glacial period. Sub-glacial streams may carry huge amounts of debris under pressure in confined tunnels at the base of the glacier. Some scientists argue that deposition occurs when the pressure is released and meltwater emerges at the glacier snout. As the glacier snout retreats, the point of deposition will gradually move backwards. This may explain why some eskers are beaded – the ridge showing significant variations in height and width – with the beads of greater size representing periods when the rate of retreat slowed or halted. However, others argue that the beads are simply the result of the greater load carried by summer meltwater (Figure 2.24). The Trim esker near Dublin is one of a group of twelve in the area. It is 14.5 km long and between 4 and 15 m high.

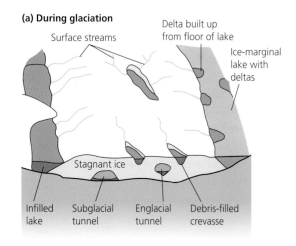

(a) During glaciation

Surface streams

Delta built up from floor of lake

Ice-marginal lake with deltas

Stagnant ice

Infilled lake

Subglacial tunnel

Englacial tunnel

Debris-filled crevasse

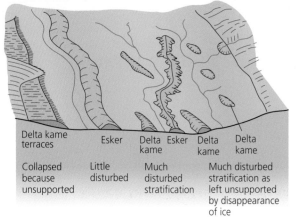

(b) After glaciation

Delta kame terraces

Esker

Delta kame

Esker

Delta kame

Delta kame

Collapsed because unsupported

Little disturbed

Much disturbed stratification

Much disturbed stratification as left unsupported by disappearance of ice

Figure 2.24 Formation of kames and eskers

61

Stretch and challenge

Research **hydrostatic pressure** and consider its role in esker formation.

Outwash plains

An **outwash plain** (also known as a sandur) is a flat expanse of sediment in the pro-glacial area. As meltwater streams gradually lose energy as they enter lowland areas beyond the ice front, they deposit their load. The largest material is deposited nearest the ice front and the finest further away. Outwash plains are typically drained by braided streams. These are river channels subdivided by numerous islets and channels.

Debris-laden braided streams lose water at the end of the melting period and so carry less material. This material is deposited in the channel, causing it to divide. Braiding begins with a mid-channel bar which grows downstream. Discharge decreases after a flood or a period of snow melt, causing the coarsest particles in the load to be deposited first. As discharge continues to decrease, finer material is then added to the bar, increasing its size. When exposed at times of low discharge, channel bars are stabilised by vegetation and become more permanent features. The river divides around the island and then re-joins. Unvegetated bars lack stability and often move, form and reform with successive flood or high-discharge events. They are very common in outwash areas due to the seasonal fluctuations in the discharge.

On the south coast of Iceland is an extensive sandur (Figure 2.25) fed by numerous meltwater streams from glaciers such as Gigjökull and Sòlheimajökull. From the edge of the upland area to the present position of the sea is a distance of some 5 km.

Modification

As with glacial deposits, glacio-fluvial deposits are often difficult to identify in the field. Again, repeated advance and retreat modify and alter the appearance of landforms which are also subject to weathering, erosion and colonisation by vegetation in post-glacial times.

As temperatures continue to rise, further melting and retreat of glaciers results in the production of more meltwater and thus a greater expanse and accumulation of outwash material in the pro-glacial zone. Kames and eskers will be exposed in greater number and of greater length during this continued retreat. Sòlheimajökull (Figure 2.26) is presently retreating about 100 m/year, and within a couple of decades (given the present rate of retreat continuing) a pro-glacial lake will probably form at its snout.

As temperatures increase, so does the growing season for vegetation. Exposed outwash material tends to become colonised over time, first by mosses and lichens and then by grasses, flowering plants and shrubs.

Figure 2.26 Pro-glacial outwash about 300 m in front of the retreating Sòlheimajökull, now colonised by vegetation

✓ Review questions

1 When do glaciers mainly produce meltwater?
2 What are the key characteristics of outwash material?
3 Distinguish between delta kames and kame terraces.
4 Explain the formation of eskers.
5 Why are glacio-fluvial landforms often difficult to recognise in a post-glacial environment?

Key idea
→ Periglacial landforms exist as a result of climate change before and/or after glacial periods

Figure 2.25 Sandur in Þórsmörk valley below Gigjökull, southern Iceland

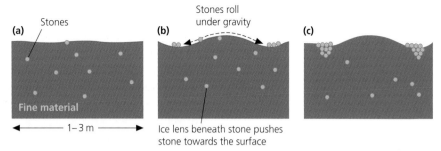

Figure 2.27 Frost heave

Periglacial environments have traditionally been referred to as being 'at or near ice sheets'. However, they are more accurately defined as areas with:

- permafrost (perennially frozen ground overlain by an **active layer**)
- seasonal temperature variations (above zero in 'summer', albeit for a short period)
- freeze-thaw cycles dominating geomorphic processes (**frost heave**, freeze-thaw).

Periglacial environments are found in high latitude areas (e.g. Alaska and northern Canada), continental interiors (e.g. Siberia) and in high mountains at lower latitudes (e.g. Plateau of Tibet, Andes and Alps). They make up 25 per cent of the Earth's land surface and it is estimated that another 25 per cent has experienced periglacial conditions in the past. Periglacial conditions dominated southern England during the most recent glacial period. At this time this part of Britain was ice-free, but experienced severe winters.

Climate change and the effects on geomorphic processes

It is difficult to establish the precise climatic conditions in southern Britain during the last glacial period. However, evidence from contemporary periglacial environments suggests that mean annual temperatures would have been less than −6°C with winter minimums much lower and only 10–30 days of summer temperatures above 0°C.

Freeze-thaw weathering is a dominant process in periglacial environments. This is due to seasonal fluctuations in temperature around freezing. In truly glacial climates (e.g. Antarctica) temperatures are invariably below freezing throughout the year.

Frost heave (Figure 2.27) is a sub-surface process that leads to a vertical sorting of material in the active layer. Stones within fine material heat up and cool down faster than their surroundings as they have a lower specific heat capacity. As temperatures fall, water beneath the stones freezes and expands, pushing the stones upwards to the surface. Ground ice also pushes overlying, finer material upwards, producing a domed surface.

The development of ground ice is also an important process. During summer melting periods, water percolates into the sub-surface geology where it accumulates below the water table. During the sub-zero winter months this water freezes and expands by between 9 and 10 per cent of its volume. As this expansion occurs, so the ground surface is pushed upwards, as it is unable to extend downwards into the permafrost below.

Periglacial landforms

The landforms of periglacial areas are very varied. Periglacial landforms are a feature of current periglacial environments, but they are also fossil features, widespread in more temperate regions today (e.g. northern Britain).

Patterned ground

Patterned ground is the collective term for a number of fairly small-scale features of periglacial environments. We have seen that as a result of frost-heave, large stones eventually reach the surface and that the ground surface is domed. The stones then move radially, under gravity, down each domed surface to form a network of stone polygons (Figure 2.28), typically

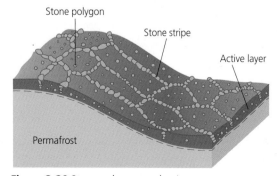

Figure 2.28 Stone polygons and stripes

1–2 m in diameter. A particularly distinctive example of patterned ground can be seen in the area around Barrow in Alaska. On slope angles of 3–50°, the larger stones move greater distances downslope and the polygons become elongated into **stone garlands**. On slopes of 60° and over, the polygons lose their shape and stone stripes develop.

Pingos

Pingos (Figure 2.29) are rounded ice-cored hills that can be as much as 90 m in height and 800 m in diameter. They grow at rates of a couple of cm/year.

Figure 2.29 Pingo in northern Canada

They are essentially formed by ground ice which develops during the winter months as temperatures fall. There are two types that are recognised (Figure 2.30):

- Open-system pingos form in valley bottoms where water from the surrounding slopes collects under gravity, freezes and expands under **artesian pressure**. The overlying surface material is forced to dome upwards. This type is common in east Greenland.
- Closed-system pingos develop beneath lake beds where the supply of water is from the immediate local area. As permafrost grows during cold periods, groundwater beneath a lake is trapped by the permafrost below and the frozen lake above. The saturated **talik**, or unfrozen ground, is compressed by the expanding ice around it and is under hydrostatic pressure. When the talik itself eventually freezes it forces up the overlying sediments. Over 1400 pingos of this type are found in the Mackenzie delta of Canada.

Modification of landforms

Patterned ground is a relatively minor and small-scale feature. As temperatures rose at the end of the periglacial period, patterned ground was often colonised by vegetation, making it hard to find and identify. Over time mass movement by creep also degrades the

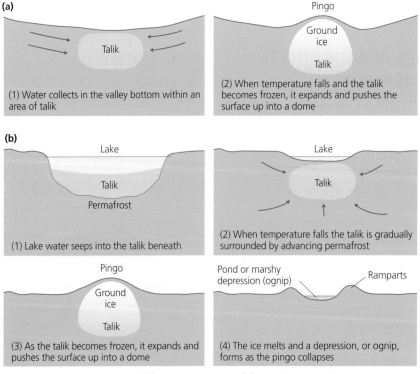

(a)

(1) Water collects in the valley bottom within an area of talik

(2) When temperature falls and the talik becomes frozen, it expands and pushes the surface up into a dome

(b)

(1) Lake water seeps into the talik beneath

(2) When temperature falls the talik is gradually surrounded by advancing permafrost

(3) As the talik becomes frozen, it expands and pushes the surface up into a dome

(4) The ice melts and a depression, or ognip, forms as the pingo collapses

Figure 2.30 The formation of (a) open-system and (b) closed-system pingos

frost-heaved domes, making the landform less obvious. Patterned ground around Leedon Tor, Dartmoor is now mainly covered by a layer of soil and grasses.

Pingos collapse when temperatures rise and the ice core thaws. When this happens the top of the dome collapses leaving a rampart surrounding a circular depression called an **ognip**. Relict ognips can be found in Britain, although due to the thawing of the permafrost only the remains of the rampart may be seen. A good example, about 15 m in diameter, can be seen at Llanberis in north Wales.

> ### ✔ Review questions
>
> 1 What are the climatic characteristics of periglacial environments?
> 2 Explain the process of frost heave.
> 3 How is patterned ground formed?
> 4 Distinguish between open-system and closed-system pingos.
> 5 How does a warming climate lead to the formation of ognips?

2.4 How does human activity cause change within glaciated and periglacial landscape systems?

Many glaciated and periglacial landscapes have opportunities for human activity. This includes the presence of raw materials, attractions for tourism and the potential for hydroelectric power. The socio-economic benefits of taking these opportunities can exceed the costs of overcoming the challenges involved. However, human activity on any significant scale can have major impacts on the often delicately balanced landscape systems in these environments.

> ### Key idea
> → Human activity causes change within periglacial landscape systems

🔍 Case study: Oil extraction in Alaska

Alaska has huge oilfields, including those around Prudhoe Bay on the north coast, with proven reserves of about 3000 million barrels. The 1300 km Alyeska pipeline, which runs from Prudhoe Bay to the ice-free port of Valdez on the south coast (Figure 2.31), transports oil up to 1.4 million barrels per day.

The need for oil

In 2014 the USA consumed 6.95 billion barrels of oil products. After a number of years of falling consumption during the global economic recession, since 2013 demand has risen again. About 40 per cent of the supply comes from imports and this is a concern for the US government, not only in terms of balance of trade, but because of the political implications of not having **energy security**.

Reserves

In order to meet as much of the demand from domestic sources as possible, exploration of potential oil fields in Alaska has been permitted by the government and it is estimated (with 95 per cent probability) that Area 1002 in the Arctic National Wildlife Refuge contains over 16 billion barrels, of which about 6 billion could be

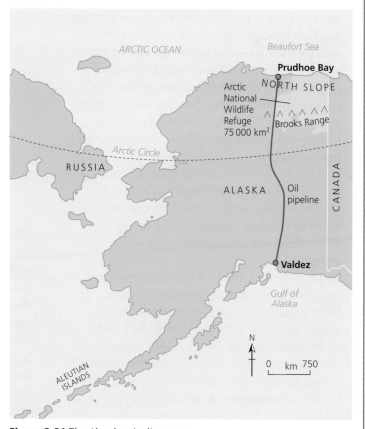

Figure 2.31 The Alyeska pipeline route

extracted using currently available technology. Extracting oil from Area 1002 means that there will be increased employment opportunities and that the existing Alyeska pipeline can be used.

The impact on the periglacial landscape system

Extraction and transportation of oil can impact upon the processes, flows of material and energy through the periglacial landscape system.

One way in which material flows are affected is through the use of gravel pads. Gravel is extracted from stream and river beds and used as an insulating base layer for road construction. The loss of gravel from the river systems alters the rate at which gravel is transported and deposited further downstream. It can also affect the equilibrium between erosional and depositional processes in the river system.

Hydrological processes are also affected. A report into gravel extraction from a glacial outwash aquifer near Palmer found that ground water levels fell by more than a metre in an area extending over 2 km from the extraction site.

Energy flows are affected in two ways:

- By the release and burning of gas during drilling. Some gases are burnt in a process called flaring, which releases mainly carbon dioxide into the atmosphere. Others, including methane, are vented without burning into the atmosphere. Both carbon dioxide and methane are significant greenhouse gases and can contribute to an enhanced greenhouse effect (see Chapter 10, Climate change, page 290), with higher levels of terrestrial radiation being trapped in the lower atmosphere, raising temperatures.
- By the production of heat from the extraction and transportation processes, as well as from the associated infrastructure. An investigation into the **urban heat island** in the small town of Barrow, Alaska found that mean temperatures were on average 2.2°C higher than in the surrounding rural area. A maximum difference of 6°C was measured on a particularly calm day. Heat from domestic heating systems in poorly insulated buildings is a major contributor to the heat island effect, and a strong correlation was found between temperature differences and oil production rates in the nearby oilfield. Energy released to the environment by human activities also affects geomorphic processes, with 9 per cent fewer days of temperature fluctuations around 0°C (i.e. freeze-thaw cycles) recorded since drilling began.

Changing landforms

Permafrost is perennially frozen ground; in other words it remains frozen from one year to the next, despite average temperatures rising above zero in the summer. Permafrost is typically overlain by a shallow active layer of surface material, much of it unconsolidated, which thaws in the summer but freezes again during the following winter. The permafrost experiences less variation in temperature than the active layer because it is not directly exposed to the seasonal differences in air temperature.

The heat released by buildings and infrastructure can lead to the thawing of permafrost, and a longer period of melting of the active layer. If the building itself is constructed directly onto the ground surface, some of the heat produced by heating systems may be transferred through the floor to the ground, melting the permafrost. This can result in subsidence and increase the mobility of the active layer, allowing a type of mass movement called **solifluction** to take place. The downslope movement of the thawed active layer results in the formation of solifluction lobes (Figure 2.32), tongues of debris, at the base of slopes when the moving material loses energy on a lower gradient.

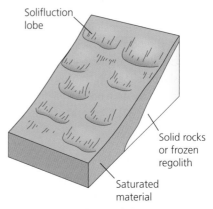

Figure 2.32 Solifluction lobes

Thermokarst

Thermokarst is a landscape dominated by surface depressions due to the thawing of ground ice. It is typified by extensive areas of hummocky ground interspersed with waterlogged hollows. Depressions may fill with water to form shallow thaw lakes, usually less than 5 m deep and 1–2 km wide. On a larger scale, **alases** are flat-floored, steep-sided depressions ranging from 5 to 50 m in depth and 100 m to 15 km in length. They develop from widespread thawing of ground ice causing large-scale subsidence. Again, these depressions may contain lakes. Where several alases combine, alas valleys may form, which can be many tens of kilometres in length.

The thawing of the ground ice can be initiated by climate change, but also by more direct human interference in the landscape system. The removal of vegetation for resource extraction or construction purposes decreases the insulation of ground ice. This leads to it thawing to greater depths, creating a thicker active layer and producing much deeper and more extensive subsidence.

On the North Slope of Alaska, research into thermokarst depressions (Figure 2.33) found subsidence rates averaging 3–4 cm per year, with a maximum rate of 12 cm/year. In this area the permafrost is 300–600 m deep and the active layer 30–80 cm thick.

Figure 2.33 Thermokarst, North Slope, Alaska

Stretch and challenge

1 Research into the impact of thawing permafrost on climate change.
   This is a good example of a positive feedback mechanism.
2 Explain the relationship between permafrost thawing and positive feedback.
   The National Snow and Ice Center has good information on its website: https://nsidc.org/cryosphere/frozenground/methane.html

Activities

1 Why was the US government so keen to extract oil from Alaska?
2 Explain how the oil industry contributes to the urban heat island effect in Barrow.
3 How can human activity be linked to the formation of solifluction lobes?
4 Explain the formation of thermokarst.

Key idea

→ Human activity causes change within glaciated landscape systems

Case study: Grande Dixence Scheme, Switzerland

Situated at the head of the Val des Dix in southwest Switzerland, Grande Dixence is the highest gravity dam in the world. It was initially constructed in the 1960s to provide Switzerland with hydroelectric power and has undergone numerous additions and alterations, most recently in 2010. Construction was a lengthy process due to both the inaccessibility of the site and the severe working conditions, which meant that construction could only take place in the summer months – it was too cold in the winter for the concrete to set. The total costs at the time were approximately 1600 million Swiss Francs.

The dam

The dam itself is 285 metres high and each year it stores over 400 million m³ of water. The dam is 200 metres wide at its base and just 15 metres wide at the top. To make the foundation soil watertight there is a deep grout curtain, which surrounds the dam extending onto each side of the valley. Aggregates for the dam were obtained locally from deposits of moraines in adjacent valleys. It has a catchment area of just over 350 km², half of which is from 35 glaciers that provide seasonal meltwater. The Lac des Dix was created behind the dam and four pumping stations send water through 100 km of tunnels into the reservoir. One of its key aims is to optimise the water level so that there is maximum availability before heavy demand periods. It is able to maximise profits by offsetting the cost of pumping water into the Lac des Dix in the summer against income from generating energy during the winter.

Figure 2.34 Grande Dixence Dam

Energy

The water stored behind the Grande Dixence dam drives the turbines in four power stations, including those at Fionnay and Nendaz (Figure 2.35), with a combined capacity of 2000 GWh annually – enough to power 400,000 Swiss households. The Grande Dixence operates by storing glacial meltwater during the summer and then using it to generate electricity during the high demand period of the winter. About one-third of Swiss electricity comes from storage power stations and another quarter from run-of-the-river schemes. The Swiss government has long had a strong environmentally aware energy policy. Less than 5 per cent of electricity production comes from fossil fuels, with nuclear providing the rest.

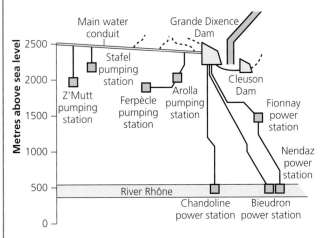

Figure 2.35 Schematic representation of the operation of the Grande Dixence Scheme

Impact on the environment

The environmental impact of the project has been minimised, partly to ensure that the area remains an attractive environment for walkers, cyclists and hikers. The pumping stations and power plants are largely built underground or are well concealed to retain the aesthetics of the location. Indeed, tourism to the locality has been enhanced and there are now guided tours and helicopter rides above the dam available to visitors.

However, the reduced flow in the Borgne River, a tributary of the Rhône, below the dam has resulted in higher concentrations of pollutants at Les Haudère, from both agricultural and domestic sources.

Impact on the glacial system

Of the water available at Grande Dixence, 85 per cent is used mainly for electricity generation (a small amount is also used to fulfil a demand for electricity caused by tourism in summer). The other 15 per cent is used to deal with the problems of sedimentation. When water is stored behind the dam, the lack of flow means a loss in energy and the deposition of sediment load behind the dam at rates of 20–40 cm/year. Sediment concentrations are >300 mg/l above the dam, 20–50 mg/l just below the dam, and <20 mg/l 3 km downstream of the dam.

To solve this problem, some of the water in the reservoir is used to purge the sediment – flushing it out and moving it downstream. At these times the water has high levels of **turbidity** and sediment concentrations of up to 20,000 mg/l.

Impact on river channels

The trapping of sediment behind the dam leads to very clear water being returned into the natural river channels below the power stations. This has excess energy as none is being used to transport sediment and results in increased channel erosion.

The lack of discharge in the below-dam rivers means that some virtually dry up in the summer. Meanwhile there has been significant contraction in the size of the channels, and the scale of contraction increases with distance downstream. The amount of sediment eventually flowing into Lake Geneva has halved since the construction of the dam.

In the Val d'Hérens, however, there is a risk of sudden and unexpected flooding when excess stored water has to be released. This has hindered both tourist use and development along the valley floor, although the local residents receive significant revenues from it and so are still strongly in favour of the scheme.

Activities

1 What is the purpose of the Grande Dixence Project?
2 How does the dam affect sediment concentrations in the river?
3 What impact has the dam had on river channels?
4 How is the problem of sedimentation in the reservoir managed?

Practice questions

A Level

Table 1 Estimated rates of glacial erosion at a location in the Rocky Mountains.

Site	Erosion rate (m/million years)
1	6.2
2	5.2
3	3.3
4	5.8
5	6.6
6	9.5
7	8.5
8	2.2
9	3.1

AS Level

Table 2 Estimated rates of glacial erosion at sites in the Rocky Mountains and Greenland

Site	Rocky Mountains Erosion rate (m/million years)	Greenland Erosion rate (m/million years)
1	6.2	1.3
2	5.2	1.4
3	3.3	0.7
4	5.8	1.1
5	6.6	0.2
6	9.5	0.4
7	8.5	1.3

1 a) Explain the influence of climatic and geomorphic processes on the formation of corries. 8 marks]

b) Study **Table 1**, which shows estimated rates of glacial erosion at a location in the Rocky Mountains.
   i) Calculate the mean erosion rate for the data in Table 1. [2 marks]
   ii) Explain why rates of glacial erosion vary spatially. [4 marks]

c) Distinguish between glacier movement by basal sliding and internal deformation. [3 marks]

d) *Assess the extent to which glaciated landscapes are influenced by present day processes rather than those of the past. [16 marks]

1 a) Study **Table 2**, which shows estimated rates of glacial erosion at sites in the Rocky Mountains and Greenland.
   i) Using evidence from Table 2, compare and contrast the two sets of data. [3 marks]
   ii) Suggest reasons for the differences between the two sets of data. [4 marks]

b) Explain the formation of corries. [8 marks]

c) *Assess the extent to which glaciated landscapes are influenced by present day processes rather than those of the past. [14 marks]

Chapter 3

Dryland landscapes

3.1 How can dryland landscapes be viewed as systems?

> **Key idea**
> → Dryland landscapes can be viewed as systems

Drylands are regions where average annual evapotranspiration is significantly higher than precipitation. As a result, for all or a large part of the year, the availability of soil moisture is limited and inhibits plant growth. According to the United Nations (UN), around 40 per cent of the land surface has an annual moisture deficit and can be classed as arid or semi-arid (Figure 3.1).

The components of dryland systems

Dryland landscapes can be viewed within a systems framework. Like other physical systems they consist of components linked by flows of energy and materials. The main components in dryland landscape systems (Figure 3.2) are:

- geology: rocks and their lithology and structure

- sediments: including boulders, screes, sand, silt and clay
- water: derived from precipitation (and condensation) and stored in rivers, lakes and aquifers
- climate: including precipitation, temperature, cloud cover and wind.

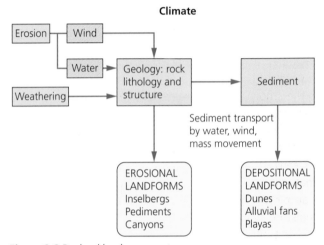

Figure 3.2 Dryland landscape system

Flows of energy and materials

The components of physical systems are linked by flows inputs, throughputs and outputs of energy and materials.

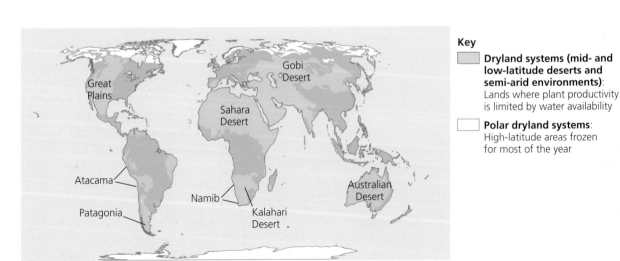

Figure 3.1 Distribution of the world's drylands

Energy and material flows

Solar radiation and precipitation are the main inputs to dryland landscape systems. They are balanced by outputs of heat (long-wave radiation from the surface), evapotranspiration and stream flow. The interaction of temperature and moisture with geology, through weathering and erosion, lead to rock disintegration and the formation of erosional landforms such as canyons, deflation hollows and pediments. Mineral materials from weathering and erosion are transported as sediments and eventually deposited by rivers and winds. Sediments are stored, often for millennia, as landforms such as screes, alluvial fans and dunes.

Aridity index

Defining drylands is complex and cannot be based on precipitation levels alone. A more accurate measure of aridity is the ratio of precipitation to evapotranspiration.

Because actual evapotranspiration is difficult to measure, calculations of aridity are based on the concept of **potential evapotranspiration (PET)**. This is the amount of evaporation that would occur if water were freely available for evaporation all year round. PET takes account of atmospheric humidity, solar radiation and wind. In drylands annual PET always exceeds annual precipitation. Therefore the moisture that could theoretically be lost through evaporation and transpiration is invariably greater than the amount of moisture available.

The United Nations Environment Programme (UNEP) aridity index is a simple way of measuring aridity:

$$AI = \frac{P}{PET}$$

where

AI = aridity index

P = mean annual precipitation

PET = mean annual potential evapotranspiration

Using this index, the UN identifies four categories of aridity: hyper-arid, arid, semi-arid and sub-humid (see Table 3.1). According to the UN classification 37.3 per cent of the Earth's surface is to some extent arid. If we include the sub-humid category, this proportion rises to 47.2 per cent (Table 3.1).

Table 3.1 UNEP classification of drylands using the aridity index (Source: UNEP)

Classification	Aridity index	% of global land surface
Hyper-arid	<0.05	7.5
Arid	0.05–0.20	12.1
Semi-arid	0.21–0.50	17.7
Sub-humid	>0.50	9.9

Table 3.2 Mean annual precipitation and mean annual potential evapotranspiration (PET)

	Mean annual precipitation (mm)	Mean annual PET (mm)
Luderitz (Namibia)	18	698
Cape Town (South Africa)	506	807
Wyndham (Australia)	703	1853
Adelaide (Australia)	288	884
Alice Springs (Australia)	252	1162
Las Vegas (USA)	460	2151

Activities

1 Calculate the aridity indices for the places in Table 3.2.
2 Using the aridity indices, classify each according to the aridity categories in Table 3.1.
3 Given your answers to Questions 1 and 2, do you think that the UNEP classifications in Table 3.1 are a reliable guide to aridity?

Key idea

→ Dryland landscapes are influenced by a range of physical factors

Potential influences on dryland landscape systems

Dryland landscapes are influenced by a range of factors including climate, geology, latitude and altitude, relief, aspect and the availability of sediment.

Climate

Most drylands have relatively small amounts of annual precipitation and all have annual values of PET which are substantially higher than precipitation. However, the thermal characteristics of drylands are more variable (see Table 3.3). Annual temperature ranges are high in continental interiors, even in the tropics. With the exception of coastal deserts like the Namib in southwest Africa (Walvis Bay) and the Atacama in Chile, temperatures are strongly influenced by latitude. Very high summer temperatures are recorded in the Sahel (Bilma), the Arabian peninsula (Baghdad) and the southwest USA (Las Vegas). Average maximum temperatures are high as far north as the Gobi Desert of central Asia (Ulaanbaatar), but decline steeply in polar regions to just 8 or 9°C.

Table 3.3 Climate characteristics of dryland environments

	Latitude	Mean max. temp. (°C) of warmest month	Mean min. temp. (°C) of coldest month	Total annual ppt (mm)
Bilma (Niger)	19°N	42	8	12
Walvis Bay (Namibia)	23°S	23	8	23
Alice Springs (Australia)	24°S	36	4	252
Baghdad (Iraq)	33°N	44	4	156
Las Vegas (USA)	33°N	41	1	106
Ulaanbaatar (Mongolia)	48°N	23	−26	216
Barrow Point (Alaska)	71°N	8	−31	134
Eureka (Canada)	80°N	9	−40	68

Weathering, the *in situ* breakdown of rocks by mechanical, chemical and biological processes, is largely determined by climate. The main processes are insolation weathering, salt weathering and freeze-thaw. These mechanical processes break rocks into smaller particles. Insolation and salt weathering dominate low- and mid-latitude drylands; in high latitudes mechanical weathering is confined to freeze-thaw. However, freeze-thaw is not uncommon in sub-tropical and mid-latitude drylands such as the American southwest and Namibia. In winter, under clear night skies, significant heat loss occurs and temperatures frequently dip below zero.

It is a paradox that despite low mean annual precipitation, flowing water is a significant geomorphic agent in many drylands. Although rainfall is often seasonal and erratic, in sub-tropical and mid-latitude drylands, it

Activities

Study the climate graphs for Las Vegas, Bilma, Barrow Point and Ulaanbaatar (Figures 3.3(a)–(d)).

1 Describe the patterns of temperature and precipitation for each place in Figures 3.3(a)–(d).
2 Explain how climate might influence the landscape at each location.

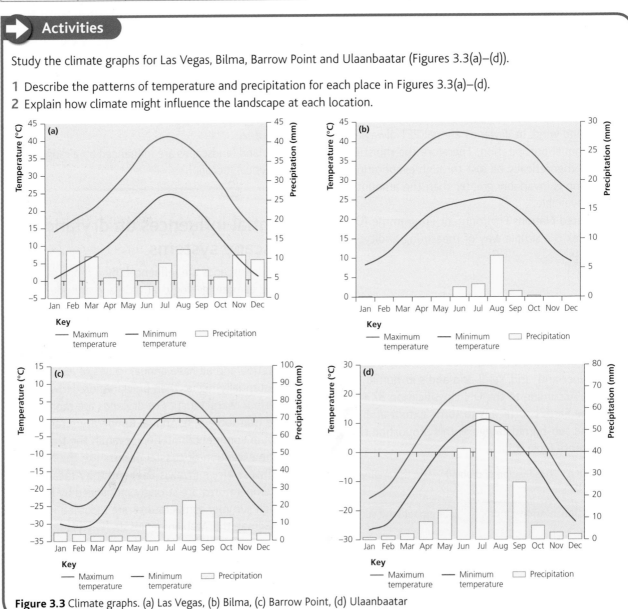

Figure 3.3 Climate graphs. (a) Las Vegas, (b) Bilma, (c) Barrow Point, (d) Ulaanbaatar

is invariably convective and highly intensive. Following heavy rainfall, ephemeral streams flow on the surface. They are short lived but have very high **peak discharges**, eroding the landscape and transporting huge volumes of sediment. Rapid run-off is enhanced by sparse vegetation cover, thin soils and ground baked hard by drought and high temperatures. In polar regions rivers only flow during the short summer. However, swollen by meltwater, for a few months they are also powerful agents of erosion and sediment transport.

In drylands wind is an active agent of landscape change, eroding rocks and transporting fine particles of sand, silt and clay. The sparse vegetation cover and dry conditions make wind action more effective in drylands than in any other environment.

Geology

Rock **lithology** and **structure** are primary controls on water transfer in drainage basins. Lithology describes the chemical composition of rocks; structure includes the physical characteristics such as jointing, bedding, folding and faulting. Both lithology and structure determine the **porosity** or **permeability** of rocks.

Dryland landscapes at a regional scale sometimes owe their character to the underlying geology as well as to climate. In the semi-arid Badlands of South Dakota in the US Midwest, water deficits are made worse by weak, poorly cemented silt and clay rocks. These rocks, which are highly impermeable, create rapid run-off and stream incision and extraordinary landscapes of canyons, ravines and gullies. In this unstable environment there is little soil development, vegetation growth and evapotranspiration (Figure 3.4). Badlands are widespread in other semi-arid regions such as the Loess Plateau in China and Almería in southern Spain.

Rocks with high levels of porosity and permeability also create drought at the surface. In central Iceland extensive plains of pumice derived from volcanic

Figure 3.4 Badlands topography in Death Valley, California

eruptions soak up heavy rainfall, leaving the surface dry and devoid of soil and vegetation. Limestone is also permeable because of its dense jointing. Precipitation quickly disappears underground leaving the surface dry and streamless. The Causses in southern France is typical: a high limestone plateau which, due to absence of water, supports only low quality grazing land.

Latitude and altitude

Latitude influences many characteristics of drylands. Tropical and sub-tropical drylands such as the Sahara, Arabian and Australian Deserts owe their existence to the atmosphere's circulation in low latitudes. These deserts lie on the poleward limb of the convective Hadley cell (Figure 3.5). Air rising near the Equator to the tropopause is then transferred polewards at heights of 10–15 km, and eventually sinks back to the surface between latitudes 20° and 30° in each hemisphere. Warmed by compression, the sinking air creates cloud-free conditions, permanent high pressure and extreme aridity at the surface. In continental interiors in Asia, North America and South America,

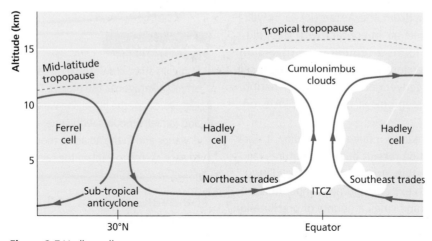

Figure 3.5 Hadley cell

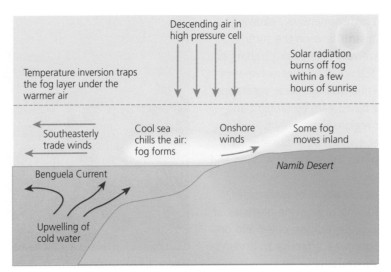

Figure 3.6 The influence of cold ocean currents on the Namib Desert

distance from the ocean and moisture-bearing air masses adds to aridity.

But the most obvious effect of latitude is to influence the amount and intensity of incident solar radiation. Thus high latitudes receive less solar radiation than low- and mid-latitudes, and thus average temperatures generally decrease polewards. Tropical, sub-tropical and even mid-latitude deserts experience intense solar radiation during summer, when the Sun is virtually overhead. Air temperatures may reach 50°C or more at this time. In contrast, polar deserts are in permanent darkness for part of the year, and even in summer the low angle of the Sun provides little heat. Low temperatures reduce humidity, and because the atmosphere contains little moisture, precipitation is sparse.

Some of the world's hyper-arid places are coastal deserts on the western side of continents, between latitudes 15° and 30°. Cold ocean currents which flow offshore play a crucial part in the formation of the Namib Desert in southwest Africa and the Atacama Desert in Chile/Peru. In Namibia the prevailing southeast **trade winds** push surface waters offshore and allow cold water (from the Antarctic) to upwell from depth. Meanwhile air in contact with the cold ocean is chilled and forms a **temperature inversion** (Figure 3.6). The inversion is spread onshore by local winds and prevents convection and cloud formation. This process has created one of the driest places on the planet.

Altitude has a more local influence on aridity. It does not affect solar radiation receipt but temperatures fall with altitude, and the thin atmosphere and cloudless skies often give rise to extreme diurnal temperature ranges, low humidity and small amounts of precipitation. The most extensive drylands at altitude are the Tibetan Plateau in central Asia and the Altiplano in Peru and

Bolivia. At an average 4500 m and 3500 m respectively, these areas are effectively highland deserts.

Relief and aspect

Several drylands such as Patagonia in South America, the Mojave Desert in the USA and the Tibetan Plateau have developed in the lee of mountain barriers. Patagonia is sheltered by the southern Andes, the Mojave Desert by the Sierra Nevada and the Tibetan Plateau by the Himalayas. While windward slopes invariably experience above-average precipitation, leeward slopes, facing downwind, are relatively dry. They lie in a so-called **rain shadow** (Figure 3.7). Generally the higher the mountain range the more pronounced the rain shadow effect. Two factors explain the rain shadow effect:

- As air masses are forced aloft by mountains, the air cools, water vapour condenses to form cloud and expends much of its moisture in the form of rain (the so-called orographic effect).
- Air descending on leeward slopes is warmed by compression which lowers its relative humidity and reduces the likelihood of rain.

Activity

Using an atlas, draw sketch maps to show the rain shadow effect in Patagonia, the Mojave Desert and the Tibetan Plateau. Show the prevailing wind direction with arrows and rainfall amounts with generalised isohyets both in the drylands and in the adjacent mountains. Add annotations to each sketch to explain the rainfall patterns.

The availability of sediment

Sediment comprises rock particles of varying size derived from weathering and erosion processes, transported

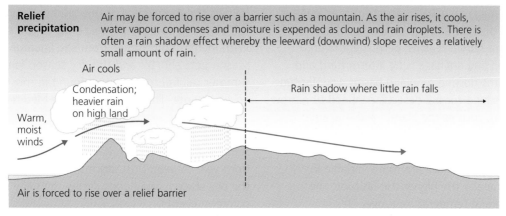

Relief precipitation | Air may be forced to rise over a barrier such as a mountain. As the air rises, it cools, water vapour condenses and moisture is expended as cloud and rain droplets. There is often a rain shadow effect whereby the leeward (downwind) slope receives a relatively small amount of rain.

Air cools

Condensation; heavier rain on high land

Warm, moist winds

Rain shadow where little rain falls

Air is forced to rise over a relief barrier

Figure 3.7 The rain shadow effect

and ultimately deposited by rivers, winds and mass movement processes. In some drylands sediments are scarce and the ground may be covered with tightly packed stones forming **desert pavements**. This rocky desert surface is known generally as **reg**. In fact, most deserts are rocky rather than sandy, suggesting that finer sediments often have been removed by wind action.

In drylands with variable relief, weathering in the mountains may provide an abundant supply of sediment. This sediment, transported by streams and rivers, is often deposited where rivers emerge from mountains onto adjacent plains. The result is **alluvial fans** and **bajadas**, which are a common feature in the deserts of the American southwest. Large sediment supplies, often derived from earlier wetter climatic phases, give rise to **sand seas**, which cover large parts of the Sahara Desert. The generic term to describe large expanses of sand desert is **erg**. Sand which forms the world's highest dunes in the Namib Desert has been transported by winds from the Orange River hundreds of kilometres to the south in South Africa. In polar drylands, sediments are largely derived from deposition by ice sheets and meltwater in the past.

> **Key idea**
> → There are different types of dryland

The characteristics of different types of dryland landscape

There are three main types of dryland landscapes: polar drylands, mid-and low latitude deserts and semi-arid environments (Figure 3.1).

Polar drylands

Polar drylands are concentrated north of the **tundra** in the high Arctic. This is a region of permafrost, free of ice cover. Almost lacking vegetation, it is essentially a cold desert (Figure 3.8). Barrow Point in Alaska

(Figure 3.3c) at latitude 71° north recorded an average precipitation of just 134 mm/year for the period 1995 to 2014. Lack of moisture in the atmosphere due to the intense cold is the main cause of low precipitation.

Figure 3.8 High Arctic, Ellesmere Island

For most of the year, water is frozen. During the brief Arctic summer, melting occurs, allowing streams to flow from snow patches. Where soil and regolith occur, the uppermost metre known as the **active layer** thaws in summer, creating mass movement known as **solifluction**. Freeze-thaw weathering is most active during spring and autumn when diurnal air temperatures fluctuate above and below freezing. Relief in the high Arctic in North America and Eurasia is generally subdued, much of it being formed of ancient pre-Cambrian rocks.

Mid- and low-latitude deserts

The world's great hot deserts straddle the tropics. The largest are the Sahara, Australian and Arabian Deserts. Although these deserts are predominantly rocky, in places vast sand sheets like the Rub' al Khali in Arabia cover the surface. In tropical deserts there is little seasonal pattern to rainfall, which is sparse, erratic and highly variable from year to year. In summer, surface temperatures may reach 80°C. Apart from exogenous rivers such as the Nile and the Colorado, which have sources in more humid regions, permanent rivers are absent, and any drainage is by intermittent and endoreic streams flowing to inland basins. Low-latitude deserts also include coastal deserts, confined to narrow coastal margins, such as the Namib and Atacama. There the proximity of the cool ocean moderates temperatures both in summer and winter.

Deserts in mid-latitudes are more variable in character. Generally the higher the latitude, the greater the annual temperature range. This is especially true of continental deserts such as the Gobi in central Asia with hot summers and very cold winters (Figure 3.3d). Rainfall is fairly evenly distributed but with a tendency towards a summer maximum. Smaller deserts in lower latitudes, such as the Mojave, Sonora and Thar, are excessively hot in summer but relatively mild in winter.

Semi-arid environments

Semi-arid environments, with an aridity index of 0.21 to 0.50, are widespread. They include the Mediterranean region; the Sahel, bordering the southern fringes of the Sahara Desert; much of central North America, including the Great Plains; large parts of eastern Australia; the Steppes in Southwest Asia; and Patagonia. Annual **potential evapotranspiration (PET)** exceeds precipitation in all of these areas but there is a seasonal pattern to rainfall, reliable enough to support agriculture and settlement. In Africa, convective rainfall occurs in summer and

follows the movement of the overhead Sun, while the period of low Sun corresponds to the dry season. With increasing distance from the Equator the rainy season gets shorter and the dry season lengthens. A summer rainfall maximum is found in most semi-arid environments. The exception is the Mediterranean, where the wet season occurs in winter.

Drought is a common and persistent feature of semi-arid environments. Rainfall amounts are variable from year to year. Dust storms and wildfires during the dry season, and floods during the rainy season, are common events.

Table 3.4 Mean annual precipitation at Tuscon, Elko and Barrow, 1995–2014 (mm)

Tuscon (32°N)	Elko (41°N)	Barrow (71°N)
387	330	81
279	295	74
415	353	131
284	283	99
179	272	121
318	188	146
164	194	132
238	206	89
227	277	131
294	224	183
246	369	139
261	382	116
232	146	62
151	197	120
273	297	153
132	195	135
382	318	144
163	159	146
194	167	218
366	267	203
258	239	194

Activity

Annual precipitation in dryland environments is generally more variable than in humid environments. In the humid northeast of the USA, Hartford in Connecticut had an average annual precipitation, 1995–2014, of 1202 mm. The inter-annual variability around the mean for this period was 17.7 per cent.

Compare this with precipitation data for three dryland locations in the USA (Table 3.4) between 1995 and 2014 – Tuscon (Arizona), Elko (Nevada), Barrow (Alaska) – by calculating the:
  a) mean annual precipitation
  b) standard deviation
  c) coefficient of variation.

Use either Microsoft Excel or Mac Numbers for your calculations and comment on your findings. You will find details on using the mean, standard deviation and coefficient of variation in Chapter 15, Geographical Skills, pp. 532–35.

Review questions

1 Explain the concept of potential evapotranspiration.
2 What is the aridity index?
3 Name two inputs and two outputs in dryland landscape systems.
4 What is:
  a) humidity
  b) a rain shadow
  c) a temperature inversion?
5 Name and briefly describe three types of dryland landscape.

3.2 How are landforms of mid- and low-latitude deserts developed?

> **Key idea**
> → Dryland landscapes develop due to a variety of interconnected climatic and geomorphic processes

Desert landforms develop through the interaction of processes such as weathering, mass movement, erosion, transport and deposition, with rock formations in dryland landscape systems.

Geomorphic processes

Weathering processes

Mechanical weathering

Weathering is the *in situ* breakdown of rocks by mechanical, chemical and biological processes. The dominant weathering processes in mid- and low-latitude deserts are mechanical, where changes in temperature and moisture cause rocks to break down into smaller particles. Mechanical breakdown is rapid because absence of soil and vegetation means that bedrock is widely exposed at the surface. Most types of weathering require moisture, and even the driest deserts receive some rain. In addition, moisture is often available from **dew** at night.

Rainwater and groundwater contain salts in solution. High temperatures draw saline groundwater to the surface and rainfall **percolates** into porous rocks. Evaporation of the water results in the growth of salt crystals, creating internal stress and rock breakdown (particularly surface layers). This process of **salt weathering** is a major cause of rock disintegration in desert areas, particularly in porous sedimentary rocks like sandstone.

Insolation weathering is due to intense solar heating of rocks at the surface. In hot deserts, surface temperatures can vary from 80°C by day, to below freezing at night. As a result, rock minerals expand and contract but at different rates. Providing some moisture is present, rocks are weakened and slowly break down.

Freeze-thaw weathering is common in winter in mid-latitude deserts, particularly where altitudes exceed 1500 m. Clear night skies, the relatively thin atmosphere and sparse vegetation cover cause significant heat loss. At sub-zero temperatures, water trapped in rock joints and pores freezes and expands (by 9 per cent of its volume) as it turns to ice. The resulting pressures break up even the most resistant rocks to form boulders and scree.

The outcome of mechanical weathering is rock particles of varying size and shape. Insolation and salt weathering often result in **granular disintegration** with rock particles no more than a few centimetres in diameter. Where rocks are massively jointed, weathering is focused on the joints (the main lines of weakness). The outcome is **block disintegration** with boulders weighing several tonnes. **Exfoliation** is the peeling of surface rock layers by insolation and salt weathering. Fluctuations in diurnal temperatures and in moisture conditions are concentrated in the surface layers of rocks. These eventually peel or flake off, giving boulders a distinctive rounded shape (Figure 3.9).

Chemical weathering

Chemical weathering processes rely on water; therefore the chemical breakdown of rocks in deserts is extremely slow. Nonetheless, chemical weathering is important: deserts are not completely dry; moisture is available from convective downpours, dew and fog. The main chemical weathering processes are **hydration**, **oxidation** and **solution**.

▶ Activities

Table 3.5 Average maximum and minimum temperatures at Ely, Nevada (1908 m)

Month	J	F	M	A	M	J	J	A	S	O	N	D
Average maximum temperature (°C)	4	6	9	14	20	26	31	29	24	18	10	5
Average minimum temperature (°C)	−13	−9	−6	−3	1	5	9	8	3	−2	−7	−12

mean annual precipitation = 257 mm

1 Draw a chart to represent the data in Table 3.5.
2 What evidence suggests that freeze-thaw weathering is active in Ely? In which months would you expect freeze-thaw weathering to occur?
3 What factor might limit the effectiveness of freeze-thaw weathering? Explain your answer.

Figure 3.9 Thermal exfoliation, Death Valley, California

Hydration occurs when rock minerals absorb water, expand and cause internal stress. This leads to granular disintegration and the surface flaking of rocks. Oxygen dissolved in water reacts with some minerals to form oxides and hydroxides. The oxidised minerals also increase in volume which weakens rocks. Some minerals dissolve in rainwater that is slightly acidic. For example, the calcite cements in sandstone may be dissolved resulting in rock disintegration.

Biological weathering

Biological weathering is limited in desert environments owing to the sparse vegetation cover. However, trees and shrubs have long root systems that penetrate and widen joints and dislodge rock particles. Lichen and algae growing on rock surfaces release CO_2; combined with water and organic matter, chemical weathering processes such as solution and **chelation** take place.

For more about weathering processes, see pages 46–48.

Mass movement

Mass movement is the downhill transfer of slope materials as a coherent body. In desert environments most mass movements are **debris flows** and rockfalls. Debris flows develop during periods of heavy rainfall, particularly on saturated slopes with sparse vegetation cover and rapid run-off. They comprise large quantities of rock fragments, mud, soil and other debris (e.g. timber) which move at speeds of up to 50 km/hour. Even on gentle slopes, debris flows can travel long distances across alluvial fans and desert basins.

Steep, angular slopes and rocky outcrops are prominent in desert landscapes. Where resistant rocks such as sandstone rest on weaker beds like shale, undercutting at the base of a slope by erosion and weathering can create rockfalls and rock slides (Figure 3.10). Mechanical weathering may also remove smaller rocks particles which fall under gravity to form **talus slopes**.

Figure 3.10 Sandstone rock debris at the base of a butte caused by mechanical weathering in Monument Valley on the Arizona–Utah border

Fluvial processes

Paradoxically many characteristic dryland landforms are the result of fluvial processes. Most dryland streams and rivers are ephemeral and only flow intermittently during and shortly after prolonged or intense rainfall events. Yet, while surface run-off is short-lived, streams and rivers in drylands have tremendous power. This is explained by:

- sparse vegetation cover, with minimal **interception flow** to slow the movement by **surface wash** to stream and river channels
- ground surfaces baked hard by the sun, which limits **infiltration**
- rainsplash on unvegetated surface which quickly fills soil pores and reduces soil **permeability**
- shallow soils, which allow little water storage and **throughflow**.

Present day fluvial erosion is highly effective in many dryland environments. Surface wash and river erosion are intensified by the sparse vegetation cover, with little to consolidate loose weathered rock debris at the surface. Consequently streams and rivers transport unusually large sediment loads. Deposition of this sediment takes place when streams and rivers lose energy. This occurs most often as discharge falls with distance (and time) from storm events (due to infiltration, percolation and evaporation); or at sudden breaks of slope such as those between lowland basins and mountain fronts.

Aeolian processes

In desert environments wind is an active agent of erosion, transport and deposition. The sparse vegetation cover and dry conditions make **aeolian** (or wind) processes more effective than in humid and sub-humid environments.

Aeolian erosion and transport

The main erosional effect of the wind in deserts is the removal of fine particles – a process called **deflation**. Tiny silt and clay particles are transported thousands of kilometres by the wind. In the UK, silt and clay exported from the Sahara is frequently washed out of the atmosphere by rain, coating windows and windscreens in red dust. It has also been suggested that Saharan dust, transported across the Atlantic, provides Amazonian soils with essential minerals that sustain the rainforest. Locally, deflation is responsible for dust storms and for surface erosion which creates shallow depressions littered with coarse lag particles. The wind transports sand and dust in three ways:

● Creep – when sand grains slide and roll across the surface. Creep is caused by drag and small differences in wind pressure on sand grains that create lift (Figure 3.11).
● Saltation – the downwind skipping motion of sand grains. It is confined to within 1 or 2 metres of the surface. When saltating grains hit a hard surface they have a ballistic effect, setting other grains moving in the direction of the wind. Saltation is the main process of wind transport in desert areas (Figure 3.12).
● Suspension – small dust particles (less than 0.15 mm in diameter) are entrained by the wind and transported aloft, sometimes vast distances and beyond desert areas.

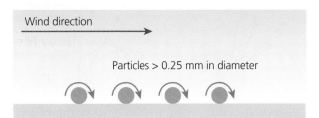

Figure 3.11 Movement of particles by surface creep

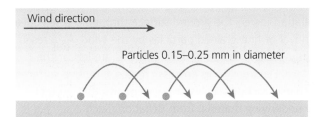

Figure 3.12 Movement of particles by saltation

The three main processes of aeolian erosion are **deflation**, **corrasion** and **attrition**. Deflation is the most important of these. Only the finest particles are removed by deflation. This selective transport often leaves a stony surface of coarse materials. Deserts with extensive stony surfaces are known as **reg**.

Aeolian corrasion is the abrasive action of wind-blown sand against rocks. Over time corrasion carves rocks into a variety of shapes. However, because sand grains are transported by saltation, the sand-blasting effect is confined to just a couple of metres above the ground. Unlike deflation, corrasion is a slow process. Depending on the strength of the rock and wind speeds, it may take a century or more to erode a layer of rock 1 mm thick.

Attrition takes place as grains of sand carried by the wind collide with each other and impact solid rock outcrops. As a result sand grains become smaller and rounder.

Aeolian deposition

As winds subside and energy levels fall, the transport of sand, silt and clay ceases and deposition occurs. Sand dunes are the most obvious evidence of deposition. Often sand accumulates in areas of reduced wind speed as vast sheets or 'sand seas'. In the Sahara these areas are known as **ergs**. Once deposited, sand attracts further deposition. This is because saltating grains are less able to rebound off soft sand compared with hard rocky surfaces.

Table 3.6 Sediment budget for a small drainage basin near Santé Fe, New Mexico (tonnes/km²/year) (Source: Leopold et al. *Fluvial Processes in Geomorphology*, 1966)

Total erosion	5452.9
Surface erosion (surface wash)	5335.2
Gully erosion	?
Mass movement	38.4
Total deposition	1205.5
Deposition in stream channels	?
Deposition in reservoir	640.6

> ### Activities
>
> Study the sediment budget in Table 3.6.
>
> 1 Calculate the amount of sediment (a) removed by gully erosion, (b) deposited in stream channels.
> 2 Calculate the annual mass balance (in tonnes/km²) for the drainage basin sediment budget.
> 3 Describe the relative importance of erosional processes in the drainage basin.

Erosional desert landforms

The erosional processes operating in mid- and low-latitude deserts give rise to distinctive landforms. Among these desert landforms are wadis, canyons, pedestal rocks, ventifacts and desert pavements.

Wadis

Wadis are stream and river channels which are dry for most of the time (Figure 3.13). Although run-off in deserts is short lived, temporary streams and rivers have abundant surplus energy for erosion. This is partly due to the nature of the rainfall, which is often convective and intense; and to rapid run-off and high peak flows. Rapid run-off is due to limited water storage because of minimal soil and vegetation cover, and ground surfaces baked hard by the Sun. The result is powerful **flash floods**.

Figure 3.13 Wadi in Death Valley, California

Canyons

Canyons are narrow river valleys with near vertical sides, cut into solid rock (Figure 3.14). They are a common feature of desert mountains and plateaux and evidence of the power of fluvial erosion in deserts. Canyons are formed by the scouring action of coarse sediment transported by rivers. Erosion is vertical rather than lateral because (a) solid rock walls allow little sideways movement of river channels (b) mass movements processes which would lower valley slopes in more humid environments (e.g. soil creep, mudlows, landslides), are absent in deserts.

Figure 3.14 The Lemoigne Canyon, Death Valley

The shape of canyons in cross profile is mainly determined by rock type. Where rocks are highly resistant and homogeneous, narrow slot canyons with vertical rock walls develop. Antelope Canyon, near Page in Arizona, is a classic example. Other canyons (including the Grand Canyon) where rocks of variable resistance crop out have stair-like sides.

Pedestal rocks

Pedestal rocks are isolated, mushroom-shaped rocks which belong to a larger group of wind-eroded rocks known as **zeugens**. Their curious shape is explained by the undercutting erosive effects of saltating sand grains confined to within 1.5 m of the ground. This process may be aided by weathering concentrated at the base of rocks where moisture is more freely available.

Ventifacts

Ventifacts are small rocks that have been abraded or shaped by wind-blown sand. The facets, formed on the upwind side, are separated from the protected lee side by sharp edges (Figure 3.15). Depending on the number of facets, ventifacts are described as einkanter (one facet), zweikanter (two facets) and dreikanter (three facets). Ventifacts with multiple facets may have been moved by the wind, or form where there are seasonal reversals in prevailing wind direction.

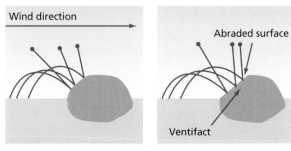

Figure 3.15 Formations of ventifacts

Desert pavements

Deflation lowers the ground surface through the selective removal of fine-grained particles by the wind. As a result tightly packed, coarse-grained particles are left at the surface. Eventually this process creates extensive surfaces of coarse, rocky particles or **lag deposits** that protect the underlying material from deflation. Such a surface is known as a desert pavement (Figure 3.16).

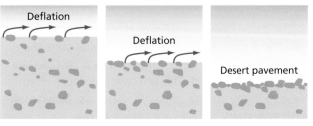

Figure 3.16 Formation of desert pavements

Depositional desert landforms

Distinctive depositional landforms comprising sediments transported and deposited by wind and water include dunes, alluvial fans and bajadas.

Dunes

Dunes are mounds and ridges of wind-blown sand. Two conditions are needed for dune formation: an adequate supply of sand and winds strong and persistent enough to transport sand. A typical dune has a windward slope of 10–15°, a sharp crest and a steeper leeward slope of 30–35° (Figure 3.17). The slip-face stands at the angle of repose, i.e. the maximum angle at which loose sand is stable. Creep and saltation transport sand up the windward slope. As sand accumulates on the crest it eventually exceeds the angle of repose, causing miniature avalanches down the slip-face which restore equilibrium. In this way, dunes advance in the direction of the prevailing wind.

There are three common types of dune: barchans, linear dunes and star dunes. However, many dunes have more complex forms than these basic types, with one form superimposed on another.

Barchans

Barchans are crescentic in planform, with two horns facing downwind (Figure 3.18). They form where winds blow predominantly from one direction. The horns move faster than the main body of the dune. The windward slope is much gentler than the leeward slope. Barchans are highly mobile and move across the desert surface at speeds of up to 30 m/year. Transverse dunes form in a similar way to barchans but develop at right angles to the prevailing wind. They have steep and less steep sides and are associated with erg environments.

Linear dunes

Linear dunes are straight or slightly curved. They are much larger features than barchans: sometimes more than 100 km long with steep slip faces on alternate sides. Again, in contrast to barchans, they form parallel to the prevailing wind direction. They occur either as isolated or parallel ridges and cover a larger area than any other dune type.

Star dunes

Star dunes are pyramidal in profile with slip faces on three or more arms that radiate from a dome-like summit (Figure 3.19). They form in areas where the wind direction is multidirectional. Because they tend to build upwards they include some of the tallest dunes in the world.

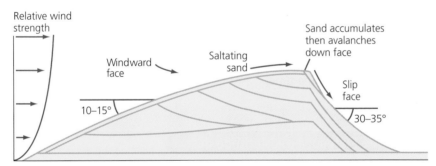

Figure 3.17 Formation of desert dunes

(a) In plan

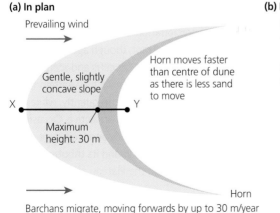

Figure 3.18 A barchan dune

(b) In profile

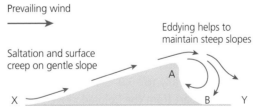

A Steep, upper slip slope of coarse grains with continual sand avalanches due to unconsolidated material (unlike a river, coarse grains are at the top)

B Gentle, basal apron with sand ripples: the finer grains. as on a beach, give a gentler gradient than coarser grains

Figure 3.19 A star dune in the Namib Desert

Alluvial fans and bajadas

Alluvial fans are cones of river-deposited sediment that accumulate at the foot of steep slopes, often along mountain fronts. They form where river channel gradients change abruptly, as a river leaves a steep mountainous course. This may occur where a river confined by narrow rock walls enters an adjacent lowland or basin. The sudden loss of energy, coupled with a huge sediment load, results in the disposition of alluvium and the main channel splitting into hundreds of smaller channels. This creates a delta-shaped mound of debris, with a concave profile in section. Where several alluvial fans develop at intervals along a mountain front they often merge to form a continuous alluvial apron called a **bajada**. Alluvial fans and bajadas are formed by ephemeral streams, so are dry features for most of the time.

> **Activity**

Study Figure 3.20 which shows average wind direction at Windhoek during the year. Explain how the wind direction at Windhoek might influence the formation and morphology of dunes in the neighbouring Namib Desert.

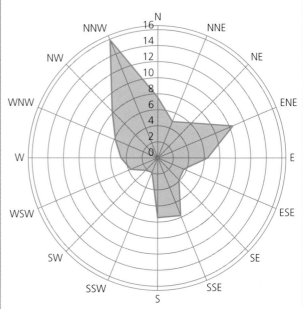

Figure 3.20 Wind rose showing average wind direction (%) at Windhoek in Namibia

Key idea
→ Dryland landforms are interrelated and together make up characteristic landscapes

Case study: Colorado Plateau: a mid-latitude desert

The Colorado Plateau in the southwest USA is situated between the Rocky Mountains in the north and east, and the Great Basin to the south and west (Figure 3.21). At latitudes 34°–40°N and with an average elevation of more than1500 m, large parts of the Plateau are a mid-latitude desert. At Monument Valley in northeast Arizona precipitation averages just 180 mm a year. Temperatures range from an average maximum of 34.4°C in July to –3.9°C in January. Drainage is sparse: apart from the Colorado River, which derives its water from snowmelt in the Rockies, and a handful of larger tributaries, permanent streams and rivers are absent on the Plateau.

Geomorphic processes

The dominant processes shaping the landscape are weathering, mass movement and fluvial action. Freeze-thaw is active in the winter months, though limited by small amounts of precipitation. Salt weathering and insolation weathering are important, though again precipitation remains a limiting factor. On cliffs and scarp slopes gravitational processes triggered by weathering, such as rockfalls and rockslides, occur frequently; while on debris slopes, landslides and slumps are active. Occasional torrential downpours create flash floods and feed huge amounts of sediment into the Colorado and its tributaries, while surface wash is widespread on the Plateau after rain.

→

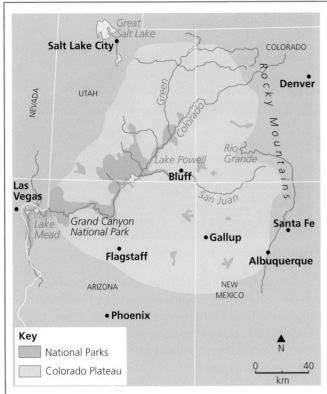

Figure 3.21 Colorado Plateau

Landforms

The principal desert landforms on the Colorado Plateau are: pediments, screes, buttes and mesas, canyons and ravines. **Pediments** are extensive, gently sloping rock platforms which in Monument Valley have formed from the parallel retreat of mesas and buttes (Figure 3.22). They are covered by a superficial layer of sediments derived from rockfall and transported by surface wash and ephemeral streams.

Over time the pediments grow at the expense of the uplands. What was once a continuous plateau of horizontally bedded shale, sandstone and conglomerate, has been reduced to a few isolated flat-topped, steep-sided remnants or **inselbergs**. The larger inselbergs are called **mesas**; the smaller ones are **buttes**. Parallel retreat

results from the undercutting of the weaker organ shale beds by weathering and lateral erosion (by ephemeral streams) leading to collapse of the overlying De Chelly Sandstone. The flat surface of inselbergs is due to a resistant **caprock**, which protects the weaker sandstone beneath. Weathering along joints in the sandstone causes rockfalls, rock slides and other types of slope failure. Scree accumulates at an angle up to 40° at the slope foot, where overloading produces sporadic rock avalanches. The sharp slope profiles of Monument Valley are typical of deserts. A striking feature of slope profiles in Monument Valley is their similarity. This is evidence that they formed by parallel retreat and remain unaltered through time (Figure 3.22).

In many places rivers and streams have cut deep canyons across the Colorado Plateau. Best known are the Grand Canyon and Glen Canyon on the Colorado River (Figure 3.23), and the Goosenecks on the San Juan River. However, many smaller rivers, many of which flow intermittently, have eroded impressive canyons. Some, like Antelope Canyon, are just a few metres wide and have vertical rock walls 20 or 30 metres high.

Canyons testify to the power of rivers as erosive agents in deserts and to the absence of **sub-aerial processes** that lower slopes in more humid environments. On the Colorado Plateau a third factor has played a part in canyon development: over the past 12 million years tectonic forces have uplifted the plateau. And as the surface has been elevated, rivers have incised their channels by downcutting to create today's canyons.

Landscape system changes

The landscapes of the Colorado Plateau have evolved over time. Some changes have occurred over immense time scales, so slowly that they are imperceptible during a human life time. Long-term changes include the uplift of the Plateau, the main factor creating the impressive canyon land scenery. Millions of years ago, during a wetter climatic phase, rivers began to dissect the Plateau surface, exposing the weaker

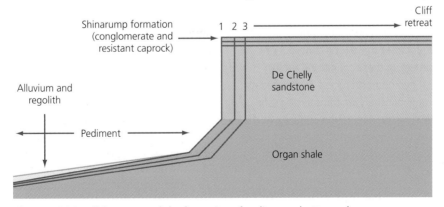

Figure 3.22 Parallel retreat and the formation of pediments, buttes and mesas

Figure 3.23 Horseshoe Bend, Glen Canyon

Figure 3.24 Alluvial fans in Glen Canyon

sandstones and shales. Sub-aerial processes and river erosion widened the valleys, and the slopes retreated parallel to themselves. Today, after millions of years of denudation, all that remains of the Plateau in Monument Valley are a few scattered inselbergs. In the course of time these too will disappear. The processes of parallel retreat can be observed today but they are slow: perhaps no more than one significant rockfall from any free-face every few centuries.

Other changes take place on much shorter time scales. Small rock particles weathered by freeze-thaw from free-faces each winter add to screes. Minor rockfalls also occur after heavy rain. Thunderstorms and flash floods transport sediment through wadis which adds to alluvial fans (Figure 3.24). Other rainfall events transport smaller particles by surface wash across pediments and debris slopes.

Case study: Namib: a low-latitude desert

The Namib Desert in southwest Africa sits astride the Tropic of Capricorn. Temperatures along the coast are moderated by the cold Benguela Current. In this hyper-arid region advection fog – a daily occurrence on the coast – is the principal source of water for plants and animals. Inland temperatures rise rapidly. South of Walvis Bay and as far as the Orange River, a vast sand desert runs parallel to the coast. Elsewhere the desert landscape consists of parched plains of gravel and bedrock, with granite inselbergs rising abruptly from the plains.

Geomorphic processes

With only 23 mm of rainfall at Walvis Bay, and mean monthly temperatures ranging from 11°C to 24°C, weathering in the Namib Desert is minimal. Also with no permanent rivers of any size and one of the driest climates on the planet, fluvial processes make little contribution to landscape development. In these conditions, aeolian action is the dominant geomorphic process. The southeast trade winds blow throughout the year, transporting sand northwards and controlling the orientation of dunes. At the same time the wind erodes fine material from desert plains by deflation.

Landforms

The Namib Desert has some of the highest dunes in the world (Figure 3.19). Dunes in the Sossusvlei region of

southern Namibia are up to 300 m high. Extensive fields of linear dunes aligned SSE to NNW run for hundreds of kilometres from the Orange River to Walvis Bay. Closer to the coast there are vast expanses of transverse dunes developed at right angles to the prevailing trade winds. Barchans and compound dune forms are also widespread. Between Lüderitz and the mouth of the Orange River large areas of streamlined rocks form linear ridges known as **yardangs**. These features form in areas of alternating outcrops of resistant and less resistant rocks. Abrasion by the wind erodes the softer rocks, creating linear depressions, leaving more resistant rocks upstanding as ridges. Like linear dunes, yardangs form parallel to the direction of the prevailing wind (Figure 3.25).

Stretch and challenge

Log in to Google Earth and zoom in on the satellite image of the Namib Desert sand sea south of Walvis Bay, between the Tropic of Capricorn and latitude 24°S.
1 Draw a generalised sketch map to show the distribution of linear dunes, transverse dunes and rocky desert.
2 Describe the orientation of the dunes in relation to the prevailing southerly trade winds and describe their morphology from photographs in the layers option.

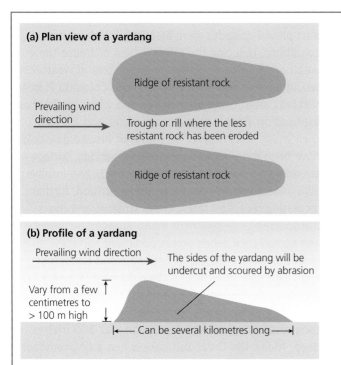

(a) Plan view of a yardang

Prevailing wind direction →

Ridge of resistant rock

Trough or rill where the less resistant rock has been eroded

Ridge of resistant rock

(b) Profile of a yardang

Prevailing wind direction →

The sides of the yardang will be undercut and scoured by abrasion

Vary from a few centimetres to > 100 m high

Can be several kilometres long

Figure 3.25 (a) Plan and (b) profile of a yardang

A large part of central and northern Namibia is rocky desert with inselbergs such as Spitzkoppe rising dramatically from low-angled pediments (Figure 3.26). Differential weathering of granite has created dome-like and conical-shaped inselbergs, gently sloping pediments and desert plains. Inselbergs have formed by deep chemical weathering along joints in the granite. These densely jointed areas have been denuded more rapidly to create extensive plains or pediments. Inselbergs are thought to have formed where jointing in the granite was more massive, creating greater resistance to deep weathering.

Landscape system changes

The Namib is the world's oldest desert. Aridity appears to have begun with the establishment of the Benguela Current around 5 million years ago. Sand which forms the dune systems dates back at least one million years. This suggests that the contemporary landscape has evolved very slowly. The Orange River, which forms the boundary between Namibia and

Figure 3.26 Spitzkoppe inselberg and pediment

South Africa, is the source of the sand. It has taken roughly 1 million years for the southerly trade winds to transport the sand over the 400 km as far north as Walvis Bay.

Even longer periods of time are needed for the formation of pediments and inselbergs. Formed by deep chemical weathering, they probably developed over millions of years when the climate was less arid.

Yet even in today's arid climate, inselbergs continue to develop. Extensive talus slopes around the base of inselbergs are evidence of recent slope failures, rockfalls and rock slides. Tafoni – rounded rock cavities or caves – occur frequently on granite inselbergs. They form by a combination of wetting and drying, hydration and wind abrasion. Together with weathering pits, formed by similar processes, they have developed on time scales of thousands of years.

On a much shorter time scale, some active geomorphic processes can be observed. The movement of sand on the windward slopes of dunes by creep and saltation takes place continuously. So too does the micro-scale avalanching of sand on the dune slip faces, and the advance of dunes, averaging 30 m/year. Although rainfall is infrequent, when it occurs surface wash and channel flow transport sediment across pediments which form desert plains.

✔ Review questions

1 What is mechanical weathering?
2 Name three sources of moisture in deserts.
3 What is exfoliation?
4 What is the difference between erg and reg deserts?

5 How do aeolian processes erode the landscape?
6 Under what circumstances do alluvial fans develop?
7 How does the formation of inselbergs and pediments in Monument Valley and the Namib Desert differ?

3.3 How do dryland landforms evolve over time as climate changes?

> **Key idea**
> → Fluvial landforms can exist in dryland landscapes as a result of earlier pluvial periods

How dryland landforms have been influenced by previous pluvial conditions

Climate change and pluvial conditions

Over the past 500,000 years the climate in drylands has oscillated between wet and dry conditions. The impact of these changes is clear in desert landscapes: many present-day landforms are relict features, left over from a period when the climate was much wetter than today.

Wetter periods in dryland environments, often lasting for thousands of years, are known as **pluvials**. In the Sahara, the world's largest desert, the most recent pluvial occurred between 10,000 and 6000 years Before Present (BP). During this time the **Inter Tropical Convergence Zone** (ITCZ) shifted north, bringing humid conditions and a savanna-type climate to much of the region. At lower altitudes tropical grasslands, similar to those in East Africa today, were widespread; and mountain environments such as the Tibesti Plateau were covered in drought-resistant mixed forests. Rainfall averaged between 100 and 600 mm/year and lakes and permanent river systems existed. Animal populations included mammalian herbivores such as giraffe, elephant and antelope and large predators. Ancient rock carvings in the Sahara depict giraffes and other animals which today are confined to the African savannas. Remarkably, small isolated populations of Nile crocodiles still survive in pools on the Tibesti Plateau, providing further evidence of past pluvial conditions.

Pluvial conditions, geomorphic processes and landforms

Pluvial conditions transformed the geomorphic environment in drylands, intensifying weathering and mass movement and increasing the importance of fluvial action. Indeed many modern desert landforms are known to be 'fossil' features – the legacy of past, wetter climatic phases.

Most weathering processes depend on water. While past pluvial episodes were associated with warmer conditions, reducing the effectiveness of freeze-thaw weathering, rates of hydration and chemical weathering would have accelerated. Today on the Colorado Plateau rockfalls often occur after heavy rain when sandstones, exposed on steep slopes, absorb water and increase their mass. Thus rates of recession of inselbergs would have been more rapid during pluvial periods. Surface streams and rivers flowing from plateaux and inselbergs also undercut steep slopes by lateral erosion, further increasing recession rates. These streams and rivers transported large sediment loads, abraded pediments and spread rock debris across pediment surfaces.

In the Mojave Desert in California and northwest Utah extensive systems of surface drainage (Figure 3.27) and lakes developed during the last pluvial. At its maximum extent, Lake Bonneville in Utah covered thousands of square kilometres and was 300 metres deep. Today, the Great Salt Lake is just a tiny remnant of Lake Bonneville. The old shorelines of Lake Bonneville are clearly visible on hillsides west of Salt Lake City and part of the former lake bed includes the famous Bonneville salt flats. To the southwest, Lake Manly extended almost the entire length of Death Valley. Similar lakes existed to the west below the eastern slopes of the Sierra Nevada. Modern salt flats and **playas** are the remains of these lakes.

A more humid climate sustained permanent rivers, and more vigorous fluvial erosion, transport and

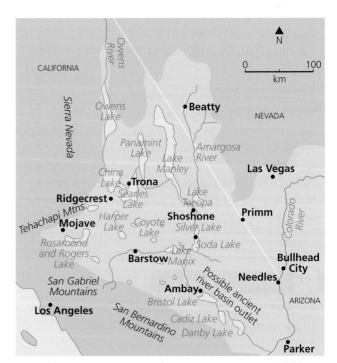

Figure 3.27 Pleistocene rivers and lakes in Mojave region

deposition. It is therefore reasonable to assume that prominent landforms in present-day dryland landscapes such as alluvial fans, bajadas and canyons, owe much of their development to wetter climate conditions in the past.

Modification of fluvial landforms by more recent and future processes associated with climate change

As dryland environments have become more arid in the past 6000 years, rates of weathering, mass movement and fluvial erosion in deserts have slowed. However, with less vegetation and soil cover, run-off may have increased in intensity. More extreme events and powerful (but sporadic) flash floods may have to some degree offset lower rainfall. Meanwhile, drier conditions will have increased the effectiveness of aeolian erosion and transport.

However, the permanent rivers that existed in the last pluvial have disappeared, replaced by intermittent drainage. Pluvial lakes have either shrunk or disappeared; with high rates of evaporation they have been replaced by old lake beds, and waters, where they survive, which are extremely saline.

Future projections of climate change in drylands are geographically variable. In hot arid environments such as Africa, Arabia and Australia further decline in rainfall (and therefore vegetation cover) is forecast; and rates of landscape change associated with sub-aerial processes might slow in these regions. However, aeolian erosion and transport, with deflation and the advance of dunes, could become more widespread. And if rainfall declines, but becomes more extreme, fluvial erosion might even increase. This scenario does not extend to mid-latitude drylands, such as the Great Plains and central Asia, where rainfall is expected to increase slightly.

➡ Activities

Log in to Google Earth and navigate to a satellite view of the Great Salt Lake in Utah.

1 Draw a sketch map to show the extent of Lake Bonneville, and the current outline of the Great Salt Lake. (Note: salt deposits show the former extent of the lake.)
2 Using the latitude and longitude grid overlay and the path function, estimate the former extent of Lake Bonneville in km².
3 What proportion of Lake Bonneville is occupied today by the Great Salt Lake?

Key idea
→ Periglacial landforms can exist as a result of earlier colder periods

How dryland landscapes have been influenced by colder climatic conditions

Climate changes in a previous time period and the resultant colder conditions

The global climate has fluctuated between colder and warmer periods on numerous occasions in the past 500,000 years. The most recent Pleistocene glacial reached its maximum around 20,000 years ago. Global temperatures at that time were 3–5°C lower than at present and large parts of Eurasia and North America were submerged by ice sheets and glaciers. However, middle latitudes were largely ice-free. There extreme cold, with average air temperatures around 6°C lower than the present day, created periglacial conditions. Their defining feature was perennially frozen ground or **permafrost**.

Today some areas that experienced periglacial conditions in the late Pleistocene are deserts and drylands. They include mid-latitude continental interiors such as central Asia and the Great Plains. In lower latitudes mountains and plateaux including the Ethiopian Highlands, the Saharan uplands (Figure 3.28) and the southern Rockies supported periglacial environments. Then between 13,000 and 10,000 BP the climate warmed and periglacial environments receded north to high latitudes.

Figure 3.28 Hoggar Plateau, in Algeria

Periglacial processes and landforms

Periglacial environments are free of ice sheets and glaciers, but apart from a shallow surface layer, the ground is permanently frozen. In these conditions a number of geomorphic processes operate such as freeze-thaw weathering, **frost heave** and **gelifluction** (Figure 3.29). Freeze-thaw weathering occurs when water, confined in rock joints, pores and crevices freezes, expands and causes the mechanical breakdown of rocks. Frost heave is the upward swelling of the ground surface due to the growth of ice crystals in soils and regolith. On slopes expansion is at right angles to the ground surface, but on melting, soil particles settle vertically. This results in a slow downslope displacement. Frost heave is key to understanding mass movements known as **solifluction** or gelifluction. Solifluction is defined as the gravitational flow of the saturated regolith. Where regolith rests on a layer of permafrost, the process is called gelifluction.

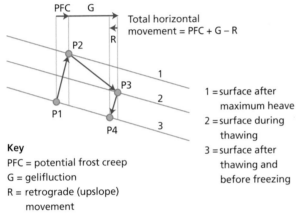

Key
PFC = potential frost creep
G = gelifluction
R = retrograde (upslope) movement

1 = surface after maximum heave
2 = surface during thawing
3 = surface after thawing and before freezing

Figure 3.29 The process of gelifluction

Periglacial processes remained active in some mid- and low-latitude drylands up to 10,000 years ago. The evidence for this is preserved in a range of periglacial landforms. In the mountains of central Arizona and New Mexico extensive **talus** slopes developed by frost shattering above 2400 m. These fossil features, inherited from cold climatic conditions, fed **rock glaciers** on their lower slopes. Rock glaciers are linear accumulations of angular rocks which have moved downslope as a result of the formation and melting of interstitial ice (i.e. ice formed between rock particles). Further north at lower altitude in Hettinger County, North Dakota, large **ognips** up to 2.6 km in diameter, are the remains of ice-cored hills or **pingos** that collapsed as the climate warmed.

In the same area, micro-scale features, known as **cryoturbation** structures (soils and sediments churned by frost heave) and **ice wedge** casts (the infilled casts of vertical, triangular sheets of ground ice) formed. Similar features also occur in the northern Gobi Desert in southern Mongolia.

In the semi-arid Semien Highlands of Ethiopia, frost shattering of bedrock is only active today at elevations above 4250 m. However, frost-shattered rubble can be found between 3100 and 3750 m, indicating colder conditions in the past. This region has other periglacial landforms: extensive boulder fields or **blockfields**, formed by frost weathering of massively bedded rocks; talus slopes and gelifluction deposits down to 3000 m. **Nivation hollows** are found where physical weathering and frost have occurred beneath long-established snow patches. The resulting debris is then removed by flowing meltwater. Once formed, nivation hollows are self-generating, trapping snow and prolonging the action of freeze-thaw, frost heave and surface wash.

The central Saharan uplands of the Tibesti Mountains and the Hoggar Plateau (Figure 3.28), surrounded by the world's largest desert, were also affected by periglacial activity in the late Pleistocene. Nivation hollows occur above 2400 m in the Hoggar Plateau, and above 3100 m in the Tibesti Mountains. Solifluction deposits comprising unsorted, fine debris that when active was water saturated are also found here.

For more on periglacial processes, see page 63.

Modification of periglacial landforms by more recent and future processes associated with climate change

The periglacial landforms in dryland landscapes are for the most part relict features, inherited from an earlier, colder climatic phase. Thus the processes that formed them are no longer operational. Present-day climate and related weathering, mass movement, fluvial and aeolian processes will modify and eventually destroy these fossil features.

In low latitudes, large boulders forming blockfields are exposed to desert weathering processes such as insolation and salt weathering. Exfoliation, as surface rock layers peel away, slowly results in granular disintegration. Rock glaciers which under periglacial conditions moved by the action of interstitial ice, become inactive. But located at the foot of slopes and often at the head of valleys, occasional flash floods transport rock glacier particles which are input as coarse sediment to dry river beds and wadis. Talus slopes may continue to develop at high altitudes where frost weathering occurs, but at lower levels rockfall from free faces is mainly caused by heavy rain saturating rocks which then fail along major joints.

In drier conditions, solifluction and gelifluction processes no longer take place. As features such as sheets, lobes and terraces dry out, these landforms are eroded and dissected by ephemeral streams and rivers, while aeolian activity removes finer particles. Meanwhile, features like nivation hollows are buried beneath fluvial- and aeolian-transported gravels, sands and silts.

1 What are pluvials?
2 What geomorphic processes operate during pluvial periods?
3 Where is periglacial activity still found today in mid- and low-latitudes?
4 What is the difference between solifluction and gelifluction?
5 What is the significance of the active layer in periglacial environments?

3.4 How does human activity cause change within dryland landscape systems?

Key idea
→ Water supply issues can cause change within dryland landscape systems

🔍 Case study: The Colorado Basin, southwest USA

The Colorado Basin in the southwest USA covers 630,000 km². It includes most of Arizona and Utah, as well as parts of Colorado, Wyoming, Nevada, New Mexico and California (Figure 3.30). The Colorado River has its source in the Rocky Mountains; its mouth is 2300 km to the southwest in the Gulf of California. It is the only major river in the southwest USA and most of the basin is desert. The climate at Moab in Utah (Table 3.7) is typical of much of the Colorado Basin.

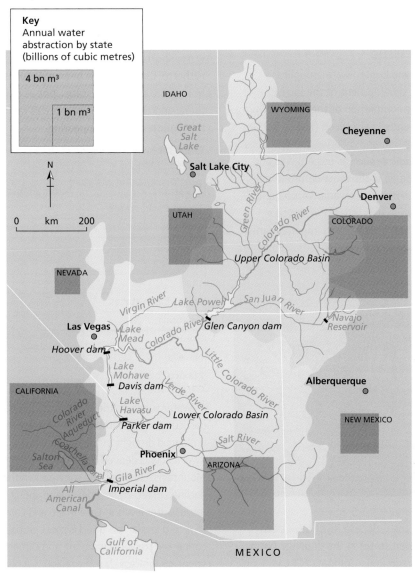

Key
Annual water abstraction by state (billions of cubic metres)

4 bn m³

1 bn m³

N

0   km   200

IDAHO

WYOMING

Cheyenne

Great Salt Lake

Salt Lake City

UTAH

Green River

Colorado River

Denver

COLORADO

Upper Colorado Basin

NEVADA

Virgin River   Lake Powell   San Juan River

Navajo Reservoir

Las Vegas   Lake Mead   Glen Canyon dam

Hoover dam   Colorado River

Lake Mohave

Little Colorado River

Davis dam   Verde River

Alberquerque

CALIFORNIA

Lake Havasu

Lower Colorado Basin

NEW MEXICO

Colorado River Aqueduct

Parker dam

Salt River

Coachella Canal

Salton Sea

Phoenix

Gila River   ARIZONA

All American Canal

Imperial dam

Gulf of California

MEXICO

Figure 3.30 Colorado Basin

Table 3.7 Climate statistics for Moab, Utah

	J	F	M	A	M	J	J	A	S	O	N	D	Total
Mean monthly temp. (°C)	−1.1	3.6	8.9	13.8	19	24.1	27.6	26.5	21.1	14.2	6.9	0.7	
Mean monthly rainfall (mm)	14	11	22	25	18	12	21	22	19	29	19	17	229
Mean monthly PET (mm)	27	45	87	130	182	219	235	202	144	90	45	27	1433

> ### Activities
>
> 1 Use the data in Table 3.7 to construct a climate graph for Moab.
> 2 Calculate the aridity index for Moab.
> 3 Comment briefly on the aridity of the climate at Moab.

The water supply issue

The problem of aridity and limited water supplies inhibited the economic development of the lower Colorado Basin until the completion of the Hoover Dam in 1936. The dam impounded Lake Mead (Figure 3.31), providing water for irrigation and meeting the demands of fast-growing urban centres such as Los Angeles and Phoenix. Today the Colorado is the most dammed river in the USA. In its lower course in Nevada, Arizona and California it is little more than a series of reservoirs created by the Hoover, Davis, Parker and Imperial dams.

Figure 3.31 Lake Mead

Water supplies from the Colorado River have been divided between the seven states in the drainage basin since 1922 by the Colorado River Compact (CRC) (Mexico was belatedly granted a water allocation in 1944). Between them the interested parties divert the river's entire annual flow: 90 per cent of the water is used for irrigation; the rest supplies municipal authorities including Los Angeles, San Diego, Phoenix, Denver, Salt Lake City and Las Vegas (Table 3.8).

The CRC splits the Colorado's water resources equally between the Upper and Lower Basin states. However, demand in the Lower Basin, due to rapid economic and population growth in the past 60 years, has led to acute water shortages, especially in California and Nevada. Hitherto, California has been able to source surplus water from the Upper Basin states, but as demand increases there, and climate change reduces rainfall and snowfall, this option is fast disappearing.

Table 3.8 Annual allocation of Colorado water between US states and Mexico

US states	Water allocations (billions m³)
Upper Basin states	
Colorado	4.79
Wyoming	1.30
New Mexico	1.04
Utah	2.13
Lower Basin states	
California	5.43
Arizona	3.70
Nevada	0.37
Mexico	0.90

Water supply in the southwest USA has acquired particular urgency in the past decade or so owing to a prolonged drought in the region. Most of the Colorado River's flow is derived from snowmelt in the Rocky Mountains. However, the snowpack in the Upper Basin has declined in recent decades due to spring warming. As a result, water levels in Lake Mead and Lake Powell have fallen to record levels. This trend is most probably long term and connected to global warming and climate change. Climate scientists forecast that the region will become more arid in the second half of the twenty-first century. ➡

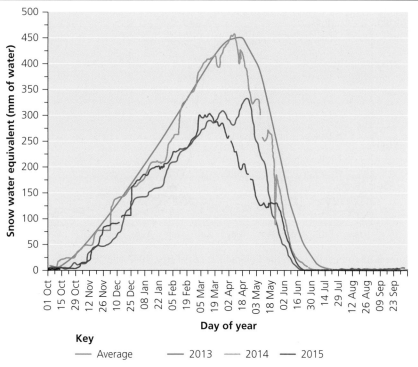

Figure 3.32 Upper Colorado Basin snowpack 2013–15

The drought, which by 2015 had lasted for fourteen years, had become the worst in the region's history (Table 3.9). A total of 64 million people were affected, with California particularly badly hit. There, failure of the winter rains and reduced snowpack in the Sierra Nevada (Figure 3.32) have contributed to water shortages. As water levels fall in Lake Mead, future water rationing is a possibility in Los Angeles and Las Vegas. Meanwhile water shortages are exacerbated by rising demand from agriculture, homes and businesses.

Table 3.9 Drought in the southwest USA, October 2015; proportion (%) of each state affected (Source: US Drought Monitor)

	California	Nevada	Arizona	New Mexico
Exceptional drought	46.0	15.93	–	–
Extreme drought	25.08	21.59	–	–
Severe drought	21.29	38.40	3.28	–
Moderate drought	4.97	18.84	59.32	7.94
Abnormally dry	2.53	5.23	26.92	34.19

Flows of water and sediment

The construction of dams on the Colorado River has fundamentally altered flows of water and sediment in the drainage basin. The biggest impact has been made by the Glen Canyon dam, completed in 1963, which is furthest upstream.

Before the Glen Canyon dam was built, the Colorado's flow rates, sediment loads and water temperatures fluctuated widely from year to year and season to season. Snowmelt in the Rocky Mountains produced peak flows in late spring and early summer, averaging between 2500 and 3000 cumecs. In contrast, low flows of less than 85 cumecs were typical of late summer, autumn and winter.

In spring and early summer, sediment loads increased sharply with discharge. Large sediment inputs also occurred in late summer following thunderstorms and flash floods in tributary catchments, while water temperatures ranged from near freezing in winter to more than 27°C in late summer.

The completion of the Glen Canyon Dam and Lake Powell resulted in huge changes to the flow regime of the Colorado River (Figure 3.33). Outflows from Glen Canyon were controlled and dominated by daily, rather than seasonal fluctuations. This reflected short-term variations in the demand for electricity generated by turbines installed in the dam. Average daily flows exceeded 850 cumecs only 3 per cent of the time (compared with 18 per cent before dam construction). At the same time low flows became less frequent: today flows of less than 140 cumecs occur around 10 per cent of the time, compared with 18 per cent before 1963.

Meanwhile the Glen Canyon Dam (and others downstream) has altered geomorphic processes. All sediments derived from the catchment upstream of Glen Canyon are now trapped in Lake Powell. Thus the sediment load of the Colorado downstream of the

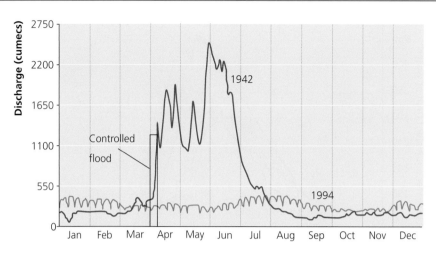

Figure 3.33 Annual flows on the Colorado at Lees Ferry, 1942 and 1994

dam is negligible. Downstream sediment input is mainly from tributary streams, which are intermittent, flowing only after heavy rain. The annual cycle of 'scour and fill' that prevailed pre-1963, based on alternating high- and low-energy conditions, no longer occurs. During peak flows the river eroded and removed sediment in the channel and from the surrounding valley floor. An equilibrium was maintained between input (deposition) and output (erosion). These processes were disrupted by the dam and this has had a direct impact on fluvial landforms.

Impact on dryland landforms and landscapes

Dryland landscapes and landforms have been affected by changes in geomorphic processes as a consequence of the Glen Canyon Dam and Lake Powell.

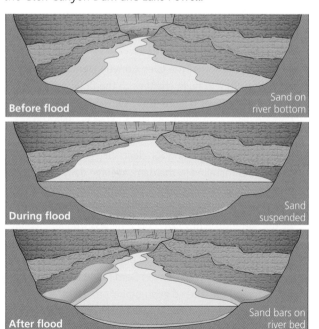

Figure 3.34 Formation of sand bars in Glen Canyon

Downstream of the dam the absence of floods and the river's minimal sediment load have caused sand bars, once a familiar landform in the channel, to disappear. Also, sand bars deposited along the margins of the valley during floods have been degraded (Figure 3.34). Debris fans, formed by tributary streams flowing into Glen Canyon, are no longer eroded by annual floods, and have encroached into the valley.

As the waters of Lake Powell rose, they flooded hundreds of tributary canyons and wadis. This created a higher base level and shortened the length of tributary wadis. The effect was to reduce wadi catchments and therefore run-off. With a decrease in stream energy, rates of erosion slowed, and **aggradation** increased. Thus many wadi channels, needing to transport smaller sediment loads and flows, have decreased in width and depth. On pediment surfaces close to Lake Powell, aggradation in multi-thread stream channels may occur as flow competence declines.

In arid environments, fluvial, aeolian and sedimentary systems are often closely linked. The Colorado River is both a source and a sink for aeolian sediment. Before the dam's construction, fluvial sand bars in the Colorado Valley were an important source of sand supply for dune building downstream in Marble Canyon and Grand Canyon. However, the loss of sand bars and the absence of river floods have starved dune fields of sediment. Deprived of sand, many dunes have become degraded – overgrown with **cryptobiotic crusts** and plants, and exposed to wind erosion.

Case study: The desert in southeast Utah, USA

Southeast Utah is a classic region of desert plateaux and canyon lands. It includes two national parks – Arches and Canyonlands – and vast wilderness areas belonging to the state and federal governments. The landscape is spectacular, with plateaux, desert plains, canyons and the world's largest concentration of natural arches. Adding to the outstanding landscape are the vivid red and orange colours of the Entrada Sandstone, sharp changes of slope and smooth sliprock surfaces.

Tourism

The region provides huge opportunities for outdoor recreation and attracts a large and growing number of visitors (Figure 3.35). Arches National Park, part of Canyonlands National Park and surrounding public lands lie within Grand County, Utah. Tourism drives employment in Grand County. The National Park Service alone provides one in three jobs and tourism accounts for nearly half of all employment in the private sector. Most of these jobs are in motels, restaurants and retailing.

Figure 3.36 Natural sandstone arches in Arches NP

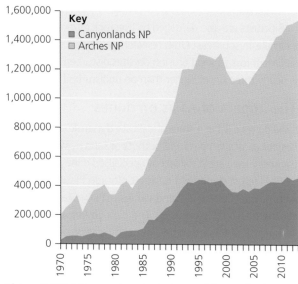

Figure 3.35 Visitor numbers to Arches and Canyonlands National Parks

Most visitors are geotourists, who come to admire the stunning scenery (Figures 3.36 and 3.37). But an increasing number are adventure recreationalists – hikers, mountain bikers and those engaged in motorised recreation, driving all terrain vehicles (ATVs), off-road vehicles (ORVs) and dune buggies.

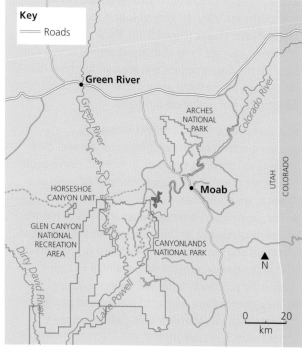

Figure 3.37 Arches and Canyonlands National Parks

Visitor impact on energy and material flows

All tourism has an impact on the environment. Its severity depends partly on the numbers involved and the types of recreational activity. Visitors to Arches and Canyonlands have increased massively in the past four decades (Table 3.10). This growth has been an important factor in increasing pressure on the fragile dryland landscape system.

Table 3.10 Visitor numbers to Arches and Canyonlands National Parks, 1979–2014

	1979	2014
Arches NP	269,840	1,284,767
Canyonlands NP	74,545	542,431

Today visitor numbers, concentrated in a relatively small area, are approaching levels of mass tourism. However, of greater concern is the growth of adventure-based recreation centred on off-road driving, mountain biking, camping and hiking. The Slickrock Trail alone, in Arches National Park, attracts 100,000 mountain bikers a year. Motorised recreation based on ATVs like quad bikes, has shown massive growth. In 1979 there were just 9000 ATVs in Utah: in 2010 this figure was 120,000. Although ATVs are banned in National Parks, they are free to operate on state and federal lands. There they cause significant erosion and other environmental damage. Such activities are especially destructive in the fragile desert environment of southeast Utah.

Cryptobiotic crust damage

In the drylands of southeast Utah and much of the southwest USA, the ground is held together by **cryptobiotic crusts** (Figure 3.38). These communities of lichens, mosses, fungi and cyanobacteria form a thin layer on the desert surface and account for more than 70 per cent of the living ground cover. Biotic crusts perform several ecological functions:

- A dense network of filaments formed by cyanobacteria excrete mucilaginous material that binds soil particles together. As a result, movements of loose mineral and organic material by wind and water are minimised.
- They absorb and store rainwater and reduce run-off.
- The rough surface of the crust reduces the energy of run-off and the shear force of wind; this is important in dryland environments where rainfall though sporadic is often intense, and winds are strong.
- They input organic matter to the soil and provide plants with essential nutrients such as nitrogen, potassium and calcium.
- They act as important seedbeds for plants.

Figure 3.38 Cryptobiotic crust

Biocrusts are thin and brittle and easily damaged by ATVs, off-trail mountain bikers, hikers and back-country campers. Moreover, recovery is extremely slow, often taking 50 years and more. Once crusts are damaged, erosion by wind and water is rapid and recovery impossible. ATVs and mountain bikes also compact soils and create artificial channels or wheelings, which accelerate water erosion and siltation of streams.

Biocrusts are also damaged by livestock grazing. For example, in southeast Utah research has shown that sediment loss to wind erosion on sites disturbed by livestock is nearly three times higher than on undisturbed sites.

The impact of ATVs on dunes

Sand dunes exist in a state of delicate **dynamic equilibrium**: despite constant inputs and outputs of sand by wind transport, they are stabilised by specialised dune plants.

The largest dune field on the Colorado Plateau is the picturesque Coral Pink Sand Dunes (CPSD) State Park in southern Utah (Figure 3.39). The park covers an area of 15 km², of which around 40 per cent is sand dune. The dunes, formed by erosion and weathering of the local Navajo Sandstone, occupy a valley between the Moquith and Moccasin Mountains (Figure 3.40). A similar area of dunes occurs across the park boundary. Administered by the Bureau of Land Management, this area has only limited environmental protection.

Despite its extraordinary beauty and its designation as a state park, 90 per cent of the area of the CPSD is open to riders of ATVs and other ORVs. The park had nearly 80,000 visitors in 2015, the majority of whom →

Figure 3.39 Coral Pink Sand Dunes State Park

were motorised recreationalists. ORVs have damaged vegetation that anchors the dunes and provides habitats for insects, reptiles and small mammals. The CPSD has special conservation significance as the habitat for two endangered species: the CPSD tiger beetle which is unique to the park; and a desert plant, Welsh's milkweed. Environmentalists have established a conservation area for the tiger beetle, though it occupies only a small area of the park.

Conservation groups in Utah in the early 1990s tried unsuccessfully to close the park to ORVs, and US Fish and Wildlife Service has attempted – so far without success – to get listed status for the tiger beetle which in practice would exclude ORVs. Not only is the ORV lobby very strong, but the state parks' mission statement is conflicting. On the one hand state parks must 'provide visitors with a range of recreational experiences' but at the same time 'preserve the natural, scenic and recreational resources'.

Dust storms and loess accumulation

Drylands like southeast Utah are sources of wind-blown silt or **loess**. Exposed dry surfaces, sparse vegetation and strong winds mean that soils are susceptible to deflation and wind erosion, especially in winter and spring. Satellite images have revealed dust plumes originating in the drylands of the American southwest, extending for hundreds of kilometres beyond the region.

Dust storms have become more frequent since the settlement of much of the American southwest in the mid-1800s. This suggests that wind erosion is closely connected to human activities such as farming and mining. The Owens dry lake bed in southern California, drained in the early 1900s to supply Los Angeles with water, is one of the largest sources of salt-bearing dust in the world.

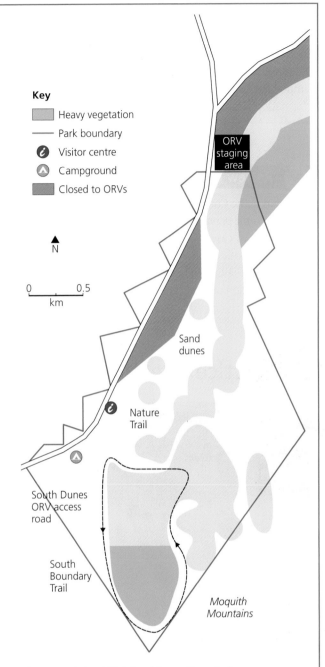

Figure 3.40 Coral Pink Sand Dunes State Park

Biocrusts stabilise soils and reduce erosion by run-off as well as wind. Studies have shown that damage to these crusts on the Colorado Plateau, caused by farm livestock and recreational activities, greatly increases soil erosion by aeolian processes. Drought is also a factor in increasing wind erosion and loess deposits.

Dust blown from the Colorado Plateau is known to accumulate on snowfields in the surrounding San Juan, Sierra Nevada and Rocky mountains. And because dust

→

is far less reflective than snow, it absorbs solar radiation and causes earlier and faster snowmelt. This increases both water loss due to evaporation and flood risks.

Wind-blown silt and dust is recognised as an important source of minerals and nutrients in soil formation. Coarser particles tend to settle out of the atmosphere within 24 hours and are more likely to be deposited within drylands like the Colorado Plateau. In contrast, fine clay particles may be transported high into the atmosphere and carried thousands of kilometres from their place of origin.

✓ Review questions

1. What is the Colorado River Compact (CRC)?
2. What is the principal source of the Colorado River's flow?
3. Name two causes of water shortage in the southwest USA.
4. How did the Colorado River's regime change after the construction of the Glen Canyon Dam?
5. What is meant by the process of 'scour and fill'?
6. What is cryptobiotic crust and why is it important?
7. What factors promote dust storms in drylands?
8. Describe two environmental effects of the transport and deposition of dust from dryland sources.

📑 Practice questions

Figure A Sand dune in the Sahara, southern Morocco

Figure B Mid-latitude dryland landscape

A Level

a) Explain how, in the past, dryland landscapes in tropical and sub-tropical regions have been influenced by wetter, pluvial conditions. [8 marks]

Table 1 Total annual precipitation at Elko, Nevada, 2000–15

222	203	188	255	295	416	290	146
205	341	325	194	202	162	265	127

b) Using the data in **Table 1**:
   i) Calculate the mean annual precipitation at Elko, 2000–15. [2 marks]
   ii) Explain why extreme variability in annual rainfall amounts is often a feature of dryland regions. [4 marks]
c) Study **Figure A** which shows a linear dune in the Sahara in southern Morocco. Describe the main transport processes operating on the dune shown in the photograph. [3 marks]

d)* To what extent is human activity the main driver of landscape change in dryland environments?

[16 marks]

AS Level

Section A

1 **a)** Study the data in **Table 2** which show rates of solifluction movement in different arctic–alpine environments in the Canadian Rockies.

    i) Compare and contrast rates of solifluction in the Canadian Rockies. [3 marks]

    ii) Suggest reasons for the variations in rates of solifluction shown in Table 2. [4 marks]

b) Study **Figure B** which shows a mid-latitude dryland landscape. Using the evidence of the photograph explain why fluvial processes are likely to be effective in this environment. [6 marks]

c)* To what extent is human activity the main driver of landscape change in dryland environments?

[14 marks]

Table 2 Solifluction movements in the Canadian Rockies (mm/year)

2.9	5.0	0.1	4.3	7.3	11.1	6.8	1.3	3.3	1.8
1.1	1.7	2.1	8.0	2.1	11.7	3.6	5.2	7.5	4.6

Chapter 4

Earth's life support systems

4.1 How important are water and carbon to life on Earth?

> **Key idea**
> → Water and carbon support life on Earth and move between the land, oceans and atmosphere

The importance of water in supporting life on the planet

Scientists believe that water is the key to understanding the evolution of life on Earth as it provides a medium that allows organic molecules to mix and form more complex structures. The ubiquity of liquid water on Earth is due to the distance of the Earth from the Sun: it lies in the so-called 'Goldilocks zone', which is 'just right' for water to exist in its liquid form.

The importance of liquid water to life can be appreciated when we compare Earth with our nearest planetary neighbour, Mars. Although water ice exists on Mars, there is new evidence of very small amounts of liquid water flowing on the Martian surface. Scientists think this liquid water greatly increases the chance of finding life forms on the planet.

Water helps to create benign thermal conditions on Earth. For example, oceans, which occupy 71 per cent of the Earth's surface, moderate temperatures by absorbing heat, storing it and releasing it slowly. Water also moderates the environment in other ways. Clouds made up of tiny water droplets and ice crystals reflect around a fifth of incoming solar radiation and lower surface temperatures. At the same time water vapour, a potent greenhouse gas, absorbs long-wave radiation from the Earth helping to maintain average global temperatures almost 15 °C higher than they would be otherwise.

The uses of water for flora, fauna and people

Water makes up to 65–95 per cent of all living organisms and is crucial to their growth, reproduction and other metabolic functions. Plants, which manufacture their own food, need water for **photosynthesis**, **respiration** and **transpiration**. Photosynthesis takes place in the leaves of plants combining CO_2, sunlight and water to make glucose and starches. Respiration in plants and animals converts glucose to energy through its reaction with oxygen, releasing water and CO_2 in the process.

Plants also require water to maintain their rigidity (plants wilt when they run out of water) and to transport mineral nutrients from the soil. In people and animals water is the medium used for all chemical reactions in the body including the circulation of oxygen and nutrients. Transpiration of water from leaf surfaces cools plants by evaporation. Sweating is a similar cooling process in humans. In fur-covered mammals, reptiles and birds, evaporative cooling is achieved by panting.

Water is also an essential resource for economic activity. It is used to generate electricity, irrigate crops, provide recreational facilities and satisfy public demand (drinking water, sewage disposal), as well as in a huge range of industries including food manufacturing, brewing, paper making and steel making.

The importance of carbon to life on Earth

Carbon is a common chemical element. It is stored in carbonate rocks such as limestone, sea floor sediments, ocean water (as dissolved CO_2), the atmosphere (as CO_2 gas), and in the biosphere. Life as we know it is carbon based: built on large molecules of carbon atoms such as proteins, carbohydrates and nucleic acids.

Apart from its biological significance, carbon is used as an economic resource. Fossil fuels such as coal, oil and natural gas power the global economy. Oil is also used as a raw material in the manufacture of products ranging from plastics to paint and synthetic fabrics. Agricultural crops and forest trees also store large amounts of carbon available for human use as food, timber, paper, textiles and many other products.

The water and carbon cycles

At the global scale, water and carbon flow in closed systems between the **atmosphere**, the oceans, land and the **biosphere**. The cycling of individual water molecules and carbon atoms occurs on time scales varying from days to millions of years.

At the macro-scale, the global water cycle consists of three main stores: the atmosphere, oceans and land. The oceans are by far the biggest store and the atmosphere is the smallest. Water moves between stores through the processes of **precipitation**, **evapotranspiration**, **run-off** and **groundwater flow** (Figure 4.1).

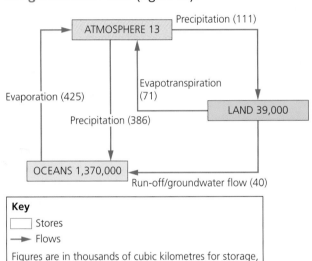

Key
☐ Stores
→ Flows
Figures are in thousands of cubic kilometres for storage, and thousands of cubic kilometres/year for flows.

Figure 4.1 The global water cycle: stores and annual flows

The global carbon cycle is similar in comprising a series of stores and flows. Long-term storage in sedimentary rocks holds 99.9 per cent of all carbon on Earth. In contrast, most of the carbon in circulation moves rapidly between the atmosphere, the oceans, soil and the biosphere. The main pathways between stores followed by carbon in this cycle include **photosynthesis**, **respiration**, **oxidation** (decomposition, combustion) and **weathering** (Figure 4.2).

The water and carbon cycles as open and closed systems

Systems are groups of objects and the relationships that bind the objects together. On a global scale the water and carbon cycles are **closed systems** driven by the Sun's energy (which is external to the Earth). Only energy (and not matter) cross the boundaries of the global water and carbon cycles – hence we refer to these systems as 'closed'.

At smaller scales (e.g. drainage basin or forest ecosystem), materials as well as the Sun's energy cross system boundaries. These systems are therefore **open systems**.

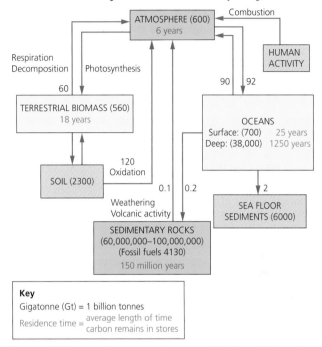

Key
Gigatonne (Gt) = 1 billion tonnes
Residence time = $\dfrac{\text{average length of time}}{\text{carbon remains in stores}}$

Figure 4.2 The global carbon cycle: stores (Gt), flows (Gt/year) and residence times

> **Key idea**
> → The carbon and water cycles are systems with inputs, outputs and stores

The global water cycle

Reservoirs and stores

The global water cycle consists of a number of reservoirs where water is stored for variable lengths of time (Table 4.1), and the linkages or pathways between these reservoirs.

Table 4.1 Global reservoirs of water

Store	Size (km³ × 10³)	% of global water
Oceans	1,370,000	97
Polar ice and glaciers	29,000	2
Groundwater (aquifers)	9,500	0.7
Lakes	125	0.01
Soils	65	0.005
Atmosphere	13	0.001
Rivers	1.7	0.0001
Biosphere	0.6	0.00004

The oceans contain 97 per cent of all water on the planet and dominate the global water cycle. Fresh water comprises only a tiny proportion of water in store and three-quarters is frozen in the ice caps of

Antarctica and Greenland. Meanwhile, water stored below ground in **permeable rocks** amounts to just one-fifth of all fresh water.

Given its pivotal role in the water cycle, it is perhaps surprising that only a minute fraction of the Earth's water is found in the atmosphere. This paradox is explained by the rapid flux of water into and out of the atmosphere: the average residence time of a water molecule in the atmosphere is just nine days.

Inputs and outputs of water

According to US Geological Survey (USGS) estimates, the global **water cycle budget** circulates around 505,000 km³ of water a year as inputs and outputs between the principal water stores.

- Inputs of water to the atmosphere include water vapour **evaporated** from the oceans, soils, lakes and rivers, and vapour **transpired** through the leaves of plants. Together these processes are known as **evapotranspiration**.
- Moisture leaves the atmosphere as precipitation (i.e. rain, snow, hail, etc.) and **condensation** (e.g. fog). Ice sheets, glaciers and snowfields release water by **ablation** (melting and **sublimation**).
- Precipitation and meltwater drain from the land surface as **run-off** into rivers. Most rivers flow to the oceans though some, in continental drylands like southwest USA, drain to inland basins. A large part of water falling as precipitation on the land reaches rivers only after **infiltrating** and flowing through the soil.

- After infiltrating the soil, water under gravity may **percolate** into permeable rocks or **aquifers**. This **groundwater** eventually reaches the surface as springs or seepages and contributes to run-off.

> ### Activities
> 1 Describe the geographical distribution of water vapour in the atmosphere in Figure 4.3.
> 2 Explain how temperature, oceans and land masses influence this distribution.

The global carbon cycle

The global carbon cycle consists of a number of stores or **sinks** connected by flows of carbon. The principal stores are: the atmosphere, the oceans, carbonate rocks, fossil fuels, plants and soils (Table 4.2). Carbon moves between these stores in an unending cycle.

Table 4.2 Principal carbon stores

Store	Carbon in store (billion tonnes)
Atmosphere	600
Oceans	38,700
Sedimentary (carbonate) rocks	60,000–100,000,000
Sea floor sediments	6,000
Fossil fuels	4,130
Land plants	560
Soils/peat	2,300

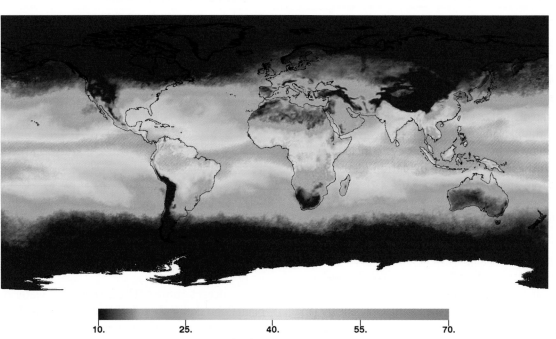

Figure 4.3 Satellite image of global water vapour (mm)

10.   25.   40.   55.   70.

Carbonate rocks, such as limestone and chalk, and deep-ocean sediments are by far the biggest carbon store. Most of the carbon that is not stored in rocks and sediments is found in the oceans as dissolved CO_2. Carbon storage in the atmosphere, plants and soils is relatively small. However, these stores play a crucial part in the carbon cycle. They also represent most of the carbon in circulation at any one time.

Inputs and outputs in the carbon cycle

There are two strands to the carbon cycle: a slow cycle and a fast cycle.

The slow carbon cycle

Carbon stored in rocks, sea-floor sediments and fossil fuels is locked away for millions of years. The total amount of carbon circulated by this slow cycle is between ten and 100 million tonnes a year. CO_2 diffuses from the atmosphere into the oceans where marine organisms, such as clams and corals, make their shells and skeletons by fixing dissolved carbon together with calcium to form calcium carbonate ($CaCO_3$). On death, the remains of these organisms sink to the ocean floor. There they accumulate and over millions of years, heat and pressure convert them to carbon-rich sedimentary rocks.

Typical residence times for carbon held in rocks are around 150 million years. Some carbon-rich sedimentary rocks, subducted into the upper mantle at tectonic plate boundaries, are vented to the atmosphere in volcanic eruptions. Others exposed at or near the surface by erosion and tectonic movements are attacked by chemical weathering (pages 106–07). Chemical weathering processes such as carbonation are the result of precipitation charged with CO_2 from the atmosphere, which forms a weak acid. The acid attacks carbonate minerals in rocks, releasing CO_2 to the atmosphere, and in dissolved form to streams, rivers and oceans.

On land, partly decomposed organic material may be buried beneath younger sediments to form carbonaceous rocks such as coal, lignite, oil and natural gas. Like deep-ocean sediments, these fossil fuels act as carbon sinks that endure for millions of years.

The fast carbon cycle

Carbon circulates most rapidly between the atmosphere, the oceans, living organisms (biosphere) and soils. These transfers are between ten and 1000 times faster than those in the slow carbon cycle. Land plants and microscopic phytoplankton in the oceans are the key components of the fast cycle. Through photosynthesis they absorb CO_2 from the atmosphere and combine it with water to make carbohydrates (sugars/glucose). Photosynthesis is a fundamental process and the foundation of the food chain. Respiration by plants and animals is the opposite process and results in the release of CO_2. Decomposition of dead organic material by microbial activity also returns CO_2 to the atmosphere.

In the fast cycle, carbon exchange also occurs between the atmosphere and the oceans. Atmospheric CO_2 dissolves in ocean surface waters while the oceans ventilate CO_2 back to the atmosphere. Through this exchange individual carbon atoms are stored (by natural sequestration) in the oceans for, on average, about 350 years.

Fieldwork ideas

Investigate differences in the soil carbon store at two contrasting sites: grassland and either deciduous or coniferous woodland.

1 Collect controlled and comparable samples (i.e. samples at the two sites should be at the same soil depth, the sites should have the same slope, aspect, altitude, characteristics, etc.).
2 Dry the samples in an oven and weigh them.
3 Use a bunsen burner to remove organic material and reweigh.
4 Calculate the percentage of organic material in each sample and test for significant differences using chi-squared, U-test or t-test (see Chapter 15, Geographical Skills, pages 536–40).

Review questions

1 Describe two ways in which water moderates global temperatures.
2 Define the term 'system'.
3 Why is water vapour described as a 'greenhouse gas'?
4 What is an aquifer?
5 State two ways in which water leaves the atmosphere.
6 What is the difference between evaporation and transpiration?
7 Outline the main differences between the fast and slow carbon cycles.
8 What are phytoplankton and why are they important in the carbon cycle?
9 What is the role of plate tectonics in the carbon cycle?

The processes of the water cycle

The water balance

The **water balance** equation summarises the flows of water in a drainage basin over time. It states that precipitation is equal to evapotranspiration and streamflow, plus or minus water entering or leaving storage:

$$\text{Precipitation (P)} = \text{Evapotranspiration (E)} + \\ \text{Streamflow (Q)} \pm \text{Storage}$$

Flows

The principal flows in the water cycle that link the various stores are: precipitation, evaporation, transpiration, run-off, infiltration, percolation and throughflow (Figure 4.4).

Precipitation

Precipitation is water and ice that falls from clouds towards the ground. It takes several forms: most commonly rain and snow, but also hail, sleet and drizzle. Precipitation forms when vapour in the atmosphere cools to its **dew point** and condenses into tiny water droplets or ice particles to form clouds. Eventually these droplets or ice particles aggregate, reach a critical size and leave the cloud as precipitation.

Precipitation also varies in character and this impacts the water cycle at the drainage basin scale.

● Most rain on reaching the ground flows quickly into streams and rivers. But in high latitudes and mountainous **catchments**, precipitation often falls as snow and may remain on the ground for several months. Thus there may be a considerable time lag between snowfall and run-off.

● Intensity is the amount of precipitation falling in a given time. High-intensity precipitation (e.g. 10–15 mm/hour) moves rapidly overland into streams and rivers.

● Duration is the length of time that a precipitation event lasts. Prolonged events, linked to depressions

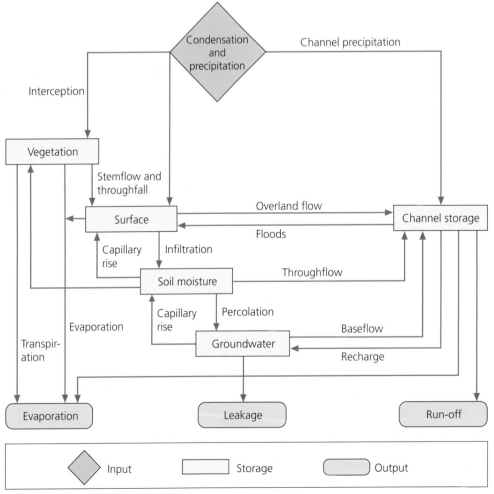

Figure 4.4 The drainage basin water cycle

and frontal systems, may deposit exceptional amounts of precipitation and cause river flooding.

- In some parts of the world (e.g. East Africa, Mediterranean) precipitation is concentrated in a rainy season. During this season **river discharge** is high and flooding is common. In the dry season rivers may cease to flow altogether.

Stretch and challenge

Cloud formation and precipitation
Clouds form when water vapour cools to its dew point. The dew point is the critical temperature when air becomes saturated (i.e. 100 per cent relative humidity) and can hold no more vapour. This results in condensation, as excess vapour changes state to form droplets of liquid water. Aggregated, these droplets (which are much smaller than rain droplets) form clouds. Rain develops either through a complex process of droplets coalescing or ice crystals growing within clouds. For a fuller explanation, research the two main processes of precipitation: (a) the collision theory, (b) the Bergeron–Findeisen theory.

Transpiration

Transpiration is the diffusion of water vapour to the atmosphere from the leaf pores (stomata) of plants. It is responsible for around 10 per cent of moisture in the atmosphere. Like evaporation, transpiration is influenced by temperature and wind speed. It is also influenced by water availability to plants. For example, deciduous trees shed their leaves in climates with either dry or cold seasons to reduce moisture loss through transpiration.

Condensation
Condensation is the phase change of vapour to liquid water. It occurs when air is cooled to its dew point. At this critical temperature air becomes saturated with vapour resulting in condensation. Clouds form through condensation in the atmosphere.

- **Cumuliform clouds**, with flat bases and considerable vertical development most often form when air is heated locally through contact with the Earth's surface. This causes heated air parcels to rise freely through the atmosphere (convection), expand (due to the fall in pressure with altitude) and cool. As cooling reaches the dew point, condensation begins and clouds form.
- **Stratiform** or layer clouds develop where an air mass moves horizontally across a cooler surface (often the ocean). This process, together with some mixing and turbulence, is known as **advection**.
- Wispy, **cirrus** clouds, which form at high altitude, consist of tiny ice crystals. Unlike cumuliform and stratiform clouds they do not produce precipitation and therefore have little influence on the water cycle.

Condensation at or near the ground produces **dew** and **fog**. Both types of condensation deposit large amounts of moisture on vegetation and other surfaces.

Activities

The conversion of snowfall to rainfall equivalent depends on temperature (Table 4.3). Generally the higher the temperature the greater the water content of snow. In the UK snow most often falls at temperatures around 0°C. The conversion rate at these temperatures is roughly 10:1. Thus a snow depth of 15 cm (150 mm) would represent approximately 15 mm of rainfall equivalent.

Table 4.3 Conversion ratios for snow to rainfall

Temperature (°C)	−1	−4	−8	−11	−15	−23
Ratio	10:1	15:1	20:1	30:1	40:1	50:1

1 On 10 November 2014, 34 cm of snow fell at St Cloud, Minnesota. The average temperature during the snow fall was −4°C. What was the equivalent rainfall for this snow event?

2 On the following day 62 cm of snow fell at Ishpeming, Michigan. The average temperature during the snowfall was −1°C. Calculate the rainfall equivalent for this event.

3 In early January 2015, 20 cm of snow was recorded at Aviemore in the Highlands of Scotland where the average temperature was −4°C. At the same time 15 cm was recorded in Glasgow, where the temperature was −1°C. Which location had the higher rainfall equivalent?

Table 4.4 Lapse rates

Environmental lapse rate (ELR)	The ELR is the vertical temperature profile of the lower atmosphere at any given time. On average the temperature falls by 6.5 °C for every kilometre of height gained.
Dry adiabatic lapse rate (DALR)	The DALR is the rate at which a parcel of dry air (i.e. less than 100 per cent humidity so that condensation is not taking place) cools. Cooling, caused by adiabatic expansion, is approximately 10 °C/km.
Saturated adiabatic lapse rate (SALR)	The SALR is the rate at which a saturated parcel of air (i.e. one in which condensation is occurring) cools as it rises through the atmosphere. Because condensation releases latent heat, the SALR, at around 7 °C/km, is lower than the DALR.

Cloud formation and lapse rates

Clouds are visible aggregates of water or ice or both that float in the free air. We have seen that they form when water vapour is cooled to its dew point. Cooling occurs when:

- air warmed by contact with the ground or sea surface, rises freely through the atmosphere. As the air rises and pressure falls it cools by expansion (**adiabatic expansion**). This vertical movement of air is known as **convection**.
- air masses move horizontally across a relatively cooler surface – a process known as advection.
- air masses rise as they cross a mountain barrier or as turbulence forces their ascent.
- a relatively warm air mass mixes with a cooler one.

Lapse rates describe the vertical distribution of temperature in the lower atmosphere, and the temperature changes that occur within an air parcel as it rises vertically away from the ground. There are three types of lapse rate (Table 4.4). Their interaction explains the formation of clouds.

Figure 4.5 illustrates how clouds form by convection. The ground heated by the Sun warms the air in contact with the surface to 18 °C. Because the air is warmer than its surroundings (13 °C) it is less dense and therefore buoyant. This situation, known as

atmospheric **instability**, results in air rising freely in a convection current. When its internal temperature reaches the dew point (8 °C) condensation occurs and cloud starts to form. The air continues to rise so long as its internal temperature is higher than the surrounding atmosphere. In this example, equilibrium is attained at 4000 m, at a temperature of −13 °C. This marks the top of the cloud. With its base at 1000 m, the cloud has a vertical height of 3000 m. Above 4000 m the atmosphere is **stable**. Air cannot rise freely in this zone because it is cooler (and therefore heavier) than its surroundings.

Stretch and challenge

Lapse rates
Study the lapse rate diagrams in Figure 4.6.

a) Explain the differences between the environmental, dry and saturated lapse rates.
b) Why is the saturated adiabatic lapse rate lower than the dry adiabatic lapse rate?
c) Give a brief explanation of atmospheric stability/instability shown in each of the diagrams in Figure 4.6.
d) Make a sketch of the diagrams in Figure 4.6 and show the likely extent of cloud development.

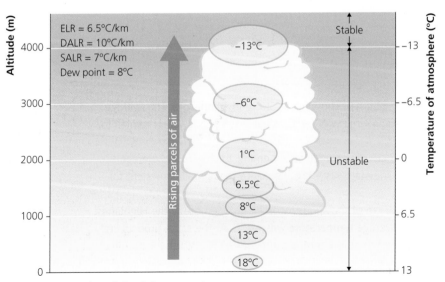

Figure 4.5 Formation of clouds by convection

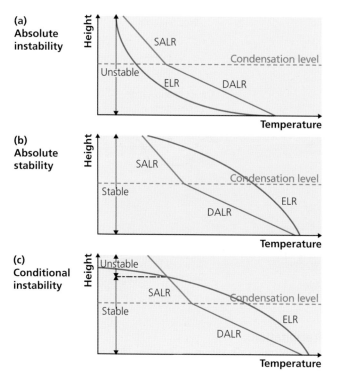

(a) Absolute instability

(b) Absolute stability

(c) Conditional instability

Figure 4.6 Lapse rates

Catchment hydrology

Evaporation

Evaporation is the phase change of liquid water to vapour and is the main pathway by which water enters the atmosphere. Heat is needed to bring about evaporation and break the molecular bonds of water. But this energy input does not produce a rise of temperature in the water. Instead the energy is absorbed as latent heat and released later in condensation. This process allows huge quantities of heat to be transferred around the planet: from the oceans to the continents; and from the tropics to the poles.

Interception

Vegetation intercepts a proportion of precipitation, storing it temporarily on branches, leaves and stems.

Eventually this moisture either evaporates (**interception loss**) (Table 4.5) or falls to the ground. Rainwater that is briefly intercepted before dripping to the ground is known as **throughfall**. During periods of prolonged or intense rainfall, intercepted rainwater may flow to the ground along branches and stems as **stemflow**.

> ### 👣 Fieldwork ideas
>
> Investigate rates of interception in adjacent areas of deciduous and coniferous woodlands. Record rainfall totals, intensity and duration for individual rainfall events for any two months from June to October in (a) deciduous woodland, (b) coniferous woodland and (c) open ground. Use chi-squared analysis (see Chapter 15, Geographical Skills, pages 539–40) to test for differences between: (a) interception rates for the two woodland types, (b) rainfall intensity and interception rates and (c) rainfall duration and interception rates.

Infiltration, throughflow, groundwater flow and run-off

Rain falling to the ground and not entering storage follows one of two flowpaths to streams and rivers (Figure 4.7):

- **Infiltration** by gravity into the soil and lateral movement or **throughflow** to stream and river channels
- **Overland flow** across the ground surface either as a sheet or as trickles and rivulets to stream and river channels

Two conflicting ideas explain the flowpaths followed by rainwater. One relates overland flow to the soil's **infiltration capacity** or the maximum rate it can absorb rain. Thus, it is argued that when rainfall intensity exceeds infiltration capacity overland flow occurs. The second idea states that rainfall, regardless of its intensity, always infiltrates the soil. Overland flow

Table 4.5 Factors affecting interception loss

Interception storage capacity	Before the onset of rainfall, vegetation surfaces are dry and their ability to retain water is at a maximum. Initially most rainfall is intercepted. However, as vegetation becomes saturated, output of water through stemflow and throughfall increases. Interception therefore depends on the duration and intensity of a rainfall event.
Wind speed	Rates of evaporation increase with wind speed. Turbulence also increases with wind speed causing additional throughfall.
Vegetation type	Interception losses are greater from grasses than from agricultural crops. Trees, which have a large surface area and aerodynamic roughness, have higher interception losses than grasses.
Tree species	Interception losses are far greater from evergreen conifers (e.g. spruce, pine) than from broad-leaved, deciduous trees (e.g. oak, ash). This is because (a) most conifers have leaves all year round, and (b) water adheres to the spaces between conifer needles (like water on a comb). This increases evaporation.

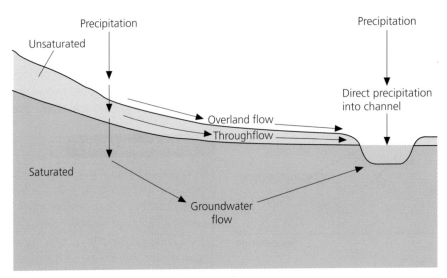

Figure 4.7 Flowpaths of the sources of stream flow

only occurs when soil becomes saturated and the water table rises to the surface. This process is known as **saturated overland flow**.

Where soils are underlain by permeable rocks, water seeps or **percolates** deep underground. This water then migrates slowly through the rock pores and joints as **groundwater flow**, eventually emerging at the surface as springs or seepages. Groundwater levels on the chalk in southern England follow a distinct seasonal pattern. By late October the water table is beginning to rise as temperatures and evapotranspiration fall. This **recharge** continues until late January. Groundwater levels then decline throughout the late winter, spring and summer, reaching their lowest point in early autumn.

Cryospheric processes

Ablation is loss of ice from snow, ice sheets and glaciers due to a combination of melting, evaporation and **sublimation**. Meltwater is an important component of river flow in high latitudes and mountain **catchments** in spring and summer. Rapid thawing of snow in upland Britain in winter is a common cause of flooding in adjacent lowlands (e.g. Welsh uplands and the Lower Severn Valley, Pennines and the Vale of York).

The processes of the carbon cycle

Carbon exchanges

The main processes involved in carbon exchanges (or **fluxes**) are: precipitation, photosynthesis, weathering, respiration, decomposition and combustion.

Precipitation

Atmospheric CO_2 dissolves in rainwater to form weak carbonic acid. This is a natural process. However, rising concentrations of CO_2 in the atmosphere, due to **anthropogenic** emissions, have increased the acidity of rainfall. This has contributed to increased acidity of ocean surface waters with potentially harmful effects on marine life.

Photosynthesis

The flux of carbon from the atmosphere to land plants and phytoplankton via photosynthesis averages around 120 gigatonnes (GT) a year. Using the Sun's energy, CO_2 from the atmosphere and water, green plants and marine phytoplankton convert light energy to chemical energy (glucose) through the process of photosynthesis.

$$\text{photosynthesis} = \underset{\substack{\text{carbon} \\ \text{dioxide}}}{6CO_2} + \underset{\text{water}}{6H_2O} \longrightarrow \underset{\text{glucose}}{C_6H_{12}O_6} + \underset{\text{oxygen}}{6O_2}$$

Plants use energy in the form of glucose to maintain growth, reproduction and other life processes. In doing so they release CO_2 to the atmosphere in respiration.

Weathering

Weathering is the *in situ* breakdown of rocks at or near the Earth's surface by chemical, physical and biological processes. Most weathering involves rainwater which contains dissolved CO_2, derived from the soil as well as the atmosphere. As we have

Figure 4.8 Limestone pavement, North Yorkshire

Figure 4.9 Norber erratics, North Yorkshire

seen, rainwater is a weak carbonic acid, which slowly dissolves limestone and chalk in a process known as **carbonation** (Figure 4.8).

$$\text{carbonation} = \underset{\substack{\text{calcium}\\\text{carbonate}}}{CaCO_3} + \underset{\substack{\text{carbonic}\\\text{acid}}}{H_2CO_3} \longrightarrow \underset{\substack{\text{calcium}\\\text{bicarbonate}}}{Ca(HCO_3)_2}$$

Carbonation releases carbon from limestones to streams, rivers, oceans and the atmosphere. The process is most effective beneath a soil cover because the higher concentration of CO_2 in the soil makes rainwater highly acidic.

It is estimated that chemical weathering transfers 0.3 billion tonnes of carbon to the atmosphere and the oceans every year. The effectiveness of solution weathering on limestone can be seen at Norber Brow in the Yorkshire Dales (Figure 4.9) where the limestone surface has been lowered by nearly half a metre over the past 13,000 years.

Skills focus

Table 4.6 shows a random sample of limestone pedestal heights which support glacial erratic boulders at Norber Brow in North Yorkshire (Figure 4.9). The erratic boulders were transported and deposited at Norber by a glacier around 13,000 years ago. Since then the whole of the limestone surface (except for the areas protected beneath the erratic blocks) has been lowered by chemical weathering.

Table 4.6 Heights of limestone pedestals at Norber Brow (cm)

22.6	34.2	28.6	41.8	32.8	41.2	31	33.4	17.6	9.9	18.2	18.1
30.6	37.2	16.5	30.6	9.5	40.6	36.8	22.4	24.6	41.8	57.6	8.8
71.2	36.6	21.6	76.4	29.7	41	37	43.6	42	56.6	59	

1 Represent the data in Table 4.6 as a histogram and a frequency distribution curve (pages 514–15).
2 Summarise the data set by calculating the mean, median and modal height of the limestone pedestals (pages 532–33).
3 Summarise the dispersion of height values by calculating the range, interquartile range and standard deviation of the data set (pages 533–35).
4 Calculate the standard error of the mean to determine the likely mean value of pedestal height with 95 per cent confidence (page 536).
5 Comment on your results and the amount and rate of lowering of the limestone surface by chemical weathering.

Physical weathering by **freeze-thaw** breaks rocks down into smaller particles but involves no chemical changes. However, it increases the surface area exposed to chemical attack. **Biological weathering** processes such as **chelation** also contribute to rock breakdown. Rainwater mixed with dead and decaying organic material in the soil forms humid acids which attack rock minerals. This process is important in humid tropical environments where decomposition is rapid and forest trees provide abundant leaf litter.

Respiration

Respiration is the process in which carbohydrates (e.g. glucose) fixed in photosynthesis are converted to CO_2 and water.

$$\text{respiration} = C_6H_{12}O_6 + 6O_2 = 6CO_2 + 6H_2O + \text{energy}$$

Plants and animals absorb oxygen which 'burns' these carbohydrates and provides the energy needed for metabolism and growth. Respiration is the reverse of photosynthesis. Whereas photosynthesis absorbs CO_2 and emits oxygen, respiration absorbs oxygen and releases CO_2. Respiration and photosynthesis are the two most important processes in the fast carbon cycle. The volume of carbon exchanged by respiration and photosynthesis each year is one thousand times greater than that moving through the slow carbon cycle.

> ### 👟 Fieldwork ideas
>
> Investigate variable rates of carbonation in an area of limestone pavement. Test the hypothesis that carbonation has been most effective where the pavement has been covered by soil longest.
>
> 1 Use a spatial sampling technique (see Chapter 15, Geographical Skills, pages 528–31) to select and measure the width of grikes (Figure 4.8) and clints (rectangular limestone blocks between grikes) (a) where the pavement has remained beneath a soil cover longest, i.e. usually at the back, (b) at the front (or scar) of the pavement, which has been exposed longest to the atmosphere.
> 2 Analyse the significance of your results using one of the following inferential statistical tests: t-test, U-test, correlation, chi-squared (see Chapter 15, Geographical Skills, pages 536–40).
> 3 Comment on your results.

Decomposition

Decomposer organisms such as bacteria and fungi breakdown dead organic matter, extracting energy and releasing CO_2 to the atmosphere and mineral nutrients to the soil. Rates of decomposition depend on climatic conditions. The fastest rates occur in warm, humid environments such as the tropical rainforest. In contrast, decomposition is slow in cold environments like the tundra or drylands such as tropical deserts.

Combustion

Combustion occurs when organic material reacts or burns in the presence of oxygen. The combustion process releases CO_2 as well as other gases such as sulphur dioxide and nitrogen oxides.

Combustion is a natural fuel use in many ecosystems. Wildfires caused by lightning strikes are essential to the health of some ecosystems such as the coniferous forests of the Rocky Mountains. Long, cold winters slow the decomposition of forest litter which builds up on the forest floor. Fire shifts this log jam, freeing carbon and nutrients previously inaccessible to forest trees. It also opens up the forest **canopy**, creating new habitats and increasing **biodiversity**.

Combustion also results from human activities such as the deliberate firing of forest and grassland in order to clear land for cultivation or improve the quality of grazing. More important is the combustion of fossil fuels. Despite international efforts to curb CO_2 emissions, oil, coal and natural gas power the global economy and their consumption continues to grow. Currently the burning of fossil fuels transfers nearly 10 GT of CO_2 a year from geological store to the atmosphere, oceans and biosphere.

Carbon sequestration in the oceans

The oceans 'take up' carbon by two mechanisms, referred to as a 'physical pump' and a 'biological pump'.

Physical (inorganic) pump

The physical carbon pump involves the mixing of surface and deep ocean waters by vertical currents, creating a more even distribution of carbon – both geographically and vertically – in the oceans. Initially CO_2 enters the oceans from the atmosphere by diffusion. Surface ocean currents then transport the water and its dissolved CO_2 polewards where it cools, becomes more dense and sinks. This **downwelling** occurs in only a handful of places in the oceans (Figure 4.10). One of these places is in the North Atlantic between Greenland and Iceland. Downwelling carries dissolved carbon to the ocean depths where individual carbon molecules may remain for centuries. Eventually deep ocean currents transport the carbon to areas of upwelling. There cold, carbon-rich water rises to the surface and CO_2 diffuses back into the atmosphere.

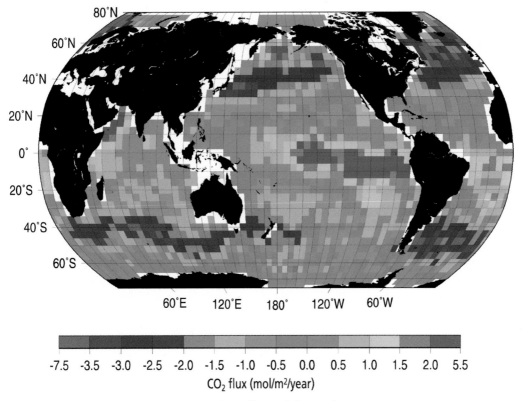

CO₂ flux (mol/m²/year)

Figure 4.10 Satellite image of ocean areas of upwelling and downwelling

---

Activities

Study Figure 4.10. The red areas on the map show the location of upwelling; the blue areas correspond with areas of downwelling.

1 Describe and explain what is happening to dissolved CO_2 at these two types of location.
2 Suggest an explanation for the global distribution of upwelling and downwelling in the oceans.

Biological (organic) pump

Carbon is also exchanged between the oceans and atmosphere through the actions of marine organisms. Globally nearly half of all carbon fixation by photosynthesis takes place in the oceans. Around 50 GT of carbon is drawn from the atmosphere by the biological pump every year.

Marine organisms drive the biological pump. Phytoplankton, floating near the ocean surface combines sunlight, water and dissolved CO_2 to produce organic material. Whether consumed by animals in the marine **food chain**, or through natural death, carbon locked in the phytoplankton either accumulates in sediments on the ocean floor or is decomposed and released into the ocean as CO_2. Other marine organisms such as tiny coccolithophores,

molluscs and crustaceans extract carbonate and calcium ions from sea water to manufacture plates, shells and skeletons of calcium carbonate. Most of this carbon-rich material eventually ends up in ocean sediments and is ultimately **lithified** to form chalk and limestone.

Vegetation

Land plants, especially trees in the rainforests and boreal forests, contain huge stores of carbon. Most of this carbon, extracted from atmospheric CO_2 through photosynthesis, is locked away for decades.

Review questions

1 What is meant by the term 'water balance'?
2 What is the dew point and why is it significant?
3 What is the difference between convection and advection?
4 What is latent heat?
5 Explain the difference between stemflow and throughflow.
6 What is saturated overland flow?
7 How do the processes of infiltration and percolation differ?
8 What is carbonation?

4.2 How do the water and carbon cycles operate in contrasting locations?

> **Key idea**
> → It is possible to identify the physical and human factors that affect the water and carbon cycles in a tropical rainforest

🔍 Case study: The Amazon rainforest

The Amazon rainforest in South America occupies an area of more than 6 million km² (Figure 4.11). The majority – 70 per cent – of the rainforest is in Brazil, but the forest also extends into parts of neighbouring Peru, Ecuador, Venezuela, Colombia, Bolivia and Guyana.

The rainforest water cycle

Amazonia is the world's largest tract of rainforest. Dominated by tall, evergreen, hardwood trees, its climatic features are (Figure 4.12):

- High average annual temperatures between 25°C and 30°C
- Small seasonal variation in temperature
- High average annual rainfall (>2000 mm) with no dry season

High average temperatures are a response to intense insolation throughout the year. However, significant cloud cover ensures that maximum temperatures do not reach the extremes of sub-tropical desert climates. Seasonal differences in temperature are small and convectional rain falls all year round, though most areas experience at least one drier period.

Between 50 and 60 per cent of precipitation in Amazonia is recycled by evapotranspiration

Figure 4.11 Amazonia

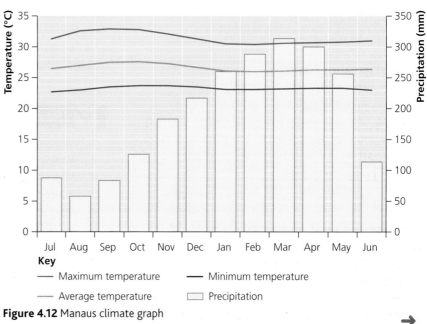

Figure 4.12 Manaus climate graph

Table 4.7 The water cycle in the Amazon rainforest

Flows and stores	Characteristics
Precipitation	High average annual rainfall (>2000 mm). Rainfall fairly evenly distributed throughout the year though short drier season occurs in some places. High-intensity, convectional rainfall. Interception by forest trees is high (around 10 per cent of precipitation). Intercepted rainfall accounts for 20–25 per cent of all evaporation.
Evapotranspiration	High rates of evaporation and transpiration due to high temperatures, abundant moisture and dense vegetation. Strong evapotranspiration–precipitation feedback loops sustain high rainfall totals. Around a half of incoming rainfall is returned to the atmosphere by evapotranspiration. Most evaporation is from intercepted moisture from leaf surfaces. Moisture lost in transpiration is derived from the soil via tree roots.
Run-off	Rapid run-off related to high rainfall, intensive rainfall events and well-drained soils. Depending on seasonal distribution of rainfall, river discharge may peak in one or two months of the year.
Atmosphere	High temperatures allow the atmosphere to store large amounts of moisture (i.e. **absolute humidity** is high). **Relative humidity** is also high.
Soil/groundwater, etc.	Abundant rainfall and deep tropical soils leads to significant water storage in soils and aquifers.
Vegetation	Rainforest trees play a crucial role in the water cycle, absorbing and storing water from the soil and releasing it through transpiration.

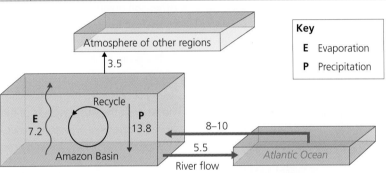

Figure 4.13 Amazonia: water balance diagram (values in gigatonnes/year)

(Figure 4.13). Water losses from the Amazon Basin result from river flow and export of atmospheric vapour to other regions. This loss is made good by an inward flux of moisture from the Atlantic Ocean.

The rainforest carbon cycle

Amazonia's humid equatorial climate creates ideal conditions for plant growth. Net primary productivity (NPP) is high, averaging 2500 grams/m²/year and the biomass is between 400 and 700 tonnes/ha. Large forest trees typically store around 180 tonnes C/ha above ground, and a further 40 tonnes C/ha in their roots. Soil carbon stores average between 90 and 200 tonnes/ha. The Amazon rainforest is a major global reservoir of stored carbon, absorbing 2.4 billion tonnes a year.

Compared to other forest ecosystems, exchanges of carbon between the atmosphere, biosphere and soil are rapid. Warm, humid conditions ensure speedy decomposition of dead organic matter and the quick release of CO_2. Meanwhile, rates of carbon fixation through photosynthesis are high.

Amazonia's leached and acidic soils contain only limited carbon and nutrient stores. The fact that such poor soils

Activities

Study the climate graph for Manaus in Amazonia (3°S, 50°W) (Figure 4.12).

1 Describe the main patterns of temperature and rainfall at Manaus.
2 Using the evidence in Figure 4.12, explain how the pattern of temperature and rainfall is typical of an equatorial location.
3 Explain how temperature and rainfall are likely to influence the carbon cycle in Amazonia.

support a biome with the highest NPP and biomass of all terrestrial ecosystems, emphasises the speed with which organic matter is broken down, mineralised and recycled.

Physical factors and stores and flows of water

Physical factors such as geology, relief and temperature affect flows and stores of water in the Amazon rainforest and other environments. These factors are described in Table 4.8.

Table 4.8 Physical factors and stores and flows in the water cycle in Amazonia

Characteristic	Effect on the flood hydrograph
Geology	Impermeable catchments (e.g. large parts of the Amazon Basin are an ancient shield area comprising impermeable, crystalline rocks) have minimal water storage capacity resulting in rapid run-off. Permeable and **porous** rocks such as limestone and sandstone store rainwater and slow run-off.
Relief (slopes)	Most of the Amazon Basin comprises extensive lowlands. In areas of gentle relief water moves across the surface (**overland flow**) or horizontally through the soil (**throughflow**) to streams and rivers. In the west the Andes create steep catchments with rapid run-off. Widespread inundation across extensive floodplains (e.g. the Pantanal) occurs annually, storing water for several months and slowing its movement into rivers.
Temperature	High temperatures throughout the year generate high rates of evapotranspiration. Convection is strong, leading to high atmospheric humidity, the development of thunderstorm clouds and intense precipitation. Water is cycled continually between the land surface, forest trees and the atmosphere by evaporation, transpiration and precipitation.

Physical factors affecting stores and flows of carbon

Forest trees dominate the biomass of the Amazon Basin and are the principal carbon store. In total approximately 100 billion tonnes of carbon is locked up in the Amazon rainforest. Absorbing around 2.4 billion tonnes of CO_2 a year and releasing 1.7 billion tonnes through decomposition, the rainforest is a carbon sink of global importance. 60 per cent of rainforest carbon is stored in the above ground biomass of tree stems, branches and leaves. The remainder is below ground, mainly as roots and soil organic matter.

Carbon cycles between the forest and other living organisms, the soil and the atmosphere. Photosynthesis connects the rainforest to the atmosphere carbon stores. High temperatures, high rainfall and intense sunlight stimulate primary production. NPP averages about 2500 grams/m²/year. Amazonia alone accounts for 15–25 per cent of all NPP in terrestrial ecosystems.

Leaf litter and other dead organic matter accumulates temporarily at the soil surface and within rainforest soils. High temperatures and humid conditions promote rapid decomposition of organic litter by bacteria, fungi and other soil organisms. Decomposition releases nutrients to the soil for immediate take-up by tree root systems, and emits CO_2 which is returned to the atmosphere.

The geology of the Amazon Basin is dominated by ancient igneous and metamorphic rocks. Carbonates are largely absent from the mineral composition of these rocks. However, in the western parts of the basin, close to the Andes, outcrops of limestone occur. In the context of the slow carbon cycle they are significant regional carbon stores.

Human factors affecting stores and flows of water

Deforestation in Amazonia averaged around 17,500 km²/year between 1970 and 2013. Since 1970 almost one-fifth of the primary forest has been destroyed

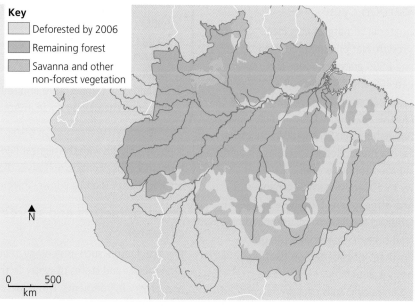

Key
- Deforested by 2006
- Remaining forest
- Savanna and other non-forest vegetation

N

0   500
km

Figure 4.14 Deforestation in the Amazon Basin

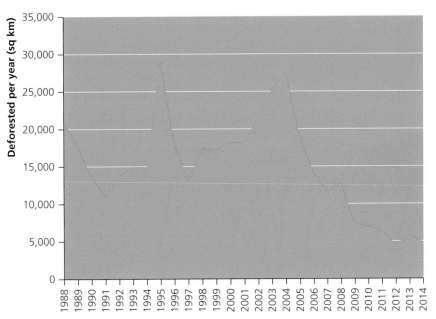

Figure 4.15 Amazonia: deforestation, 1988–2014

or degraded (Figure 4.14), though in recent years rates of deforestation have slowed (Figure 4.15).

In April 2014 devastating floods occurred on the Madeira River, the largest tributary of the Amazon River (Figure 4.16). At Porto Velho the river reached record levels of 19.68 m above normal. Vast expanses of floodplain were inundated; 60 people died; 68,000 families were evacuated; and there were outbreaks of cholera and leptospirosis.

In the Upper Madeira drainage basin human activity has modified stores and flows in the water cycle. Deforestation has reduced water storage in forest trees, soils (which have been eroded), permeable rocks (due to more rapid run-off) and in the atmosphere. At the same time fewer trees mean less evapotranspiration and therefore less precipitation. Meanwhile, total run-off and run-off speeds have increased, raising flood risks throughout the basin.

Despite torrential rains in the upper basin of the Madeira River, the main driver of the floods was deforestation in Bolivia and Peru. Between 2000 an d 2012, 30,000 km² of Bolivian rainforest was cleared for subsistence farming and cattle ranching. Much of this deforestation occurred on steep lower slopes of the Andes. The result was a massive reduction in water storage and accelerated run-off.

Deforestation has a huge impact on the water cycle and has the potential to change the climate at local and regional scales. Converting rainforest to grassland increases run-off by a factor of 27, and half of all rain falling on grassland goes directly into rivers. Rainforest trees are a crucial part of the water cycle, extracting moisture from the soil, intercepting rainfall and releasing it to the atmosphere through transpiration, as well as

stabilising forest **albedo** and ground temperatures. This cycle sustains high atmospheric humidity which is responsible for cloud formation and heavy conventional rainfall. Deforestation breaks this cycle and can lead to permanent climate change (Figure 4.17).

However, the impact of deforestation on water cycles is not just local. Projections of future deforestation in Amazonia predict a 20 per cent decline in regional rainfall as the rainforest dries out and forest trees are gradually replaced by grassland. Nor is it just deforested areas that experience a reduction in rainfall: disruption of the regional water cycle means that forests hundreds of kilometres downwind of degraded sites are affected too.

Figure 4.16 Madeira drainage basin

113

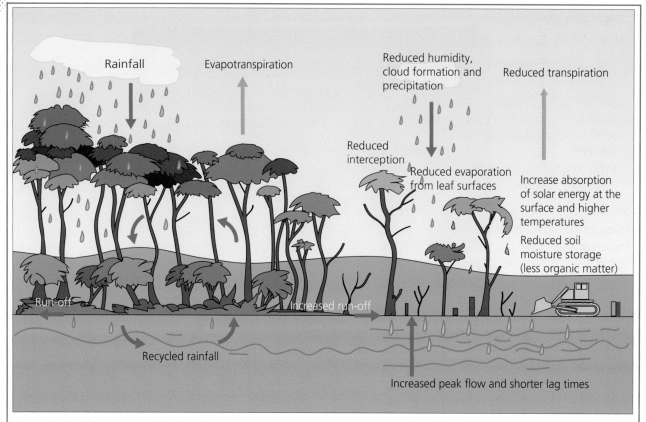

Figure 4.17 Deforestation and the water cycle

Human factors affecting carbon and nutrient flows and stores

Present-day deforestation is most severe in the tropical rainforest. In primary rainforest, unaffected by human activity, the biomass of trees represents about 60 per cent of all the carbon in the ecosystem. The above ground carbon biomass in the rainforest is approximately 180 tonnes/ha. Most of the remaining carbon is found in the soil as roots and dead organic material.

Deforestation exhausts the carbon biomass store. Croplands and pasture contain only a small amount of carbon compared to forest trees. For example the biomass of grasslands in areas of former rainforest is 16.2 tonnes/ha; and for soya cultivation it is just 2.7 tonnes/ha. At the same time deforestation drastically reduces inputs of organic material to the soil. Soils, depleted of carbon and exposed to strong sunlight, support fewer decomposer organisms, thus reducing the flow of carbon from the soil to the atmosphere.

In tropical rainforests, the principal store of plant nutrients such as calcium, potassium and magnesium is forest trees. Rainforest soils contain only a small reservoir of essential nutrients and the forest is only sustained by a rapid nutrient cycle. Deforestation destroys the main nutrient store – the forest trees – and removes most nutrients from the ecosystem. Nutrients no longer taken up by the root systems of trees are washed out of soils by rainwater; and soils, without the protective cover of trees, are quickly eroded by run-off.

Strategies to manage tropical rainforests: the positive effects on the water and carbon cycles

The degrading or outright destruction of large areas of Amazon rainforest is an issue of international as well as national concern. This is because deforestation has implications for global climate change (see Chapter 10). Brazil is committed to restoring 120,000 km² of rainforest by 2030.

Indigenous people have lived sustainably in the rainforest for thousands of years, maintaining the water balance, carbon cycle and the forest's biodiversity. These people survived as hunter-gatherers and **shifting cultivators**. In stark contrast to exploitative commercial farming, logging and mining of the past 50 years, indigenous people pursued a way of life perfectly adapted to the limited resources and fragility of the rainforest. →

Modern strategies to manage the Amazon rainforest sustainably fall into three categories:

- Protection through legislation of large expanses of primary forest so far unaffected by commercial developments
- Projects to reforest areas degraded or destroyed by subsistence farming, cattle ranching, logging and mining
- Improving agricultural techniques to make permanent cultivation possible

Since 1998, the Brazilian government has established many forest conservation areas. These Amazon Regional Protected Areas now cover an area twenty times the size of Belgium. By 2015, 44 per cent of the Brazilian Amazon comprised national parks, wildlife reserves and indigenous reserves where farming is banned.

Several reforestation projects, sponsored by local authorities, non-governmental organisations (NGOs) and businesses are underway, but so far progress has been slow. One such example is the Parica project in Rondônia in the western Amazon. This sustainable forestry scheme aims to develop a 1000 km² commercial timber plantation on government-owned, deforested land. The plan is for 20 million fast-growing, tropical hardwood seedlings, planted on 4000 smallholdings, to mature over a period of 25 years. Financial assistance is given to smallholders for land preparation, planting and the maintenance of plots. Tree nurseries provide them with seedlings. Timber will be exported along the Amazon and its tributaries through Manaus or Port Velho.

Although this project is a **monoculture** and cannot replicate the biodiversity of the primary rainforest, it is sustainable. It also sequesters carbon in the trees and soil; reduces CO_2 emissions from deforestation; re-establishes water and carbon cycles; and reduces run-off and the loss of plant nutrients and carbon from the soil.

Also in Rondônia, the indigenous Suruí people participate in a scheme that aims to protect primary rainforest on tribal lands from further illegal logging, and reforest areas degraded by deforestation in the past 40 years. The Suruí plant seedlings bred in local nurseries in deforested areas around their villages. The native species planted are chosen to provide them with timber for construction, food crops and, through logging, a sustainable source of income.

In 2009 the Suruí were the first indigenous group in Amazonia to join the UN's Reducing Emissions

from Deforestation and Degradation (REDD) scheme. This scheme provides payment to the tribe for protecting the rainforest and abandoning logging. It is a market-based approach involving granting of carbon credits to the Suruí. These credits can be purchased by international companies which have exceeded their annual carbon emissions quotas. In 2013, Natura, a large cosmetics transnational corporation (TNC), purchased 120,000 tonnes of carbon credits from the Suruí. This was the first carbon credit sale by indigenous people in Amazonia.

 Activity

Investigate in more detail the UN's REDD+ scheme: its purpose, how it operates, its benefits and its possible weaknesses. Start by logging in to: www.un-redd.org and www.carbonplanet.com

Improved agricultural techniques

Farming has been the main cause of deforestation in Amazonia. However, the low fertility of soils meant that permanent cultivation proved unsustainable. After a few years, smallholders abandoned their plots which were then converted to low quality grassland. Extensive ranching enterprises could scarcely support stocking levels of one head of cattle per hectare.

One response to improve agriculture has been diversification. Soil fertility can be maintained by rotational cropping and combining livestock and arable operations. Integrating crops and livestock could allow a fivefold increase in ranching productivity and help slow rates of deforestation.

European explorers observed that the Amazon rainforest, as late as the sixteenth century, supported high population densities, and many large urban centres. This appears to contradict the view that natural resources for farming in the region are too poor to support settled, permanent cultivation. The explanation is thought to be human-engineered soils: so-called dark soils made from inputs of charcoal, waste and human manure. Charcoal in these soils attracts micro-organisms and fungi and allows the soils to retain their fertility long-term. Scientists are currently investigating these dark soils. If they can be successfully recreated they would allow intensive and permanent cultivation which would drastically reduce deforestation and carbon emissions.

Case study: The Arctic tundra

The Arctic tundra occupies some 8 million km² in northern Canada, Alaska and Siberia. It extends from the northern edge of the boreal coniferous forest to the Arctic Ocean and its southern limit approximates the 10 °C July isotherm (i.e. climatic limit of the **tree line**). Climatic conditions in the tundra are severe and become more extreme with latitude. For eight or nine months a year the tundra has a negative **heat balance** with average monthly temperatures below freezing. As a result the ground is permanently frozen with only the top metre or so thawing during the Arctic summer (Figure 4.18).

Permafrost underlies much of the tundra and is an important feature of the region's water cycle (Figure 4.19). In winter, when for several weeks the Sun remains below the horizon, temperatures can plunge below −40 °C. Long hours of daylight in summer provide some compensation for brevity of the growing season. Mean annual precipitation is low.

Few plants and animals have adapted to this extreme environment; biodiversity is low and apart from a few dwarf species, the ecosystem is treeless. In the southern areas – the Low Arctic – conditions are less severe, and vegetation provides a continuous ground cover. Further north in the High Arctic, plant cover is discontinuous with extensive areas of bare ground.

Water cycle in the tundra

The main features of the water cycle in the tundra are:

- Low annual precipitation (50–350 mm) with most precipitation falling as snow
- Small stores of moisture in the atmosphere owing to low temperatures which reduce absolute humidity

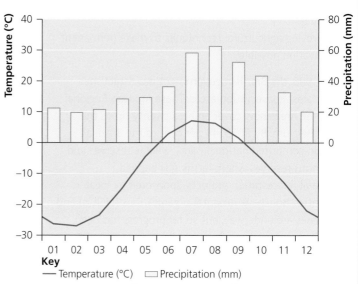

Figure 4.18 Iqaluit (Baffin Island, Canada) climate chart (Source: Climate-Data.org)

- Limited transpiration because of the sparseness of the vegetation cover and the short growing season
- Low rates of evaporation. Much of the Sun's energy in summer is expended melting snow so that ground temperatures remain low and inhibit convection. Also, surface and soil water are frozen for most of the year.
- Limited groundwater and soil moisture stores. Permafrost is a barrier to infiltration, percolation, recharge and groundwater flow.
- Accumulation of snow and river/lake ice during the winter months. Melting of snow, river and lake ice, and the uppermost **active layer** of the permafrost in spring and early summer, results in a sharp increase in river flow.
- Extensive wetlands, ponds and lakes on the tundra during summer (Figure 4.20). This temporary store of liquid water is due to permafrost which impedes drainage.

| Continuous | Discontinuous | Sporadic |

Figure 4.19 Distribution of permafrost in the northern hemisphere

Figure 4.20 Wetlands in the Arctic tundra

Carbon cycle in the tundra

The permafrost is a vast carbon sink. Globally it is estimated to contain 1600 GT of carbon. The accumulation of carbon is due to low temperatures which slow decomposition of dead plant material. Overall, the amount of carbon in tundra soils is five times greater than in the above-ground biomass.

The flux of carbon is concentrated in the summer months when the active layer thaws. Plants grow rapidly in the short summer. Long hours of daylight allow them to flower and fruit within just a few weeks. Nonetheless, net primary productivity (NPP) is less than 200 grams/m²/year. Consequently the tundra biomass is small, ranging between 4 and 29 tonnes/ha depending on the density of vegetation cover.

During the growing season tundra plants input carbon-rich litter to the soil. The activity of micro-organisms increases, releasing CO_2 to the atmosphere through respiration. However, CO_2 (and methane (CH_4)) emissions are not just confined to the summer. Even in winter, pockets of unfrozen soil and water in the permafrost act as sources of CO_2 and CH_4. Meanwhile snow cover may insulate microbial organisms and allow some decomposition despite the low temperatures.

In the past the permafrost functioned as a **carbon sink**. But today, global warming has raised concerns that it is becoming a **carbon source**. At the moment the evidence is unclear. While outputs of carbon from the permafrost have increased in recent decades, higher temperatures have stimulated plant growth in the tundra and greater uptake of CO_2. This in turn has increased the amount of plant litter entering store. It is possible therefore, that despite the warming Arctic climate, the carbon budget in the tundra today remains in balance.

Stretch and challenge

Table 4.9 Mean monthly flow (cumecs) and air temperatures (°C), Meade River, Atqasuk, Alaska
(Source: USGS Water Resources)

	J	F	M	A	M	J	J	A	S	O	N	D
Flow	0	0	0	0	43.6	104	66.7	115.2	21.7	7.4	0.3	0
Temp.	−28	−25	−27	−14	−7	6.7	12.4	10.3	3.6	−3.6	−16	−20

The river flow and temperature data (Table 4.9) were recorded on the North Slope of Alaska (70°30′N). The Meade River drains a catchment of 4618 km².

1 Plot the monthly flows and temperature values as a two-axis (two y axes) chart. Flows should be plotted as a bar chart; temperatures as a line chart.
2 Summarise the relationship between the flow regime of the Meade River and monthly air temperatures.
3 Suggest a possible explanation for two peaks in river flow.

Physical factors, seasonal changes and stores and flows of water and carbon

Water cycle

Flow and stores of water in the tundra are influenced by temperature, relief and rock permeability.

● Average temperatures are well below freezing for most of the year so that water is stored as ground ice in the permafrost layer. During the short summer the shallow active layer (top metre) thaws and liquid water flows on the surface. Meltwater forms millions of pools and shallow lakes which stud the tundra landscape. Drainage is poor: water cannot infiltrate the soil because of the permafrost at depth. In winter, sub-zero temperatures prevent evapotranspiration. In summer some evapotranspiration occurs from standing water, saturated soils and vegetation. Humidity is low all year round and precipitation is sparse.

→

- Permeability is low owing to the permafrost and the crystalline rocks which dominate the geology of the tundra in Arctic and sub-Arctic Canada.
- The ancient rock surface which underlies the tundra has been reduced to a gently undulating plain by hundreds of millions of years of erosion and weathering. Minimal relief and chaotic glacial deposits impede drainage and contribute to waterlogging during the summer months.

Carbon cycle

- Carbon is mainly stored as partly decomposed plant remains frozen in the permafrost. Most of this carbon has been locked away for at least the past 500,000 years.
- Low temperatures, the unavailability of liquid water for most of the year and parent rocks containing few nutrients, limit plant growth. Thus the total carbon store of the biomass is relatively small. Averaged over the year, photosynthesis and NPP are low, with the growing season lasting for barely three months. However, there is some compensation for the short growing season in the long hours of daylight in summer.
- Low temperatures and waterlogging slow decomposition and respiration and the flow of CO_2 to the atmosphere.
- Owing to the impermeability of the permafrost, rock permeability, porosity and the mineral composition of rocks exert little influence on the water (and carbon) cycle.

Oil and gas production and the carbon and water cycles in Alaska

The North Slope of Alaska, between the Brooks Range in the south and the Arctic Ocean in the north, is a vast wilderness of Arctic tundra (Figure 4.21). Oil and gas were discovered here at Prudhoe Bay in 1968. From the start the development of the oil and gas industries on the North Slope presented major challenges: a harsh climate with extreme cold and long periods of darkness in winter; permafrost, and the melting of the active layer in summer; remoteness and poor accessibility; and a fragile wilderness of great ecological value.

Despite the challenges, production went ahead, driven by high global energy prices and the US government's policy to reduce dependence on oil imports. Massive fixed investments in pipelines, roads, oil production plants, gas processing facilities, power lines, power generators and gravel quarries were completed in the 1970s and 1980s. By the early 1990s, the North Slope accounted for nearly a quarter of the USA's domestic oil production. Today the proportion is 6 per cent though Alaska remains an important oil and gas province (Table 4.10). Decline in recent years reflects two things: high production costs on the North Slope and the massive growth of the oil shale industry in the USA.

Table 4.10 Oil production on the North Slope and in the USA, 1980–2014 (thousand/barrels/day) (Source: US Energy Information Administration)

Year	North Slope	USA
1980	1524	8597
1985	1779	8971
1990	1743	7355
1995	1441	6560
2000	942	5822
2005	845	5181
2010	589	5482
2014	479	8653

Impact on the water and carbon cycles

Oil and gas exploitation on Alaska's North Slope has had significant impacts on the permafrost and on local water and carbon cycles. Permafrost, the major carbon store in

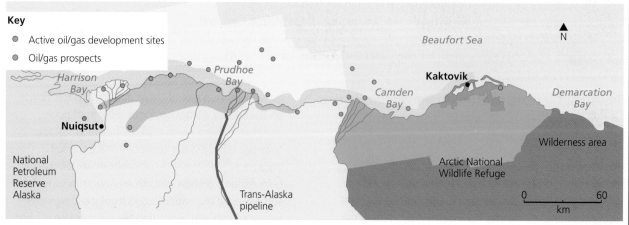

Figure 4.21 North Slope of Alaska

the tundra, is highly sensitive to changes in the thermal balance. In many areas, this balance has been disrupted by the activities of oil and gas companies which have caused localised melting of the permafrost. Melting is associated with:

- construction and operation of oil and gas installations, settlements and infrastructure diffusing heat directly to the environment
- dust deposition along roadsides creating darkened snow surfaces, thus increasing absorption of sunlight
- removal of the vegetation cover which insulates the permafrost

Permafrost melting releases CO_2 and methane (CH_4). On the North Slope, estimated CO_2 losses from the permafrost vary from 7 to 40 million/tonnes/year, while CH_4 losses range from 24,000 to 114,000 tonnes/year. Gas flaring and oil spillages also input CO_2 to the atmosphere. Other changes to the local carbon cycle are linked to industrial development. For example the destruction or degrading of tundra vegetation reduces photosynthesis and the uptake of CO_2 from the atmosphere; and the thawing of soil increases microbial activity, decomposition and emissions of CO_2. Moreover, the slow-growing nature of tundra vegetation means that regeneration and recovery from damage takes decades.

Similar changes have occurred to the water cycle. Melting of the permafrost and snow cover increases run-off and river discharge making flooding more likely. This means that in summer, wetlands, ponds and lakes have become more extensive, increasing evaporation. Strip mining of aggregates (sand and gravel) for construction creates artificial lakes which disrupt drainage and also expose the permafrost to further melting. In addition, drainage networks are disrupted by road construction and by seismic explosions used to prospect for oil and gas. Finally, water abstracted from creeks and rivers for industrial use and for the building of ice roads in winter reduce localised run-off.

Management strategies to moderate the impact on the water and carbon cycles

Development on the North Slope has often involved the deliberate destruction of the permafrost. Today the emphasis is on protecting the permafrost (Table 4.11), thus minimising disruption to the water and carbon cycles and wildlife. However, the purpose of these strategies is also pragmatic: melting permafrost causes widespread damage to buildings and roads as well as increased maintenance costs for pipelines and other infrastructure.

Table 4.11 Strategies to reduce the impact of development on the water and carbon cycles

Insulated ice and gravel pads	Roads and other infrastructural features can be constructed on insulating ice or gravel pads, thus protecting the permafrost from melting.
Buildings and pipelines elevated on piles	Constructing buildings, oil/gas pipelines and other infrastructure on piles allows cold air to circulate beneath these structures. This provides insulation against heat-generating buildings, pipework, etc. which would otherwise melt the permafrost.
Drilling laterally beyond drilling platforms	New drilling techniques allow oil and gas to be accessed several kilometres from the drilling site. With fewer sites needed for drilling rigs, the impact on vegetation and the permafrost due to construction (access roads, pipelines, production facilities, etc.) is greatly reduced.
More powerful computers can detect oil- and gas-bearing geological structures remotely	Fewer exploration wells are needed thus reducing the impact on the environment.
Refrigerated supports	Refrigerated supports are used on the Trans-Alaska Pipeline to stabilise the temperature of the permafrost. Similar supports are widely used to conserve the permafrost beneath buildings and other infrastructure.

✔ Review questions

1. State three features of temperature that define the climate of the tropical rainforest.
2. Describe the main carbon stores in the rainforest.
3. Describe the impact of deforestation on the water cycle.
4. In what ways does shifting cultivation mimic the ecology of the natural rainforest?
5. What is permafrost?
6. How does the distribution of carbon in the tundra differ from the rainforest?
7. What is the active layer of the permafrost?

4.3 How much change occurs over time in the water and carbon cycles?

> **Key idea**
> → Human factors can disturb and enhance the natural processes and stores in the water and carbon cycles

Dynamic equilibrium and the water and carbon cycles

Most natural systems, unaffected by human activity, exist in a state of **dynamic equilibrium**. They are dynamic in the sense that they have continuous inputs, throughputs, outputs and variable stores of energy and materials. In the short term, inputs, outputs and stores of water or carbon will fluctuate from year to year.

In the long term, however, flows and stores usually maintain a balance, allowing a system to retain its stability. **Negative feedback loops** within systems restore balance. In a drainage basin unusually heavy rainfall will increase the amount of water stored in aquifers. This in turn will raise the water table, increasing flow from springs until the water table reverts to normal levels. In the carbon cycle, burning fossil fuels increases atmospheric CO_2 but at the same time stimulates photosynthesis. This negative feedback response should remove excess CO_2 from the atmosphere and restore equilibrium.

Land-use changes

Urbanisation

Urbanisation is the conversion of land use from rural to urban. Farmland and woodland are replaced by housing, offices, factories and roads; natural surfaces such as vegetation and soil give way to concrete, brick or tarmac. These artificial surfaces are largely impermeable so they allow little or no infiltration and provide minimal water storage capacity to buffer run-off. Urban areas also have drainage systems designed to remove surface water rapidly (e.g. pitched roofs, gutters, sewerage systems). As a result a high proportion of water from precipitation flows quickly into streams and rivers, leading to a rapid rise in water level.

In addition to changing land use, urbanisation also encroaches on floodplains. Floodplains are natural storage areas for water. Urban development on floodplains reduces water storage capacity in drainage basins, increasing river flow and flood risks.

Farming

Farming brings changes to vegetation and soils which have implications for the carbon and water cycles.

The clearance of forest for farming reduces carbon storage in both the above- and below-ground biomass. Soil carbon storage is also reduced by ploughing and the exposure of soil organic matter to oxidation. Further losses occur through the harvesting of crops with only small amounts of organic matter returned to soils. Soil erosion invariably accompanies arable farming. Erosion by wind and water is most severe when crops have been lifted and soils have little protective cover.

Changes to the carbon cycle are less apparent on pasture land or where farming replaces natural grasslands. For instance in North America, the net primary production of annual crops such as wheat on the Great Plains exceeds that of the original Prairie grasslands. However, carbon exchanges through photosynthesis are generally lower than in natural ecosystems. In part this is explained by a lack of biodiversity in farmed systems, and the growth cycle of crops often compressed into just four or five months.

Farming also modifies the natural water cycle. Crop irrigation diverts surface water from rivers and groundwater to cultivated land. Some of this water is extracted by crops from soil storage and released by transpiration; but most is lost to evaporation and in soil drainage.

Interception of rainfall by annual crops is less than in forest and grassland ecosystems (Figure 4.22). So too is evaporation and transpiration from leaf surfaces. Ploughing increases evaporation and soil moisture loss, and furrows ploughed downslope act as drainage channels, accelerating run-off and soil erosion. Infiltration due to ploughing is usually greater in farming systems, while artificial underdrainage increases the rate of water transfer to streams and rivers. Surface run-off increases where heavy machinery compacts soils. Thus peak flows on streams draining farmland are generally higher than in natural ecosystems.

Forestry

Forest management in plantations modifies the local water and carbon cycles. Changes to the water cycle include:

● Higher rates of rainfall interception in plantations in natural forests. In eastern England, interception

Figure 4.22 Arable farming in Dorset

Figure 4.23 Hill farming in the Lake District

rates for Sitka spruce are as high as 60 per cent. In upland Britain, where temperatures and evaporation are lower, interception is about half this figure. In the UK, preferred plantation species are conifers. The needle-like structure of conifer leaves, their evergreen habit and high density of planting all contribute to high rates of interception.

- Increased evaporation. A large proportion of intercepted rainfall is stored on leaf surfaces and is evaporated directly to the atmosphere.
- Reduced run-off and stream discharge. With high interception and evaporation rates and the absorption of water by tree roots, drainage basin hydrology is altered. Streams draining plantations typically have relatively long **lag times**, low peak flows and low total discharge. The effect of conifer plantations in upland catchments is often to reduce water yield for public supply.

- Compared to farmland and moorland, transpiration rates are increased. Typical transpiration rates for Sitka spruce in the Pennines are around 350 mm/year of rainfall equivalent.
- Clear felling to harvest timber creates sudden but temporary changes to the local water cycle, increasing run-off, reducing evapotranspiration and increasing stream discharge.

Changing land use from farmland, moorland and heath to forestry increases carbon stores. In a typical plantation in the UK, mature forest trees contain on average 170–200 tonnes C/ha. This is ten times higher than grassland, and 20 times higher than heathland. The soil represents an even larger carbon pool. In England measurements of forest soil carbon are around 500 tonnes C/ha.

Forest trees extract CO_2 from the atmosphere and sequester it for hundreds of years. Most of the carbon is stored in the wood of the tree stem. However, forest trees only become an active carbon sink (i.e. absorbing more carbon than they release) for the first 100 years or so after planting. Thereafter, the amount of carbon captured levels off and is balanced by inputs of litter to the soil, the release of CO_2 in respiration and by the activities of soil decomposers. In consequence, forestry plantations usually have a rotation period of 80–100 years. After this time the trees are felled and reforestation begins afresh.

➡ Activity

With reference to the two different types of farming shown in Figures 4.22 and 4.23, compare and contrast the features of the water cycle you would expect in each area.

👣 Fieldwork ideas

Investigate the influence of local climate (microclimate) on one or more features of the water cycle:

1 Contrasts in soil moisture between north- and south-facing slopes in a valley.
2 Contrasts in rainfall and wind speed between a woodland and an adjacent open site.
3 Temperature contrasts in urban and surrounding rural areas at night, under still air conditions.

Water extraction

Water is extracted from surface and groundwater to meet public, industrial and agricultural demand. Direct human intervention in the water cycle changes the dynamics of river flow and groundwater storage.

Water extraction on the River Kennet catchment

The River Kennet in southern England drains an area of around 1200 km² in Wiltshire and Berkshire (Figure 4.24). The upper catchment mainly comprises chalk which is highly permeable. Thus groundwater contributes most of the Kennet's flow. As a chalk stream, the river supports a diverse range of habitats and wildlife. Its water, filtered through the chalk, has exceptional clarity, high oxygen levels and is fast-flowing. Among the native fauna are Atlantic salmon, brown trout, water voles, otters and white-clawed crayfish.

Within and close to the catchment, several urban areas rely on water from the Kennet basin to meet public supply. Swindon, the largest, has a population of over 200,000. The Kennet also supplies water for local industries, agriculture and public use. Thames Water abstracts groundwater from the upper catchment from boreholes. None of this water is returned to the river as waste water.

Water extraction from the Kennet and its catchment has had a significant impact on the regional water cycle:

- Rates of groundwater extraction have exceeded rates of recharge, and the falling water table has reduced flows in the River Kennet by 10–14 per cent.
- During the 2003 drought flows fell by 20 per cent, and in the dry conditions of the early 1990s by up to 40 per cent.
- Lower flows have reduced flooding and temporary areas of standing water and wetlands on the Kennet's floodplain.
- Lower groundwater levels have caused springs and seepages to dry up and reduced the incidence of saturated overland flow on the chalk.

Aquifers and artesian basins

Aquifers are permeable or porous water-bearing rocks such as chalk and New Red Sandstone. Groundwater is abstracted for public supply from aquifers by wells and boreholes. Emerging in springs and seepages, groundwater feeds rivers and makes a major contribution to their base flow. Within an aquifer the upper surface of saturation is known as the water table. Its height fluctuates seasonally and is also affected by periods of exceptional rainfall, drought and abstraction. In normal years in southern England the water table falls between

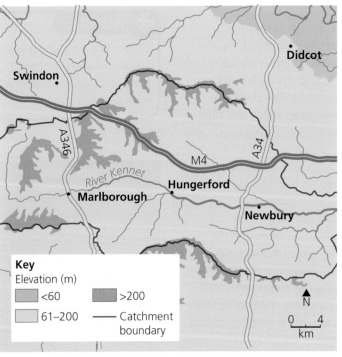

Figure 4.24 Kennet drainage basin (Source: UK National River Flow Archive)

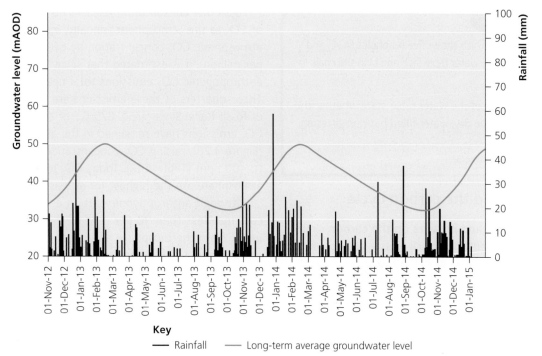

Figure 4.25 Rainfall and groundwater levels at Crompton, Sussex

March and September as rising temperatures increase evapotranspiration losses. **Recharge** resumes in the late autumn (Figure 4.25).

Artesian basins

When sedimentary rocks form a **syncline** or basin-like structure, an aquifer confined between impermeable rock layers may contain groundwater which is under **artesian pressure**. If this groundwater is tapped by a well or borehole, water will flow to the surface under its own pressure. This is known as an **artesian aquifer**. The level to which the water will rise – the **potentiometric surface** – is determined by the height of the water table in areas of recharge on the edges of the basin.

London is located at the centre of a synclinal structure which forms an artesian basin. Groundwater in the chalk aquifer is trapped between impermeable London Clay and Gault Clay. Rainwater enters the

chalk aquifer where it outcrops on the edge of the basin in the North Downs and Chilterns (Figure 4.26). Groundwater then flows by gravity through the chalk towards the centre of the basin. Thus under natural conditions the wells and boreholes in the London area are under artesian pressure.

Groundwater from the chalk is an important source of water for the capital. However, overexploitation in the nineteenth century and in the first half of the twentieth century caused a drastic fall in the water table. In central London it fell by nearly 90 m. In the past 50 years declining demand for water by industry in London and reduced rates of abstraction have allowed the water table to recover. By the early 1990s it was rising at a rate of 3 m/year and began to threaten buildings and underground tunnels. Since 1992 Thames Water has been granted abstraction licenses to slow the rise of the water table which is now stable.

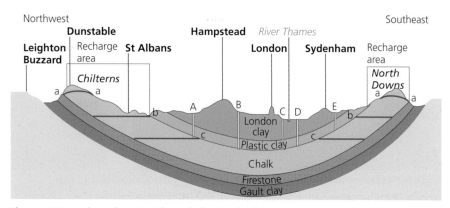

Figure 4.26 Geological section through the London Basin

Fossil fuels and the carbon cycle

Use of fossil fuels and impacts on the carbon cycle

For the past two centuries, fossil fuels – coal, oil and natural gas – have driven global industrialisation and urbanisation. Despite the development of nuclear power and renewable energy, the global economy remains overwhelmingly dependent on fossil fuels. In 2013 they accounted for 87 per cent of global energy consumption (Figure 4.27).

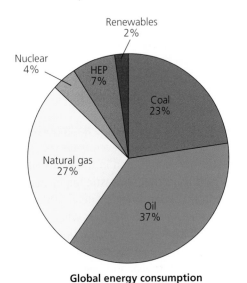

Global energy consumption

Figure 4.27 Global energy consumption, 2013

Fossil fuel consumption releases 10 billion tonnes of CO_2 to the atmosphere annually, increasing atmospheric CO_2 concentration by over 1 ppm (parts per million). It is estimated that since 1750 cumulative anthropogenic CO_2 emissions total nearly 2000 GT. Three-quarters of these emissions are from the burning of fossil fuels. Since 1750, 879 GT of anthropogenic CO_2 emissions have remained in the atmosphere (Figure 4.28), raising CO_2 concentrations from 280 ppm to 400 ppm (Table 4.12). Today CO_2 levels in the atmosphere are the highest for at least 800,000 years.

Although anthropogenic carbon emissions comprise less than 10 per cent of the natural influx from the biosphere and oceans to the atmosphere, they impact significantly on the size of the atmosphere, ocean and biosphere carbon stores. Despite international efforts to limit human carbon emissions, in the period 2000–09 they grew faster than in any previous decade. Without increased absorption of anthropogenic carbon by the oceans and biosphere, today's atmospheric CO_2 concentrations would exceed 500 ppm.

Sequestration of waste carbon

The combustion of fossil fuels and the transfer of carbon from geological store to the atmosphere and oceans is the main driver of present-day global warming. One possible solution to this problem is to capture and store CO_2 released by power plants and industry. This new technology of carbon sequestration is known as **carbon capture and storage** (CCS).

So far the technology has been piloted at just a handful of coal-fired power stations. It involves three stages. First the CO_2 is separated from power station emissions. The CO_2 is then compressed and transported by pipeline to storage areas. And finally it is injected into porous rocks deep underground where it is stored permanently (Figure 4.29).

Where our carbon emissions have come from: carbon emission sources 1750–2012 (Gt CO_2)

Where our carbon emissions have gone: carbon emission sinks 1750–2012 (Gt CO_2)

Figure 4.28 Carbon emissions and sinks (Sources: IPCC (2007) WG1, Global Carbon Project, CDIAC, NOAA)

Table 4.12 Human CO_2 emissions, 1750–2012: flows and stores (Source: IPPC)

		1750	2012
Atmospheric CO_2 concentration (ppm)		280	393
Influx CO_2 to atmosphere (ppm)	Coal	+86	
	Oil	+64	
	Gas	+26	
	Cement	+5	
	Land use	+76	
	Total	**257**	
Absorption of CO_2 from atmosphere (ppm)	Biosphere sink	−68	
	Ocean sink	−76	
	Total	**−144**	
Net gain CO_2 in atmosphere		257 − 144 = 113 ppm	

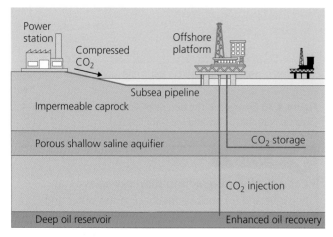

Figure 4.29 Carbon capture and storage

CCS could eventually play an important part in reducing CO_2 and other greenhouse gas emissions. For instance, in the USA 40 per cent of all CO_2 emissions are from coal- and gas-fired power stations and CCS has the potential to reduce these emissions by 80–90 per cent. In the UK, a CCS pilot project is underway at Peterhead in north-east Scotland. However, due to rising costs, the Drax project in North Yorkshire, designed to capture 2 million tonnes of CO_2 per year, was axed in 2016. The plan was for the carbon to be transported by pipeline to the North Sea and stored in depleted gas reservoirs. CO_2 gas can also be pumped into 'mature' oilfields to extract oil that would otherwise be uneconomic to recover.

Although the technology of CCS is feasible, its effectiveness is limited by economic and geological factors. This is because CCS:

- involves big capital costs – the Drax and Peterhead projects will cost at least £1 billion

- uses large amounts of energy – typically 20 per cent of a power plant's output is needed to separate the CO_2 and compress it
- requires storage reservoirs with specific geological conditions, i.e. porous rocks overlain by impermeable strata.

Positive and negative feedback loops in the water and carbon cycles

Feedback is an automatic response to changes which disturb a system's balance or equilibrium. Change in natural systems can produce either **positive** or **negative feedback** responses. Positive feedback occurs when an initial change causes further change (a kind of 'snowball' effect). Negative feedback is the opposite: it counters system change and restores equilibrium.

Feedback in the water cycle

Rising temperatures affect the water cycle at the global scale. In a warmer world, evaporation increases and the atmosphere holds more vapour. The result is greater cloud cover and more precipitation. These changes create a positive feedback effect. Because water vapour is a **greenhouse gas**, more vapour in the atmosphere increases absorption of long-wave radiation from the Earth causing further rises in temperature.

Alternatively, more atmospheric vapour can induce negative feedback. This works as follows: more vapour creates greater cloud cover which reflects more solar radiation back into space. And as smaller amounts of solar radiation are absorbed by the atmosphere, oceans and land, average global temperatures fall.

In drainage basins in the longer term, inputs and outputs of water are in equilibrium. The main input, precipitation, is balanced by outputs of evapotranspiration and run-off. However, this balance varies from year to year. The system responds to above average precipitation by increasing river flow and evaporation; and excess water recharges aquifers, increasing water storage in permeable rocks. Both responses are examples of negative feedback. During droughts the system adjusts to lower precipitation by reducing run-off and evapotranspiration. Meanwhile, as the water table falls, springs and seepages dry up, helping to conserve groundwater stores.

Feedback in the water cycle also takes place at the smallest scale. In most years precipitation is sufficient to satisfy an individual tree's demand for water. However, in drought years, shallow-rooted trees like silver birch become stressed: water lost in transpiration is not replaced by a similar uptake of water from the soil. The tree responds, reducing transpiration losses by shedding some or all of its leaves. This negative feedback loop restores the water balance and ensures the tree's survival.

Feedback in the carbon cycle

The global carbon cycle is currently in a state of disequilibrium. Human activity, primarily through burning fossil fuels, has increased the concentration of CO_2 in the atmosphere, the acidity of the oceans and the flux of carbon between the major stores. Within the carbon cycle there are feedback loops which could either restore equilibrium or induce further disequilibrium.

Negative feedback could neutralise rising levels of atmospheric CO_2 by stimulating photosynthesis. This process is called **carbon fertilisation**. In this way excess CO_2 is extracted from the atmosphere and stored in the biosphere. Eventually much of this carbon would find its way into long-term storage in soils and ocean sediments, allowing the system to return to a steady state.

However, increased primary production through carbon fertilisation is conditional on the availability of other requirements for photosynthesis such as sunlight, soil nutrients, nitrogen and water. So, although there is evidence that primary production has indeed increased in recent years it is not possible to say with certainty that this is due to increased atmospheric CO_2. For instance, recently significant increases in primary production have been observed in the Amazon rainforest but this could be explained by lower rainfall, with less cloud cover and more sunlight rather than an increase in CO_2.

Positive feedback could tip the carbon cycle into greater disequilibrium. For instance, global warming will intensify the carbon cycle, speed up decomposition and release more CO_2 to the atmosphere, thus amplifying the greenhouse effect. Another positive feedback effect is seen in Arctic tundra where global warming is occurring faster than in any other region (1.5°–2.5°C in the past 30 years). As the Arctic sea ice and snow cover shrinks, large expanses of sea and land are exposed (Figure 4.30). This means that more sunlight is absorbed, warming the tundra and melting the permafrost. This is significant for the global carbon cycle because the tundra stores an estimated 1600 GT of organic carbon in the permafrost.

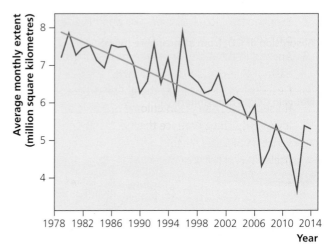

Figure 4.30 Changes in the extent of Arctic sea ice, 1979–2014 (Source: National Snow and Ice Data Center)

> **Activity**
>
> Construct flow diagrams to show examples of (a) negative and (b) positive feedback in the water and carbon cycles.

> **Key idea**
> → The pathways and processes which control the cycling of water and carbon vary over time

Monitoring changes to the global water and carbon cycles

Given the potentially damaging impact of climate change, accurate monitoring of changes in global air temperatures, sea surface temperatures (SST), sea ice thickness and rates of deforestation is essential. Because ground-based measurements of environmental change at the global scale are impractical, monitoring relies heavily on satellite technology and remote sensing (Table 4.13). Continuous monitoring by satellite on a day-to-day, month-to-month or year-to-year basis allows changes to be observed on various time scales. Using Geographic Information Systems (GIS) techniques these data can then be mapped and analysed to show areas of anomalies and trends, and regions of greatest change.

Activities

Investigate global maps and GIS data which relate to the water and carbon cycles at www.climate.gov/data/maps-and-data (access the Data Set Gallery). With reference to global air temperatures and sea surface temperatures (SSTs):

1 Describe the global patterns of temperature of the past seven days.
2 Examine the global temperature anomalies of the past seven days and describe the global pattern.
3 Analyse in detail the temperature anomalies in eastern Greenland in January 2015.
4 Describe and explain (a) the global distribution of, (b) seasonal changes in chlorophyll in the oceans.
5 Investigate changes in ice thickness and front variation of three glaciers: Aletsch (Alps), Kotarjökull (Iceland) and Athabasca (Canada).

Flows of carbon vary both diurnally and seasonally. During the daytime CO_2 flows from the atmosphere to vegetation. At night the flux is reversed. Without sunlight, photosynthesis switches off, and vegetation loses CO_2 to the atmosphere. The same diurnal pattern is observed with phytoplankton in the oceans.

Seasonal changes

Ultimately the seasons are controlled by variations in the intensity of solar radiation. In the UK, solar radiation intensity peaks in mid-June. A typical solar input in June in southern England is around 800 W/m^2; in December the input falls to little more than 150 W/m^2 (Figure 4.31). As a result, evapotranspiration is highest in the summer months and lowest in winter (Table 4.14). In the driest parts of lowland England up to 80 per cent of precipitation may be lost to evapotranspiration. With

Diurnal changes

Significant changes occur within a 24-hour period in the water cycle. Lower temperatures at night reduce evaporation and transpiration. Convectional precipitation, dependent on direct heating of the ground surface by the Sun, is a daytime phenomenon often falling in the afternoon when temperatures reach a maximum. This is particularly significant in climatic regions in the tropics where the bulk of precipitation is from convectional storms.

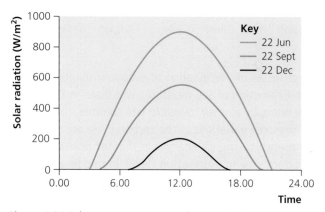

Figure 4.31 Solar energy input in southern England

Table 4.13 Remote sensing: satellite technology to monitor changes to the water and carbon cycles

Feature	Satellite	Analysis
Arctic sea ice	NASA's Earth Observing System(EOS) satellites have monitored sea ice growth and retreat since 1978 (http://earthobservatory.nasa.gov/Features/WorldOfChange/sea_ice.php)	Measures microwave energy radiated from Earth's surface. Comparison of time series images to show changes.
Ice caps/glaciers	As well as ground-based estimates of mass balance, satellite technology, e.g. ICESat-2 (http://icesat.gsfc.nasa.gov/icesat2/)	Measures surface height of ice sheet and glaciers using laser technology. Shows extent and volume of ice and changes.
Sea surface temperatures (SSTs)	NOAA satellites (www.ospo.noaa.gov/Products/ocean/sst.html)	Radiometers measure the wave band of radiation emitted from the ocean surface. Changes in global SSTs and areas of upwelling and downwelling.
Water vapour	NOAA polar orbiters (www.ospo.noaa.gov/Products/atmosphere/mspps/noaa18prd.html)	Measures cloud liquid water, total precipitable water, etc. Long-term trends in cloud cover and water vapour in the atmosphere.
Deforestation	ESA albedo (reflectivity) images from various satellites (www.globalbedo.org)	Measurements of reflectivity of Earth's surface and land use changes.
Atmospheric CO_2	NASA's Orbiting Carbon Observatory-2 (OCO-2) (http://oco.jpl.nasa.gov/). Ground-based measurements at Mauna Loa, Hawaii, since 1958.	New satellite measurements of global atmospheric CO_2 from NASA's Orbiting Carbon Observatory-2 (OCO-2). The satellite also measures the effectiveness of absorption of CO_2 by plants.
Primary production in oceans	NASA's MODIS/AQUA (http://modis.gsfc.nasa.gov/data/dataprod/mod17.php)	Measures net primary production in oceans and on land.

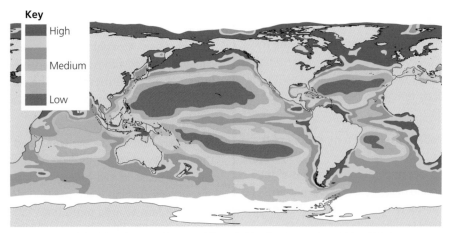

Figure 4.32 Global distribution of ocean chlorophyll, July–September, 1997–2000

Table 4.14 Average monthly evapotranspiration in lowland England (mm)

April	May	June	July	August	September
54	79	92	96	79	49

large losses of precipitation to evapotranspiration and the exhaustion of soil moisture, river flows in England are normally at their lowest in late summer.

Seasonal variations in the carbon cycle are shown by month-to-month changes in the **net primary productivity of vegetation** (NPP). In middle and high latitudes, day length or **photoperiod**, and temperature drive seasonal changes in NPP. Similar seasonal variations also occur in the tropics, though there the main cause is water availability.

During the northern hemisphere summer, when trees are in full foliage, there is a net global flow of

CO_2 from the atmosphere to the biosphere. This causes atmospheric CO_2 levels to fall by 2 ppm. At the end of summer, as photosynthesis ends, the flow is reversed with natural decomposition releasing CO_2 back to the atmosphere. Seasonal fluctuations in the global CO_2 flux are explained by the concentration of continental land masses in the northern hemisphere. During the growing season ecosystems such as the boreal and temperate forests extract huge amounts of CO_2 from the atmosphere which has a global impact.

In the oceans phytoplankton are stimulated into photosynthetic activity by rising water temperatures, more intense sunlight and the lengthening photoperiod. Every year in the North Atlantic there is an explosion of microscopic oceanic plant life which starts in March and peaks in mid-summer. The resulting algal blooms are so extensive, they are visible from space (Figure 4.32).

Skills focus

Table 4.15 River Ribble at Salmesbury: monthly flows, rainfall totals and sunshine hours, 2011

	J	F	M	A	M	J	J	A	S	O	N	D
Average flow (cumecs)	34	23.5	22	12.1	4.12	4.36	15.35	11.7	51.7	28	59	17.7
Rainfall (mm)	59.7	64.2	94.2	21.2	20.9	43.2	87.4	54.9	88.4	85.7	159	28.3
Sunshine (hours)	46.6	44.7	127.1	159.1	179.1	204.4	104.4	149.5	122.1	114.1	68.1	41.6

1 Using Table 4.15 analyse the seasonal relationship between (a) river flow and rainfall, (b) river flow and sunshine (evapotranspiration) by calculating the Spearman rank correlation coefficient (r_s) for the data (see Chapter 15, Geographical Skills, pages 541–42).

2 Calculate the coefficient of determination ($r_s \times 100$) to find the percentage of variation in river flow explained by rainfall and sunshine (page 541). Comment on the results with reference to the possible influence on river flow of factors other than rainfall and sunshine totals.

3 With reference to evaporation, transpiration, infiltration, saturated overland flow and rainfall, give possible explanations for differences in river flow between November and February, and May and August.

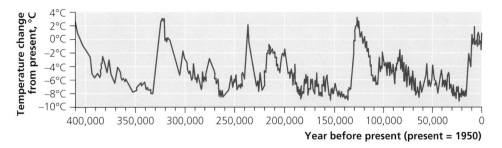

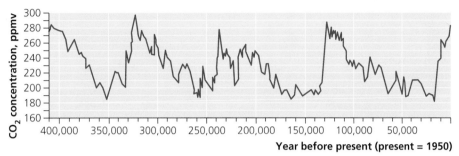

Figure 4.33 Temperature and CO_2 changes in the past 400,000 years (Source: J.R. Petit, J. Jouzel, et al. Climate and atmospheric history of the past 420 000 years from the Vostok ice core in Antartica, *Nature* 399 (3 June), pp. 420–36, 1999)

Long-term changes

The climate record over the last million years shows the Earth's climate has been highly unstable, with large fluctuations in global temperatures occurring at regular intervals. In the past 400,000 years there have been four major glacial cycles with cold **glacials** followed by warmer **inter-glacials** (Figure 4.33). Each cycle lasted around 100,000 years. At the height of the last glacial, 20,000 years ago, average annual temperatures in the British Isles were 5 °C lower than today, and Scotland, Wales and most of northern England and Ireland were submerged by ice up to 1 km thick.

During the warm inter-glacial periods, temperatures were similar to those of today. However, on much longer time scales global temperatures have been even more extreme. For example, 250 million years ago average global temperatures reached 22 °C – at least 7–8 °C higher than today's (Figure 4.33). These climatic shifts had a major impact on the water and carbon cycles.

Water cycle

During glacial periods the water cycle undergoes a number of changes. The most obvious is the net transfer of water from the ocean reservoir to storage in ice sheets, glaciers and permafrost. As a result, in glacials the sea level worldwide falls by 100–130 m; and ice sheets and glaciers expand to cover around one-third of the continental land mass. As ice sheets advance equatorwards they destroy extensive tracts of forest and grassland. The area covered by

vegetation and water stored in the biosphere shrinks. Meanwhile, in the tropics, the climate becomes drier and deserts and grasslands displace large areas of rainforest.

Lower rates of evapotranspiration during glacial phases reduce exchanges of water between the atmosphere and the oceans, biosphere and soils. This, together with so much freshwater stored as snow and ice, slows the water cycle appreciably.

Carbon cycle

The most striking feature of the carbon cycle during glacial periods is the dramatic reduction in CO_2 in the atmosphere. Figure 4.33 shows the close correlation between temperature and atmospheric CO_2 over the past 400,000 years. At times of glacial maxima CO_2 concentrations fall to around 180 ppm, while in warmer inter-glacial periods they are 100 ppm higher.

No clear explanation exists for the drop in atmospheric CO_2 during glacial periods. It is, however, possible that excess CO_2 finds its way from the atmosphere to the deep ocean. One mechanism is changes in ocean circulation during glacials that bring nutrients to the surface and stimulate phytoplankton growth. Phytoplankton fix large amounts of CO_2 by photosynthesis before dying and sinking to the deep ocean where the carbon is stored. Lower ocean temperatures also make CO_2 more soluble in surface waters.

Other changes occur in the terrestrial biosphere. The carbon pool in vegetation shrinks during glacials

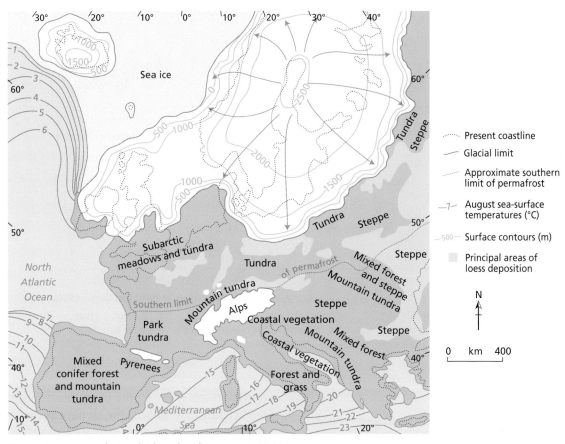

Figure 4.34 Europe during the last glacial maximum, 20,000 years ago

as ice sheets advance and occupy large areas of the continents (Figure 4.34). In this process deserts expand, tundra replaces temperate forests and grasslands encroach on tropical rainforests. With much of the land surface buried by ice, carbon stored in soils will no longer be exchanged with the atmosphere. Meanwhile expanses of tundra beyond the ice-limit sequester huge amounts of carbon in permafrost. With less vegetation cover, fewer forests, lower temperatures and lower precipitation, NPP and the total volume of carbon fixed in photosynthesis will decline. The implications are an overall slowing of the carbon flux and smaller amounts of CO_2 returned to the atmosphere through decomposition.

✔ Review questions

1 What is meant by the term 'dynamic equilibrium'?
2 What are the differences between positive and negative feedback?
3 What factors influence the flow characteristics of rivers draining urban areas?
4 State two ways in which forestry modifies the water cycle.
5 Under what conditions does groundwater become artesian?

6 What happens to CO_2 emitted through the combustion of fossil fuels?
7 Why has progress of CCS been limited?
8 What is meant by the term 'carbon fertilisation'?
9 Why is global warming fastest in the Arctic?
10 Why do global atmospheric CO_2 levels fall between April and September?
11 Outline two factors responsible for contemporary **eustatic** sea level rise.

4.4 To what extent are the water and carbon cycles linked?

> **Key idea**
> → The water and carbon cycles are linked and interdependent

Increasing levels of CO_2 (and other greenhouse gases) in the atmosphere drive global warming and focus attention on the linkages between the two cycles. The linkages between the atmosphere, oceans, vegetation, soils and cryosphere are complex: the principal ones are summarised in Table 4.16.

Human activities cause changes in water and carbon stores

Rapid population and economic growth, deforestation and urbanisation in the past 100 years have modified the size of water and carbon stores and rates of transfer between stores in the water and carbon cycles. The impact and scale of these changes is most apparent at regional and local scales.

The human impact on the water cycle is most evident in rivers and aquifers. Rising demand for water for irrigation, agriculture and public supply, especially in arid and semi-arid environments, has created acute shortages. In the Colorado Basin in the southwest USA, surface supplies have diminished as more water is abstracted from rivers, and huge amounts are evaporated from reservoirs like Lake Mead and Lake Powell. Elsewhere, the quality of fresh water resources has declined. Overpumping of aquifers in the coastal regions of Bangladesh has led to incursions of salt water, often making the water unfit for irrigation and drinking.

Compared to natural ecosystems, human activities such as deforestation and urbanisation reduce evapotranspiration and therefore precipitation; increase surface run-off; decrease throughflow (or interflow); and lower water tables (Figure 4.35). In Amazonia, forest trees are a key component of the water cycle, transferring water to the atmosphere by evapotranspiration which is then returned through precipitation. In places, extensive deforestation has broken this cycle, causing climates to dry out and preventing regeneration of the forest.

Human activity is also altering the carbon cycle, depleting some carbon stores and increasing others. The world relies on fossil fuels for 87 per cent of its primary energy consumption. The exploitation of coal, oil and natural gas has removed billions of tonnes of carbon from geological store – a process that has gathered momentum in the past 30 years with the rapid industrialisation of the Chinese and Indian economies. Currently around 8 billion tonnes of carbon a year are transferred to the atmosphere by burning fossil fuels (Figure 4.36). In addition, land use change (mainly deforestation) transfers approximately 1 billion tonnes of carbon to the atmosphere annually. The additional carbon is stored primarily as atmospheric CO_2 where its concentration increases year-by-year. Around 2.5 million tonnes is absorbed by the oceans, and a similar amount by the biosphere.

Massive deforestation has reduced the planet's forest cover in historic times by nearly 50 per cent. Thus the amount of carbon stored in the biosphere, and fixed by photosynthesis, has declined steeply (Figure 4.37). Even

Table 4.16 How the water and carbon cycles are inter-linked and inter-dependent

	Interlinkages
Atmosphere	Atmospheric CO_2 has a greenhouse effect. CO_2 plays a vital role in photosynthesis by terrestrial plants and phytoplankton. Plants, which are important carbon stores, extract water from the soil and transpire it as part of the water cycle. Water is evaporated from the oceans to the atmosphere, and CO_2 is exchanged between the two stores.
Oceans	Ocean acidity increases when exchanges of CO_2 are not in balance (i.e. inputs to the oceans from the atmosphere exceed outputs). The solubility of CO_2 in the oceans increases with lower SSTs. Atmospheric CO_2 levels influence: SSTs and the thermal expansion of the oceans; air temperatures; the melting of ice sheets and glaciers; and sea level.
Vegetation and soil	Water availability influences rates of photosynthesis, NPP, inputs of organic litter to soils and transpiration. The water-storage capacity of soils increases with organic content. Temperatures and rainfall affect decomposition rates and the release of CO_2 to the atmosphere.
Cryosphere	CO_2 levels in the atmosphere determine the intensity of the greenhouse effect and melting of ice sheets, glaciers, sea ice and permafrost. Melting exposes land and sea surfaces which absorb more solar radiation and raise temperatures further. Permafrost melting exposes organic material to oxidation and decomposition which releases CO_2 and CH_4. Run-off, river flow and evaporation respond to temperature change.

Before

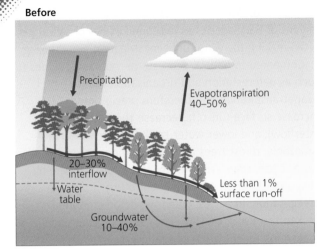

Precipitation

Evapotranspiration
40–50%

20–30%
interflow

Water
table

Less than 1%
surface run-off

Groundwater
10–40%

After

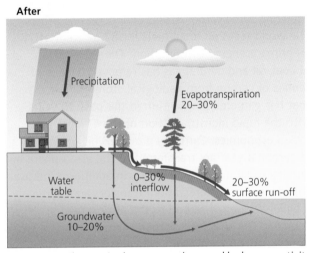

Precipitation

Evapotranspiration
20–30%

Water
table

0–30%
interflow

20–30%
surface run-off

Groundwater
10–20%

Figure 4.35 Changes in the water cycle caused by human activity

more significant is photosynthesis by phytoplankton in the oceans. Ultimately phytoplankton absorb more than half the CO_2 from burning fossil fuels – significantly more than the tropical forests. Acidification of the oceans threatens this vital biological carbon store as well as adversely affecting marine life. Soil is another important carbon store which is being degraded by erosion caused by deforestation and agricultural mismanagement. Carbon stores in wetlands, drained for cultivation and urban development, have also been depleted as they dry out and are oxidised.

> ### ➡ Activities
>
> 1 Study Figure 4.35, which shows the impact of human activity on the water cycle.
>
>   a) Summarise the main changes shown.
>   b) Explain why deforestation has caused a fall in the water table.
>   c) Why does surface run-off increase as a consequence of deforestation.
>
> 2 Study Figure 4.37, which shows the impact of deforestation on stores and flows of carbon in a small area of tropical rainforest:
>
>   a) Calculate the mass balance of carbon in (i) undisturbed forest, (ii) areas deforested for 10 years.
>   b) Calculate the impact of these changes in mass balance on major carbon stores.

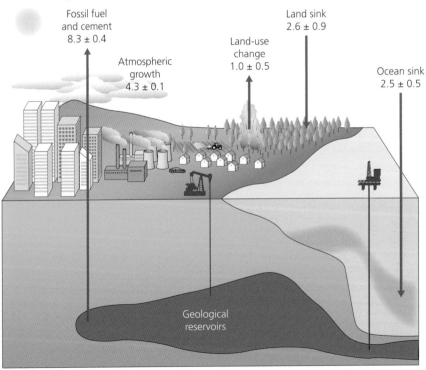

Fossil fuel
and cement
8.3 ± 0.4

Land sink
2.6 ± 0.9

Land-use
change
1.0 ± 0.5

Atmospheric
growth
4.3 ± 0.1

Ocean sink
2.5 ± 0.5

Geological
reservoirs

Figure 4.36 Human impact on the global carbon cycle (Gt of carbon/year)(Source: US Carbon Cycle Program)

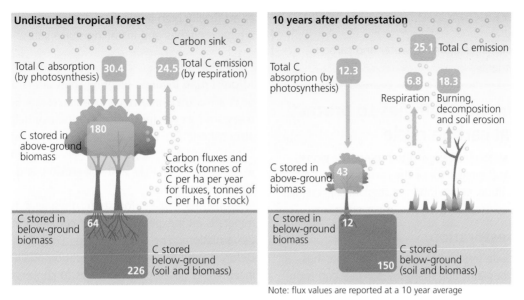

Figure 4.37 The impact of deforestation on the carbon cycle: (a) undisturbed tropical forest, (b) 10 years after deforestation

The impact of long-term climate change on the water and carbon cycles

Water cycle

Climate change is already modifying the global water cycle. Global warming has increased evaporation and therefore the amount of water vapour in the atmosphere. More vapour, which is a natural GHG, has a feedback effect, helping to raise global temperatures, increase evaporation and precipitation. Meanwhile increased precipitation will result in higher run-off in the water cycle and greater flood risks. Water vapour is also a source of energy in the atmosphere releasing latent heat on condensation. With more energy in the atmosphere extreme weather events such as hurricanes and mid-latitude storms become more powerful and more frequent.

Global warming is accelerating the melting of glaciers, ice sheets like Greenland and permafrost in the Arctic tundra. Thus water storage in the cryosphere shrinks, as water is transferred to the oceans and atmosphere.

Carbon cycle

The impact of global climate change on the carbon cycle is complex. It depends not just on rising temperatures, but also on geographical differences in rainfall amounts. Higher global temperatures will in general increase rates of decomposition and accelerate transfers of carbon from the biosphere and soil to the atmosphere. However, in the humid tropics climate change may increase aridity and threaten the extent of forests. As forests are replaced by grasslands the amount of carbon stored in tropical biomes will diminish. In contrast in high latitudes, global warming will allow the boreal forests of Siberia and Canada and Alaska to expand polewards.

Carbon frozen in the permafrost of the tundra is being released as temperatures rise above freezing and allow oxidation and decomposition of vast peat stores. Meanwhile, acidification of the oceans through the absorption of excess CO_2 from the atmosphere reduces photosynthesis by phytoplankton, limiting the capacity of the oceans to store carbon. Thus long-term climate change will probably see an increase in carbon stored in the atmosphere, a decrease in carbon stored in the biosphere and possibly a similar decrease in the ocean carbon stores. Movement of carbon into and out of the atmosphere will vary regionally, depending on changes in rates of photosynthesis, decomposition and respiration.

Management strategies to protect the global carbon cycle

Management strategies designed to protect the global carbon cycle as the regulator of the Earth's climate include wetland restoration, afforestation, sustainable agricultural practices and controls on greenhouse gas emissions.

Wetland restoration

Wetlands include freshwater marshes, salt marshes, peatlands, floodplains and mangroves. Their common feature is a water table at or near the surface causing the ground to be permanently saturated. Wetlands are important in the carbon cycle: they occupy 6–9 per cent of the Earth's land surface and contain 35 per cent of the terrestrial carbon pool.

Population growth, economic development and urbanisation have placed huge pressure on wetland environments. In the lower 48 US states the wetland area has halved since 1600. Apart from loss of biodiversity and wildlife habitats, destruction of wetlands transfers huge amounts of stored CO_2 and CH_4 to the atmosphere.

However, climate change and the need to reduce CO_2 emissions have led to a re-evaluation of the importance of wetlands as carbon sinks. In the twentieth century, Canada's prairie provinces lost 70 per cent of their wetlands. Restoration programmes in this area have shown that wetlands can store on average 3.25 tonnes C/ha/year. Now 112,000 ha have been targeted for restoration in the Canadian prairies which should eventually sequester 364,000 tonnes C/year.

The need for protection of wetlands as wildlife habitats as well as carbon stores is reflected in management initiatives such as the International Convention on Wetlands (Ramsar) and European Union Habitats Directive. In the UK up to 400 ha of grade 1 farmland in east Cambridgeshire is currently being converted back to wetland. This project will assist the UK government to meet its target to restore 500 ha of wetland by 2020. A similar scheme is underway in Somerset.

Restoration focuses on raising local water tables to re-create waterlogged conditions. Wetlands on floodplains for example can be reconnected to rivers by the removal of flood embankments and controlled floods. Coastal areas of reclaimed marshland used for farming can be restored by breaching sea defences (Figure 4.38). Elsewhere water levels can be maintained at artificially high levels by diverting or blocking drainage ditches and installing sluice gates.

Afforestation

Afforestation involves planting trees in deforested areas or in areas that have never been forested. Because trees are carbon sinks, afforestation can help reduce atmospheric CO_2 levels in the medium to long term and combat climate change. It also has other benefits such as reducing flood risks and soil erosion, and increasing biodiversity.

Protecting tropical forests from loggers, farmers and miners is an inexpensive way of curbing greenhouse gas emissions. The UN's Reducing Emissions from Deforestation and Forest Degradation (REDD) scheme incentivises developing countries to conserve their rainforests by placing a monetary value on forest conservation. Several projects are already well established such as those in Amazonia (Puras, Russas-Valparaiso) and the Lower Mississippi.

In China a massive government-sponsored afforestation project began in 1978. It aims to afforest 400,000 km^2 (an area roughly the size of Spain) by 2050. In the decade 2000–09, 30,000 km^2 were successfully planted with non-native, fast-growing species such as poplar and birch. However, the project has a wider purpose: to combat **desertification** and **land degradation** in the vast semi-arid expanses of northern China.

Agricultural practices

Unsustainable agricultural practices such as **overcultivation**, **overgrazing** and excessive intensification often result in soil erosion and the release of large quantities of carbon to the atmosphere. Intensive livestock farming produces 100 million tonnes/year of CH_4, a potent greenhouse gas. Almost as important are CH_4 emissions

Figure 4.38 Freiston Shore, Lincolnshire – managed coastal realignment where salt marsh has replaced arable land

from flooded (padi) rice fields and from the uncontrolled decomposition of manure. Agricultural practices to reduce greenhouse gas emissions are shown in Table 4.17.

International agreements to reduce carbon emissions

Climate change affects all countries. Solving the problem therefore requires international co-operation. So far, co-operation has been patchy. For a variety of economic and political reasons some of the world's largest greenhouse-gas emitters have opted to pursue narrow self-interest.

Until recently the only significant international agreement to tackle climate change has been the Kyoto Protocol (1997). Under Kyoto most rich countries agreed to legally binding reductions in their CO_2 emissions, though controversially, developing countries, and some of the biggest polluters (e.g. China and India), were exempted. Also, several rich countries, notably the USA and Australia, refused to ratify the treaty. Kyoto expired in 2012. After several rounds of negotiation, a new international agreement was finally reached at the Paris Climate Convention in 2015 for implementation in 2020. The Paris Agreement aims to reduce global CO_2 emissions below 60 per cent of 2010 levels by 2050, and keep global warming below 2°C. However, countries will set their own voluntary targets. These are not legally binding and a timetable for implementing them has yet to be agreed. Meanwhile rich countries will transfer significant funds and technologies to assist poorer countries to achieve their targets. Major CO_2 emitters such as China and India argue that global reductions in CO_2 emissions are the responsibility of rich countries because:

- countries such as China and India are still relatively poor and industrialisation, based on

fossil fuels energy, is essential to raise living standards to levels comparable with those in the developed world
- historically, Europe and North America through their own industrialisation and economic development are largely to blame for contemporary global warming and climate change.

Cap and trade

Cap and trade offers an alternative, international market-based approach to limit CO_2 emissions. Under this scheme businesses are allocated an annual quota for their CO_2 emissions. If they emit less than their quota they receive **carbon credits** which can be traded on international markets. Businesses that exceed their quotas must purchase additional credits or incur financial penalties. **Carbon offsets** are credits awarded to countries and companies for schemes such as afforestation, renewable energy and wetland restoration. They can be bought to compensate for excessive emissions elsewhere.

Management strategies to protect the global water cycle

Forestry

The crucial role of forests in the global water cycle is recognised by multilateral agencies such as the United Nations (UN) and World Bank (WB). They, together with other organisations and governments, fund programmes to protect tropical forests. The UN's Reducing Emissions from Deforestation and Forest Degradation (REDD) programme and the World Bank's Forest Carbon Partnership Facility (FCPF) fund over 50 partner countries in Africa, Asia-Pacific and South

Table 4.17 Reducing emissions from agriculture

Type	How emissions of greenhouse gases are reduced
Land and crop management	<ul><li>Zero tillage – growing crops without ploughing the soil. This conserves the soil's organic content, reducing oxidation and the risk of erosion by wind and water.</li><li>Polyculture – growing annual crops interspersed with trees. Trees provide year-round ground cover and protect soils from erosion.</li><li>Crop residues – leaving crop residues (stems, leaves, etc.) on fields after the harvest, to provide ground cover and protection against soil erosion and drying out.</li><li>Avoiding the use of heavy farm machinery on wet soils, which leads to compaction and the risk of erosion by surface run-off.</li><li>Contour ploughing and terracing on slopes to reduce run-off and erosion.</li><li>Introducing new strains of rice that grow in drier conditions and therefore produce less CH_4. Applying chemicals such as ammonium sulphate which inhibit microbial activities that produce CH_4.</li></ul>
Livestock management	<ul><li>Improving the quality of animal feed to reduce enteric fermentation so that less feed is converted to CH_4; mixing methane inhibitors with livestock feed.</li></ul>
Manure management	<ul><li>Controlling the way manure decomposes to reduce CH_4 emissions. Storing manure in anaerobic containers and capturing CH_4 as a source of renewable energy.</li></ul>

America. Financial incentives to protect and restore forests are a combination of carbon offsets and direct funding.

Brazil has received support from the UN, World Bank, World Wildlife Fund (WWF) and the German Development Bank to protect its forests. The Amazon Regional Protected Areas (ARPA) programme now covers nearly 10 per cent of the Amazon Basin. Areas included in the programme are strictly protected. The benefits are significant: stabilising the regional water cycle; offsetting 430 million tonnes of carbon a year; supporting indigenous forest communities; promoting ecotourism; and protecting the genetic bank provided by thousands of plant species in the forests.

Water allocations

In countries of water scarcity governments have to make difficult decisions on the allocation of water resources. Agriculture is by far the biggest consumer. Globally it accounts for 70 per cent of water withdrawals and 90 per cent of consumption (Figure 4.39). Wastage of water occurs through evaporation and seepage through inefficient water management (e.g. over-irrigating crops). Improved management techniques which minimise water losses to evaporation include mulching, zero soil disturbance and drip irrigation. Losses to run-off on slopes can be reduced by terracing, contour ploughing and the insertion of vegetative strips.

Meanwhile, better water harvesting, with storage in ponds and reservoirs, provides farmers with extra water resources. Recovery and recycling of waste water from agriculture, industry and urban populations is technically feasible, but as yet little used outside the developed world.

In semi-arid regions of water scarcity, such as the Lower Indus Valley in Pakistan, and the US Colorado Basin, water agreements divide up resources between downstream states. In Pakistan the Punjab and Sindh receive 92 per cent of the Indus's flow; in the Colorado Basin water resources are allocated to California, Arizona, Nevada, Utah and New Mexico. In both regions, the vast bulk of water is used for irrigation.

Drainage basin planning

The management of water resources is most effective at the drainage basin scale. At this scale it is feasible to adopt an integrated or holistic management approach to accommodate the often conflicting demands of different water users. Agriculture, industry, domestic use, wildlife and recreation and leisure generate demands that impact on water quality, river flow, groundwater levels, wildlife habitats, biodiversity and so on.

Specific targets for drainage basin planning include run-off, surface water storage and groundwater. Rapid run-off is controlled by reforestation programmes in upland catchments, reducing artificial drainage and extending permeable surfaces (e.g. gardens, green roofs) in urban areas.

Surface water storage is improved by conserving and restoring wetlands, including temporary storage on floodplains. Groundwater levels are maintained by limiting abstraction (e.g. for public supply, farming and industry) and by artificial recharge, where water is injected into aquifers through boreholes.

In England and Wales drainage basin management is well advanced. Under the EU's Water Directive Framework, ten river basin districts have been defined. The districts comprise major catchments, such as the Severn, Thames and Humber. Each district has its own River Basin Management Plan published jointly by the Environment Agency and Defra (Department for Environment, Food and Rural Affairs). The plan sets targets in relation to, for example, water quality, abstraction rates, groundwater levels, flood control, floodplain development and the status of habitats and wildlife.

✔ Review questions

1 Why has ocean acidity increased in the past 100 years? What are its effects?
2 State two factors connected to climate change that increase rates and frequency of coastal erosion and flooding.
3 Why are montane plants and animals most at risk from climate change?
4 How can afforestation help to combat problems of desertification and land degradation?
5 What are (a) cap and trade, (b) carbon credits and (c) carbon offsets?

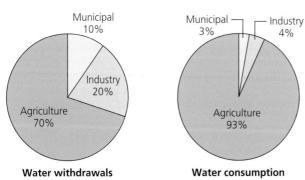

Figure 4.39 Global water withdrawal and consumption

📑 Practice questions

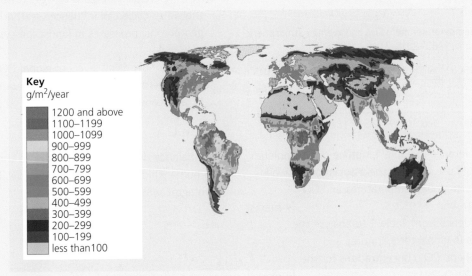

Figure A Global annual net primary productivity

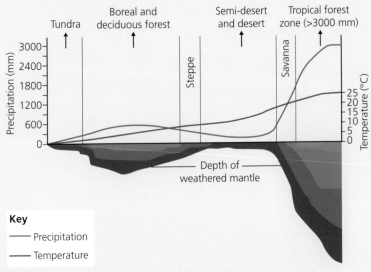

Figure B Latitudinal variations in average annual precipitation and temperature, and the depth of the weathered mantle

A Level

1 Study **Figure A** which shows global net primary productivity (NPP).

a) Outline the factors responsible for geographical differences in NPP at the global scale. [4 marks]

b) With reference to Figure 4.18 on page 116 which describes the climate at Iqaluit, explain **three** limitations of climate graphs as representations of the climate of a place. [3 marks]

c) With reference to a **case study** of the tropical rainforest, examine the significance of forest trees in the carbon cycle. [10 marks]

d) 'The damaging impact of human activities on the carbon cycle is of greater significance than on the water cycle.' Discuss. [16 marks]

AS Level

Section A

1 a) Explain how volcanic activity can influence climate change. [4 marks]

b) Suggest why anthropogenic greenhouse gas emissions have increased since the pre-industrial era. [6 marks]

c) Study **Table 1** which shows global temperature anomalies and greenhouse gas emissions between 1960 and 2015.

Table 1 Global temperature anomalies and average atmospheric CO_2 concentrations, 1960–2015

	1960	1970	1980	1990	2000	2010	2015
Temperature (°C)	−0.03	0.03	0.28	0.44	0.42	0.72	0.87
CO_2 (ppm)	317	326	339	354	370	390	401

i) With reference to Table 1, outline one technique you could use to show the association between changes in global temperatures and CO_2 concentration. [4 marks]

ii) Use evidence from Table 1 to analyse the changes in global temperatures and atmospheric CO_2 concentrations for the period 1960–2015. [6 marks]

Section B Synoptic

2 a) With reference to **Figure B** suggest how changes to the water cycle can influence weathering and other geomorphic processes in landscape systems. [8 marks]

b) Examine how rising levels of global CO_2 in the atmosphere can affect people, economy and society in landscape systems. [8 marks]

Section C

3 'On balance the impact of changes to the carbon cycle on humankind will be strongly negative'. How far do you agree with this statement? [20 marks]

Part 2

Human interactions

Chapter 5

Changing spaces; making places

5.1 What's in a place?

Geographers are interested in how and why places differ from one another. We also look at the similarities and interconnections between places. Globalisation has meant that increasing numbers of places are linked with each other through complex webs of connections.

At the heart of places are people. What makes a place different from, or the same as, other places is in large part down to people. As they live, work and play, space is changed into place. All of us are involved in changing spaces and making places. As the world's population climbs towards 9 or 10 billion by 2050, how we make places will become ever more important to the quality of people's lives.

Key idea
→ Places are defined by a combination of characteristics which change over time

What characteristics make up the identity of a place?

Using Google Earth to view the place where you live allows different characteristics of that place to be recognised. The view from space identifies the outlines of the land areas. It is then possible to zoom closer and closer in and as this happens more and more characteristics such as the settlement pattern, transport routes and eventually individual buildings become successively visible. Adding various layers to the image gives more information about the places shown on the Google Earth image.

Maps, such as those produced by the Ordnance Survey, for example at the scales of 1:50,000 and 1:25,000, are also made up of layers of information (pages 507–08).

Activities

1 Source an Ordnance Survey 1:25,000 scale map of the area in which you live (Digimap for Schools or a hard copy).
2 Stick a copy of the couple of kilometre grid squares immediately around your home in the centre of a piece of paper.
3 Around these squares construct a spider diagram of the characteristics **shown on the map** which indicate something of what the place is like.
4 Compare the map's portrayal of your home area with that on Google Earth.
5 To what extent do the map and Google Earth indicate the identity of your home location as you would describe it? What characteristics are not shown on the map nor on Google Earth?

Stretch and challenge

Source at least one other OS map but at a different scale, for example the 1:50,000 scale. What does this map tell you about the place which the other map did not? How important is the scale at which a place is studied to an understanding of the identity of that place?

A number of characteristics interact to make the identity of a place at a local scale. They include:

- Physical geography, e.g. altitude, slope angle, aspect, drainage, geology
- Demography, e.g. number of inhabitants, their ages, gender and ethnicity
- Socio-economic, e.g. employment, income and family status, education
- Cultural, e.g. religion, local traditions, local clubs and societies
- Political, e.g. local, regional and national government, local groups such as resident associations
- Built environment, e.g. age and style of buildings including building materials, density of housing

These characteristics are investigated in detail by the study of two contrasting localities: Lympstone in Devon, and Toxteth in Liverpool.

Lympstone, East Devon and Toxteth, Liverpool

Lympstone is a small settlement on the east bank of the River Exe estuary, approximately 15 km south of Exeter (Figure 5.1a). Toxteth is part of inner Liverpool, about 1.5 km south of the city centre (Figure 5.1b).

Natural characteristics

Lympstone occupies a small valley cut by Wotton Brook through the red breccia cliffs which mark the edge of the Exe estuary. Extensive tidal mudflats extend out into the estuary which is about 1.5 km wide at Lympstone. A small beach of pebbles and gravel runs along the foot of the cliff.

Toxteth occupies undulating land rising up from the banks of the River Mersey. A stream flows from the northeast, dividing into two before discharging into the river. The Mersey is fast flowing past Toxteth as its channel narrows before passing into Liverpool Bay, so there is little foreshore at Toxteth.

Past characteristics – Lympstone

The Saxons established Lympstone, having forced out the original Celtic inhabitants from the region. Connections with the continent continued with the Norman Conquest when ownership of the area

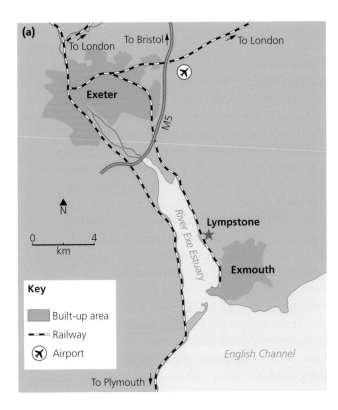

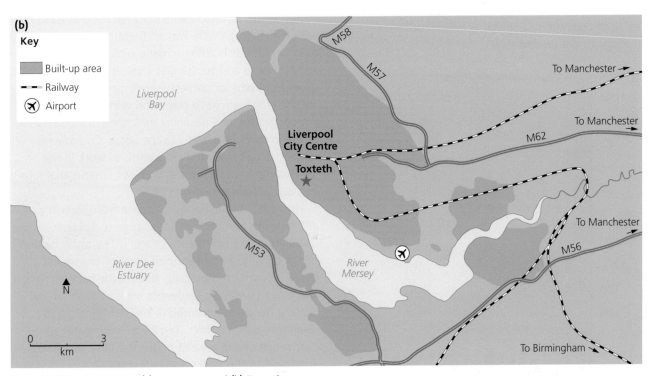

Figure 5.1 The locations of (a) Lympstone and (b) Toxteth

transferred to a succession of French families such as the de Traceys and the de Courtlands.

By the thirteenth century, Lympstone was one of a number of small ports operating under the administration of Exeter with coastal and cross-Channel trading links.

Lympstone maintained its strong links with the sea as shipbuilding thrived into the nineteenth century. Eventually, as the scale of ships increased, this industry ceased. The estuary was used for fishing and the shellfish industry, in particular mussels. Boats also went out into the North Atlantic from Lympstone to catch cod and hunt for whales.

The beginnings of tourism in the 1840s saw Lympstone attract considerable numbers of wealthy families from Exeter and East Devon who enjoyed the novelty of being 'by the sea'. Lympstone began to change from the mid-nineteenth century with the expansion of the built environment to accommodate visitors. However, Lympstone remained essentially a small village (Figure 5.2).

The arrival of the railway in 1861 improved Lympstone's connectivity both regionally and nationally. Time–space compression allowed the shellfish industry access to a wider market and local residents too could travel more easily to Exeter, the main regional centre.

Lympstone underwent further change in the twentieth century when it became a dormitory settlement for Exeter. However, it has retained a strong sense of community, something which continues through to the present day.

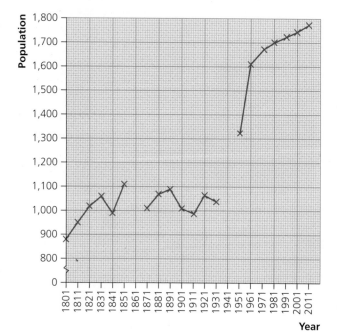

Figure 5.2 Population change, Lympstone, 1801–2011 (data for 1861 and 1941 missing from records)

Past characteristics – Toxteth

The Saxons established a settlement in Toxteth, and the Domesday survey of 1086 records it as one of just a handful of coastal villages along the banks of the Mersey. It was given to a knight, Roger of Pictou, after the Norman Conquest.

King John took the area into his control in the early thirteenth century when it became part of a large royal hunting forest. It remained as a fenced-off forest for around three hundred years.

Towards the end of the sixteenth century the area's status changed and it began to be opened up for farming. Small-scale industry was a growing characteristic during the seventeenth century, making use of water power from the dammed stream.

As the changes associated with the Industrial Revolution gathered pace, Liverpool began to emerge as a major port with many associated industries. Toxteth took on a more urban and industrial nature with activities such as several forges, a copper works established in 1772 and later a ceramics factory. Industries such as flour milling and brewing developed to serve the rapidly growing population. The river bank became lined with docks, ship-building yards and associated industries such as rope walks. This section of docks tended to specialise in the handling of timber, much of which was imported from Scandinavia.

Residential developments occurred hand in hand with industrial growth. Part of Toxteth was given over to an ambitious housing scheme with wide streets lined by large and substantial villas. It was an attractive greenfield site for property developers who built for the growing middle class who wanted to escape the congestion and declining environmental quality of the innermost suburbs to the north. Commuters journeyed daily into the city centre to work in the developing service sector as well as in managerial jobs in manufacturing.

However, the demands for space from industry and housing brought further change to Toxteth throughout the nineteenth century. The areas behind the large villas had been used for very cheap and poorly constructed housing, much of it back-to-back and court dwellings. Epidemics such as typhoid and cholera frequently erupted in the unhygienic and insanitary surroundings. Terraced housing spread over much of the area to house the families of those employed in the docks and industries.

Meanwhile there was an exodus of middle-class residents. As their disposable income increased and urban transport improved (trams and suburban

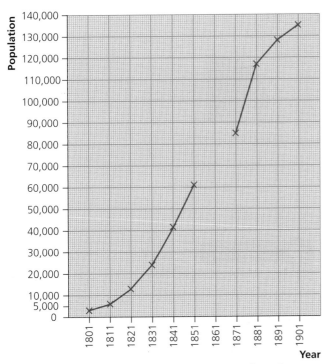

Figure 5.3 Population change, Toxteth, 1801–1901 (Data for 1861 missing from records. After 1901, Toxteth data included within Liverpool.)

railways) they relocated to greenfield sites on the edge of the built area. Thus, over a period of about 150 years, Toxteth had changed from a small rural community to an inner city suburb in a large metropolitan area (Figure 5.3). Original rural features such as fields and hedgerows were replaced by the built environment with just the rise and fall of the ground evident in the streets.

Toxteth's growth had much to do with increasing connections both nationally and internationally as goods passing through the docks were traded all over the country and beyond the UK.

Present-day characteristics: Lympstone and Toxteth compared

Just as a diamond has several facets or faces, so do places. Individually they indicate something about the place and when put together they give a comprehensive place profile.

The demographic face of a local place

The demographic profiles of the communities living in Lympstone and Toxteth are quite different (Table 5.1). Lympstone has a top-heavy age structure whereas Toxteth has fewer elderly residents, with the majority of its residents aged between 16 and 64. It has significantly fewer children than Lympstone.

The ethnic profile of the two places (Figure 5.4) helps us to understand how migrations have altered each place.

The West Country is relatively homogeneous ethnically and even a city such as Exeter does not have the ethnic diversity seen in Liverpool and other large urban centres. Some locations in some large cities have a much higher proportion of ethnic groups than Toxteth. Devon and the West Country, dominated by agriculture and tourism, provided few job opportunities for immigrants in the 1960s and 1970s. That is beginning to change, especially in larger urban centres like Exeter and Plymouth as second and third generation migrants assimilate socially and economically.

However, large industrial cities such as Liverpool have always attracted migrants, with large influxes of Irish during the nineteenth century. Following the Second World War, the employment and housing opportunities for migrants in inner cities such as Toxteth led to significant changes in place profiles. Toxteth became

Table 5.1 Some key demographic characteristics of Lympstone and Toxteth (All figures from the 2011 census. Lympstone – East Devon LSOA 014C; Toxteth – various census sources)

Variable	Lympstone	Toxteth	England
Population density (persons per hectare)	16.6	87.8	4.1
0–15 years old	19.8	13.9	18.9
16–64 years old	55.6	75.5	64.8
≥ 65 years old	24.6	10.6	16.3

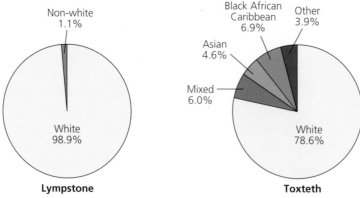

Non-white
1.1%

White
98.9%

Lympstone

Black African
Caribbean
6.9%

Other
3.9%

Asian
4.6%

Mixed
6.0%

White
78.6%

Toxteth

Figure 5.4 Ethnic composition of Lympstone and Toxteth

a destination for migrants from parts of Asia such as Hong Kong, West Africa and the Caribbean. Liverpool's port function had created links with many countries, especially in the tropics.

The socio-economic profile of a local place

As with demography, the socio-economic profiles of Lympstone and Toxteth are very different (Table 5.2).

Although the census does not ask direct questions about income, the data it records relating to factors such as **housing tenure**, car ownership and education give indications of relative wealth and poverty. A household's economic status has a direct effect on social factors such as health.

> **Activities**

1 Using the data in Table 5.2, compare and contrast the socio-economic profiles of Lympstone and Toxteth.
2 Suggest reasons for the differences between Lympstone and Toxteth.

The cultural face of a local place

Cultural change is seen in Toxteth around the time of major Muslim religious festivals such as Eid and Ramadan. As well as marking these festivals, the importance of Friday prayers in the Muslim community is different from the Christian tradition of Sunday

Skills focus

A key source of authoritative information in many ACs (Advanced Countries) is a census. In the UK a national census is held every ten years when all **households** have to complete a detailed form about both the people living in the household and its accommodation.

The website www.neighbourhood.statistics.gov.uk contains data packaged into spatial units of various sizes. When researching places at the local scale, data at the Lower Layer Super Output Area (LSOA) are very helpful as they include a wide range of information presented in absolute and percentage figures, which helps compare one place with another.

However, in cases such as Toxteth that have become absorbed into a larger urban settlement, it makes sense to use ward data. A ward is an amalgamation of several LSOAs and gives a more accurate picture of the area. The Riverside ward in Liverpool is largely made up of LSOAs from Toxteth.

It is helpful to know the postcode of the local area you are researching as this is a quick way of accessing relevant data on the National Statistics website. The Topics

section on the home page gives access to a wide variety of information about local areas such as demographic data, housing, economic indicators (for example occupation groups), education, skills and training, and the physical environment in terms of factors such as air quality.

The Map Viewer allows you to identify the boundaries of the census units making up your local area.

The Neighbourhood Summary generates a report for the local area around whichever postcode is entered. This is a useful compilation of data bringing together a great deal of information into one location on the website.

As with all census data, you need to be aware of its limitations. A key one for any census is that it offers a snapshot of a place at a particular moment in time. As the years advance from when the last census was held, its data become more and more out of date, for example the building of a major housing development soon after the last census. Another limitation can be that the boundaries of units used by a census change over time. This can make it difficult to compare local places through time.

Table 5.2 Some socio-economic characteristics of Lympstone and Toxteth

Variable	Lympstone	Toxteth	England
Average household size (number of people)	2.3	1.9	2.4
Owner-occupiers	66.1	24.0	64.5
Rent from social landlord	12.2	34.2	17.6
Rent from private landlord	20.6	39.3	16.7
Car availability (% with no access to a car or van)	12.7	54.4	25.6
% people with bad or very bad health	4.1	9.4	5.6
% aged sixteen and over with no formal qualifications	14.1	27.2	22.5

services. Many from the Black African Caribbean community have distinctive forms of Christian worship which are culturally different from traditional English styles. Toxteth's cultural diversity stretches back into the nineteenth century when a Greek Orthodox church was built there and a synagogue is evidence of Jewish culture.

In Lympstone, the rhythm of the cultural year is a long-established one based upon the Christian year with Christmas and Easter as key times in the calendar (Figure 5.5).

The political characteristics of a local place

All places in the UK have a hierarchy of political authorities from the local to the national scale. Some elements are the same for all places, for example every place is part of the constituency of a Member of Parliament who sits in the House of Commons at Westminster. Other elements are different.

Lympstone has a parish council with eleven elected people serving on this local government body. The parish council has various powers and duties all focused on local matters. Examples of the sort of community affairs they deal with are lighting local roads, and providing and equipping community facilities such as a village hall, playground or local sports field. They express views on any planning applications in the place such as house building and extensions. Their powers are relatively limited as both a district (East Devon) and county council (Devon) exist above the parish council. Lympstone is part of an area that elects two district councillors and one county councillor. The parliamentary constituency that includes Lympstone (East Devon) has an electorate of just over 72,000 and returns one MP.

The urban place of Toxteth is part of Riverside Ward, one of 30 wards making up the Liverpool City Council; each ward returns three councillors. The city council has powers and duties similar to the county council for Lympstone, such as education and children's services, and regeneration, housing and sustainability.

Toxteth is part of the Liverpool Riverside parliamentary constituency with an electorate of about 73,000.

Places can also have local groups which can be said to have a political influence on the place profile. For example, residents' associations who make their voice heard on planning matters and groups focused on protecting a place's heritage.

The built environment of a local place

In Toxteth, the built environment has been adapted by migrant communities. Mosques and ethnic retailers are visual indicators of a changing place. Many buildings have changed their use and new ones have been constructed.

Lympstone has undergone some change in its built environment. Former low-order shops have closed and been converted to residences. Some new housing has been constructed, including large houses on the cliff-top on the village's periphery. The heart of the village is, however, subject to strict planning rules and local residents protect the architecture.

> **Activities**
>
> Think about how local media might reflect the cultural profile of a local place.
>
> 1 What types of media might be available to indicate the cultural profile of the place?
> 2 What sources of information in each type of media might indicate the cultural profile of the place?
> 3 Consider different techniques to obtain details about the cultural profile of the place such as questionnaires, counts of news stories or advertisements or photographs past and present.
> 4 How might you present your findings? Think of a variety of media which might help give as complete a representation as possible of the place's profile such as prose, e.g. newspaper reports; spoken word, e.g. local radio, music; photographs.

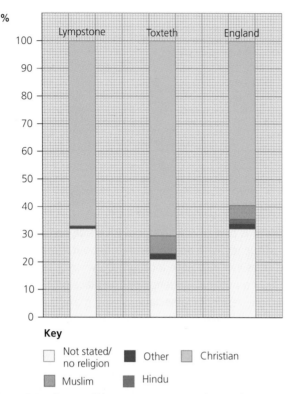

Key

☐ Not stated/ no religion   ■ Other   ▨ Christian

▨ Muslim   ▨ Hindu

Figure 5.5 Religious affiliation, Lympstone and Toxteth

Shifting flows of resources and ideas change local places

Toxteth has direct experience of the loss of employment in the docks as well as manufacturing jobs, which affected the whole of Merseyside. Many in Toxteth worked either in the docks or in industries linked with the processing of goods imported through the docks, such as sugar refining and flour milling. Containerisation and the migration of docks further to the north and the mouth of the river at Seaforth led to substantial unemployment. The consequent poverty was a significant factor contributing to the Toxteth riots in 1981. This was a period of serious civil disturbance in the area and probably represents the low point in Toxteth's history as a changing place.

The shift in the national economy towards services as well as the growing significance of the knowledge economy has had a greater positive effect on Lympstone than on Toxteth. Just over a third of its employed residents are classified as being managers, directors, senior officials or in professional occupations. They are a part of the flow into the Exeter region of employment in services and knowledge-based firms, which has been considerable. The Meteorological Office moved from Reading to Exeter in 2003 and at the time was the single largest move of IT equipment in Europe. It has provided a strong economic stimulus to the region, bringing in many knowledge-based jobs.

Additionally, Exeter's tertiary employment has seen a growing higher education sector including a university of about 3000 staff (academic and administrative), a major hospital and the regional headquarters of many firms and organisations in both services and high-tech.

Since the early 1980s, successive initiatives have attempted to remake Toxteth as a thriving place. The Merseyside Development Corporation of the 1980s set about trying to regenerate some of the former dock areas. In 1988 a major art gallery, Tate Liverpool, opened and in 2008 Liverpool was the European Capital of Culture, both of which aimed to boost tertiary employment through cultural tourism. Some of the waterfront has now been recognised as having international importance with its designation as a UNESCO World Heritage Site, offering more employment opportunities. The Liverpool Science Park was started in 2003 to support the growth and development of the knowledge economy in Liverpool.

EU funding via its Structural and Investment Funds Strategy has been attempting to help regenerate Liverpool's profile. For example, the city's Low Carbon sector has been expanding with employment created in offshore wind technology. Europe's largest offshore wind farm is in Liverpool Bay.

However, in 2015, a survey of local business in Liverpool highlighted that their main issue was in recruitment as many applicants lack the necessary technical or job-specific skills. Many of Toxteth's residents do not yet possess the skills and qualifications which would allow them to access the service and knowledge economy of Liverpool. Due to the multiplier effect, some of the residents of Toxteth will be hoping to gain from the substantial inward flow of investment. However, it will take sustained local and regional efforts to bring about socio-economic change to Toxteth's place profile.

▶ Activities

1 Using the study of Lympstone and Toxteth as a guide, investigate the present-day characteristics of your local place (read pages 520–24 before you start). Present your findings under the following headings:
   ● Demographic profile
   ● Socio-economic profile
   ● Cultural profile
   ● Built characteristics
   ● Political characteristics
2 Explain the present-day identity of your local place, referring to:
   ● Natural characteristics
   ● Past and present connections with other places locally, regionally, nationally and internationally
   ● Changing flows of people and money

Data sources for investigating present-day characteristics of local places include:

- Demographic and socio-economic data at www.neighbourhood.statistics.gov.uk. You can access data at various scales, e.g. local authority, ward, super output areas (Middle Layer Super Output Area (MSOA) represents 5000–15,000 people and Lower Layer Super Output Area (LSOA) represents 1000–3000 people). Demographic data (numbers of people, age and sex, ethnicity) as well as socio-economic data (employment, unemployment and economic activity, educational qualifications, occupations, car availability, health status, disability, types of housing, housing tenure) are available. There are also data concerning environmental quality such as air quality.
- Most local authorities also have reports on the wards within their areas which bring together census information. Both urban authorities such as Liverpool and county administrations such as Devon have links to their reports on their websites (www.liverpool.gov.uk and https:new.devon.gov.uk).
- Local crime data is available at www.police.uk.
- Local fieldwork would give data about environmental quality and land-use; the numbers and locations of cultural facilities; numbers and locations of services such as schools and colleges, health care facilities such as doctors and dentists; access to public transport.

✔ Review questions

1 Name the six characteristics that make up the identity of a place.
2 What are the differences between demographic and socio-economic characteristics?
3 How have the physical characteristics of Lympstone influenced its place identity?
4 What have been the major flows of people affecting Toxteth?
5 What effects have the contrasting flows of people had on Lympstone and Toxteth?
6 In what ways can the built environment of a place indicate its identity?
7 What is meant by the term 'knowledge economy'?
8 Draw up a table comparing the demography and socio-economic profile of your local place with both Lympstone and Toxteth.
9 Write a short summary of the key differences and similarities between the profile of your local place and those of Lympstone and Toxteth.

5.2 How do we understand place?

> **Key idea**
> → People see, experience and understand place in different ways and this can change over time

Defining what is meant by 'place'

Place is more than a term used in an academic way. Think of the ways 'place' can be used in how we speak. 'Shall we meet at your place or mine?', 'That picture looks out of place in this room', 'I was caught between a rock and a hard place'.

A place can have an objective meaning: something that just is, such as an address, or a set of map co-ordinates. But it can also have a subjective meaning: some aspect of a place that humans have added for whatever reason. For example, 10 Downing Street is a fixed location that can be plotted exactly on a map. However, it is also the official residence of the Prime Minister of the UK. As such it represents the focus of political power so has meaning in terms of how this country is governed. It also has a political meaning internationally as heads of governments from overseas visit the Prime Minister there.

'Space' is different from 'place'. Space exists between places and does not have the meanings that places do. Think of two places, one where you live and the other a major city, such as New York. Where you live has significant meaning for you in a variety of different ways. It is where you live, study and enjoy recreation. On the other hand your image of a place like New York might be associated with buildings such as the Statue of Liberty, the Empire State Building, the Twin Towers destroyed in the 9/11 attacks or with the city's lively, cosmopolitan culture. But what of the North Atlantic Ocean between your home and New York? It is likely that, for you, this is simply an enormous space to cross in order to travel from one place to another.

But one person's 'space' might be another person's 'place'. The North Atlantic Ocean is a workplace to deep-sea fishermen. It is a corridor along which large numbers of cargo ships sail. The crews of these ships have an understanding of the currents, winds and wave patterns which they routinely encounter. The navies of the North Atlantic Treaty Organization (NATO) and Russia see the North Atlantic as full of meaning as they patrol its waters keeping an eye on each other. For these people, the North Atlantic is a place.

The perception of place

We do not see the world around us in the same way as others. How we experience the world influences our perceptions of it.

> ### ➡ Activities
>
> 1 Find one photograph of a place that you think of positively, and one of a place that you regard negatively.
> 2 Write a description of each place, stating what it is about each place that leads to your positive or negative perceptions.
> 3 Show the photographs to others in your class and find out what they feel about the places. Do not reveal your feelings about the places.
> 4 Compare your perceptions. Discuss what factors might be responsible for the similarities and differences in your perceptions.

Factors influencing perceptions

Because perception is an individual thing, people's personal characteristics influence how they see the world. Key factors influencing perceptions can include age, gender, sexuality, religion and role in society. This last factor is closely related to education and socio-economic status.

Age

People's perceptions change as they get older. Think of how perceptions of the same place, a local park for example, might alter through time. A five year old might see the park as a place to have fun by playing on the swings or riding a bike. A few years later, the place might be used for different recreational activities such as tennis and skateboarding. How might older teenagers and those in their early twenties perceive a park?

Have you ever revisited a location about which you had very positive feelings only to be rather disappointed on your return? This might apply to a holiday venue because the geographic features (sandy beach) and facilities (playground) which were so appealing when you were a child are perceived differently as a teenager.

Many people move through a **life cycle** that involves changing their residence and therefore where they might live. Such moves are often associated with changes to income or family size. When a young person leaves home to set up their first independent household they usually have limited income and do not need much living space so they often live in rented accommodation close to a city centre. This allows easy access to employment and services such as shops, bars and clubs. On the other hand, a couple with children might buy a larger house with more space and a garden in the suburbs. In retirement the need for accommodation is reduced and people may regard peace and quiet as a priority. As a result many retired people may downsize their accommodation or migrate to smaller centres or quieter suburbs.

Gender

In different societies, the roles men and women have are reflected in the way the two groups can move around and the types of places they can use.

Traditionally, many places have been defined as being 'male' or 'female'. A division along gender lines has been most apparent in the separation of public and private places. The phrase 'A woman's place is in the home' represents a stereotypical image of women that was widespread until the later twentieth century in many Western societies. The female private place of the home contrasted with male public spaces, such as factories, offices and many places of recreation. Past photographs of sporting events show overwhelmingly male crowds, for example.

Such divisions among places reflect the way society sees male and female roles. That females were more or less excluded from certain spaces was seen as a key element of how males dominated and controlled society. It is not that long ago that if a woman married, she often had to resign from her job. She was expected to stay at home, run the household and bring up children. In the UK, for example, married women were not employed in the civil service nor by many local government offices until the Second World War.

The question of safety is a significant one in giving meaning to places. For many people fear can influence their mental maps and therefore the decisions they make about where, when and how they go to certain places. Certain places can be perceived as 'unsafe' and therefore some people avoid these routes, neighbourhoods or places on the basis of their gender. Locations that a person would go to during the day might be avoided at other times. Places which are isolated or dark and late night public transport can represent a 'geography of fear' for some people and may restrict their personal geography, especially if on their own. Urban geographies of fear are also influenced by a person's age and sexuality. Groups such as local politicians, architects and planners are giving the issue of safety a higher priority when making decisions about the layout of places, especially of city centres.

Sexuality

Sexuality can influence the way in which people use places. As the acceptance of different sexual orientations becomes more widespread, some places acquire a meaning because they are where lesbian, gay, bisexual and transgender groups (LGBT) tend to cluster. In some cities, LGBT 'zones' have been identified and mapped. These areas centre on concentrations of restaurants, bars and clubs which are 'gay friendly' such as the Castro District in San Francisco and the 'Gay Village' in Manchester. The south coast resort of Brighton has acquired an image as the LGBT capital of the UK. The LGBT community is a large one and Brighton has developed into a cosmopolitan place, accepting of a wide diversity of people.

Some researchers have seen similarities between the emergence of these places and the emergence of **ghettos**. People cluster together for a sense of security and a place where they can 'be themselves'. The idea of strength in numbers is well known in urban residential patterns but has mainly been seen in terms of ethnicity. LGBT neighbourhoods in a predominantly heterosexual society allow people to express themselves and in the case of San Francisco, to win political power and thus to gain an influence over decision-making. The election of LGBT local councillors is seen as important in creating a strong sense of place for San Francisco's LGBT community.

There is an economic aspect to the emergence of places defined on the basis of sexuality. The 'pink' pound, euro or dollar is important in some locations in helping regeneration and the rebranding of places. Manchester, Brighton and San Francisco benefit from LGBT tourism as people seek out the places to visit where they can relax and have a sense of security in being openly themselves and enjoy personal geographies not restricted by fears and anxieties.

Religion

People have given locations spiritual meanings for millennia. Some natural landscape features are sacred to certain human groups. The giant mass of sandstone Uluru (Ayers Rock) (Figure 5.6) in the centre of Australia has a major role in Aboriginal creation stories. These were disregarded by colonising

Figure 5.6 Uluru (Ayers Rock), Australia

Europeans and the name Ayers Rock came from the Premier of South Australia in the late nineteenth century. More recently, a greater sensitivity towards Aboriginal culture has remade the place into one of spiritual significance.

Activities

1 Research other examples of natural physical features which have acquired spiritual meanings, for example, Mount Fuji, Japan; the River Ganges, India; Kilauea, Hawaii and Lake Titicaca, Bolivia. Plot their locations on a world map and explain their spiritual meaning.
2 Identify the physical features at each location which people have given spiritual meanings to and indicate how these link with the beliefs of the people.

Humans have long given locations religious meaning through buildings. Ancient stone circles and megaliths (literally 'large stones') are found all over the world. Stonehenge was originally built as a wooden structure around 5000 years ago, which was later replaced by stone, much of which survives today. Earlier still are the cave paintings at Lascaux, southwest France. The images of animals, birds and human figures drawn on the cave walls 17,000 years ago have been interpreted as having magical and religious significance.

Religions such as Judaism, Christianity and Islam have given meanings to many places through the building of synagogues, churches and mosques. There is one particular location which has come to represent a place of very great religious significance for all three of these religions: Jerusalem (Figure 5.7).

149

Figure 5.7 Jerusalem

- Judaism – capital of the united kingdom of the tribes of Israel in about 3000 BC. The First Temple was built on the Temple Mount by Solomon and housed the Ark of the Covenant, containing the stone tablets given to Moses by God on which were written the Ten Commandments. Jerusalem came to represent Judaism's most sacred site and the ancient capital of the Jewish state.
- Christianity – site of Christ's crucifixion and resurrection. Jerusalem represents Christianity's most holy site.
- Islam – site of Muhammad's 'night journey' and Islam's third most sacred shrine; the Dome of the Rock is located in Jerusalem.

Jerusalem, the focus of spiritual meaning for three different but linked religions has become 'contested space'. The city has been captured eleven times in the past 1500 years, and destroyed and rebuilt on five occasions.

Many religious places are associated with refuge, peace and healing. People go to practise their religion at various shrines, wells and buildings. One of the most famous places associated with healing is Lourdes, in the foothills of the Pyrenees in the south of France. Some 6 million pilgrims, mostly Roman Catholics, visit Lourdes each year. In the mid-nineteenth century a local young woman experienced visions of the Virgin Mary, the mother of Jesus Christ. Lourdes became a place where people travel to bathe and/or drink the spring water flowing from the cave where the visions took place, as the water is claimed to have curative powers. Cures are examined by the Church to make sure they are authentic and have occurred solely on the basis of the healing power of the water.

Role

Each of us performs a variety of roles at different times. For example, as a sixth former, you are a student for much of the time. Within your school or college you may have a position of responsibility such as a prefect, member of a sports team, band or play cast. At home you are a son or daughter but perhaps also a brother or sister. You may have a job and work as an employee.

The role we have at any one time can influence our perceptions of a location and how we behave. For example, you are likely to act in a different way in your local shopping centre when you are with your friends compared to when you are with your parents or even grandparents!

As we go through life we gain and lose roles. And as we change, so do our attitudes and our perceptions of places. An independent twenty year old is likely to view locations differently from the parent of young children. As a parent your perception of potential threats such as traffic may be heightened. In older age your perception of accessibility may be more acute.

Role influences perceptions of fear, insecurity and anxiety, which are then reflected in the ways boundaries are used to include or exclude people and activities. As British towns and cities grew rapidly during the nineteenth century, many residential developments tried to exclude certain types of people from living in them. Most people rented their accommodation and leases contained rules about who could and could not rent certain properties. In this way, some landowners and developers tried to sustain a high socio-economic status for their developments.

During the past 40 years, gated communities have become more common in many countries. They tend to be high value properties which are 'defended' by secure boundaries and controlled access points.

The influence of emotional attachment to place

People remember places in many different ways. What are your first thoughts as you look at the photographs in Figure 5.8? These four places are likely to have stirred a variety of emotions as you thought about what each place means to you.

Memory is a personal thing because our experiences are unique to us. Our memories are also highly selective: we remember some things and forget others.

If we have positive experiences of a place we are likely to have strong emotional attachments to that place and vice versa. And it is not just our personal experiences that influence how we feel about places. Memory and feelings are also social, that is we receive them as part of a group. Think of the very strong emotional attachments some sports fans feel to their team's home ground. People often have a similar, perhaps deeper attachment to nations. This is especially true for people exiled from their homeland for whatever reason.

Figure 5.8a Buckingham Palace

Figure 5.8b Glastonbury

Figure 5.8c Auschwitz

Figure 5.8d Twickenham

 Activities

1 Think back to your time at a previous school. What feelings do you have about that time? How do your experiences during that time affect your feelings about that place?

2 If others in your current geography class also attended your previous school, share your feelings about this place. What different and similar feelings do you have about the place? What reasons might help explain these?

People without their own state: the Kurds

The Kurds are an ethnic group spread across a number of Middle Eastern countries. The present 'heartland' of the Kurdish nation, known as Kurdistan, spreads across Iran, Iraq, Syria and Turkey (Figure 5.9).

The Kurds are estimated to number about 28 to 30 million in this heartland, with another 2–3 million living as a **diaspora**, away from this region. For example, Germany has a sizeable Kurdish community.

The Kurds have long wanted their own independent state. They have suffered persecution, notably under the Iraqi regime of Saddam Hussein, and there is also a history of armed conflict with the Turkish police and military. The Kurdistan Workers' Party (PKK), an organisation using armed conflict to bring about a Kurdish nation, is regarded as a terrorist organisation by the USA and EU countries.

Activities

1 Research the origins and history of the Kurds (or a similar stateless group such as the Basques that has a strong emotional attachment to a place). News sites such as Reuters and newspapers such as *The Times*, *Guardian*, *Independent* and *Washington Post* have material on such groups.
2 Describe the physical and human geography of their 'homeland' including the landscape and economic activities.
3 Plot the distribution of your chosen group, including clusters located away from the homeland region.
4 Make a presentation using a medium such as PowerPoint or Prezi outlining the origins and current situation of the group, making clear the role of the group's emotional attachment to their homeland.

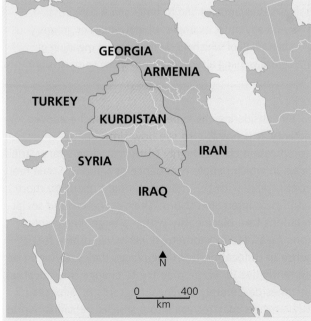

Figure 5.9 The location of Kurdistan

How globalisation and time–space compression can influence a sense of place

Globalisation is the increasing interconnectedness and interdependence of the world, economically, socially, politically and culturally. In terms of a sense of place, it is a set of forces that are changing the ways in which people experience and understand places, both familiar and unfamiliar. As with many changes, different places and people are affected in different ways by globalisation.

The term 'global village' is used to convey the idea that the world has become smaller, not in a physical sense but because of its greater inter-connectedness. Communications and the flow of goods are quicker and more reliable than ever before. Capital moves around the globe at the press of a key. More people travel further and with increasing frequency. This view of the world in the twenty-first century is summarised in the phrase 'time–space compression'. In other words, space is no longer the barrier it once was to communication and movement of people, goods and ideas.

In the UK one way we can understand the significance of time–space compression is in the local supermarket. Much of our food is sourced from overseas and the availability of foods such as strawberries or melons no longer depends on the season. Similarly, many of the goods we buy, including our clothes, come from countries on the other side of the globe.

As a result of these changes, how we perceive places alters.

Activities

1 Choose a local shopping cluster of about 20+ outlets.
2 Survey the cluster, noting the number of outlets that represent a good or service which originated overseas, e.g. Indian or Chinese restaurants, kebab shops, foreign banks, stores specialising in international foods (read pages 523–24 before you start).
3 Assess the impact of such retail outlets on different people living in the area, such as a long-time local resident, a teenager who had grown up in the area and a recently arrived foreign migrant.

For some people the changes to places as a result of time–space compression are easily accommodated and they benefit from them. For others, the changes can be disturbing and mean that they no longer 'feel at home' in a location where they may have lived all their lives.

Winners and losers

Advantages from time–space compression are not available equally to all people. For those who are able to manipulate time–space compression to their own advantage, their sense of place probably does fit the idea of the 'global village'. However, there are those who are not able to gain much from the shrinking of the world and who are more controlled than controlling.

Activities

1 Consider the following types of people:
   - A currency trader in a bank in New York or London or Frankfurt or Tokyo
   - An unemployed Mexican waiting to the south of the USA–Mexico border for a chance to cross illegally
   - An elite sportsperson from a Low-Income Developing Country (LIDC)
   - A mother from an LIDC or an Emerging and Developing Country (EDC) working away from her family in an Advanced Country (AC)
   - A teenager in an AC travelling during a gap year
   - An ex-steel worker, now 52 years old in an AC whose place of work was shut and the business transferred to an EDC
   - A homeless person in a city in an AC
   - A homeless person in a city in an LIDC
2 Draw up a table with two columns, headed Advantages from time–space compression and Disadvantages from time–space compression.
3 For each person listed above, note ways in which the individual's sense of place has been given advantages or disadvantages due to time–space compression.
4 Add any other examples of individuals or groups you can think of which illustrate how the meaning of place is being altered by time–space compression.

By considering the types of people in the above activity, it is possible to see how time–space compression creates highly complicated and very varied ways in which people experience geography. Some feel comfortable with the changes brought about by more and faster interconnections. Their economic and social relations take place easily within a larger linear space (one measured in kilometres). However, others feel a sense of dislocation from the places they grew up in or currently live in. Some may try to change their location, at times desperately as illegal migrants or refugees, while others may retreat into a smaller individual world which has few interactions beyond it.

Ways of representing places

The way in which a place is represented by the media influences how we feel about that place. Media agencies can be divided into formal and informal categories. Perhaps the best-known formal agency is a census such as those held in the UK, USA and India every ten years. Other formal agencies include any data that has clear locational positioning such as a road network, the location of victims of Ebola or the distribution of different soil types. Very often such representations of places are closely linked with statistics which describe data associated with particular places such as crime figures or rainfall totals.

The informal category includes a great diversity of media such as television, film, music, art, photography, literature, graffiti and blogs.

Informal ways of representing places

Television and film play major roles in representing places offering sounds as well as sights. The lens of the camera can give wide-angled views of places perhaps showing their geographical context. It can also zoom in on the detail of a place.

Television soaps are interesting because they represent places through the lives of local people. These programmes, partly through their continuity over the years, build up a strong fictional representation of places. *EastEnders* and *Coronation Street* represent two inner city places and *Emmerdale* a rural area in the north of England.

Activities

1 Choose one television programme. Write an analysis of the ways in which the programme portrays key elements of the place it is located in. Consider the following aspects:
   a) The physical geography of the location such as landscape, street layout, architecture, age of buildings, weather and climate
   b) The human geography of the location such as age and sex structure, occupations, ethnic origins, religion and occupations
2 What representation of the programme's location are the programme-makers trying to convey? How successful are they?

Many films rely on their representation of place to tell their story to the viewer. Both fictional and real places play important roles in films and sometimes the two are combined. Tolkien's *Lord of the Rings* fictional trilogy was set in the actual landscape of New Zealand (Figure 5.10). New Zealand then took the image to promote its tourism industry.

Figure 5.10 New Zealand South Island, one of the film locations for the *Lord of the Rings* trilogy

It is not just the visual media that represent places. Across the centuries, composers and authors have used their media to depict places. This continues today and has been joined by twentieth- and twenty-first-century media such as photography, blogs and even graffiti.

Activities

1 Choose a place such as a major city or a distinctive landscape such as the Lake District or Antarctica.
2 Research how your chosen place has been represented by a range of media, for example in literature, poetry, films, art or blogs. You could seek advice from fellow students or staff in departments other than geography, for example those studying music, art history or art.
3 Analyse the similarities and differences that the media highlight when representing your chosen place.

Formal ways of representing places

Vast quantities of data are collected and stored, much of it spatial. Such data are invaluable in representing places, allowing them to be described and investigated.

Census

In 1801 England and Wales held its first census of population. Thereafter the census was completed every ten years, continuing without a break (apart

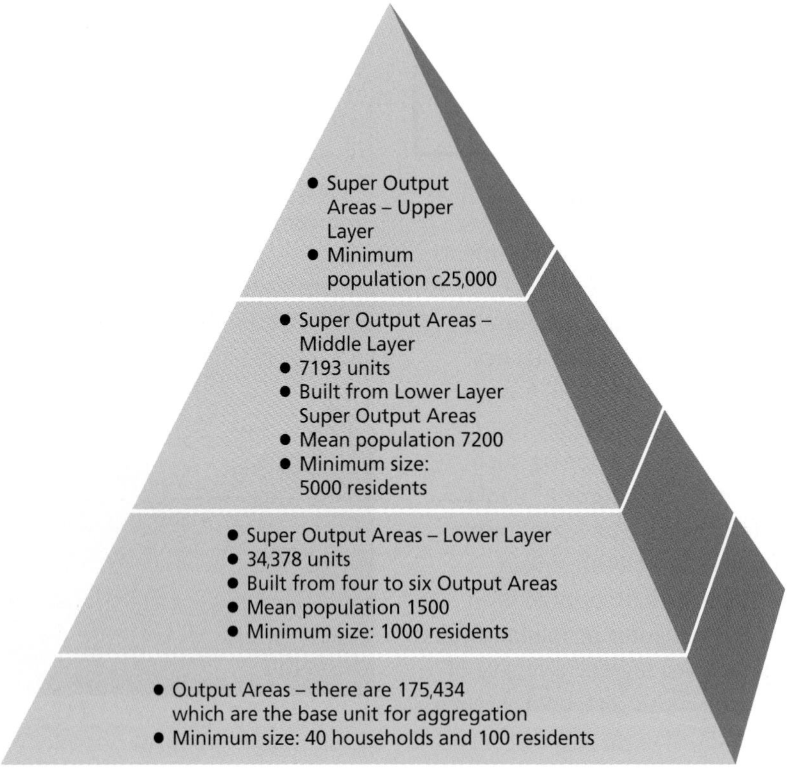
Super Output Areas – Upper Layer
Minimum population c25,000
Super Output Areas – Middle Layer
7193 units
Built from Lower Layer Super Output Areas
Mean population 7200
Minimum size: 5000 residents
Super Output Areas – Lower Layer
34,378 units
Built from four to six Output Areas
Mean population 1500
Minimum size: 1000 residents
Output Areas – there are 175,434 which are the base unit for aggregation
Minimum size: 40 households and 100 residents

Figure 5.11 Geography of the 2011 census for neighbourhood statistics

from in 1941). Initially a simple counting of heads, over time the census has become increasingly sophisticated and detailed. Today it includes personal information such as date of birth, gender, educational qualifications, ethnicity, religion, health, welfare, housing and employment.

The basic source of data is the **household**. A household is defined as: one person living alone, or a group of people (not necessarily related) living at the same address who share cooking facilities and share a living room or sitting room or dining area. Short-term residents, such as students at university, do not count as a household. Household data are then added together into various sizes of areas (Figure 5.11).

Up-to-date census data are an essential element for government planning and the allocation of resources to areas, such as schools, health care facilities and housing.

Representing rural places

Numerical data are often used to present formal and objective representations of places. Nonetheless, they almost always include some degree of subjectivity and bias. A key question for collecting any spatial data is where the boundaries should be drawn. For example, a village surrounded by agricultural land and looking in every part like a typical rural settlement (Figure 5.12) may be inhabited by a majority of people who work, shop and spend their leisure time outside the village. The census may record this place as rural. But do the residents think of themselves as rural dwellers? And do others see this place as rural?

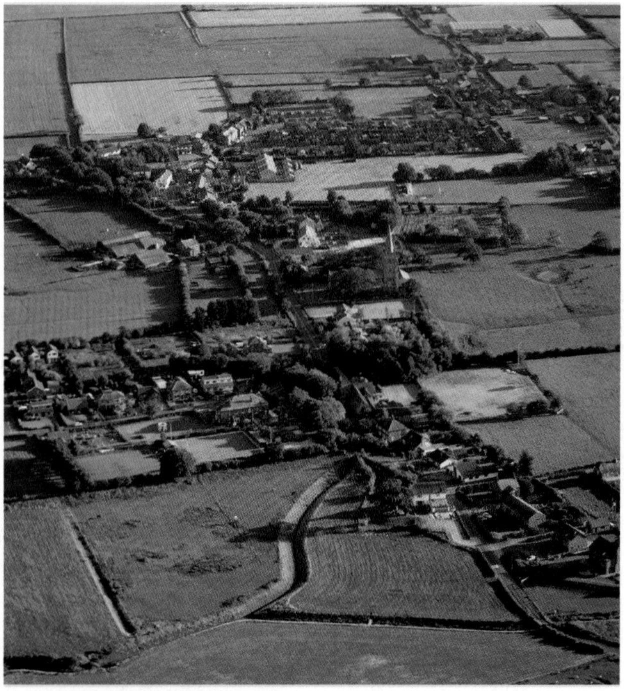

Figure 5.12 A 'typical' village

The conventional view is that rural communities possess a number of characteristics that distinguish them from urban places:

- Closely knit, supportive community where everyone knows everyone else
- More conservative and traditional in views
- More homogeneous ethnically
- Less mobility, both spatially and socially

➡ Activities

1 Search the web for a rural place's website. Many villages host their own site.
2 Describe the location of the village in its regional context, including its physical setting.
3 Using pages from the website, list the activities and or characteristics you think mark this place out as 'rural'. Give reasons why these indicate a rural place.
4 Use Google Earth to view images of the place. Do these suggest a rural place?
5 Use the website www.neighbourhood.statistics.gov.uk to investigate the census data for your chosen place, such as age structure, population density, ethnicity, occupations and commuting flows. Do these data suggest a rural place?

✔ Review questions

1 Using examples, distinguish between place and space.
2 How does the stage in a person's life cycle affect their perception of place?
3 How does gender influence a geography of fear for some people?
4 Suggest how sexuality can influence how people use places.
5 Identify two places with particular religious meanings and suggest why they have come to have such meaning.
6 What is meant by the term 'time–space compression'?
7 Outline how time–space compression changes a person's experience of place.
8 Name two informal and two formal representations of place.
9 Show how different informal agencies present contrasting representations of one particular place.
10 Assess the relative advantages and disadvantages of formal and informal representations of places.

5.3 How does economic change influence patterns of social inequality in places?

Key idea
→ The distribution of resources, wealth and opportunities are not evenly spread within and between places

What is social inequality and how can it be measured?

Differences, based on factors such as age, ethnicity, gender, religion, education and wealth exist in all societies. Such differences often raise moral issues concerning inequality.

Because social differences vary from place to place, geographers can make significant contributions to debates about spatial inequality.

The terms 'quality of life' and 'standard of living' (Figure 5.13) are frequently used when discussing differences between places. It is important to distinguish between them as they are not the same.

- Quality of life – the extent to which people's needs and desires (social, psychological or physical) are met. This can be seen in areas such as the treatment of people. Are all people treated with equal dignity and do they have equal rights? Does everyone have reasonable access to services such as health care, education and leisure? Are all opinions heard and respected?
- Standard of living – the ability to access services and goods. This includes basics such as food and water, clothes, housing and personal mobility.

Clearly, income and wealth are significant factors in determining both standard of living and quality of life. Higher incomes tend to offer people greater choice of housing, education and diet. But a rise in income may not always lead to improved quality of life. Longer hours at work, a longer daily commute, migration away from family and friends sometimes to a foreign country, poor air quality and pollution are examples of factors that can lead to a higher income and standard of living but can result in a lowering in the quality of life. In other words quality of life may be sacrificed for a higher income.

When social inequalities lead to very great differences between groups of people, the term

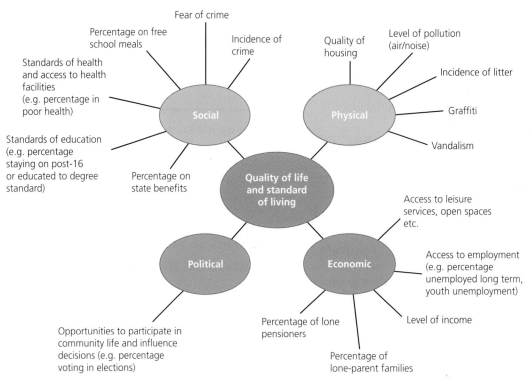

Figure 5.13 Indicators of quality of life and standard of living

deprivation is commonly used to refer to a context when quality of life and standard of living are low. Deprivation is more than just poverty. Poverty is not having enough money to support a decent standard of living whereas deprivation refers to a general lack of resources and opportunities (Figure 5.14).

The UK government uses an Index of Multiple Deprivation to assess relative levels of deprivation. Seven factors are combined to give an overall measure: income, employment, health, education, crime, access to housing and services, and living environment such as air quality.

All the smallest scale LSOA census units in the census can be ranked according to their Index of Multiple Deprivation score. This allows us to identify the most and least deprived areas in England and Wales and compare levels of deprivation between areas. It is important to understand that such indices do not tell us by how much one place is more or less deprived than another. For example, if a place has a rank of 80, it is not necessarily twice as deprived as a place with a rank of 160.

Measuring social inequality

In order to understand social inequalities we first need to access appropriate data. In addition to census data a whole range of social information has become available in the past 50 or 60 years, including data on employment, education and health care.

Income

At the global scale the role of income in measuring social inequality is seen in the World Bank's definition of absolute poverty: US$1.25/day PPP (purchasing power parity). Below this level of income, a person

Figure 5.14 The cycle of deprivation

- Poverty
  low wages or unemployment
- Poor living conditions
  poor accommodation overcrowding run-down area
- Ill-health
  stress and strain
- Poor education
  old schools
- Poor skills
  poor occupational skills

Activities

1 Access the Multiple Deprivation Index at www.gov.uk/government/statistics/english-indices-of-deprivation-2015.

2 Click on the link to the 'Infographic' which is an introduction to the Index. The 'Statistical Release – Main Findings' also gives useful summaries.

3 Look up the reference name and number of your local area, starting with the smallest spatial unit, Lower Super Output Area (LSOA). You can also use ward data which combines several LSOAs. (See Chapter 15, Geographical Skills, page 506.)

4 Use the links to access data on multiple deprivation in your local area. The data are available at a variety of scales of local authority such as county and district council. Note the scores for the larger units where your local area is located.

5 a) Compare your local area with those around it.

   b) Choose a different type of area from where you live (urban/rural for example) and compare the scores.

   c) Draw up a table of comparison of multiple deprivation scores at LSOA or ward scale and local authority scale.

6 It is also possible to access data for the seven domains (factors) used to construct the Index of Multiple Deprivation such as income, health or living environment. Choose one of these domains to compare the same areas.

7 Use the indices of deprivation 2015 explorer (http://dclgapps.communities.gov.uk/imd/idmap.html) to access the mapping tool. Generate maps to show the pattern of Multiple Deprivation in your city, town or rural area.

8 Comment on the pattern shown on the map and in your table.

---

cannot afford to purchase the minimum amount of food and non-food essentials such as clothes and shelter. The use of purchasing power is important as the cost of obtaining a particular good or service can vary greatly from one country to another.

Relative poverty is a useful measure as it relates the level of poverty to the distribution of income across the whole population. For the UK and throughout the EU, the relative poverty level is 60 per cent of the median household income. About 13 million inhabitants of the UK, including some 3.5 million children, currently fall below this threshold.

The USA also uses a poverty threshold to differentiate between poor and non-poor. It is a monetary value adjusted each year to take account of changes to inflation.

The **Gini coefficient** (Figure 5.15) is a technique that can be used to measure levels of income inequality within countries. It is defined as a ratio with values between 0 and 1.0. The lower the value the more equal is income distribution. A Gini coefficient of 1.0 would mean that all the income in a country was in the hands of one person while a value of 0 indicates that everyone in a country has equal income.

Housing

Being able to afford accommodation of an adequate standard is closely related to income. At all scales social inequality is evident in the type and quality of housing people occupy.

Housing tenure is an important indicator of social inequality. Owner-occupiers own their house outright. In many ACs this is achieved through borrowing money in the form of a mortgage which is normally paid back over 25 to 30 years. Some people rent from private landlords while others rent from a local authority (council). Charities and housing associations also provide subsidised accommodation for rent. In many LIDCs, housing tenure is complex, especially in the slum areas. In many such areas a well organised system of landlords and tenants exists. The term 'squatter settlement' is often misleading and should only be used where people have no legal right to the land they occupy.

Education

Formal education is provided by schools, colleges, apprenticeships and universities. Informal education can be gained from doing something in the home or workplace, for example learning how to cook at home or watching an older brother or sister milk a cow or repair some machinery. The acquisition of skills can be underestimated if only formal qualifications are measured. This is particularly true when studying societies in EDCs and LIDCs.

Contrasts in literacy levels give an indication of inequality in education. Literacy is a measure of the

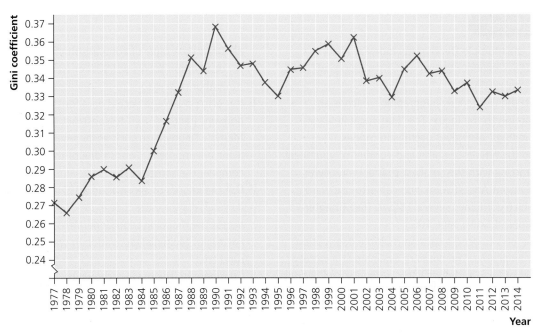

Figure 5.15 Changes in UK Gini coefficient 1977–2012

ability to read and write to a basic level. Globally there are clear contrasts among countries in levels of literacy and especially in terms of gender equality (Figure 5.16).

Health care

Access to health care and levels of ill-health are closely associated with social inequality. The association between poverty and ill-health is very strong and reflects a number of influences. These influences include variables which are clearly health-focused such as number of health care professionals. The measure of number of doctors per 1000 people is often used to describe health inequality between places at the global scale.

In the UK, increasing attention is paid to unequal access to health care, the so-called 'postcode lottery'. Depending on where you live, the level of medical provision through the National Health Service varies.

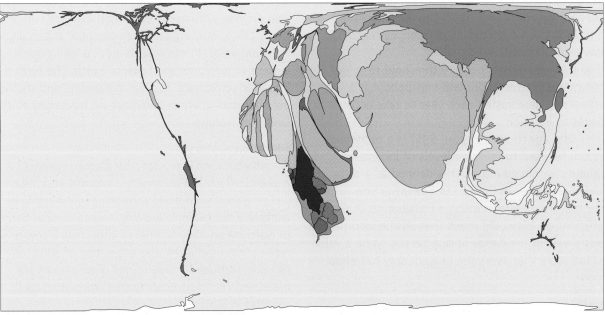

Figure 5.16 Global distribution of girls not at secondary school. The size of each country represents the proportion of all girls in the world who are not attending secondary school that live in that country. 'Slim' territories (e.g. the Americas) indicate the smallest difference between male/female enrolments in secondary schools and 'fat' territories indicate the largest differences. Colours are used to distinguish between different countries

In part these differences may reflect the priorities of Health Care Trusts and differences in morbidity and life expectancy.

But health is not just a matter of medical provision. Access to clean water, effective sanitation, quality and quantity of diet, type of housing and air quality, for example, have significant influences. It is also a matter of social behaviour and lifestyle, for example attitudes towards vaccination and tobacco and alcohol consumption. HIV/AIDS has had a serious impact in places such as sub-Saharan Africa. In part this is due to the attitudes towards male–female relationships and ignorance.

Employment

Whether a household includes someone who is in receipt of regular income has a profound impact on the standard of living and quality of life. Unemployment is perhaps the obvious measure but it is not always straightforward to assess. Not all countries take counts of employment and defining what counts as employment and what does not can be difficult.

Someone may be in employment but receiving only low wages. In ACs, it is generally the case that rural places have average incomes below those in towns and cities. In urban areas in LIDCs and EDCs millions of people make a living by working in the **informal sector** of the economy. This offers a relatively easy way into employment but has drawbacks. For example, someone selling fruit or brooms on a street corner might work for ten hours or more in a day and yet still be unable to afford to live anywhere except in a slum settlement.

Social inequality exists at all scales. It is clear that an individual's life chances are closely related to where they live.

Human Development Index (HDI) – a composite measure of inequality

From 1990 the UN has produced an annual assessment of the level of development of each of its member states. It is based on economic and social indicators:

- Income adjusted to take account of purchasing power in the country
- Life expectancy at birth
- Education using the adult literacy rate and the average number of years spent in school

The index ranges from 1 (most developed) to 0 (least developed). A high index equates to 0.8 and above; medium index from 0.5 to 0.79; and a low index is less than 0.5. The HDI highlights the great inequalities that exist between countries in both economic and social terms.

Stretch and challenge

1 Find out exactly how the HDI is calculated.
2 Critically assess its value as a measure of inequality.
3 Research other composite measures and assess their value in measuring spatial inequality.
4 Is income the same as wealth? Discuss this in small groups and then debate within your group.

How and why spatial patterns of social inequalities vary

Inequalities exist at all scales from the global to the local. There are inequalities between urban places and rural places and there are also contrasts at the intra-urban scale. Several factors influence the relative level of social inequality between places. A single factor is unlikely to explain inequalities at any scale. It is the interaction of several factors which tends to lead to spatial patterns of inequality.

Wealth

The ability to purchase goods and services is fundamental to social well-being. Everywhere low incomes are linked to factors such as ill-health, lower educational attainment and poor access to services. The lack of formal qualifications and low skill sets are major obstacles to raising income and thereby reducing social inequality.

The cost of living is an important consideration when discussing the role of wealth. If a person's income rises, but increases in the cost of food, housing, clothes and fuel outstrip the additional income, then that person is relatively less well-off. A key factor here is **disposable income**: the amount left over after the essentials of life (food, housing and clothing) have been bought.

Housing

Quality of accommodation is a significant influence on social inequality. The smaller the income of a household, the less choice of housing they have. Poor quality housing and overcrowded conditions often create ill-health. Such inequality in access to the housing market often occurs when demand exceeds supply. In LIDCs and some EDCs, millions of people have little choice but to live in slum housing. Often because of rapid urbanization, the municipal authorities are simply overwhelmed by the scale of demand and lack the resources to increase the supply of decent housing.

Homelessness is a growing problem among urban populations in many ACs. This group often exists on the margins of society and may resort to squatting illegally in derelict or empty buildings.

Also in ACs the affordability of housing contributes to social inequality. When the cost of housing, either through purchase price or rent, inflates at a faster rate than wages and other prices, those with low or irregular incomes can find themselves excluded from the housing market. In the UK, traditionally low-cost social housing was provided by local authorities but since the 1980s, the availability of this type of accommodation has decreased dramatically. This is also a cause of inequality in rural regions. The rise in second-home ownership and the migration of wealthy people into the villages and small towns of the countryside have raised property prices beyond the reach of many young families.

Health

In all societies there is a clear link between ill-health and deprivation. Sub-standard housing, poor diet, unhealthy lifestyles and the additional stress of day-to-day living in poverty take their toll on human health. Access to medical services also plays a part: often, and at all scales, the distribution of health care services is uneven. Within a local area for example, some groups such as the elderly have limited mobility which restricts their access to GPs and primary health care. In rural areas where health facilities are widely dispersed, accessing medical care can be an issue for households without access to a car or public transport.

Education

Differing access to educational opportunities is recognised as a significant element in creating and maintaining inequalities. Achieving universal primary education was one of the **Millenium Development Goals** and most governments invest in education to raise standards of living and quality of life. Illiteracy excludes people from accessing education and skills training and therefore reduces employment opportunities. Accessing even basic education can be a major issue in rural regions in LIDCs.

Access to services

It is often the case that how accessible services are to people greatly affects both their quality of life and standard of living. It is a significant disadvantage to people if they find accessing services difficult.

At the global scale there are stark inequalities between societies in ACs, EDCs and LIDCs in terms of access to services. For example, one measure of access to medical services is the number of doctors per thousand people. In Norway (AC) there are just over four, in Brazil (EDC) there are just under two while in Kenya (LIDC) the figure is well below one. The picture for access to education is likewise one of clear inequality among the three categories of country.

At the national scale, most countries display inequalities between regions. People living in core regions, such as most capital cities, where wealth and investment are high, tend to have good access to services whereas the more peripheral regions, those usually furthest away from the core, suffer from limited access.

Access to services is influenced by three factors:

- number of services
- how easy it is to get to the service, e.g. quantity and quality of transport links and geographical distance
- social and economic factors, e.g. factors such as age, gender, income.

There is often a clear urban–rural divide in access to services. On average urban dwellers have better access to services than their rural counterparts. However, within both urban and rural locations, those with higher incomes are nearly always advantaged as regards their access to services. People living in low status housing districts, whether located in ACs, EDCs or LIDCs, struggle to access services such as retailing, public transport or banking.

One recent service which shows distinct differences among places is access to the internet. A digital divide exists in terms both of possessing the means to be online, for example owning equipment such as a mobile phone or laptop, as well as the quality (speed and bandwidth) of a connection. In the UK, for example, there are significant contrasts between urban–rural areas close to urban centres and remote rural areas in terms of broadband speeds. Although in nearly all countries investments in broadband are growing, in general the faster speed areas tend to get faster and faster with slow speed locations lagging behind.

It is interesting how in some EDCs and LIDCS mobile phone technology is beginning to transform lives and reduce inequalities. The growth in satellite technologies removes the need to set up fixed copper cables and with solar-powered recharging equipment, even very remote places can become linked in.

One way in which social inequality can persist is in those places where the authorities restrict access to internet services. In both China and the Democratic People's Republic of Korea (North Korea) internet access is severely controlled by the governments.

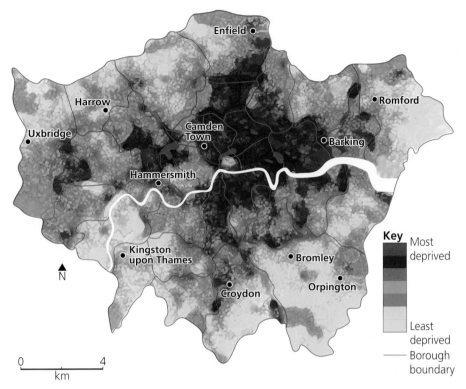

Figure 5.17 The pattern of multiple deprivation across the London boroughs

---

Activities

1 Study Figure 5.17 The pattern of multiple deprivation in London.
2 Describe the pattern of multiple deprivation as shown in Figure 5.17.
3 Suggest reasons for this pattern.

Key idea

→ Processes of economic change can create opportunities for some while creating and exacerbating social inequality for others

---

The role of globalisation in economic change

In economic terms, globalisation has led to increasing flows of ideas, capital, goods and services and people. The global economy has become knitted together in ways that have never happened before. Globalisation is driven by economic changes across all scales from the global to the local (Figure 5.18).

Geographers identify transnational corporations (TNCs) and nation states as key players in the global economy. Relationships among TNCs and between TNCs and states drive changes that impact the lives of billions of people.

One major consequence is **global shift**. This refers to the relocation of manufacturing production on a global scale. Fifty years ago most manufacturing was concentrated in western Europe and North America. Raw materials such as copper and coffee were exported by countries like Zambia and Brazil, which had limited manufacturing bases of their own. From the 1980s the **New International Division of Labour (NIDL)** gathered pace. European, North American and Japanese TNCs created labour-intensive factories in what were called Newly Industrialising Countries (NICs), mainly in East Asia and Latin America. Containerisation and bulk handling brought down relative costs dramatically and so contributed to the locational changes.

With **economic restructuring** came the loss of employment in the **primary** and **secondary sectors** as the comparative advantages of ACs in primary and secondary activities declined. ACs transformed into **post-industrial** societies in which most people worked in the **tertiary** and **quaternary sectors**.

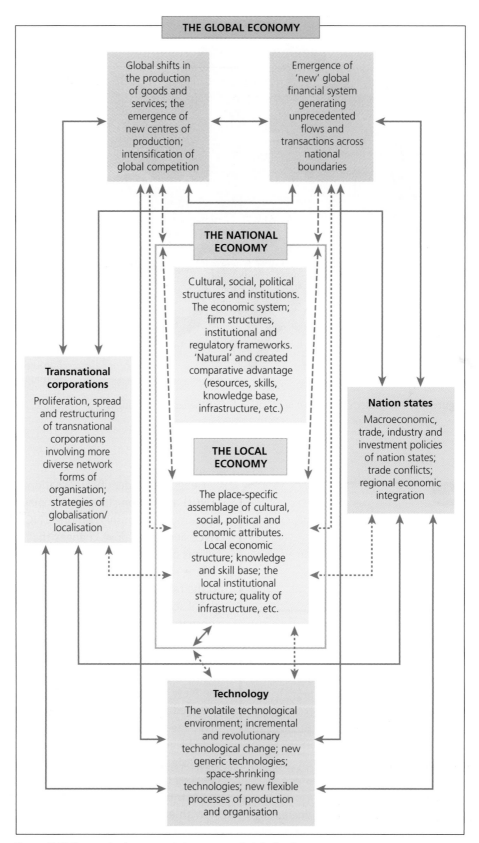

Figure 5.18 Economic changes and the process of globalisation

Comparing employment structures – the use of triangular graphs

Study the outline graph in Figure 5.19. Note how the three axes relate to each of the primary, secondary and tertiary sectors (all quaternary employment is included in the tertiary data). (See Chapter 15, Geographical Skills, page 517.)

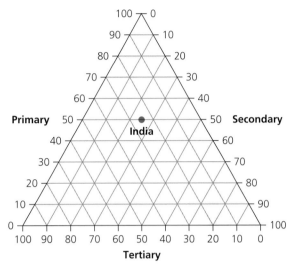

Figure 5.19 Employment structures in selected countries

1 Plot the data in Table 5.3 as a triangular graph.

Table 5.3 Employment structures in selected countries

Country	% employed in primary	% employed in secondary	% employed in tertiary
Australia	5	20	75
Chile	13	23	64
Denmark	2.5	20	77.5
Ethiopia	85	5	10
Hungary	7	30	63
India	48	22	30
Indonesia	39	15	46
Japan	3	26	71
Mozambique	81	6	13
Romania	28	28	44
United Kingdom	1.5	15	83.5

2 Using your graph attempt to classify the data into groupings of countries with similar employment structures.

3 Describe and justify the groupings you have made. Are the terms ACs, EDCs and LIDCs useful when distinguishing different types of countries?

The impacts of structural economic change on people and places

Inevitably economic restructuring led to mine and factory closures and job losses in ACs, a process known as **deindustrialisation**. Some places which relied heavily on a narrow range of traditional economic activities such as mining, iron and steel making, shipbuilding or textiles were badly affected by deindustrialisation. Unemployment and associated problems such as ill-health increased significantly and were often concentrated in inner city neighbourhoods or on local authority housing estates in the suburbs. The skills required by traditional heavy industries were not easily transferable to the growing service sector. Moreover, the physical environment of deindustrialised regions was often poor, with a legacy of abandoned and derelict buildings and polluted land and waterways. However, some places hardest hit by deindustrialisation and economic change, such as Essen in Germany, Pittsburgh in the US and Swansea in south Wales, have been rebranded and have witnessed significant regeneration.

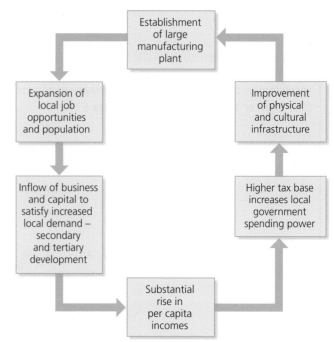

Figure 5.20 The effect of positive economic change on a place

Table 5.4 Positive and negative impacts of economic change on people and places (Source: Figure 4.36 The positive and negative impacts of global shift from Nagel, G. and Guinness, P. (2011) *Geography for Cambridge International A and AS Level* Hodder p. 453)

	Positive	Negative
In ACs	• Cheaper imports of all relatively labour-intensive products can keep cost of living down and lead to a buoyant retailing sector. • Greater efficiency apparent in surviving outlets. This can release labour for higher productivity sectors (this assumes low unemployment). • Growth in LIDCs may lead to a demand for exports from ACs. • Promotion of labour market flexibility and efficiency, greater worker mobility to area with relative scarcities of labour should be good for the country. • Greater industrial efficiency should lead to development of new technologies, promotion of entrepreneurship and should attract foreign investment. • Loss of mining and manufacturing industries can lead to improved environmental quality.	• Rising job exports leads to inevitable job losses. Competition-driven changes in technology add to this. • Job losses are often of unskilled workers. • Big gaps develop between skilled and unskilled workers who may experience extreme redeployment differences. • Employment gains from new efficiencies will only occur if industrialised countries can keep their wage demands down. • Job losses are invariably concentrated in certain areas and certain industries. This can lead to deindustrialisation and structural unemployment in certain regions. • Branch plants are particularly vulnerable as in times of economic recession they are the first to close, often with large numbers of job losses.
In EDCs and LIDCs	• Higher export-generated income promotes export-led growth – thus promotes investment in productive capacity. Potentially lead to a multiplier effect on national economy. • Can trickle down to local areas with many new highly paid jobs. • Can reduce negative trade balances. • Can lead to exposure to new technology, improvement of skills and labour productivity. • Employment growth in relatively labour-intensive manufacturing spreads wealth, and does redress global injustice (development gap).	• Unlikely to decrease inequality – as jobs tend to be concentrated in core region of urban areas. May promote in-migration. • Disruptive social impacts, e.g. role of TNCs potentially exploitative and may lead to sweatshops. Also branch plants may move on in LIDCs too, leading to instability (e.g. in Philippines). • Can lead to overdependence on a narrow economic base. • Can destabilise food supplies, as people give up agriculture. • Environmental issue associated with over-rapid industrialisation. • Health and safety issues because of tax legislation.

During the 1970s and 1980s there was significant investment by foreign-owned TNCs in the EU and in particular the UK (see page 178).

Training and employment grew and a positive multiplier effect was created. Although the number of people employed in such enterprises was fewer than in the traditional industries they replaced, they represented a significant opportunity for individuals and places.

Globalisation has led to greater international opportunities. Firms have specialised in areas where they have a comparative advantage. In manufacturing this has often meant specialist high-tech industries, aerospace, pharmaceuticals and biotechnology. Highly qualified workers and cutting edge research, design and development are required. Some places have built on an existing high-tech reputation, such as Cambridge, while others have developed this. Bangalore in south central India, for example, has become a centre for aerospace engineering and IT development. In west Cumbria, a remote rural area, the largest concentration of high-tech employment in the nuclear industry in the EU is located at Sellafield.

Birmingham Research Park

The traditional focus of Birmingham's economic activity has been 'metal bashing', the manufacture of a wide variety of metal products from screws and nuts and bolts to jewellery, firearms and vehicles. Many trades have declined and the city has faced problems of deindustrialisation. However, several initiatives are currently underway to provide job opportunities as part of the restructuring of Birmingham to include a greater emphasis on the knowledge economy.

Birmingham Research Park (Figure 5.21) is a joint venture between the University of Birmingham and Birmingham City Council. It is designed to attract research-led companies wishing to work in partnership with academic staff. In particular the Park is developing its BioHub and is at the heart of the Edgbaston Medical Quarter, which includes the large Queen Elizabeth Hospital and the University's College of Medical and Dental Science.

The nature of firms developing innovative products means that close links with academic research are required. Accommodation for new businesses needs to be flexible – for example, small-sized units for start-up firms and room for expansion as a business grows. Biotechnology also requires very specific laboratory conditions such as sterile manufacturing environments.

Figure 5.21 Birmingham Research Park

How booms and recessions impact on people and places

The economic health of a place is rarely static. Over time places grow and decline and this impacts on social opportunities and inequalities. Within countries, places often experience in different ways the effects of boom and recession. This is because different types of economic activities are not distributed evenly in space.

It has been claimed that the capitalist economic system operates in a series of interconnected cycles. Notably the Russian economist Kondratieff concluded that roughly 50-year cycles of growth and decline have characterised the capitalist world since 1750 (Figure 5.22). These cycles of growth and stagnation have been linked with technological innovation with new industries providing the basis for a boom. Once the technology is no longer 'new', fewer opportunities for growth exist and boom is followed by recession.

Technical innovation is not evenly distributed. Centres of innovation and their inhabitants often benefit from above average economic growth. Within these centres or core regions the multiplier effect is strong. The greater economic opportunities available in core regions help to explain the higher standards of living found there. It is a matter of debate as to why some places are able to support technological innovation when others

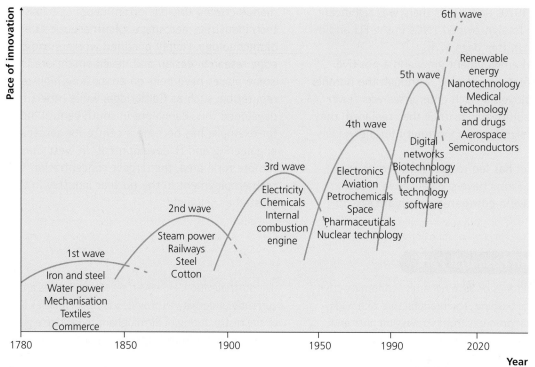

Figure 5.22 Waves of innovation

do not. Explanations tend to focus on the role education, government and social organisations have in encouraging or discouraging enterprise and change.

Recessions are general slowdowns in economic activity. Macroeconomic indicators such as gross domestic product (GDP), investment spending, household income, business profits and inflation fall, while bankruptcies and unemployment rise. Some people are more able to cope with a recession than others. In general, the more skilled someone is the more employment opportunities they are likely to have. Households tend to cut back on spending on non-essentials such as leisure and entertainment when there is pressure on incomes. This can result in fewer jobs in service activities such as bars and restaurants. But recessions affect people across the socio-economic spectrum.

Silicon Valley

Silicon Valley is a by-word for technological innovation, enterprise and high standards of living. It is the area around the southern part of San Francisco Bay, California (Figure 5.23).

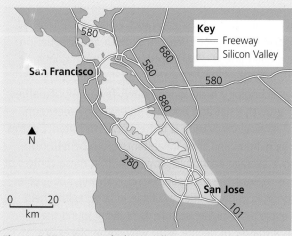

Figure 5.23 Location of Silicon Valley

Centred on Santa Clara valley and the town of San Jose, Silicon Valley is home to many of the world's largest high-tech corporations and thousands of start-up businesses all hoping to be the next Intel, Apple or Facebook. Stanford University has close links with much of the research being carried out and offers a regular supply of high-quality graduates. Just as important is the availability of **venture capital**, which funds the development of risky innovations. Nearly half of all the venture capital in the USA is spent in Silicon Valley and it has the most millionaires and billionaires in relation to the population of the region. In addition law firms specialising in patent and copyright law have clustered in Silicon Valley also offering opportunities for those with relevant skills.

However, many of the production line workers are not paid more than the state's minimum wage and the manufacture of some computer components involves exposure to toxic chemicals which pose health risks. The majority of these workers are female migrants from Asia or Latin America.

The roles governments can play in patterns of social inequality

Governments operate at different geographical scales. There are transnational governments such as the EU, national governments such as the UK and local bodies such as county, city and parish councils. In many countries, governments play an important role in decision-making and the allocation of resources. Most governments are motivated by ideals of social justice and political cohesion and seek to reduce the extremes of poverty and inequality (Table 5.5).

UK government – measures for tackling social and economic inequalities

Table 5.5 Methods used to tackle social and economic inequalities

Taxation	Income tax is often used by governments to redistribute wealth from more prosperous to less prosperous groups, and so create a fairer society. Most governments have progressive tax systems where the better-off pay a larger proportion of their incomes in tax. Essential items such as food may be exempt from tax. This benefits poorer groups that spend a larger percentage of their income on food. This information can be found at www.investorwords.com
Subsidies	Governments also try to reduce inequality by giving subsidies to poorer groups. Children in poor families may get free school meals, clothing allowances and help with university fees. Pensioners may get subsidies for fuel and transport. Other subsidies may include free child care for single parents. Low wage earners, unemployed workers and those with long-term disability are entitled to benefits.
Planning	Governments, charities and housing agencies often give priority to upgrading housing and services in the poorest areas. Planning is often organised geographically and is targeted at the most deprived areas which vary in scale from neighbourhoods to entire regions.
Law	Legislation exists which outlaws discrimination on racial, ethnic, gender and age criteria and aims to give equal opportunities to all groups. The poorest groups of workers are protected by minimum-wage legislation.
Education	Governments often provide funding for training and upgrading skills in order to raise skill levels and qualifications, improve employment prospects and boost economic growth. Education programmes designed to improve personal health (e.g. diet, obesity, smoking) are often targeted at the poorest groups in society.

In 2016 it is estimated that government spending in the UK will be £760 billion, which is split approximately three-quarters by central government and one-quarter by local government. The pie chart indicates how this sum is allocated (Figure 5.24).

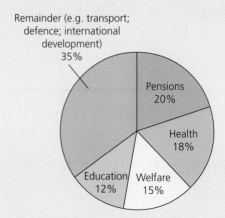

Figure 5.24 UK government spending, 2016 (estimated)

Pensions

Spending on pensions in the UK has almost doubled in the past dozen years. This largely reflects increasing life expectancy and an ageing of the population. As a result the age at which people can draw their state pension is rising. Some of the poorest members of society are those relying on the state pension although some are entitled to benefits. There tends to be a disproportionate number of these people living in inner cities and this contributes to high levels of multiple deprivation in these areas. It is important though to distinguish between this group and pensioners who are relatively wealthy. Large numbers of people now retiring have occupational pensions which provide an additional income to the state pension. Many of these retirees also own their homes outright and have benefited from house price inflation, which increases their wealth.

Health care

Health care in many ACs is provided by a combination of government and private organisations. In the UK, the National Health Service is 'free at the point of delivery'. People pay for the NHS through the taxation system. However, the provision of health services varies, for example, inner cities and remote rural areas can be short of GPs and other health care workers. Others, such as affluent suburban areas, are well served.

Language and cultural barriers and the role and status of women have been obstacles to recently arrived migrants accessing health care services. Within some groups, such as Somali and South Asian, fears about immunisation and a lack of resistance to childhood diseases have meant that children are unprotected against common diseases such as measles and whooping cough. Literature is produced in ethnic minority languages to inform parents of the benefits of immunisation and as literacy has improved and more assimilation has taken place, overall health care has improved.

There is a hierarchy in the provision of health care from the GP surgery through to specialist units treating rare diseases in large teaching hospitals. As medical care has become more technological, it has become concentrated in fewer but larger facilities. Meanwhile in rural areas and many small towns, local 'cottage' hospitals have closed. For lower-income families and the elderly with lower levels of personal mobility, access to health services in these areas can be a problem.

Activities

1 Use the table of data and map you generated in the Activity on page 157 to remind yourself of the pattern of inequality in your area.

2 Using the data available at www.gov.uk/government/statistics/english-indices-of-deprivation-2015 identify the pattern of inequality in health and disability. Use the mapping tool to generate a map showing the pattern of inequality (see Chapter 15, Geographical Skills, page 510).

3 Plot the location of health care services such as GP surgeries, health centres and dentists. Use internet searches, local directories and personal fieldwork to obtain the data (read pages 523–24 before you start).

4 Compare and contrast the location of health care services with the pattern of health and disability inequality.

Rural services

For many decades a cornerstone of rural planning in the UK has been to support rural areas (especially remote ones) through the **key settlement** policy. Services such as education and health care, employment and housing have been concentrated in large villages and small towns. These places act as hubs for people living in surrounding smaller settlements. The idea behind this policy was that if a service is supported by a critical mass of people or **threshold**, then it would be sustainable. However, as improvements in personal mobility have taken place many rural residents no longer rely exclusively on their nearest key settlement. They often combine trips for employment and shopping; and access a range of destinations such as the supermarkets and retail parks on the outskirts of urban places. This behaviour has been helped by the extended opening hours. The availability of home delivery services by food retailers also provides other options.

Key idea
→ Social inequality impacts people and places in different ways

The contrasts between two places in social inequality

Case study: Jembatan Besi, Jakarta, Indonesia

Jakarta is the capital of Indonesia, a country of 256 million people, making it the fifth largest in the world. Of these, some 10 million live in the capital, which, as in other EDCs, is a city of extremes. In the country as a whole, the wealthiest 10 per cent control nearly 30 per cent of household income while the poorest 10 per cent have access to 3.4 per cent. No official figures exist for Jakarta but it is likely that the distribution of income is very similar to the country as a whole. Just over a quarter of Jakarta's inhabitants live in slum settlements but many others live in districts which ACs would call slums.

Figure 5.25 The location of Jembatan Besi within Jakarta

Jembatan Besi is a slum in Jakarta about 4 km northwest of the city centre (Figure 5.25). It has developed organically over the past 40 years as Jakarta's population has grown. The settlement is hemmed in on all sides by other built-up areas and the Ciliwung River. With a population of about 4000 it is one of Jakarta's most densely populated districts. The inhabitants include people whose families have lived there for several generations but also migrant workers who may only stay for a few months.

Social and economic conditions

The reason why people live in slums like Jembatan Besi (Figure 5.26) is that demand for affordable housing greatly exceeds the supply. Neither the government nor the private sector has the resources to cope with the increasing numbers of people wanting to live in Jakarta. Also, the people themselves do not have the resources to afford more expensive formal housing.

Most people in Jembatan Besi struggle to make ends meet. The average income of residents is about US$4/day but for many this is not a regular income. Employment is often insecure with most residents able to provide only unskilled and casual labour. There is much self-employment with many families running their own small business. Selling food or second-hand goods, some of which are salvaged from waste tips, is common. Such activities require very little start-up capital and many are run from home. Even jobs in more formal employment come with little security. Jakarta has a significant garment industry and there are many small-scale producers operating in and around the slum areas. There is little protection for those employed in these small factories and few health and safety precautions.

Health is a major concern. Sanitation hardly exists in Jembatan Besi. Few homes have a toilet and although there are toilets in the slum, they are poorly built and run for profit by local businesses. The toilets tend to flush out into open sewers in the street. There is no clean running water. Groundwater supplies are available but are polluted because Jembatan is built on a former waste tip. This also means that if a family is able to afford a water pump to raise groundwater, it is likely to be highly polluted.

Epidemics of water-borne diseases such as cholera and typhoid are common. The tropical hot and humid climate means that malaria is an issue, as is hepatitis A. The very young and elderly are at risk from dehydration due to diarrhoea caused by poor hygiene.

Air pollution is at very high levels. The use of kerosene for cooking as well as high levels of emissions drifting over the city pose significant health risks for residents.

The nutrition of most slum dwellers is dominated by rice with little fresh protein or fruit and vegetables.

There are schools but most are poorly equipped. Too often families simply cannot allow their children to complete their formal education because they need to earn money to supplement family incomes. The garment industry is a major source of employment for many young females.

Housing conditions

Jembatan Besi is one of the most densely populated places in Indonesia. Most homes consist of a relatively well built ground floor using timber and brick. This represents the original house but as pressures on space have grown, extra stories have been added. Consequently the construction is increasingly makeshift with height. Residents make use of any materials they can find, scrap wood and metal being common. Fire is a constant risk due to overcrowding, use of kerosene and the improvised nature of electrical wiring, most of which suffers from serious over-loading. Due to the narrow alleys and tall buildings, houses have virtually no direct sunlight. Electric lighting is by neon tubes and bare light bulbs.

Figure 5.26 Slum housing, Jakarta

The future?

Slums such as Jembatan Besi will not be disappearing in the forseeable future. The urban authorities are trying to make inroads in the worst areas but planning is difficult and non-existent in most slums. The Jakarta Housing and Administrative Buildings Agency has identified 392 'community units' that are slum areas planned for improvement. However, slum clearance to allow rebuilding has tended to result in people relocating to other slum areas and making the situation worse there.

The Ciliwung River slums are notorious in Jakarta for the very poor living conditions and overcrowding.

However, there is often a strong sense of community within the slums. Their inhabitants are remarkably resilient; they just about make ends meet and look to build a better future for themselves and their children. Nonetheless, the inequalities between the residents of Jembatan Besi and others in Jakarta, not to mention those living in ACs, are stark.

Case study: Northwood, Irvine, southern California

Northwood is a community in the northern part of the City of Irvine in Orange County, California (Figure 5.27). Irvine was developed as a fully planned city, beginning in the 1960s and built on the former Irvine Ranch. It is a classic edge city. The city's vision statement is 'to create and maintain a community where people can live, work, and play in an environment that is safe, vibrant, and aesthetically pleasing'. The city was developed around a series of communities called 'villages' of which Northwood is one.

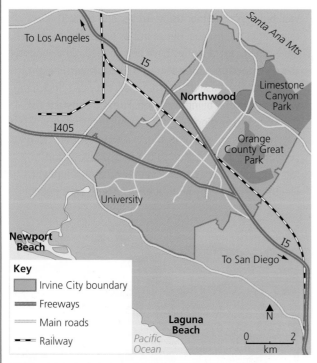

Figure 5.27 The location of Northwood within Irvine

Northwood is located on the southern edge of the Los Angeles conurbation about 10 km east of Newport Beach on the Pacific Ocean. To the east are the low Santa Ana Mountains. Some 22,000 people live in Northwood in just over 8200 households. A third of the households are families with children of school age.

Social and economic conditions

In general the residents of Northwood are well-off financially. Their median income is about US$86,500 a year compared with the national average of about US$52,250. One of the reasons for choosing to live in Northwood is access to employment in Irvine. The University of California, Irvine is the city's single largest employer and offers a range of opportunities. There are a number of well-established high-tech companies such as Blizzard Entertainment (video games), Broadcom (semiconductors) and several medical, pharmaceutical and aerospace firms. Several TNCs have their USA headquarters in Irvine such as Kia Motors, Mazda Corporation and Toshiba. The area is home to many new business ventures and offers opportunities for start-up companies such as small and well-serviced premises and venture capital.

Health care for Northwood's residents is excellent, even by the high standards of an AC. Air pollution is relatively low as Northwood is on the edge of the Los Angeles conurbation and therefore avoids the smog common in places towards the centre of the conurbation.

The schools in Irvine are regularly assessed as being among the best-achieving in the USA. There are five high schools and several tertiary education facilities. The educational standards of the residents is high, many having first degrees (68.5% of residents aged 25+) with 20.5% possessing masters or doctorates.

The overall crime rate in Northwood is 70% lower than the national average. Irvine as a whole is safer than 96% of the cities in California. Violent crime, for example, stands at about 50 incidents per 100,000 people in Irvine compared with 366 per 100,000 nationally. The figures for vehicle theft are 52 per 100,00 in Irvine but 220 per 100,000 nationally.

Ethnically, about half the residents are white with the second most numerous group being Asian, mainly from Vietnam (after the USA left Vietnam following the success of the Viet Cong in the Vietnam War, many former South Vietnamese migrated to the west coast of the USA).

Figure 5.28 Typical housing in Northwood

Housing conditions

Northwood is characterised by single-family houses on relative large 'lots' (Figure 5.28). Most households (91 per cent) own their own home and of these two-thirds have lived in Northwood for more than ten years. The average household size is 2.8 persons. Many of the streets are lined with trees, mostly eucalyptus, which are a legacy of the windbreaks established when the area was farmland (see www.statisticalatlas.com/neighborhood/California/Irvine for data based on the US census).

The future?

Irvine in general regularly features among the highest ranked cities within the USA for factors such as safety, management and 'best place to live'. Even during downturns in the economy, the area retains its reputation for high-paid employment. With its combination of high-quality housing, transport infrastructure, education, facilities such as retailing and Mediterranean-style climate along with the easy access to the beaches along the Pacific, Northwood and Irvine represent the opposite end of the economic spectrum to Jembatan Besi.

✓ **Review questions**

1   What is meant by the term 'spatial inequality'?
2   Distinguish between 'quality of life' and 'standard of living'.
3   Draw a diagram(s) to show how factors such as income, housing, education and health care combine to lead to multiple deprivation.
4   Why do so many people in LIDCs live in slum housing?
5   What is meant by 'global shift' in economic change?
6   Describe and explain how the employment structure of an AC, an EDC and an LIDC have changed over the past 50 years.
7   Suggest how the concept of 'comparative advantage' has led to opportunities in ACs.
8   Outline the reasons why some places benefit from booms while others suffer from recessions.
9   How do governments influence the spatial pattern of inequality?

5.4 Who are the players that influence economic change in places?

Key idea
→ Places are influenced by a range of players operating at different scales

The players involved in driving economic change

Economic change is a complex process that affects places at a variety of scales, from local and neighbourhood through to an entire country. At any scale, change is brought about by the interaction of a considerable number of **players** or stakeholders. Players are individuals, groups of people or formal organisations who can influence, or can be influenced by, the processes of change (Figure 5.29). Some players have more 'power' or influence than others, especially in economic terms.

Public players include government. The EU is a trans-national government which can influence economic change via grants for infrastructure development, for example. National government has departments and agencies responsible for strategic planning such as education and training, major transport links and environmental management. Local government has similar responsibilities and carries out the planning and implementation at the local scale such as a county or city or town. Government tries to stimulate economic growth, sustain existing employment and create new jobs and improve the environment.

Private players include a very wide range of different people and organisations. Businesses range from TNCs to those who are self-employed and across all sectors in the economy. The primary aim of business players is to generate money to make a profit on their investment.

Local communities are concerned about their immediate area. They are interested in economic change, such as employment, and also social and environmental matters,

Figure 5.29 Possible players involved in the construction of a by-pass around a market town

for example the construction of new housing on the edge of a town or the redevelopment of a brownfield site.

A large number of non-government organisations exist which tend to have a particular focus. Some are small local groups while others cover the whole country. For example, the National Trust is a private organisation with over a million members interested in the conservation of historic buildings and landscape.

Activities

1 Using Figure 5.29, suggest how each of the players has a stake in the change.
2 Assess the relative 'power' of each player to influence the change.

Case study: Structural economic change in Birmingham Metropolitan Region

Birmingham is a large metropolitan region (Figure 5.30) at the heart of the West Midlands **conurbation**.

In mid-2014 there were 1.1 million residents, making it the second largest city after London.

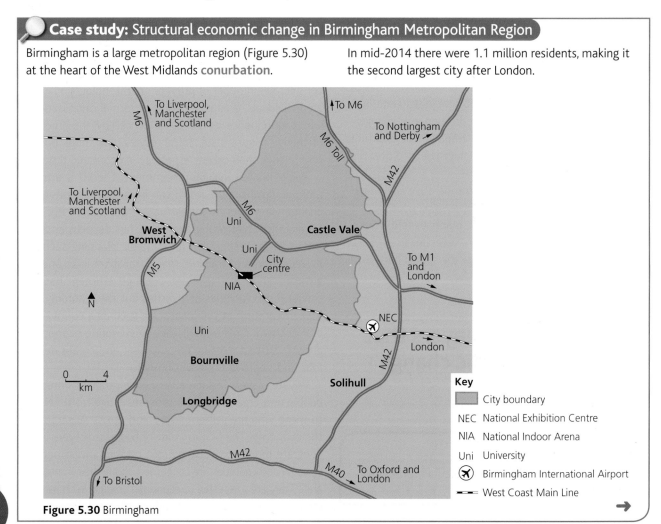

Figure 5.30 Birmingham

Various machines for processing cotton were invented in Birmingham in the 1730s and 1740s, and electro-plating and the pneumatic tyre were either invented or first used in the city. Boulton and Watt began their partnership in 1775 in Birmingham, revolutionising the production of more efficient steam engines, and the first medical X-ray was taken there in 1896. During the Second World War, pioneering work on radar development took place in the city. The majority of inventions receiving British patents in the nineteenth and twentieth centuries originated within 50 km of the city centre and today about 25 per cent of British manufactured exports originate in Birmingham.

Birmingham's development to the 1960s: the making of a 'city of a thousand trades'

Birmingham is a product of the Industrial Revolution. Although mentioned in the Domesday survey (1086) the place at that time was a relatively poor agricultural manor.

The first players to influence change were the de Bermingham family who purchased a royal charter in 1166 that allowed them to hold a market. The place began to grow as a result of trade, a theme that continues right through to the present day.

Medieval Birmingham consisted of about half a dozen streets focused on the parish church and the market. Metal-working was already established and this was a sign of change to come. The main area of metal smelting, the Black Country (for example, Dudley and Wolverhampton), lay to the west and supplied the raw material for Birmingham's metal-working trades.

In 1563 William Camden, the English historian and topographer, reported that the town was 'swarming with inhabitants and echoing with the noise of anvils (for here are great numbers of smithes)'. Power came from watermills, with the products increasingly being non-agricultural such as blades for swords.

The Industrial Revolution

By the early 1700s Birmingham's population had grown to 15,000. Many of the town's inhabitants had migrated from rural areas in search of employment. Among the inhabitants, the middle classes were beginning to increase in number as service sectors such as law and banking developed. At this time Birmingham also developed clearly differentiated housing areas based upon socio-economic status.

Matthew Boulton was a key player in moving the industrial base of the town forward. An entrepreneur and engineer, he established the first factory in the world in 1761. His 'Soho Manufactory' brought 700 employees under the one roof along with complete industrial processes.

The nineteenth century witnessed extraordinary growth. The gun, jewellery, button and brass industries dominated. Other metal manufacturing industries and all the industries needed to supply a growing population such as food processing were present. The Cadbury family set up their Bournville factory and model village for the workers on the southeast outskirts of the city. Industry and trade required financing, and banks, insurance and legal firms developed. Lloyds and Midland banks were founded in Birmingham in the mid-nineteenth century.

Transport infrastructure developed hand in hand with industrial growth. Birmingham lay at the heart of the national canal network and the Midland terminus of the London to Birmingham Railway was opened in 1838.

1900 to the 1950s

Birmingham continued to grow during the first half of the twentieth century. New engineering industries developed, such as the Austin car plant which opened in 1906 at Longbridge. As the vehicle industry expanded, hundreds of small firms supplying the industry with vehicle components grew up in and around the city. For example, in 1917 the Dunlop tyre company, founded in Birmingham, established a large factory, which employed 10,000 people by the 1950s. A chemical industry developed, with Bakelite being manufactured in the city. Birmingham survived the Great Depression of the inter-war years relatively well, largely due to the diversity of its metal-working industries.

Throughout the first half of the twentieth century Birmingham sustained economic growth (Figure 5.31). This was accompanied by continuous population growth from both natural increase within the city as well as immigration from rural areas and other parts of the British Isles such as Ireland.

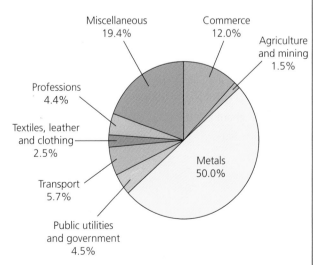

Figure 5.31 Employment in the Birmingham conurbation, early 1950s

Large areas of terraced housing built for workers in the many industries were squeezed into spaces between factories. With limited personal mobility, most people lived within walking distance of their work. The development of a middle class who could afford to commute helped drive urban expansion. Transport innovations such as trams, suburban railways and later buses and cars facilitated urban expansion and the outward growth of the built area. A ring pattern of land-use centred on Birmingham's Central Business District emerged, interrupted by sectors along major arterial routeways.

Until the 1950s Birmingham's population was overwhelmingly white in ethnic origin. Employment was dominated by males; 60 per cent had skilled jobs such as lathe operators and precision engineers. The inter-war suburbs such as Northfield and Marston Green were classic areas of semi-detached and detached housing but also included some of the largest local authority housing estates in the country, for example Kingstanding.

Meanwhile, inner city areas mainly comprised poor-quality housing at high density such as in Aston and Handsworth. In addition decades of industrial activity had left land sites, canals and rivers with high levels of pollution. Air pollution reached high levels with controls on emissions (smoke, sulfur dioxide (SO_2)) almost non-existent.

Post-war Birmingham

Industrial decline

Birmingham was a prosperous city for most of the 1950s and 1960s with unemployment below 1 per cent. Yet between 1970 and 1983, earnings fell from being the highest in the UK to almost the lowest of any region. In 1982 unemployment reached 19.4 per cent.

The industrial geography of the city changed dramatically (Table 5.6).

Table 5.6 Changing employment structure in Birmingham, 1978–2000

Type of employment	% of work force employed		% change
	1978	2000	
Energy and water	1.0	0.2	−0.8
Metal, mineral and chemical manufacture	5.8	4.7	
Metal goods and vehicles	28.4	5.8	
Other manufacturing	9.6	7.2	
Construction	4.6	3.4	
Distribution and catering	15.7	19.8	+ 4.1
Transport and communication	5.2	6.4	
Finance and business services	7.3	21.4	
Other services, e.g. law, health and education	22.4	31.1	

Activities

1 Using the data in Table 5.6, calculate the percentage change for each type of employment between 1978 and 2000.
2 Plot these data using the outline on the right as a guide.
3 Describe the pattern of employment change in Birmingham in the last quarter of the twentieth century.

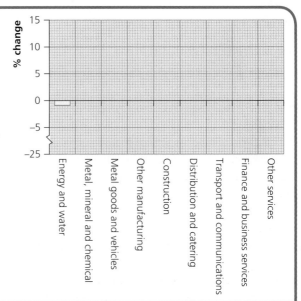

Table 5.7 Car ownership levels in the UK (%)

Number of cars per household	1970	1980	1990	2000	2010
No car	49	43	33	28	21
One car	44	45	45	44	44
Two cars	5	12	18	22	28
Three or more cars	1	2	4	6	7

Economic change occurred through the interaction of several players. Birmingham was caught up in the global recession of the 1970s in which the oil crisis of 1973 was most significant. The fragile geopolitics of the Middle East had resulted in another Arab–Israeli war. Most western countries supported Israel, and the Organisation of Petroleum Exporting Countries (OPEC), dominated by Arab states such as Saudi Arabia, used oil supplies as a 'weapon'. An embargo on supplies followed and the oil price increased about tenfold. This was a catastrophic shock to the global economic system as until then all planning and economic policy assumed that energy would be cheap for the foreseeable future. Birmingham's traditional industries were already beginning to suffer from increasing overseas competition from TNCs based in countries with lower production costs. This was part of the global shift in economic structure to which the oil crisis gave added impetus.

By the 1970s the British vehicle industry was in decline, despite car ownership levels rising (Table 5.7).

Foreign-based TNCs and in particular Japanese car manufacturers began to make significant in-roads into the British car market. Their products were seen as more reliable and better value for money. Japanese manufacturers gained wide acceptance from the consumers along with makers such as VW, Renault and Peugeot. Some overseas car manufacturers established factories in the UK but interestingly not one located in the West Midlands. Their locations were strongly influenced by grants from central government, another key player in economic change. The aim was to attract investment and create employment in regions that had even greater economic problems than Birmingham.

Within the car industry labour relations were far from smooth. Strikes were frequent during the 1970s and both management and the unions were players involved in the decline affecting the industry. This made Birmingham less attractive to potential investors – another group of influential players bringing about change.

Birmingham's manufacturing industries had been to a large extent made up of small and medium-sized enterprises (SMEs). Most occupied small inner city sites. As a result of the local authority's desire to bring about **comprehensive redevelopment** as part of a slum clearance programme, many small industrial premises were demolished. As a result, SMEs had difficulty in finding suitable premises. Those purpose built by the local authority often charged rents that were unaffordable to start-up businesses.

Housing

During the Second World War 5000 houses were destroyed in Birmingham and many more were damaged. In the immediate post-war years Birmingham had 110,000 sub-standard houses. The local government had resisted large-scale redevelopment based on flats but the scale of the housing need was such that 400 tower blocks were built in the 1950s and 1960s. In total, between 1945 and 1970, over 81,000 new dwellings were constructed. Not all were high-rise and the accommodation was significantly better than the old slums.

One consequence of redevelopment was the redistribution of people. The central zone was to a large extent cleared of residential land-use and people relocated to peripheral estates such as Castle Vale to the northeast. People also moved to new towns such as Redditch to the south. As the economy became more service orientated, flows of commuters increased from outer areas into the centre where the majority of services were located.

Both national and local governments were involved in the establishment of a **green belt** around the city to restrict outward expansion. This had the effect of increasing the value of land adjacent to the belt, especially in places close to principal road and rail routes to the city centre, such as Solihull and Knowle.

Demography

From the 1950s onwards there was significant international in-migration in inner city areas. Most immigrants were from the Caribbean, South Asia and the Far East (Table 5.8).

Table 5.8 Ethnic composition of Birmingham's population, 2011 (Source: 2011 census)

Ethnic group	Birmingham		England
	Number	%	%
White British	570,217	53.1	79.8
Pakistani	144,627	13.5	2.1
Indian	64,621	6.0	2.6
White other	51,419	4.8	5.7
Caribbean	47,641	4.4	1.1
Mixed	47,605	4.4	2.3
Bangladeshi	32,532	3.0	0.8
African	29,991	2.8	1.8
Chinese	12,712	1.2	0.7
Other	71,680	6.7	3.1

Immigrants tended to cluster in areas of cheap housing that had good access to employment. The growing service sector required a whole range of low-skilled jobs such as office cleaners, hospital porters and taxi-drivers. Birmingham became a cosmopolitan city and this was soon reflected in the religious landscape as mosques and temples were built. A great diversity of ethnic food shops and restaurants appeared. Ethnic clothing and fabric shops and financial services such as banks based in the home countries of the migrants were another addition to the city's increasingly diverse service functions.

Activities

1 Using the data in Table 5.8, select an appropriate technique and construct a chart to represent the ethnic structure of Birmingham and England (page 516).
2 Compare and contrast the ethnic composition of Birmingham with England.
3 Describe the impacts of Birmingham's ethnic composition on Birmingham's:
   a) urban landscape
   b) culture.
4 Assess the reasons why many migrants are attracted to a place such as Birmingham.

The city has a relatively youthful population compared with England as a whole (Figure 5.32): 38 per cent are 24 years old or younger whereas the equivalent figure for England is 31 per cent. Birmingham's elderly (more than 65 years old) account for 13 per cent; England's for over 16 per cent.

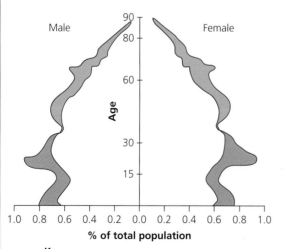

Key
■ Proportion greater in Birmingham
■ Proportion greater in England

Figure 5.32 Birmingham's age structure compared with England, 2014

Recent regeneration

Birmingham's socio-economic profile bore many of the characteristics of industrial decline and urban decay well into the late twentieth century. However, a variety of players are actively involved in trying to reposition Birmingham as a modern, forward-looking city with key roles to play at the local, regional, national and international scales.

The role of government

Local government is playing a vital role in promoting change. It actively promoted the city region, attracting inward investment to bring about the construction of the National Exhibition Centre (NEC) and the expansion of the adjoining Birmingham International Airport, both to the west of the city. Both developments attract the international conference and exhibition market which brings significant money into the region. Birmingham's improved international accessibility is another factor attracting investment.

From the late twentieth century onwards, use was made of national government schemes such as City Challenge and the Single Regeneration Budget which were designed to help places access funding for redevelopment. The Birmingham Heartlands Development Corporation was established in 1992 to bring regeneration to old industrial areas such as Nechells. In 1999 the New Deal for Communities (NDCs) scheme provided broadband access in parts of the inner city and set up work-experience programmes to try to improve employment opportunities.

The city council has a European and International Affairs team who seek to promote Birmingham beyond the UK and to gain funding from various EU departments. £235 million has been secured from the EU Social Investment Fund for the period 2014–20. Money from the European Regional Development Fund (ERDF) has gone into projects such as improving insulation in run-down social housing and building small premises for start-up SMEs.

The role of planning and flagship developments

Planning was important in bringing about physical change which it was hoped would act as a catalyst for socio-economic changes. More public spaces were developed in the centre such as Centenary Square. Victoria Square and the buildings surrounding it, such as the Council House and the Town Hall (both nineteenth-century 'statement' buildings) have been refurbished. At the centre of Victoria Square is a spectacular water cascade with large pieces of public art surrounding it (Figure 5.33).

A flagship development, funded in part by the EU, was the International Convention Centre and Symphony

Figure 5.33 Victoria Square, Birmingham

Hall opened in 1991. The latter is home to the City of Birmingham Symphony Orchestra, an example of a cultural player helping bring about economic change. The Convention Centre, now called the Barclaycard Centre, attracts conferences internationally as well as nationally. It is a good example of the multiplier effect as hotels, restaurants and leisure facilities benefit from visitors and their spending. The National Indoor Arena attracts large sporting events such as the 2003 World Indoor Athletics Championships and in 2010 the World Wheelchair Basketball Championships.

Three other important players are the city's universities which have a combined student population of some 50,000. Demographically this contributes to the city's relatively youthful profile. Economically, such institutions bring very significant wealth to an area. They employ large numbers of people who then have income to spend. This is in addition to the spending of students themselves. Students provide a market for a whole host of goods and services.

As a post-industrial city, Birmingham has been improving its service provision. The Bull Ring redevelopment in the 1960s was Europe's first major indoor shopping centre. It has been refurbished and along with flagship buildings, such as the Selfridges department store with its radical design, contributes to the city's status as one of the country's leading retail centres.

The environment of the city centre has also been significantly upgraded. Streets have been pedestrianised

and provided with high-quality street furniture and trees. The city's canals and their surroundings, largely ignored for most of the twentieth century, are playing a key role in economic revitalisation. Players such as The Canal and River Trust, a charity set up in 2012 to oversee the canal system, as well as local government and the private sector, have contributed to the regeneration of locations such as the Gas Street Basin and Brindley Place.

The role of transport

Birmingham's transport infrastructure has seen some of the most dramatic changes during the past twenty years. The main railway station, New Street, has been transformed by a £700 million investment. This involved Network Rail, private developers and retailers such as John Lewis creating an area called Grand Central. Shops, cafes, restaurants and offices have created 1000 jobs. As with many cities, Birmingham scrapped its tram system several decades ago, but tram and light rail systems have made a comeback. The Midland Metro links central Birmingham with places to the west such as West Bromwich and Wolverhampton. Further expansion of the system is currently taking place.

HS2 is the national government's flagship transport development which will impact directly on Birmingham. The £20 billion Phase 1 links London to Birmingham via a high-speed railway. Although still in the early planning stages, the project will involve a whole host of players such as national and local government, TNCs involved in major engineering works, banks, local community groups and a wide range of non-governmental organisations such as the National Trust. It is projected to bring further socio-economic change to Birmingham. It should stimulate employment growth in the city and make Birmingham, just 50 minutes from London by HS2, a more attractive proposition for investment.

A transport legacy of the 1970s, 1980s and 1990s is the West Midlands' motorway network. Birmingham is the hub of the national network. The M6 linking to the M1 just to the east, and the M5 and M40 are major spokes, while the M42 provides a ring-road from the southwest to the southeast of the city. These have been developed by national government, though private investment was responsible for the toll motorway that helps reduce congestion on the M6.

The Birmingham Development Plan (BDP) guides decisions on development and regeneration in the city up to 2031. It deals with how and where new homes, jobs, services and infrastructure will be delivered and the type of place Birmingham is likely to become over the next few decades.

1   What is meant by the term 'player' in the context of economic change?

2   Describe and explain the role of *three* players who were important in bringing economic change to Birmingham between the twelfth and nineteenth centuries.

3   Use a flow diagram to show how the development of industry can lead to a multiplier effect in a place.

4   In what ways did the economic changes of the Industrial Revolution change Birmingham's demographic profile?

5   Describe and explain how external forces influenced change in Birmingham in the second half of the twentieth century.

6   Identify three players influencing recent changes in Birmingham. State what changes they are helping bring about.

5.5 How are places created through placemaking processes?

> **Key idea**
> → Place is produced in a variety of ways at different scales

How places are produced by a range of people

The world's population reached 7 billion in 2011 and is expected to climb to between 9 and 10 billion by 2050. Urban places are at the forefront of this growth but increasing interconnection between places means that many rural places will also experience significant growth. The various challenges that need to be faced in the next few decades must have place and placemaking as a key element. For example, where will people live? Where will people work? What types of communities will people live in? How will people and goods move around?

The role of governments and other organisations in placemaking

It has been suggested that between 60 and 80 per cent of all data now includes a locational component. The explosion in the use of **Geographic Information Systems (GIS)** has meant that more and more people and governments are using geography as part of their work. Governments at all scales are becoming more aware of the need to consider placemaking as part of their operations.

In the UK, there is a hierarchy of government stretching from the national to local scale. National government represents a country as a place on the international stage. The UK is represented by an ambassador in the majority of the world's countries. As a representative of the Head of State (the Queen), the ambassador is independent of the elected government but nevertheless is closely linked with it. Government departments also engage with foreign countries. For example, trade links are supported and military co-operation is organised. The British Council is an organisation that specialises in promoting educational and cultural links abroad.

The attraction of Foreign Direct Investment

With the growth of transnational corporations (TNCs) in all sectors of the economy, governments around the globe have been keen to encourage inward investment by TNCs. Many TNCs have considerable choice when identifying locations for investment. Sometimes described as 'placeless', TNCs and their operations are widely distributed across several countries.

Foreign Direct Investment (FDI) has increased enormously since 1980 but with ebbs and flows depending on the health of the global, regional (e.g. EU) or national economies. Most FDI flows of capital are from TNCs headquartered in ACs, such as Barclays, Sony and Nestlé. Over 60 per cent of their investments are in other ACs. TNCs originating from EDCs and even LIDCs are having an increasing presence regionally, and in some cases globally. For example Tata (an Indian-based TNC) owns Jaguar Land Rover, and China is set to become one of the world's biggest overseas investors by 2020. In 2015 some US$100 billion of FDI was invested by Chinese companies.

Hitachi Rail

In September 2015, Hitachi Rail Europe (HRE) opened a new manufacturing plant at Newton Aycliffe (Figure 5.34), in northeast England. This plant represents an £85 million flow of FDI from Japan into the UK. Since the 1980s Japanese TNCs have invested heavily in new car assembly plants in the UK, notably at Sunderland (Nissan), Swindon (Honda) and Burnaston (Toyota). This FDI has been accompanied by many parts manufacturers, which have relocated from Japan to the UK.

Successive UK governments have been keen to attract inward investment to help raise the economic status of places undergoing major structural change and the loss

Figure 5.34 Prototype train at Newton Aycliffe

of traditional industries such as coal mining, iron and steel and heavy engineering. The UK's membership of the EU was a positive factor as Japanese TNCs were exempt from import tariffs on cars exported to mainland Europe. Another factor favouring the UK as a location for FDI was the English language – the preferred foreign language for Japanese business people. The British government also awarded a contract worth £5.7 billion to Hitachi to design and build the next generation of inter-city passenger express trains for the UK.

The plant at Newton Aycliffe is not just an assembly plant; it also houses research and development and design departments. Some 700 people will be employed and it is hoped that the multiplier effect will add up to another 6000 jobs in the factory's supply chain.

The Hitachi plant is intended to help bring some positive social and economic improvements to Newton Aycliffe and the surrounding region, which in recent decades has experienced significant employment losses. Railway engineering has strong historic connections with the northeast region. The world's first regular passenger service in the 1820s ran just a few miles away from the new Hitachi plant, and the North Eastern Railway had major workshops at nearby Darlington.

How planners and architects make places

Architecture can make an important contribution to placemaking through the design of individual buildings. This has been true across the centuries and continues today. 'An efficient planning system and a good spatial plan are essential to achieving high-quality places and good design.' This statement was made by the Commission for Architecture and the Built Environment (CABE) in 2009. CABE is part of the Design Council and provides advice on architecture, urban design and public space.

Local authorities in the UK maintain their own planning departments. They develop a Local Plan for their own local areas which includes elements of place such as industrial and housing developments, transport and amenities such as parks. This strategy is important in placemaking as it sets the framework for new buildings or uses of land. The Royal Town Planning Institute guides all professional planners working in local authorities, architecture firms and property consultancies.

Activities

1 Research the Local Plan for the local authority in which you live. Entering 'Local Plan Exeter', for example, into a search engine brings up links such as www.exeter.gov.uk and from there you should be able to find the relevant documents.

2 Within your class, work in small groups to produce a short presentation focused on one of the following elements of the Local Plan:
   - The history of the place and its regional setting (e.g. Exeter's place within southwest England)
   - Employment
   - Housing
   - Retail
   - Transport
   - Climate change

3 Share your presentations to build up an overall picture of the vision planners and architects have for the place in which you live.

4 What are the views of your class regarding the making of a place that is 'high-quality' and has 'good design' as set out in the Local Plan for your home area?

Individual buildings and public spaces are designed by architects. Through their design, buildings and spaces, such as parks and squares, reflect the history and culture of a place. They also influence how our lives are lived. As a result architects and the architecture they design make major contributions to placemaking. Design that pleases people and works well tends to be valued and cared for. However, some places become liabilities both to individuals and to the wider community. For example, such places can lead to crime, vandalism, high maintenance costs, poor health and a feeling of isolation for those living in them.

In the period between 1950 and 1980 the UK faced a huge problem of a growing population but too few houses. Also, in major urban areas, much of the inner city housing stock was either destroyed or badly damaged due to aerial bombing during the war. This housing crisis resulted in architects designing cheap, system-built housing. Much of the new build comprised tower blocks of flats. These structures created places that often had a negative image such as Hulme in inner Manchester and Orchard Park, Hull.

Since the 1980s, many tower blocks have been demolished and replaced with low-rise housing. However, this has not solved the issues associated with such places. Concerns such as unemployment, poverty, education and health remain significant.

The 24-hour city

The idea of the 24-hour city can be found around the world with many cities in EDCs and LIDCs, such as Cairo and Mumbai, having long histories of being cities that never sleep. Large urban places can be transformed into different places depending on the time of day.

Planners and architects are developing ideas that support and promote the 24-hour city (Figure 5.35). In London, night bus routes doubled between 1999 and 2013 and passenger numbers tripled. From September 2015, five underground lines have been operating 24 hours during the weekend. Expansion of the night-time service is planned for other lines as modernisation programmes are completed. In 2006 just six McDonald's opened overnight; by 2015 the number was nearly 50. Increasingly gyms and hairdressers remain open throughout the night. Much of this change is led by licensed premises being open until well after midnight. In the City, London's financial heart, some 250 licensed premises remain open after midnight. Yet just a few years ago, there were none. Art exhibitions and theatres also open into the early hours when particularly popular events are being held.

This growth of around the clock activity is partly due to population change. London's population has been rising since the mid-1980s, in particular in central areas where large numbers of young professional people have taken

---

Activities

1 Research images of the following buildings:
- The Shard, London
- Gateshead Millennium Bridge, Gateshead/ Newcastle upon Tyne
- Sydney Opera House, Sydney
- Petronas Towers, Kuala Lumpur

2 What contribution do these structures make to the identity of the place in which they are located?

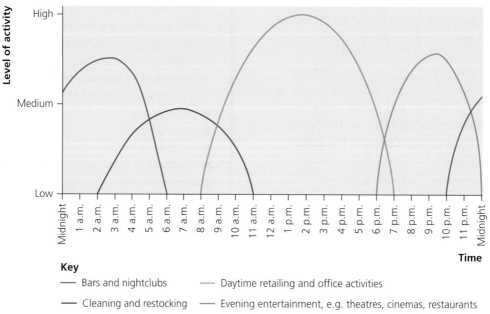

Key

— Bars and nightclubs   — Daytime retailing and office activities

— Cleaning and restocking   — Evening entertainment, e.g. theatres, cinemas, restaurants

Figure 5.35 Changes to an urban place through a 24-hour period

up residence. This trend is replicated in other cities with a 24-hour culture such as Berlin, Madrid and New York.

The growth in international tourism also contributes to changing places into 24-hour cities. Places such as New Orleans and Paris promote themselves as having attractions for tourists around the clock. But it is also an increasing number of people working shifts that is propelling places to change. Transport for London estimates that half the people using its night services are commuting to or from work. Retailers such as the larger supermarkets have stores open 24 hours which allows deliveries to be made overnight.

Common perceptions are of central places being deserted, threatening and unsafe at night. The rebranding into a 24-hour city is attempting to alter such perceptions.

How local community groups shape the place they live in

As has been identified in both Lympstone and Toxteth (see pages 141–47), and your own local place, local communities can have a significant influence in shaping places. As well as local political organisations such as councils, there are a whole host of groups whose activities go towards local place profiles.

In some locations, residents' associations have been in existence since the end of the nineteenth century. However, the majority came into existence when urban areas were expanding rapidly in the 1920s and 1930s. As these were new developments, often on what were then greenfield sites on the edge of the built-up area, they were focused on the generation of a community. This continues into the twenty-first century on some new housing developments. For example, Newcourt is an urban extension to the east of Exeter which will provide around 3700 dwellings and twenty hectares of employment land when complete. A community organisation has been established which influences the making of this new place from the point of view of local residents.

Residents' associations tend to be concerned with housing, community and environmental matters. These are important at a local scale, for example issues such as traffic speeds through a neighbourhood, on-road parking, footpath maintenance, grass cutting of public spaces and the use of a community centre.

Perhaps the most active associations are those where a place can be clearly identified as having a distinctive character, such as a former village which is now part of a larger urban settlement. Such places might have a cluster of local shops and services, a couple of places of religious observance and even an area of open space such as a former village green.

Similarly, heritage associations can be active in the placemaking process when there is a distinguishing character based on the survival of past characteristics such as architecture. Most of these associations are non-governmental organisations (NGOs) and range in scale from national bodies such as the National Trust to local groups focused on a particular building or a feature such as a stretch of preserved canal or railway. Their contribution to placemaking is seen in the ways in which both physical elements such as buildings, machinery or clothes and human characteristics such as ways of living and working are preserved. Together these allow people to see and experience places as they were in times past.

Digital placemaking is increasingly playing a role in the evolution of places. As social media becomes more integrated in the lives of more and more people, it is being used to encourage public participation and collaboration in processes such as planning and decision making about land-use in local neighbourhoods. One concept used to help individuals participate in placemaking is the 'Power of 10+'. The idea behind this concept is that places thrive when those who use them have several reasons, 10+, to be out and about in that place. Social media allows people to share their individual reasons and an overall pattern can be readily built up. The multiple ideas can be brought together, mapped and a design for a place adopted which serves the needs of the community. In Baltimore, USA a plan aimed at creating a network of open spaces through the city's central area was generated. This used not only professional expertise such as engineers, planners, architects and landscape architects but also employed an online crowdsourcing application which allowed local people to identify open spaces in their city that mattered most to them. An overall picture thereby emerged which guided the final plan.

> **Key idea**
> → Rebranding changes places through regeneration and giving a place a new image

Why places rebrand

All places have an image. This image depends upon how people perceive a place. A number of separate images then come together to form a collective view of a particular place.

A place's brand is the popular image the place has acquired and by which it is generally recognised. It includes both objective aspects, such as its location but also subjective ones such as its atmosphere, safety or level of economic activity. If a place has acquired a negative brand then **rebranding** can be attempted.

Rebranding is not a new phenomenon. After the Great Fire of London in 1666, Sir Christopher Wren, the foremost architect of the day, drew up a master plan which involved ridding the city of its medieval narrow streets and alleys. His vision was of a number of large piazzas (squares) linked in a geometric manner by wide, long boulevards. As is so often the case with rebranding, the plans were contested, mainly by owners of buildings that would have been demolished. As a result, little of the original plan was completed.

In the 1960s and 1970s, the UK attempted to rebalance the country's spatial economy by relocating public and private sector employment from London and the southeast to peripheral regions such as south Wales and Merseyside. Most peripheral regions suffered from a negative image due to declining heavy industry, large areas of slum housing, multiple deprivation and polluted environments. Advertising campaigns were run to try to **reimage** and rebrand these places.

In today's increasingly globalised world, places are competing not just regionally, but nationally and internationally for investment. This investment might be from the private sector such as a TNC or from a government body which might include trans-national governments such as the EU. Places seek to rebrand because their current brand is failing to attract sufficient investment.

Rebranding involves three key elements (Figure 5.36):

1 Brand artefact – the physical environment, such as individual buildings, the built environment (cityscape) or features in a rural area such as dry stone walls or evidence of former industry such as waste tips
2 Brand essence – people's experiences of the place
3 Brandscape – how the place positions itself in relation to other competitor places

Urban and rural places tend to possess many identities so when considering rebranding it can be difficult to decide exactly what elements of a place to promote. Sometimes existing identities are reworked but new identities can also be created. Places offer different experiences to different people so when rebranding, one issue to consider is avoiding alienating groups such as existing residents. There can be elements which everyone agrees need rebranding but the process can often be contested; in other words, there are divided opinions about what and how to rebrand.

Strategies for rebranding a place

There are several rebranding strategies: usually a place uses a combination of them.

- Market-led – involves private investors aiming to make a profit. Typically includes property developers, builders and business owners, for example those running restaurants, wine bars or retailing.

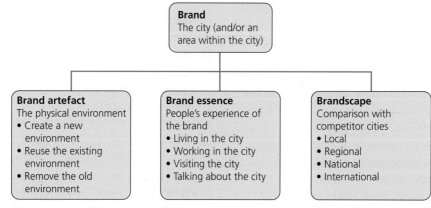

Figure 5.36 Key elements of rebranding

Gentrification is typical of this strategy such as in Islington, London or Le Marais, Paris.

- Top-down – involves large-scale organisations such as local authorities, especially their planning departments, development agencies and private investors such as insurance and pension fund managers. Several former dockland areas such as Salford Quays, Manchester and Inner Harbor, Baltimore are examples.
- Flagship development – large scale, one-off property projects with distinctive architecture. They act as a catalyst to attract further investment and regeneration. The Millennium Stadium, Cardiff and The Waterfront, Belfast are examples.
- Legacy – following international sporting events which brought investment and regeneration to a place. Examples include the Olympics in Barcelona (1992) and London (2012) and the Commonwealth Games, Manchester (2002).
- Events or themes – major festivals such as those associated with the European Capital of Culture, Liverpool (2008) and Riga, Latvia (2014). This serves as a catalyst for the cultural development and the transformation of the city. Consequently, the beneficial socio-economic development and impact for the chosen city are now also considered in determining the chosen cities.

Whatever the strategy, various elements can be involved in the rebranding process.

- Architecture – nearly always plays a role and can be used to reinforce a particular heritage look or to promote the place as modern and forward looking. Examples include the Covent Garden area in central London for the re-use of existing buildings or the Pompidou Centre in Paris for the transforming effect of a radical design. Architecture is not just about the look of a place, it can also alter how people use a place and thus change its image.
- Heritage use – nearly always plays a role and can revitalise a place. The Wessex Tourist Board has based much of its rebranding of this rural region (covering parts of Dorset, Somerset and Wiltshire) on the rich heritage of the region: ancient monuments such as Avebury and Stonehenge, Saxon history including Alfred the Great, the myths and legends surrounding King Arthur and Glastonbury, and the setting of Thomas Hardy's novels in Wessex. (Hardy lived much of his life in Wessex and used local places as the basis for his fictional places – Casterbridge is based on Dorchester and Sherton Abbas on Sherborne.)
- Retail – with the growth in the importance of consumer spending and the increased emphasis given to the 'shopping experience', rebranding can be aided by retail developments. The flagship Selfridges store opened in 2003 has become an icon of the rebranding of central Birmingham. Further developments have followed in the area such as the rebuilding of New Street Station and the Bull Ring. Dubai has sought to raise its international profile through the shopping experience it offers.
- Art – both through art galleries and art events. Galleries such as the Guggenheim Museum in Bilbao and the Tate Gallery in St Ives, Cornwall have been pivotal in the rebranding of these places. Art events such as the Edinburgh and Glastonbury Festivals contribute both economically and culturally to the image of the place where they are based.
- Sport – major international sporting events, such as the Olympic Games, a World Cup or a Formula 1 Grand Prix, can be the catalyst that helps kick-start rebranding. Bahrain has been very keen to establish itself on the F1 list of races as part of the rebranding of the place as a major hub at the global scale.
- Food – some places have developed a reputation of high-quality food to help in their rebranding. Ludlow, a small market town in Shropshire, has become known as 'the food town'. It has several restaurants with international reputations for fine dining, many specialist food shops and food festivals.

People and groups involved in rebranding

Rebranding is such a diverse process that many different people, organisations and occupations can be involved. The terms players and stakeholders are used to summarise the whole range of people involved in, and affected by, a process such as rebranding.

Key players are those involved in funding it. Governments of various scales and their directly funded organisations, such as tourist boards and planning departments, play significant roles. The EU's European Regional Development Fund (ERDF) gives grants to assist projects to aid places that fall well below the average income levels found in the EU. Many infrastructure projects have been part-funded by the ERDF, such as the improvements to the A55 along the north Wales coast. This has played a part in the rebranding of this rural region and the development of its tourist industry and has helped to offset social and economic decline.

Corporate bodies, such as banks, insurance companies, pension funds and development companies, not only help fund rebranding strategies, they also carry out actual physical developments. Investing in and building large shopping malls is used to generate long-term regular flows of income for pension providers, for example.

Not-for-profit organisations, such as bodies responsible for the arts, the National Trust and local community groups like residents' associations, are involved in rebranding, sometimes invoking change but also being affected by it.

Rebranding can be a contested process

Rebranding often brings changes to the character of a place, something that is not always welcomed by certain groups, particularly local residents, due to different perspectives on what should be changed and how change should proceed.

Change in the character of a place

Gentrification brings about a socio-economic change as wealthier people move into a neighbourhood. Their relatively large disposable incomes lead to changes in the types of local services available. Newsagents, corner shops and hardware shops can be displaced by restaurants, wine bars and specialist shops, such as boutique clothes and shoe shops. Rising property prices invariably accompany gentrification, forcing poorer residents, who cannot afford the increased rents and prices of goods and services, to move out.

Favouring one group over another

Some players/stakeholders benefit more than others from rebranding. When Liverpool One (a large shopping centre in inner Liverpool) was developed, many local residents felt that regeneration would be of little benefit to them. The rebranding was felt to be more suitable for entrepreneurs and those living in the suburbs or outside Liverpool.

Differences in priorities

Development agencies may have different priorities from local residents. Liverpool Vision, the urban regeneration body working in inner Liverpool, employed cheaper foreign labour rather than local people. It wanted to attract organisations and individuals from outside the local area, believing that in this way it would help change Liverpool's negative image. However, established residents and existing owners of shops, offices and leisure facilities felt they were being ignored in the rebranding process.

> **Key idea**
> → Making a successful place requires planning and design

🔍 **Case study:** Barcelona

The City of Barcelona is the principal place in the Greater Barcelona Metropolitan Region. It lies on the Mediterranean coast in northeast Spain, about 125 km from the French border (Figure 5.37). Barcelona is the capital of Barcelona province and the Spanish region of Catalonia, and has approximately 1.6 million inhabitants (Greater Barcelona has 3.8 million residents). It is Spain's second largest city and its principal industrial and commercial centre. One-quarter of Spain's exports originate from Barcelona. The main manufactured products are textiles, precision instruments, machinery, railway equipment, paper, glass and plastics. Barcelona is also a major Mediterranean port, a financial and publishing centre and a tourist destination receiving some 7 to 8 million visitors a year.

The historical context of Barcelona's rebranding

The site of Barcelona has been occupied for over 2000 years. Originally a Roman settlement, it became a tightly packed medieval city, with a dense network of streets and alleys contained within city walls. Barcelona became the principal city in northeast Spain and was

Figure 5.37 Barcelona's location in the western Mediterranean

the capital of the Kingdom of Aragon in the twelfth to fourteenth centuries. Barcelona's maritime heritage was well established by the fifteenth century and the city received and sent goods throughout the known world. Nineteenth-century industrialisation added another dimension to the city's geography, with cotton, cork, iron and steel, shipbuilding and wine all becoming important. Accompanying these developments were cramped, densely populated tracts of housing which contributed to the poor health of their inhabitants. Eventually, the medieval walls were knocked down and growth began to spread out into the surrounding countryside.

The city's population grew rapidly in the nineteenth century from 115,000 to over 500,000 and reached 1 million by 1930. The city became noted as a centre for culture with close links with Paris and movements such as Art Nouveau in the late nineteenth and early twentieth centuries.

The second half of the nineteenth century saw a resurgence of support for all things Catalan including the Catalan language. Various political events crystallised support for a separate Catalan identity. When the Spanish Civil War broke out in 1936, Barcelona became a volatile centre of Republican opposition to the Nationalists led by General Franco. The city was taken by the Nationalists in 1939 and in the years that followed, Catalan culture was suppressed. The city's architecture was not well maintained and Barcelona lost its vibrant spirit, acquiring the image of a run-down and rather weary, ageing port city. With Franco's death in 1975, the scene was set for Barcelona's renaissance.

Barcelona's rebranding

Barcelona has a history of trying to reinvent itself. Major events were held to raise the city's international profile. In 1888 the Universal Exhibition was held and in 1929 the World Exhibition showcased Barcelona, attracting exhibitors and visitors from Europe, the USA, Latin America and the Far East. Both events were designed to show off the city's contemporary architecture and the products of Catalan industry.

Following the end of the Franco regime, Barcelona gradually reclaimed more democratic rights which included the ability to make decisions about the management of the city. The local authority began regeneration with a seven-year plan starting in 1980. This focused on 140 small projects providing more piazzas (public squares) and better housing, transport routes, schools and hospitals.

The role of sport

Winning the right to stage the 1992 Olympic Games was a major moment in Barcelona's rebranding. The athletes' village was located on the waterfront, with the main stadium a little way inland. The city used the games to generate city-wide redevelopment and to renovate run-down areas such as the harbour and beaches. A riverside park was created along with business and media parks and an international conference centre.

Rebranding using the Olympics as its catalyst was presented to Barcelona's inhabitants as a 'one city' exercise. It intended to offer something to all the residents and tried to unite the city around a public project. Decisions about Olympic developments were taken at all levels of the planning and design process, not just top-down. The project was key in helping to reassert Catalan pride and identity.

Barcelona Football Club is another key element to strengthening the city's self-belief and confidence. The development of the club's stadium, the Nou Camp, into one of the world's great sporting venues capable of holding 98,000 spectators has created another flagship location in the city (Figure 5.38).

The role of culture

Barcelona has a long tradition of artistic culture. The distinctive architecture of Antoni Gaudi gave Barcelona buildings that are recognised as World Heritage Sites by the United Nations Educational, Scientific and Cultural Organization (UNESCO). The cathedral of Sagrada Familia and the Casa Batlló for example are visited by thousands of tourists each year. The city used the Universal Forum of Cultures in 2004 to promote the regeneration and reimaging of the city. Refurbishment of public spaces and galleries gave further momentum to rebranding.

Art galleries, museums, restaurants, cafés, architecture and public spaces are all used to promote the city as one of the most vibrant and creative places in Europe. La Rambla is a tree-lined boulevard designed to attract locals and tourists to its shops, galleries, restaurants and bars.

The role of business

Players such as the business community and the municipal government have built Barcelona's reputation for services, innovation, the knowledge economy and entrepreneurship. In November 2010 a new 'Strategic Metropolitan Plan of Barcelona – Vision 2020' was presented by the municipal authority. Its key aim was to position the city as 'one of the most attractive and influential European regions for innovative global talent and as the best setting for economic and business growth.' It encouraged people to see Barcelona as not only an attractive place to live and study in and visit but as a centre of twenty-first-century business innovation. It looks to develop a lead in sustainability and adapting to climate change in areas such as managing water supply.

In 2014 the city was given the title of European Capital of Innovation by the EU. It aims to establish itself as a leading centre for mobile technologies and one of the top ten 'smart' cities in the world.

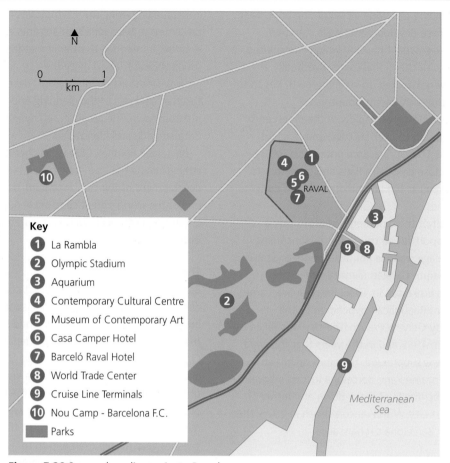

Figure 5.38 Some rebranding projects, Barcelona

Key
1 La Rambla
2 Olympic Stadium
3 Aquarium
4 Contemporary Cultural Centre
5 Museum of Contemporary Art
6 Casa Camper Hotel
7 Barceló Raval Hotel
8 World Trade Center
9 Cruise Line Terminals
10 Nou Camp - Barcelona F.C.
⬛ Parks

Activities

1 Research the architecture of the rebranding projects in Figure 5.38.
2 How well do you think they reflect the rebranding strategies outlined in the case study?

Small-scale rebranding within Barcelona: the example of the Raval

While there has been a rebranding of Barcelona *in toto*, neighbourhoods within the city have had their own specific rebranding (Figure 5.38). One of these is the Raval.

Why the Raval required rebranding

The Raval district in Barcelona's inner city has been the focus of a major rebranding exercise. Until recently this was one of the most densely populated urban areas in the world. It was where the textile industry, brick making, many abattoirs and the tanning industry were based. Mixed in and around these polluting industries were tenement blocks which housed workers in the nearby factories. The area closest to the port, southern Raval, contained Barrio Chino or Chinatown where immigrants first settled on arrival in Barcelona. Immigrants took advantage of the very cheap lodgings in the run-down tenements and the easy access to informal employment in and around the docks. It was also an area with a reputation for illicit activities such as organised and petty crime, prostitution and drug dealing.

Rebranding projects in the Raval

Within the Raval, the nature of rebranding has differed between its northern and southern areas. In the north is a zone with numerous flagship cultural buildings such as the Museum of Contemporary Art and the Contemporary Cultural Centre, both of which have international reputations. A private university has been constructed and some streets house art galleries, coffee and wine bars, restaurants and high-quality food shops. Much of the change is associated with gentrification as high socio-economic status people have moved in.

→

In southern Raval, overcrowded residential areas, and old and derelict factories remain while its inhabitants are among the more deprived in the city. However, efforts to change the district are underway. In 1995, in an area known as the Raval Rambla, significant physical regeneration was undertaken. Nearly 1700 residential and commercial properties were demolished to clear space for a large, tree-lined pedestrianised space. This space has been kept virtually free from street furniture to allow flexible use for events such as street markets and small local festivals. New housing has been constructed for some of the residents displaced by the scheme. Student accommodation occupies refurbished buildings and a modern pneumatic refuse collection service has been introduced. The scheme cost €5 million of which 80 per cent came from EU funding.

Private investment can be seen in two flagship hotel projects. The Barceló Raval Hotel built in 2008 has an 'avant garde' design. This building rises high above its surroundings and represents a significant financial commitment to the area's rebranding. The Casa Camper Hotel has refurbished an old stone building and claims that its 'Old age shines on the outside, youthful good looks blush within.'

Contested rebranding in the Raval

As with much rebranding, not everyone agrees with what has been done. The in-migration of wealthy residents with different perspectives on the way the area should be managed and the new employment opportunities more in demand from the better-off, has generated social tensions. Wealthy tourists also make poorer residents feel excluded from many of the new facilities. Property speculation has raised prices and some landlords are keen to see existing residents leave to allow them to cash in. Stories have emerged of pressure being applied, especially to long-standing elderly residents, to move out. Protest graffiti is common on the properties where such conflict is occurring.

But much has improved in the city and the Raval (Figure 5.39). Indices of employment, education, health and crime all show both physical and socio-economic gains for locals as well as visitors. The key question is whether the rebranding is sustainable in the medium to long term.

Activities

1 Research the rebranding of another major urban location, such as Liverpool, Salford Quays, Berlin or Baltimore.
2 Compare and contrast the strategies used in your chosen location with what took place in Barcelona.
3 Research the rebranding of a rural location such as Cornwall or Snowdonia.
4 In what ways is rural rebranding (a) similar to and (b) different from urban rebranding?

Assessing the success of rebranding

Most of the strategies used in rebranding fall into one or more of the following categories:

● Economic – to improve wealth creation, employment and incomes
● Social – to improve the quality of life of residents
● Environmental – to improve the physical characteristics of the place

Whatever the context of the rebranding (for example rural or urban), criteria are needed against which to measure its success.

Because rebranding is a multi-faceted process, there are a variety of ways in which it can be assessed (Figure 5.40).

Quantitative analysis uses data that can be expressed numerically. This type of data can be statistically analysed and represented visually on maps or in charts and graphs. The Office for National Statistics (ONS) and local authorities publish suitable data for, for example, investigating changes in types of employment and levels of unemployment; comparing the extent of multiple deprivation from one year to another; comparing crime statistics and looking at the availability of affordable housing. Assessments of the quality of the physical environment in regenerated areas compared with neighbouring areas which have not been rebranded can be quantified through a scoring system and plotted on a map. Pedestrian counts in regenerated retail areas can be compared with areas not improved. Qualitative data is non-numerical information that can be collected from sources such as interviews, websites, photographs, brochures, television, film, paintings, books, music and cartoons.

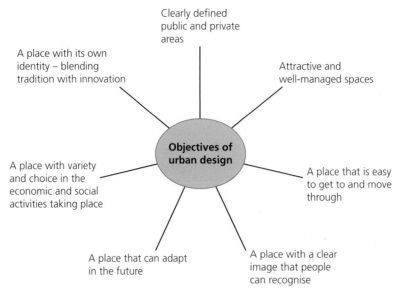

Figure 5.39 Core objectives in the rebranding process

Figure 5.40 Research ideas for assessing rebranding success

Review questions

1 What is meant by Foreign Direct Investment (FDI)?
2 Give two examples of how architecture can lead to negative placemaking and two examples of where it has led to good placemaking.
3 What is meant by the term 'the 24-hour city'?
4 What are the three elements that make up the rebranding process?

5 Give an example of where each of the following has been used in rebranding: architecture, heritage, retail, art, sport, food.
6 Outline why rebranding is often a contested process.
7 Why did Barcelona need rebranding?

Practice questions

Figure A Part of Salford Quays, Manchester

A Level

1 Study **Table 1**.

Table 1 Selected data for Borrowdale, a remote upland rural area in the Lake District, northwest England

% population change, 1991–2015 (est.)	−16.0%
% change in number of elderly >65 years, 1991–2015 (est.)	+10.0%
% change in number of children <15 years, 1991–2015 (est.)	−15.0%
Number of post offices in 1980	2
Number of post offices in 2015	0

Explain how **one** piece of evidence from Table 1 indicates that this place's identity has changed over time. [3 marks]

2 Study **Figure A**.

Using evidence from Figure A, explain how rebranding contributes to constructing a new place image. [8 marks]

3 Explain how informal representations of place influence how people understand place. [6 marks]

4 With reference to **one** country or region you have studied, assess the roles a range of players can have in influencing economic change. [16 marks]

AS Level

1 a) Explain how time–space compression can alter a person's sense of place. [4 marks]

b) **Figure A** shows part of an area that has undergone rebranding.

   i) Using Figure A state **one** piece of evidence which shows that this area has been rebranded. [1 mark]

   ii) With reference to Figure A, suggest **two** ways in which rebranding can alter perceptions of a place. [4 marks]

c) Using evidence from **Table 2**, suggest how social inequality can have impacts on people's daily lives in different ways. [6 marks]

Table 2 Selected census statistics for two wards in Birmingham, Nechells and Edgbaston, 2011

	Nechells	Edgbaston
Unemployment rate (16–64 year olds)	11%	5%
Long term sick or disabled	6%	3%
AS/A Level or equivalent qualifications	18%	32%
Living in social or local authority housing	51%	23%
Living in detached housing	5%	20%

d)* Discuss the relative advantages and disadvantages of informal representations of places. [14 marks]

Chapter 6

Trade in the contemporary world

6.1 What are the contemporary patterns of international trade?

Patterns of international trade can be measured in different ways. These include national statistics for merchandise, services and capital, such as value of exports or percentage share of world imports. International organisations involved in regulating trade at the global scale such as the World Trade Organization (WTO), the United Nations Conference on Trade and Development (UNCTAD), the International Monetary Fund (IMF) and the Organisation for Economic Co-operation and Development (OECD) provide a wealth of trade and trade-related data. In addition many international bodies for major trade commodities monitor and assess flows of merchandise, services and capital. They have vested economic and professional interests in producing databases for their particular trade. Examples include the International Bar Association which regulates trade in legal services, the International Sugar Organization and the Organization of the Petroleum Exporting Countries (OPEC).

An extremely wide range of products is traded internationally (see the India, USA and Sierra Leone case studies on pages 207–09, 210–11 and 212–13). However, a geographical feature they all have in common is the unevenness of their global patterns of trade. In nearly every instance trade is dominated by the more advanced and the rapidly emerging economies which have the economic wealth and political control to reinforce their trade position. It is the least developed countries that have limited access to global markets, and which in many instances are still exploited and have much weaker terms of trade.

> **Key idea**
> → International trade involves flows of merchandise, services and capital which vary spatially

An understanding of the terms 'merchandise', 'services' and 'capital' as components of international trade

Flows of merchandise, services and capital occur at different scales within the global trade system, between countries, regions and continents. There are spatial variations in the international trade of these items in terms of direction of flow and in volume, composition and value.

Merchandise

The WTO defines the international trade of merchandise in simple terms as all inward and outward movement of goods through a country. Table 6.1 gives an overview of the uneven pattern of this trade in terms of value of merchandise exports for each global region. Figures are

Table 6.1 Merchandise exports by region, 2013 (US$ billion) (Source: WTO International Trade Statistics, 2014)

World region	Manufactured goods	Fuels and mining products	Agricultural products	Total
Europe	4,910	812	708	6,430
Asia	4,566	690	390	5,646
North America	1,616	408	266	2,290
Middle East	276	880	33	1,189
Commonwealth of Independent States (CIS)	174	514	69	757
South and Central America	194	297	217	708
Africa	112	397	62	571

Figure 6.1 The garments sector of Bangladesh has grown impressively and captured an increasing share of the world market. Garments now make up 75 per cent of Bangladesh's merchandise exports.

based on customs records which reflect the movement of goods across international borders. Categories shown include primary products (agriculture, mining and fuels) and secondary products (manufactured goods such as iron and steel, chemical, machinery, textiles and clothing – Figure 6.1 – and automobiles).

Table 6.1 demonstrates not only the geographical differences between world regions but also the scale of the inequalities. The totals show the very high value of merchandise exports in the economies of Europe (US$6430 billion) and Asia compared with the very low values of Africa (US$571 billion). Asia has nearly ten times the value of merchandise exports than Africa. Moreover, values for Europe are nine times those of South and Central America. Of course these figures conceal variations between countries within each region. And, not showing rates of change, they provide only a snapshot of what is a rapidly changing and complex situation.

There is also considerable variation in the components of merchandise exports between the regions. Export values of fuel and mining products are greatest in the Middle East with the combined oil production of Saudi Arabia, UAE, Iran, Iraq and Kuwait. Agricultural products are a very strong feature of exports from European countries. Exports of agricultural products from Europe (US$708 billion) are more than ten times those of both Africa (US$62 billion) and the Commonwealth of Independent States (CIS) (US$69 billion) and more than twenty times those of the Middle East (US$33 billion). Another huge inequality is that exports of manufactured goods from Asia are 40 times greater than those of Africa.

Patterns of international trade are shown in more detail in Figure 6.2. The size of merchandise trade is shown for each country within the global regions. Economic, social, environmental and political factors which explain the extent of the inequalities are outlined in Figure 6.9 on page 196. Detailed reasons are exemplified by the trade patterns of the contrasting case studies for the USA, India and Sierra Leone later in this chapter.

Skills focus

1 Describe the pattern of global merchandise trade shown in Figure 6.2.
2 With reference to Table 6.14 on page 210 and Table 6.17 on page 212 suggest reasons for the contrasts in merchandise trade between USA and Sierra Leone.
3 Discuss advantages and disadvantages of the choropleth map used in Figure 6.2 to represent these data.

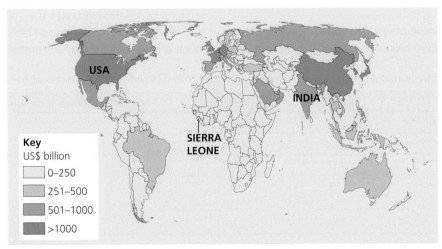

Figure 6.2 Size of merchandise trade by country, 2013 (Source: WTO International Trade Statistics, 2014)

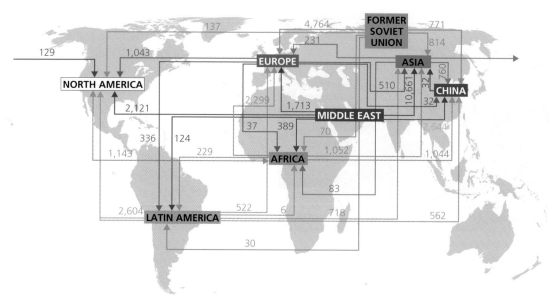

Figure 6.3 World crude oil exports by destination, 2013 (1000 b/d) (Source: OPEC Annual Statistical Bulletin, 2014)

The idea of flow within the international system of merchandise trade is perhaps best represented by oil, being one of the most important and valuable of commodities. Figure 6.3 demonstrates the complexity and density of the network of flows of crude oil. There are distinct regions of supply where export is the dominant feature of the trade. And these are clearly linked to regions of heavy demand. The precise volume of flow is also related to demand and price, which changes frequently. For example, over 10 million barrels per day (b/d) moved between the Middle East and Asia excluding China. China imported over 1 million barrels of crude oil per day from Africa alone.

> **Activity**
>
> Using Figure 6.3, summarise the 2013 pattern of global flows of crude oil.

Services

In 2013 the total value of world exports of commercial services was US$2550 billion. In comparison with the world total of merchandise exports, which amounted to US$17,591 billion, this is a relatively small figure. But it is nonetheless a very significant element of the global trade system and it is growing very rapidly. Furthermore the global pattern has become increasingly complex especially with the continued growth of **outsourcing** and the increasing connectivity of global supply chains, in particular between countries within regional trading blocs such as the Association of Southeast Asian Nations (ASEAN).

Commercial services include transport and travel services and a range of other services in the fields of communications, construction, insurance, finance and ICT plus government services. At the global scale the broad pattern of trade in services replicates that of merchandise. It is dominated by the more advanced economies, and the less developed countries tend to be net importers. The figures for export and import of these services are derived from transactions in the **balance of payments** statistics.

Europe is the largest net exporter of commercial services, which demonstrates the importance of its trade with other global regions (Table 6.2). However, much of this type of trade in Europe is intra-regional, in

Table 6.2 World trade in commercial services by region, 2013 (US$ billion) (Source: WTO International Trade Statistics, 2014)

Region	Exports	Imports	Net trade balance in commercial services
Europe	1301	976	+325
Asia	611	517	+94
North America	457	295	+162
South and Central America	63	92	−29
Middle East	48	63	−15
CIS	47	75	−28
Africa	24	68	−44

particular between the 28 countries of the EU. At the other extreme, African countries, especially those in sub-Saharan Africa, are overall net importers of commercial services, mainly through trade with **Advanced Countries** (ACs) and **Emerging and Developing Countries** (EDCs).

Maps showing indices for global trade, including Figure 6.4, can be found on the WTO website: www.wto.org. These are interactive, giving trade figures for individual countries plus links to national profiles in international trade.

Almost every country exports commercial services. Even a relatively poor country such as Sierra Leone exported commercial services to the value of US$0.18 billion in 2013, mostly derived from transport and travel services.

As with merchandise, the global pattern of commercial services exports is very uneven. The most important exporters of commercial services are the countries of the EU, the emerging economies of Asia, especially India and China, North America and Japan. This is in stark contrast to nearly all of Africa, especially the sub-Saharan countries where the value of exports is very low. These contrasts in ability to supply commercial services depend on factors such as skill and education levels of the work force, government and private investment, for example in communications and transport, and the strength and reliability of financial and legal institutions.

This global pattern illustrates the vast differences in this trade between advanced, emerging and least developed economies (Table 6.3).

Table 6.3 Value of exports of commercial services, 2013

Country	Value of exports of commercial services (US$ billion)
Advanced Country: USA	662
Emerging and Developing Country: Brazil	37
Low-Income Developing Country: Equatorial Guinea	0.03

Capital

International flows of capital are the result of purchases and sales of real and financial assets across national borders. Real assets include physical or tangible items that are traded, such as commodities, minerals, land and real estate. Financial assets are the so-called intangibles such as currency, stocks and bonds, which can be traded on the world's financial markets. In effect the term 'capital' refers to the financial side of international trade. One way of considering this is that finance flows in the opposite direction to the merchandise and services that are purchased.

Flows of capital in the global trade system are more complex than this suggests. This is because the term 'capital' includes so many diverse elements. Moreover the growing interconnectivity of global trade has increased globalisation of capital flows. This has been rapid and far reaching especially in the last decade or so. Most of these flows are between small numbers of advanced economies, but the network is widening as a growing number of low-income developing countries have become integrated into the global trade system.

Another significant feature of global capital flow is that most flows are intra-firm, especially within the **multinational corporations** (MNCs). In these very large enterprises, goods, materials and semi-finished products flow between parent companies and their subsidiaries through **global supply chains**. And the consequent financial transactions are often completed by accounting within the company.

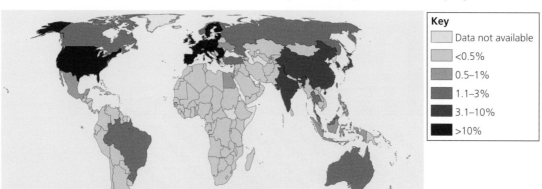

Figure 6.4 World share of exports of commercial services, 2013 (Source: WTO Maps)

In addition, the term 'capital' also includes cross-border investments. These include investment in securities, the buying and selling of international reserves which pass between the central banks of different countries and **Foreign Direct Investment** (FDI).

Foreign Direct Investment

FDI is a key element of international economic integration; it creates direct, stable and long-lasting links between economies, and is one of the driving forces of economic globalisation. Figure 6.5 shows the differences in global FDI inflows between world regions. There are significant inward flows of capital into Africa and Asia. This reflects the very high level of interest in investment, largely by multinational corporations, in these two regions.

Inward flow of FDI to India is derived from a large number and wide variety of countries (Figure 6.6). Figure 6.7 shows the range of sectors in which the investments have been made in this rapidly emerging economy.

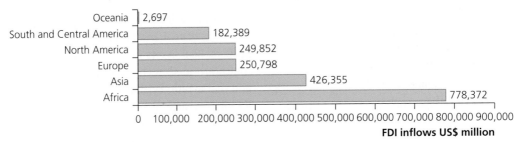

Figure 6.5 Inflows of FDI by world region, 2013 (Source: UNCTAD statistics)

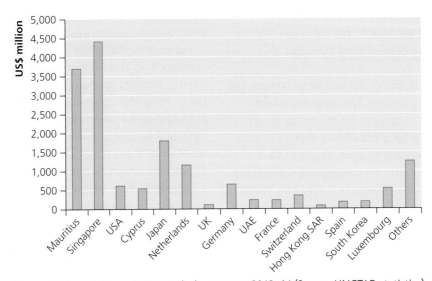

Figure 6.6 Inward flows of FDI to India by country, 2013–14 (Source: UNCTAD statistics)

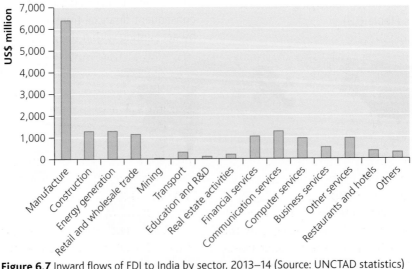

Figure 6.7 Inward flows of FDI to India by sector, 2013–14 (Source: UNCTAD statistics)

Inward FDI in India is mainly the response to active encouragement by the Indian government:

- Treaties between India and Mauritius and Singapore enable companies located in these countries to avoid high taxation, for example under the Double Tax Avoidance Agreement. The high value of investment from Mauritius (Figure 6.6), a relatively poor country, is explained by large companies based there which invest in business and services in India. These include Cairn UK Holdings, Oracle Global, TMI and Vodafone.
- India needs FDI to help overhaul its infrastructure of ports, airports and highways. FDI norms including rules and constrictions regarding inward investment have been relaxed in sectors such as defence, oil refining, telecom, retail, airline and railway infrastructure companies and pharmaceuticals.
- The Foreign Investment Board has raised inward investment ceilings. This has increased FDI in the service sector for private insurance companies, financial investment companies and communication, business and computer services, mostly by outsourcing.
- The Reserve Bank of India has facilitated transactions required for mergers, takeovers and new investment.

Activities

1 Describe and explain the pattern of FDI inflows by global region, 2013 (Figure 6.5).
2 Summarise the characteristic features of FDI by geographical source and sector in India (Figures 6.6 and 6.7).
3 Discuss possible reasons why so much FDI has been directed towards India in recent years.

Current spatial patterns in the direction and components of international trade

In addition to the global patterns of trade in merchandise, services and capital, international trade patterns can be illustrated and explained at inter-regional and intra-regional scales.

Inter-regional trade between Europe and North America

Inter-regional trade is the flow of international trade among major world regions such as Europe, North America and Asia. For example, there is a long-established transatlantic trade relationship between Europe and North America. The EU and the USA are the world's largest trading partners by value of merchandise, services and capital. In 2014, total trade between the

two accounted for over 30 per cent of global trade flows. Reciprocal investments between the UK and USA involve very high flows of capital. Overall this trade contributes to growth and jobs on both sides of the Atlantic.

Table 6.4 identifies the top ten merchandise imports and exports between the UK and USA. The high value of the goods involved and the degree of economic interdependence between the two countries is explained by the principle of **comparative advantage**.

Table 6.4 Top ten exports: USA to UK and UK to USA, 2014

Rank	US exports to UK	Value US$ billion	UK exports to USA	Value US$ billion
1	Aircraft, spacecraft	9.1	Machines, engines, pumps	9.3
2	Machines, engines, pumps	6.3	Vehicles	6.6
3	Gems, precious metals, coins	5.7	Oil	5.9
4	Electronic equipment	3.8	Pharmaceuticals	3.9
5	Collector items, art, antiques	3.0	Medical, technical equipment	3.3
6	Medical, technical equipment	2.9	Electronic equipment	2.7
7	Vehicles	2.7	Organic chemicals	2.5
8	Pharmaceuticals	2.5	Alcoholic beverages	2.0
9	Oil	2.5	Collector items, art, antiques	1.7
10	Plastics	1.3	Aircraft, spacecraft	1.5

Intra-regional trade within the EU

Intra-regional trade is the flow of international trade within one of the major world regions such as Europe or Asia. Most international trade is intra-regional. The intra-regional trade within the EU is complex. This complexity is the outcome of changing patterns and trends of trade flows in many diverse merchandise, services and capital products between 28 member states.

In order to simplify and illustrate this trade we can consider the international trade of the UK and its EU trading partners based on data from the Food and Drink Federation. This is a large and significant sector of intra-regional trade.

In 2014 the UK relied on the EU for 75 per cent of its £12.8 billion food and non-alcoholic drinks exports. Figure 6.8 shows the relative values of these

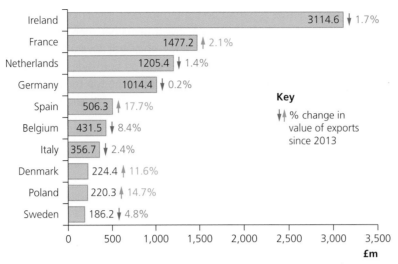

Figure 6.8 The top ten EU markets for UK food and drink exports, 2014 (Source: Food and Drink Federation)

UK exports to other EU countries. Ireland was by far the biggest importer of UK food and drink, worth over £3.11 billion. There are strong bilateral trade links between Ireland and the UK, and goods from the UK are exempt from import duties.

Table 6.5 shows specific UK food and drink products, their export destinations in the EU, and their growth in value between 2013 and 2014. Exports of milk to Ireland were of greatest value (£198 million) followed by the value of salmon exports to France (£134 million). Demand for quality salmon in France has soared and this has been met by production in Scotland supported by the export drive of the Scottish government.

Table 6.5 The top gains in value for UK food and drink exports to EU markets, 2013–14 (Source: Food and Drink Federation)

1	Wheat to Spain	+£41m to £55m
2	Salmon to France	+£24m to £134m
3	Barley to Spain	+£21m to £26m
4	Milk to Ireland	+£20m to £198m
5	Processed milk to Ireland	+£15.6m to £57m
6	Soybeans to Poland	+£23m to £23m
7	Confectionery to Netherlands	+£11.5m to £100m
8	Butter to France	+£11m to £18m
9	Wheat to Portugal	+£10.9m to £17m
10	Confectionery to Belgium	+£9.6m to £52.8m

Factors that influence patterns of international trade

The economic, political, social and environmental factors that influence patterns of global trade are shown in Figure 6.9; these are varied but interrelated. For example, trade in salmon depends on environmental conditions for producing Scottish salmon, government and private

ECONOMIC
- Physical infrastructure including transport
- Technology including communications
- Transport costs
- Cost of production
- Foreign Direct Investment
- Speed of border formalities, e.g. documentation/use of ICT

POLITICAL
- Supranational organisations
- Regional trading blocs
- National government policy
- Trade agreements/market integration
- Tariff/non-tariff barriers
- Free trade/Free Trade Areas
- Governance/transparency of customs authorities

SOCIAL
- Demographic factors which affect labour force and import demand including age structure and migration
- Stage in demographic transition
- Female empowerment/women in the labour force
- Levels of education

ENVIRONMENTAL
- Distribution of natural resources including oil and mineral ores
- Climate/soils/water scarcity (affecting food and agricultural products in international trade)
- Deep-water ports
- Natural hazards

Figure 6.9 Factors influencing contemporary patterns of international trade

investment in its marketing, free trade within the EU and technology in transport.

In simple terms international trade began because a country was able to produce a surplus and supply a commodity not produced by another. In some instances this continues to apply, for example in the production and export of tropical crops to more temperate regions.

However, in the current global system many Advanced Countries such as the USA and UK trade with each other in the same kinds of goods which they are both capable of manufacturing. This is partly explained by comparative advantage. The UK and USA both export similar high-value items of merchandise to each other such as vehicles (Table 6.4). These overall figures conceal the fact that each country tends to specialise in producing goods for which it has the greatest relative or comparative advantage. This might include particular components of a product such as vehicle parts or a particular type of vehicle. Specialisation and international trade are illustrated by the modern automobile industry. Specific vehicle parts can be produced efficiently in one country and assembled in another, often by an MNC which benefits from **economies of scale**.

Comparative advantage in merchandise trade is influenced by economic, political, social and environmental factors (Figure 6.9).

> **Key idea**
> → Current patterns of international trade are related to global patterns of socio-economic development

The relationship between patterns of international trade and socio-economic development

There is a close relationship between trade and development. Measures such as value of exports and Human Development Index (HDI) (Table 6.6) demonstrate a positive relationship. China's HDI is lower than might be expected for a country with such high value of exports, but overall value of exports is high for more developed and lower for less developed countries.

There are also close relationships between trade indices such as 'percentage share in world trade' or 'percentage share of services imported' and other indicators of development such as infant or maternal mortality rates. Specific links

between trade and socio-economic development are examined in the case studies of India, USA and Sierra Leone.

> **Skills focus**
>
> 1 Draw a scatter graph to represent the data shown in Table 6.6 and describe the association between exports and level of development (measured by the Human Development Index).
> 2 Calculate Spearman's rank correlation coefficient between exports and HDI (see Chapter 15, Geographical Skills, pages 541–42), and comment on the outcome.

How international trade can promote stability, growth and development within and between countries through flows of people, money, ideas and technology

The significance of the statistical links between trade and socio-economic indices lies in the importance of trade in the development process. The WTO recognises the link between trade and development, as do many other institutions and national governments. It launched the Aid for Trade Initiative in 2005 to help developing countries expand their trade, and in 2013 this was reinforced at the Bali Ministerial Conference. This has been designed to assist the least developed countries integrate better into a multilateral free trading system.

The positive effects of international trade on stability, economic growth and development are closely related to flows or transfer of people, money, ideas and technology (Table 6.7). The points made in Tables 6.7 and 6.8 are illustrated in the India, USA and Sierra Leone case studies at the end of this chapter (see pages 207, 210 and 212).

International migration of highly skilled workers has an innovative impact. This flow of people is

Table 6.6 Value of exports and HDI for selected countries, 2013

Country	Exports (US$ billion)	HDI	Country	Exports (US$ billion)	HDI	Country	Exports (US$ billion)	HDI
China	2,210	0.719	Czech Republic	137	0.861	New Zealand	37.8	0.910
USA	1,575	0.914	Vietnam	129	0.638	Bangladesh	26.9	0.558
Germany	1,493	0.911	Ireland	113	0.899	Tunisia	17.4	0.721
UK	813.2	0.892	Portugal	61	0.822	Nicaragua	4.2	0.614
Japan	697	0.890	Israel	60.6	0.888	Sudan	4.1	0.473
Italy	474	0.872	Colombia	58.7	0.711	Liberia	0.92	0.412
Spain	458	0.869	Oman	56.2	0.783	Haiti	0.87	0.471
Brazil	244	0.744	Philippines	47.5	0.660	Tajikistan	0.82	0.607

Table 6.7 How international trade can promote stability, growth and development within and between countries

Stability	• International trade can contribute to stability and international peace, especially if countries trade under the WTO's 'most-favoured-nation' principle, i.e. by the same rules. • Trade encourages states to co-operate; multilateral trade agreements can contribute to economic and political stability. • Some bilateral agreements extend beyond trade to assistance in political issues such as strengthening democratic processes and human rights, which create a more stable environment for foreign investors.
Economic growth	• Trade stimulates production, contributes to GDP growth and further investment, including FDI. • Employment opportunities are created, incomes are raised and poverty levels reduced. • The economic multiplier can be enhanced by international trade at national, regional and local scales.
Development	• Removal of tariffs and other obstacles to LIDC trade help generate foreign exchange which can be invested to reduce internal inequalities in poverty, health, education, infrastructure and transport. • Corporate responsibility of MNCs can be of economic and social benefit. • Membership of trade and political unions with a common purpose can help socio-economic development.

an important channel for circulation of ideas and information especially within North America and Europe and it is also significant for many EDCs. For example, migrant scientists and engineers from China and India employed in Silicon Valley, California share information on technology development with colleagues in their native countries. Also the 'brain drain' from LIDCs has an effect on flows of ideas especially when migrants return to their country of origin and disseminate acquired knowledge. For these reasons, flows of people are important factors in economic growth and development at all scales.

Monetary flow or foreign exchange is generated through international trade, and this stimulates further investment. In many LIDCs, domestic and foreign investment in communications, health, education, transport and infrastructure are important for development (Sierra Leone, Table 6.18 on page 213).

Furthermore, trade agreements may extend beyond the economics of trade, and social and political relationships develop between countries. For example, this can lead to the spread of democracy and acceptance of **human rights norms**. And, in many agreements, trade is encouraged if no child labour is used in the supply chain or if programmes to combat child labour are negotiated in Free Trade Agreements.

Diffusion of technology related to international trade includes the increasing availability of ICT around the world. Speed of movement of merchandise through customs has been significantly increased, and corruption has been reduced under border controls which benefit from electronic messaging and data storage and retrieval systems. Exporting has been made simpler, quicker and cheaper by use of online currency trading platforms which allow financial balances to be held in many different currencies. The logistics of tracking of products is easier using cloud-based platforms which are capable of monitoring incidents in supply chains around the world. Improved

cyber security helps to protect information systems from theft or disruption. And the prevalence of mobile phones has allowed the increasing trend for mobile money transfers.

How international trade causes inequalities, conflicts and injustices for people and places through uneven flows of people, money, ideas and technology

The relationship between trade and development is not a straightforward causal link. One difficulty lies in knowing whether or not the level of trade is the cause of development or the consequence. The emergence of the BRICS countries (Brazil, Russia, India, China and South Africa) illustrates the importance of trade for socio-economic development, but not all countries would necessarily benefit from the WTO aim of free trade in a multilateral system. The global trade system is still dominated by a few rich and powerful countries which retain protectionist policies and barriers to trade at the expense of poorer and weaker economies.

The points made in Tables 6.7 and 6.8 are illustrated in the USA and Sierra Leone case studies at the end of this chapter (see pages 210 and 212).

International trade is spatially uneven. Inequalities, conflicts and injustices are related to unequal flows of people, money, ideas and technology associated with the global trade system (Table 6.8).

Technology brings many benefits for international trade, but access to it is unequal across the world. For example, the huge investment in new technologies for handling large-scale shipping at ports in rich countries, and the limited access to these facilities for poorer countries, reinforces global inequalities in international trade.

Computer and mobile phone ownership and access to signals are still very limited in many low- and

Table 6.8 How international trade causes inequalities, conflicts and injustices for people and places

Inequalities	• Many LIDCs have limited access to global markets; this widens the **development gap** between developed and developing countries. • Skilled workers, especially men, tend to benefit the most from employment opportunities created by trade. • In many LIDCs internal inequalities are exacerbated by trade activity, often spatially concentrated in ports.
Conflicts	• Trade disputes can arise over tariffs, prices of commodities and changes in trade agreements. • Border and customs authorities can be subject to corruption and breaches of security. • Port development, mining and deforestation linked to trade create environmental conflicts.
Injustices	• Displacement of communities can result from land grabbing by investments in industry and agri-business. • Use of child labour and other forms of modern slavery in attempts to secure cheap labour. • Unequal power relations, unfair trade rules such as tariffs and other trade barriers and opening up to free trade can adversely affect businesses such as small-scale farmers or fishermen in LIDCs .

middle-income countries. Without this technology, border control and customs authorities, for example, are more easily subject to breaches of security and corruption in some developing countries.

Flows of capital such as FDI in mining operations can have a significant negative impact on indigenous populations and natural environmental systems.

Concentration of economic activity by TNC investment in ports or other large urban areas creates employment opportunities, but this also leads to internal migrant flows and the widening of socio-economic inequalities, especially within LIDCs and EDCs.

✔ Review questions

1. What do you understand by capital as a component of international trade?
2. What factors account for global patterns of trade in merchandise?
3. How would you assess the relationship between international trade and socio-economic development?
4. What is meant by the term 'comparative advantage'?
5. How can international trade promote economic growth?
6. In what ways can international trade cause conflicts?

6.2 Why has trade become increasingly complex?

> **Key idea**
> → Access to markets is influenced by a multitude of interrelated factors

International trade has increased connectivity due to changes in the twenty-first century

International trade has been characterised by increasing connectivity and complexity in the twenty-first century. The most recent and significant developments within this system are the driving forces of the globalisation process.

Technology, transport and communications have increased connectivity of global supply chains

Supply chains

Global supply chains involve flows of materials, information and finance in a network of customers, suppliers, manufacturers and distributors. Increasingly LIDCs are becoming integrated into global supply chains. This has increased the connectivity of countries of all economic status and their corporations within the global trade system. Supply chains are therefore of fundamental importance in international trade.

The World Economic Forum (WEF) and the OECD recognise specific 'pillars' which ease market integration at every stage of the supply chain. These include: enabling use of ICT, investing in transport and trade-related infrastructure and improving information availability.

A simple depiction of a supply chain in the oil industry is shown in Figure 6.10. Where value is added to the product by the processes at each stage, often in different parts of the world, they are known as **global value chains**.

But there are many risks for companies involved in global supply chains. These include ensuring the quality of the product itself, political and economic instability, natural hazards, customs issues, terrorism and piracy, cyber-crime and ethical issues such as use of child labour. Improving information availability, transport and communications can help to mitigate these issues. Modern digital techniques involve the use of ICT in

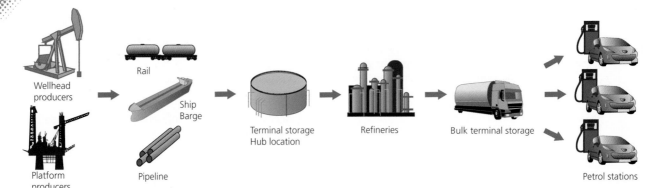

Figure 6.10 The oil supply chain (Source: OPEC)

tracking global shipments, logistics competency in reducing shipping time, plus the application of so-called Internet of Things (IoT) and Big Data (see Chapter 15, Geographical Skills, page 520) to increase efficiency of supply chains.

Communications and technology

Digital connectivity is important for businesses helping to connect producers and customers along supply chains quickly and easily. And it facilitates delivery of basic services including finance. The use of ICT by border and customs agencies is essential for governments to ensure accountability, administration and more transparent governance of corruption.

However, use of ICT and quality of access to broadband is not universal. ICT is still very limited in the developing world and this clearly inhibits access to global markets for some countries. In 2012 less than 1 in 100 of the population of sub-Saharan Africa had a broadband internet subscription. The figure was only slightly better in the developing countries of Asia.

Therefore, it is perhaps no coincidence that global patterns of internet users and smartphone subscriptions are very similar to global trade patterns (Figure 6.11).

Stretch and challenge

1 Identify and suggest reasons for similarities and differences in the global patterns shown in Figures 6.2 and 6.11.
2 To what extent is there a link between internet usage and merchandise imports at global scale?

Transport and technology

Recent developments in transport infrastructure and transport technology have also helped to improve connectivity within supply chains. This is an aspect of global trade in which gaps between the advanced economies and the rest of the world are very wide. Investment in domestic road, rail and air infrastructure and in connectivity of sea lines is essential to ease the

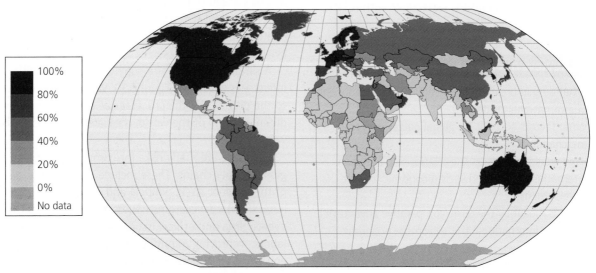

Figure 6.11 Internet users as a percentage of a country's population (Source: International Telecommunication Union)

movement of goods within a country and across its borders. This is therefore fundamental to the efficiency of global trade.

The construction of large container vessels and the development of deep-water ports is another obvious effect of technology on transport. The Port of Felixstowe has invested heavily in:

- new berthing facilities and gantries capable of handling the largest of container ships
- a Logistics Park with space for warehousing and logistics companies
- modernisation of computer systems for transhipment of containers
- road links to the UK motorway network
- new rail links.

Felixstowe is now one of only a handful of ports in the world that can accommodate vessels such as the CSCL (China Shipping Container Lines) Globe (Figure 6.12). This container ship was built with the intention of sailing the trade loop between China, Malaysia and Europe with a capacity of 19,100 TEUs (twenty foot equivalent unit).

Figure 6.12 The CSCL Globe at Felixstowe in 2015

👟 Fieldwork ideas

Investigate the economic, social and environmental impacts of port development at Felixstowe or London Gateway.

These developments of technology in transport and communications along supply chains require considerable investment. World Economic Forum (WEF) rankings in its Enabling Trade Index, 2014 are closely related to levels of development. Most of the top six countries are OECD members and most of the lowest six are in sub-Saharan Africa (Table 6.9).

Table 6.9 WEF national rankings based on the Enabling Trade Index (ETI)

Top six countries by ETI		Lowest six countries by ETI	
1	Singapore	133	Burkina Faso
2	Hong Kong SAR	134	Zimbabwe
3	Netherlands	135	Guinea
4	New Zealand	136	Angola
5	Finland	137	Venezuela
6	UK	138	Chad

The ETI is a composite index of 56 indices which measure basic attributes that govern a country's ability to benefit from trade; see www.weforum.org/reports/global-enabling-trade-report-2014

The increasing influence of MNCs in EDCs including outsourcing

International trade is dominated by the multinational corporations (MNCs). The top 500 MNCs account for over 70 per cent of flows of goods, services and capital. They are a powerful economic force and are major drivers of the global trade system and of globalisation.

Most have headquarters in ACs, especially in the USA and Europe (Table 6.10) with many factories and businesses in EDCs. One example is JCB, a British MNC, which has recently invested in India. This choice of EDC location is explained on page 207 and by government policies in the India case study.

Investment by MNCs in emerging economies such as India, China and Brazil has continued to increase in the twenty-first century. In addition, new inroads are

Table 6.10 The top twelve companies in the Fortune Global 500, 2014

Company/MNC		Revenue US$ billion	Country of origin	Main product
1	Walmart	476	USA	Retail
2	Royal Dutch Shell	459	Netherlands/UK	Petroleum
3	Sinopec Group	457	China	Oil and gas
4	China National Petroleum	432	China	Petroleum
5	Exxon Mobile	407	USA	Petroleum
6	BP	396	UK	Petroleum
7	State Group	333	China	Energy
8	VW	261	Germany	Automobiles
9	Toyota Motors	256	Japan	Automobiles
10	Glencore	232	Switzerland	Commodities
11	Total	227	France	Petroleum
12	Chevron	220	USA	Petroleum

being made in sub-Saharan Africa, Southeast Asia and Latin America. These large companies are responsible for developing the global value chains which are increasingly significant and add to the complexity of international trade.

The effects of MNCs' investment in host nations are economic, social and environmental (Table 6.11), often at a local scale.

While recognising the importance of MNCs in global trade, the OECD and the UN have attempted to reinforce norms of Corporate Social Responsibility (CSR) among MNCs. The guidelines encourage MNCs to enhance their contribution to sustainable development and hence minimise the disadvantages outlined in Table 6.11.

Stretch and challenge

Unilever, which has invested heavily in EDCs such as India for many years, has initiated the Unilever Foundation.

1 Use the information and links in www.unilever.co.uk/aboutus/foundation-2014 to investigate the different ways in which Unilever aims to achieve sustainable development in India.
2 Discuss the extent to which Unilever's policies for CSR meet OECD and UN codes of conduct for MNCs.

Table 6.11 The impact of MNC operations in host countries

Advantages	Disadvantages
● Provide inward investment and create jobs for local people ● Increase incomes and raise living standards among employees ● Boost exports and improve the trade balance ● Develop and improve skills levels and expertise among the work force, and technology and process systems among local firms ● Increase spending and create a multiplier effect in local economies ● May attract related investment by suppliers and create clusters of economic activity	● Exploitation of work force especially in LIDCs, with poor working conditions and low wages ● Environmental pollution which governments tolerate to attract investment ● Lack of security, with closure of operations as lower-cost locations attract investment elsewhere ● In many LIDCs and EDCs jobs are mainly low skilled in labour-intensive industries (e.g. electronics and clothing) ● Lack of control, with key decisions that have important economic implications for a country taken overseas at company HQ ● Competition could lead to closure of domestic firms

Outsourcing

Outsourcing is a long-established form of investment and trade used by most large corporations. In 2015 European businesses outsourced services by significant amounts (UK 17%, Spain 17%, Finland 20%, Germany 15%). India is the world's major destination of outsourcing contracts (Figure 6.13), especially in the provision of services for large MNCs.

India dominates the global supply of IT services. A geographical outcome is the development of IT clusters in Indian cities. India has six of the top ten cities in the global list of outsourcing destinations and overall employs 10 million people in Information Technology Outsourcing (ITO).

Bangalore is the largest of these cities in India, with 40 per cent of India's IT industry, including over 500 companies and accounting for around US$85 billion

Figure 6.13 Location of top twenty outsourcing destinations, 2015 (Source: Tholons)

in exports annually. Reasons for this concentration, include:

- many nearby educational institutions which have provided a skilled, professional work force in IT
- the ability to rapidly interchange information and knowledge about best practice and market opportunities
- lower costs of providing infrastructure in one locality rather than it being dispersed
- the ability to raise the market profile from a recognised specialist location
- positive government incentives for investment
- the widespread fluency of the English language among the work force
- relatively low labour costs.

A significant trend which is developing rapidly is the rise of emerging economy MNCs. In particular this involves outward FDI by Chinese (Figure 6.14) and Indian companies.

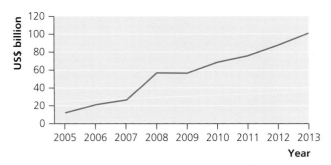

Figure 6.14 China – outward FDI, 2005–13 (Source: UNCTAD Stats)

In fact outward investment by MNCs from each of the BRICS countries is growing rapidly, especially in sub-Saharan Africa. There is a new breed of MNCs arising from the emerging economies, not just for exploitation of primary products but in an increasingly diversified range of products in manufacturing and service sectors. This includes telecommunications, financial services, food processing and infrastructure.

This represents a phase of globalisation characterised by the growing economic power of the emerging nations. Moreover, this new investment is not just from the BRICS countries but also MNCs from South Korea, Mexico, Malaysia, Poland, Saudi Arabia, Thailand and Turkey. In addition to Africa, much of this FDI is in their neighbouring countries of Latin America and the Caribbean, East Asia and South Asia. Of the top 500 MNCs, 40 emanated from EDCs in 2007 and by 2014 the figure had grown to 122.

The role of regional trading blocs

The largest regional trade blocs are the European Union (EU), the North American Free Trade Agreement (NAFTA), the Association of Southeast Asian Nations (ASEAN) and Mercado Común del Sur (Mercosur). Others include EFTA, CARICOM, CIS, COMESA, ECOWAS and GAFTA (Figure 6.15). They are all groups of countries which are relatively close geographically, and may be classified as Free Trade Areas, Customs Unions or Economic Unions.

Figure 6.15 Membership of regional trade blocs (2016)

Their purpose is to achieve economic benefits for their member states through trade policies including:

- encouragement of intra-regional free trade between member states by removing tariffs
- protection of manufacturing and service industries from foreign competition by using tariffs and subsidies
- entering into inter-regional trade agreements with other countries/trade blocs.

The effect of these powerful organisations on global trade is very significant. The volume of intra-regional free trade has ensured that trade blocs are responsible for most of global trade. And their negotiating power, trade agreements and defence of trade have been strong influences on the direction, composition and volume of inter-regional trade.

The EU is the largest economic and political union with 28 full member states and a total population of 520 million. In 2013, it was alone responsible for over 16 per cent of world trade. Germany, Netherlands, France, Belgium, Italy, UK and Spain are the leading countries to have gained from EU membership in terms of value of trade. Together, in 2013, they accounted for 72.3 per cent of intra-EU trade and 79.9 per cent of EU trade with the rest of the world (Figure 6.16).

In 2013, the EU was the biggest trade partner of 59 other countries. For China and the USA the figures were 36 and 24 respectively.

(a) Merchandise

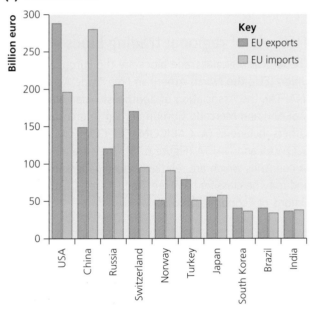

(b) Services

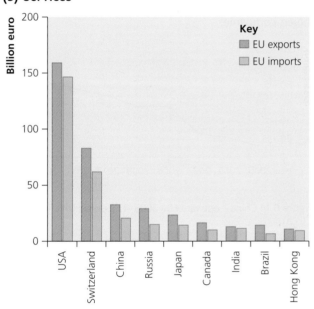

(c) Foreign Direct Investment

Outflow of FDI from the EU

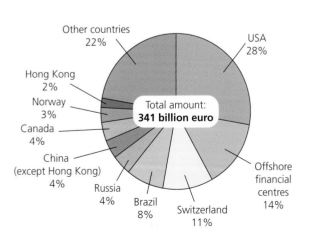

Inflow of FDI into the EU

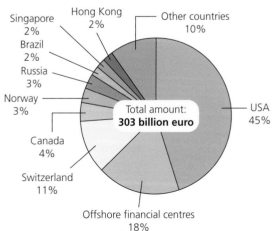

Figure 6.16 EU's largest trade partners in (a) merchandise, (b) services and (c) Foreign Direct Investment (FDI), 2013

The EU operates a **single market** economy allowing free flow of goods within the EU. In recent years it has initiated a number of specific measures which reinforce its powerful position in global trade. These include:

- Trade defence policy. The opening up of global markets to free trade is recognised as a means of achieving economic growth and development. But when 'price-dumping' occurs, i.e. very low prices for imported goods, the EU imposes import duty on those goods. This protects EU manufacturers, allowing them and their subsidiaries to stay in business, maintain jobs and compete fairly.
- Trade and development policy. This is designed to improve trade links between the EU and LIDCs. Many of the poorest countries are being held back in their development by lack of access to global markets. Since 2014 the EU has adopted a system of preferences for developing countries and specific imports from them by tariff reductions.
- Free Trade Agreements (FTAs). The EU is a strong negotiator of trade agreements throughout the world. For example, it is in the process of establishing FTAs with some of its eastern neighbours such as Ukraine. There are agreements with a range of countries in Latin America, the southern Mediterranean, Africa, the Pacific and Southeast Asia including Singapore and South Korea. The EU–South Korea FTA established in 2011 helped to increase trade between the two by 35 per cent in the first year from €30.6 to €41.5 billion.
- Development of trade partnerships. The EU–USA Transatlantic Trade and Investment Partnership (TTIP) is an important arrangement under negotiation between the EU and its biggest trading partner. This removes trade barriers such as customs duties and restrictions on investment on each side of the Atlantic in order to further boost the economies, create jobs and lower prices.

> ### Activities
>
> 1 Describe the flows of merchandise, services and capital shown in Figure 6.16.
> 2 Suggest possible reasons for these patterns.

Furthermore, EU expansion policy has led to integration of more countries into the trading bloc. Croatia was the latest in 2013 and before that Romania and Bulgaria joined in 2007. Membership brings the prospect of unfettered access to the EU markets, plus the benefits of free flow of labour, capital, technology and information. It also gives access to the EU Structural and Cohesion funds with the chance to improve quality of life. For Romania, after eight years of membership, economic growth has begun to accelerate. Exports increased from US$57 billion in 2012 to US$66 billion in 2013 as industrial output, connectivity and trade relations strengthened.

The growth of South–South trade between developing countries

The proportion of world trade between developing countries (LIDCs and EDCs), the South–South trade, has more than doubled in the last decade (Figure 6.17).

This rapid increase in trade has become a vital driver of economic growth for many emerging and developing countries. Merchandise, services and capital flow along new commercial corridors linking the emerging markets of Asia, Africa and Latin America. The total value of this South–South trade in 2013 was US$4.7 trillion. South–South trade accounted for 50 per cent of the total

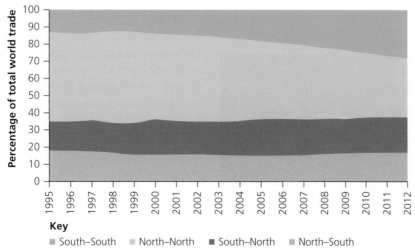

Figure 6.17 Percentage growth in South–South trade (Each region pair represents exports from the first region to the second one.) (Source: UNCTAD statistics)

205

trade of China and 60 per cent of India's total trade in 2013. This trend has been spurred on by:

- rising demand in fast-growing China and India for raw materials and energy to fuel their development
- the vast size of the potential customer market in Asia and Latin America
- increasing demand from the growing middle class of countries such as Brazil, China and India
- intra-regional trade between emerging nations and their immediate neighbours often within a trading bloc
- growth in FDI, especially from China and India into other developing countries of Latin America and Africa.

The least developed countries have remained marginalised, held back by their low productive capacity, limited economic diversification, limited infrastructure and poor governance. Problems caused by being landlocked or remote can also inhibit their trade performance. Most countries in sub-Saharan Africa have lagged behind because their trade is mainly in unprocessed primary products, they have limited access to modern technology and achieve very little value addition in their supply chains. But even in this region exports are increasingly heading off to other developing countries, especially China. And intra-regional trade within Africa is rising despite many barriers such as infrastructure and border restrictions.

Growth of services in the global economy

Since 2000, international trade in commercial services has expanded more rapidly than trade in merchandise. And this is a continuing trend. In 2013, the value of world merchandise exports grew by 2 per cent while export of commercial services grew by 6 per cent to a total of US$4.65 trillion.

Absolute growth in services has occurred in all regions. Europe was the highest exporter in 2013 with 47.2 per cent, but like North America (16.4 per cent) its percentage share has been dropping in recent years with the rise of Asia as a service exporter (26.2 per cent).

Figure 6.18 Canary Wharf – an important hub of the global service industry

In the last decade, exports of commercial services from the least developed countries have been expanding rapidly, albeit from a low base of 0.7 per cent global share in 2013. This growth has been fuelled by the expansion of travel services in new tourist destinations in Asia and Africa such as Cambodia and Uganda and by the emergence of ICT services in new locations such as Bangladesh.

Economic growth in countries such as the BRICS has led to significant growth in the commercial services market. Financial services, ICT and communication services sectors are all increasing in importance as growth in BRICS' merchandise trade and capital investments also continue to grow.

For all of the developing countries, growth in services is important in the development process. It creates employment, adds to GDP, provides funds for investment and enhances global value chains. It is estimated that the value created by services adds over 30 per cent to manufactured goods. For the poorest countries therefore income produced by service growth is important in addressing poverty. Even São Tomé and Principe, the country with the lowest value of services exports in 2014, had receipts of US$18 million from this trade.

For these reasons, according to the IMF, growth in this sector of international trade is essential for sustained economic development. Income from services is therefore important to all economies. It is hardly surprising that the most wealthy economies are in a position to drive this type of trade (Figure 6.18). In 2014, the USA was the largest exporter of commercial services, to the value of US$583.2 billion. And in Europe, the Netherlands, France and Germany produced the largest growth in service exports in 2013.

Increasing labour mobility and New International Division of Labour

Both labour mobility and the New International Division of Labour continue to be significant influences on the global trade system.

New International Division of Labour (NIDL) is the term used to describe global reorganisation of production in the last 40 years. It is associated with the global advance of MNCs and deindustrialisation in advanced economies. This has produced a broad pattern of higher paid managerial jobs and research and development (R&D) in the Advanced Countries and lower paid labourers involved in manufacturing in LIDCs. MNC investment remains strong in many parts of the world and the organisation and structure of these large corporations perpetuates this pattern. In fact, this basic model has been reinforced with specialisation of production in particular countries, for example in the automobile industry. And it is being replicated as firms gain a foothold in sub-Saharan Africa, Mexico, Central America and the Andean states.

A recent example of this type of reorganisation in 2014 was the opening of JCB's biggest factory in Jaipur, India. This investment takes advantage of the Indian government's plans for construction and road building to help boost its domestic manufacturing. Furthermore, located in India, JCB will have easy access to the Middle East and Southeast Asian markets. The success of this new venture is based partly on the plentiful labour supply with a range of differing skill levels and the relatively low costs.

International labour mobility is complex, like the trade on which it is based. It is influenced by both positive and restrictive factors such as:

- government policy
- effectiveness of border control
- demand for labour such as in construction in the Middle East

- prospects of higher and regular wages which can be sent home as a migrant remittance
- ease of access and/or ability to pay air/rail fares
- modern slavery in which labour is exploited and often restricted in movement
- relative ease of labour migration within trade blocs.

> **Key idea**
> → There is interdependence between countries and their trading partners

The rapid integration of emerging and developing economies into the global trade system is a significant feature of the twenty-first century. During this recent period of intensified globalisation the growing connectivity and **interdependence** of trading nations has been a major part of this integration. The interdependent relationships which are developing between countries extend beyond the economic links and include aspects of the political, social and environmental geography of the countries involved. This can be illustrated by the interdependence between India and its trading partners.

🔍 **Case study:** India, an EDC

Direction and components of its current international trade patterns

We have already seen the importance of FDI and outsourced services in India. In addition, the main merchandise exports include refined petroleum, engineering goods, chemicals, pharmaceuticals, gems, jewellery, agricultural products and clothing. And imports include crude oil (34 per cent of imports), gold and silver, machinery and electricals. Figure 6.19 shows the direction of movement of these products and economic interdependence between India and its main trading partners. A useful website is www.tradingeconomics.com/india/indicators which updates frequently.

Changes in its international trade patterns over time

Until the early 1990s India adopted a trade policy of **import substitution** with high tariffs on imported goods plus restrictions on foreign investment to protect its domestic agriculture and manufacturing industry. Since that time India's policies have been more liberal and it is forging strong relationships with its trading partners. It is now one of the world's largest exporters of merchandise, services and capital

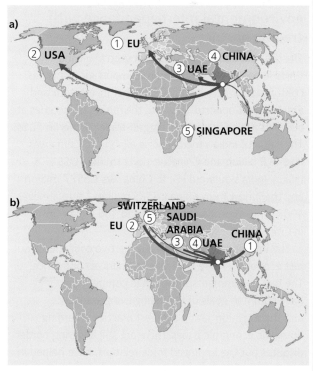

Figure 6.19 India's main trading partners in merchandise exports (a) and imports (b), 2014 (Source: WTO Maps)

(US$313 billion, 2014) and one of the fastest growing economies (average GDP growth of 7 per cent, 2000–14). There have been significant changes in the economy since reforms began in the early 1990s, illustrated by the shift in sector contribution to GDP (Table 6.12). During this period, India doubled its global share of merchandise exports, and share of services exports increased fivefold.

Table 6.12 Changing sector contribution to GDP in India

Sector	1990 %	2014 %
Agriculture	32	18
Industry	27	26
Services	41	56

 Activity

Investigate changes in the patterns and composition of India's global trade. Useful statistics can be found at https://atlas.media.mit.edu/en/profile/country/ind/ and on the websites for the World Bank, WTO and Reserve Bank of India.

Economic, political, social and environmental interdependence with trading partners

The bilateral trade relationship between India and China is one of growing economic importance. This has contributed to growth in GDP, further investment and employment opportunities and the raising of incomes at all levels. China was India's biggest trade partner in 2014. The value of India's merchandise exports to China was US$18.8 billion and China's exports to India US$47.8 billion. India's outward FDI to China was US$27 million and the inflow from China US$25 million. There are over 100 Chinese companies in India and 165 Indian firms in China.

The interdependence between the two countries is not just economic. There is increasing political confidence. The two governments, although in competition, have common goals of improving living standards, and sharing information and best practice regarding domestic economic issues. There are still lingering border disputes but the improved trade relations have helped in stabilising political conflicts.

Despite border conflicts with China, Bangladesh and Pakistan, the Indian government has led economic reform in the last two decades. India is said to be the world's largest democracy and as a member of the G20 it is in a position to have significant influence in global socio-economic, political and environmental affairs.

In the past India and China have been in conflict over environmental issues such as water supply, deforestation and land degradation in Himalayan border areas. But recently there has been joint monitoring and investigation of these problems. For example, trans-border rivers such as the Brahmaputra are important to the socio-economic development of both countries. China now provides India with flood-season hydrological data and there are the beginnings of co-operation over emergency flood management in this basin.

Moreover, criticised in the past for air pollution, China and India have signed climate change treaties.

The UK is a long-standing trade partner of India. Nearly 70 years after independence, India and the UK have a strong interdependent relationship. The UK Trade and Investment government body is currently working with the Indian government to improve business links so that UK companies can succeed in India and Indian companies can invest in the UK. In addition, in 2015, political leaders of the UK and India have agreed to collaborate on issues including finance, defence, nuclear power and climate change. Furthermore there is a strong social relationship; the UK has one of the largest Indian communities outside Asia, which makes a significant contribution to the UK economy and society.

Impacts of trade on India including economic development, political stability and social equality

India's socio-economic development corresponds to growth in its share of the world's merchandise exports. Human Development Index (HDI) has increased from 0.483 in 2000 to 0.586 in 2013; share of merchandise exports has more than doubled in the same time period from 0.7 per cent to 1.7 per cent.

The impact of trade on economic development can be illustrated at different scales from local multiplier effects to larger scale agglomeration of industry. There are many examples of growth stimulated by international trade around the large ports in India, including the automobile industry at Chennai. The multinationals GM and Ford, and major Indian companies such as Ashok Leyland (commercial vehicles) and Tractors and Farm Equipment Ltd, →

are all able to trade in vehicles and components by shipping at this port. Ancillary industries have been attracted to the region and overall this has had the effect of contributing to GDP and stimulating the economic multiplier throughout the state of Tamil Nadu.

The success of India's trade and domestic policies is partly responsible for its political stability in the last decade. The growth in India's share of global trade and the development of its trade partnerships have been driven by government policies:

- Creation of an open market economy including privatisation of state-owned enterprises
- **Trade liberalisation** with reduced controls on foreign trade and investment
- Growth in inward and outward FDI
- Investment in education, producing a skilled and educated work force
- Investment in infrastructure and technology for transport and communications
- Development of global trade agreements

Inequality is often the result of rapid economic growth. But for India, the Gini coefficient has come down from 36.8 to 33.6 in the last seven years and these figures compare favourably with other emerging economies (Brazil 51.9, Mexico 48.3, China 47.3, and Russia 42). Nevertheless there are still great inequalities within India's urban areas, between urban and rural areas, and between regions in terms of health, nutrition, education, poverty and gender.

 Activities

1 Examine further the inequality in India using http://openindia.worldbankgroup.org/#!overview. This provides socio-economic data for each state plus information on the issues of gender inequality, urbanisation and poverty.
2 To what extent is there a relationship between internal inequalities in a country and its international trade?

 Review questions

1 What is meant by the term 'global value chain'?
2 How do developments in communication technology affect patterns of global trade?
3 Why is Bangalore one of the most important locations for the outsourcing of IT services?

4 Why does the EU have such a strong position in global trade?
5 Why has India developed one of the world's fastest growing economies?

6.3 What are the issues associated with unequal flows of international trade?

Key idea
→ International trade creates opportunities and challenges which reflect unequal power relations between countries

Issues which arise from the global inequalities in international trade are illustrated by case studies

of the USA and Sierra Leone. These exemplify both the challenges and the opportunities in each case. For example, the economic and political strength of ACs enables them to drive international trade which creates many economic benefits but it can also lead to problems such as trade disputes, border control and the effects of trade deficit and environmental impacts. And while international trade contributes to socio-economic development in LIDCs, their limited influence in the global trade system leaves them with the challenges of integration into global supply chains, attracting investment and dealing with the problems of their internal inequalities.

This case study shows how core economies have a strong influence and drive trade in the global trade system to their own advantage. This is illustrated through economic, political and social factors.

Advantages for trade, including patterns, partners, negotiations and agreements

The USA is the most advanced and powerful of all nation states, economically and politically. In 2015 the population of 325 million included a labour force of some 158 million. Per capita GNP was US$54,800. The service sector contributed 77.7 per cent to overall GDP (industry 20.7 per cent and agriculture 1.6 per cent). Of the top 500 global largest companies, 128 have their headquarters in the USA. It is the world's second largest trading nation after China. It has strong trade relationships with Canada, China, the EU, Mexico and Japan, and for many economies in Central and South America and Southeast Asia, the USA is their top trading partner. Table 6.13 shows the high value of USA trade in three main components.

The USA's top trading partner is Canada. This is the world's largest bilateral trade relationship, in which US$1.4 million in goods and services are traded every minute. Trade in energy (crude oil, refined petroleum

products, natural gas, electricity and uranium) is just one important category of USA imports from Canada. The USA is the most important destination for Canadian FDI.

The USA's second largest trade partner is China. It has a large and growing trade deficit with China for merchandise but it has a trade surplus with China for commercial services, especially in education, professional services, travel, licence fees and royalties. USA FDI in China in 2014 was US$51.4 billion, mainly in manufacturing, wholesale, finance, banking and insurance, while China's FDI in the USA was relatively small at US$5.2 billion, mainly in banking.

A range of interrelated factors contribute to USA strength in world trade (Table 6.14). The Office of the US Trade Representative is responsible for negotiating and implementing USA trade agreements in order to create opportunities for its economic growth. Details of the work of this government body are found at https://ustr.gov/. Recent examples include the Trans-Pacific Partnership (TPP) Agreement which is a regional Asia-Pacific agreement, and the Transatlantic Trade and Investment Partnership currently being negotiated with the EU.

The advantages of the USA which explain its strong position in the global trade system include: the negotiating strength of the Office of the US Trade Representative; and practical factors such as availability and use of ICT, including the number of mobile phone and mobile broadband internet subscriptions; the availability and quality of its transport services and transport infrastructure; and the efficiency and transparency of its border administration. Further contributory factors are shown in Table 6.14.

Table 6.13 USA trade, 2014

USA trade 2014	Exports/outward, US$	Imports/inward, US$
Merchandise	1.61 trillion	2.33 trillion
Commercial services	662 million	432 million
FDI	5.26 trillion	3.25 trillion

Table 6.14 Factors contributing to the strength of USA trade

Economic	<ul><li>Investment in domestic transport infrastructure such as freeways, railroads, airports, pipelines, ports and vehicles</li><li>High levels of productivity in agriculture, manufacturing and service industries, taking advantage of economies of scale</li><li>Inward and outward FDI often by large MNCs</li><li>Technology in communications including high level of broadband connectivity</li><li>Strong demand from large domestic market</li></ul>
Environmental	<ul><li>Ability to exploit its many natural resources such as minerals, ores, water and timber</li><li>Wide range of climatic conditions, soils and relief which benefit agriculture</li><li>Extensive coastline with natural harbours and access to rich fishing grounds</li></ul>
Political	<ul><li>Stable, democratic government, which has ability to negotiate trade agreements providing easy access to global markets</li><li>Membership of global organisations such as the WTO, OECD and regional bodies including NAFTA and Trans-Pacific Partnership which promote trade liberalisation</li></ul>
Social	<ul><li>Large population/work force with high levels of entrepreneurship</li><li>Diverse multi-cultural migrant labour force</li><li>Highly skilled and educated work force</li></ul>

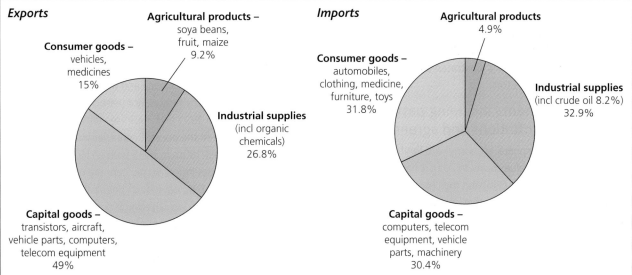

Figure 6.20 The main categories of USA merchandise imports and exports, 2014

The main types of merchandise traded by the USA are shown in Figure 6.20.

Opportunities and challenges

The growth of trade in the USA has created many economic, political, social and environmental opportunities and challenges at different scales (Table 6.15).

Table 6.15 USA – opportunities and challenges created by international trade

Opportunities	Challenges
● Job creation – it is estimated that 9 million jobs depend on the trade and investment between Canada and the USA. This includes employees in transport, customs and border security, finance and ICT. ● Investment leads to cumulative causation described by the economic multiplier effect. This leads to economic growth at a variety of scales and, in time, to overall sustained socio-economic development. ● Relations between countries can develop beyond economic to include improved political relations, cultural understanding, co-operation over security and defence, and stewardship of the environment. This benefits countries such as Honduras under the Central American Regional Security Initiative. ● Advanced Countries such as the USA can have a global influence by encouraging the economic integration of LIDCs into global supply chains – for example, the projects of USAID in economic growth and trade in sub-Saharan Africa.	● The rapid growth of container ports can lead to local environmental issues. The San Pedro Bay ports of Los Angeles and Long Beach and their freight corridors have experienced major expansion in the last 40 years. While they employ over 1500 workers locally and create 1.4 million jobs in California, they are the biggest single source of air pollution in southern California. Emission of diesel particulates, nitrogen oxides and sulphur dioxide contribute to asthma, cardiovascular disease and lung cancer. The University of Southern California estimates 3700 premature deaths/year in this area. Moreover the economic cost of these deaths, medical care, missed work and school days is US$30 billion/year. ● Recent trade disputes with Mexico include sugar and tomato disputes with Mexico accused of 'price-dumping'. ● Border control is an increasing issue in terms of illegal migration and illicit trade. ● The growing trade deficit with China (US$318 billion, 2013). Since establishing the US–China Strategic Economic Dialogue, China has opened up more of its markets to US industries but China still produces many goods at a lower cost and US companies cannot compete.

Case study: Sierra Leone, an LIDC

This case study shows how peripheral economies exert limited influence and can only respond to change in the global trade system. This is illustrated through economic, political and social factors.

Trade components including patterns, partners, negotiations and agreements

Sierra Leone is a peripheral economy with limited access to global markets and value chains. In 2014 it had a trade deficit in every category (Table 6.16) and merchandise trade in total was 1500 times smaller than that of the USA. It is heavily dependent on aid from the USA, UK and increasingly China.

Table 6.16 Sierra Leone trade, 2014

Sierra Leone trade 2014	Exports/ outward, US$	Imports/ inward, US$
Merchandise	0.76 billion	1.27 billion
Commercial services	177 million	524 million
FDI	400 thousand	2.7 billion

Sierra Leone is a small West African country with an Atlantic coastline. It has many natural resources, especially metallic ores, and the potential for a wide range of agricultural produce. But its trade and development have been held back by many factors, not least over a decade of civil war between 1991 and 2002 (see Table 6.17). Sierra Leone is one of the poorest countries in the world (Figure 6.21), ranked 180th in 2015 with an HDI of 0.374. Of its 6.1 million population, 60 per cent are under 25.

Bilateral partnerships have been negotiated with China and the UK and Sierra Leone is a member of the Economic Community of West African States (ECOWAS). The main exports are ores, especially iron and titanium, and agricultural products, such as cocoa beans and coffee; imports include rice, chemicals, and vehicles and machinery for construction. The main export trade partners are China, Japan and Belgium, and imports come from China, India, the UK and other EU countries. The effects of the trade agreements and integration into supply chains with its trade partners are beginning to provide opportunities for Sierra Leone despite the many challenges that remain (Figure 6.21).

Limited access to global markets

A range of interrelated factors are responsible for Sierra Leone's peripherality in the global trade system (Table 6.17).

Figure 6.21 Sierra Leone per capita GDP, 1990–2013 (Source: World Economic Forum)

GDP (PPP) per capital (int'l $) — vertical axis: 500, 1,000, 1,500, 2,000, 2,500, 3,000. Horizontal axis: 1990, 1992, 1994, 1996, 1998, 2000, 2002, 2004, 2006, 2008, 2010, 2012. Key: Sierra Leone; Sub-Saharan Africa

Table 6.17 Sierra Leone – factors contributing to limited access to global markets

Economic	• Overdependence on primary product exports, such as iron and titanium ores, cocoa beans, coffee and palm oil, makes trade vulnerable to sudden shocks as global demand and prices fluctuate • Limited access to finance and secure banking sector • Inadequate infrastructure, including water and energy provision, unreliable road, rail, sea and air transport, limited networks • Poor communications networks with very limited access to broadband internet • High costs of production and sub-optimal productivity • Low levels of technology
Political	• Slow recovery from the consequences of civil war (1991–2002). Previous political instability has affected confidence of investors and international trade • Inefficient government bureaucracy and prevalence of corruption • Inability to control crime including illegal trading
Social	• High level of unemployment – 70% of under 25s unemployed • Inadequately educated work force with high percentage (91% women, 80% men) not reaching secondary education • Gender inequality presents serious barriers to the contribution of women • Other human rights abuses such as the use of child labour • Other socio-economic inequalities in health, poverty and life expectancy
Environmental	• Debilitating effects on the population/work force of prevalent diseases such as malaria, hepatitis A, yellow fever, typhoid, dengue fever, Lassa fever and ebola • Limiting effects of environmental damage to water, soil and forest reserves as a result of large-scale mining operations and agri-business

Opportunities and challenges

Sierra Leone is poorer than most sub-Saharan countries in Africa. Its limited international trade, weak economy and lack of political strength have contributed to its socio-economic problems. Recent growth in trade and investment has created opportunities for development (Table 6.18).

Table 6.18 Sierra Leone – opportunities and challenges created by international trade

Opportunities	Challenges
● Sierra Leone has the support of bilateral trade partners to strengthen trade and socio-economic development. Chinese companies, exploitative at first in demand for iron ore, have now much greater corporate social responsibility. They have financed medical teams, a new hospital in Freetown, and a new foreign ministry, airport and rail network. The UK government aims to strengthen social and economic development, prosperity, democracy, human rights, and beat organised crime and prevent conflict – by enhancing bilateral trade. And the USA has similar aims. Unfortunately Sierra Leone's legal and financial institutions are too weak to make effective progress. ● Membership of ECOWAS has benefited Sierra Leone by abolishing tariff and non-tariff barriers between member states. These include other sub-Saharan LIDCs within this regional group of 15 West African countries such as Togo, Benin, Burkina Faso and Ivory Coast. This has increased trade, boosted economic activity and increased competitiveness in global markets. ● Membership of the Mano River Union (MRU) has involved Sierra Leone in social and economic development programmes. ● The Sierra Leone government in combination with multi-lateral donors has set up the Small Holder Commercialisation Programme, providing tractors, processing equipment, reclaiming swamps and improving roads to markets. ● Political stability and good trade relations have improved since the civil war. There have been three successful democratic elections, the government is modernising customs to ease trade and rule of law is strengthening in dealing with corruption and new gender laws.	● Reducing barriers which inhibit participation in global value chains. Sierra Leone is close to a point of becoming much more integrated but is restricted by poor infrastructure in marketing cash crops and tourism. Reduction of illegal practices is a priority in the diamond trade and fishing. ● Broadening the economic base by attracting investment in manufacturing industries and in services. In 2014, the percentage contributions to GDP were: agriculture 42.5%, industry 26.8% and services 30.7%. ● Managing conflicts between indiscriminate development and environmental degradation. Unsustainable mining practices and agri-business have caused soil erosion, silting of rivers, deforestation and displacement of communities. ● Reducing severe inequalities in life expectancy, gender, education and income (Figure 6.22). Generation of GDP, government funding and employment opportunities are insufficient for investment in services, infrastructure and industry. Rural areas are particularly deprived with 66% living in poverty. Gender inequality is an injustice (Sierra Leone is ranked 139th on the Gender Inequality Index) which is both a cause and consequence of limited trade and development.  **Figure 6.22** Mabella, a squatter settlement in Freetown, Sierra Leone's major port

---

✓ Review questions

1 What are the factors that contribute to the strength of USA trade?

2 In what ways does international trade present challenges for the USA?

3 What are political factors that have restricted Sierra Leone's access to global markets?

4 How does membership of ECOWAS benefit Sierra Leone trade?

5 What are the social and economic challenges for Sierra Leone in gaining access to global markets?

Practice questions

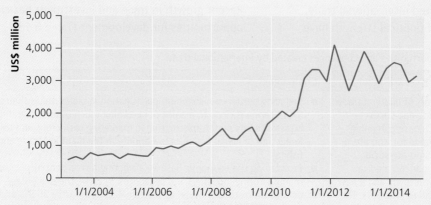

Figure A Ghana, value of exports by month, 2004–15 (Source: www.tradingeconomics.com/Bank of Ghana)

Structured/data response

1 a) Study **Figure A**, which shows the value of Ghana's exports between 2003 and 2015. Ghana is an LIDC in West Africa with an Atlantic coastline.

i) Suggest **two** ways in which changes in the value of Ghana's exports, shown in Figure A, can be influenced by economic factors. [2 marks]

ii) Explain **one** political factor that might account for the variations in value of exports over time in Figure A. [3 marks]

iii) Examine the effectiveness of the data presentation technique in Figure A for showing the value of Ghana's exports between 2003 and 2015. [4 marks]

b) With reference to a **case study**, explain how international trade creates economic benefits for low-income developing countries (LIDCs). [8 marks]

Essay

2*To what extent do physical factors influence contemporary patterns of international trade? [16 marks]

Chapter 7

Global migration

7.1 What are the contemporary patterns of global migration?

Key idea
→ Global migration involves dynamic flows of people between countries, regions and continents

There has been significant growth in the numbers of people migrating across international borders in the twenty-first century. In 2015, according to the United Nations Population Fund (UNFPA), 244 million people – 3.3 per cent of the world's population – were living outside their country of origin.

Migration is inextricably linked to globalisation processes. Places are increasingly interconnected and it is not surprising that the magnitude, complexity and impact of global migration make it a priority issue for almost all nations. Migration policies, border control and migrant safety have become increasingly important issues.

The global migration system is dynamic; flows of people are constantly changing in number, direction of movement and in demographic and ethnic composition. International migration occurs at differing scales: between neighbouring countries in Europe such as France and Germany; across the globe between the UK and Australia; and between major global regions such as Africa and Europe (Figure 7.1).

The reasons for migration are diverse and the decisions of potential international migrants depend on many factors. Globally the majority of migrants are **economic migrants**, seeking work and social opportunities, often sending money back to their family, known as **migrant remittances**. There are also a growing number of **refugees**, fleeing conflict zones and persecution, and **asylum seekers**.

The impact of migration has affected every country either as a place of origin, transit or destination. The consequences are demographic, economic, social, cultural, environmental and political. The vast majority of migrants make meaningful contributions to host countries. But at the same time international migration entails loss of human resources from countries of origin and may lead to tensions in the country of destination.

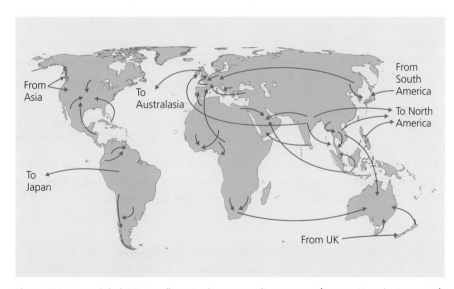

Figure 7.1 Major global migrant flows in the twenty-first century (Source: iRevolutions 2012)

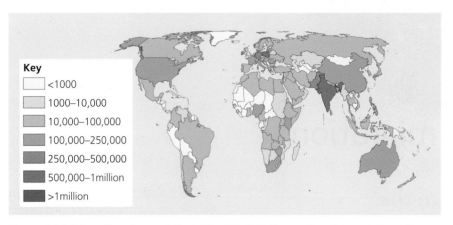

Figure 7.2 Origins of immigrants living in the UK, 2013 (Source: Pew Research Centre)

Current spatial patterns in international migrant flows

International migration

International migration is one element of total population change in a country. This can be expressed in a simple equation and exemplified by UK statistics (Table 7.1):

$$\text{Population change} = (\text{births} - \text{deaths}) \pm \text{international migration}$$

The fact that the migration figures are estimates points to the issue of defining an international migrant. Lack of uniformity among countries creates difficulty in obtaining accurate, reliable and comparable statistics. This is compounded by the many undocumented, illegal migrations.

The UN defines a long-term migrant as a person who moves to a country other than his or her usual residence for a period of at least a year. And a short-term migrant is a person who moves for at least three months but less than a year.

The term **net migration** refers to the difference between numbers of immigrants and emigrants for a particular country. In 2014, the UK had an estimated net migration gain of 318,000 (Table 7.1); in 2015, it was 333,000.

According to Eurostat, EU countries with the largest net migration gain in 2013 were Italy, 1.18 million, and Germany, 466,254. Those with greatest net migration loss were Spain, 265,849, and Greece, 52,000.

Numbers, composition and direction

The scale and direction of international migration and the composition of migrant flows are complex and can be demonstrated by the **immigration** and **emigration** patterns for the UK (Figure 7.2, Table 7.2).

Table 7.2 Main countries of origin of UK immigrants and destination of UK emigrants, 2013

Country of origin of UK immigrants	Number of immigrants resident in UK (million)	Country of destination of UK emigrants	Number of UK emigrants resident in destination country (million)
India	0.76	Australia	1.28
Poland	0.66	USA	0.76
Pakistan	0.48	Canada	0.67
Ireland	0.41	Spain	0.38
Germany	0.31	New Zealand	0.31
Bangladesh	0.24	South Africa	0.31
USA	0.22	Ireland	0.25

The figures in Table 7.2 and Figure 7.2 represent cumulative numbers living in the country in 2013.

In total 5.1 million people born in the UK lived abroad in 2013 (Figure 7.3). The main reasons were:

● Employment opportunities – the majority were of working age, including a high proportion in managerial occupations

Table 7.1 UK population change, 2014

Total population	Births	Deaths	Natural change	Immigration (estimated)	Emigration (estimated)	Net migration	Total population change
64,596,000	695,233	570,341	+124,892	641,000	323,000	+318,000	+442,892

- Retirement – a large number were of retirement age. High UK house prices enable many to sell up and live more cheaply abroad, often in a locality with a warm climate and good quality of life
- Family reunification – moving to join relatives overseas

Figure 7.3 Australia – main destination for UK migrants

> **Activities**
>
> 1 a) Use Figure 7.2 to describe the global pattern of origin of immigrants living in the UK, 2013.
>   b) Suggest reasons for this pattern.
> 2 Discuss the advantages and disadvantages of using choropleth maps to represent spatial patterns of international migration.

There were 7.8 million foreign-born people living in the UK in 2013. The composition of this immigrant population can be analysed according to gender and ethnic origin.

Figure 7.4 shows the overall growth and male–female trends, with women representing just over half the immigrant population in 2013.

There is also a wide range of ethnic diversity among immigrants. London has the largest number of immigrants of all regions in the UK (36.2 per cent). And Figure 7.5 shows that Asian countries have the largest representation, with India more than double that of any other nationality.

Of the 641,000 immigrants moving to the UK in 2014, 178,000 had secured an employment contract before arrival, and 106,000 were looking for work; 193,000 were students in full-time education and 91,000 were seeking family reunification.

Inter-regional migrant flows

Many thousands of people have risked their lives fleeing conflict and instability in Africa and the Middle East in the last decade in order to reach European territory. Total numbers, demographic composition and major countries of origin for those reaching Italy in 2014 are shown in Table 7.3. Often having travelled overland across desert areas, the migrants are transported and sometimes set adrift in small vessels in the Mediterranean by the traffickers. According to the International Organisation for Migration (IOM), 3279 died at sea in the crossings in 2014.

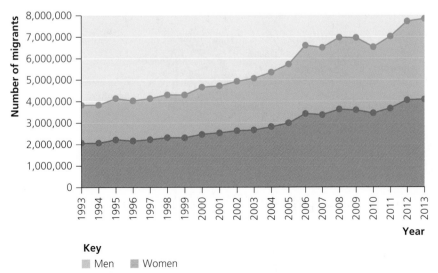

Key
■ Men ■ Women

Figure 7.4 Foreign-born population in the UK, 1993–2013 (Source: Labour Force Survey Quarter 4 and Oxford University migration observatory)

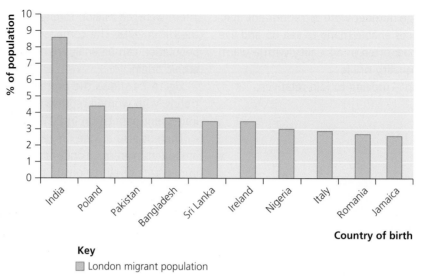

Figure 7.5 Top ten countries of birth of migrants in London, 2013 (Source: Labour Force Survey Quarter 4 and Oxford University migration observatory)

Table 7.3 Italy: arrivals by sea, 2014 (Source: IOM)

Country	Men	Women	Minors	Total
Syria	25,155	6,203	10,965	42,323
Eritrea	24,061	6,076	4,192	34,329
Mali	9,382	27	529	9,938
Nigeria	6,989	1,454	557	9,000
Gambia	7,409	28	1,270	8,707
Palestine	3,413	1,035	1,634	6,082
Somalia	3,010	1,104	1,642	5,756

One migration route is in the central Mediterranean from Libyan ports to Italy's most southerly point, the island of Lampedusa. Others include West African routes to Spain via its North African territories of Ceuta and Melilla and the Canary Islands. Large numbers of migrants have crossed between Turkey and Greece (Figure 7.6). Numbers of migrants increased significantly in 2015; this became a major issue of rescue and border control for the Italian coastguard and Frontex the EU's border management agency as well as the UN's Refugee Agency, UNHCR and many NGOs concerned with migrant welfare.

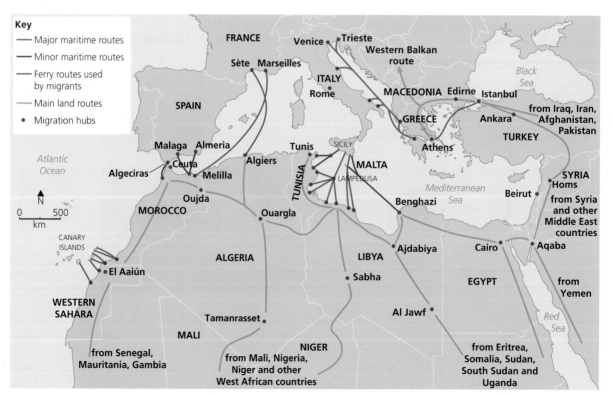

Figure 7.6 Inter-regional migration routes from Africa and Asia to Europe

The Frontex interactive Migratory Routes Map is a useful, detailed source of information, updated for each of the inter-regional routes from Africa and Asia to Europe: http://frontex.europa.eu/trends-and-routes/migratory-routes-map/. Further details and updates are also available on the IOM website at www.iom.int.

The Lee migration model

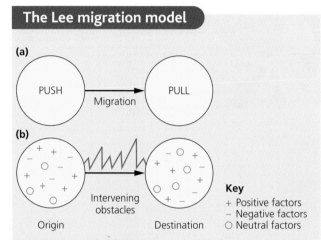

Figure 7.7 The push–pull mechanism (a) and the Lee model of migration (b)

Migrations are caused by:

- **push factors**; these are negative factors which operate in a migrant's current location
- **pull factors**; these are the perceived advantages of a potential destination which attract migrants.

In many instances, the decision to migrate is influenced by a combination of factors; these can be economic, social, political and environmental.

This model provides a useful framework for the understanding of international migration. Places of origin and destination possess attributes which each potential migrant perceives differently according to their personal characteristics and circumstances.

The Lee model also incorporates the idea of **intervening obstacles**. These could occur at any point from origin to destination, and include: costs; physical features such as oceans, rivers, mountain ranges; climatic factors; health; transport; and cultural factors such as language.

Intra-regional migrant flows

International migration within the EU-28 is complex. In any one year not only is there significant movement between the EU member states but also between many non-EU countries and the EU.

In 2012, 1.7 million people resident in an EU country migrated to another EU country. And a further 1.7 million moved to the EU from countries outside. Furthermore 2.7 million people migrated from the EU to a non-EU state.

Activities

The basic mechanism for international migration identified in the Lee model is a development of the simpler 'push–pull' model incorporating 'intervening obstacles' (Figure 7.7).

1 a) Investigate and outline the specific push and pull factors which have caused migration for any one of the routes between Africa or the Middle East and Europe shown in Figure 7.6.
  b) Identify the intervening obstacles for migrants along your chosen route.
2 With reference to Table 7.3, suggest reasons for both gender and ethnic differences in the composition of migrant flows from African and Middle Eastern countries to Italy.

EU intra-regional migration is represented in Figure 7.8. This shows:

- by choropleth, the percentage of foreigners in the population of any one country
- by pie chart, the percentage of a country's foreign population which is of EU origin (green) or non-EU origin (yellow).

Table 7.4 Intra-regional immigration in the EU, 2013

EU-28 country	Immigrants from other EU states 2013	EU-28 country	Immigrants from other EU states 2013
Austria	56,485	Italy	75,710
Belgium	55,807	Latvia	1,517
Bulgaria	1,801	Lithuania	1,728
Croatia	1,966	Luxembourg	13,479
Cyprus	6,259	Malta	2,999
Czech Rep.	11,870	Netherlands	50,537
Denmark	21,158	Poland	45,555
Estonia	385	Portugal	2,523
Finland	9,761	Romania	6,460
France	94,393	Slovakia	5,400
Germany	345,692	Slovenia	3,120
Greece	13,903	Spain	85,020
Hungary	13,574	Sweden	26,175
Ireland	22,211	UK	192,495

The size of these migrations (Table 7.4) is explained partly by the Schengen Agreement which, although not applicable to every EU state, allows freedom of movement within most of the EU across its internal national borders. Details can be found under *Policies* on the *Migration and Home Affairs* page of the European Commission website: http://ec.europa.eu/index_en.htm.

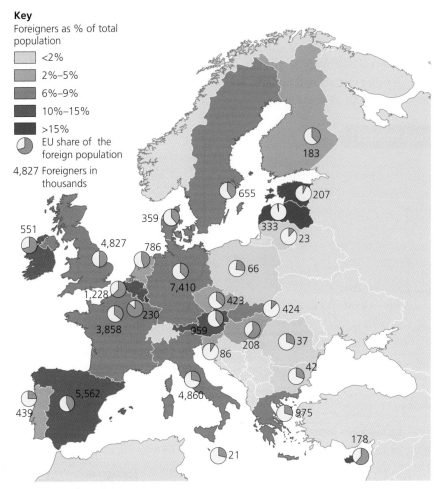

Key

Foreigners as % of total population

- <2%
- 2%–5%
- 6%–9%
- 10%–15%
- >15%
- EU share of the foreign population

4,827 Foreigners in thousands

Figure 7.8 EU intra-regional migration, 2012 (Source: http://one-europe.info/eu-migration-and-eu-welfare-state)

It can also be explained by recent EU expansion which has increased the potential number of migrants. The Czech Republic, Estonia, Latvia, Lithuania, Malta, Poland, Cyprus, Hungary, Slovakia and Slovenia all joined the EU in 2004; Romania and Bulgaria in 2007; and Croatia in 2013. Most of these countries have a relatively high proportion of population in working age groups seeking employment. They are also attracted by higher wages elsewhere within the EU.

The overriding motive for migration is economic. For example, Polish accession to the EU in 2004 stimulated a new wave of immigration from Poland to the UK. Polish migrants were attracted by greater employment opportunities, higher wages, better living standards and ease of return, intended after perhaps two or three years of earning. In 2013 there were approximately 660,000 Poles living in the UK.

In the same year there were also 40,000 UK migrants living in Poland. Many of these were business professionals; the potential business market in Poland

was attractive and open to innovation and inward investment. Some were employed by MNCs, and all were benefiting from the quality of life in a country where the cost of living is lower than the UK.

Intra-regional migration within the EU is also explained by education, retirement, joining family and return flows.

Activities

1 a) Using the data in Table 7.4, describe the spatial pattern of intra-regional immigration within the EU.
  b) Suggest reasons for this pattern.
2 a) Outline one statistical mapping technique to represent the spatial pattern of EU intra-regional immigration, using the data in Table 7.4, and justify your choice.
  b) Suggest an alternative technique and discuss its advantages and disadvantages for showing these data.

Key idea
→ Current patterns of international migration are related to global patterns of socio-economic development

The relationship between patterns of international migration and socio-economic development

There is a close relationship between migration and development.

● Migration can contribute significantly to development; it can be a positive process for stability, economic growth and socio-economic change (Table 7.6).
● Inequalities in levels of development can be a cause of migration; this has a major influence on the direction and scale of global migrant flows.

One statistical measure of international migration which is linked to development is the value of migrant remittances. These are private funds sent by migrants usually to the non-migrant members of their family.

The financial framework for the international transfer of this money is being strengthened. Therefore official figures are increasingly accurate, but in reality the amounts are much higher.

Remittances are of considerable importance in the development process and are shown in Table 7.5 as a percentage of the recipient country's GDP. Development is measured according to the UN's Human Development Index (HDI), a composite index incorporating social and economic indices for life expectancy, education and GDP per capita.

The figures in Table 7.5 show that migrant remittances to the more advanced countries, with higher HDI, are a lower percentage of GDP. For LIDCs and EDCs, they represent a higher proportion of GDP, and therefore are of greater significance to their economy and development. For example, in 2013, US$6.69 billion was sent by migrants as remittances to the USA, less than 0.1 per cent of GDP; India received more than ten times this amount, US$69.97 billion, 3.7 per cent of GDP; and Haiti received much less in absolute terms, US$1.78 billion, but this was 21.1 per cent of GDP.

Not all poor countries receive large migrant remittances. Potential migrants from sub-Saharan Africa countries such as Sudan (HDI 0.473, remittances only 0.6 per cent of GDP) are affected by restrictive immigration policies of developed countries and costs of travel, including payments to traffickers.

Table 7.5 Migrant remittances as percentage of GDP and HDI for selected countries, 2013 (Source: World Bank)

Country	Remittance (% of GDP)	HDI	Country	Remittance (% of GDP)	HDI	Country	Remittance (% of GDP)	HDI
Nepal	28.8	0.540	Mali Rep.	8.2	0.407	Sudan	0.6	0.473
Haiti	21.1	0.471	Togo	7.8	0.473	China	0.6	0.719
The Gambia	20.0	0.441	Pakistan	6.3	0.537	Germany	0.4	0.911
Liberia	19.7	0.412	Nigeria	4.0	0.504	South Africa	0.3	0.658
Honduras	16.9	0.617	India	3.7	0.586	Russia	0.3	0.778
Jamaica	15.0	0.715	Kenya	2.4	0.535	Australia	0.2	0.933
Uzbekistan	11.7	0.661	Romania	1.9	0.785	Norway	0.2	0.944
Montenegro	9.6	0.789	Mexico	1.8	0.756	Argentina	0.1	0.808
Sri Lanka	9.6	0.750	Poland	1.3	0.834	UK	0.1	0.892
Bangladesh	9.2	0.558	France	0.8	0.884	USA	<0.1	0.914

How global migration can promote stability, growth and development within and between countries through flows of people, money, ideas and technology

International organisations such as the United Nations and the International Organisation for Migration (IOM) recognise the importance of migration as a key factor in development. The Joint Migration and Development Initiative, implemented as part of the United Nations Development Programme (UNDP) post-2015 sustainable development agenda, is one strategy designed to harness the potential of migration for development – details at www.migration4development.org/content/about-jmdi.

Table 7.6 outlines some of the positive effects of international migration on stability, growth and development, for countries of origin and destination. Global migrations are intrinsically related to flows of money, ideas and technology. This transfer of resources is significant in promoting stability, growth and development.

Monetary transfers are most evident in the billions of dollars sent worldwide as migrant remittances. These flows of money have been made easier, more efficient and more secure by the use of technology. This includes mobile money transfers made more reliable by cash transfer programming and the use of smartphones, which have become more prevalent.

Global migration also leads to the geographical diffusion of ideas, information and values which can be transmitted back to place of origin. This includes ideas on family size, education and marriage, referred to as social remittances. Information on migrant reception and progress at the destination can be useful for prospective migrants. And values such as democracy and other norms of behaviour can flow from one country to another through use of social media or when migrants return.

Technology is increasingly important. It is used, for example, by international humanitarian organisations including NGOs, to assess crises so that response can be where needs are greatest. Human mobility analysis is conducted through use of 'big data' including mobile phone records and credit card transactions. These data, in addition to existing cartography, satellite imagery, field reports, including conventional media reports, and crowd-sourced data from text messages, emails and tweets are used in 'crisis mapping' (see Chapter 15, Geographical Skills, pages 520–21). An example is the use of smartphones (Chapter 15, page 523) and digital data collection to give access to numbers of urban internally displaced persons (IDPs) in Pakistan. Many IDPs flee to urban areas rather than the established camps due to perception of increased economic and social opportunities. The IDP Vulnerability Assessment and Profiling Project (IVAP) has been set up jointly with the Pakistan government and fourteen humanitarian organisations. Data collected by surveys using smartphones are uploaded to an online database designed for automated analysis of around 400,000 urban IDPs so that the most vulnerable can be targeted for assistance.

How global migration causes inequalities, conflicts and injustices for people and places through unequal flows of people, money, ideas and technology

Problems created by international migration may be economic, social, cultural or political. And their impacts may exist in countries of origin and destination.

Table 7.6 How global migration can promote stability, economic growth and development within and between countries

Stability	• Migrant remittances are a source of foreign exchange which can contribute to economic stability of the recipient country. • Returning migrants, having acquired new ideas and values including democracy and equality, can contribute to peacebuilding and conflict resolution. • Where there is ageing population, youthful migrant working populations contribute to a more balanced age structure and population growth.
Economic growth	• The GDP and tax base of the host nation can be boosted by working migrants. • Migrants as consumers themselves can stimulate local economies in a host country, even opening up new markets in demand for food, clothing, music, etc. • Migrants can fill skills gaps and shortages in the labour market of a host country at local and national scales. • Migrant remittances can supplement household income, stimulate consumption, provide funds for local investment and stimulate local **multiplier effects** in the country of origin of the migrants.
Development	• Skills and knowledge acquired by returning migrants can be of benefit to countries of origin. • Migrants can create networks which ease flows of skills, financial resources, values and ideas through their links to **diaspora** associations, including professional, business, social and religious networks. • UN 'migration and development' projects between partner countries are involving families, local authorities, and public and private service providers in effective 'bottom-up' approaches to development.

Table 7.7 How global migration can cause inequalities, conflicts and injustices

Inequalities	• Countries of origin lose a proportion of the young, vibrant and fittest element of the labour force; this may contribute to downward economic spiral at local, regional and national scales. • Often it is the better educated that migrate; this represents a 'brain drain' and loss of human resources in the country of origin. • The demographic selectiveness of international migration causes redistribution of population of reproductive age; this influences crude birth rates in countries of origin (decline) and destination (growth). • Migrant remittances can increase inequality between families who receive them and those who do not.
Conflicts	• Social conflict can develop between host communities and 'newcomers'; people of a particular culture or ethnic origin may find difficulty integrating perhaps because of language. • Immigrant populations, especially if concentrated in specific areas, can place pressure on service provision such as education, health and housing in the host country. • International borders can be areas of conflict for border control authorities, traffickers and illegal migrants.
Injustices	• Migrants are vulnerable to violation of their human rights as a result of forced labour, exploitation of women and children, and human trafficking. • Treatment of asylum seekers can include being held in detention centres, not being allowed to work, and being supported on meagre financial resources for food, sanitation and clothing for the duration of application. • The plight of refugees in terms of shelter, food, water, medicines and safety, including possibility of return to country of origin where risks are high.

Areas of potential conflict and injustice are exemplified by:

● reported human rights violations in detention camps for Myanmar refugees on the Thai–Malaysian border
● the new fence and high-tech surveillance at the Bulgaria–Turkey border (Figure 7.9)
● the refugee camp and Channel Tunnel at Calais (Figure 7.10). Another camp, 'the Jungle', outside Calais contained over 3000 migrants in February 2016 – mostly unaccompanied young males.

Inequalities, conflicts and injustices are related to unequal flows of people, money, ideas and technology within the global migration system (Table 7.7).

Migrant flows are spatially uneven. Globally the South–North and South–South migrations are dominant. The transfer of ideas and money are closely related to these patterns of inequality in migration,

and tend to flow in the opposite direction, back to the place of migrant origin. For example, the talent-based immigration policies of countries such as Canada and the USA encourage flows of highly skilled migrants from LIDCs and EDCs, and this gives rise to consequent contra-flow of social and financial remittances.

Access to technology in some countries is limited and reflects low levels of socio-economic development and infrastructure for ICT. For example, many low- and middle-income countries lack the skilled human resources and computer forensic tools to analyse digital data or download and use maps and satellite imagery.

In some countries, government or military control of information and limited broadband speed affect the work of civilian relief organisations. These inequalities restrict the effective management of conflict and injustices linked to migration.

Figure 7.9 Boundary of 'Fortress Europe' at the Bulgaria–Turkey border

Figure 7.10 Refugee camp near the port of Calais

223

1 What is the UN definition of an international migrant?
2 What is meant by the term 'net migration'?
3 What is meant by the term 'intervening obstacle' referred to in the Lee model of migration?
4 What are the specific reasons for increased intra-regional migration following expansion of the EU since 2004?
5 What are migrant remittances and how can they contribute to development?
6 How can international migration lead to increased socio-economic inequalities?

7.2 Why has migration become increasingly complex?

Key idea
→ Global migration patterns are influenced by a multitude of interrelated factors

Changes in the twenty-first century have increased the complexity of global migration.

Economic globalisation leading to the emergence of new source areas and host destinations

An increasing number of countries and their economies have become more interdependent. And the increasing complexity of global migration can be linked to this intensification of the globalisation process. Major bilateral corridors and traditional migration partnerships have remained strong, but in addition new places of origin and new destinations for migrants have emerged as the effects of globalisation have spread.

Examples of contemporary migration patterns which reflect these recent changing global economic trends include:

Inter-regional

● Migration of highly skilled workers from China, India and Brazil to the USA – including graduates, especially in science, mathematics and technology, and those in professional and business services, attracted by high salaries and the quality of life.

● Migration of workers from India, Bangladesh, Pakistan, Egypt, the Philippines and Indonesia to oil-producing Gulf States and Saudi Arabia – attracted by increased demand for labour, relatively high wages, ease of returning formal remittances, accommodation, and improved transport and communications.

Intra-regional

● Rapid increase of **international migrant stock** (6.5 million, 2013) among ASEAN member states – the fast-growing economies of Singapore, Malaysia and Thailand are the main destinations (Figure 7.11); and Myanmar, Lao PDR and Cambodia, the main sources. Most migrants are low-skilled, many undocumented, seeking employment and higher wages via cyclical migration to countries of higher socio-economic development.

● Increased migration streams within South America, especially to the 'southern cone' of Argentina, Chile, Paraguay and Uruguay. The main drivers are disparities in wages and labour opportunities. Regional integration (Mercosur, Andean Community trading blocs) has also eased immigration through free movement of labour.

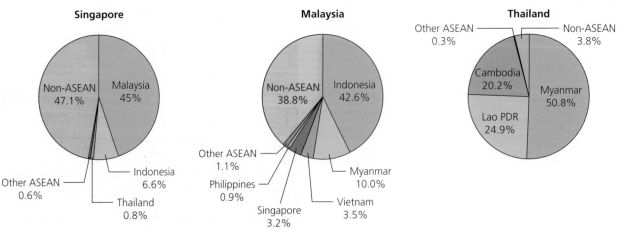

Figure 7.11 Origins of international migrant stock in Singapore, Malaysia and Thailand, 2013 (Source: Trends in International Migrant Stock. The 2013 Revision Database)

- Return migration has been a high proportion of migrant flows within the EU – Romania (93% of its total migration), Lithuania (88%), Latvia (72%), Portugal (64%), Poland (63%) and Estonia (58%). This includes young workers having achieved their pre-planned economic goals after two or three years – often taking low-skilled jobs abroad before returning to more prestigious positions in their home country.

Internal

- Internal migrant flows within EDCs such as India, China, Mexico and Brazil, driven by FDI which has created agglomerations of economic activity near large urban centres. Rural–urban migration is not new but it has been reinforced and is a major element of the global migration system.

High concentration of young workers and female migrants

Young workers

The main reason for international migration by the younger elements of the labour force is economic – greater employment opportunities, higher wages and the possibility of remittance.

Figure 7.12 shows the population structure of all migrant populations in Asia, 2013. There is clearly male dominance and the largest groups are young, aged 25–39.

Demand for workers in the oil-producing countries of the Middle East such as UAE, Qatar and Saudi Arabia is an example of young labour driven migration. Numbers of foreign-born residents in UAE have increased

dramatically in the twenty-first century from 2.45 million in 2000 to 7.83 million in 2013.

Typical of the patterns in this area, in Saudi Arabia in 2013 the migrant population included 2.85 million born in India, 1.09 million Bangladeshis and 0.95 million originating in Pakistan. The flows were dominated by young males, with a high proportion working in construction. The majority were low skilled, many not educated beyond primary level. Only 3.6 per cent of migrants were employed in health, and 2.6 per cent in education sectors.

Female migrants

Globally there has been an increase in the number of women and girl migrants in the twenty-first century. In 2013, 52 per cent of all migrants in developed countries were female; in developing countries the figure was 46 per cent. Regionally, in Europe, Latin America and the Caribbean, North America and Oceania, female participation in international migration exceeded that

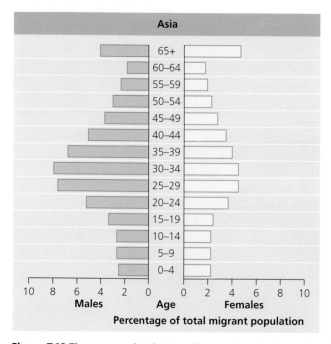

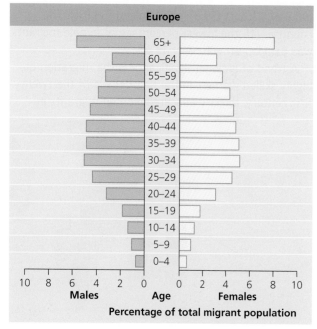

Figure 7.12 The age–sex distribution of international migrants in Asia and Europe 2013 (%) (Source: UN Department of Economic and Social Affairs, Population Division. International Wallchart 2013)

of men. The opposite was the case for Africa and Asia (Figure 7.13), although the percentage of female migrants in sub-Saharan Africa increased from 40.6 per cent in 1990 to 46.4 per cent in 2013.

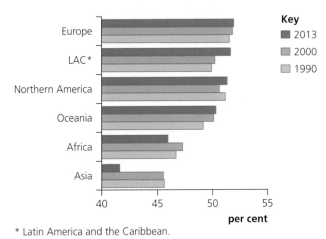

* Latin America and the Caribbean.

Figure 7.13 Percentage of women among all international migrants, 1990, 2000 and 2013 (Source: unmigration.org factsheet 2013/2)

Reasons for the differences centre on the regulations governing admission and departure of migrants in different countries of destination and origin – many of which are linked to the status of women in these countries.

In the last two decades growth in the number of female migrants can be accounted for by their greater independence, status, freedom and increasing importance as main income earners.

In 2013 there were 101 countries in which the female international migrant stock was greater than that of men (Figure 7.14). The highest were Latvia (60.8%), Estonia (59.8%) and Poland (58.8%). The country with the lowest percentage of female migrants was Bangladesh with only 13.4%.

A trend of growing significance is the migration of highly skilled women. In the first decade of the twenty-first century, tertiary educated women migrants in OECD countries increased by 80 per cent. During this time the emigration of female graduates and other highly skilled women was higher than that of highly skilled/educated men for African and Latin American countries. This was also true of women in India, China and the Philippines. The main destinations include Canada, USA, UK and Israel, countries in which there is less discrimination in the labour market and where in general women's rights are better respected.

Flows in South–South corridors are now equal in magnitude to those in South–North corridors

UN statistics for 2013 show that South–South international migrant stock now outnumbers that of the South–North flows (Table 7.8).

Table 7.8 Numbers of international migrants in the main global corridors, 2013

Migration corridor	Number of migrants (million)	Share of global migrant stock (%)
South–South	82.3	36
South–North	81.9	35
North–North	53.7	23
North–South	13.7	6

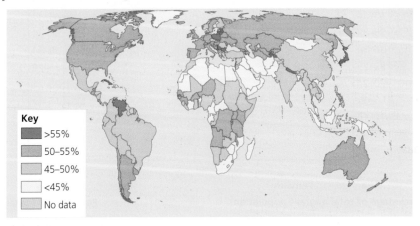

Figure 7.14 Percentage of females among all international migrants, 2013 (Source: unmigration.org factsheet 2013/2)

Until the second decade of the twenty-first century the largest international migrant flows had been from the poorer less developed countries of the South to the wealthier more developed countries of the North. But as global economies have become more interconnected, South–South flows have increased significantly. In 2015, South–South flows were expected to be equal to all other flows combined.

South–South flows are explained mainly by labour migration. Evidence for this is that two-thirds of migrant remittances were sent between countries of the South in 2013. An example is the Bangladesh–India corridor which had the world's third largest bilateral flow of 3.3 million migrants.

Another factor is the increase in refugees fleeing persecution or conflict, such as the 2.3 million having migrated from Afghanistan to Pakistan.

The main reasons for the magnitude of South–South flows include:

- restrictive administrative barriers for migrants from the South attempting to enter the North; often in response they redirect their migration to another South country
- the number of fast-growing economies in the South which offer employment opportunities and are increasingly accessible
- increased awareness of opportunities in the South resulting from improved communications and developing social and business networks
- preventative costs of moving to more distant richer countries.

Two examples of South–South migration corridors illustrate specific factors:

Burkina Faso to Ivory Coast
West Africa has high levels of intra-regional international migration. Burkina Faso is a landlocked, low-income country with GDP of US$684 per head. It is bordered to the south by the Ivory Coast, a lower-middle-income country, which is the world's largest exporter of cocoa, and in 2013 had a GDP of US$1529 per head.

In 2013, there were 560,000 Ivorians living in Burkina Faso. Many of these settled in Burkina Faso having escaped conflict in the Ivory Coast. There were also 1.46 million born in Burkina Faso living in the Ivory Coast; this excess movement of migrants from Burkina Faso to the Ivory Coast is explained by:

- employment opportunities and higher wages available in the Ivory Coast's cocoa and coffee plantations. The income disparity between the two countries is relatively small but sufficient to encourage significant flows.

- opportunities for migrant farmers in the more fertile lands of the Ivory Coast
- former French colonial administration in both countries has led to shared language, currency and a cultural system which has made it easier for those of Burkinabé descent to travel to the more prosperous Ivory Coast.

Myanmar to Thailand
The largest ASEAN migrant corridor is the flow from Myanmar to Thailand, both 'South' countries, involving 1.9 million migrants. Thailand is Southeast Asia's fastest growing economy and many migrants who live below the poverty line in the lower-income country, Myanmar, are attracted for economic reasons. Thailand needs to resolve labour shortages in agriculture, fisheries, manufacturing, construction and domestic services. Moreover it has recently introduced a legal daily minimum wage of 300 baht (US$9), some ten times that of Myanmar. The geographical proximity of the two countries and the freer flows of labour possible within the newly formed ASEAN Economic Community (AEC) are further contributory factors. In addition many are refugees from the Myanmar government, escaping forced labour in government development projects such as railway construction as part of Myanmar's economic reforms.

Conflict and persecution have increased numbers of refugees
A refugee is someone who has moved outside the country of his nationality or usual domicile because of genuine fear of persecution or death. According to the United Nations High Commission for Refugees (UNHCR) the number of refugees worldwide increased from 15.7 million in 2012 to 19.5 million in 2014. In 2015 Syria had become the largest source of refugees, overtaking Afghanistan which had held this position for three decades; Turkey was the largest recipient. Of all refugees 87.2 per cent live in the global south (10.2 million in Asia).

An asylum seeker is a person who seeks entry to another country by claiming to be a refugee. Those judged not to be refugees nor requiring international protection can be sent back to their home countries. Globally in 2014, 1.66 million asylum applications were submitted; the Russian Federation, Germany and USA were the largest recipients of these applications.

Statistical updates for the rapidly changing global situation are available at www.unhcr.org/pages/49c3646c11.html.

An indication of the complexity and global reach of international refugee migration is shown in Table 7.9.

Table 7.9 The 32 countries with the highest ratio of refugees to total population, 2014

Country of asylum residence	Refugees per 1000 inhabitants	Country of asylum residence	Refugees per 1000 inhabitants	Country of asylum residence	Refugees per 1000 inhabitants	Country of asylum residence	Refugees per 1000 inhabitants
Afghanistan	5.04	Gambia	6.8	Liberia	9.65	Sudan	6.75
Austria	6.62	Guinea-Bissau	5.4	Malta	32.31	Sweden	12.17
Burundi	5.18	Iran	13.19	Mauritania	22.16	Switzerland	7.38
Cameroon	10.98	Iraq	8.21	Montenegro	11.55	Syria	6.94
Chad	38.81	Israel	6.5	Norway	9.43	Turkey	11.43
Congo Rep.	11.96	Jordan	114.11	Pakistan	9.30	Uganda	10.55
Djibouti	24.81	Kenya	13.13	Rwanda	6.71	Venezuela	7.05
Ethiopia	6.75	Lebanon	257.08	S. Sudan	24.21	Yemen	54.69

Skills focus

1 a) Suggest and justify an appropriate statistical mapping technique to represent the global pattern of refugees per 1000 inhabitants shown in Table 7.9.

   b) Outline the advantages and disadvantages of one alternative technique.

2 Using the link www.therefugeeproject.org evaluate the mapping techniques used in 'The Refugee Project' to show the global pattern of refugees and refugee connections for each individual country between 1975 and 2012.

The crisis in Syria in the second decade of the twenty-first century has been a major factor in the recent increase of refugees. The civil war, which began in March 2011, has led to internal displacement of 7.6 million people and a further 4.7 million international refugees.

A high percentage of the refugees have moved relatively short distances to countries which share a border with Syria (Figure 7.15); Turkey (1.9 million) and Lebanon (1.5 million) being the biggest recipients (January 2016). Residents of Kobani, Syria, for example, have been living in tented camps near Turkish border towns. The intention is to return to their neighbourhoods and farmlands as soon as the border crossing is reopened.

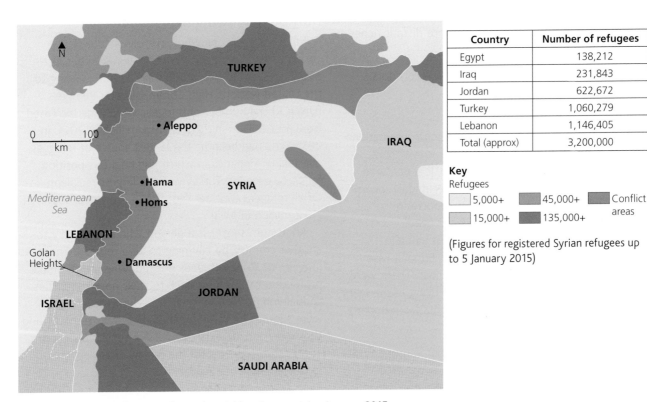

Country	Number of refugees
Egypt	138,212
Iraq	231,843
Jordan	622,672
Turkey	1,060,279
Lebanon	1,146,405
Total (approx)	3,200,000

Key
Refugees

5,000+   45,000+   Conflict areas
15,000+   135,000+

(Figures for registered Syrian refugees up to 5 January 2015)

Figure 7.15 Location of Syrian refugees in neighbouring countries, January 2015

Lebanon has been so overwhelmed by Syrian refugees that in January 2015 the government imposed the need for an entry visa whereas before this date movement between the two countries was largely unrestricted. Syrians seeking work must be sponsored by a Lebanese individual or company.

Other refugees have embarked on longer and more complex journeys to reach their planned destination in the EU. For example, many Syrians reach Aksaray, an Istanbul neighbourhood, where they live temporarily before making contact with the agents of the traffickers. The route to Europe depends to some extent on their wealth: there is the less costly sea route with greater risk; or the more expensive but safer arrangement for a fake passport and direct flight.

The main reasons for the large number of refugees globally include:

- the effects of conflict, including personal safety, loss of homes, access to services, damage to other infrastructure including communications
- political persecution, discrimination and violation of human rights
- economic hardship including forced labour and modern slavery
- the impacts of natural hazards.

Changes in national immigration and emigration policies

National migration policies are designed to meet the economic, social and political needs of a country. For example, some ACs such as the UK, Australia and Canada use a points-based system to satisfy labour shortages in particular sectors. In the developing world some countries actively encourage emigration, largely to assist in the development process through the financial benefits of migrant remittances, and the skills, ideas and business contacts brought by returning migrants.

Differing effects of recent policy changes are illustrated by Pakistan and Canada.

Emigration policy – Pakistan

Pakistan is a lower-middle-income country. In 2014 its population was 196.1 million, 45 per cent of whom were under 20. It had a GDP per head of US$4736 and an HDI of 0.537.

The Pakistan government is pro-emigration. There are 7 million Pakistanis working abroad and 96 per cent of these are in the Gulf Cooperation Council countries (Table 7.10). In 2013, migrant remittances amounted to US$11.5 billion (the fifth largest in the world) which is important to the socio-economic development of the country. This has been recognised formally in Pakistan's new emigration policy.

The Pakistan National Emigration Policy has been drafted by the Ministry of Overseas Pakistanis and Human Resource Development jointly with the International Labour Organization (ILO). It aims to promote emigration and safeguard migrants, and includes the following requirements:

- Ratification of ILO and UN Conventions regarding rights of workers and protection of basic human rights
- Promotion of the export of Pakistani manpower abroad
- Positive steps to encourage female participation in overseas employment (currently only 0.12 per cent)
- Support for social networks and associations abroad (Pakistani diaspora)

Table 7.10 Total emigrants from Pakistan to the Gulf Cooperation Council countries, 2008–13 (Source: Pakistan Bureau of Emigration and Overseas Employment)

Country of destination	2008	2009	2010	2011	2012	2013	Total
Saudi Arabia	138,283	201,816	189,888	222,247	358,560	270,502	1,381,296
UAE	221,765	140,889	113,312	156,353	182,630	273,234	1,088,183
Oman	37,441	34,089	37,878	53,525	69,407	47,794	280,134
Bahrain	5,932	7,087	5,877	10,641	10,530	9,600	49,667
Qatar	10,171	4,061	3,039	5,121	7,320	8,119	37,831
Kuwait	6,250	1,542	153	173	5	229	8,352
Total	419,842	389,484	350,147	448,060	628,452	629,478	2,845,463

- Establishment of training institutions to help Pakistani youth in preparation for working abroad
- Enhancement of the impact of economic remittances and skills of returning migrants for development

Skills focus

1 Construct an appropriate chart to show trends in Pakistan emigration, 2008–13 (Table 7.10).
2 Discuss the advantages and disadvantages of the type of chart you have chosen to construct.

Immigration policy – Canada

Canada is a high-income country. In 2013 it had a GDP per capita of US$44,843, a GDP growth rate of 2.5 per cent and an HDI of 0.902. Its total population was 35.1 million. Changes were made to its immigration policy in January 2015 to address the skills gap in the labour market.

The new policy is aimed at the country's long-term requirements for engineers, IT specialists and health care workers. Potential migrants are ranked on a 1200-point system which enables young, highly skilled immigrants to be fast-tracked. Every effort is made to employ a Canadian citizen first, but half the necessary points can be awarded to migrant applicants with a permanent job offer from an employer – especially if the employer is located in one of the provincial employment schemes away from the big cities of Toronto, Vancouver or Montreal. Applicants in their twenties receive maximum points for age, and graduates are also favoured.

In the short term, Canada has agreed to take 10,000 Syrian refugees over a three-year period from 2015.

Development of distinct corridors of bilateral flows

Bilateral migration is simply the migrant flow between two countries. The number of migrants, their composition and the directions of flow are important characteristics of bilateral migration. Some bilateral corridors are very large and long-standing such as that between Mexico and USA. Others cannot match this in scale, but new and significant flows have been recorded in the last decade. For example, migration between Sudan and South Sudan includes large numbers of refugees (Figure 7.16).

As we have seen, the decision to migrate depends on many factors, real or perceived, in both place of origin and potential destination. Explanation of these strong

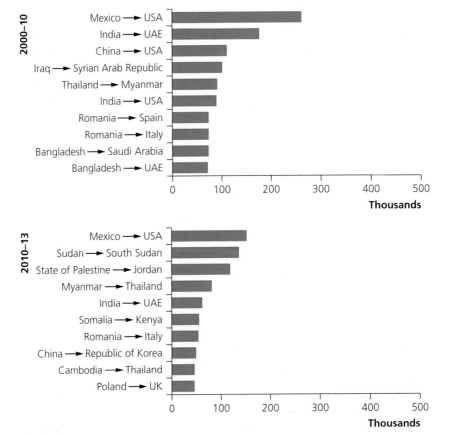

Figure 7.16 Ten bilateral migration corridors with the largest number of international migrants per annum, 2000–10 and 2010–13 (Source: UN Department of Economic and Social Affairs, Global Migration Report 2013)

bilateral migrant corridors includes economic, social and political influences:

- Costs of travel which may be related to proximity
- Ease of access and communication between the two countries
- Efficiency and cost of sending remittances
- Employment opportunities and wage differentials
- Established diaspora communities and networks
- Effects of conflict and persecution
- Migration policy, including accession to economic unions and policy on refugees
- Former colonial influence, such as language

Key idea
→ Corridors of migrant flows create interdependence between countries

> **Activity**
>
> Referring to Figure 7.16, describe and explain how the scale and pattern of migrant corridors have changed between 2000–10 and 2010–13.

Global integration and interdependence have proceeded rapidly. Flows of trade, capital, technology, information, ideas and people across international boundaries have not only increased but have become more complex in their geographical pattern. International migration is a very important part of these globalisation processes, contributing significantly to social and economic interdependence.

This has brought benefits and opportunities for many, but others have been excluded from the process or have experienced the effects of inequalities and injustice. These issues can be illustrated by a case study of an EDC such as Brazil.

Case study: Brazil, an EDC

Brazil is the seventh largest economy in the world and the leading economic power in Latin America. As an emerging economy it experienced a significant increase in GDP per capita from US$4874 in 2007 to US$5823 in 2014. The sectoral contribution to GDP is: services 69 per cent, industry 25 per cent and agriculture 6 per cent.

Brazil is moving rapidly through the demographic transition with declining crude birth rate and ageing population (Figure 7.17). Migration has contributed to the economic growth and development of Brazil over a long period.

Current patterns of immigration and emigration

Features of Brazil's current migration patterns include:

- A net migration loss of half a million in each of the four-year periods 2000–04 and 2005–09, but this slowed to 190,000 between 2010 and 2014 (source: World Bank)
- Increased migration between Brazil and its neighbouring countries, especially Mercosur members, but also Chile and the Andean states

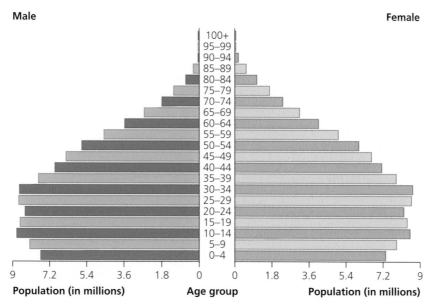

Figure 7.17 Population structure of Brazil, 2014 (Source: http://www.indexmundi.com/brazil/age_structure.html)

231

- A slowing down of emigration of lower skilled economic migrants to the USA
- Increased emigration of highly skilled workers to Europe, USA and Japan
- An influx of migrants from Haiti and increasingly from African countries using Ecuador and Chile as transit countries
- A rise in the number of international labour migrants attracted by the construction industry for the 2014 football World Cup and the 2016 Olympics
- Strong and continuing internal migration especially from the northeast to the cities of the southeast

Stretch and challenge

1 Using information at http://populationpyramid.net/brazil/2015/ describe the projected change in Brazil's population structure in the next fifteen years and suggest reasons for the changes.
2 Suggest an alternative statistical technique to represent the global migration pattern of the current Brazilian squad.

Changes in immigration and emigration over time

During the late nineteenth and twentieth century, Brazil was a net recipient of migrants. There were periods in which Europeans were attracted to work in the agricultural sector, especially coffee cultivation – particularly Italians, Germans and Portuguese. Also Japanese migrants have long been drawn towards agricultural and industrial sectors. Economic migration between near neighbours Paraguay and Argentina has always been relatively high. And political crises at various times have led to migrations from Bolivia, Angola and Lebanon.

Immigration into Brazil has slowed in the last fifteen years. There are 80,000 fewer immigrants living in Brazil than at the start of the century. And during the same period emigration has increased; in 2013, 1.77 million Brazilians lived abroad compared with 0.98 million in 2000. The USA has Brazil's largest population overseas mainly for economic reasons. And many Brazilians of Japanese descent, having strong cultural links, have emigrated to Japan, encouraged by the employment opportunities.

Table 7.11 illustrates Brazil's main international migration links. These estimates are useful indicators of Brazil's migration patterns, past and present.

Table 7.11 Estimated numbers of immigrant populations living in Brazil, and Brazilians living abroad, 2013 (Source: www.pewglobal.org/2014/09/02/global-migrant-stocks/)

Country of birth	Number living in Brazil	Country of destination	Number of Brazilian-born
Portugal	140,000	USA	370,000
Japan	50,000	Japan	370,000
Paraguay	40,000	Portugal	140,000
Bolivia	40,000	Spain	130,000
Italy	40,000	China	120,000
Spain	30,000	Italy	110,000
Argentina	30,000	Paraguay	80,000
Uruguay	20,000	France	60,000
USA	20,000	UK	50,000
China	20,000	Germany	40,000

Economic, political, social and environmental interdependence with countries connected to Brazil by migrant flows

Interdependence between Brazil and countries connected to it by international migration is illustrated by the following three examples.

Portugal

Brazil and Portugal have a long-standing bilateral relationship on a political, social and economic basis. Brazil was a former colony of Portugal and today the Portuguese government still gives special status to Brazilian migrants. For economic migrants, Portugal has become a gateway for entry to the EU. The shared language, ancestry and family ties contribute to the ease with which migrants in both countries can be integrated. Reciprocal migration is supported by the well-developed social diaspora networks in both countries. Meanwhile migrant remittances are an important economic factor for many families.

USA

The links between the USA and Brazil are important in political, socio-economic and environmental terms. The many thousands of low-skilled economic migrants working in the USA are able to remit significant monies, while returning migrants having acquired skills and knowledge are able to contribute to Brazilian development. Highly skilled Brazilians are increasingly finding opportunities to work in the USA especially in the service sector. There are strong links in education and teacher training. The USA has negotiated agreements with Brazil regarding agriculture, trade, finance, education

→

and defence. In addition USAID gives support to Brazil in many environmental projects. These range from practical help such as training Xavante indigenous people to protect their tribal lands from forest fire, to assisting the Brazilian government in designing and implementing laws concerning forest governance and sustainable forest management.

Haiti

Brazil has developed a political, economic and humanitarian relationship with Haiti. The National Immigration Council for Brazil enables Haitian immigrants to obtain visas relatively easily in Haiti (Figure 7.18), and thereby reduce their vulnerability to trafficking networks. This is of great benefit to Haitians who have found it difficult to recover from the devastating earthquake of 2010, which displaced 1.5 million, and the effects of Hurricane Sandy in 2012. The number of immigrants grew from 1681 in 2010 to 11,072 in 2013 and this continues to rise as Haitians attempt to escape the political instability, unemployment, poverty, poor access to education and the country's appalling human rights record, especially gender-based violence. Many intend to join friends and relatives in the southeast of Brazil where low-skilled jobs are available in agriculture and the factories of Rio Grande do Sul and Santa Catarina.

Figure 7.18 Haitians and West Africans receiving visas at the state immigration centre, Rio Branca, Acre, Brazil

The impact of migration on Brazil's economic development, political stability and social equality

Migration has had significant direct and indirect impacts on the development of Brazil.

Economic development

- Waves of immigration in different periods from Japan, Portugal and other European countries including Italy, Spain and Germany have contributed to the growth in agriculture and manufacturing sectors.
- Recent arrivals of highly skilled professionals with employment contracts have contributed to entrepreneurship, innovation and reducing gaps in the labour market.
- Emigration to the USA, Japan, Portugal and other European countries has resulted in migrant remittances to Brazil, used by families in housing improvements, education and general consumption, which has contributed to development at all scales (US$2.4 billion, 2014 – 0.1 per cent of GDP).

Political stability

- Brazil has a stable and democratic political system; it is also a leading member of Mercosur, an important member of G20 and OECD, and is one of the so-called BRICS group of emerging economies.
- Membership of Mercosur, primarily a trading bloc in which there is free flow of trade, capital and labour migration, has helped South American integration and promotion of political stability.
- There are stable political relationships between Brazil and the countries with which it has significant bilateral migrant flows, especially USA, Japan and Portugal.
- Brazil is an important receiver of environmental and political refugees and as a stable government accepts responsibility for their welfare and employment prospects by providing visas and work permits.

Social equality

- According to UNESCO, there are inequalities in Brazilian society between different ethnic groups. Inequalities exist in housing provision, access to services, educational attainment and income; Brazilians of African descent are most affected.
- Inequalities have a spatial perspective with poverty concentrated in rural areas or in the favelas to which the poor migrate.
- There is prejudice and discrimination in the labour market, especially against black and indigenous populations, and this impedes their full economic, political and social development.

Review questions

1 What are the main factors influencing international migration within trade blocs such as ASEAN and Mercosur?

2 Why has there been growth in the total number of females in international migration?

3 Why has there been growth in the number of migrants in South–South corridors?

4 What are the differences between refugees, asylum seekers and economic migrants?

5 What are the main purposes of the new Pakistan National Emigration Policy?

7.3 What are the issues associated with unequal flows of global migration?

Key idea

→ Global migration creates opportunities and challenges which reflect the unequal power relations between countries

The issues associated with the unequal flows of global migration are illustrated by contrasting case studies of the USA and Lao PDR.

Case study: The USA

This case study shows how the USA influences and drives change in the global migration system.

Patterns of emigration and immigration

The USA has a strong influence on global migration. In 2013, there were 41.3 million immigrants living in the USA, 13% of the total population. This included 78 nationalities with over 50,000 people each, and nine with over 1 million. The main contributors to the immigrant population in 2013 were Mexico (28% of all foreign-born residents in the USA), India (5%), China (5%), the Philippines (4%), Puerto Rico (3.5%), and Vietnam, El Salvador, Cuba and South Korea (all 3%).

There has been rapid growth in the number of immigrants entering the USA in the twenty-first century but there are recent signs that the rate of increase is slowing (Figure 7.19).

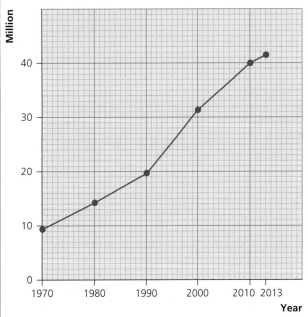

Figure 7.19 USA – size of immigrant population (Source: Migration Policy Institute)

Reasons for the attractiveness of the USA to immigrants from such a wide range of countries include:

● The positive immigration policy and possibility of obtaining a Green Card (becoming a permanent resident)
● Employment opportunities for both low- and high-skilled workers
● Wage differentials and the opportunity to send remittances
● Educational opportunities and access to other services such as health
● The importance of family reunification within the migration policy
● The policy on refugee admission

The USA is a country of net migration gain, but in 2013 there were 2.98 million US citizens living abroad and the number is growing. The main destinations were Mexico, Canada, UK, Puerto Rico, Germany, Australia and Israel. This is explained to a large extent by the return of migrants such as Mexican and other Latin American families and their American-born children. In addition a number of highly skilled workers in education, IT and communications have migrated to countries with political, economic and historical ties such as Canada, UK and other EU countries.

Migration policies

The Immigration and Naturalisation Act is the body of law which governs current immigration policy. (There are periodic amendments including the Immigration and Nationality Acts.) This allows an annual worldwide limit of 675,000 permanent immigrants. There are also a separate number of refugee admissions.

US immigration policy is based on the following principles:

● Reunification of families – 480,000 visas are available per year for family members to join US citizens/legal residents. →

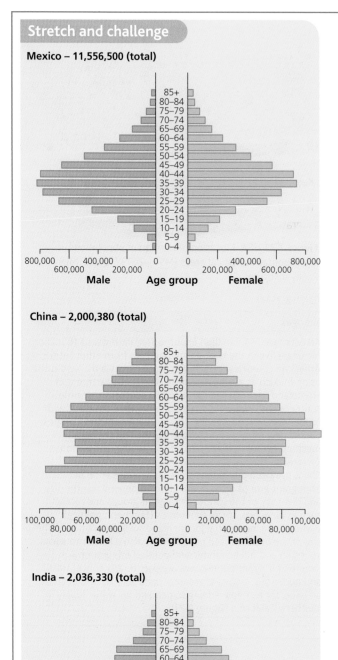

Mexico – 11,556,500 (total)

China – 2,000,380 (total)

India – 2,036,330 (total)

Figure 7.20 Age–sex distribution of US immigrants, 2013 – Mexico, China and India (Source: Migration Policy Institute Data Hub)

Compare the three population pyramids in Figure 7.20. Suggest reasons for the differences and similarities.

- Admission of migrants with skills valuable to the US economy – there are 140,000 visas available (with limits for any one country) for highly skilled workers, normally sponsored by an employer for a specific job offer.
- Protecting refugees – the number of admissions is decided each year; in 2013 the ceiling was 70,000 apportioned by major world region.
- Promoting diversity – the Diversity Immigrant Visa Program makes available 50,000 immigrant visas annually, drawn by random selection of individuals, from countries with low rates of immigration to the USA. In 2014 these went mostly to African countries.
- Humanitarian relief – temporary visas are available each year for relief from natural disasters or ongoing armed conflict.

Details of the 2014 policy are available in a fact sheet produced by the American Immigration Council at www.immigrationpolicy.org (search: How the US immigration system works).

Interdependence with countries linked to the USA by migration

Political, economic, social and environmental interdependence can be illustrated by connections in the world's largest bilateral migrant corridor, between USA and Mexico.

- In 2013, over 11.5 million Mexicans lived in the USA and 1 million Americans in Mexico. Each of the two countries has its largest diaspora living in the other; there is growing social and cultural connectivity.
- Low-skilled Mexicans, many illegal, contribute to the US economy by working in agriculture, construction and low-paid services. Wages are much higher than in Mexico, providing opportunity for remittances – which, via formal channels, amounted to US$22 billion in 2013, 2 per cent of Mexico's GDP.
- Since the formation of NAFTA, bilateral trade between the USA and Mexico has grown significantly. Reciprocal merchandise trade alone accounts for US$1.4 billion per day. Mexican industry has benefited, for example development of aerospace and IT sectors has been boosted mainly by FDI from the USA.
- Political power relations remain imbalanced but increasingly there has been co-ordination and co-operation over issues in common. These include border security, the drug trade, human trafficking and environmental issues such as water scarcity. There is joint management of the Colorado River Basin, and the ecology of the Sonoran desert (see http://sonoranjv.org/programs-projects/)

Opportunities and challenges created by USA international migration

There are far more immigrants in the USA of Hispanic origin than any other group. But the pattern is changing as a result of deceleration of the Mexican influx, tightening of border control, increasing numbers of Mexican returnees, and growth in the numbers of Asian migrants especially from China and India. In 2013, more immigrants entered the USA from each of China and India than Mexico. This trend provides new opportunities and new challenges for the USA (Table 7.12).

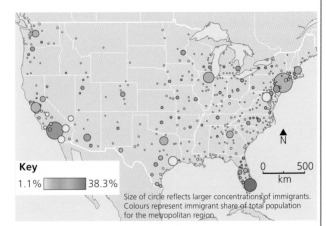

Key

1.1% ▓▓▓▓ 38.3%

Size of circle reflects larger concentrations of immigrants. Colours represent immigrant share of total population for the metropolitan region.

Figure 7.21 USA – location of immigrant populations in metropolitan statistical areas (MSA), 2013 (Source: Migration Policy Institute)

Table 7.12 USA – opportunities and challenges created by international migration

Opportunities	Challenges
• Immigrant populations take many of the low-paid jobs which native-born Americans find unattractive. In 2013, the main employment of Mexican immigrants was in services (31%), natural resources, construction and manufacturing (25%), production and transport (22%) – providing significant contributions to GDP and US economic growth. • US immigration policy also aims to attract highly skilled and well-qualified professionals. This includes 'Persons of extraordinary ability' in the arts, science, education, business or athletics; outstanding professors, researchers and some multinational executives. In 2013, the main employment of immigrants from both China (51%) and India (73%) was in business and science. • Most immigrants to the USA are in the young working age groups (Figure 7.20). This is of economic (e.g. contributing to the tax base) and social benefit in a country where crude birth rate has declined and population is ageing. • Immigrants are consumers themselves which helps job creation and business start-ups, and generates wealth.	• An estimated 11.5 million unauthorised immigrants (6 million Mexican) live in the USA. Many migrants from other Latin American countries are attempting to cross via the land route using Mexico as a transit country. The US Department of Homeland Security has tightened its southern maritime, southern land and west coast borders (www.dhs.gov/) with more Border Patrol Agents. And in December 2015 Congress passed new immigration bills strengthening the legal mechanisms for Refugee Security Screening, and Counter-Terrorism. • There has been uneven progress in the integration of immigrant groups into US society. This depends on the size and diversity of the various immigrant groups and factors such as language, socio-economic attainment, political participation, residential location and social interactions. The size of the unauthorised population is also a barrier to social cohesion and full economic and political integration. • Where immigrant populations are particularly concentrated and numerous, supply of resources and services has become an issue, for example adding to demand for water supply in southern California (Figure 7.21).

🔍 Case study: Laos (Lao PDR)

This case study shows how Laos (officially, the Lao People's Democratic Republic) has limited influence and restricted ability to respond to change within the global migration system.

Laos is a landlocked LIDC of 6.8 million people, located in Southeast Asia (Figure 7.22). It is a poor country (GDP per capita US$1,660, 2014) with 73 per cent employed in agriculture. Although a communist state, it has been a member of ASEAN since 1997 and increasingly has encouraged private enterprise. Net migration loss is high (1.1 migrants per 1000 population, 2014). Remittances bring much needed income, but human trafficking is a major problem for the government.

Patterns of emigration and immigration

In 2013, 1.29 million Laos-born emigrants lived abroad, mainly in Thailand, and only 20,000 foreign-born immigrants, mainly Vietnamese, lived in Laos (Table 7.13).

Table 7.13 Laos migration patterns, 2013

Destination countries	Number of resident Laos-born emigrants	Source countries	Number of foreign-born immigrants
Thailand	930,000	Vietnam	10,000
USA	200,000	China	<10,000
Bangladesh	90,000	Thailand	<10,000
France	40,000	Cambodia	<10,000

Figure 7.22 Laos – main urban areas, River Mekong and neighbouring countries

Figure 7.23 Land grabs drive subsistence farmers into deeper poverty in Laos (Source: Farmlandgrab.org)

The main reasons for emigration to Thailand include:

- Many of those working on the land are subsistence farmers so farming holds no promise of financial gain or personal independence (Figure 7.23).
- There is a lack of alternative occupations in rural areas.
- There is insufficient land available for farming and periodic droughts which lead to food insecurity.
- There is a strong motivation to follow others who have returned from financially successful migrations.
- The daily minimum wage in Thailand is 300 baht compared with 80 baht in Laos; average monthly earnings for Laos migrants in Thailand is approximately 6800 baht.
- For many families in Laos, migrant remittances are the main source of income.
- Low levels of education in Laos mean that many are suited only to unskilled jobs; there is insatiable demand for unskilled labour in Thailand's rapidly growing economy.

And the migrations have been made easier by:

- familiarity with Thai culture and language
- improved access across the Mekong
- use of 'brokers' to reach the Thai border and access employment in Thailand.

Migration to Laos from neighbouring countries, especially Vietnam, is largely the result of employment opportunities in the government-driven, World Bank-funded programmes linking the countries in the region by highways, bridges and tunnels. Many Vietnamese immigrants work in construction and mining.

Migration policies

Laos is a source country for human trafficking. In 2013, 36 per cent of the population were under fifteen, therefore there are an increasing number of entrants into the work force in a country where opportunities are limited. The vulnerability of young migrants to forced labour and sex exploitation in Thailand is of major concern to the government and international organisations. A recent development is the 'trafficking' of under-age footballers from Liberia to Laos to play for Champasak United (www.bbc.co.uk/news/world-africa-33595804).

Government policies have been ineffective; many young migrants do not obtain the required passport and risk fines on return. However, the Lao PDR Ministry of Labour and Social Welfare, and the Ministry of Public Security now work in co-operation with the IOM, UN agencies such as ILO, UNICEF and UN Women and international NGOs such as CARE International (Cooperative for Assistance and Relief Everywhere), Save the Children and World Vision to implement a more stringent anti-trafficking policy.

The National Plan of Action for Human Trafficking led by the Lao PDR government has three strands:

- Prevention – awareness campaigns, education, child protection, alleviation of poverty reducing the need to migrate
- Protection – repatriation and reintegration of returning migrants, including shelters for women who may need counselling
- Prosecution – investigation of trafficking networks, training border officials, strengthening legal framework

As a member of ASEAN, Laos is also subject to its laws on migration. The newly formed ASEAN Economic Community (AEC) aims to allow freer movement of skilled labour from 2015. Mutual Recognition Agreements allow professionals employed in nursing, medicine, dentistry, architecture and tourism, after five years of working in their country of origin, to have greater freedom of movement between ASEAN countries.

Interdependence with countries linked to Laos by migration

ASEAN countries are becoming increasingly interdependent, illustrated by the Laos–Thailand relationship:

- The Laos–Thailand migration corridor is dominated numerically by the outward flow of unskilled Laotians to work in Thailand. Their contribution to the Thai economy is in construction, agriculture, fisheries and factory work; remittances assist development in Laos.

- The Laos–Thailand Cooperation Committee has been established; completion of the latest Friendship Bridge has helped to strengthen communication and trade – Thailand is the principal access to the sea for Laos. Thailand has funded a large health service development and drugs treatment centre in Laos.
- Laos and Thailand work together as members of the Mekong River Commission to manage flooding and economic activities in the basin and the Don Sahong hydro-power project (www.mrcmekong.org).
- Laos signed agreements to build rail links between Thailand and Vietnam in 2012, with which it also has special relations, including a high-speed rail link to China – all of which will open up Laos to development.
- Laos and Thailand are members of the Coordinated Mekong Ministerial Initiative against Trafficking (COMMIT).

Opportunities and challenges

Table 7.14 Laos – opportunities and challenges created by migration

Opportunities	Challenges
<ul><li>The migration corridor between Laos and Thailand is one of the largest within ASEAN. It has helped to stimulate political and economic co-operation in terms of trade, investment, development projects and security.</li><li>Bilateral relations with Vietnam also extend beyond reciprocal labour migration. There is economic co-operation, with Vietnam involved in over 400 investment projects in Laos.</li><li>Migrant remittances are very important to the life of returnees and their families. In Laos 22% of families live below the poverty line and the economic impact of this money on local and national development ranges from purchase of simple domestic appliances to agricultural machinery.</li><li>Political stability is improving between Laos, Thailand and Vietnam.</li></ul>	<ul><li>Most economic migrants from Laos are low skilled, of limited education and under 18 at their first migration. Many travel illegally and are vulnerable to human trafficking, forced labour and exploitation. Laos government policy has been difficult to implement and, for its success, depends on transnational governance by organisations such as IOM, ILO and Civil Society.</li><li>There is loss of skilled labour such as carpenters and mechanics to Thailand. This is set to increase if wage differentials remain high and there is freer movement in the ASEAN Economic Community.</li><li>The Laos garment industry is the largest sector of manufacturing employment, but its growth depends on improved working conditions in the factories of the Laos capital, Vientiane, and retention of the many young female workers who may leave seeking higher wages in Thailand.</li></ul>

 Review questions

1 What are the basic principles of USA immigration policy?
2 What is the basis of interdependence between the USA and Mexico?
3 What are the challenges of international migration for the USA?

4 How does the Lao PDR government attempt to tackle human trafficking?
5 What opportunities does international migration create for Laos?

Practice questions

Structured/data response

1 a) Study **Table 1** which shows the largest immigrant populations of France and New Zealand, 2013.

Table 1 The largest immigrant populations of France and New Zealand, 2013 (Source: www.pewglobal.org/2014/09/02/global-migrant-stocks)

Country of birth	Numbers living in France (million)	Country of birth	Numbers living in New Zealand (million)
Algeria	1.46	UK	0.31
Morocco	0.93	China	0.11
Portugal	0.64	Australia	0.08
Tunisia	0.39	Samoa	0.07
Italy	0.38	India	0.06
Spain	0.32	South Africa	0.05

i) Suggest **two** ways in which the immigrant populations, shown in Table 1, might influence flows of ideas. [2 marks]

ii) Explain **one** economic benefit of immigration for the destination countries in Table 1. [3 marks]

iii) Evaluate **one** alternative presentation technique that could be used to illustrate the data in Table 1. [4 marks]

b) With reference to a **case study**, examine the impacts of migration on social equality in an emerging and developing country (EDC). [8 marks]

Essay

2 'International migration creates opportunities for development.' To what extent do you agree with this statement? [16 marks]

Chapter 8

Human rights

8.1 What is meant by human rights?

> **Key idea**
> → There is global variation in human rights norms

Understanding what is meant by human rights

Human rights are the basic rights and freedoms to which all human beings are entitled. They are applicable at all times and in all places and they protect everyone equally, without discrimination.

The United Nations Office of the High Commissioner for Human Rights states:

> *Human rights are rights inherent to all human beings, whatever our nationality, place of residence, sex, national or ethnic origin, colour, religion, language, or any other status. We are all equally entitled to our human rights without discrimination.*

Definitions and understanding of human rights and the issues that surround them in the twenty-first century are derived from The Universal Declaration of Human Rights (UDHR). This was one of the most significant events in human rights history when it was adopted by the United Nations General Assembly in 1948.

Two examples of the statements found within the Declaration are:

- Article 5: No one shall be subjected to torture or to cruel, inhuman or degrading treatment or punishment.
- Article 9: No one shall be subjected to arbitrary arrest, detention or exile.

Other examples, and there are 30 in all, may be found at www.un.org/en/documents/udhr.

Since that time it has become evident that many of the principles set out in 1948 have not been adhered to uniformly. From a geographical perspective these violations of human rights have occurred in many different parts of the world, on every continent, in advanced countries as well as in developing countries (LIDCs and EDCs), and at different scales from individuals to large-scale groups. Examples include use of child labour, people trafficking, genocide and modern slavery.

Globalisation has contradictory impacts on human rights. Transnational integration and increased mobility have had the effect of simultaneously strengthening and diminishing the protection of human rights.

- On the one hand this has enhanced the ability of **civil society** to work across borders and to promote human rights.
- On the other it has enabled some organisations to gain power and perpetrate violations.

Geographical patterns of socio-economic inequality are closely associated with inequalities in respect for human rights. Many development programmes and the steps towards achieving the UN's Millennium Development Goals (MDGs) have been human rights led.

The website of the UN Office of the High Commissioner for Human Rights www.ohchr.org/EN/Issues/Pages/ListOfIssues.aspx is a useful resource for further investigation of human rights issues.

Human rights norms

Human rights norms represent ways of living that have been inculcated into the culture of a country or area over long periods of time. They are the foundation of human rights. It was on the basis of established customs and norms drawn from all cultures, religions and philosophies across the world that the UDHR was devised. These norms are based on the moral principles that underpin the universally accepted standards of human behaviour.

The statements set out in the UDHR are generally accepted as international human rights norms. And although this is a non-binding resolution, human rights are in fact protected by international law.

International human rights law sets out the obligations of state governments. By signing **international treaties**, it is the duty of states to respect, protect and fulfil international human rights. Governments that ratify or sign treaties therefore have to put in practice domestic measures and legislation which are compatible with that treaty.

There are growing numbers of human rights norms, laws and treaties or conventions. The most widely ratified of all international human rights is the UN Convention on the Rights of the Child (UNCRC). This Convention, signed by governments worldwide, has been designed to change the ways in which children are viewed and treated. The rights embedded within it describe what a child needs in order to survive, grow and achieve full potential. It explains the responsibilities of adults and governments to ensure that children everywhere can enjoy all their rights. This forms the basis of UNICEF's work today.

Nevertheless, there is still significant global variation in deaths of young children. The **infant mortality rate** (IMR) is defined as the annual number of deaths of infants under the age of one per 1000 live births. And the 2013 figures exemplify the global range, from Mali 106.5 and Chad 91.9 to the UK 4.5, Czech Republic 3.7 and Italy 3.3 (Figure 8.1). Afghanistan also has a particularly high IMR.

Most of these infant deaths could be prevented. The UN view is that if a country is not doing what it can to prevent these deaths it is not meeting its legal and moral obligations. It is not upholding the rights of its most vulnerable people. Therefore infant mortality is not just a health matter but a human rights concern.

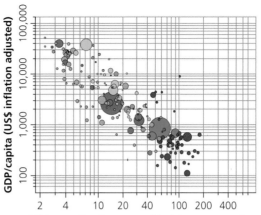

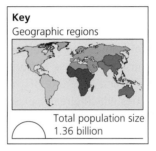

Figure 8.2 The relationship between child mortality and GDP per capita, 2011

> ### → Activities
>
> 1 Describe the global pattern of infant mortality rates shown in Figure 8.1.
> 2 Use Figure 8.2 to describe the relationship between child mortality and GDP/capita.
> 3 Suggest reasons why the right to life for infants is not upheld in many parts of the world.

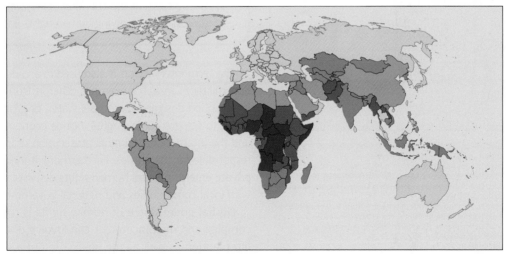

Key
Number of deaths of infants under age 1 per 1000 live births.
■ 75+ ■ 50–74 ■ 25–49 ■ 15–24 ☐ <15

Figure 8.1 Global pattern of infant mortality rates, 2013

241

Intervention

Intervention includes the use of military force by a state or group of states in a foreign territory in order to end gross violation of fundamental human rights of its citizens. This type of intervention in pursuit of humanitarian objectives is often referred to as **humanitarian intervention**.

The UN Security Council is the only body that can legally authorise use of force. Irrespective of this, the entire process of military intervention is controversial. It can be effective in stopping the violations. This can have immediate benefits for local communities and contribute to longer term socio-economic development and political stability. But it can have unintended negative impacts. These include injuries and deaths of civilians, loss of homes and population displacement. It may also cause an increase in human rights abuses, further injustices and widening of the socio-economic inequalities which already exist within the country.

United Nations involvement takes many forms and its peacekeeping, political and peacebuilding missions (see Figure 8.3) serve many purposes, not least concerning human rights violation. If the international community is called upon, the UN Security Council establishes a mandate so that its workers and troops can be authorised and drawn from a wide range of member states. Usually military presence helps protect citizens from human rights abuse, with non-use of force except in self-defence. In addition a UN human rights team works in the area to protect and promote human rights. Its task is to monitor the situation, attempt to empower populations to assert their human rights, enable governments to implement their human rights obligations and strengthen rule of law.

The UN also co-ordinates the input of a wide range of agencies and organisations in the area affected. These include:

- regional organisations such as the North Atlantic Treaty Organisation (NATO), the Organization for Security and Co-operation in Europe (OSCE) and the ASEAN Intergovernmental Commission on Human Rights
- non-governmental organisations such as the International Committee of the Red Cross (ICRC) and Oxfam
- public-private partnerships such as the Gavi Alliance (Global Alliance for Vaccines and Immunisation).

> ### Activities
>
> The North Atlantic Treaty Organization (NATO) was established in 1949 with twelve member countries from Europe and North America; in 2015 there were 28 member states.
>
> 1 Access the NATO website www.nato.int and investigate the purpose of the organisation and the different ways in which it attempts to achieve its aims.
> 2 a) Use NATO's interactive map (Figure 8.4) to examine specific details of its work in any one area (current operations, videos, statistics, partners): www.nato.int/nato-on-duty/index.html.
>   b) To what extent has this kind of intervention been successful in resolving human rights issues?

The term intervention is also used in a wider non-military sense. For example, other instruments of intervention designed to compel states, or groups within them, to respect human rights include economic sanctions and the international criminal prosecution of individuals responsible for the abuses. Furthermore, it includes NGOs, private enterprises and human rights activists working with local communities and national governments.

Global governance of human rights is therefore complex and multifaceted: it can involve direct physical intervention as well as the application of a growing number of human rights norms, laws and treaties or conventions, plus the work of civil society. Effective intervention depends on their interaction and co-ordination at all scales.

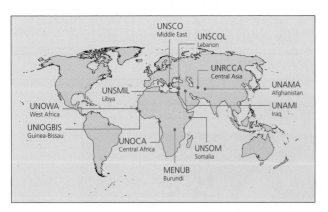

MENUB	United Nations Electoral Observation Mission in Burundi (established 2015)
UNAMA	United Nations Assistance Mission in Afghanistan (established 2002)
UNAMI	United Nations Assistance Mission for Iraq (established 2003)
UNIOGBIS	United Nations Integrated Peacebuilding Office in Guinea-Bissau (established 2010)
UNOCA	United Nations Regional Office for Central Africa (established 2011)
UNOWA	United Nations Office for West Africa (established 2001)
UNRCCA	United Nations Regional Centre for Preventive Diplomacy for Central Asia (established 2008)
UNSCO	Office of the United Nations Special Coordinator for the Middle East Peace Process (established 1999)
UNSCOL	Office of the United Nations Special Coordinator for Lebanon (established 2007)
UNSMIL	United Nations Support Mission in Libya (established 2011)
UNSOM	United Nations Assistance Mission in Somalia (established 2013)

Figure 8.3 Ongoing political and peacebuilding UN missions

Figure 8.4 Interactive map of NATO operations, 2015. Access similar maps on the NATO website at www.nato.int/nato-on-duty

Geopolitics

The term **geopolitics** refers to the global balance of political power and international relations. The pattern of political power is closely related to economic power especially in terms of the relative wealth and international trade strength of nations and groups of nations.

Historically there have been a number of 'geopolitical transitions' in which geopolitical world order or power has shifted. The most recent has been the ending of the Cold War. This situation existed from 1946 to 1989 in which period the USA and the USSR were the two dominant superpowers.

Contemporary geopolitical power has a very uneven spatial distribution and is viewed from different perspectives:

- the USA is the only superpower. It may have lost its place to China as the world's leading trading nation but it remains dominant militarily and politically
- there are inequalities in power between individual states depending on wealth, political strength and development. According to the International Monetary Fund (IMF) there are the powerful **advanced countries** (ACs), the increasingly influential **emerging and developing countries** (EDCs) and the peripheral economies of the **low-income developing countries** (LIDCs)
- there are supranational political and economic organisations such as the UN, EU, ASEAN and OPEC, which exert greater geopolitical influence than their individual member states

- there are the effects of globalisation in which trans-state organisations such as multinational corporations (MNCs) have considerable influence on the countries in which they invest.

The geopolitics of intervention in human rights issues requires an understanding of the:

- political composition of the groups of countries and organisations that are involved in intervention
- nature of the intervention itself
- reasons why intervention has been deemed necessary
- characteristic features of the country, government and peoples affected
- possible political, socio-economic and environmental consequences of intervention/global governance
- complexity of human rights issues and their spatial patterns.

> **Key idea**
> → Patterns of human rights violations are influenced by a range of factors

Current spatial patterns of human rights issues including forced labour, maternal mortality rates and capital punishment

Article 3 of the UDHR states that 'everyone has the right to life, liberty and security'. Forced labour, maternal mortality rates and capital punishment are all connected to this most basic of human rights. However, maps and statistics of their global patterns illustrate significant spatial variation in their prevalence.

Forced labour

The term **forced labour**, as described by the International Labour Organization (ILO), refers to:

situations in which persons are coerced to work through the use of violence or intimidation, or by more subtle means such as accumulation of debt, retention of identity papers or threats of denunciation to immigration authorities.

It is estimated that globally 21 million people are victims of forced labour – 11.4 million women and girls and 9.5 million men and boys. Nineteen million of these are exploited by private individuals or enterprises and 2 million by state or rebel groups. In the private economy this generates US$150 billion in illegal profits annually.

Forced labour includes:

- children who are denied education because they are forced to work
- men unable to leave work because of debts owed to recruitment agents
- women and girls exploited as unpaid, abused domestic workers.

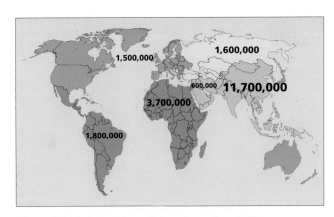

Figure 8.5 Victims of forced labour by region, 2012

The global distribution of these estimates is very uneven (Figure 8.5). Southeast Asia has the highest overall incidence but significantly no region is unaffected.

Activities

Forced labour is a major element of modern slavery (Figure 8.5).

1 Describe the global pattern of modern slavery shown in Figure 8.6.
2 Use the information in Figure 8.7 to suggest reasons for this pattern.

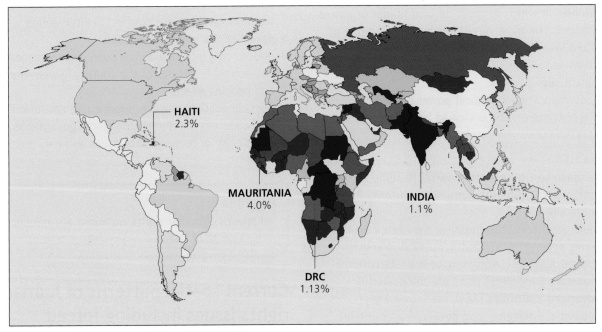

Key

Percentage of the population that is enslaved (2014)

| 0.04 | 0.2 | 0.3 | 0.4 | 0.7 | 0.75 |

Figure 8.6 Percentage of each country's population that is enslaved, 2014

Factors that influence global variations of forced labour

ECONOMIC	POLITICAL
• Poverty • Lack of economic opportunities and unemployment • Low wages • Subsistence farming • Migration and seeking work	• Political instability • Conflict • Breakdown of rule of law • Corruption • State sponsorship of modern slavery, e.g. cotton harvest in Uzbekistan • High levels of discrimination and prejudice

SOCIAL	ENVIRONMENTAL
• Gender inequality • Age, especially children • Entire families enslaved through bonded labour, e.g. construction, agriculture, brick making, garment factories in India and Pakistan • Women and children trafficking for sexual exploitation, e.g. through organised crime in Europe from Nigeria • Indigenous people	• Escaping climate-related disasters including food and water shortages • Hazardous working conditions in open mines

Figure 8.7 Factors that contribute to vulnerability to forced labour

Stretch and challenge

Investigate the factors that contribute to the prevalence of modern slavery in any one country and assess their relative significance. The Walk Free Foundation Global Slavery Index is a useful source: www.globalslaveryindex.org.

Maternal mortality rate (MMR)

The definition of maternal mortality used by the World Health Organization (WHO) states: 'the death of a woman while pregnant or within 42 days of termination of pregnancy ... from any cause related to or aggravated by the pregnancy or its management'. The maternal mortality rate (MMR) is the annual number of these deaths per 100,000 live births.

In 2013, globally 289,000 women died during and following pregnancy and childbirth. Most of these deaths occurred in developing countries (Figure 8.8). The worst affected countries were all in sub-Saharan Africa – Sierra Leone (1100 per 100,000 live births), Chad (980) and the Central African Republic (880). The lowest figures were in the more developed countries in Europe, such as Belarus (1) and Italy (4), and in North America.

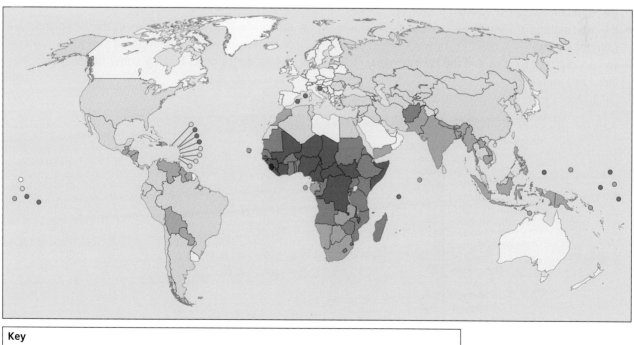

Key

Annual number of female deaths per 100,000 live births

☐ <20 ☐ 20–99 ☐ 100–299 ☐ 300–499 ☐ 500–999 ■ ≥1,000 ☐ Data not available

☐ Not applicable ☐ Population <100,000 not included in the assessment

Figure 8.8 Maternal mortality rates, 2013 (Source: WHO)

Factors that influence global variations of MMR

The global inequalities in MMR can be explained by:

- access to treatments for pregnancy and birth complications, especially emergency care
- quality of medical services, especially provision of skilled attendance at birth
- level of political commitment and government investment
- availability of information and education
- cultural barriers which affect discrimination
- poverty.

The vast majority of these deaths are preventable therefore this is not just a matter of development but of human rights. These rights are all legally protected by international human rights treaties including:

- the Convention on the Elimination of All Forms of Discrimination Against Women (CEDAW)
- the International Covenant on Economic, Social and Cultural Rights
- various regional treaties and the laws of many individual states.

Capital punishment

The death penalty is a denial of the most basic of human rights, i.e. that states must recognise the right to life. The UN General Assembly has called for an end to the death penalty. Human rights organisations such as Amnesty International and Human Rights Watch campaign against its imposition as a fundamental breach of human rights norms.

Nevertheless, according to Amnesty International, in 2014 there were at least 607 executions globally and 2466 people were sentenced to death in 55 countries. See: www.viewsoftheworld.net/?tag=death. There are significant global inequalities in capital punishment (Figure 8.9).

Factors that influence global variations of capital punishment

The inequalities can be explained by factors such as:

- differences between countries in the range and type of crimes for which it is imposed
- the incidence of its legality under national law

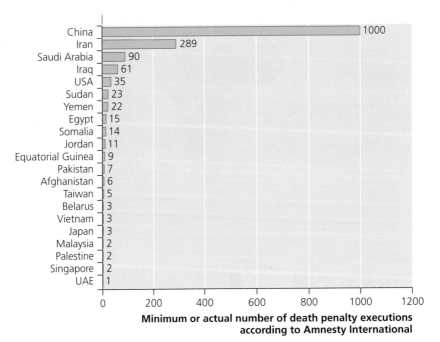

Figure 8.9 Death penalty executions, 2014

- the increase in the number of countries in which it is being abolished
- its reinstatement in some countries for threats to state security and public safety posed by terrorism
- the number of commutations and pardons (granted in 28 countries in 2014).

✔ Review questions

1 Identify two international human rights norms stated in the UDHR.
2 What do you understand by the term human rights norm?
3 What is meant by intervention in human rights issues?
4 What is meant by forced labour?
5 What factors account for variation in the global pattern of maternal mortality rates?
6 Why do numbers of executions vary spatially throughout the world?

8.2 What are the variations in women's rights?

Key idea
→ The geography of gender inequality is complex and contested

Gender inequality is the unequal treatment of individuals based on their gender. It is a situation in which women and men do not enjoy the same rights and opportunities across all sectors of society specifically because they are a woman or a man.

The extent of the rights denied to women or men can be measured using a range of indices. The statistics show that in many instances it is females that do suffer the most. This is a major obstacle to development. Yet this is not just an issue for women; increasingly international organisations and civil society are involving men and boys as well as women in their education programmes concerning the roles of both genders.

Global patterns of gender inequality are closely related to disparities in respect for the rights of women. This is demonstrated by the Global Gender Gap Index (GGGI) devised by the World Economic Forum (WEF) (Figure 8.10).

Skills focus

1 Using Figure 8.10, describe the global pattern of the gender gap.
2 Evaluate the effectiveness of the WEF's choropleth map used to show the Global Gender Gap Index.
3 To what extent is there a relationship between gender inequality and development? Use the data in Table 8.1 to calculate Spearman's rank correlation coefficient (see Chapter 15, Geographical Skills, pages 541–42). Evaluate the outcome.

There have been great improvements in protecting and promoting human rights in the twenty-first century but women and girls continue to experience harmful gender-based discrimination and exploitation. This is true of all countries but it is particularly prevalent in the poorer economies.

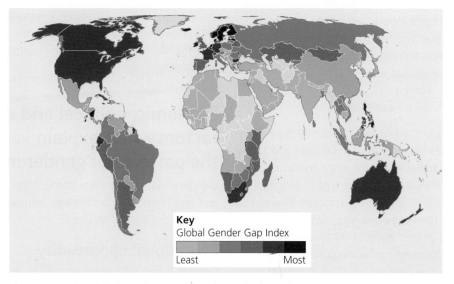

Key
Global Gender Gap Index

Least — Most

Figure 8.10 The Global Gender Gap Index 2014. This map shows national gender gaps based on economic, political, education and health criteria. In the key: Most = highest index score, i.e. least inequality; Least = lowest index score, i.e. greatest inequality (see Table 8.1). Grey = no data (Source: WEF)

Table 8.1 HDI (Human Development Index) and GGGI for selected countries, 2014

Country	GGGI	HDI	Country	GGGI	HDI
Iceland	0.859	0.895	Honduras	0.693	0.617
Finland	0.845	0.879	Montenegro	0.693	0.789
Norway	0.837	0.944	Russian Federation	0.692	0.778
Sweden	0.816	0.898	Vietnam	0.691	0.683
Denmark	0.802	0.900	Senegal	0.691	0.485
Nicaragua	0.789	0.614	China	0.683	0.719
Rwanda	0.785	0.506	India	0.645	0.586
Ireland	0.785	0.899	Guinea	0.600	0.392
Philippines	0.781	0.660	Morocco	0.598	0.617
Belgium	0.780	0.881	Jordan	0.596	0.745
Switzerland	0.779	0.917	Lebanon	0.592	0.765
Germany	0.778	0.911	Ivory Coast	0.587	0.452
UK	0.738	0.892	Iran	0.581	0.749
Chile	0.697	0.622	Mali	0.577	0.407
Bangladesh	0.697	0.558	Syria	0.577	0.658
Italy	0.697	0.872	Chad	0.576	0.372
Brazil	0.694	0.744	Pakistan	0.552.	0.537
Romania	0.693	0.785	Yemen	0.514	0.500

The complexity and contested nature of gender inequality is demonstrated by the remaining challenges:

- Forced marriage, often involving children
- Trafficking into forced labour, including sex slavery
- Access to education and health care
- Employment opportunities and political participation
- Wage equality for similar work to men
- Violence against women
- Access to reproductive health services

The denial of rights in these areas is often derived from entrenched attitudes towards women by men and the deeply rooted patriarchal norms of some countries or regions. It is hardly surprising that two Millennium Development Goals (MDGs) were specifically directed towards improving the lives of women, female empowerment and gender equality.

The UN has established conventions such as the Convention on the Elimination of all Forms of Discrimination Against Women (CEDAW) and is doing much work to strengthen the rule of law and

to reinforce norms to outlaw gender discrimination. International treaties, recommendations and declarations set out the obligations of national governments to protect the rights of women. And many NGOs such as the International Center for Research on Women (ICRW) are working within local communities to resolve these issues. Yet women and girls continue to be at risk in many areas.

Economic, political and social factors which explain variation in the patterns of gender inequality

Gender inequality is seen in the areas of educational and employment opportunities and access to reproductive health services.

Educational opportunity

Gender inequality in education in many parts of the developing world tends to favour males (Figure 8.12). There have been significant improvements in

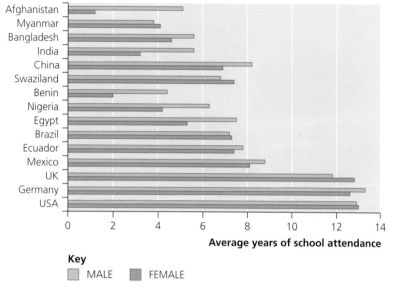

Figure 8.12 Average years of school attendance for population aged 25 years and over, 2013 (Source UNDP)

female enrolment, especially under the MDG to achieve universal primary education. But challenges remain, and girls still suffer severe disadvantages and exclusion in education in the poorer countries, in particular in rural areas and among the rural poor. The obstacles to female secondary school participation are greatest in sub-Saharan Africa and South Asia (Table 8.2).

Table 8.2 Average secondary school attendance, 2008–13

Region	% of female population	% of male population
Sub-Saharan Africa	30	34
South Asia	47	56

Female education is the key to empowering women and achieving gender equality in all respects. It helps women to move into the labour market and to increase the production capacity of the labour force. In countries where there has been increasing parity in education, total fertility rates, population growth rates and infant mortality rates have fallen. Family health and child nutrition have improved and there has been a significant effect on poverty reduction.

The UN has established the Girls' Education Initiative, for which UNICEF is the lead agency. Many NGOs are involved in education partnerships in poorer countries and increasingly MNCs are assuming a role in education as part of their

Corporate Social Responsibility. But there are still many challenges and barriers which marginalise women and girls in education in the developing world (Table 8.3).

Table 8.3 Factors influencing female educational participation in developing countries

- Costs may prohibit all the children in a family from continuing in secondary education; it is usually the girls that suffer.
- Household obligations often fall on the eldest girl when the family burden of work increases because of male out-migration.
- In patriarchal systems, female education may only be of benefit to the family into which a daughter marries and not to her own family.
- Negative classroom environments in which girls face violence, exploitation and corporal punishment
- Inadequate sanitation in schools which do not offer private or separate latrines
- Insufficient numbers of female teachers
- The impact of girls being exploited for child labour
- The prevalence of child marriage
- The differing levels of support for education by different religions
- Early pregnancy
- Inadequate legislation
- Insufficient government investment

Activities

1 Attempt to classify the factors outlined in Table 8.3 as political, social or economic.
2 What are the difficulties you encounter in attempting this classification?

Figure 8.13 Factors affecting female reproductive health in developing countries

Access to reproductive health services

Factors that affect female reproductive health in developing countries are outlined in Figure 8.13. These are mostly social factors, often related to cultural practices such as female genital mutilation (FGM) which are discriminatory and closely linked to human rights. Female reproductive health rights are violated when women and girls are denied access to health care services. Girls and young women living in poor communities in the developing world are most at risk. Economically and socially disadvantaged women are less likely to gain access to health services, information and education, and to become empowered to negotiate safer sex and decide on the number and spacing of their children.

Girls in poor communities face the additional obstacles of early marriage and early child bearing. In developing countries, one in three teenagers marry before they are eighteen and one in nine before fifteen years of age. Every day there are approximately 20,000 births to girls under eighteen. This has a cascading effect on their lives. Education ends, job prospects diminish, they become vulnerable to poverty and exclusion, health suffers and maternal mortality rates in this age category are among the highest.

A number of international organisations and charitable institutions are involved in resolving these issues – for example, the Office of the High Commissioner for Human Rights; and the 'on the ground' work of NGOs such as ICRW, Amref Health Africa and Womankind (Figure 8.14), which are working closely with the communities worst affected.

Figure 8.14 Women For Change and Womankind is working with local communities to empower women in rural Zambia. Detailed statistics are available at www.unicef.org/infobycountry/zambia_statistics.html (Source: Womankind)

Employment opportunity

It is the right of both women and men to have equal access to employment opportunities. The Labour Force Participation Rate is an index of equality used in the UNDP. This is the ratio of females to males within a country's working population (fifteen years and over) that engages in the labour market either by working or actively looking for work.

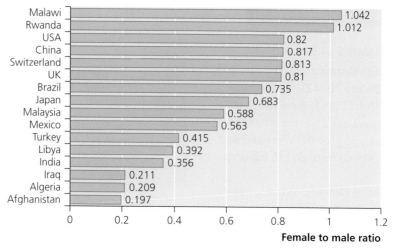

Figure 8.15 Global labour force participation rates (Source: UNDP Data)

There is significant global variation in this index (Figure 8.15), ranging from Malawi, where the value is higher than one, indicating greater female participation than male, to Afghanistan where 0.197 represents very limited female access to the labour market relative to men. Countries with a high HDI such as USA, UK and Switzerland have a high Labour Force Participation Ratio, but none have achieved female–male employment parity. India has a relatively low ratio compared with other rapidly emerging economies. North African and Middle East states have the lowest ratios.

The global pattern is therefore complex but factors that affect the spatial variation include differing:

- social norms, for example where primary responsibility in securing household income through employment is attributed to men and where women are expected to devote time to unpaid domestic care
- cultural beliefs and practices of religious or social groups
- levels of governmental and company support for child care
- degrees to which equal opportunity is safeguarded by law
- social acceptance of women as contributors to household income
- gender-based norms that shape the educational and job decisions of women and men
- levels of discrimination by employers
- sectoral structure of the labour market.

Changes in the ratio of females to males in the labour force of selected countries are shown in Figure 8.16.

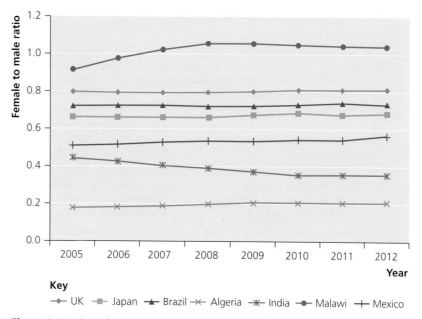

Key

UK — Japan — Brazil — Algeria — India — Malawi — Mexico

Figure 8.16 Labour force participation rates, 2005–12 (Source: UNDP Data)

Case study: Women's rights in India

Gender inequality issues in India

India has one of the world's fastest growing economies. Between 2008 and 2014 GDP per capita increased from US$863 to US$1165; there has been significant change in sectoral employment structure with decline in agriculture and growth in services; and total value of exports increased from US$15 billion to US$28 billion.

However, in the WEF Global Gender Gap Index, India is ranked 114th out of 142 countries and in the UN's Gender Inequality Index, 135th out of 187. Gender-based inequalities place women at a disadvantage in all aspects of Indian society, economically, socially and politically (Table 8.4).

Activities

1 Use Figure 8.17 to describe the pattern of the crime rate against women in India.
2 Suggest possible reasons to explain the spatial inequalities.

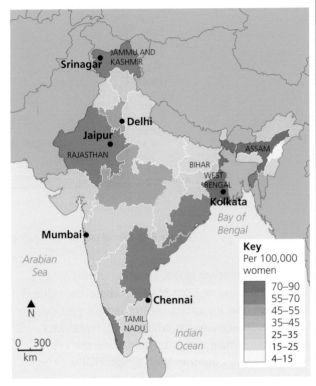

Figure 8.17 Crime rate against women, India, 2012 (Source: M. Tracey Hunter in Wikimedia)

Table 8.4 Gender inequality issues in India

Violence against women	This is underpinned by persistent social norms. According to UNICEF, 52% of women in India think it justifiable for a man to beat his wife. UNICEF research also indicates that domestic violence is tolerated by communities, and to some extent by the state. There are deeply entrenched patriarchal and customary practices which are perpetrated by husbands, in-laws and other family members. There have been increases in dowry killings and increases in rape and violence outside the home, for example against women using public transport.
Modern slavery	It is estimated that 14.3 million people, mostly women and girls, were subject to modern slavery in India in 2014. This includes trafficking for sexual exploitation, early forced marriage and forced labour.
Property ownership	Women have very few rights in ownership of land and property and, in practice, inheritance is invariably patriarchal.
Employment opportunity	There is gender inequality in the labour market in India. Women have limited access to employment opportunities and are often expected to remain at home, raise children, conduct domestic chores and work in subsistence farming. This is especially true of the rural poor. Even women who have received a full secondary or tertiary education still do not enter the work force, finding it hard not to conform to social norms of marriage and immediate motherhood.
Discrimination in the workplace	Discrimination in the workplace is common practice. Maternity benefits are denied by many employers and most women do not return to work after childbirth. In Delhi only 25% of married women returned to work after childbirth, including those who can afford to pay for child care. The social conditioning is that it is their responsibility to bring up children.
Political participation	Gender inequality is perpetuated by the lack of women in government at national, provincial and local government level. Women have poor representation in India's parliament with only 11% in Lok Sabha (lower house) and 10.6% in Rajya Sabha (upper house).
Access to health care	Gender discrimination in health care is closely related to the cultural norms in Indian society in which women have little influence. According to CARE International, nearly a third of all households in Bihar do not access government health services.
Access to education	Nationally 70% of girls attend primary school, but the figure is much lower at secondary. Strong opposition from families and communities, poverty and cultural beliefs are restrictive factors.

The consequences of gender inequality on society in India

Women have been subject to murder and disfigurement, mostly by burning, when their family cannot meet the demands for a dowry by the husband. Often marriage is used by the husband to obtain property and other assets from his wife and her family. In 2012 there were 8233 dowry-related deaths in India, affecting almost every state, with highest rates in the north and west in Uttar Pradesh and Orissa (between 1 and 1.4 per 100,000 women) and least in the south and east such as in Gujarat, Karnataka and Kerala. The effect is disproportionate on the lower castes and tribes, including Dalit and Adivasi women.

Women can be subject to honour killing by their family members for not agreeing to arranged marriage or for not conforming with other gender norms. Many women are beaten in the domestic home, are subjected to sexual violence and lead a life of servitude and harassment.

Women's health is at risk during and after pregnancy; there is high incidence of maternal mortality and morbidity and this places existing children at further risk. Limited education and poverty especially among the rural poor has adverse effects on maternal and child nutrition and contributes to India's high infant mortality rate (43.2 under age one per 1000 live births, 2014). Furthermore, in this patriarchal society, women have been subjected to sex-selective abortions in the desire for male offspring. And limited access to the work force further adds to dependency on their husbands. Added to this, many women have been coerced into sterilisation schemes, sometimes with disastrous consequences (Figure 8.18).

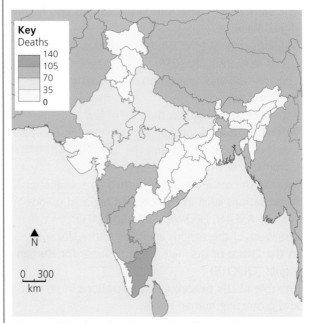

Figure 8.18 India – distribution of sterilisation deaths since 2008 (Source: Quartz India qz.com)

Evidence of changing norms or strategies to address gender inequality issues in India

The problems faced by women in India are the result of deeply entrenched gender-based norms which favour men. The problems persist, but slowly change is occurring as a result of intervention by strengthening the rule of law, increasing numbers of laws and treaties, and by the work of NGOs in local communities, improving education and through the influence of the media.

India has ratified a number of international human rights treaties. This means that the government agrees to incorporate the principles and international laws bound by these treaties into its own national laws and to uphold and implement them. Examples include the International Convention on Civil and Political Rights and the International Convention on Economic, Social and Cultural Rights. Together with the Universal Declaration of Human Rights (UDHR), these constitute the International Bill of Human Rights.

The Indian government has also joined UN Treaty Bodies such as the Committee on the Elimination of Discrimination Against Women (CEDAW). And in consultation with the Indian government, the UN Special Rapporteur on Violence against Women visited the country in 2013. Other examples can be found in the UN's India section of the OHCHR at www.ohchr.org/EN/countries/AsiaRegion/Pages/INIndex.aspx.

At national and state level the Indian government has passed many Acts of Parliament designed to address women's rights. For example:

- The Prohibition of Child Marriage Act, 2006
- The Dowry Prohibition Act, 2008
- Protection of Women from Domestic Violence Act, 2005
- Sexual Harassment of Women at Workplace Act and Rules, 2013

However, long-term shifting of norms and effective application of the law requires a more practical, 'on the ground' approach:

- In 2014 the Indian Ministry of Home Affairs launched an anti-trafficking portal. This involves expansion of anti-trafficking police units, specific training of police units, more accurate reporting of crime, a victim support programme including temporary accommodation and co-operation with Bangladesh at the border.
- Some large companies are beginning to provide child-care facilities and to organise flexi-work options.
- Interventions by NGOs in implementing development projects are taking a gendered approach. For example, the International Center for Research on Women (ICRW) is working in Delhi neighbourhoods. ICRW has set up →

a 'Safe Cities' project working in partnership with UN Women, the Indian government and a New Delhi-based organisation, Jagori. This approach has led to women slowly gaining the confidence to report more crimes and to speak up for their rights in their communities. Details of this project are available at www.icrw.org, especially in the links to 'What we do' and 'Where we work'. In addition, see the Jagori website (which means 'awaken women!') at www.jagori.org and UN Women at www.unwomen.org/en.

● In 2014 the Delhi police increased the number of women police officers and overall police force in the outer districts of the city. This was in response to the findings of their project to map crimes against women. Crimes such as rape and molestation were reported as almost absent in Central Delhi but they were most prevalent in West Delhi and South Delhi (Figure 8.19). The shortfall in policing had arisen as a result of rapid urbanisation with unregulated urban growth in the rural–urban fringe of the city.

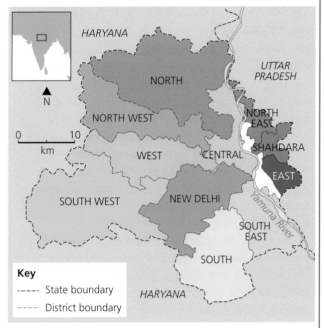

Figure 8.19 Districts of the National Capital Territory of Delhi

Conduct research to investigate the effects of recent change on the lives of women in India in the last decade:

1 quantitatively, by tracing statistical change in indices such as female life expectancy, maternal and infant mortality rates and total fertility rate
2 qualitatively, by searching for evidence, for example in media reports, of changing roles of women in the home and the workplace, their increasing empowerment such as in family size and attitudes of men towards women.

Review questions

1 What is meant by the term 'gender inequality'?
2 What are the challenges faced by women as a result of gender inequality?
3 Explain the low levels of female participation in education in many developing countries.
4 Identify three factors which limit female access to reproductive health services in developing countries.
5 State three factors which restrict female participation in the labour force.
6 What do the letters CEDAW represent?

8.3 What are the strategies for global governance of human rights?

Strategies for global governance of human rights include the following:

● Attempts to change and modernise norms
● The work of NGOs, private organisations and human rights activists
● The influence of MNCs in terms of their Corporate Social Responsibility (CSR)
● The creation and application of international and national laws
● Attempts to strengthen the rule of law
● Reference to legal mechanisms such as the European Court of Human Rights and the International Criminal Court (ICC)
● Treaties or conventions established by supranational organisations such as the UN and regional unions such as ASEAN, ECOWAS, EU and NATO
● The work of the various agencies of the UN such as the Office of the High Commissioner for Human Rights (OHCHR)
● The role of UN peacekeeping operations in promoting and protecting human rights
● Humanitarian (military) intervention and humanitarian relief assistance

Human rights violations are involved in every conflict, either as a root cause or as a consequence, or both. There are many examples, current and in recent history, with high concentration in sub-Saharan Africa and the Middle East.

How the violation of human rights can be a cause of conflict

The causes of conflict through violation of human rights are complex and interrelated. It is difficult to attribute any one cause; often a combination of factors is involved which interact and reinforce each other. The following human rights violations are either the initial cause of a conflict or a major contributory factor:

- Denial of human needs such as food, housing, employment opportunity and limited access to education over long periods of time
- Discrimination and denial of freedom in an undemocratic society
- Unrepresentative government in which people are treated unequally or unjustly and excluded from decision-making
- Oppressive governments that do not respect the needs of all groups, including preservation of ethnic identity or discrimination on the basis of gender
- Genocide and torture

Stretch and challenge

Investigate the violation of human rights in South Africa between 1948 and 1991 under the apartheid system. The ensuing conflict, demonstrations, violence and outcome are well documented in a series of television programmes archived by the BBC at www.bbc.co.uk/archive/apartheid/.

How the violation of human rights can be a consequence of conflict and how this can be addressed through geopolitical intervention

Conflict may have the effect of further human rights violation, with outcomes such as the following:

- High incidence of mortality, of military involved in fighting and civilians in local communities nearby
- Damage to homes and other property
- Damage to infrastructure, including transport systems, hospitals, schools and tele-communications
- The impact on food and water supply
- Displacement of populations including the effects of internal and international refugee migration
- Exploitation of women
- Further so-called 'ethnic cleansing'

Intervention in emergency situations may involve military action with forces acting under UN mandate. There have been a number of instances of humanitarian intervention given authority by the UN in order to prevent humanitarian crises. For example, the 2011 military intervention in Libya was given UN mandate to prevent further violation of human rights such as bombing and starvation of civilians by the dictatorship. This intervention was led by France and the UK, with seventeen other countries involved, until NATO forces took over. Invariably continued military presence is required to support all other forms of global governance in the ensuing period, as illustrated by the Afghanistan case study (pages 257–59).

The role of flows of people, money, ideas and technology in geopolitical intervention

The exchange of ideas and information, at the planning stage and during its operation, is important for effective intervention. In places where human rights abuses are perpetrated the UNHRC employs experts, rapporteurs, special representatives and working groups to promote ideas and values. In addition the OHCHR provides education and training for civilian law enforcement and judicial officials to strengthen legal frameworks. Many NGOs such as Amnesty help by publishing information on the human rights abuses in each country to increase awareness. And international organisations at their conferences share ideas and information on courses of action.

Technologies such as social media and other online tools are being used increasingly to help the flow of ideas and information in the intervention process. The widespread use of ICT including the internet and mobile phones are important in communications. And technology such as remote sensing using satellite imagery and unmanned aircraft are used for surveillance in areas that are too inaccessible or too dangerous for conventional observation.

Intervention involves flows of people and money to areas affected. This includes aid workers such as NGO staff and the many uniformed personnel required for peacekeeping missions of the UN and regional organisations such as NATO. Funds to support these operations, often amounting to many millions of dollars, are provided by member states; overall the USA is the greatest contributor. Haiti has benefited from the influx of personnel, money, technology and ideas since the early 1990s. The current UN stabilising mission has over 4500 uniformed personnel and access to over US$500 million donated from 51 countries around the world. There are many hundreds of NGOs working in Haiti, funded by US$5 billion in foreign aid. The IMF has provided debt relief and financial support.

Figure 8.20 UN peacekeepers patrol Golan Heights

> **Key idea**
> → Global governance of human rights involves co-operation between organisations at scales from global to local, often in partnership

How human rights are promoted and protected by institutions, treaties, laws and norms

Strategies to promote and protect human rights are put in place by organisations at supranational scale, such as the UN, and at regional scale, such as OSCE (the Organization for Security and Co-operation in Europe) or ASEAN, by national governments and by NGOs. Many NGOs are international in structure like the UN but operate at local scale, working closely with small communities affected by human rights violations.

The United Nations

The United Nations is an intergovernmental organisation with 193 member states, each of which agrees to accept the obligations of the UN Charter. Human rights are at the core, one of the main aims of the UN being to 'reaffirm faith in fundamental human rights, in the dignity and worth of the human person, in the equal rights of men and women and of nations large and small.' Many of its agencies are involved in promoting and protecting human rights. The Office of the High Commissioner for Human Rights has the lead responsibility. In addition the Human Rights Council and the Human Rights Treaty Bodies work with the legal backing of the International Bill of Rights. The Security Council deals with grave human rights violations, often in conflict areas (Figure 8.20). Further information on how the UN promotes and protects human rights can be found at www.un.org/en/sections/what-we-do/protect-human-rights/index.html.

NGOs

The many non-governmental organisations (NGOs), part of civil society, play a vital role in human rights. In zones of conflict their work 'on the ground' can involve monitoring and providing early warning of new violence, modification of social norms through education, training in practical skills of agriculture (Figure 8.21), water conservation and improved sanitation, provision of medicines, medical assistance and health education, and implementing local strategies to support women and children's rights.

Figure 8.21 Farmers being trained by the International Development Enterprises (iDE) NGO advisers in Kabwe, Zambia. iDe helps to create income and livelihoods for poor rural households

Treaties and laws

Treaties are formal, written agreements between groups of countries which are binding in international and national law. Often drawn up by the UN or a regional organisation, one example is the Convention on the Rights of the Child (*continued on page 259*).

Case study: Strategies for global governance of human rights in Afghanistan, an area of conflict

Contributions and interactions of different organisations at a range of scales from global to local, including the United Nations, a national government and an NGO

Afghanistan is a landlocked, largely mountainous country in Southwest Asia. It is a poor country and its development has been held back by a long period of political instability. In 2013, HDI was 0.468, 169th out of 187 countries; IMR (infant mortality rate) was 117 per 1000 live births; only 5.8 per cent of women over 25 had received secondary education; and 65 per cent of the 30.5 million population had to live on less than US$2 per day – most employed in agriculture (Table 8.5).

Table 8.5 Afghanistan – employment structure, 2014

Afghanistan	% of the labour force	% contribution to GDP
Agriculture	80	31
Industry	10	26
Services	10	43 (mostly banking)

Although not recognised as a legitimate government, the Taliban were in control of 90 per cent of the country before 2001. After that time, their fundamentalist government was overthrown by the USA and its allies. But resurgence of the Taliban, especially in the south, has led to continued violation of many human rights.

By the end of 2014, NATO-led combat troops officially completed their mission in Afghanistan – although some 10,000 advisory forces remain to train the Afghan military (Figure 8.22). Steps towards a negotiated peace have begun with Taliban representatives meeting Afghan officials in May 2015. Nevertheless, in 2016, attacks were continuing and the Taliban insist they will not stop until all foreign troops have left the country.

Figure 8.22 Afghan troops being trained by NATO at one of the military training centres in Kabul

According to Human Rights Watch, this period of renewed insurgency has led to further decline in respect for human rights. They cite:

- increased casualties among Afghan security forces and civilians
- domestic violence towards women and continued inequality in access to employment, health services and education
- disruption of the 2014 Presidential election
- attacks on journalists and hence freedom of expression
- extra-judicial executions
- kidnapping, detentions and torture
- the issue of food security partly fuelled by poppy cultivation, heroin production and the illicit drugs trade.

The UN in Afghanistan

The UN Assistance Mission in Afghanistan (UNAMA) was established by the Security Council in 2002 in order to help achieve sustainable peace and development. More information including mandate details and 2015 renewal are available at http://unama.unmissions.org/.

The UNAMA mission and the UN High Commissioner for Human Rights are interacting with the Afghan government and NGOs to strengthen the work of the Afghanistan Independent Human Rights Commission. Their aims are to:

- promote respect for international humanitarian and human rights laws
- co-ordinate the efforts of all organisations and communities to ensure protection
- promote accountability
- implement the freedoms and human rights provisions in the Afghan constitution and the treaties to which it is party
- achieve full enjoyment of their rights, for women, displaced persons and returning refugees (an estimated 750,000 are displaced).

The Afghan government

The Afghan government has joined the Economic Cooperation Organisation and the South Asian Association for Regional Cooperation to help promote economic growth. It has passed laws such as the Law on the Duties and Structures of the Independent Elections Commission to help improve the democratic process, and the Elimination of Violence Against Women Law of 2009. Also, attempts are being made to pass a law to remove the quota for the number of women in the Afghan Parliament. These measures are all related to human rights issues and contribute to the process of socio-economic development in Aghanistan.

Afghan Aid, an NGO

Hundreds of NGOs have registered to work in Afghanistan. Afghan Aid is just one example, which is involved in sustainable rural development strategies. It co-ordinates its project work with that of the UN and the Afghan government. Details can be found at www. afghanaid.org.uk along with other examples such as Action Aid, CARE International, the International Rescue Committee and the International Committee of the Red Cross (ICRC) – each of which tends to specialise in particular aspects of rights which have been neglected or violated during this period of instability.

Table 8.6 Consequences of global governance of human rights for rural and urban communities

Rural projects in the Chaghcharan district, Ghor	Urban neighbourhood projects in Kabul
<ul><li>Ghor is one of the most geographically inhospitable regions; politically it has become increasingly hostile and insecure (Figure 8.23).</li><li>Poverty in the Chaghcharan district has led to families having to sell assets such as livestock.</li><li>Basic rights have been neglected by the effects of the conflict, including: serious gender inequality; the selling of daughters; limited, if any, access to services such as education, sanitation and infrastructure; health; and nutrition.</li><li>Afghan Aid is one example of many NGOs working in the field to co-ordinate donor funding, the input of the Afghan government and the work of local communities.</li><li>The work of Afghan Aid has transformed the lives of individuals and local communities.</li><li>Economic, social and political rights are promoted by local groups brought together to design, implement and realise their own projects. For example, EU funding and Afghan Aid training have introduced more effective agricultural practices, reduced risk of disease by securing safe water supplies and improved hygiene.</li><li>Local people now have greater freedom, women are more integrated in society and local democratic practices have been strengthened by the election of community groups.</li></ul>	 **Figure 8.24** Afghanistan – urban areas (Source: World Factbook CIA) <ul><li>UN Habitat is working to co-ordinate the Afghan government, local governments, community councils and funding from the Japanese government to upgrade neighbourhoods in the 33 provincial capitals and Kabul (Figure 8.24).</li><li>The basis of these projects is the election of Community Development Councils (CDCs), each of which includes about 200–250 households. CDC 11 in District 5, Jalalabad has 2657 residents living in 296 households</li><li>Denial of basic human rights plus rapid urban growth in Kabul (economic and security motivated rural–urban migration) has deprived local communities of many services.</li><li>The CDCs are locally elected (some of mixed gender) and plans are submitted which reflect specific needs of the area.</li><li>In most CDC areas plans include upgrading of housing, infrastructure, electricity, sanitation, schools and health care.</li><li>Other benefits are improved engagement of women in the projects, employment opportunities, security of land tenure for informal settlements, improved roads and drainage and greater provision of shops.</li></ul>

Figure 8.23 Inhospitable landscape. Hari Rud river flowing through its narrow valley at the base of red rock mountains between Jam and Chist-I-Sharif, Ghor province, Afghanistan

Consequences of global governance of human rights for local communities

The Afghan government is supported by international organisations to engage in local community projects. This approach is proving to be much more effective than 'top-down' programmes since local residents are elected and form their own plans and priorities for the areas in which they live.

As a result by 2014: nearly 6 million children were attending school (up from 1 million in 2001) and nearly 40 per cent of these were girls (unthinkable under Taliban control); access to primary health care had increased to over 50 per cent of the population (up from 9 per cent in 2003); MMR (maternal mortality rate) had halved since 2001 and average life expectancy increased from 55 (2000) to 61 (2013).

The effects of global governance are not uniform across the country but work is being undertaken to improve people's lives through protection of human rights not only in the cities but also remote rural areas (Table 8.6).

(continued from page 256) Examples of laws drawn up by national governments can be found in the India and Afghanistan case studies (see pages 252 and 257). A particularly important role of many NGOs is their ability to reinforce and strengthen the rule of law through education.

Therefore a combination of legal and practical instruments is used for protection of human rights and it is the co-ordination of strategies by different organisations operating at different scales which provides the most effective global governance.

✓ Review questions

1 Identify three strategies for global governance of human rights issues.
2 What is meant by the term 'civil society'?
3 Give one example of a treaty for the protection of a specific human right.
4 State two examples of national laws to help resolve human rights issues.
5 Name one agency of the UN involved in the promotion and protection of human rights.
6 Outline two ways in which rural communities have benefited from global governance of human rights.

8.4 To what extent has intervention in human rights contributed to development?

'The litmus test of development is the degree to which any strategies and interventions satisfy the legitimate demands of the people for freedom from fear and want, for a voice in their own societies and for a life of dignity.'
(Navi Pillay, UN High Commissioner for Human Rights)

Human rights are essential for achieving and sustaining development. The links between human rights and development are embodied in the UN's Millennium Development Goals (MDGs) and Sustainable Development Goals (SDGs) in its 2030 Agenda for Sustainable Development. For example, achieving MDG 3, concerning the rights of women, depends directly on the rights to gender equality especially in education, employment and in political participation. Sustainable development in any country depends on equal opportunities for women and men.

Stretch and challenge

Specific targets for each of the eight MDGs, and the statistical indicators for their success by 2013 and continuation beyond 2015 are set out at www.un.org/millenniumgoals.

1 Use the Action 2015 link to investigate how the UN and its partners aim to improve the lives of people beyond 2015 by tackling human rights issues.
2 Use the UNDP link http://hdr.undp.org/en/countries to access data for change in HDI for three contrasting countries, such as the UK, Mexico and Benin. Discuss reasons for the differing HDI levels and rates of change.

Key idea
→ Global governance of human rights has consequences for citizens and places

How the global governance of human rights issues has consequences for citizens and places, including short-term effects and longer-term effects

The desired effect of global governance of human rights is positive, long-term, sustainable social and economic development for all nations. Shorter-term effects, especially in cases of military intervention are more varied. The impact on individuals and local communities can be of immediate relief or conversely further strife (Table 8.7).

259

Table 8.7 Examples of possible effects of global governance on human rights

Short-term effects	Longer-term effects
Benefits of intervention: • Medical assistance and provision of medicines – NGOs such as Médecins Sans Frontières • Provision of shelter, sanitation, food and water – NGOs such as ICRC, Oxfam, Save the Children • Military protection preventing further casualties and providing protected areas to live and safety for aid workers – UN peacekeeping operations Negative impact of military intervention: • Damage to property and infrastructure • Population displacement • Further disrespect for human rights • Civilian casualties • Disruption of education • Tensions can be fuelled over aid and conflict prolonged into longer term • Military action and ensuing dependence on aid can undermine the local agricultural economy	Positive impact on development: • Improvement in health and life expectancy including IMR and MMR • Education equality, increased enrolment for girls and boys • Improved transport systems – physical access to services • Development of infrastructure networks • Internalisation of accepted societal norms • Freedom from abuse of women and children • Democratic elections, democratic government and political stability • Strengthened judicial system including new national laws and stronger rule of law • Employment opportunities and reduction of poverty • Development of local agricultural systems including skills training/education

 Case study: The impact of global governance of human rights in Honduras, an LIDC

Honduras (Figure 8.25) is beset by economic, political, social and environmental problems. These are exacerbated by human rights issues which further hold back Honduran development.

- Honduras is one of the least developed countries in Central America. In 2014, GDP per capita was US$4700; life expectancy was 69 (male), 73 (female); total fertility rate was 2.8; MMR was 120 per 100,000 live births; IMR was 17.8 per 1000 live births; the employment structure was made up of 39 per cent agriculture, 21 per cent industry, 40 per cent services; and there was 85 per cent literacy.
- The country has been politically unstable, with military control until 1982, but since that time the civil authorities of elected governments have failed to control the security forces and to deal with human rights.
- The country was devastated by Hurricane Mitch in 1998 from which it is only slowly recovering.

The human rights issues

Honduras has one of the highest murder rates in the world. Institutions for public safety are ineffective. The effects of unresolved human rights issues include:

- Unlawful use of force and corruption by the police
- Killings in rural areas over land disputes – often agro-industrial firms pitted against local indigenous organisations over rightful ownership
- Discrimination against indigenous populations
- Gang culture and drug-related violence
- Organised crime which includes trafficking of child labour and child prostitution
- Pervasive societal violence, including harassment of and violence against women

- Intimidation, threats and killings of journalists
- Poverty (44 per cent of the population living on under US$2 per day)
- Limited access to primary health care
- Limited access to education

Global governance strategies used

The UN has sent a Human Rights Adviser to Honduras at the request of its government. The aims are to strengthen government institutions working in human rights, to build a stronger human rights culture, to help implement the Honduran National Human Rights Action Plan, and to co-ordinate the work of civil society and the Honduran government.

The USA is also a key international actor, providing US$50 million in security aid between 2010 and 2014. And it continues to provide assistance through the Central American Regional Security Initiative. Military and police aid is also available if the Honduran government meets specific human rights stipulations. Details of the work of USAID in Honduras can be found on the website of the US Agency for International Development at www.usaid.gov/.

The Honduran government itself has set up a Ministry for Justice and Human Rights, a Ministry for Security and various commissions to reform citizen security and prevent torture. It also uses and relies heavily on NGOs to address issues such as provision of primary health care in rural areas. Details of each current NGO project in Honduras can be found using the interactive map at: www.ngoaidmap.org/location/gn_3608932.

Care International is just one example of an NGO which in several projects in rural Honduras is tackling human rights issues such as poverty, education,

➜

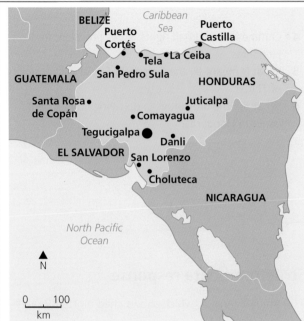

Figure 8.25 Location of urban settlements in Honduras

employment opportunity, health care, water provision and food production – see www.careinternational.org.uk/where-we-work/honduras.

In addition, the corporate responsibility of large companies such as Kenco is evident in its education scheme in Honduras, Coffee vs Gangs with links at www.mykenco.com/charity/1/.

Opportunities for stability, growth and development, and challenges of inequality and injustice

The effects of global governance of human rights are beginning to have an impact on Honduran socio-economic development. HDI has increased from 0.461 in 1980 under military control to 0.617 in 2013. Nevertheless many problems remain including the challenges of urban and rural poverty and discrimination (Table 8.8).

Table 8.8 Opportunities for stability, growth and development, and challenges of inequality and injustice in Honduras

Opportunities	Challenges
Stability: ● The bilateral links with the USA are helping the economy and national security. ● Political stability is being achieved with US support for anti-corruption and free and fair elections. ● Under the Central America Regional Security Initiative, local governance is being strengthened. ● Community-based efforts to prevent crime and gang activity include education for at-risk youths. **Economic growth:** ● The USA, UN and civil society are effectively implementing new food security programmes, promoting economic diversification and training citizens in emergency response to natural disasters. ● Under the Dominican Republic – Central America Free Trade Agreement (CAFTA-DR), small farmers and other enterprises are being assisted to increase their trade opportunities. ● USAID (lead US government agency) is working in local areas to end poverty and enable communities to realise their potential. **Socio-economic development:** ● 40% of the population is under 15: the Honduras Ministry of Education has reformed its policies by decentralising to local authorities in rural and urban areas in order to meet MDG targets for school enrolment. ● Efforts to improve maternal and child health and nutrition, and to prevent HIV/AIDS are providing opportunities for further development.	**Inequality:**  **Figure 8.26** Squatter settlement, San Pedro Sula (Source: Creation Care Crossroads) ● Inequalities between rich and poor are evident in urban areas such as the industrial centre of San Pedro Sula. Rural–urban migration leads to housing shortages, and inevitable problems of water supply and sanitation (Figure 8.26). ● In both urban and rural areas there remains unequal access to education and health care, not only between girls and boys but also by discrimination against HIV-positive people. **Injustice:** ● Judges face acts of intimidation. ● The criminal justice system needs modernising with greater protection of human rights. ● Attacks on journalists continue. ● Violence against children continues. ● Discrimination against indigenous populations ● There is a high incidence of violence related to drug trafficking and urban gangs.

1 Explain why the work of NGOs is particularly effective in tackling human rights issues.

2 State three possible short-term negative impacts on people and/or places of intervention in human rights issues.

3 Identify three long-term effects of global governance of human rights on development.

4 Outline three human rights issues prevalent in Honduras.

5 What is meant by the Corporate Social Responsibility of TNCs?

6 What are the difficulties in resolving human rights issues in Honduras?

📖 Practice questions

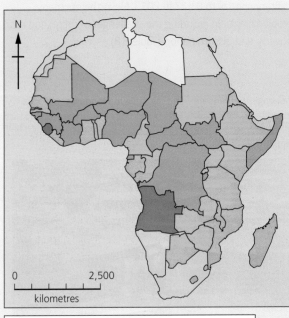

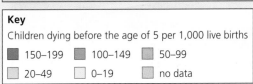

Key

Children dying before the age of 5 per 1,000 live births

- 150–199
- 100–149
- 50–99
- 20–49
- 0–19
- no data

Figure A Child mortality rates in Africa, 2012

Structured/data response

1 a) Study **Figure A**, which shows child mortality rates in Africa, 2012.

   i) Suggest **two** ways in which political factors influence child mortality rates in Figure A. [2 marks]

   ii) Explain **one** socio-economic factor that might account for the spatial variation in child mortality rates in Figure A. [3 marks]

   iii) For the data presentation technique used in Figure A, suggest how effective the technique is for showing spatial patterns of child mortality rates. [4 marks]

b) With reference to a **case study**, explain strategies used to address gender inequality issues in a country. [8 marks]

Essay

2 'Global governance to protect and promote human rights has only positive consequences.' Discuss. [16 marks]

Chapter 9

Power and borders

9.1 What is meant by sovereignty and territorial integrity?

> **Key idea**
> → The world political map of sovereign nation-states is dynamic

The world political map shows territories of sovereign nation-states. These are spatially bounded areas of land, which physically define independent, self-governing countries. These political units are the dominant entity in the global political system and are considered to be the most important form of spatial governance.

The dynamic nature of the map is demonstrated by the formation of new countries since 1990. Examples include:

- South Sudan seceded from Sudan in 2011, following protracted civil war; border regions such as Abyei remain contested (see South Sudan case study, page 281).

- Eritrea gained **independence** from Ethiopia in 1993 following years of armed conflict.
- Fifteen new countries in eastern Europe and central Asia were formed by **secession** from the USSR following the demise of the communist regime there in 1991 (Figure 9.1).
- Germany achieved unity in 1990 as a result of political, economic and social reunification of East (GDR) and West (FRG) Germany.
- The area of former Yugoslavia has seven countries today; the division, in 1990, was based on political, economic and ethnic factors. These Balkan states include Kosovo (Figure 9.2), recognised as independent by most other countries.
- Namibia gained independence from South Africa in 1990 following the Namibian War of Independence.
- The Republic of Yemen replaced the Yemen Arab Republic (North) and the People's Democratic Republic of Yemen (South) in 1990; this was partly to exploit oil fields either side of the former border.
- The Czech Republic and Slovakia became independent states after the dissolution of Czechoslovakia in 1993, following the 'Velvet Revolution'.

Figure 9.1 Political map of eastern Europe and central Asia following the breakdown of the former USSR

Figure 9.2 The countries of former Yugoslavia

The significance of these border changes is far more than a matter of **territory**. They affect sovereignty over populations and physical resources; they influence the economy and social geography of each area, including ethnic groups. And they influence global patterns of trade, and migration, internal and international, for each new country.

The world political map also shows disputed **international borders**. Examples in South Asia include the claims of India, Pakistan and China in the Jammu and Kashmir area, and the Arunachal Pradesh border between India and China (Figure 9.3).

Stretch and challenge

For any one example of change in the global political map since 1990, investigate:

a) specific reasons for the change
b) factors affecting the location of new international borders
c) effects on the economic and social geography of the countries involved.

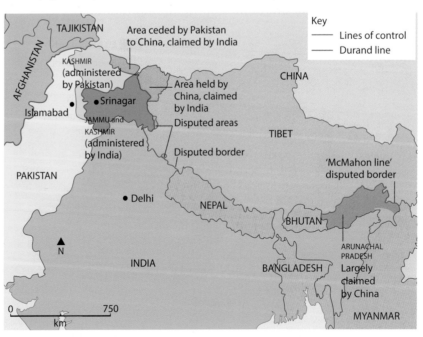

Figure 9.3 Disputed borders in South Asia

Furthermore, changes to the world political map, through processes of integration, have created politically strong and economically important groups of countries, for example:

- regional trade blocs such as the EU – its recent expansion includes accessions of Croatia in 2013, Bulgaria and Romania in 2007 and ten other states in 2004
- global organisations such as the UN, IMF, World Bank and WTO
- G20 and G7.

Definitions of state, nation, sovereignty and territorial integrity and how they are fundamental in understanding the world political map

The following terms are important in understanding the global political system and the challenges it faces in the twenty-first century.

State

The term **state** refers to the area of land, of an independent country, with well-defined boundaries, within which there is a politically organised body of people under a single government. States therefore are political entities that have territories over which the body politic exercises **sovereignty**.

States have the following characteristics:

- Defined territory: which is internationally recognised.
- Sovereignty: in which the political authority is effective and strong enough to assert itself throughout the bounded territory.
- Government recognised by other states: often achieved through UN elected membership.
- Capacity to engage in formal relations with other states.
- Independence: self-governing.
- Permanent population which has the right to **self-determination**.

The term **state apparatus** refers to the set of institutions and organisations through which state power is achieved.

Globally there is inequality in the power and influence of states. Some states have the ability to dominate and drive global systems and have significant influence on geopolitical events. Others have little influence and can only react or respond to global change.

Economic power can be measured in terms of trade and wealth generated over long periods (Figure 9.4).

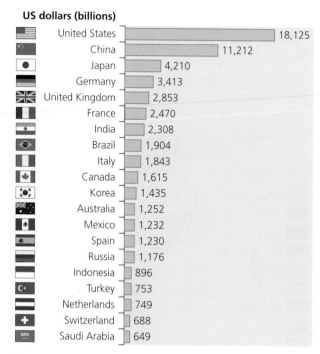

US dollars (billions)

United States	18,125
China	11,212
Japan	4,210
Germany	3,413
United Kingdom	2,853
France	2,470
India	2,308
Brazil	1,904
Italy	1,843
Canada	1,615
Korea	1,435
Australia	1,252
Mexico	1,232
Spain	1,230
Russia	1,176
Indonesia	896
Turkey	753
Netherlands	749
Switzerland	688
Saudi Arabia	649

Figure 9.4 States with highest GDP in US$ (billions) – a measure of economic wealth and power, 2015 (Source: IMF World Economic Outlook (WEO), April 2015, http://knoema.com/nwnfkne/world-gdp-ranking-2015-data-and-charts)

Military power may also depend on wealth and government policy (Figure 9.5).

A state may also be influential in the global spread of its cultural attributes, such as so-called 'Americanisation'.

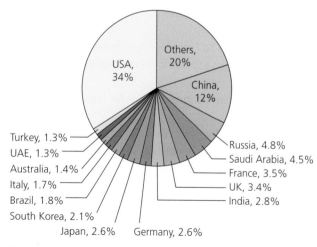

Figure 9.5 Share of world military expenditure, 2014 (Source: www.sipri.org)

State power depends on diverse economic, social, political and physical factors, including ability to exploit natural resources, geographical location, human resources, demographic structure, industrial development, ability to regulate its economy, trade strength, wealth, internal political organisation, international relations, government policy, and events in its history.

The degree of **resilience** of a state is measured by the Fund for Peace (FFP) Fragile States Index. The global pattern of state fragility/resilience (Figure 9.6) is based on a wide range of social, economic, political and military indices – see http://fsi.fundforpeace.org/indicators. Examples of specific indices include refugees per capita, fatalities from conflict and political prisoners.

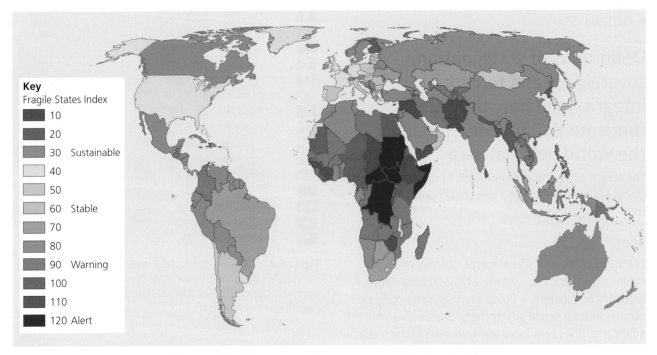

Key
Fragile States Index

- 10
- 20
- 30 Sustainable
- 40
- 50
- 60 Stable
- 70
- 80
- 90 Warning
- 100
- 110
- 120 Alert

Figure 9.6 Global pattern of state fragility, 2014 (Source: http://fsi.fundforpeace.org/rankings-2014)

Activities

1 Describe the global pattern of state fragility shown in Figure 9.6.
2 With reference to Fund for Peace indices:
   a) suggest two economic reasons for high levels of fragility
   b) suggest two political reasons for high levels of fragility.
3 Discuss advantages and disadvantages of the choropleth technique used to show levels of state fragility in Figure 9.6.

Nation

The term **nation** refers to a large group of people with strong bonds of identity – they are united by shared descent, history, traditions, culture and language.

A nation is different to a state. Spatially, a nation may be confined to one country, or its people may live in an area across adjoining countries, and some may be scattered globally in the diaspora. Many states contain several national groups.

An example is the Kurdish nation. Kurds are a non-Arab, Middle Eastern population inhabiting the region known as Kurdistan. This is an extensive plateau and mountain area in Southwest Asia, including parts of eastern Turkey, northeast Iraq, northwest Iran, and small areas of Armenia and northeast Syria (Figure 9.7). Traditionally a nation of nomadic herders, Kurds are now mostly semi-nomadic or sedentary. Other examples of nations include Maori, Basque and Sioux nations.

Stretch and challenge

Table 9.1 Fragile States Index and Fund for Peace rankings, selected countries, 2015

FFP classification	Global rank	State	Fragile States Index	FFP classification	Global rank	State	Fragile States Index
Very high alert	1	South Sudan	114.5	Warning	72	Nicaragua	79
	2	Somalia	114		79	Bosnia	77.4
	3	Central African Republic	111.9		84	Ukraine	76.3
	4	Sudan	110.8		100	Mexico	71.8
High alert	5	Congo DR	109.7	Low warning	112	Cuba	67.4
	9	Syria	107.9		113	South Africa	67
	12	Iraq	104.5		123	Brazil	62.6
Alert	15	Ivory Coast	100	Less stable	130	Bulgaria	55.4
	19	Niger	97.8		134	Greece	52.6
	24	Eritrea	96.9	Stable	141	Argentina	47.6
	30	Mali	93.1		147	Italy	43.2
	34	Sri Lanka	98.6		151	Spain	40.9
High warning	41	Angola	88.1	More stable	158	USA	35.3
	51	Uzbekistan	85.4		161	UK	33.4
	55	Laos	84.5	Sustainable	165	Germany	28.1
	60	Senegal	83		173	Switzerland	22.3
	65	Russia	80	Very sustainable	178	Finland	17.8

1 Outline and justify a statistical technique to represent the data shown in Table 9.1.

2 Evaluate the criteria used by the Fund for Peace in devising the Fragile States Index.

3 To what extent is there a relationship between state fragility and GDP?

Figure 9.7 Location and extent of Kurdistan (Source: Wars in the World)

Nations do not have sovereignty; the Kurds are a nation united by culture but without a state or sovereign power.

When a nation has an independent state of its own it is referred to as a **nation-state**. A nation-state is a state that has sovereignty over a single nation. Japan and France are examples. The boundaries of the state coincide with the geographical area inhabited by the nation.

In modern global politics nearly all states refer to themselves as nation-states, even though many contain citizens of more than one nationality. This is partly because every government attempts to build a sense of national identity among its citizens. In many nation-states the government actively promotes nationality, often through its education system and media, in a process of **nation-building**.

Sovereignty

The term 'sovereignty' refers to the absolute authority which independent states exercise in the government of the land and people in their territories. The global system of states is therefore based on territorial sovereignty; this is fundamental in understanding current political geography.

Sovereignty is sustained by two political processes:

- Internal sovereignty: a state has exclusive authority within its bounded territory and no other state can intervene in its domestic politics.
- External sovereignty: a state cannot simply proclaim sovereignty, there has to be mutual recognition among other sovereign states. This ensures its territorial integrity and enables it to enter into international relations with other states.

Sovereign authority exists beyond the dimension of spatial land area – it applies also to:

- rocks, soils, minerals and space beneath the surface
- agreed areas of sea and sea-bed resources
- agreed air space.

These are important principles because, as we shall see later, the concepts of sovereignty and territorial integrity are being challenged by processes developing as a result of globalisation which are a potential source of conflict.

Territorial integrity

Territorial integrity and sovereignty are interrelated. States exercise their sovereignty within a specific territory, the boundaries of which have been established by international law. This provides the framework for the current international political structure based on territorial division and sovereign states.

The Charter of the United Nations, which aims to maintain international peace and security and develop friendly relations among nations, specifically refers to the importance of territorial integrity, viz:

- *Article 2.4: All Members shall refrain in their international relations from the threat or use of force against the territorial integrity or political independence of any state, or in any other manner inconsistent with the Purposes of the United Nations.*

The preservation of territorial integrity and sovereignty is therefore important in achieving and maintaining international security and stability in the world.

Understand the terms, norms, intervention and geopolitics, and how they are fundamental in appreciating that sovereignty and territorial integrity are complex issues

Norms

Norms are derived from moral principles, customs and behaviours which have developed over time throughout the world. They are embedded in international law to be upheld by state governments and their citizens.

These norms are based on principles set out in the Charter of the United Nations which outline the universally accepted understanding of sovereignty and territorial integrity. They refer to customary, internationally accepted behaviour of state governments. This includes state responsibilities of maintaining the global system of sovereign states with bounded territories, and protecting their citizens.

Examples of principles set out in the UN Charter which have become important norms for sovereignty and territorial integrity include:

- *Article 2.1: The Organization (the UN) is based on the principle of the sovereign equality of all its Members.*
  This means that all member states have equal right to determine their own form of government, which they can choose without outside influence. This chosen government is able to make authoritative decisions with regard to the people and resources within its defined territory. But it also has the responsibility to respect the sovereignty of other states.
- Article 2.4 (above) which makes specific reference to preservation of the territorial integrity and political independence of a state.
- *Article 4.1: Membership in the United Nations is open to all other peace-loving states which accept the obligations contained in the present Charter and, in the judgment of the Organization, are able and willing to carry out these obligations.*
  The obligations of states referred to in this norm include promoting and developing friendly external relations between nations. And internally, the obligation of a state is to protect its citizens. This includes: to respect, protect and fulfil human rights; to allow citizens to be involved in government; and to allow the freedom and opportunity to have a role in which they can contribute to society.

Governments are expected to put in place domestic measures and laws compatible with the UN Charter and any other treaty obligations which they have signed.

Some states are said to be fragile (Table 9.1) because the sovereign government has been unable to fulfil these obligations or responsibilities, often because state apparatus is ineffective.

There are increasing numbers of norms and principles of accepted behaviour which are established not only by the UN but also in the charters of regional organisations such as the EU, AU and ASEAN. Norms regarding sovereignty and territorial integrity are reinforced by treaties to which member states sign and ratify.

Intervention

The term **intervention** encompasses actions of international organisations in resolving conflicts or humanitarian crises arising from challenges to sovereignty and territorial integrity. Examples of different types of intervention include:

- economic sanctions, as outlined in Article 41 of the UN Charter
- military intervention authorised by the UN, in Article 42
- missions of regional organisations such as NATO
- humanitarian assistance by Civil Society organisations, including many NGOs and aid agencies.

Intervention is deemed necessary in circumstances such as a state government failing to protect its citizens from violation of human rights, or a direct act of aggression by another state, perhaps over territorial claims. It is also applied where, for example, civil war is the result of poor or corrupt governance, where there is conflict between ethnic groups, or where religious fundamentalism and terrorist activity have serious effects. And there is intervention where TNCs have negative economic, social or environmental impacts on countries in which they invest.

Intervention is controversial. It is argued that the principle of sovereignty, promoted by the UN, is undermined by the very act of intervention even though sanctioned by the UN Security Council.

Geopolitics

Geopolitics involves the global balance of political power and international relations. Geopolitical power

Stretch and challenge

Responsibility to Protect is gaining ground as a relatively new norm in the twenty-first century. Refer to details of organisations involved and the aims of the International Coalition for R2P at www.responsibilitytoprotect.org/index.php/about-coalition.

- What are the arguments for intervention under the notion of R2P?
- To what extent does R2P challenge the longer held norms that all states have equal sovereignty and that non-intervention is inherent in sovereignty?

is very uneven throughout the world. As we have seen, there are inequalities in power between states explained by their wealth, the political strength of their governments, and their level of development. For example, there are: powerful **Advanced Countries** (ACs), including the USA superpower; **Emerging and Developing Countries** (EDCs), which are increasingly important economically and politically; and **Low-Income Developing Countries** (LIDCs), which are less powerful peripheral economies. In addition, supranational political and economic organisations such as the UN, EU, MERCOSUR, ASEAN and OPEC exert strong geopolitical influence. And trans-state organisations such as TNCs have increasing influence on countries in which they locate as globalisation continues to spread.

The **geopolitics of intervention** in sovereignty and territorial integrity issues is therefore important. When the **international community** is called upon to intervene, it requires consideration of: reasons why intervention is necessary; appropriate type of intervention; political composition of groups of countries and organisations involved; characteristic features of the country, government and peoples affected; and potential socio-economic, environmental and political effects (see Mali case study, page 285).

Global governance of sovereignty and territorial integrity issues is therefore complex and multifaceted; it can involve economic intervention, military intervention, and the humanitarian work of CSOs including reinforcement of a growing number of human rights norms, laws and treaties. Effective global governance depends on interaction and co-ordination at every scale (see South Sudan case study, page 281).

Review questions

1 Give examples of one sovereign state formed in each of three different continents since 1990.
2 What is the difference between the terms 'state' and 'nation'?
3 What is meant by the terms 'sovereignty' and 'territorial integrity'?
4 What do you understand by the term 'norm' in the context of sovereignty and territorial integrity?
5 What are the different ways in which intervention might be imposed by the international community?
6 Why is intervention controversial?

9.2 What are the contemporary challenges to sovereign state authority?

Key idea
→ A multitude of factors pose challenges to sovereignty and territorial integrity

Erosion of sovereignty and loss of territorial integrity are influenced by economic, political, social and environmental factors

Current political boundaries

The current system of nation-states with clearly defined political boundaries is based on the **Westphalian model**. This was established on the principles of sovereignty, territorial integrity and the sovereign equality of all states. A state cannot violate the sovereignty or territory of another state since, in this respect, all are equal.

These principles are reinforced in the UN Charter today. But this system has been challenged by many current threats. Disruptive and destabilising forces interact to erode the long-established Westphalian concepts. And as a result, control of territory and its borders has been increasingly contested in the last two decades both within and between sovereign states. This includes:

- contested territory: for example, Russia's annexation of Crimea and support for separatists in Ukraine; contested islands in South and East China Seas; UK and Argentinian claims over the Falkland Islands
- **separatism**: for example, claims for secession by Basque and Catalan national groups in Spain and France; and Scottish nationalists in the UK
- factional or sectarian tensions: for example, in the Middle East and North Africa where political and ethnic conflict challenge sovereignty and territorial integrity
- transnational movement of terrorist and extremist activity: for example, across the Turkey–Syria border, where smuggling of foreign fighters, oil, weapons and other military supplies threatens territorial integrity and sovereign control of the two countries (Figure 9.8)
- contested maritime boundaries: many are disputed over resources and exploration rights; one example is the boundary dispute in Atlantic waters off the Ivory Coast and Ghana where oil reserves are being exploited
- the legacy of colonialism: for example, the 'scramble for Africa' where arbitrary political boundaries and European administration of territory caused ethnic partitioning (see Mali case study, page 285).

Activity

The International Boundary Research Unit, Durham University is an institution dealing with global boundary matters and related geographical issues including the changing nature of sovereignty, territory, citizenship and political organisation of space. Use www.dur.ac.uk/ibru/ to conduct research:

a) Identify one current boundary dispute.
b) Outline factors that explain the challenges to territorial integrity.
c) Suggest a possible solution.

Stretch and challenge

Sovereignty of Arctic Ocean waters and sea bed is currently contested and unresolved. The possibility of resource exploitation is the reason for claims by countries with an Arctic coastline. With reference to the map and notes at www.dur.ac.uk/resources/ibru/resources/ibru_arctic_map_27-02-15.pdf:

a) Describe the geographical pattern of established territorial seas in the Arctic Ocean.
b) Examine difficulties in delimiting areas of territorial sea and Exclusive Economic Zones in the Arctic Ocean.

Figure 9.8 Despite barriers, the boundary between Turkey and Syria at Kilis is still permeable

Transnational corporations

Transnational corporations (TNCs) are large corporate enterprises which operate in more than one country. For example, Nike, the US sports equipment and clothing company, has 692 factories in 42 countries and employs over a million people worldwide.

There are growing numbers of TNCs, not only emanating from advanced countries but increasingly from the outward FDI of emerging and developing countries. TNCs have become a driving force of global economic integration. Many countries especially LIDCs have become reliant on TNCs to integrate their economy into the global economy and to encourage development. And in so doing, TNCs have brought many benefits to poorer countries.

But TNCs have brought disadvantages too, many of which present challenges to government control and in effect to state sovereignty. In the last two decades many TNCs have expanded their operations regardless of state boundaries. And such is their economic power that some nation-states have in part lost control of territory, work force, environment, and in some instances their own political decision-making. These decisions have been strongly influenced by the large

corporations on which they depend for economic growth and development.

Many TNCs have been criticised for pursuing their own profit-making interests at a cost to the countries in which they have invested:

- Business decisions to invest or disinvest, affecting the lives of many people, have been taken outside the host country which has little or no involvement.
- Disrespect for human rights by some TNCs is another challenge to sovereignty; exploitation of working populations, payment of low wages, demands for overtime and long hours, poor working conditions and use of child labour are all abuses of human rights levelled at some large businesses.

These are issues over which the state government may have little control, affecting its ability to protect citizens within its own territory. The dilemma for small, peripheral economies is that they are dependent on the economic and social benefits derived from TNC investment such as local multiplier effects, the lifting of population above poverty levels and the stimulation of other affiliated industries.

It is not surprising that these challenges to sovereignty are a concern for international

271

organisations such as the UN and OECD. Both have established guidelines, setting out TNC and state government responsibilities.

Nike has long been criticised for its negative impacts. Its stated aim now is to shed this reputation and conform to the guidelines. Other TNCs such as Nestlé, Toyota and Royal Dutch Shell also have clear policies to achieve their corporate social responsibility and conform to the UN Global Compact.

Supranational institutions such as large regional trading blocs

Supranational institutions represent a tier of governance above that of the individual state. Examples include international organisations at global scale such as the UN and its agencies including the IMF and WTO, and at regional scale such as NATO and trading blocs including the EU, ASEAN, MERCOSUR and ECOWAS.

Within supranational institutions member states retain their sovereignty: they are independent countries, have equal rights, and exercise exclusive control over, and responsibility for, their citizens. But having achieved membership they are also bound to the requirements of the supranational body, including any treaties they sign. In this respect, member states are said to 'surrender' some aspects of their sovereignty since they must comply with the international or regional laws of these institutions.

The European Union

The EU is a trading bloc which is not only an economic union but also a political union with its own parliament and a monetary 'Eurozone'.

There are 28 sovereign member states (Figure 9.9), but their integration in the EU can bring challenges to their individual power and autonomy.

● On the one hand there are benefits of integration which protect states' interests. These include the ability to address transnational issues such as air and water pollution and international crime; and economic and trade advantages such as protection of industry by common tariff and access to a large European market.

● However, these same states are also required to implement EU laws and decisions even if they did not vote for them. They cannot pass laws in the interests of their own state if they conflict with those of the EU. And the 19 members of the Eurozone have additional financial restrictions such as being unable to set their own interest rates, and being forced to accept harsh austerity measures and contribute to large bail-out funds, as for Greece's debt crisis in 2015.

Other aspects of EU membership which limit sovereign power of individual member states include enforced fishing quotas in the Common Fisheries Policy, and compliance with EU regulations governing Basic

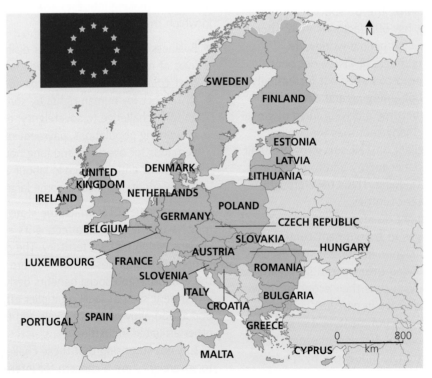

Figure 9.9 The 28 independent states of the European Union, 2016

Payments Schemes and Countryside Stewardship under Common Agricultural Policy.

United Nations

Each of the 193 sovereign states in the UN is a member of the General Assembly. As we have seen in the UN Charter, sovereignty and territorial integrity are important norms (Articles 2.1, 2.4) which underpin the global political system. And despite the effects of globalisation, the system of nation-states is still the basic framework for international law; it is the state which has responsibility for its bounded territory and its citizens as well as relations with other states (Article 4.1).

Nevertheless, what appears to conflict with these norms is that the UN, with the backing of the Security Council, has the right to sanction intervention as in Articles 41 and 42. These norms are increasingly applied by the international community when a state fails to protect its citizens.

At the General Assembly World Summit in 2005, a UN resolution reaffirmed that the primary responsibility to protect its citizens still lies with the individual state, but intervention should apply in instances of genocide, war crimes, crimes against humanity and ethnic cleansing. The international community will intervene in a state without its consent only when that state is allowing violation of human rights to occur or if it commits them itself.

Activities

Globally there are conflicting opinions over the right to intervention and the principle of state sovereignty. Investigate these views using the report of the UN Security Council Meeting in February 2015. www.un.org/press/en/2015/sc11793.doc.htm

1 Compare and contrast views of representatives of China, the Russian Federation, the USA, India, Venezuela and Nigeria.
2 Explain what is meant by UN Secretary-General Ban Ki-moon's statement: 'Sovereignty remains a bedrock of international order. But in today's world the less sovereignty is viewed as a wall or a shield, the better our prospects will be for protecting people and solving our shared problems.'

Political dominance of ethnic groups

The geographical distribution of ethnic groups does not always coincide with current political borders.

- A sovereign state may include more than one ethnic group within its territory: South Sudan is estimated to include 60 different ethnic groups or indigenous tribes, of which the Dinka and Nuer peoples are the largest.
- A single ethnic group may be partitioned by modern state borders: as we have seen, Kurdistan extends across five states in Southwest Asia (Figure 9.7), and the Tuareg homeland stretches across five states in North and West Africa (see Mali case study, Figure 9.22, page 285). In each instance, national borders bear little resemblance to these areas of common culture and ethnicity.

Challenges to the territorial integrity of a state come from ethnic groups that have strong identity, culture and political organisation and demand full independence to create a new state. This is the case for the Tuareg in Mali, who claim independence and the right to self-determination in Azawad.

Challenges to the sovereignty of a state occur where internal conflict between ethnic groups results in a government being unable to protect all its citizens. Conflict between the Dinka and Nuer has been fuelled by political differences between leaders and politicians that originate from these different ethnic groups in South Sudan (see case study, page 281).

The Basque Nation

Challenges to territorial integrity and sovereignty have arisen in Spain and France where the Basque nation has long struggled for independence.

The Basque country straddles the international border of northern Spain and southwest France (Figure 9.10). The area is centred on the western end of the Pyrenees and includes seven provinces, four in Spain and three in France, with a population of more than 3.1 million.

Basque people have a distinctive culture, including a unique language (Euskara) which is taught in schools. Many traditions and cultural features are maintained in the architecture, food and dance of the area. The Basques have a strong tradition of independence and they have achieved political autonomy, including the Basque parliament.

Basque nationalists have demanded the right to self-determination and even full independence from France and Spain. To this end the separatist movement, ETA, has carried out violent acts in the past. This has been in the face of a history of suppression of Basque identity by the French and Spanish governments, who see this as a threat to their sovereign power and potentially their territorial integrity.

Recent issues have been the dispersal of political prisoners among French and Spanish prisons. These include members of ETA, pro-independence politicians, members of the Basque youth movement and journalists. Following the ETA ceasefire in 2011, the Basque

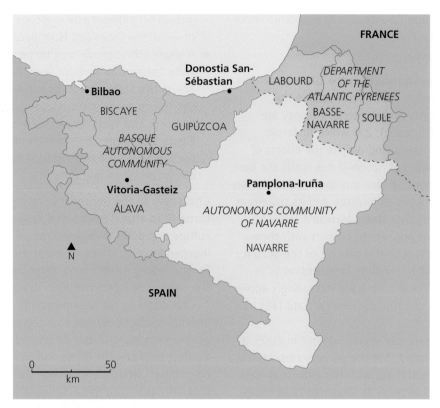

Figure 9.10 The seven historic provinces of the Basque Country are divided into three administrative regions

separatists have undertaken peaceful protests. They include a human chain from Durango to Pamplona in 2014 supporting the Basque right to self-determination; and marches in the streets of Bilbao have highlighted the lack of respect for the human rights of the political prisoners (Figure 9.11). The Basque pro-sovereignty movement still supports the right to self-determination and territorial unity following the ETA ceasefire.

Figure 9.11 Tens of thousands march in the city of Bilbao in January 2015 to protest against the 'policy of dispersal' of Basque prisoners and call for their repatriation

Case study: Ukraine: a country in which sovereignty has been challenged

The area of modern-day Ukraine has been contested politically throughout its history. It is a large European country, stretching from the Carpathians in the west to the Donets Basin in the east. The southern boundary is the long Black Sea and Sea of Azov coastline including the Crimean Peninsula (Figure 9.12).

Ukraine is classified by the World Bank as a lower-middle-income economy. Even though it is a major producer of food, especially grain on its extensive fertile plains, and it produces heavy manufacturing products based on large coal, iron ore and other mineral deposits, GDP per capita has been falling in recent years. Moreover, its population

of 44.5 million in 2015 is also in decline as a result of net migration loss and natural decrease.

Causes and challenges to the government

Ukraine gained independence following the collapse of the Soviet Union in 1991. Since then democracy, civil liberty, economic reform and prosperity have been difficult to achieve. This is explained by a combination of factors which reflect its geography, culture and political diversity:

- Geographical position: between Russia and the countries of the EU.

Figure 9.12 Ukraine and neighbouring countries (Source: CIA World Factbook)

Activities

1 Use the BBC News Country profile of Ukraine at www.bbc.co.uk/news/world-europe-18018002 and your own research to produce a timeline of the main events in Ukraine's political history in the twenty-first century.

2 Identify the main factors which have challenged the Ukraine government's ability to exercise full sovereign control and to sustain its territorial integrity as recognised by the UN in international law.

Figure 9.13 The east–west linguistic division in Ukraine

- Internal political division: Russian-speaking provinces in the east and the Ukrainian-speaking west (Figure 9.13).
- Ethnic disparities: the two main ethnic groups are Ukrainian (77.8 per cent) and Russian (17.3 per cent).
- Inability to build a common national identity and develop strong state mechanisms: there has been endemic corruption, attempts to rig an election, unpopular government policies, and demonstrations, some of which have been extremely violent.

Recent political events have challenged both the state sovereignty and the territorial integrity of Ukraine. These include the serious civil unrest in 2013. This resulted from the Yanukovych government's failure to sign the EU Association Agreement on trade and co-operation with the EU in favour of closer economic ties with Russia. Also the invasion and annexation of Crimea, an autonomous republic of Ukraine, in 2014 by the Russian Federation followed shortly after the election of the pro-West Poroshenko government. And there has been military conflict in the Donbass region of eastern Ukraine, where pro-Russian separatists have been supported by the Russian Federation providing manpower, weaponry and finance.

There has been assistance from NATO and the Organization for Security and Co-operation in Europe (OSCE), but the Ukraine government needs to reform the state apparatus. This includes development of an electoral system which is fair, transparent and reliable. The judicial system needs to be more capable of dealing with criminal activity such as smuggling and corruption. And this requires more effective support from repressive systems, including civilian police and armed forces. Health, education and welfare systems need to be improved. Currently there are human rights issues to be resolved, especially concerning internally displaced people (IDP) as a result of the military conflict. Ukraine needs to become more energy independent, reducing energy dependence on Russia. And greater treasury involvement is required to improve the deteriorating business and investment climate and restore international confidence in trade.

Impacts on people and places

Conflict between Ukrainian and pro-Russian separatists in the Donbass region has serious social, economic and environmental impacts (Table 9.2).

Table 9.2 Impacts of conflict on people and places in Ukraine

People	Places
<ul><li>Approximately 1.47 million of the 5.2 million inhabitants of the Lugansk and Donetsk oblasts are IDPs, having fled their homes since April 2014; a further 600,000 moved to neighbouring countries, mostly to Russia, as residential areas came under fire.</li><li>Evacuees have moved to dormitories in summer camps, disused huts in pinewoods and villages, and a sanatorium at Svetagorsk. Their lack of income, poor quality of shelter, and poor access to health care, medicines and food have added to the vulnerability of the many old, young and disabled.</li><li>There have been 7000 deaths and 13,900 injuries, including 298 people shot down in a civilian aircraft.</li><li>The town of Debaltseve has been bombed, causing damage to housing, services, places of work, communications and livelihoods. Located on the highway linking other rebel strongholds of Donetsk and Lugansk, it is a strategically important rail link for goods from Russia. Those who stay, sheltering in basements, have no power or heating.</li></ul>	Within Ukraine:<ul><li>Donetsk airport has been the scene of heavy fighting. Rockets were fired on the Ukraine military HQ at Kromatorsk.</li><li>Donbass industrial and residential areas have suffered loss of power, water supply and gas.</li><li>Industrial plants have been damaged, including the coal mine at Zasyodko, the Makiyvka chemical works, an oil refinery and the explosives factory at Petrovske.</li><li>Areas of steppes and forests, usually prone to fire in the dry summer, have been burning more than usual. Movement of heavy vehicles has damaged nature reserves.</li></ul>Impacts on other countries:<ul><li>Economic sanctions by the EU and the USA have been effective in leading to increased prices and a drop in value of the rouble in Russia.</li><li>NATO has increased its strength in potentially vulnerable former Soviet states on the Baltic.</li></ul>

1 Identify three ways in which state sovereignty may be challenged.
2 Give three located examples of challenges to the territorial integrity of a state.
3 Give three examples where ethnic groups have claimed independence and right to self-determination.

4 In what ways do supranational organisations challenge the sovereign power of their member states?
5 In what ways can TNCs challenge the sovereignty of a state?
6 Identify two social, two economic and two environmental impacts of conflict in eastern Ukraine.

9.3 What is the role of global governance in conflict?

Key idea
→ Global governance provides a framework to regulate the challenge of conflict

How challenges to sovereignty and territorial integrity can be a cause of conflict

Challenges to sovereignty and/or territorial integrity, which lead to conflict, arise where:

● citizens are treated unjustly or groups of citizens have limited representation in government
● there is competition for the same, or scarce, resources such as water supplies, agricultural land or oil
● people seek autonomy, independence and self-determination if suppressed or marginalised by a state government

● a government fails to protect its citizens from violation of human rights
● people's religious or political beliefs, on which their identity depends, conflict or they are persecuted
● there are differing ethnic identities and ethnic conflict within a state
● a government, through poor management or deliberate policy, fails to supply people's basic human needs.

Patterns of conflict throughout the world are represented by the Global Peace Index (Figure 9.14). This interactive map which can be used to show changes in global patterns of peace over time is produced by the Institute for Economics and Peace. The index is composite, based on 23 indicators, grouped into three main categories:

● level of safety and security in a society
● number of international and domestic conflicts
● degree of militarisation.

Numbers are allocated for each criterion, the highest total score being the least peaceful country.

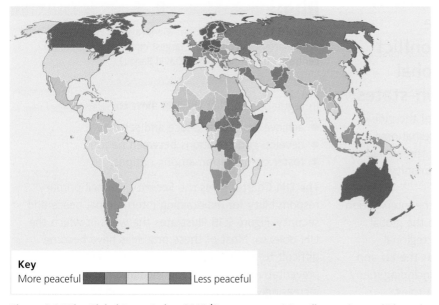

Key
More peaceful ▬▬▬▬▬ Less peaceful

Figure 9.14 The Global Peace Index, 2015 (Source: www.visionofhumanity.org/#/page/indexes/global-peace-index)

1 With reference to Figure 9.14, describe the global pattern of peace.
2 What are the similarities and differences in the global pattern of peace (Figure 9.14) and the global pattern of state fragility (Figure 9.6)? With reference to selected criteria for each index, suggest reasons for your answer.

Natural resources: water supply conflict, Kashmir

Sovereignty over Kashmir has been contested between Pakistan and India since the partition of India in 1947 (Figure 9.3). Periodic firing across the border by both countries has caused deaths of military personnel and internal displacement of thousands of poor farmers and their families. Troops are even stationed high in the Karakoram Range at over 6000 m to control the territory of the Siachen Glacier, a major source of the River Indus.

Although there are ethnic, cultural and religious differences, water insecurity is at the heart of the Kashmir dispute. The Indus is a very important natural resource to both countries for irrigation and hydro-electric power. Mediated by the World Bank, the Indus Water Treaty of 1960 shared the waters of the Indus and is still in force. But Pakistan, occupying the lower part of the Indus Basin, complains that India adversely affects its water supplies by damming the upper tributaries which flow through that part of Kashmir under Indian control.

This dispute is one of escalating importance for both countries owing to:

- the rapid growth of their populations, increasing demand for water
- the water resource itself depleting as global warming causes Himalayan glaciers to retreat.

The dispute remains unresolved, requiring greater co-operation between the two countries.

The role of institutions, treaties, laws and norms which are significant in regulating conflict and in reproducing the global system of sovereign nation-states

The sovereign nation-state is the basis of the global political system. Intervention by the international community is used when sovereignty is threatened by genocide, war crimes, crimes against humanity and ethnic cleansing.

Examples of institutions involved in the intervention process in order to sustain or reproduce the global political system include the UN, NATO, regional economic and political groupings such as the EU and ASEAN, and civil society organisations including many international NGOs. They have different roles.

The United Nations

The UN, founded in 1945, is an international organisation of elected member states. Its headquarters is in New York City, but its historic home is the Palais des Nations, Geneva. Formerly housing the League of Nations, overlooking Lake Geneva in Switzerland (Figure 9.15), today it houses the Offices of the High Commissioner of Human Rights (OHCHR) and the UN Refugee Agency (UNHCR).

Figure 9.15 Palais des Nations, Geneva – UN European headquarters and one of the largest diplomatic conference centres in the world (Source: David Barker)

In regulating conflict the UN aims to:

- achieve worldwide peace and security
- develop good relations between nations
- foster co-operation among nations.

The UN Charter gives the Security Council primary responsibility for maintaining international peace and security. Figure 9.16 illustrates the ways in which the UN does so. Most of these practices have become difficult to achieve in the twenty-first century; preventative diplomacy and mediation are therefore increasingly important. Monitoring and observation

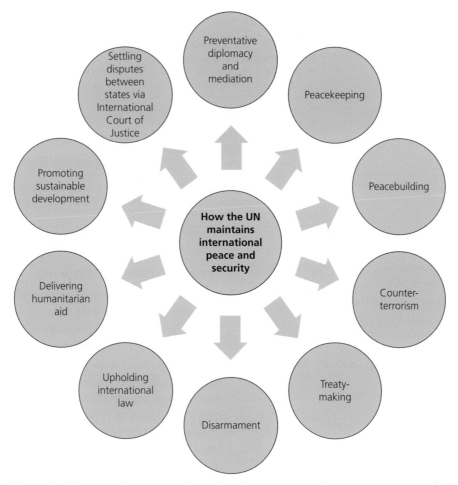

Figure 9.16 How the UN maintains international peace and security

have also taken on greater significance, enabling peacekeeping operations to be deployed earlier.

The UN operates all its policies as part of a global partnership. This helps to give its interventions legitimacy, sustainability and global reach. Currently there are 16 peacekeeping missions (see South Sudan and Mali case studies, pages 281 and 285). These are backed by the legal and political authority of the UN Security Council, the experience of the Secretariat departments, which carry out the day-to-day responsibilities of the UN, the support of the host country, and the use of personnel and finance contributed by member states.

The North Atlantic Treaty Organization

NATO is an alliance of European and North American countries. Its purpose is to safeguard the freedom and security of its 28 members through political and military means.

- Politically: NATO promotes democratic values and encourages consultation and co-operation

on defence and security issues to help prevent conflict.
- Militarily: NATO attempts to achieve peaceful resolution of disputes. If diplomatic measures are ineffective, it has the military capacity and mandate needed to undertake crisis management operations either alone or in co-operation with other countries and international organisations.

The European Union

After the Second World War, the aim of the EU was to foster economic co-operation on the understanding that international trade between member countries would enhance economic interdependence, making them more likely to avoid conflict.

In addition, today the EU has an important security role, providing forces from member states on an ad hoc basis for rapid response operations. It also has many institutional bodies through which it aims to regulate conflict by its policies. Three examples are relevant to the issue of conflict – see Table 9.3.

Table 9.3 EU policies specifically linked to conflict

Foreign Affairs and Security Policy	● Preserve peace and international security ● Promote international co-operation ● Develop and consolidate respect for human rights and fundamental freedoms, rule of law and democracy
Common Security and Defence Policy	● Joint disarmament operations ● Humanitarian and rescue tasks ● Military advice and assistance ● Conflict prevention and peacekeeping ● Crisis management, including peace-making and post-conflict stabilisation
European Neighbourhood Policy	The Eastern Partnership is a key element of EU foreign relations in which the EU aims to co-operate with its close neighbour states in eastern Europe in terms of security, stability and prosperity

Stretch and challenge

ASEAN has a similar approach to the EU in terms of conflict prevention. Use the information in the Political–Security Community section of its website at www.asean.org/communities/asean-political-security-community to investigate ways in which ASEAN aims to prevent conflict in its region.

Non-government organisations

International civil society organisations (CSOs), including NGOs, intervene in many different ways in conflict zones. They work in co-operation with global institutions, national and local governments and citizens resident in local communities within the conflict zones. They provide humanitarian relief including health care, medicines, education, food and water. And their work involves monitoring and early warning of new violence; direct mediation and open dialogue between adversarial parties; strengthening local institutions, rule of law and democratic processes; and reinforcing 'in the field' an increasing number of norms, treaties and laws.

Examples include Amnesty International, the International Committee of the Red Cross (ICRC), Oxfam and ACCORD, the African Centre for the Constructive Resolution of Disputes (see also NGOs in the South Sudan and Mali case studies, pages 281 and 285).

Activity

ACCORD is a South African-based CSO specialising in conflict management, analysis and prevention in Africa.

Examine specific ways in which ACCORD works to resolve conflict in locations such as Somalia and South Sudan, using www.accord.org.za/home/accords-work.

Treaties, laws and norms

International organisations such as the UN and the EU are involved in formulating treaties, laws and norms. These govern legal and generally accepted practices which regulate conflict and maintain peace.

A **treaty**, or convention, is a written international agreement between two or more states and/or international organisations. States that sign and ratify a treaty are bound to it by **international law**.

Many multilateral treaties have been adopted by the UN. One example relating to conflict is 'The Convention on the prohibition of the use, stockpiling, production and transfer of anti-personnel mines and on their destruction'. By 2015, 162 countries had signed up to this legally binding agreement. It remains open to ratification by the others, which include the USA, Russia, China and India. The Broken Chair sculpture in Place des Nations, Geneva is a monument to many thousands of victims of land mines, and it serves as a reminder to all countries to sign the treaty (Figure 9.17).

A primary goal of the UN is to develop international law based on multilateral treaties it has adopted. International law defines responsibilities of states in their conduct with each other and treatment of their citizens. In terms of conflict, there are laws relating to human rights, disarmament, refugees, nationality issues, treatment of prisoners, use of force and conduct of war.

International law also regulates conflict over **global commons**. These are resource domains or areas which lie outside political reach of any one nation-state, including the high seas, the atmosphere, Antarctica and outer space. There are legal and institutional frameworks which address environmental issues of the global commons, including for example the UN Convention on the Law of the Sea (UNCOL).

Treaties and laws are derived from norms. Norms are long-established, common practices in many countries

Figure 9.17 The Broken Chair, Place des Nations, Geneva
(Source: David Barker)

set out in the UN Charter; they in turn are formalised and reinforced by treaty and legal requirements.

Six new norms relating to **cyber conflict** have been created by Microsoft. Cyber conflict is an increasing threat to sovereignty. Microsoft, the US multinational technology corporation and fifth largest TNC in the world, has assumed responsibility for this issue. By creating cybersecurity norms for state governments it has attempted to limit potential conflict in cyberspace. This is expected to bring stability and security to the international environment in our increasingly globally connected society.

The role of flows of people, money, ideas and technology in geopolitical intervention

The flows of people and money are an integral part of the global governance of conflict. Attempts by the

international community to intervene and provide assistance in conflict zones include UN missions and the involvement of regional organisations plus the work of NGOs. These all require movement of personnel into conflict zones and the transfer of finances, donated by member states, sometimes amounting to more than 10,000 people and over US$1 billion per mission (see South Sudan and Mali case studies on pages 281 and 285).

Planning and executing intervention involves exchange of ideas. The sharing of good practice, co-ordination of strategies and flows of intelligence are essential for effective governance of conflict. Flows of ideas and information are a feature of bilateral meetings of governments, regional council meetings, UN conferences and discussion at the General Assembly.

There is increasing dependence on technology in peacekeeping. The advanced technology of the military such as satellite imagery, remotely controlled drones and weaponry is used for surveillance and air strikes. Also the growth of modern ICT enables information to be supplied via the internet, international databases and the media. Communications via mobile telephony and web-based social network services are indispensable for transnational networking in the monitoring of behaviour and in conflict management.

> **Key idea**
> → Global governance involves co-operation between organisations at scales from global to local, often in partnership

---

🔍 **Case study:** Strategies for global governance in South Sudan, an area of conflict

South Sudan is the world's newest sovereign state, gaining independence from Sudan in 2011. A landlocked country in east-central Africa (Figure 9.18) with a population of just over 12 million in 2015, it has many natural resources, including oil which generates 98 per cent of its income. But after years of internal conflict, industry and infrastructure are severely undeveloped and poverty is widespread. Most of the population depends on subsistence agriculture. The Dinka and Nuer are the largest ethnic groups.

Since independence South Sudan has experienced poor governance and difficulty in nation-building. In 2013 the political infighting between President Kiir, a Dinka, and Vice-President Machar, a Nuer, turned into serious armed conflict with an ethnic component. This has led to thousands of deaths, 1.5 million internally displaced

persons (IDPs) and 730,000 refugees in neighbouring countries. A further 8 million are at risk of food insecurity, with one in three children suffering malnutrition. Today South Sudan is a very fragile state; its power and sovereignty have been weakened by government failure to meet its responsibility to protect its citizens.

Interventions and interactions of organisations at a range of scales, including the United Nations, a national government and an NGO

Intervention by the international community has been necessary to resolve the conflict and ease suffering. Effective intervention in South Sudan depends on co-operation between the UN, the Inter-Governmental Authority on Development for Eastern Africa (IGAD) →

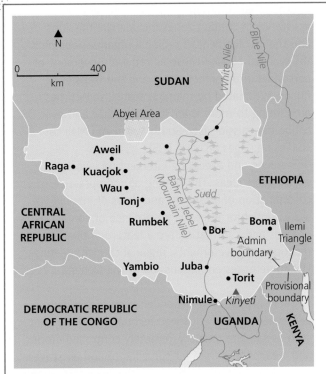

Figure 9.18 Political boundaries of Sudan and South Sudan (Source: www.cia.gov/library/publications/the-world-factbook/geos/od.html)

and its partners, and many NGOs. But this is difficult because of the obstructive and intransigent attitude of the government.

The UN

The UN Peacekeeping Mission (UNMISS), was reinforced in 2014 with a new mandate. Originally, after four years of independence, the mandate was to support the South Sudanese government in peace consolidation, state-building and economic development. But following the December 2013 crisis, military and police presence has been increased. This involves 12,500 military personnel, 1000 police plus 2500 others contributing from countries across the globe. The new emphasis has been to consolidate peace and security, protect civilians, monitor human rights, implement the Cessation of Hostilities Agreement and support delivery of humanitarian assistance.

Other UN agencies working in the country include UNICEF, important in a country where total fertility rate is 5.3 and 45 per cent of the population is under 15; WHO, operating an oral cholera vaccination campaign; and OCHA, the UN Office for the Coordination of Humanitarian Affairs, concerned with hygiene and provision of safe drinking water. In addition, UNHCR has urged the government to sign international conventions for refugee protection and it co-ordinates partnerships

between UN agencies and South Sudan government agencies to assist displaced people, especially unaccompanied and separated children.

The South Sudanese government

The UN Security Council is highly critical of the South Sudan government for failing to protect its citizens. It has called upon the government to put the good of the country and its peoples before the personal ambitions of the leaders, and to find a political rather than a military solution to the conflict.

UNHCR has attempted to engage government agencies as partners, including the Commissioner for Refugee Affairs, the Directorate of Nationality, and South Sudan AIDS Commission. But human rights violations in Unity and Upper Nile Provinces have continued, including attacks on UN peacekeepers, humanitarian personnel and on IDPs under UN protection at their sites near Juba, where 20,000 people are sheltered.

IGAD is attempting to include the South Sudan government in mediation, but the government has prevented representatives from travelling to Addis Ababa in order to participate in the IGAD Plus Peace Process.

South Sudan has signed the following treaties:

- Convention against Torture
- African Charter on Human and Peoples' Rights
- Convention on the Elimination of all Forms of Discrimination Against Women
- Convention on the Rights of the Child.

Ratification of these treaties and a peace agreement signed in August 2015 have brought hope, but there are major obstacles to be overcome. Lack of government co-operation leaves long-term challenges such as alleviating poverty, improving the business environment, negotiating water and grazing rights on the Central African Republic border, and the trafficking of women and children.

NGOs in South Sudan

Many NGOs work with the UN assisting local communities. South Sudan NGO Forum organises more than 300 NGOs addressing humanitarian and development needs. But there has been harassment and violence against NGO workers, including their forced evacuation of Upper Nile state, leaving populations vulnerable and blocking use of the Nile for the delivery of food relief. Examples of NGOs working in South Sudan include Médecins Sans Frontières (MSF), Amref, Save the Children, Oxfam (Figure 9.19), Terre des Hommes and Care South Sudan.

Activity

Research one NGO, such as Oxfam, operating in South Sudan to investigate the precise nature of its relief work in local communities and the problems faced by the aid workers.

Figure 9.19 Vaccination programme for children in Rumbek, Lakes State, South Sudan

Consequences of global governance of the conflict for local communities

Intervention in South Sudan has benefits at the local scale. Positive effects of global governance for local communities in South Sudan include:

- aid agencies have co-ordinated their efforts to position essential supplies during the dry season, enabling easier access to emergency food during the wet season when roads are difficult
- villagers have received training in maintaining livestock health and use of fishing equipment to improve longer-term food security
- vulnerable children have received treatment for acute malnutrition, for example via the work of MSF
- aid agencies have negotiated access in areas where fighting is ongoing
- civilian protection camps have been expanded for IDPs, providing shelter and food during the rainy season and protection from the fighting, which tends to escalate in the drier season
- WHO and its health partners have set up cholera treatment centres and provide advice on hygiene and access to clean drinking water
- co-ordinated efforts have secured funds from various European governments.

Many inhabitants of Upper Nile and Unity states have suffered because transport routes have been blocked and aid organisations on which they depend have been extorted for their resources such as supplies of food and medicine. They are faced with problems such as having to abandon their homes, starvation and illness, and children being forced into military training camps.

Review questions

1 Give three ways in which the sovereign power of a state can be weakened.
2 Give three ways in which the UN attempts to maintain peace and security.
3 How does NATO attempt to resolve conflict?
4 What is the role of NGOs in conflict zones?
5 Why are treaties important in regulating conflict?
6 Identify three ways in which intervention in South Sudan has helped local communities.

9.4 How effective is global governance of sovereignty and territorial integrity?

Key idea

→ Global governance of sovereignty and territorial integrity has consequences for citizens and places

How the global governance of sovereignty issues has consequences for citizens and places

Where sovereignty has been threatened, global governance by the international community aims to re-establish effective political control by the state and restore the social and economic life of the country. Such interventions have consequences, not only for the state government but also for its citizens and the communities in which they live. Intended positive effects are outlined in Table 9.4.

Activity

Haiti is a very fragile state. Chronic political instability has been compounded by human rights violations, criminal activity including misappropriation of financial aid, ineffective rule of law and the impact of environmental hazards.

Investigate the aims, strategies and effectiveness of:

a) MINUSTAH, the UN's stabilising mission – www.un.org/en/peacekeeping/missions/minustah/
b) an NGO working in Haiti, such as Haven – www.havenpartnership.com/about-haven

How the global governance of territorial integrity issues has consequences for citizens and places

Sovereignty and territorial integrity are interrelated. Table 9.5 shows benefits of global governance where territorial integrity has been threatened.

Unintended effects of military intervention which may exacerbate already existing inequalities and injustices include:

- increased numbers of civilian casualties
- displacement of population (IDPs and refugees)
- damage to housing/residential districts
- damage to infrastructure, including roads, limiting access to services such as clean water, health care, medicines
- food insecurity as farmers are unable to grow crops/tend livestock
- disruption to education
- escalation of violence and human rights violation, including ethnically targeted attacks and attacks on aid workers.

Table 9.4 Intended benefits of global governance for citizens and places where sovereignty has been threatened

Short-term effects	Longer-term effects
Humanitarian aid, via aid agencies and donated funds	Agricultural training to improve food security
Supply of food	Education programmes, which aim to prevent further conflict, help post-conflict rehabilitation and support minority groups
Access to supplies of clean drinking water	Building of democratic institutions
Supply of medicines and medical treatment	Technical assistance to improve legislative and administrative frameworks
Provision of shelter/safe havens for IDPs	Support democratic and fair electoral processes
Assistance for vulnerable refugees and returnees	Upholding of human rights, including training of police and military, and reinforcement of norms, treaties, laws
Maintain peace, strengthen rule of law and protect civilians	Integrate gender equality into policies and practices

Table 9.5 Intended benefits of global governance for citizens and places where territorial integrity has been threatened

Short-term effects	Longer-term effects
Security and protection of civilians in conflict zones	Mediation and fostering of co-operation
Periods of ceasefire negotiated and monitored	Development of sustainable food and water supplies
Border control to facilitate movement of people and goods	Improving trade relationships to help reduce effects of economic shock
Human rights of minority groups/women and children/civilians monitored for further response	Improve business environment by counter-corruption and diminishing terrorism financing
Early warning of renewed/potential conflict	Cyber defence
Reduce forced conscription of child soldiers	Restore territory according to international law
Rule of law strengthened	Support transition to democracy/fair elections and representation
Assistance for IDPs and their return	Re-establish state authority and state apparatus

Case study: The impact of global governance of sovereignty/territorial integrity in Mali

There are significant socio-economic and environmental contrasts in Mali. The north is a vast area of desert and semi-desert in which the Tuareg are the dominant ethnic group (Figure 9.20). The south has the most economic activity as well as the capital, Bamako, located on the Niger (Figure 9.21).

Figure 9.20 Tuareg at the Festival au Désert near Timbuktu, Mali

Figure 9.21 Dyeing and rinsing cotton cloth on the bank of the Niger river near Bamako, Mali

Gold, cotton and agricultural exports generate income but overall this is a very poor, landlocked country which depends heavily on foreign aid and migrant remittances.

Sovereignty/territorial integrity issues

In 2013 Malian interim authorities requested the assistance of France 'to defend Mali's sovereignty and restore its territorial integrity'. This was in response to a military coup d'état in 2012 and subsequent insurgency, including continued terrorism.

These challenges to sovereignty and territorial integrity can be explained by the following:

● International boundaries delineated by European colonial powers in the early twentieth century had little regard for tribal lands, resulting in arbitrary division of the Tuareg ethnic group (Figure 9.22).

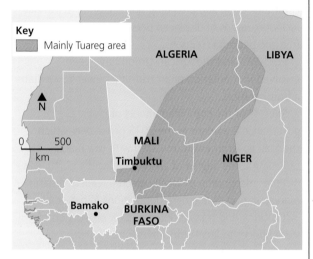

Figure 9.22 Tribal lands of the Tuareg in Sahelian and Saharan Africa (Source: http://trama-e-ordito.blogspot.co.uk/2012/04/tuareg-lo-spirito-degli-uomini-blu.html)

● The Tuareg declared independence for Awazad, an area of northeast Mali over which they claim territorial and cultural rights.
● There was fighting in the north to control routes for both legal trade and illicit smuggling, for example the Tilemsi valley.
● There has been ineffective state governance of the north, which has been marginalised and neglected from Bamako.
● There is not just a centre–periphery divide in Mali; in addition to the Tuareg there are many other significant ethnic groups, for example the Songhai are prominent in the Gao area.

Global governance strategies

Global and regional institutions have intervened to resolve the sovereignty and territorial integrity issues. Their overarching aim is to sustain the global system of sovereign nation-states by restoring and maintaining Malian sovereignty and territorial integrity.

➜

The United Nations Multidimensional Integrated Stabilization Mission in Mali (MINUSMA) was established by the UN Security Council in 2013. It aims to support the political process and stabilise Mali, ensure security, protect civilians, assist re-establishment of state authority and promote and protect human rights.

Currently just over 9000 military personnel, around 1000 police and more than 1300 other international and local staff are involved. They operate in the main population centres, keeping open important lines of communication and providing humanitarian assistance, including return of displaced persons and preparation for free and peaceful elections.

ECOWAS and the African Union have been involved in mediation and returning power to civilian administration.

Success of the combined effects of global governance is evident in the 2015 peace deal which the Mali government formulated with the Tuareg, providing some degree of autonomy for the north. This includes recognition of locally elected leaders, greater representation of northern populations in national institutions, and transferring a greater proportion of state budget to local authorities in the north.

NGOs provide assistance to local communities. Examples include:
- Population Services International: reproductive health projects
- Care: food insecurity and poverty alleviation
- World Education Mali: addressing educational barriers to literacy
- Solidarités International: water, hygiene, sanitation and food security in the northern settlements of Timbuktu, Kidal and Gao, and in Koulikoro in the south (Figure 9.23).

Despite the presence of MINUSMA forces, this is increasingly difficult work in areas of armed bandits involved in smuggling, car-jacking, kidnapping (of tourists, NGO workers and diplomats for ransom money) and land mines.

Opportunities for stability, growth and development and challenges of inequality and injustice

In Mali the effects of global governance have been limited, there is continuing instability, and the huge inequalities, from which many of the problems stem, remain a challenge. Nevertheless, HDI figures show slow improvement overall (Table 9.6).

Figure 9.23 Location of Azawad in northeast Mali

Table 9.6 Opportunities for stability, growth and development, and the challenges of inequality and injustice in Mali

Opportunities	Challenges
Stability: • MINUSMA is establishing mechanisms for political, social and economic stability. Provision of military force and strengthening of police are designed to minimise terrorist activity, and to support the government in providing more effective legislation and rule of law, and to set up democratic elections. • Protection of human rights is a priority, reinforcing international norms of behaviour. • Re-establishing sovereign state control of the north and territorial integrity within internationally recognised boundaries is also a UN aim.	Inequality: • Socio-economic inequalities between south and north remain; the north is undeveloped and under-represented politically. • Deep-rooted cultural and linguistic divisions within and between Tuareg, Arab, Songhai and other ethnic groups in the north are a major challenge for peace. The international community has mistakenly attempted to deal with the north as one area; not all its peoples see Azawad as common territory.

Opportunities	Challenges
Growth: ● GDP per capita is recovering after it fell in the two years following the coup (US$495 in 2015). ● Reduction of import dependency and increased economic diversity are essential for Mali to reduce its trade deficit; main exports are gold, 72%, and cotton, 10%. The World Bank supports smallholder farmers by enhancing supply chains for farming and fishery products, for which Mali has a strong comparative advantage, through the Agricultural Competitiveness and Diversification Project. Development: ● Development indices for Mali, 2015 include:	● Like much of sub-Saharan Africa, inequalities between urban and rural areas in Mali demonstrate the problem of limited infrastructure and service provision; many villagers feel abandoned and disconnected from the state.

HDI	Male 0.455, Female 0.350
Mean years of schooling	2 years
Secondary education	10.9% of population
Literacy	33.4% of population
Life expectancy	55 years
Living on < US$1.25/day	50.43% of population

% population with access to services	Rural	Urban
Electricity	8.5%	86%
Water pipes	33%	92.5%
Paved roads	11.5%	84%
Sewage system	1%	43%
Health clinics	48%	89%

● The UN mission is paving the way for stability to precede the development process. Other international organisations in co-operation with the Mali government include Water Aid, working with local communities in urban and rural areas to secure sustainable water supplies and sanitation – a vital opportunity for citizens to become healthier, better educated and more food secure.

Injustice:
● The government's inability to police its own country has led to high levels of human trafficking, drug smuggling, kidnapping, embezzlement and corruption.
● The government is unable to protect citizens from human rights abuse. In particular, children and women are subject to a disproportionate amount of domestic and agricultural work, early marriage, female genital mutilation (FGM), military conscription and unsafe conditions in gold mines.
● There are abductions, killings, bombings and the problems of land mines.
● There are high rates of maternal and child mortality.

Activities

Georgia's territorial integrity is threatened by the declarations of independence of Abkhazia and South Ossetia. These are supported by Russia with which they both share a border (Figure 9.1, page 263). Tension between these states and Georgia came to a head in 2008 when there was serious military conflict between Russia and Georgia.

1 Investigate:
   a) the reasons why South Ossetia and Abkhazia have claimed independence
   b) the impact of conflict on the citizens of these two areas
   c) the reasons why the UNOMIG mandate has not been extended.
2 Identify the positive effects of intervention by International Alert (www.international-alert.org/what-we-do/where-we-work), a peacebuilding organisation, in the two areas.
3 Examine the developing relations between NATO and Georgia.

Review questions

1 Give three short-term benefits of global governance in areas where sovereignty has been threatened.
2 Give three possible longer-term benefits of global governance for areas where sovereignty has been challenged.
3 Identify three unintended effects of intervention in areas where sovereignty is threatened.
4 Explain the challenges to the territorial integrity of Mali.
5 What are the global governance strategies to resolve sovereignty issues in Mali?
6 Explain how intervention has created opportunities for stability in Mali.

Practice questions

Photograph 1

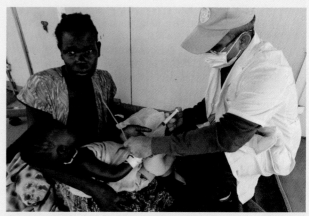

Photograph 2

Figure A Different types of assistance by the UN in areas of conflict (Source: www.huffingtonpost.com/hilde-johnson/the-un-in-south-sudan-imp_b_4920176.html)

Structured/data response

1 a) Study **Figure A**, which shows different types of assistance by the UN in areas of conflict.

   i) Suggest **two** ways in which global governance of conflict influences flows of people in Figure A.
                                     [2 marks]

   ii) Explain **one** consequence of UN intervention in areas of conflict for local communities in Figure A.
                                     [3 marks]

   iii) Evaluate the use of the photographs in Figure A for showing UN intervention in areas of conflict.
                                     [4 marks]

b) With reference to a **case study**, explain the contemporary challenges to the sovereignty of a country. [8 marks]

Essay

2 'The impact of global governance to resolve threats to territorial integrity is always beneficial.' Discuss. [16 marks]

Part 3

Geographical debates

Chapter 10

Climate change

10.1 How and why has climate changed in the geological past?

> **Key idea**
> → The Earth's climate is dynamic

The Earth's climate has undergone huge changes in the past 100 million years. These changes have been particularly rapid in the last 2.5 million years, known as the Quaternary period, with numerous transitions from cold to warm conditions. Just 20,000 years ago, ice sheets covered most of northwest Europe; today, ice sheets and glaciers are shrinking and global temperatures are rising at an unprecedented rate.

Methods used to reconstruct past climates

Our knowledge of the Earth's past climates comes largely from sediments in marine, lake and peat bog environments, and from ice cores. Other evidence is obtained from tree rings and fossils (Table 10.1).

Fluctuations in the Earth's climate in the past

Over hundreds of millions of years, the Earth's climate has fluctuated between greenhouse and icehouse conditions. During greenhouse conditions atmospheric CO_2 concentrations, global temperatures and sea levels are high. Icehouse conditions are the opposite: low CO_2, low global temperatures and large parts of the continental surface submerged by ice. Glacial and inter-glacial periods occur in icehouse phases but on a much smaller timescale. The most recent, the Pleistocene, lasted for around 1 million years.

Long-term changes

During the mid-Cretaceous period around 100 million years ago, average global temperatures were 6–8°C higher than today. Sub-tropical conditions extended from Antarctica to Alaska and there were no polar ice caps. This exceptional warm phase, which coincided with atmospheric CO_2 levels five times higher than today's, lasted for tens of millions of years. Also at this time the continents had a very different configuration

Table 10.1 Methods used to reconstruct the Earth's past climate

Sea-floor sediments	The fossil shells of tiny sea creatures called foraminifera, which accumulate in sea-floor sediments, can be used to reconstruct past climates. The chemical composition of foraminifera shells indicates the ocean temperatures in which they formed.
Ice cores	Ice cores from the polar regions contain tiny bubbles of air – records of the gaseous composition of the atmosphere in the past. Scientists can measure the relative frequency of hydrogen and oxygen atoms with stable isotopes. The colder the climate, the lower the frequency of these isotopes.
Lake sediments	Past climates can be reconstructed from pollen grains, spores, diatoms and varves in lake sediments. Pollen analysis identifies past vegetation types and from this infers palaeoclimatic conditions. Pollen diagrams (Figure 10.1) show the number of identified pollen types in the different sediment layers. Diatoms are single-celled algae found in lakes with cell walls made of silica. They record evidence of past climates in their shells (see foraminifera above). Varves are tiny layers of lake sediment comprising alternating light and dark bands. The light bands, formed from coarser sediments, indicate high energy, meltwater run-off in spring and summer. The darker bands, made up of fine sediment, show deposition during the winter months.
Tree rings	The study of tree rings or dendrochronology is the dating of past events such as climate change through study of tree ring (annule) growth. Annules vary in width each year depending on temperature conditions and moisture availability.
Fossils	Plants and animals require specific environmental conditions to thrive. Some, such as coral reefs, are highly sensitive to temperature, water depth and sunlight. Where they exist in the fossil record they can be used as proxies for climate. Animals are more adaptable. However, some herbivorous dinosaur species only survived in sub-tropical habitats.

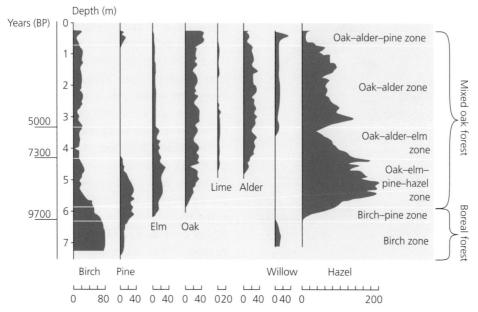

Pollen (except hazel) is expressed as a percentage of the total tree pollen.
The hazel scale relates to pollen count per unit of peat.
Age is shown on the left; equivalent modern forest types are on the right.

Figure 10.1 Pollen diagrams

(Figure 10.2), which affected ocean circulation and the Earth's energy budget. A further, but short-lived, spike in global temperatures occurred 55 million years ago. This is known as the Palaeocene–Eocene thermal maximum, when average global temperatures peaked around 23°C. At the start of the Oligocene period, 35 million years ago, there was a rapid transition to colder conditions, which have continued to the present day. This change was related to a major reduction in atmospheric greenhouse gases (especially CO_2).

Glaciation of Antarctica

Today the entire continent of Antarctica is covered by a vast ice cap that extends to sea level. This is the largest glacial system on the planet, containing $25–30 \times 10^6$ km³ of glacial ice. The ice cap is so thick that only the tops of the highest mountains appear above the ice. Yet, 40 million years ago, the fossil record shows that the continent experienced sub-tropical conditions. The descent of Antarctica into a permanent icehouse state occurred rapidly around 35 million years ago.

This transition to icehouse conditions has been explained by changes in atmospheric CO_2 and tectonic processes.

- CO_2 levels dropped abruptly 35 million years ago, from 1000–1200 parts per million (ppm) to 600–700 ppm.

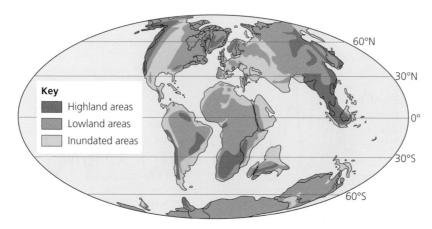

Figure 10.2 Earth 100 million years ago

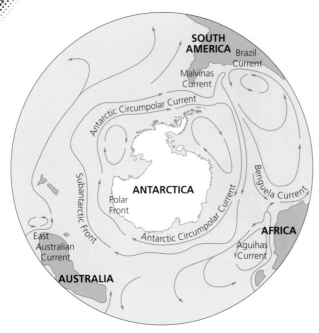

Figure 10.3 Antarctica and the Antarctic Circumpolar Current

- Continental drift: the movement of Antarctica towards the south pole, and away from South America and Australia, isolated the continent. This allowed the Antarctic Circumpolar Current to insulate Antarctica from the warmer water further north (Figure 10.3).
- The growth of the South Sandwich Islands' submerged volcanic arc disrupted deep-water ocean currents around Antarctica, isolating the continent from warmer water from the South Pacific, Atlantic and Indian Oceans.

Quaternary glaciation

The Quaternary period spans the last 2.6 million years. The main feature of the Quaternary has been cyclic changes of climate, with long cold periods or **glacials** interrupted by shorter, warmer **inter-glacials**. Typically, glacials have lasted for around 100,000 years, inter-glacials

for 10,000–15,000 years. During glacial periods, average annual temperatures in northwest Europe remained well below zero. In the past 450,000 years there have been four major glacial episodes and four inter-glacials (Figure 10.4). The most recent glacial, the Devensian, reached its maximum around 20,000 years ago when approximately one-third of the continental surface was covered by snow and ice.

In Scotland, ice sheets were up to 1 km deep, and ice sheets covered most of northern England, the West Midlands, Wales and Ireland. As the climate warmed, the ice thinned and retreated, though glaciers remained in the mountains until c13,000 before present (BP). A brief cold spell or **stadial** (Loch Lomond Stadial or Younger Dryas) returned for around 1000 years until 11,700 BP. Confined to cirque hollows and U-shaped valleys in the uplands, this was the last time that glaciers occupied the British Isles.

> **Activities**
>
> Study Figure 10.4, which shows a reconstruction of global temperature change in the past 450,000 years.
>
> 1 Describe the main characteristics of the long-term trend of the global climate in the past 450,000 years.
> 2 How many glacials and inter-glacials have occurred during this period?
> 3 Compare the average length of time of glacials with that of inter-glacials.

The Holocene

The current geological period is known as the Holocene. It began at the end of the last glacial, 11,700 years ago. The Holocene is, therefore, an inter-glacial period: a brief interlude separating glacials which have dominated 90 per cent of the Quaternary. During the Holocene ice sheets and glaciers have shrunk and sea level has risen

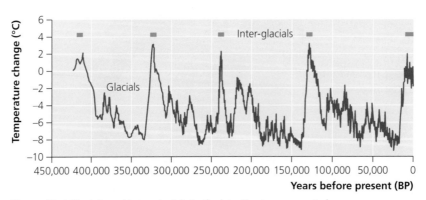

Figure 10.4 Glacials and inter-glacials in the late Quaternary period

by over 100 m, and with the exception of Antarctica and Greenland, ice sheets have disappeared from the continental surface. Their remnants – ice fields and valley glaciers – survive only in high mountains such as the Himalayas and Andes, and in high-latitude regions like Alaska.

Although the global climate has warmed steadily throughout the Holocene, warming has not been continuous. There have been several relatively short-lived episodes of warming and cooling – 6000 years ago global temperatures were 1–2°C higher than today's and an early medieval warm period occurred between 1100 and 1300. This was followed by the 'little Ice Age' (1550 to 1850) when average global temperatures fell by 1°C and sent winters in Europe into deep freeze. The vast majority of climate scientists believe that climate change in the past 200 years has been driven by human activity. Indeed, the influence of people on the global climate has become dominant, so much so that many scientists believe we have entered a new geological period, called the Anthropocene.

Natural forcing and climate change in the geological past

Natural forcings, which induce change in the climate system, may be external or internal. External forcings include astronomical shifts in the Earth's orbit and axial tilt, the precession of the **equinoxes** and fluctuations in solar output. These forcings control the incidence and distribution of solar radiation on the planet's surface. However, external forcing alone does not explain the massive climatic shifts

of the past million years; internal forcings play an important role in global climate change. Among these forcings are volcanic eruptions, continental drift, changes in ocean circulation and fluctuations in atmospheric CO_2.

Milankovitch cycles

Milutin Milankovitch argued that long-term climatic shifts such as glacial cycles are caused by astronomical events such as changes in the Earth's axis and orbit and the precession of the equinoxes (Table 10.2). These external forcing mechanisms affect the amount of solar radiation reaching the planet's surface and its spatial and temporal distribution. They operate on timescales that vary from 10,000 to 100,000 years. Milankovitch identified climate cycles at 100,000, 43,000, 24,000 and 19,000 years, with long glacial periods followed by shorter inter-glacials.

Volcanic eruptions

Explosive eruptions pump huge amounts of volcanic ash and sulphur dioxide into the stratosphere. Such eruptions have the potential to change the global climate, at least in the short term. Although volcanic ash is quickly removed, sulphur dioxide is more persistent and has a cooling effect. In the atmosphere, sulphur dioxide is converted to sulphuric acid, which forms sulphate aerosols. They reflect solar radiation back into space and lower temperature in the **troposphere**. The Mount Pinatubo eruption in the Philippines on 15 June 1991, one of the largest in the twentieth century, injected 20 million tonnes of sulphur dioxide into the stratosphere and, over a period of three years, cooled the Earth's climate by 1.3°C.

Table 10.2 Climate change: external forcing mechanisms hypothesised by Milankovitch

Forcing mechanism	Effect
Obliquity (tilt of the Earth's axis)	Over a period of around 40,000 years, the Earth's axial tilt (perpendicular to its orbital plane) varies from 22 degrees to 24.5 degrees (current tilt is 23.4 degrees). When the tilt is close to 22 degrees, seasonal temperature differences are reduced, i.e. summers are cooler and winters are warmer. As a result, snow and ice, accumulated during winter, do not melt during the summer, allowing glaciers and ice sheets to expand. This has a positive feedback effect, increasing the reflection of incoming solar radiation and lowering temperatures even further.
Eccentricity of the Earth's orbit	The Earth's orbit around the Sun follows an elliptical path. The eccentricity of the orbit varies from near circular to markedly elliptical over periodicities of 96,000 and 413,000 years. With maximum eccentricity, differences in solar radiation receipt of around 30 per cent occur between perihelion (when the Earth is closest to the Sun) and aphelion (when the Earth is furthest from the Sun). Ice ages correspond to periods of maximum orbital eccentricity.
Precession of the equinoxes	The Earth gyrates on its axis like an enormous spinning top, so that the point in the Earth's orbit when the planet is closest to the Sun (perihelion) changes over time. This shift or precession which occurs with a periodicity of around 22,000 years is due to the gravitational influence of the Moon and Jupiter and affects the intensity of the seasons. If perihelion occurs during the northern hemisphere's winter, winters will be warmer and summers cooler. Snow and ice accumulating in winter will not melt completely in the following summer, so that ice and snow cover expands, eventually triggering a glacial period.

Plate tectonics and continental drift

Driven by plate tectonics and sea-floor spreading, the global distribution and configuration of the continents have changed dramatically over geological time – 250 million years ago the continents formed a single huge land mass: Pangaea. Since then the contents have drifted apart to their present-day position.

Tectonic changes on this scale have a direct impact on the global climate. In part they explain the periodic extreme shifts of the Earth's climate between greenhouse and icehouse conditions. For example, as a larger continental area occupies higher latitudes, the land area with permanent ice cover expands. This in turn increases global albedo and positive feedback, which forces global cooling.

Ocean circulation

Ocean currents are a vital component of the global energy budget, transferring surplus energy (i.e. warm water) from the tropics to the poles. Continental drift can modify ocean circulation and energy transfer. This happened around 5 million years ago with the formation of the Isthmus of Panama, which joined the North and South American continents, and closed the 'gateway' between the Pacific and Atlantic Oceans.

It is believed that this event intensified the Gulf Stream, conveying warm surface water from the Caribbean to the North Atlantic. As a result, evaporation and precipitation increased and the prevailing westerly winds deposited more precipitation in the North Atlantic, Europe and Siberia, diluting the salinity of the North Atlantic and Arctic Oceans. Meanwhile, reduced salinity weakened the downwelling of water in the North Atlantic, which acts as a pump driving not only the Gulf Stream conveyor but the entire global thermo-haline circulation (Figure 10.5).

The combination of less saline surface waters and the reduction in heat transferred by the Gulf Stream led to the expansion of sea ice in the North Atlantic and Arctic. This change was amplified by positive feedback effects, with increased reflection of solar radiation, and the insulation of warmer ocean waters sealed beneath the sea ice. This was the prelude to the onset of glaciation around 3 million years ago.

More recent shifts in the pattern of ocean currents in the North Atlantic are thought to have been responsible for climate change in Europe. Around 12,000 years towards the end of the last glacial, the melting of ice sheets in the Arctic reduced ocean salinity, weakening downwelling and the North Atlantic conveyor. This event heralded a return to glacial conditions in northern Europe. The resulting cooldown lasted for nearly a millennium and led to a resumption of valley glaciation in the Scottish Highlands, the Lake District and north Wales.

Natural greenhouse gases

We have seen that there is a close relationship between atmospheric CO_2 levels and average global temperatures (Figure 4.33 on page 129). Periods of icehouse Earth, such as the past 3 million years, correspond with low levels of CO_2 in the atmosphere which reduce the Earth's natural greenhouse effect. During the past 800,000 years, when the Earth's climate has been dominated by numerous glacial periods, CO_2 levels have fluctuated between just 170 ppm and 300 ppm. Yet in the Pliocene, 3–5 million years ago, atmospheric CO_2 concentrations were around 400 ppm and temperatures 2–3.5°C above today's average. Fifty million years ago CO_2 levels reached 1000 ppm; average global temperatures were 10°C higher, the poles were ice-free and sea level was 60 m higher than it is today.

Such changes beg the question of how CO_2 is removed from the atmosphere. Plate tectonics and continental drift offer an explanation. In the Tertiary era tectonic plate movements created extensive fold mountain ranges such as the Himalayas, Andes and Rockies. Uplift of these mountains increased rainfall, erosion and chemical weathering by rainwater

Key

→ Warm, less dense, surface current

→ Cold, dense, deep water current

Figure 10.5 North Atlantic circulation

charged with CO_2 (carbonic acid). Thus large volumes of CO_2 were removed from the atmosphere and transferred to storage in carbonate sediments in the oceans (see page 101, Chapter 4). At the same time, the increase in nutrients in the oceans stimulated vast blooms of phytoplankton, which also extracted CO_2 from the atmosphere. When these organisms died, the CO_2 was trapped in deep-ocean sediments.

Solar output

The Sun's output is not constant but varies markedly over time. On timescales measured in millennia or centuries, these variations may be shown to contribute to climate change. Observations of sunspot activity have been made for at least 400 years and used as a proxy for solar output. There is a positive correlation between the number of sunspots and solar energy output. Only in the past 30 years or so have satellites been able to measure solar irradiance more accurately.

Solar output follows an 11-year cycle. However, the difference in energy output between maximum and minimum sunspot activity is only 0.1 per cent: not enough to impact global climate significantly.

On longer timescales solar output is more variable. Near the end of the seventeenth century the number of sunspots declined to almost zero for several decades. This was the so-called Maunder Minimum and corresponded to the severe winters in Europe known as the 'Little Ice Age'. A similar period of low activity, the Dalton Minimum, occurred in the early nineteenth century (Figure 10.6). In the past 50 years, sunspot activity has been relatively high and coincides with a warming of the global climate. Nonetheless, isolating the impact of variations in solar output from other influences, such as volcanic eruptions, human activity and feedback effects, is difficult.

10.2 How and why has the era of industrialisation affected global climate?

Key idea
→ Humans have influenced the climate system, leading to a new epoch, the Anthropocene

In recent decades, humankind, due to population growth and technological advances, has become the dominant influence forcing environmental change.

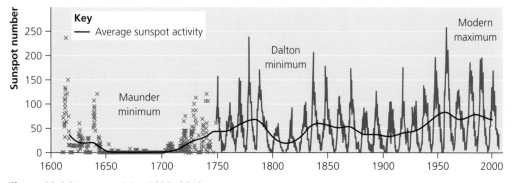

Figure 10.6 Sunspot activity 1600–2010

This change, which is both deliberate and inadvertent, marks the end of the Holocene period: today the Earth has entered a new geological epoch known as the Anthropocene.

Evidence that the world has warmed since the late nineteenth century

Strong evidence of worldwide warming in the past 100–150 years comes from several different sources: increases in global surface and atmospheric temperatures, shrinking of valley glaciers and ice sheets, rising sea level, increases in atmospheric water vapour, and declining snow cover and sea ice.

Increases in global temperatures

The consistent rise of global land and ocean temperatures in the twentieth century, and in particular in the past 40 years, is powerful evidence of climate change. Accurate records of global and ocean temperatures began in 1880. While rising temperatures have been the dominant trend of the past 135 years, several anomalies occurred in the period 1880–2014 (Figure 10.7). For example, average land and ocean temperatures for the month of June were below the twentieth century mean for every year between 1880 and 1937. In complete contrast, every year since 1977 has recorded above-average June temperatures. A similar pattern emerges when land and ocean temperatures are considered separately.

More remarkable is the steep rise of global temperatures in the twenty-first century – 2014 was the warmest year for mean global land and ocean temperatures since records began, with land and ocean temperatures 1°C and 0.57°C respectively above average. Perhaps more significant, 2014 was the 38th consecutive year that the annual global temperature rise had been above average, while nine of the ten warmest years since 1880 have occurred in the twenty-first century.

Activity

Log on to the NOAA website at www.ncdc.noaa.gov/cag/time-series/us.

a) Plot the time trend for average July temperatures in the USA 1901–2014.
b) Insert the best-fit trend line (see pages 543–44).
c) Describe the short-term temperature changes and the long-term trend in July for the period 1901–2014.
d) Create a second time trend for December temperatures. Compare the fluctuations and trend with July's.

Shrinking valley glaciers and ice sheets

Everywhere valley glaciers in the past century or so have retreated, and in some instances have disappeared completely. This trend is set to continue: in the Alps, valley glaciers may shrink by 80–96 per cent by the end of the century. Both satellite imagery and ground photographs document this change (Figure 10.8). Between 1961 and 2005 the thickness of small glaciers worldwide decreased on average by 12 m, the equivalent of 9000 km³ of water.

The Antarctica and Greenland ice sheets contain 97 per cent of the global ice store and in places are up to 4 km thick. Antarctic ice covers 14 million km² in three

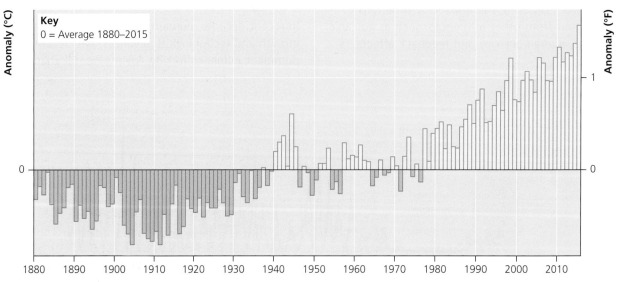

Figure 10.7 Global temperature anomalies June 1880–2014

Figure 10.8 Retreat of the Muir glacier, Alaska, 1941–2004

> **Activity**

Sea level is measured globally by satellites and locally by tidal-gauging stations. Local measurements can be highly variable. They are influenced by a range of factors, such as tectonic and isostatic land movements, ocean currents, configuration of the coastline and so on. Investigate the variability of local sea level change by logging on to the NOAA website: http://tidesandcurrents.noaa.gov/sltrends/sltrends_global.htm

a) Analyse the average sea level change at the following tidal-gauging stations around the UK coast: Lerwick, Stornoway, Aberdeen, North Shields, Sheerness and Newlyn.

b) Comment on your findings.

c) Data from the tidal-gauge station at Lagos in Portugal show that since 1900, the mean sea-level rise has been 1.50 mm/year ±0.24 mm/year.

   i) Refer to Chapter 15 (pages 535–36) and explain what this statistic means.

   ii) At the 68 per cent and 95 per cent confidence levels, estimate the possible range of mean sea-level rise at Lagos.

separate ice sheets: West Antarctica, East Antarctica and the Antarctic Peninsula. The Greenland ice sheet is smaller, covering 1.7 million km². At current rates of melting it would take centuries for the polar ice sheets to disappear, but like valley glaciers, they are shrinking fast. Present-day melting of polar ice sheets adds approximately 1 mm to sea level every year. Losses of ice are due to:

- warming of the atmosphere, which melts the ice surface
- warming, which produces meltwater that penetrates the ice and increases the velocity of glacier flow
- ocean warming, accelerating melting of ice sheets in coastal areas (in 2002 the Larsen B **ice shelf** in Antarctica covering 3250 km² broke up in less than a month).

Rising sea level

Sea level began to rise in the mid-nineteenth century. Since 1900 the average rise has been 1.0–2.5 mm/year. Recent evidence from satellite altimetry suggests that sea level is currently rising at a faster rate – 3 mm/year – and that this trend will continue throughout the twenty-first century. Two processes account for global sea level rise: the thermal expansion of the oceans as the world's climate warms and the melting of land-based ice sheets and glaciers.

Increasing atmospheric water vapour

Water vapour is the most important greenhouse gas (GHG). It traps huge amounts of energy radiated from the Earth's surface (Figure 10.9) and creates a natural greenhouse effect. The amount of water vapour in the atmosphere is directly related to temperature and rates of evaporation. This means that in a warmer world there will be more atmospheric moisture.

It is estimated that for every 1°C increase in temperature caused by enhanced CO_2 levels, rising levels of water vapour will double the warming. Moreover, large concentrations of water vapour will amplify the effects of warming by positive feedback; hence more vapour intensifies the greenhouse effect, increases evaporation, which in turn leads to more atmospheric vapour, more evaporation, more warming and so on.

Decreasing snow cover and sea ice

Satellite measurements reveal a decline in spring snow cover of 2 per cent per decade since 1966 in the northern hemisphere. This change has an impact on climate. Snow has a high albedo, reflecting 70–80 per cent of incoming solar radiation compared with just 10–20 per cent for soil and vegetation surfaces.

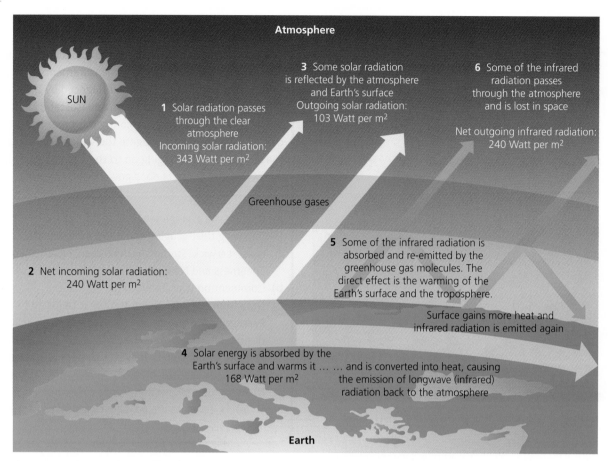

Figure 10.9 The greenhouse effect (Sources: Okanagan University College in Canada, Department of Geography; University of Oxford, School of Geography; United States Environmental Protection Agency (EPA), Washington; Climate change 1995, The science of climate change, contribution of Working Group 1 to the Second Assessment Report of the Intergovernmental Panel on Climate Change, UNEP and WMO, Cambridge University Press, 1996)

Diminishing snow cover therefore increases the absorption of solar radiation. With the Sun's energy used to warm the ground rather than melt the snow, air temperatures rise. Again, this creates a positive feedback, which explains the rapid warming currently taking place in the Arctic.

Sea ice is frozen sea water that floats on the ocean surface. It is found only in the polar regions and forms and melts with the seasons. In winter, Arctic sea ice covers on average 17–20 million km²; in summer this shrinks to between 4 million km² and 6 million km² (Figure 10.10).

Since 1979 when satellite monitoring began, the extent and volume of Arctic sea ice has declined dramatically (Figure 10.11). In summer the area of sea ice has decreased on average by nearly 8 per cent per decade, and by 3–4 per cent in winter. Arctic sea ice has also declined in thickness, from an average 3.6 m in 1980 to 1.9 m 30 years later. In September 2012 the area of Arctic sea ice shrank to a new minimum

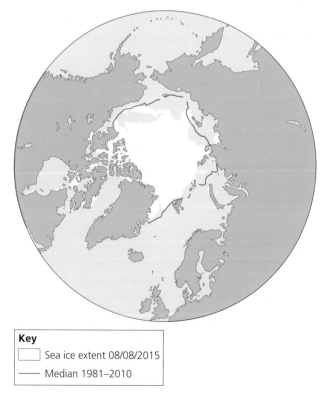

Key

☐ Sea ice extent 08/08/2015

— Median 1981–2010

Figure 10.10 Arctic sea ice extent, 18 August 2015

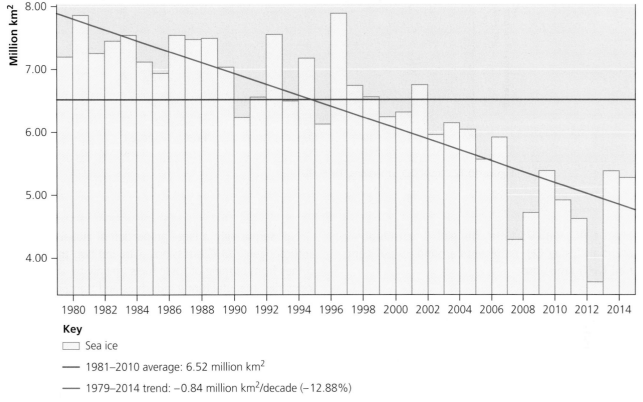

Figure 10.11 Arctic sea ice extent in September, 1979–2014

of just 3.63 million km² (Table 10.3). Less sea ice results in more solar energy absorption by the open sea surface, raising air temperatures and causing even more rapid melting.

In contrast to the situation in the Arctic, sea ice in the Antarctic has actually expanded by just over 1 per cent per decade (compared with the average for 1981–2010).

Table 10.3 Arctic sea ice extent at the September minimum (million km²)

2002	5.96	2009	5.39
2003	6.15	2010	4.93
2004	6.05	2011	4.63
2005	5.57	2012	3.63
2006	5.92	2013	5.35
2007	4.3	2014	5.28
2008	4.73	2015	5.79 (Aug 16)

> **Activities**
>
> 1 Summarise the changes in sea ice extent in the northern hemisphere 1979–2014 (Figure 10.11).
> 2 a) Calculate the Spearman rank correlation for the data in Table 10.3. (See Chapter 15, pages 541–42.)
>   b) What conclusions can you draw from your analysis?

Anthropogenic greenhouse gas emissions

Rising emissions of greenhouse gases

Greenhouse gases warm the Earth and its atmosphere by intercepting outgoing, terrestrial radiation and re-radiating it back to the surface and to the atmosphere. The principal GHGs are water vapour, carbon dioxide (CO_2) and methane (CH_4). For the past 200 years or so these gases have been increasing in the atmosphere. This trend, directly or indirectly, is the result of human activities.

CO_2 is the second most important GHG and accounts for more than three-quarters of all anthropogenic GHG emissions. Before 1800 its concentration in the atmosphere was fairly stable at around 280 ppm. In 2015 it passed 400 ppm for the first time: an increase of over 40 per cent in 200 years (Table 10.4 and Figure 10.12). Moreover, the increase in atmospheric CO_2 is accelerating: nearly half of the increase has occurred since 1960. Atmospheric concentrations of CH_4 are much smaller and are measured in parts per billion. Direct measurements of CH_4 began only in 1984. In that year, at sea level, they were 1735 ppb; by 2009 this figure had increased to 1890 ppb. Although a much slower rate of increase than CO_2, CH_4 is a more

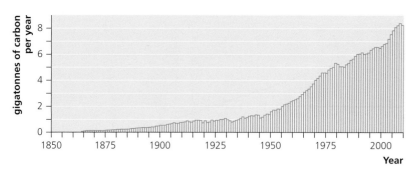

Figure 10.12 Global CO$_2$ emissions, 1850–2010

potent GHG, with 25 times the global warming potential. Today CH$_4$ represents around 15 per cent of all GHG emissions.

Table 10.4 Increases in global atmospheric CO$_2$, 1960–2015

	CO$_2$ concentrations (ppm)	Decadal increase (ppm)
1960	316	–
1970	326	10
1980	339	13
1990	354	15
2000	370	16
2010	390	20
2015	401	23 (estimated 2010–2020)

Reasons for rising GHG emissions since the pre-industrial era

Over the past two centuries GHG emissions have grown rapidly and at an accelerating rate. Three sets of factors explain this increase:

- The huge surge in demand for energy due to industrialisation and technological advances, particularly in manufacturing and transport.
- Massive global population growth, from 1 billion in 1800 to 7.4 billion in 2015, together with rising living standards.
- Land-use change, especially deforestation and the draining of wetlands for food production and urban development.

Since the first industrial revolution in Europe in the early nineteenth century, the world has relied heavily on energy from fossil fuels. Despite the development of nuclear power and renewables, this dependence on fossil fuels remains. Today fossil fuels still supply 87 per cent of the world's energy.

Not surprisingly, a large proportion of global anthropogenic GHG emissions – around two-thirds – comes from burning of fossil fuels (mainly by energy industries, manufacturing industries and transport), which release 10 billion tonnes of CO$_2$ annually (Figure 10.13). Since 1750, cumulative anthropogenic CO$_2$ emissions total 2000 gigatonnes (GT); three-quarters of these emissions are from burning fossil fuels.

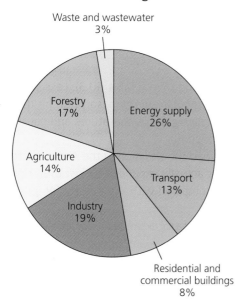

Figure 10.13 GHG emissions by source (Source: IPCC (2007))

Since the mid-twentieth century, the demand for fossil fuels has grown hugely in both scale and intensity. Although coal is no longer the leading fossil fuel, coal production in the early twenty-first century reached record levels (Figure 10.14). Two of the world's largest economies, China and India, are powered largely by coal. Unfortunately, coal is also the dirtiest fossil fuel: it emits nearly twice as much CO$_2$ as natural gas and 20 per cent more than oil. Expansion of the world economy between 1990 and 2012, stimulated by globalisation and the growth of EDCs (especially China),

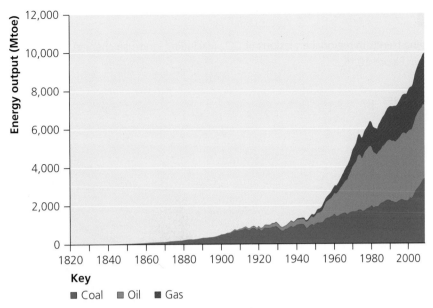

Figure 10.14 Historical production of fossil fuels, 1820–2009

increased the consumption of fossil fuels and according to the International Energy Agency led to a 51 per cent increase in CO_2 emissions.

Fossil fuels are only one, albeit the major, source of GHG emissions. Around one-third of GHG emissions comes from land-use changes and the emission of carbon from deforestation, drained wetlands and cultivated soils. These processes transfer carbon from the biosphere to the atmosphere. Today 40 per cent of the planet's land surface is used for agriculture compared with just 7 per cent in 1700. Estimates of the global loss of forest cover since 1700 vary from 15 per cent to 25 per cent. Today forests clothe only 31 per cent of the land surface. Meanwhile, deforestation, mostly in the tropics and sub-tropics, continues apace, with 5.2 million ha cleared between 2000 and 2010.

The changing balance of anthropogenic emissions around the world

Most CO_2 and other GHG emissions in the period 1850 to 1960 originated from the industrialised economies of North America and Europe. Historically the USA has dominated CO_2 emissions. Its historic emissions since 1850 (28 per cent) are almost equal to the combined emissions of China, Russia, Germany and the UK.

Since 1960, however, significant regional shifts have occurred in emissions of CO_2 and other GHGs. Asia's emissions have increased massively while those of North America and Europe have stabilised and in the case of Germany and the UK have actually declined. Early in the twenty-first century, China, with its

reliance on coal and its insatiable demand for energy to sustain its explosive economic growth, overtook the USA as the world's leading emitter of CO_2.

Global emissions remain highly uneven: the top ten CO_2-emitting countries account for nearly 80 per cent of all emissions. However, when CO_2 emissions are measured in per capita terms, ACs such as the USA, Australia, Germany and the UK are well ahead of the emerging economies of China and India. For example, the USA and Australia both emit an average of 17 tonnes/year per person of CO_2, compared with 5.4 tonnes by China and just 1.4 tonnes by India.

When emissions of other GHGs such as CH_4 from land-use changes are added to CO_2 from fossil fuel combustion, a slightly different regional picture emerges. Although China and the USA remain in first and second place, Brazil and Indonesia currently rank as the third and fourth largest emitters of GHGs. This is due to the effects of large-scale deforestation in Amazonia and Indo-Malaysia.

The enhanced greenhouse effect

We have seen that GHGs such as water vapour, CO_2 and CH_4 occur naturally in the atmosphere. There they have a warming effect, being transparent to short-wave radiation from the Sun but absorbing the Earth's long-wave radiation. Without the natural greenhouse effect the average temperature at the Earth's surface would be around 34°C lower than it is today.

Since the early nineteenth century the volume of GHGs in the Earth's atmosphere has increased rapidly. The most striking change has been an

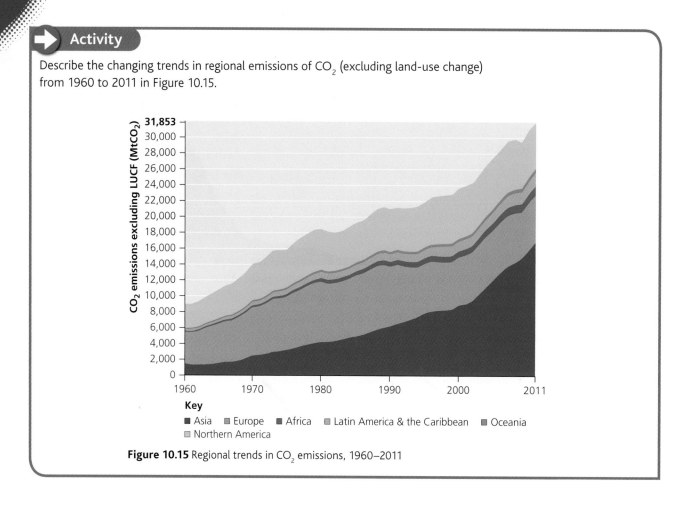

Activity

Describe the changing trends in regional emissions of CO_2 (excluding land-use change) from 1960 to 2011 in Figure 10.15.

Key
- Asia
- Europe
- Africa
- Latin America & the Caribbean
- Oceania
- Northern America

Figure 10.15 Regional trends in CO_2 emissions, 1960–2011

increase in CO_2 by one-third, from 280 ppm in 1800 to just over 400 ppm today. This has created an **enhanced greenhouse effect** due almost entirely to the consumption of fossil fuels, deforestation and other land-use change. As the Earth's climate has warmed, change has been amplified by increases in evaporation and therefore atmospheric water vapour, and the melting of permafrost in the Arctic releasing CH_4 and CO_2 (see page 116, Chapter 4).

The enhanced greenhouse effect increases absorption of long-wave radiation in the atmosphere and raises global temperatures. This trend is consistent with the history of climate change over the past 1 million years, which clearly shows that warming is associated with unusually high concentrations of GHGs in the atmosphere.

The global energy balance and human activity

The Earth and its atmosphere system is a closed system, with inputs of solar radiation and outputs of terrestrial radiation. When inputs and outputs balance, the system is stable, with no long-term changes in global temperature.

Figure 10.16 shows that only about 45 per cent of incoming solar radiation at the top of the atmosphere reaches and warms the Earth's surface. A small amount of insolation is reflected from surfaces such as snow and ice, around a third is reflected from clouds, and about one-fifth is absorbed by water droplets, ozone, CO_2 and dust in the atmosphere.

About two-thirds of outgoing, long-wave radiation from the Earth's surface is lost to space. The rest is absorbed by GHGs, especially water vapour, CO_2 and CH_4. Meanwhile, energy is also exported from the surface to the atmosphere by convection currents or thermals, and by latent heat, released when water vapour cools and condenses. Thus 84 per cent of all the heat energy leaving the Earth's surface returns as back radiation. The global energy balance underlines the important role of GHGs in stabilising and moderating global temperatures.

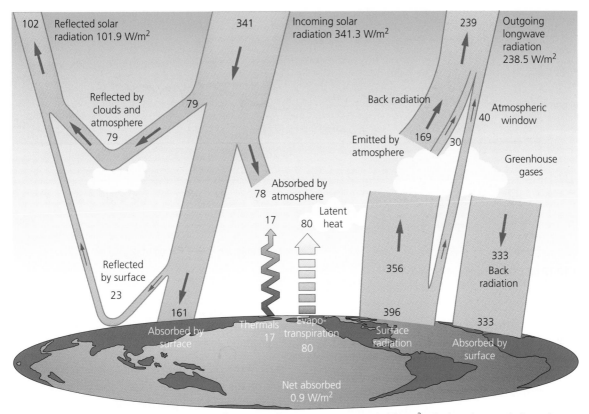

The global annual mean Earth's energy budget for the March 2000–May 2004 period (W/m²). The broad arrows indicate the schematic flow of energy in proportion to their importance.

Figure 10.16 Global energy balance (Source: Trenberth *et al.* 2009)

However, human activity in the past 200 years, through the consumption of fossil fuels and land-use change, has upset this delicate balance.

● Increasing concentrations of GHGs absorb a larger proportion of terrestrial radiation. This is returned to the surface as back radiation, raising average global temperatures.
● Rising global temperatures increase evaporation, transfer more latent heat to the atmosphere and increase concentrations of water vapour, the most important GHG.
● Rising global temperatures melt snow, glaciers and sea ice in the Arctic, reducing albedo and the reflection of incoming solar radiation, adding heat energy to the Earth–atmosphere system. Deforestation also reduces albedos, increasing absorption and the warming effect.

Case study: The contribution of the UK to anthropogenic GHG emissions

Historic data on the UK's carbon emissions have been reconstructed as far back as 1751, before the start of the Industrial Revolution (Table 10.5).

Table 10.5 UK's CO_2 emissions from burning fossil fuels: 1751–1900 (1000s tonnes of carbon) (Source: CDIAC)

1751	1800	1850	1900
2,252	7,269	33,468	114,558

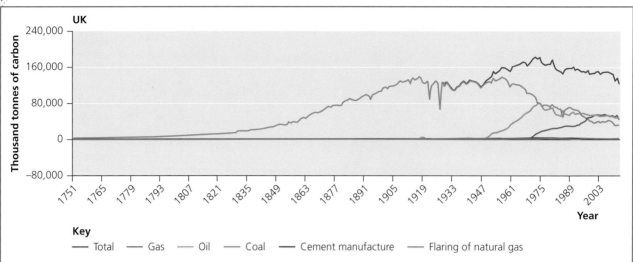

Figure 10.17 UK CO$_2$ emissions, 1751–2011

It was not until the early nineteenth century that industrialisation in the UK, powered by coal, took off. Significant emissions of CO$_2$ were the result, and in the course of the century these emissions increased sixteenfold (Figure 10.17). The UK remained dependent on coal as its main energy source for much of the twentieth century. Although coal production peaked in 1916 (137 million tonnes), as late as 1961 output was still more than 120 million tonnes. Until the early 1970s, reliance on coal together with an expanding economy meant that carbon emissions increased year-on-year. However, this trend was occasionally interrupted, notably when GDP and economic activity fell. An abrupt decline in emissions occurred for example in 1921 during the miners' strike and also in the general strike of 1926. There was also a marked decline in the early 1980s, associated with economic recession and widespread unemployment. More recently, the global financial crisis of 2009 caused a sudden reduction in CO$_2$ emissions from fossil fuels.

The UK's CO$_2$ emissions peaked in 1971. Since then, annual emissions have fallen by around one-third. There are several reasons for this:

- The shift away from coal as the primary fuel for electricity generation (and for space heating) to cheaper natural gas (Figure 10.18).
- The development of nuclear power stations (particularly in the 1970s and 1980s) and renewable sources of energy since the 1990s.
- Improvements in energy conservation through more energy-efficient homes, offices and factories.

- International obligations and legally binding reductions in carbon emissions.

In 2014 the UK's carbon emissions fell by a record 9 per cent on the previous year. In part this reflected a fall in energy demand owing to the mild winter and spring. Coal consumption was the lowest since the 1850s, with two major coal-fired power stations decommissioned and the largest, Drax in North Yorkshire, switching more of its capacity to biomass fuels. In addition, an unprecedented 15 per cent of electricity generation came from renewables. The UK government's drive to **decarbonise** the British economy is likely to lead to further reductions in carbon emissions in the future.

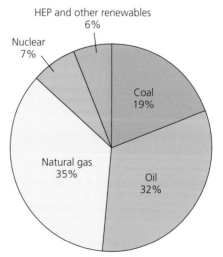

Figure 10.18 UK energy mix

Case study: The contribution of China to anthropogenic GHG emissions

Table 10.6 China's CO_2 emissions from burning fossil fuels: 1899–2013 (1000s tonnes of carbon)

1899	1950	1970	1990	2000	2010	2013*
26	21,465	210,422	671,051	928,601	2,459,645	2,490,000

Source: CDIAC *estimates revised down by 14 per cent due to low carbon content of Chinese coal

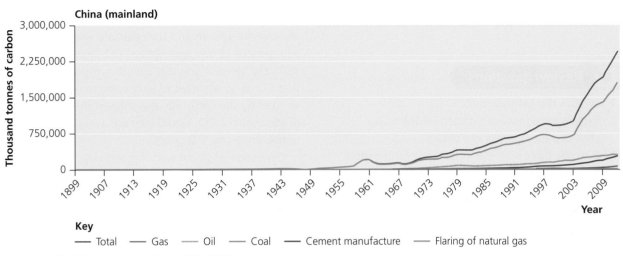

Key

— Total — Gas — Oil — Coal — Cement manufacture — Flaring of natural gas

Figure 10.19 China's CO_2 emissions, 1899–2009

Despite China's huge population, until the 1970s the country's economy was overwhelmingly rural. With an economy largely dependent on biofuels (timber, animal waste, etc.), total emissions of carbon from burning fossil fuels were modest. After 1978 the situation changed dramatically (Figure 10.19). An abrupt U-turn of policy by China's leaders moved the country away from a strict command economy and the country embraced the free market. Economic liberalisation simulated international trade and foreign direct investment. Thus China emerged as a major player in the global economy, with economic growth based on export-led manufacturing. Today China controls 12 per cent of the world's exports, compared with just 1 per cent in 1970.

Industrialisation was accompanied by spectacular urbanisation; involving hundreds of millions of people, the migration from rural to urban areas in China in the past 40 years was the largest population movement in history. Economic progress has raised average incomes to unprecedented levels. GDP per capita rose from $US299 in 1980 to $US12,763 in 2012.

China's economic development and industrialisation were made possible by massive energy consumption. Most of this energy came from the country's huge indigenous reserves of coal (Figure 10.20). Inevitably, this produced large increases in carbon emissions (Table 10.6).

Unhampered by international protocols like Kyoto, carbon emissions rose two and a half times in the period 2000 to 2014. Today China consumes almost as much coal – 1845 million tonnes in 2014 – as the rest of the world together and in 2006 overtook the USA as the world's largest emitter of carbon. Although emission rates have slowed recently, coal still provides nearly two-thirds of China's energy. China's aim to reduce CO_2 emissions by 40 per cent in the period 2005–20 appears to have little chance of success.

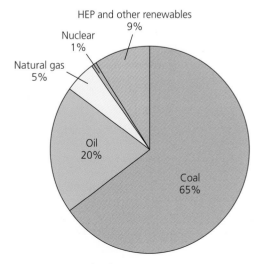

Figure 10.20 China's energy mix, 2014

Log on to the CDIAC database at http://cdiac.ornl.gov/ CO₂_Emission/timeseries/regional.

a) Plot time-series charts (for at least 50 years) for total CO_2 emissions for an AC, EDC and LIDC of your choice (Chapter 15, Geographical Skills, pages 516–17).

b) Justify your choice of chart type.

Review questions

1 What is the Anthropocene?
2 Explain the difference between ice sheets, ice shelves, valley glaciers and sea ice.
3 Describe three processes responsible for the rapid melting of the ice sheets and glaciers.
4 Describe the feedback effect causing the rapid rise in temperatures in the Arctic.
5 Describe the relative importance of water vapour, CO_2 and CH_4 as GHGs.
6 What is meant by:
   a) the enhanced greenhouse effect?
   b) the global energy balance?

10.3 Why is there a debate over climate change?

Key idea
→ Debates on climate change are shaped by a variety of agendas

Historical background to the global warming debate

The historical evidence for global warming

The idea of global warming goes back nearly two centuries. The greenhouse effect was discovered as early as 1824 by the French physicist Joseph Fourier. In 1862 John Tyndall first suggested that certain gases – water vapour and CO_2 – trapped heat escaping from the Earth–atmosphere system and that **glacials** were associated with periods of low atmospheric CO_2. Meanwhile, in 1896 the Swedish scientist Svente Arrhenius observed that CO_2 is an absorber of long-wave radiation emitted by the Earth. He also showed that a doubling of atmospheric CO_2 would increase average global temperatures by 5°C or 6°C.

Discussion of global warming then lapsed for several decades until 1938 when Guy Callender, an amateur scientist, linked global warming in the nineteenth century to emissions of CO_2 from burning fossil fuels. At the time his assertion aroused little interest because it was widely believed that any excess CO_2 would be absorbed immediately by the oceans. But this view changed dramatically in 1957. Hans Suess and Roger Revelle discovered a complex chemical process which limited the capacity of the oceans to absorb CO_2. This meant that the residence time of CO_2 in the atmosphere was much longer than previously thought. It was from this time that the debate on global warming finally took off.

In 1958 accurate measurements of global atmospheric CO_2 began at the Mauna Loa observatory on Hawaii. This database provided incontrovertible proof that CO_2 concentrations in the atmosphere were increasing year by year (Figure 10.21). This trend, known as the Keeling curve, shows that today, atmospheric CO_2 is at its highest level for 700,000 years.

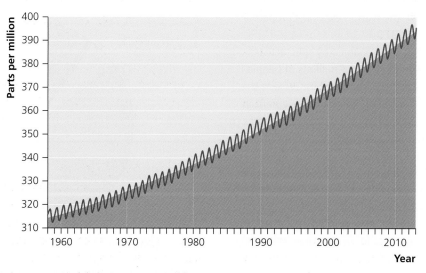

Figure 10.21 Global CO_2 mean monthly concentrations in the atmosphere

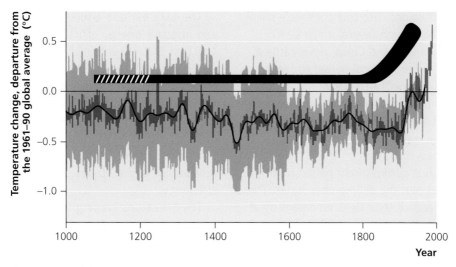

Figure 10.22 Global temperature changes 1000–2000 AD

Meanwhile, more evidence of global warming began to accumulate. Michael Mann's famous 'hockey stick' chart shows average global temperatures over the last millennium compared with the 1961–90 average, with a sharp rise since the onset of industrialisation (Figure 10.22).

The global warming debate

A long-running debate took place in the 1970s between scientists who argued that global warming was a reality and those who favoured global cooling and an imminent new glacial period. However, as satellite imaging and computer modelling advanced, the argument swung in favour of global warming. Moreover, computer models revealed that human activity, by releasing CO_2 and other GHGs, was probably the main forcing agent of global warming and climate change. Such was the concern that in 1988 the International Panel on Climate Change (IPCC) was set up.

In the past two decades a scientific consensus has emerged: although there remains some dissent, 97 per cent of climate scientists support the view that global warming is taking place. Yet while there is little doubt that the planet's climate has warmed, there is still debate about its causes. Dissenters argue that rising temperatures in the first half of the twentieth century were due to increased **solar output** and exceptional volcanic activity. But this argument has been undermined by the continued rise in temperatures since 1950, despite reductions in both solar and volcanic activity.

The potential impact of global warming is sufficiently alarming to convince even the most sceptical politicians to take action. For example, once global temperatures increase above the 2°C threshold (compared with pre-industrial levels), it is widely held that abrupt and irreversible climate change will begin to occur.

Already there is clear evidence that the Greenland ice cap is melting. Other forecasts suggest fundamental changes in global oceanic circulation and the release of massive amounts of carbon from tropical forests, soils, **permafrost** and undersea stores (as **methane hydrates**). In a nightmare scenario, these changes could accelerate average global temperatures by up to 10°C by the end of the century. Even in the USA, until recently highly sceptical of climate change, two-thirds of the population now accept the causal linkage between global warming and anthropogenic CO_2 emissions.

> **Activity**
>
> Log on to www.climatechange.procon.org. Outline and evaluate the arguments that global warming and climate change are explained by natural processes and not by anthropogenic carbon emissions.

Climate change and the role of governments and international organisations

Climate change is a global problem. Its solution requires a co-ordinated response by all countries. A number of international organisations, notably the UN and the EU, have taken the lead in the battle against global climate change.

The United Nations

In 1992 41 countries joined an international treaty, the UN Framework Convention on Climate Change (UNFCCC), to consider what action should be taken to limit global warming. The efforts of the UNFCCC culminated in the Kyoto Protocol of 1997. For the first time it set legally binding targets on countries to reduce their GHG emissions. 192 countries were parties to the Kyoto Protocol, though several major GHG emitters, including the USA and China, never ratified the treaty. The first Kyoto Protocol period ended in 2012; a second runs from 2013 to 2020.

Kyoto recognised that ACs, through their cumulative GHG emissions over the past 150 years, were primarily responsible for global warming and climate change. It therefore placed a heavier burden on these countries to take action. Kyoto's first commitment period (2008–12) involved 37 countries plus the EU member states. These countries agreed to reduce their GHG emissions by at least 5 per cent below 1990 levels by 2012. The second commitment set a more ambitious target: 18 per cent below 1990 levels by 2020. A number of countries withdrew from Kyoto at this stage, including Canada, Japan and Russia.

International climate change conferences under the auspices of the UN, with delegates from all countries, take place every year. Their aim is to achieve legally binding agreement to combat climate change. The priority is to reduce GHG emissions and limit global temperature increases to no more than 2°C above the pre-industrial average. Unfortunately, the conferences held between 2009 and 2014 failed to reach agreement. Meanwhile, global carbon emissions have continued to rise.

The European Union

The EU leads the world in its commitment to tackling climate change (Figure 10.23). In the past 25 years it has compiled a comprehensive package of measures to reduce GHG emissions through its European Climate Change Programme (ECCP). Each EU state also has in place its own policies which build on and complement the ECCP.

The EU's earliest efforts to limit carbon emissions and improve energy efficiency date back to 1991. The first ECCP was launched in 2000, the second in 2005. Through the ECCP the EU has set targets for reducing GHG emissions up to 2050. The aim is to transform the EU into a low-carbon economy. Taking 1990 as the base year, legally binding targets commit member states by 2020 to a 20 per cent cut in GHG emissions, 20 per cent of electricity generated from renewables and a 20 per cent improvement in energy efficiency.

The Emissions Trading System (ETS) is the cornerstone of the EU's climate change policy (see page 135) and covers 45 per cent of emissions from the EU. This **cap-and-trade** scheme targets a 21 per cent reduction in emissions from power stations, industry and aviation by 2020 (compared with 2005 emissions). National Emissions Reduction Targets cover the remaining 55 per cent of GHG emissions, mainly from agriculture, housing, waste and transport. Each EU country has binding annual targets up to 2020 for cutting emissions in these four sectors. Member states also have targets for expanding the contribution of renewables to national energy production and for improvements in energy efficiency.

The UK

Within the international framework of EU policies and the Kyoto Protocol, the UK has freedom to develop its own policies to tackle GHG emissions. The Climate Change Act (2008) commits the government to reduce emissions by at least 80 per cent by 2050, compared with 1990, and to develop a more energy-efficient, low-carbon economy.

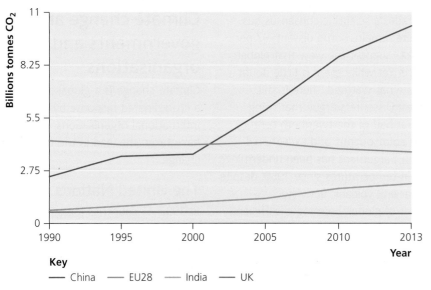

Figure 10.23 CO_2 emissions from China, the EU, India and the UK, 1990–2013

The UK strategy includes:

- setting carbon budgets (five-year periods) as stepping stones towards the 2050 target – for example, the reduction targets for the periods 2013–17 and 2018–22 are 29 per cent and 35 per cent respectively
- reducing the demand for energy using domestic smart meters, promoting energy efficiency (e.g. insulation in buildings) and reducing GHG emissions from transport and agriculture
- investing in low-carbon technologies such as carbon capture and storage (see page 321) and encouraging the growth of renewables (e.g. wind, solar, biomass) by offering subsidies, and reducing GHG emissions from power stations
- carbon taxes to switch electricity generation from coal to greener fuels. In 2014 there were just ten coal-fired power stations in the UK, but they were responsible for one-fifth of UK carbon emissions. Introduced in 2013, taxed carbon emissions started at £16/tonne, with a planned rise to £70/tonne by 2030. Although in 2015 the tax was frozen at just over £18/tonne amid concerns about cost of electricity to consumers, three of the UK's remaining coal-fired stations – Ferrybridge, Eggborough and Longannet – closed in 2015/16. Closures were due in part to the carbon tax and in part to low global energy prices. Carbon taxes have been levied on vehicle CO_2 emissions since 2002.

India

India did not ratify the Kyoto Protocol. Like other middle- and low-income countries, the Indian government argued that rich countries should shoulder the costs of reducing carbon emissions. This view prevented the USA – at the time the world's biggest GHG emitter – from supporting Kyoto: the US government believed that without the participation of India and China the Kyoto Protocol would be meaningless.

India's case was that:

- its per capita energy consumption (1.7 tonnes) was well below the global average of 5 tonnes
- its priorities were alleviating poverty and expanding access to electricity in India, rather than reducing emissions
- current high concentrations of CO_2 in the atmosphere were largely the result of industrialisation and economic growth in ACs over the previous 150 years.

At the 2014 climate conference India's environment minister stated that India would not consider reducing its emissions for at least 30 years. Nonetheless, India is concerned about climate change and has its own National Action Plan on Climate Change (NAPCC). But so far it has only decided to reduce its GHG emissions as a ratio of its GDP. Using this measure it targets a 20–25 per cent reduction in emissions (based on 2005 levels) by 2020. This commitment is voluntary and will do little to stem the country's huge absolute rise in GHG emissions.

On the domestic front, India's NAPCC aims to improve energy efficiency and develop renewables, especially solar power. A sort of carbon tax is currently levied on coal, both imported and domestic. Given its current policies GHG emissions are expected to be in the range 4.0–7.3 billion tonnes by 2030. This compares with 2.43 billion tonnes in 2010 (Figure 10.23).

The public image of climate change: the role of the media and interest groups

Given that the overwhelming majority of climate scientists believe in human-caused climate change, the issue remains remarkably controversial. It is argued that this situation is partly due to media misrepresentation (Figure 10.24).

The Climate-Change Religion

Climate Science Is Not Settled

Schoolroom Climate Change Indoctrination

The real 'deniers' in the climate change debate are the warmists

HOW ARCTIC ICE HAS MADE FOOLS OF ALL THOSE POOR WARMISTS

Can Southern Company Shareholders Persuade Its Board to Stop Supporting Climate Change Denial?

Bias Bash: Liberal media push global warming fears

Is climate change really that dangerous? Predictions are 'very greatly exaggerated', claims study

Fossil fuel firms are still bankrolling climate denial lobby groups

Exxon Exposed for Spending Millions on Climate Change Denial

The danger of ideology-based newspaper coverage of climate change

What a shower! The more money the Met Office gets, the more ludicrously inaccurate its doom-mongering on climate change

Figure 10.24 Newspaper headlines on climate change

Most people do not read scientific papers, reports, blogs and specialist websites on environmental issues. Instead they rely on popular media reporting through newspapers, magazines and television. These media outlets therefore play a crucial role in forming public perception and opinion. However, depending on the political leanings of media organisations, reporting may be deliberately slanted. Thus right-leaning newspapers such as *The Times* and *Sunday Telegraph* are more likely to report sceptical opinions on climate change than left-leaning newspapers like the *Guardian*. This makes the climate change issue appear more open to question than it really is.

Also, in an attempt to provide a balanced discussion and accommodate dissenting views, some media organisations, including the BBC, have been accused of false balance. By giving disproportionate coverage to contrarians and sceptics, they suggest that climate change is far more controversial than experts believe. Instead of balance, climate experts argue that reporting should reflect the huge weight of scientific research in favour of climate change.

The strongest opponents of climate change have been energy industries, such as major oil and gas corporations and mining companies. Clearly such groups have a vested interest in maintaining the pre-eminence of fossil fuels and protecting their profits. They see restrictions on the consumption of fossil fuels as weakening economic growth, creating unemployment and even as a political ploy, to redistribute wealth. Moreover, these groups have the resources and influence to manipulate the media and elected politicians. Some energy companies even recruit dissident scientists to support their case. At the same time the media have often weakened the climate change arguments in the public mind by simplistic and sensational reporting. All of this has tended to confuse the public and add doubt to the credibility of climate change and its causes.

> **Activity**
>
> Log on to the website www.wsj.com/articles/the-climate-change-religion-1429832149.
>
> Read the article by Lamar Smith, a Republican Congressman, in *The Wall Street Journal*, published on 23 April 2015. Summarise and critically assess the views presented in the article 'The climate-change religion'.

> **Review questions**
>
> 1 What contribution did Suess and Revelle make to the climate change debate?
> 2 What is significant about a 2°C rise in average global temperature above pre-industrial levels?
> 3 What was the first international treaty on climate change and what did it propose?
> 4 What are low-carbon technologies?
> 5 What arguments were presented by some countries for not ratifying the Kyoto Protocol?
> 6 Describe the ways in which the media have influenced the climate change debate.

10.4 In what ways can humans respond to climate change?

> **Key idea**
> → An effective human response relies on knowing what the future will hold

Importance of the carbon cycle

At its most basic level the carbon cycle consists of a number of carbon stores and pathways along which carbon moves into and out of store. The principal stores are the oceans, the atmosphere, carbonate rocks, the soil and the biosphere. The main pathways include **photosynthesis**, **respiration**, decomposition and chemical weathering. A detailed description of the carbon cycle can be found in Chapter 4, pages 100–01.

The carbon cycle is crucial to life on the planet:

- All living organisms depend on carbon, which is a fundamental building block of life.
- Green plants and **phytoplankton** extract carbon from the atmosphere in the process of photosynthesis. They are **primary producers** in **ecosystems**, converting sunlight and CO_2 to carbohydrates, which support all consumer organisms, including humans.
- Carbon stores such as ocean sediments and carbonate rocks lock away carbon for millions of years, helping to maintain atmospheric CO_2 at levels conducive to life on the planet.
- Decomposition and oxidation ensure that CO_2 is recycled rapidly, replenishing stores of CO_2 in the atmosphere for photosynthesis.

- CO_2 and CH_4 in the atmosphere are important GHGs, absorbing long-wave radiation from the Earth's surface and contributing to the natural greenhouse effect.

The carbon cycle operates in a state of dynamic equilibrium, with carbon moving continuously between stores. Over millions of years a balance exists between flows of carbon into and out of stores, but on shorter timescales, the amount of carbon held in the atmosphere can vary. So, for example, concentrations of CO_2 in the atmosphere have varied from 200 ppm to more than 900 ppm.

Positive and negative feedback in the carbon cycle and its climatic impact

Feedback is a response to change in a system: **positive feedback** amplifies change and increases disequilibrium (see Table 10.7); **negative feedback** restores balance in a system (see Table 10.8). The carbon cycle, which is undergoing rapid change, contains a number of feedback loops. These feedback loops have the potential to alter the global climate radically. Unfortunately, most are poorly understood and some, such as the response of the global climate to increasing cloud cover, are conflicting.

Table 10.8 Potential negative feedback related to rising atmospheric CO_2 levels and global warming

Change	Negative feedback effect
Expansion of forests	As temperatures rise, the tree line advances polewards. The expansion of forests will absorb more CO_2 from the atmosphere.
Increased cloudiness	Higher rates of evaporation and levels of atmospheric water vapour will increase cloud cover. This in turn will increase reflection of incoming solar radiation (albedo) back into space and lower temperatures. Increased reflection is associated with stratocumulus clouds.
Increased aerosols in the atmosphere	Burning fossil fuels releases tiny airborne particles (aerosols) of smoke, dust and sulphur to the atmosphere. They reflect incoming solar radiation back into space, which lowers global temperatures. This process is known as **global dimming**.

Future emissions scenarios and their impact on global temperatures and sea levels

Because of uncertainties about future emissions of GHGs and the complexity of the Earth–atmosphere system, the Intergovernmental Panel on Climate Change (IPCC) provides a range of forecasts of global temperature and sea level rises during the twenty-first century. Forecasts

Table 10.7 Positive feedback related to rising atmospheric CO_2 levels and global warming

Change	Positive feedback effect
Increased evaporation	Global warming intensifies evaporation from ocean and land surfaces. As a result, atmospheric water vapour, a potent GHG, increases, which raises global temperatures, creating further evaporation and so on.
Reduced albedo	Melting of glaciers, sea ice and snow fields reduces albedos, so that more solar radiation is absorbed at the surface. Temperatures increase, with further melting, etc.
Declining forest cover	As global temperatures rise, tropical forest trees become stressed and die, releasing CO_2 from store, which results in higher temperatures, threatening the future of forests.
Increased cloudiness	More water in the atmosphere increases the cover of high-altitude cirrus clouds. They help retain radiated heat from the Earth and so contribute to further temperature rises, more evaporation, etc.
Release of methane hydrates	Methane hydrates are locked away in ocean sediments. The stability of this carbon store depends on temperature. As oceans warm there is the potential for massive release of methane, a GHG 20 times more potent than CO_2.
Melting of **permafrost**	CH_4 and CO_2 are stored in vast quantities in the Arctic and sub-Arctic permafrost. Thawing of the permafrost will release these GHGs, creating an enhanced greenhouse effect, which results in more melting.
Increased ocean acidity	As the oceans absorb more CO_2 they become more acidic. Increased acidity reduces the oceans' capacity to absorb CO_2. In the long term this increases CO_2 in the atmosphere, which enhances the greenhouse effect.

Geographical debates

311

range from worst- to best-case scenarios. Table 10.9 shows the IPCC's four scenarios for changes in the global energy balance 2010–2100.

Table 10.9 The IPCC's representative concentration pathways (RCPs)*

RCP	Description
+2.6	GHG emissions peak 2010–20 and decline thereafter (most optimistic)
+4.5	GHG emissions peak around 2040 and then decline
+6.0	GHG emissions peak around 2080 and then decline
+8.5	GHG emissions rise throughout the twenty-first century (most pessimistic)

*Global energy balance by 2100 in watts/m².

Mean global temperature changes

All of the IPCC projections for the next 70–80 years show significant rises in mean global temperatures during the twenty-first century. Depending on the trajectory of GHG emissions, temperature increases range from 0.3°C to 4.8°C (Figure 10.25). However, a mean global temperature rise of around 2°C is the most likely scenario.

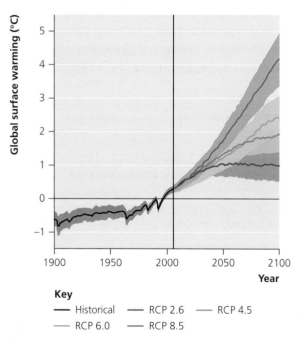

Key
— Historical   — RCP 2.6   — RCP 4.5
— RCP 6.0   — RCP 8.5

Figure 10.25 IPCC global temperature forecasts

Mean global sea level changes

Rising sea level is the result of the thermal expansion of the oceans due to warming, and the loss of ice by ice sheets and glaciers on land. Ice losses are caused by surface melting and increased rates of glacier flow. Sea level has been rising since the onset of industrialisation in the mid-nineteenth century. Between 1901 and 1990

mean global sea level (MGSL) rose by 1.5 mm/year. However, between 1990 and 2010 it rose twice as fast, averaging just over 3.0 mm/year. IPCC forecasts indicate an even more rapid rise during the rest of the twenty-first century. Also, after 2100 it is almost certain that MGSL will continue to rise for several centuries.

Like mean global temperatures, forecasts of MGSL are subject to uncertainty, depending on the future trend of GHG emissions. Current projections suggest a minimum rise (RCP 2.6) of 0.28 m by the end of the century, and a maximum of 0.98 m (RCP 8.5). However, the most pessimistic forecast does not include the possible collapse of marine-based sections of the West Antarctic ice sheet, which could increase the 0.98 m figure substantially.

> **Key idea**
> → The impacts of climate change are global and dynamic

Implications of climate change for people and the environment

Climate change has important implications for the environment, including ecosystems and weather. Changes to the environment also impact people, not least in areas of human health and well-being.

Ecosystems

Ecosystems are communities of plants, animals and other organisms and the physical environment. Within ecosystems, living organisms interact with each other and with the physical environment through flows of energy and the cycling of nutrients. Climate is a key element of ecosystems, shaping habitats that support plants and animals. Some plant and animal species will adapt to change; others, more highly specialised or geographically isolated, face decline and ultimately extinction.

Marine ecosystems

Global warming affects marine ecosystems by raising sea surface temperatures (SSTs). Coral reefs, one of the most biodiverse and productive ecosystems on the planet, are currently threatened by bleaching, caused by higher SSTs. Bleaching and the death of corals are possible with only a small rise in water temperature (1–2°C). In the last 30–40 years, Indonesia has lost half of its reefs to bleaching. In the Caribbean the proportion has reached 80 per cent.

Massive changes to marine ecosystems are currently taking place in the Arctic. Warming of the Arctic Ocean and shrinking of the sea ice have decimated ice algae, which are the base of the marine food chain. Sea ice is also crucial to the survival of marine mammals such as walrus and seals. Walrus use the coastal sea ice as diving platforms for foraging on the sea bed and travel long distances on floating ice. Ring, harp and other seal species use the sea ice to rest, give birth, raise pups and moult. Exposed on the ice, they are hunted by polar bears, the Arctic's top predator. The disappearance of sea ice is the main reason for a projected two-thirds decline in the polar bear population by mid-century.

Few human groups rely directly on natural ecosystems. One exception is the indigenous Inuit hunters of the Arctic (Figure 10.26). Their economy and culture depend on hunting marine mammals, especially seals, walrus and whales. But thinning and melting of the sea ice makes hunting hazardous. And with more open water in the Arctic, the number of killer whales is increasing. As natural predators of bowhead whales, narwhals and seals, killer whales are in direct competition with Inuit hunters.

Figure 10.26 Inuit hunters

Around the UK, average sea surface temperatures have risen by 1.6°C since 1980. Warming seas limit the food supplies, growth rates and spawning for many fish species. As a result, some indigenous cold-water species such as cod, haddock and mackerel have moved northwards towards Iceland and the Faroe Isles. Meanwhile, warm-water species like sea bass, hake and red mullet have migrated into UK waters. This movement has implications for the UK fishing industry. Not only will commercial fishing have to switch to new species but the fishing effort for cold-water fish like cod will have to shift northwards.

Terrestrial ecosystems

Temperatures are rising faster in the tundra than in any other ecosystem. One result is habitat change. As the permafrost thaws, wetland areas expand, attracting more migratory birds, especially wild fowl and waders. Influxes of spring migrants from the south will occur earlier, the breeding season will last longer and higher temperatures will ensure an abundance of insects. Meanwhile, the southern fringes of the tundra will lose their open aspect as the tree line advances north. As forests replace the tundra, changes in habitat will affect indigenous plant and animal species. Migration patterns among caribou, which spend the summer on the tundra, will be disrupted. As open tundra habitats shrink, predators such as snowy owls and arctic foxes, which rely on the lemming population, will be forced northwards.

The Cairngorms is a high-level plateau (averaging around 1000 m) in northeast Scotland. It supports the most extensive mountain tundra in the British Isles. Several animal species are restricted to this type of habitat. They include the arctic hare, ptarmigan and snow bunting. Arctic hares and arctic-alpine plants such as purple saxifrage and moss campion are essentially stranded on this island plateau. It is estimated that a 1°C rise in temperature requires an uphill movement of 200–275 m to maintain the same habitat. Animals living at 600 m would have to move to 800 m. Even with this modest warming, Scotland would lose 90 per cent of its arctic-alpine habitats. Ultimately, if warming continued, specialist arctic-alpine species would eventually run out of suitable habitat and face extinction.

Phenology, the study of changes in the timing of spring and other natural seasonal events, is an indicator of global warming. According to the IPCC, in the past 30 years spring has occurred earlier by 2.3–5.2 days per decade on average. This creates a loss of synchronisation between species: animals awaken from hibernation or start to breed before the emergence of food resources such as leaves, insects, larvae and caterpillars.

In continental Europe, southern species of birds on the northern limit of their ranges, such as cirl buntings and little egrets, have colonised southern Britain. A similar movement has been noted with butterfly species such as orange tips and commas, and plants native to southern Europe like tongue orchids. Meanwhile, species on the southern edge of their range have been pushed northwards by warming and more competition. Examples include birds such as lapwings and arctic skuas, and butterflies such as the scotch argus and large heath. But warmer winters have allowed several bird species that previously migrated in the autumn (e.g. blackcap, Dartford warbler) to overwinter in the UK.

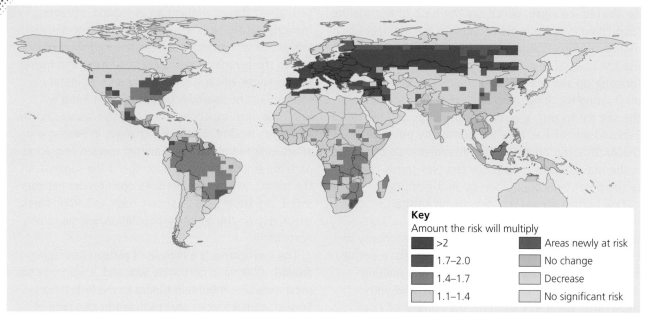

Figure 10.27 Predicted change in risk of transmission of malaria, 2020 prediction compared with average risk in 1961–90 (Source: Pim Martens, Maastricht University)

Key
Amount the risk will multiply

- >2
- 1.7–2.0
- 1.4–1.7
- 1.1–1.4
- Areas newly at risk
- No change
- Decrease
- No significant risk

Human health

Overall, climate change is likely to have a negative effect on human health. The World Health Organization (WHO) forecasts an additional 250,000 deaths a year worldwide between 2030 and 2050 linked to climate change and the spread of infectious disease, malnutrition and diarrhoea.

Climatic change will in general stimulate the transmission of vector-borne diseases and their geographic range (see Disease dilemmas, page 341). Dengue fever, a disease spread by the *Aedes* mosquito and formerly confined to the tropics and sub-tropics, is today found in 28 US states. Between 1995 and 2005, 4000 cases were reported. Rising temperatures and increased rainfall have favoured the spread of mosquitoes carrying the disease. Climate change is also responsible for the spread of Lyme disease in the USA. The disease is transmitted by ticks, which thrive in warmer conditions. Currently the distribution of Lyme disease is expanding northwards; it is likely that the disease-carrying ticks will eventually colonise Canada.

Malaria is a leading cause of mortality in the developing world (mainly in Africa), claiming around 800,000 lives a year. An infectious disease, malaria is spread by *Anopheles* mosquitoes, which thrive in warm, wet conditions. The disease is seasonal throughout tropical Africa (see Chapter 11, pages 350–51). However, in a warmer and wetter world, the disease could spread to malaria-free parts of the world, including southern Europe and the Mediterranean (Figure 10.27).

Higher temperatures also increase the risk of food contamination by salmonella and other bacteria which cause food poisoning; and heavier rainfall increases flood frequencies, and the probability of water supplies polluted by human waste. Bacteria in drinking water will multiply the risk of diarrhoea, already a major cause of death among children in LIDCs. Nor are people in ACs immune: heavy rainfall can increase the contamination of drinking water with parasites like cryptosporidium.

Human health could be compromised by droughts and floods reducing crop yields and food production. This could threaten food security and human health with widespread malnutrition and undernutrition in LIDCs. Global food supply has already been reduced by climate change and forecasts by the IPCC suggest significant reductions in staple cereal crops by 2030 (Figure 10.28).

Activity

Study Figure 10.27. The map compares the risk of malaria transmission in 2020 with the risk in 1961–90. It assumes an increase in average global temperatures of 2°C and no attempts made to contain the spread of the disease.

a) Describe the global distribution of transmission risk.
b) Suggest possible reasons for geographical variations in the risk of transmission (see Disease dilemmas, pages 339–41).

The future of food and farming: 2030s
In the 2030s, climate change will adversely affect food production, particularly among small farmers in poor countries
Crop and pasture yields are likely to decline in many places

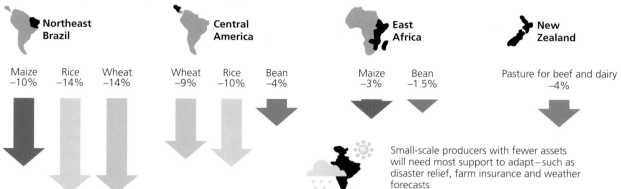

Northeast Brazil

Maize −10%   Rice −14%   Wheat −14%

Central America

Wheat −9%   Rice −10%   Bean −4%

East Africa

Maize −3%   Bean −1.5%

New Zealand

Pasture for beef and dairy −4%

Small-scale producers with fewer assets will need most support to adapt—such as disaster relief, farm insurance and weather forecasts

Figure 10.28 The future of farming and food

Extreme weather

Global warming is predicted to disrupt regional weather patterns. Extreme weather events such as storms, droughts and heatwaves (Table 10.10) are forecast to occur with greater frequency and intensity.

The connection between global warming and more extreme weather is complex and not fully understood. However, rising temperatures are known to disrupt the middle latitude and sub-tropical jet streams, increase evaporation and humidity, and modify pressure patterns both in the atmosphere and at the surface.

Since the mid-1970s the frequency, intensity and duration of extreme weather have increased. For example, the 2005 hurricane season in the North Atlantic Caribbean area produced a record-breaking number of storms, including five category 5 hurricanes;

Table 10.10 Nature and distribution of extreme weather phenomena

Extreme weather event	Nature	Distribution
Tropical cyclones (hurricanes, typhoons)	Large, violent revolving storms generating hurricane-force winds and torrential rain.	The tropics and sub-tropics, mainly in the North Atlantic and North Pacific Oceans, and the western Pacific.
Mid-latitude depressions	Large, mobile low-pressure systems which form on the jet stream and bring spells of heavy rain and gale force winds.	Middle to high latitudes, between 35° and 70°, e.g. northwest Europe, South Island New Zealand.
Tornadoes	Violent, rotating, funnel-shaped columns of air in contact with the ground.	Continental interiors in mid-latitudes, e.g. Midwest USA between the Rockies and the Gulf of Mexico.
Heavy rainfall	Either intense, short-lived thunderstorms or prolonged periods of rainfall associated with cyclones and depressions. The outcome is often river floods or groundwater flooding.	Widespread and most extreme in climates with wet and dry seasons, e.g. monsoon Asia.
Blizzards	Heavy snowfall driven by strong winds.	High latitudes and mountainous regions.
Severe cold spells	Prolonged spells of abnormally low temperature and/or snow cover, e.g. below zero.	Middle to high latitudes. Continental rather than oceanic locations.
Heatwaves	Prolonged spells of abnormally high temperature, e.g. > 30°C.	Middle latitudes in summer.
Droughts	Prolonged spells of abnormally low rainfall.	Widespread but most common in the sub-tropics (e.g. the Sahel), continental interiors (e.g. Midwest USA) and the Mediterranean.

and in August 2015, for the first time, three large hurricanes were active simultaneously in the Pacific. The Sahel region of Africa has endured numerous severe droughts in the past 40 years, which have accelerated **desertification**, while in 2015 California entered its fourth successive year of drought. In the UK, temperature and rainfall records have been broken in recent years. The highest temperature ever in the UK, 38.5°C, was recorded in August 2003; 1 July 2015 was the hottest July day (36.5°C) on record; the winter of 2014 was the wettest for 250 years; and in southern England, the period 2010–12 was the driest since records began in 1910.

Although extreme weather is part of the natural variability of climate, there is mounting evidence that the trend towards more extreme weather is driven by global climate change.

Stretch and challenge

Investigate extreme weather in the UK in the past fifteen years. For each event consider:

a) the type of event
b) its intensity
c) its frequency
d) its duration
e) its human impact.

The Meteorological Office website is a good place to start: www.metoffice.gov.uk/. Also browse the MetLink website at: www.metlink.org/extreme-weather/. Most national daily newspapers (especially *The Times*, *Guardian*, *Telegraph* and *Independent*) report exceptional weather events in detail.

Extreme weather has wide-ranging implications for people, economy and society:

- Heavier and more frequent rainfall associated with depressions and thunderstorms will cause severe river floods. Also groundwater floods, such as those that devastated the Somerset Levels in winter 2014, will become more common. Possible responses include engineering to strengthen hard flood defences, managed land-use change in catchments to delay run-off, and restrictions on housing and commercial development on active floodplains.
- More powerful tropical cyclones which generate huge **storm surges** threaten populations in low-lying coastal areas. Hurricane Katrina caused massive loss of life and economic damage in New Orleans and along the Gulf coast in August 2005. Even more vulnerable is the densely populated Ganges-Brahmaputra Delta in Bangladesh and India. While rich countries like the USA can afford to build costly coastal defences, poor countries such as Bangladesh and Myanmar may ultimately have to abandon areas of greatest flood risk.
- More frequent and more powerful depressions will accelerate rates of coastal erosion in mid-latitudes. Given the costs of defending coastlines, coastal management will involve difficult decisions. The options are hard defences to give complete protection (expensive), coastal realignment, allowing the coastline to retreat inland and coastal processes to operate and form natural defences such as mudflats and salt marshes, and non-intervention allowing erosion and deposition to establish a new equilibrium.
- Heatwaves are expected to occur with greater frequency in future. They pose a direct threat to human health, especially to vulnerable groups like the elderly in large urban areas where the **heat island effect** is most intense. An exceptional and prolonged heatwave struck much of Europe in summer 2003. It began in June and continued until mid-August, causing an estimated 35,000 excess deaths. Most of the victims were aged 75 years and over. The health impact of heatwaves in future will be mitigated by early warning of heatwaves, advice on the precautions that vulnerable groups should take, increasing the extent of reflective surfaces in urban areas, and greening urban areas to promote evaporation and cooling.
- Droughts are gradual weather events which ultimately lead to water shortages. The impact of drought is first felt by activities that depend heavily on water, such as crop growing, ranching, river transport and power supplies. Severe droughts may threaten public water supplies. As global weather patterns become less predictable, droughts will occur more frequently (Figure 10.29). The failure of the Asian monsoon could lead to severe regional and even global food shortages.

The vulnerability of people and the environment to climate change

People

The vulnerability of people to climate change depends largely on two factors: where they live and their ability to cope.

Most rural communities in the developing world are subsistence or semi-subsistence farmers. The majority

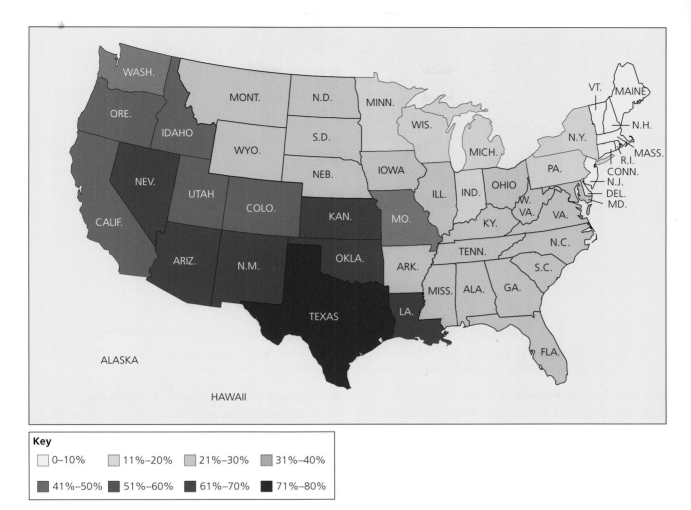

Key

☐ 0–10%	☐ 11%–20%	☐ 21%–30%	☐ 31%–40%
■ 41%–50%	■ 51%–60%	■ 61%–70%	■ 71%–80%

Figure 10.29 Risk of drought in the USA in the next 100 years

depend on direct rainfall for the successful cultivation of crops and raising of livestock. But climate change will make rainfall more erratic and floods and droughts will become more frequent. These problems will hardest hit farmers in marginal farming environments, where rainfall is only just sufficient to support agriculture.

Since the 1970s extensive areas of crop and grazing land in the Sahel in northern Africa have been abandoned due to severe **land degradation** and desertification. In part this results from overcultivation and excessive exploitation of soil, water and pasture resources. But the problem has been exacerbated by prolonged droughts linked to climate change. Other farming regions likely to be affected by drier conditions in future include prime agricultural areas such as the Prairies in North America and the Pampas in Argentina. Declining cereal yields in these regions could eventually lead to global food shortages.

Meanwhile, with 98 per cent of the world's glaciers currently retreating, regions that rely on glacial

meltwater for irrigation will almost certainly experience water shortages in future. Some of the rural regions affected, such as northern India and eastern China, are among the most densely populated places in the world (Figure 10.30).

Populations in low-lying coastal regions in the tropics and sub-tropics are vulnerable to flooding caused by a combination of rising sea level and more powerful tropical storms. Once again people in poorer countries are at most risk: a storm surge in Bangladesh in 1991 killed 138,000 people; this compared with a death toll from Hurricane Katrina in the USA (one of the most powerful storms on record) in 2005 of just over 1400. Storm surges not only cause loss of life but destroy crops and livestock, and leave a legacy of salinised soils and contaminated water supplies.

We have seen that global warming is most rapid in the Arctic, where the thawing of permafrost and the melting of sea ice threaten the livelihoods of

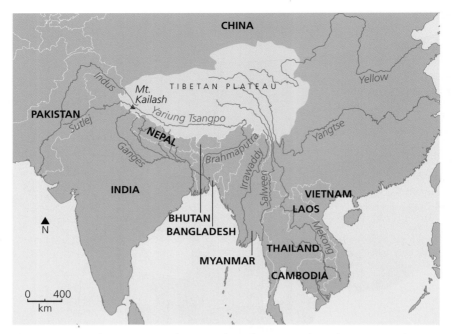

Figure 10.30 Major rivers in South, East and Southeast Asia that rely on seasonal mountain streams and meltwater from the Himalayas and the Tibetan Plateau

native Inuit hunters. Meanwhile, urban populations, in both the developed and the developing world, are increasingly at risk from more frequent heatwaves and soaring summer temperatures.

The ability of people and groups to cope with disasters also determines the human impact of climate change. Inequality and poverty at a global scale mean those at greatest risk are concentrated in the world's poorest countries. People living in poverty have fewest **entitlements** to protect themselves and their families against natural climatic hazards and to cope with losses of crops, property and steep rises in food prices.

> ### Activity
>
> Vulnerability to climate change refers to a country's or society's predisposition to be adversely affected. It includes the type and severity of climate-related changes and the preparedness to counter hazards. Study Figure 10.31. Summarise and suggest explanations for the geographical distribution of vulnerability on a global scale.
>
>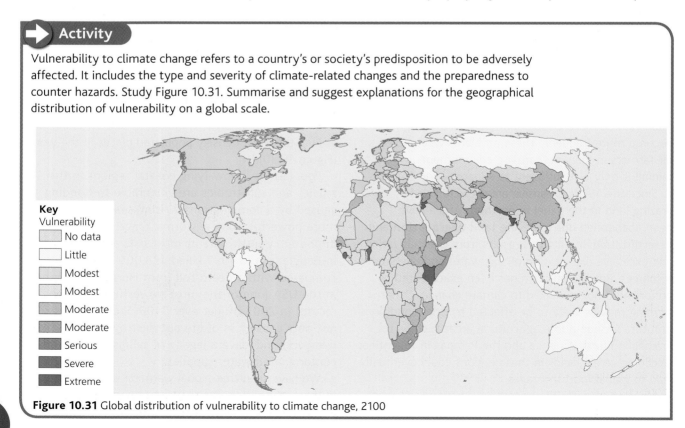
>
> **Key**
> Vulnerability
> - No data
> - Little
> - Modest
> - Modest
> - Moderate
> - Moderate
> - Serious
> - Severe
> - Extreme
>
> **Figure 10.31** Global distribution of vulnerability to climate change, 2100

Table 10.11 Environments most vulnerable to climate change

Environment	Change
Tundra	Rising temperatures melt the permafrost, disrupting vegetation, creating extensive thaw lakes and wetlands, and initiating mass movement. Forests will invade the southern margins of the tundra.
Mountains	Glaciers will retreat in warmer conditions. Thawing and glacier retreat will make slopes less stable and trigger more frequent mass movements (e.g. landslides). The snow line will recede upslope and the winter snowpack will thin. These changes will reduce meltwater inputs to rivers.
Hot semi-arid environments	Rainfall is likely to become more erratic, the rainy season will shorten and droughts will become more frequent and prolonged. With less vegetation cover and drier conditions, wind erosion (deflation) will increase and dust storms will become more frequent.
Rainforest	Computer models predict that the Amazon rainforest will become warmer and drier by the mid-twenty-first century. As deforestation increases (due partly to climate change and partly to clearance for agriculture, ranching and logging), the water cycle will weaken, creating positive feedback and accelerating forest loss – 30–60 per cent of the Amazon rainforest could become dry savanna grassland by the end of the century.
Coasts	Higher sea levels and more powerful storms will increase rates of erosion on both upland and lowland coasts. Shorelines will retreat inland. Coastal environments at particular risk include dunes, salt marshes and mudflats.

The elderly, the young and the chronically ill are among those groups least able to cope with the effects of climate change. This is illustrated by the impact of heatwaves in densely populated urban areas. Prolonged spells of extreme temperature like the heatwave that struck Europe in summer 2003 caused 35,000 deaths, the majority of them in elderly people aged 75 years and over.

Environment

Among the environments that are most vulnerable to climate change are fragile ecosystems like tundra, mountains and deserts, the Amazon rainforest, and coastal regions (Table 10.11). In each environment climate change will adversely affect biodiversity, plant productivity and food webs.

> **Key idea**
> → Mitigation and adaptation are complementary strategies for reducing and managing the risks of climate change

Mitigation strategies aim to reduce GHG emissions and tackle the causes of global warming and climate change. This approach is long term and may be unachievable before the end of the twenty-first century. In the meantime, the threats posed to people and the environment require immediate action. The response is adaptation: a strategy to minimise the impact of climate change today. Both mitigation and adaptation strategies are complementary. They represent a two-pronged approach to the challenge of climate change.

Mitigation strategies to cut global emissions of GHGs

Energy efficiency and conservation

Improvements in energy efficiency and conservation which reduce energy consumption are a first step to cutting GHG emissions and moving to a low-carbon economy.

In the UK, domestic demand accounts for nearly one-third of primary energy consumption. Service activities based in offices account for another 16 per cent. Already in the UK building regulations ensure that new homes and offices conform to minimum standards of heat insulation and limits to the ratio of window/door space to floor area. Energy performance certificates are required for all new buildings completed since 2008. Government, local authorities and energy companies provide financial incentives to eligible householders to insulate lofts and cavity walls. Figure 10.32 shows a home designed for maximum energy efficiency. Apart from generating its own electricity from solar power and heat exchanges, and conserving energy with effective insulation, the house is oriented to achieve maximum solar gain with large south-facing windows and is equipped with smart energy meters.

Fuel shifts and low-carbon energy sources

Three important changes have taken place in the UK's energy economy in recent years. First, there has been a steady decline in overall energy consumption since 2005. Indeed, energy use fell by 6.6 per cent in 2014, despite economic growth of nearly 3 per cent. Second, there was a marked reduction in the use of coal and oil between 1990 and 2014: by 2014 coal consumption had fallen to levels not seen since the nineteenth century. Third, the contribution of renewables has expanded, albeit from a very low base.

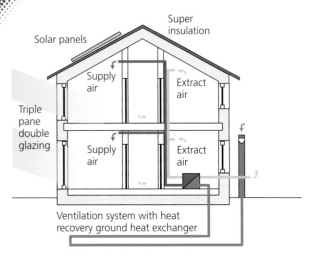

Figure 10.32 House designed for maximum energy efficiency

These fuel shifts can be explained by advances in energy conservation and energy efficiency, and government policies to decarbonise the UK economy. In common with other EU countries, the UK has to conform to the EU's Renewable Energy Directive. Under the Directive the UK has to achieve a 15 per cent target for electricity generation from renewables by 2020. Its strategy includes expanding renewables, especially wind power in offshore locations, closing several large coal-fired power stations (e.g. Ferrybridge 2015, Eggborough 2016) and converting some coal-fired power stations to biofuels. The latter is already taking place at Drax in North Yorkshire. By 2018, three of its six units will burn

biomass fuel (wood pellets imported from the USA) instead of coal. Wind power has been promoted by offering generous subsidies. Solar energy has also been promoted by green subsidies. In 2016, the feasibility of building a power station to harness tidal energy in Swansea Bay was being investigated.

Nuclear-generated electricity has been an important part of the UK's energy mix since the 1960s. Although a low-carbon energy source, the contribution of nuclear power has fallen in the past decade as a number of older stations have been decommissioned. Construction of the Hinkley Point C nuclear station in Somerset, the first to be built in the UK for more than 25 years, is under consideration, though its completion is unlikely before 2024.

Despite significant progress in the past ten years, the UK still relies on fossil fuels for 86 per cent of its energy supply (Figure 10.33). It remains a long way from a decarbonised economy and lags well behind other EU countries such as Denmark, Sweden and Germany.

Activity

Study Figure 10.33.

a) Suggest reasons why renewable energy targets for 2020 vary between individual countries.

b) Find out why some countries have already exceeded their 2020 target.

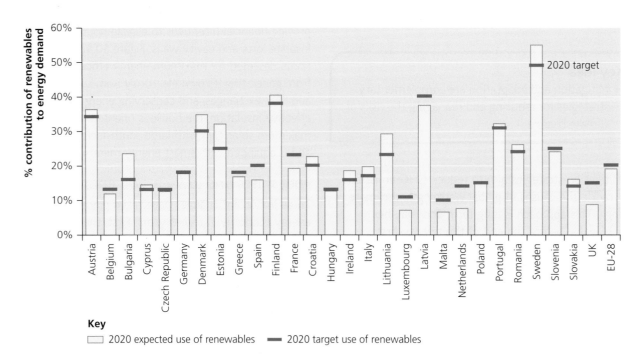

Key

☐ 2020 expected use of renewables ▬ 2020 target use of renewables

Figure 10.33 EU renewable energy projections and targets, 2020

Which is the best way to control climate change?
Evaluating geoengineering techniques for temperature and carbon

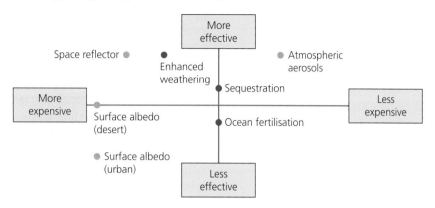

Figure 10.34 Geoengineering evaluation (Source: Royal Society)

Carbon capture and storage

Carbon capture and storage (CCS) is a new technology that extracts CO_2 emitted by coal-burning power stations and transfers it to long-term storage underground. It has the potential to cut drastically anthropogenic emissions of CO_2. However, even when the technology has been perfected, the use of CCS may be limited by costs and by a shortage of suitable storage sites, such as old oil- and gasfields. Refer to pages 124–125, Chapter 4 for more details on CCS.

Geoengineering

Some scientists believe that technology can be the answer to climate change. The use of technology to modify the environment on a large scale is known as geoengineering. Figure 10.34 shows the relative feasibility and costs of some geoengineering solutions. So far geoengineering has focused on two strategies: reducing the amount of insolation absorbed by the Earth and its atmosphere, and removing CO_2 from the atmosphere.

Increasing reflection of incoming solar radiation could reduce the amount of energy absorbed and converted to heat. In theory this could be done either by placing huge reflecting plates in orbit or by siting them on Earth. Alternatively, it might be possible to seed the stratosphere with aerosols (i.e. tiny particles) that will scatter insolation back into space.

Removing CO_2 from the atmosphere offers another solution. One suggestion is to fertilise the oceans with nutrients such as iron. Shortage of nutrients, especially in the open ocean, is a factor limiting phytoplankton growth. The availability of iron and other nutrients would stimulate the growth of phytoplankton and extract CO_2 from the air via photosynthesis.

Enhanced weathering is another avenue of research. Some abundant silicate minerals such as olivine react on exposure to air and absorb CO_2 to form carbonates. These would eventually find their way to long-term storage on the ocean floor. Enhanced weathering is based on the idea of crushing huge amounts of rock to increase their surface area and thus speed up weathering processes.

Extracting CO_2 from the atmosphere using 'artificial trees' appears fanciful, but the science suggests that it is feasible. A plastic resin is used to capture CO_2, which is subsequently compressed and stored. It is estimated that a single sheet of resin the size of a door could extract up to 700 kg/day of CO_2.

Reforestation and forest conservation

Deforestation, mainly in the humid tropics, is responsible for around one-fifth of global carbon emissions. Reforestation and forest conservation are the cheapest and most effective strategies to combat and slow climate change. To this end the UN has taken a lead in promoting its Reducing Emissions from Deforestation and Forest Degradation (REDD) programme. This scheme gives a financial value to the carbon stored in forests by providing incentives to developing countries to conserve forests and reduce carbon emissions.

> **Activity**
>
> Investigate one of the geoengineering strategies in Figure 10.34.
>
> a) Explain its principles.
> b) Examine its feasibility and practicality.

Adaptation strategies to reduce the vulnerability of human populations at risk

So far we have investigated mitigation strategies which cut GHG emissions. Adaptation is an additional rather than an alternative strategy. It involves anticipating the adverse effects of climate change and taking appropriate action to prevent or minimise potential damage. The UK government's National Adaptation Programme is a long-term strategy which addresses the risks and opportunities of climate change.

Most adaptations to climate change fall into one of three categories: retreat, accommodate and protect.

Retreat strategies

Populations living in coastal zones and river valleys face increasing risks from natural hazards caused by extreme weather events, floods and erosion. In coastal zones, rising sea level adds to risk. One response is 'managed realignment': vulnerable coastlines, with few settlements, are set back inland, where the risks of flooding and erosion are less. In England and Wales this policy, supported by the DoE, Defra and local authorities, is applied in Shoreline Management Plans.

River valleys are also at increased risk of flooding due to climate change. Retreat policies operate in this environment in the UK and many other ACs. Land-use zoning can prevent housing and businesses from locating on floodplains or even enforce their relocation. But retreat strategies may not be an option for coastal and valley communities in the world's poorest countries. Poverty dictates that nearly 140 million people live in the Ganges–Brahmaputra Delta, a region of high risk, exposed to storm surges and violent river floods. Despite the storm surge disasters of 1970 and 1991, which caused huge loss of life, the population of the delta has continued to grow rapidly.

Accommodation strategies

Climate change has major implications for agriculture. It will affect growing conditions, crop types, cultivation practices, crop yields and so on. More than any other economic activity, agriculture will have to adjust and accommodate climate change. New crop strains will be developed, adapted to shorter or longer growing seasons. Irrigation will be extended in regions of water shortage, and there will be more efficient irrigation systems such as lined irrigation canals and drip irrigation. In areas where rainfall is deficient, livestock farming or tree crops could replace cereal cultivation.

Other methods to conserve soil moisture include **zero tillage**, rotating crops, growing drought-resistant crops, mulching, using lower plant densities, planting hedgerows to act as wind breaks, and using smaller fields to reduce evapotranspiration. While farmers in rich countries can accommodate change, those in poor countries with limited access to credit and capital will struggle to adjust.

Accommodation strategies will also be important in the water supply industry. Action is already taken to make water usage more efficient by reducing losses to leakage, recycling waste water and using grey water (from baths, showers, washing machines, etc.) for gardens and flushing toilets. More costly options include increasing reservoir capacity, the desalination of sea water, and the construction of pipelines and canals for inter-basin transfers of water from areas of surplus to areas of deficit. In the developing world, water consumption per capita is relatively small and capital is in short supply. Even so, water could be conserved by more efficient irrigation systems, the construction of ponds and reservoirs, and by reducing losses to run-off. Water supplies could be increased by drilling inexpensive tube wells.

Improving education and public awareness of hazards such as storms and heatwaves can also help to accommodate climate change. Early warning of extreme weather events allows the public in developed countries to prepare for floods, storm surges, gales and blizzards. Thanks to modern satellite communications, smart phones and the internet, this is also beginning to happen in the developing world. Meanwhile, new building codes applied to houses, offices and factories can help combat the effects of extreme heat or extreme cold, while improvements in health education in the developing world (e.g. boiling drinking water, washing hands, using pit latrines) could limit the spread of water-borne disease.

Protection strategies

Hard engineering structures such as sea walls and dykes are used to protect coastal communities from flooding and erosion. Storm surge barriers such as those on the Thames and the Eastern Schelde (Netherlands) protect low-lying coasts at risk of flooding from violent storms and rising sea level. Hard defences such as storage reservoirs and levées protect communities at risk from river floods.

While large-scale, capital-intensive flood defences are beyond the means of most countries in the developing world, soft engineering offers a cheaper and more sustainable option. Conservation of beaches, salt marshes and mudflats provides natural barriers against flooding and erosion in coastal areas. In river catchments, natural water storage can be increased by expanding areas of wetland, and run-off can be controlled by afforestation and other land-use changes. Halting deforestation in developing countries in the tropics and sub-tropics would help control river floods and provide much-needed protection for vulnerable communities. In cities, urban heat islands can be moderated by increasing reflection from urban surfaces, planting trees, creating areas of open water and reducing energy consumption.

Finally, climate change is a threat to human health: directly through natural hazards and indirectly through infectious diseases and food shortages. In warmer and wetter conditions vector- and water-borne diseases will become more widespread. Communities can be protected from vector-borne diseases like malaria and dengue through control of mosquitoes with pesticides, as well as drugs, new vaccines and the use of bed nets. Improved water treatment and sanitation, together with screening for pathogens, could protect millions of people from the threat of water-borne disease such as typhoid, cholera and bilharzia.

Future adaptations to buildings, cities, transport and economies

In future, the design of housing, offices and cities, and the functioning of transport and economic systems, will need to adapt to rising temperatures and more frequent floods and droughts associated with climate change.

Housing and offices

Look at Table 10.12.

Transport

Transport systems will also need to adapt to more extreme weather and sea-level change in future.

The winter of 2014 in the UK showed how parts of the rail network are vulnerable to extreme weather. Storm waves destroyed part of the sea wall at Dawlish in Devon (Figure 10.35), severing the only rail link between Cornwall and the rest of the country. After six weeks the sea wall was repaired and the route reopened. This example showed that railways on the coast are vulnerable to rising sea level and powerful storms. Future strengthening of sea walls and dykes will be needed and in some cases rail routes diverted inland. Also in winter 2014 the Somerset Levels experienced severe flooding, interrupting rail services to the West Country. Low-lying parts of the rail network in river valleys or close to sea level will require protection, either by building flood barriers or elevating sections of track.

Warmer summers increase the problems of rails buckling in extreme heat: this problem can be overcome by improved sleepers and rail fastenings.

The road network is vulnerable to high summer temperatures which melt tarmac and cause rutting. Although winters in the UK and the rest of Europe are likely to become milder, extreme cold spells cause freeze-thaw action, leading to the formation of potholes and damage to road surfaces. Changes in the grades of asphalt can help to counteract both problems. Meanwhile, heavier rainfall and higher river flows could damage bridges. Faster currents scour the piers and abutments and can induce bridge collapse (Figure 10.36). The solution is to improve maintenance and protect piers and abutments either with rip-rap or by concreting around foundations.

Table 10.12 Adaptive strategies for housing, commercial buildings and cities

	Housing and commercial buildings	Cities
High temperatures	Heat-adaptive strategies: air conditioning and fans; limit solar gain by reducing window areas and south-facing windows; insulation to reduce external heat gain; increase albedo with reflective roofing (e.g. white roofs) and white walls; improved ventilation, e.g. night-time cooling with open windows; reduce high levels of glazing which trap heat; install sun shades over windows; high-efficiency lighting to avoid emitting waste heat; air-tight buildings to prevent heat entering from outside.	Altered heat exchanges create an urban heat island effect. This will be addressed by: creating green infrastructure and replacing where possible concrete and tarmac with trees, shrubs and grass; greening cities increases evapotranspiration and cooling and trees also provide shade; rooftop gardens reduce albedo; cool roofs (white) and light coloured roofs, walls and walkways increase reflection. Future policies will aim to protect existing green space and enhance it wherever possible. Increased reliance on renewable energy rather than combustion, which adds to urban heat generation.
Floods	Relocating more important and valuable services from basements and ground floors. In office blocks using these spaces for parking. Green roofs to increase interception, storage and evapotranspiration; houses built on stilts. In coastal regions, floating houses to counter sea-level rise.	By favouring changes in land use that conserve natural ecosystem principles, evaporation can be increased and run-off reduced and slowed. Increasing the green cover, with greater interception of rain, more evapotranspiration and soil moisture storage. Replacing impermeable roads and walkways (concrete, tarmac) with permeable paving.
Droughts	Rainwater harvesting (collecting run-off from roofs and gutters) and using grey-water systems, e.g. in flushing toilets, watering gardens.	Develop storage systems to capture run-off and recycle water for domestic, commercial and industrial users.

Figure 10.35 Damage to seawall and railway line at Dawlish

More frequent and prolonged droughts will disrupt traffic on major European waterways such as the Rhine and the Danube. In low flow conditions, when water depth falls below 1.6 m, traffic comes to a standstill. Fleet modification and river engineering are adaptations that can maintain traffic in times of low flow. Fleet modification would involve the use of barges with reduced draught, buoyancy aids allowing vessels to sit higher in the water, and coupled convoys. River engineering includes dredging and the construction of groynes. Groynes encourage sediment deposition, narrowing and deepening river channels.

Climate change has negative implications for airlines. Strong headwinds reduce landing rates when aircraft on their final approach are separated by distance. At Heathrow airport, a new system based on time separation allows more aircraft to land within a given time frame during stiff headwinds. This system is likely to be widely adopted. Among other problems facing air transport are airport flooding, heat damage to runway surfaces, sea-level rise threatening airports on the coast, increases in the strength of the jet stream and more convective thunderstorm weather. As convective thunderstorms occur more often in future, flights will be delayed and flight paths changed, increasing costs and flight duration. Higher temperatures also mean less dense air and aircraft needing longer runways for take-off.

> **Activity**
>
> Investigate how well adapted your own neighbourhood (postcode unit) is to climate change and increases in temperature, rainfall, high winds and drought.
>
> Write a report recommending and justifying modifications that would reduce the impact of climate change on your neighbourhood.

Figure 10.36 Damage to Workington Bridge on the River Derwent, Cumbria, following the 2009 floods

Economies

It is difficult to predict how a country's economy will adapt to climate change in the course of the twenty-first century. Most ACs will probably accommodate the worst effects; others, mainly in the developing world, will be harder hit. In all countries adapting to climate change will be costly, and spending in this area will limit investment elsewhere and weaken economic growth.

Most LIDCs are situated in the tropics and sub-tropics, where a rise in global temperature of 2°C or more will have much greater significance than in cooler climates. Moreover, many LIDCs depend heavily on climate-sensitive activities such as agriculture and tourism. Unable to adapt to drought and shifting rainfall patterns, tens of millions of African farmers will become climate refugees. Most will migrate to towns and cities. In Asia, declining flows of major rivers fed by glaciers could reduce crop yields at the same time as population and demand are growing. Problems of food production will be compounded if rising sea level forces the abandonment of farming in densely populated river deltas like the Ganges–Brahmaputra in India and Bangladesh and the Mekong in Vietnam. The outcome for many economies will be rising prices, inflation, unemployment, food insecurity and declining exports.

Case study: Bangladesh

Bangladesh is a low-income country, situated at the head of the Bay of Bengal in South Asia. Most of the country occupies the Ganges–Brahmaputra Delta, which supports the highest rural populations densities in the world. Exposure to climatic hazards is already high, but the impact of future cyclones, storm surges, river floods and droughts is expected to worsen with climate change. Unfortunately, Bangladesh's ability to adapt to the challenge of climate change, both on national and community scales, is limited by poverty.

Impacts of climate change

The future environmental and socio-economic impact of climate change will almost certainly be negative, for the following reasons:

● Most of the country is low and flat. Average elevation above sea level is 4–5 m and 10 per cent of the country is just 1 m above sea level. Exposure to storm surges and river floods is high (Figure 10.37) – on average, 70 per cent of the country floods every year.

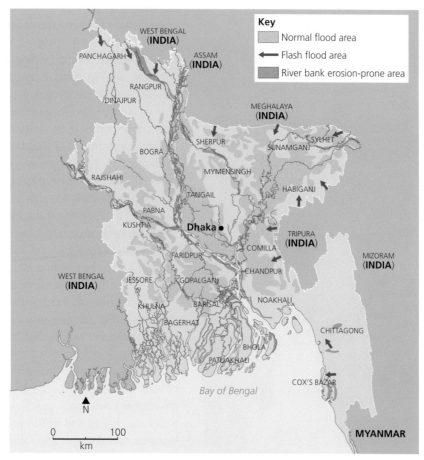

Figure 10.37 Bangladesh – flood-prone areas

- In the past 25 years, 60 per cent of all deaths worldwide from tropical cyclones have occurred in Bangladesh (Figure 10.38). The country is still recovering from cyclones in 2007 and 2009. Rising sea level, more intense cyclones and a rapidly growing population increase exposure to cyclone disasters.
- Around 20 million people live in areas less than 1 m above sea level; before the end of the century these people could be forced to abandon their homes and one-sixth of the country's land area could be lost to the sea.
- Rising sea level will increase the contamination of groundwater and soils by salt water intrusion. By 2100, 600,000 ha of agricultural land could be too highly salinised for cultivation.
- Coastal mangrove forests, which provide protection against storm surges and valuable ecological services, are threatened by rising sea level and human activity (clearance for agriculture and fish farming).
- Most communities exposed to environmental hazards are poor and are least able to adapt. According to the World Bank, around 60 million people in Bangladesh live in poverty.
- More frequent flooding will increase the disease burden, with inadequate sanitation likely to spread water-borne disease and flooded areas providing breeding sites for insect vectors such as mosquitoes.
- The ability of the country to adapt to climate change is constrained by its economic situation: GDP per capita was only US$2364 in 2014, and on the UN's Human Development Index, Bangladesh ranked 142nd out of 187 countries.

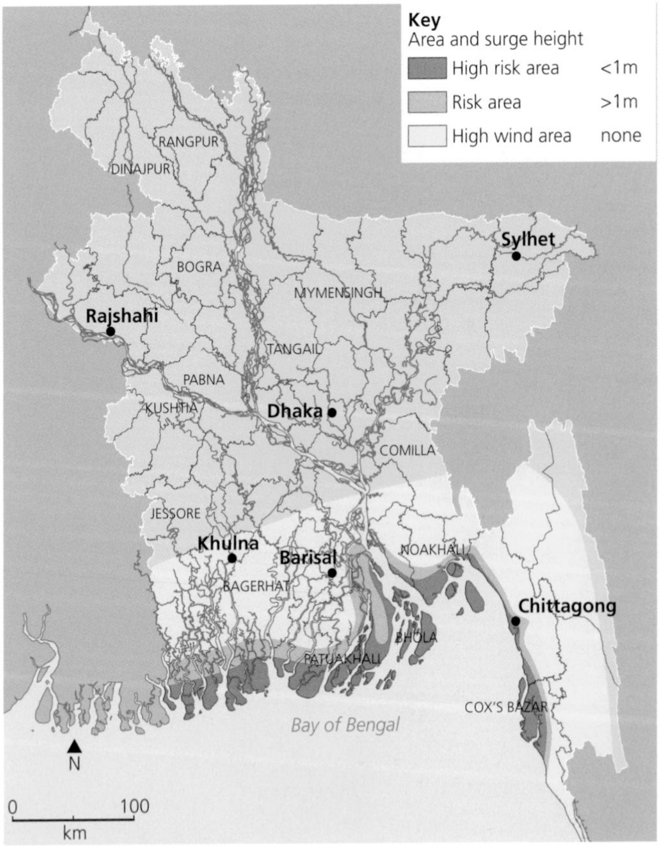

Mitigation and adaptation

Actions taken by the Bangladesh government to address climate change have focused exclusively on adaptation. Like other developing countries, Bangladesh regards mitigation, and the reduction in GHG emissions, as the responsibility of the developed world. The more urgent need in Bangladesh is to adapt to climate change that is already taking place.

In 2009 the government published its strategy for action against climate change. Adaptation includes investment in major engineering projects and community-based action. Lack of financial resources is an obstacle, and the government has set up a trust fund to raise capital from overseas donations. Currently a US$2 billion flood embankment project is under way. It will improve and strengthen some of the country's 3500 km of dykes, which provide protection against floods and erosion. The project is financed by a loan from the World Bank. Warning systems of storm surges are being upgraded, and cyclone and flood shelters built.

Protection of the coastal mangrove forests is a priority. Threatened by rising sea level and clearance for farmland and shrimp farming, local communities are encouraged to conserve the mangroves, which provide a natural barrier against storm surges.

In a country with a huge population and limited financial resources, community-based solutions are an important part of Bangladesh's adaptive strategy. As Himalayan glaciers recede and river flows decline, farmers need to use techniques that conserve water, including drip irrigation, rainwater harvesting and building storage reservoirs. Floating vegetable gardens on rafts of water hyacinth or bamboo mesh are a low-cost solution to crop damage caused by flooding. Houses have been raised on platforms by a metre or so above flood levels and similar platforms built as refuges for livestock. Other adaptive responses include the development and introduction of new cropping systems for waterlogged and saline soils.

Finally, flooding encourages the spread of water- and vector-borne diseases. The government has been proactive in establishing a climate change and health promotion unit to research new diseases and inform and educate populations about the risks.

Key
Area and surge height

High risk area	<1m
Risk area	>1m
High wind area	none

Figure 10.38 Bangladesh – cyclone-prone areas

Australia is a developed country with an advanced economy. In 2014 it ranked second in the world on the UN's HDI and had a per capita gross national income of US$41,500.

Impacts of climate change

Despite its wealth, Australia is highly vulnerable to climate change. Natural disasters cost the Australian economy US$4.5 billion/year, a figure that is expected to double by 2030. With most of its population, major cities and infrastructure on the coast, the country is at high risk from coastal flooding (Figure 10.39). Severe floods, caused by tropical cyclones, have devastated the Queensland and New South Wales coasts in the past 15 years. Cyclone Yasi in 2011 and its storm surge reduced Queensland's GDP by US$4 billion and badly affected tourism and coal exports.

Rising sea level has also affected tourism and ecosystems. Accelerated coastal erosion threatens beaches and infrastructure at the Gold Coast, Australia's main tourism centre. Salt-water intrusion along estuaries has damaged freshwater ecosystems in sensitive areas like Queensland's Kakadu National Park. Rising sea level, together with bleaching and ocean acidification, has damaged parts of the Great Barrier Reef along the country's east coast. Meanwhile, delicate sea grass and salt marsh ecosystems, trapped between advancing shorelines and sea walls and dykes, are exposed to wave attack and erosion, a process known as **coastal squeeze**.

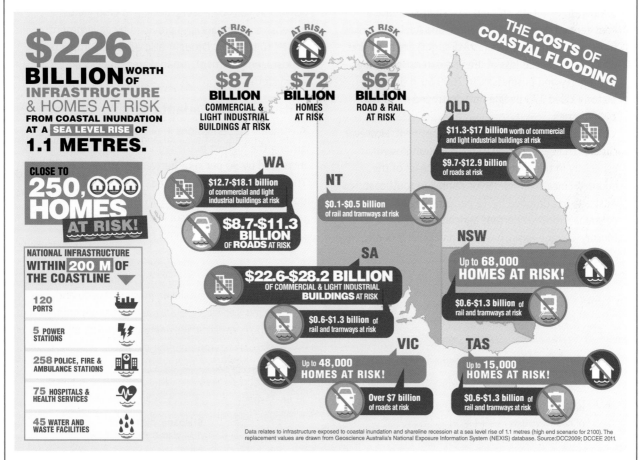

Figure 10.39 Australia – costs of coastal flooding

1 Select a suitable graphical technique and represent the data in Table 10.13 as a chart.

Table 10.13 Changes in the number of heatwave days in Australia's largest cities, 1950–2011

	Sydney	Melbourne	Brisbane	Perth	Adelaide	Hobart	Darwin	Canberra
1950–80	6	5	10	6	5	4	3	6
1981–2011	9	6	10	9	9	5	7	13

2 Describe the changes in the frequency of heatwave days in Australia's largest cities between 1950 and 2011.

3 Choose and justify an appropriate statistical test to analyse the significance in differences in the number of heatwave days in 1950–80 and 1981–2011 (pages 536–40).

Extreme heat creates human health problems. The 2009 heatwave in southeast Australia, when temperatures hit 45.1°C, led to a sharp increase in heat-related deaths in Melbourne and Adelaide. Linked to extreme temperatures (and drought) are bush fires, which have caused loss of life and extensive damage to property and businesses. In 2009 bush fires in Victoria killed 173 people and destroyed more than 2000 homes.

Australia is the driest inhabited continent. However, climate change has increased the risk of drought, especially in the southwest and southeast of the country. Since the mid-1990s rainfall in southeast Australia has dropped by 15 per cent (Figure 10.40). Australia's most important agricultural region, the Murray–Darling Basin, has been particularly badly affected. The so-called Millennium Drought, which lasted from 1996 to 2010, was the longest on record. As well as causing a steep decline in crop yields, expensive relief packages provided by government added to the cost. Drought also adversely impacts tourism, employment, urban water supplies and ecosystems.

Mitigation and adaptation

Australia's CO_2 emissions in per capita terms are among the highest in the world and its GHG emissions increased by 30 per cent between 1990 and 2010. The government's response has been to develop both mitigation and adaptation strategies. Australia ratified the Kyoto Protocol in 2007 and following the Copenhagen climate summit of 2013 agreed a

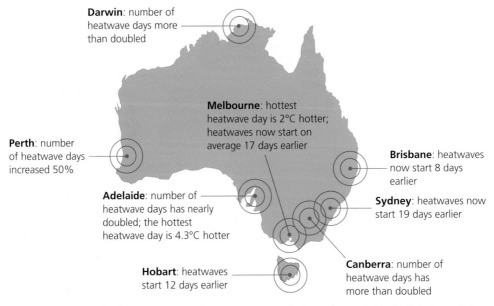

Figure 10.40 Comparison of heatwaves during 1950–80 with those during 1980–2011 in Australia's capital cities

5–25 per cent reduction in CO_2 emissions by 2020 (based on 2000 emissions). This target, however, has been criticised as being well below that of other ACs. As part of its mitigation strategy Australia has also developed a cap-and-trade scheme linked to the EU's trading scheme and has introduced targets to increase production of renewable energy by 2020.

Mitigation is a long-term strategy. Even if effective, climate change between now and the end of the century is a certainty. Thus more urgent action is needed in the short term. Central, state and local governments have been proactive, commissioning research to improve understanding, target priority areas and draw up action

plans. Adaptation on the coast will include protection by building new sea walls, and flood gates and dams will alleviate river flooding. In areas of high flood risk, some relocation of houses and businesses and stricter planning laws enforced to prevent development will be inevitable. Australia currently has a $US9 billion investment programme to develop new water resources, use existing resources more efficiently, and protect river and wetland ecosystems. In order to reduce agriculture's vulnerability, crop strains suited to hotter and drier conditions will be developed. Dams will be constructed to retain and manage water, and early warning and disaster-response management will be upgraded.

Review questions

1 Give three examples of positive feedback caused by rising GHG emissions.
2 State two reasons for the contemporary rise in sea level.
3 State two ways in which climate change threatens human health.
4 What is the heat island effect and how can it be mitigated?
5 With reference to climate change, what is the difference between mitigation and adaptation?

10.5 Can an international response to climate change ever work?

Key idea
→ Effective implementation depends on policies and co-operation at all scales

Geopolitics associated with the human response to climate change

The role of the Intergovernmental Panel on Climate Change (IPCC) in shaping policy making

The IPCC was created in 1988 by the United Nations Environment Programme (UNEP) and the World Meteorological Organisation (WMO). Its mission was to provide:

- objective, science-based reports on climate change and its impacts
- understanding of possible risks associated with human-induced climate change
- options for mitigation and adaptation.

The IPCC's reports, designed to inform policy makers, are neutral with respect to policy. Since 1988 the IPCC has delivered five reports, the most recent in 2013. Reports have provided increasingly strong evidence that climate change is due to rising concentrations of CO_2 and other GHGs in the atmosphere linked to human activity.

Supported by a huge body of research, and consensus within the scientific community, the IPCC is highly influential in shaping climate change policies at international, national and sub-national levels. Warnings by the IPCC that continued GHG emissions could have catastrophic and irreversible consequences have underlined the need for policy makers to take urgent action. Thus the UN Climate Change Conference in Paris in 2015, informed by IPCC's latest report, has driven discussions towards a new legally binding, universal international agreement.

International directives

The Kyoto Protocol (1997–2012) was the first legally binding international agreement on limiting carbon emissions. The main drivers of Kyoto were the EU countries. A voluntary second commitment period to Kyoto operates from 2013 to 2020. Although it is supported by many ACs, together they account for only 14 per cent of carbon emissions. This is because the USA, Russia, Japan, Canada and developing countries are not party to the second commitment.

Kyoto set an average target of 5 per cent reduction in carbon emissions relative to 1990 by 2012. Many countries, particularly in Europe, achieved their targets. Others fell well short and some even increased their emissions substantially (Figure 10.41).

In a sense Kyoto was flawed because the largest carbon-emitting countries, notably the USA, China, Russia and India, failed to ratify the treaty. The USA never ratified Kyoto; Russia only belatedly in 2007. EDCs like China and India, and LIDCs were exempt. China and India prioritised economic development over climate change. They also argued that ACs had a moral responsibility to deal with a problem caused by their

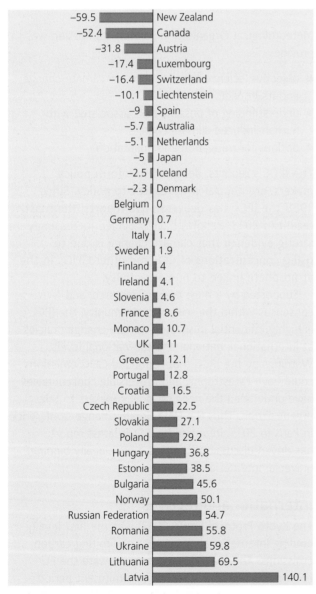

Figure 10.41 Percentage changes in carbon budgets 1990–2010 (including carbon emissions, sinks and land use). Countries in blue surpassed their Kyoto targets; those in red failed to achieve their targets (Source: *Guardian*)

historic carbon emissions over the past 150 years. The USA used the opt-out of China and India as a reason not to ratify Kyoto. It argued that the treaty would make little or no difference to climate change if China and India refused to limit their carbon emissions.

At the Paris Conference in December 2015, 195 countries adopted a new, legally binding climate deal (see page 135). It will come into force in 2020 and aims to limit global warming to less than 2°C by the end of the century.

EU Climate Directive

The EU has been at the forefront of international efforts to mitigate GHG emissions and climate change. It has set tough, mandatory targets for emission reductions in EU countries, as well as for renewable energy, vehicle emissions and carbon trading. For the most part the EU has made substantial progress in a number of key areas, which would not have been the case if countries had simply been left to pursue their own national policies. Thanks to the Renewables Directive, the UK increased its renewable energy output by nearly 90 per cent between 2007 and 2011.

The EU's carbon trading scheme is the world's largest and has been widely copied, and tough emissions and fuel efficiency standards for motor vehicles have been successful. However, there have been disappointments. Emissions targets have lacked ambition. The 20 per cent reduction by 2020 had been achieved by 2011. Progress in improving energy efficiency has been limited while some member states have been very slow to implement the directive's policies.

Significance of carbon trading and carbon credits

The EU Emissions Trading System (EUETS) is the world's largest GHGs trading scheme. Introduced in 2003, it includes more than 11,000 power stations and industrial plants across the EU, such as oil refineries, iron and steel works, cement and chemical works. Aviation operations are also covered by the scheme. EUETS does not include agriculture or transport, but in all around 50 per cent of emitters are covered by the scheme.

The EUETS is a so-called 'cap and trade' system. Essentially this is a market-based solution to climate change, where polluters either cut their emissions or incur extra costs. It sets a cap or limit on GHG emissions and this cap is converted into tradable allowances. Participants are allocated a tradable emissions allowance or credits, with one allowance giving the holder the right to emit 1 tonne of CO_2 or its equivalent. Emissions are monitored and reported each year and participants must surrender enough allowances to cover their emissions. If emissions exceed allowances, then participants can purchase extra allowances from those holding allowances that have not been used. The number of allowances issued is reduced each year. Participants with insufficient credits to cover their emissions are fined €100 for every tonne of CO_2 in excess. They can also be named and shamed.

Overall, EUETS has achieved real, though relatively small, reductions in carbon emissions. However, it has been criticised for issuing too many credits and in its early stages not imposing sanctions. Industries criticise the scheme because it imposes extra costs, giving non-EU competitors an unfair advantage. It is also suggested that some energy-intensive industries might relocate overseas, though there is little evidence of this so far.

National and sub-national policies

Denmark

Although Denmark is a member of the EU, its climate policy is only in part determined by the EU's Climate Directive. Denmark has its own ambitious targets to reduce its GHG emissions: an interim 40 per cent reduction on 1990 levels by 2020, and by 2050 all energy consumption is expected to be from renewables. To achieve this target it will phase out fossil fuels and increase investment in wind and solar power. Demand for fossil fuels will be reduced by carbon taxes and tax exemptions for hydrogen cell and electric cars, cheaper public transport, railway electrification and the promotion of cycling. In the agricultural sector, methane (CH_4) produced by domestic livestock and fertilisers will be controlled. Prior to 2010, the main thrust of Danish policy was towards mitigation. More recently the government has focused on adapting to climate change that is already taking place, such as increases in winter rainfall, higher temperatures, more extreme weather and rising sea level.

However, adaptation policies have mainly devolved to regional and local levels. Copenhagen, the capital and largest city in Denmark, developed its own adaptation strategy in 2011. The aim is to create 'climate-proof neighbourhoods'. Detailed plans have been drawn in the event of heavy rain overwhelming the city's drainage systems. It is hoped to pre-empt flooding by improving the city's storm water drains.

Because Copenhagen is located at sea level, rising sea levels increase the risk of flooding from storm surges. Storm barriers are planned and dykes will be raised along the most vulnerable shorelines, new buildings elevated 1 m or so on platforms, and land use on the ground floors of buildings converted to those less susceptible to flood damage. Large areas of Copenhagen are also predicted to be a risk from flooding by rainfall by 2100.

The threat of extreme temperatures in summer, exacerbated by Copenhagen's heat island effect, has prompted the authorities to create more areas of open water, vegetation and shade, and improve air circulation. In adaptation strategy the priority is to prevent a disaster. This is the purpose, for example, of building hard flood defences. If flooding still occurs, interventions such as sandbagging can help reduce the impact. In the worst-case scenario, pumping and disaster management plans to provide emergency relief can be implemented.

California

The state of California is a world leader in formulating and implementing both mitigation and adaptation and policies to tackle human-induced global climate change. Its policies are far more radical and far-reaching than those of the US government.

Legislation passed in 2006 – the Global Warming Solutions Act – set absolute state-wide limits on GHG emissions and committed the state to decarbonisation and a clean energy economy. This is the first pillar of California's climate change policy. The target is to reduce GHG emissions by 80 per cent below 1990 levels by 2050. California's long-term mitigation strategy involves phasing out fossil fuels for electricity generation, a cap-and-trade scheme, reducing emissions from motor vehicles and promoting renewable energy (Table 10.14).

Table 10.14 California's climate change mitigation strategy

Fossil fuel-generated electricity	Retirement of power stations fuelled by fossil fuels in the long term. No new coal-fired plants will be built. Currently about 60 per cent of electricity is produced in gas-fired power stations.
Cap-and-trade	State scheme introduced in 2013. A market-based system to reduce carbon emissions. California's scheme is already the second largest in the world after the EUETS. It covers 85 per cent of carbon polluters in the state and despite strong economic growth in 2013–15 is proving a success.
Vehicle emissions	Stringent laws introduced to limit carbon emissions. The aim is to have 15 per cent of cars powered by electricity, compressed natural gas and hydrogen fuel cells (as well as hybrids) by 2020. Currently nearly 3 per cent of vehicles are powered by batteries.
Renewables	California aims to generate 33 per cent of its electricity from renewables by 2020 and 50 per cent by 2050. Subsidies are available for renewables. 1940 MW of new solar energy was installed 2012–15.

Adaptation is the second pillar of California's climate change policy. Its purpose is to tackle the effects of climate change that are already occurring. Among these effects are increases in average temperatures, sea-level rise, decreased winter rainfall and snowfall, more frequent droughts, and increases in the frequency and severity of wildfires. Guided by the state's climate adaptation strategy, the priorities are to improve water management, avoid new development in areas with inadequate protection against flooding, wildfires and erosion, safeguard existing areas of economic, social and cultural value against floods, fires and erosion, and protect aquatic habitats and ecosystems.

Review questions

1 What is the IPCC and what is its role?
2 Why were EDCs and LIDCs exempted from the Kyoto Protocol?
3 Explain how cap-and-trade is a market-based approach to climate change mitigation.
4 What are the main features of Denmark's policies for tackling climate change?
5 How do California's policies towards climate change differ from those of the US federal government?

Practice questions

A Level

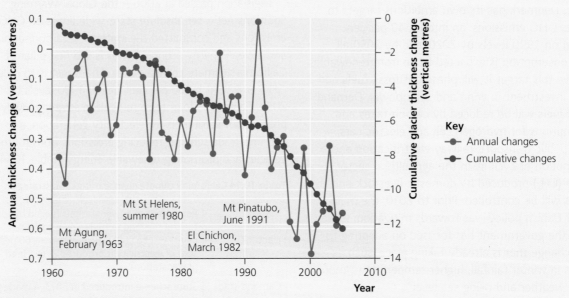

Figure A Changes in glacier thickness, 1960–2010

Section A

1 Study **Figure A**.

a) Give **three** reasons why data showing recent trends in glacier thickness provide evidence of climate change. [3 marks]
b) Explain how the shrinking of glaciers and other ice stores might impact human activity. [6 marks]

Section B Synoptic

2 Examine how climate change may be impacting the water cycle in the Arctic tundra. [12 marks]

Section C

3 Assess the relative merits of mitigation and adaptation strategies to deal with the problems caused by global climate change. [33 marks]

AS Level

Table 1 Global temperature anomalies and average atmospheric CO_2 concentrations, 1960–2015

	1960	1970	1980	1990	2000	2010	2015
Temperature (°C)	−0.03	0.03	0.28	0.44	0.42	0.72	0.87
CO_2 (ppm)	317	326	339	354	370	390	401

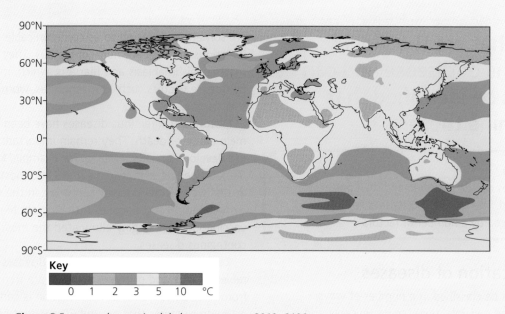

Figure B Forecast changes in global temperatures, 2010–2100

Section A

1 a) Explain how volcanic activity can influence climate change. [4 marks]

b) Suggest why anthropogenic greenhouse gas emissions have increased since the pre-industrial era. [6 marks]

c) Study the data in **Table 1**.

i) Outline and justify **one** statistical method you could use to analyse the relationship between global temperature anomalies and atmospheric CO_2 concentrations between 1960 and 2015. [4 marks]

ii) Using evidence from Table 1 comment on the trends in global temperatures and atmospheric CO_2 concentrations for the period 1960–2015. [6 marks]

Section B Synoptic

2 a) With reference to **Figure B** suggest how temperature change can influence geomorphic processes in landscape systems. [8 marks]

b) Examine the role of other dimensions of climate change to processes operating in physical landscape systems. [8 marks]

Section C

3 Assess the effectiveness of international efforts to tackle the problems of climate change. [20 marks]

Chapter 11

Disease dilemmas

11.1 What are the global patterns of disease and can factors be identified that determine these?

> **Key idea**
> → Diseases can be classified and their patterns mapped. The spread of disease is complex and influenced by a number of factors

Classification of diseases

Diseases can be classified in a number of ways: infectious and non-infectious, communicable and non-communicable, contagious and non-contagious, epidemic, endemic and pandemic.

Infectious diseases are spread by pathogens such as bacteria, viruses, parasites and fungi. Most, but not all, can be transmitted from one person to another. Contagious diseases are a class of infectious disease easily spread by direct or indirect contact between people. They include bacterial infections such as typhoid and plague, and viral diseases like yellow fever and Ebola. Infectious diseases which spread from host to host, but which do not require quarantine, are sometimes referred to as communicable. Some infectious diseases are non-contagious. They include malaria, leishmaniasis and filariasis, which are spread by disease vectors such as mosquitoes, worms and their larvae.

Historically, infectious diseases have been the main cause of death. They remain important in the developing world. Malaria, for example, killed 583,000 people worldwide in 2013 and most of these deaths were in the poorest countries. In the developed world, medical technologies including antibiotics and vaccination have largely eliminated the most dangerous contagious diseases.

Zoonotic diseases are infectious diseases such as rabies, plague and psittacosis, which are transmitted from animals to humans. Transmission is common: one estimate suggests that around 60 per cent of infectious diseases are spread from animals.

Non-infectious diseases are not communicable. They have a variety of causes, including nutritional deficiencies (e.g. rickets), lifestyle (e.g. diabetes, heart disease, cancer) and genetic inheritance (e.g. heart disease, stroke, cancer). Non-infectious diseases are the main cause of death in developed countries and increasingly in developing countries. For example, cardiovascular disease and cancer account for nearly two-thirds of all premature deaths in the UK (see Table 11.1).

Table 11.1 Top ten disease-related causes of death: UK and Malawi, 2011

	UK		Malawi	
1	Coronary heart disease	Non-communicable	HIV/AIDS	Infectious
2	Stroke	Non-communicable	Influenza and pneumonia	Infectious
3	Lung cancer	Non-communicable	Stroke	Non-communicable
4	Influenza	Infectious	Coronary heart disease	Non-communicable
5	Breast cancer	Non-communicable	Diarrhoeal disease	Infectious
6	Lung disease	Non-communicable	Diabetes mellitus	Non-communicable
7	Alzheimer's/dementia	Non-communicable	Malaria	Infectious
8	Colon–rectum cancer	Non-communicable	TB	Infectious
9	Liver disease	Non-communicable	Kidney disease	Non-communicable
10	Prostate cancer	Non-communicable	Lung disease	Non-communicable

Compare and contrast the causes of disease-related deaths in the UK and Malawi (East Africa).

Stretch and challenge

1 Research the following diseases online: influenza, filariasis, Ebola, HIV/AIDS, leishmaniasis, diabetes, cancer, cholera, smallpox, typhoid, CVD, anthrax, Lyme disease, measles, West Nile virus, yellow fever.
2 Classify the diseases as infectious (contagious), infectious (non-contagious) and non-communicable.
3 Select one infectious disease that is contagious and one that is non-contagious and research their epidemiology.

Endemic, epidemic and pandemic

Endemic diseases exist permanently in a geographical area or population group. Examples include sleeping sickness, confined to rural areas in sub-Saharan Africa (Figure 11.1), and Chagas disease, found in Central and South America. Sleeping sickness is caused by a parasite transmitted to humans by the bite of an infected tsetse fly. Chagas disease is also caused by tiny parasites transmitted by blood-sucking insects. It affects 7.6 million people.

An **epidemic** is an outbreak of a disease that attacks many people at the same time and spreads through a population in a restricted geographical area. An outbreak of Ebola disease in West Africa in March 2014 quickly led to an epidemic. A year later nearly 25,000 people had been infected, with 10,500 deaths, mainly in Liberia, Sierra Leone and Guinea.

An epidemic that has spread worldwide is known as a **pandemic**. Historic examples of pandemics are the Black Death (bubonic plague) of the mid-fourteenth century and Spanish flu in 1918–19. The Black Death killed an estimated 75–200 million people worldwide, and possibly half of Europe's population in just four years. Spanish flu affected one-third of the global population. One in five people infected died, with a worldwide death toll of around 50 million. More recent pandemic flu outbreaks include Asian flu (1957–58) and Mexican flu (2009).

Degenerative diseases and lifestyle

Currently 60 per cent of deaths worldwide are due to degenerative diseases, reflecting the ageing of the global population and increasingly unhealthy lifestyles. Common examples of degenerative disease are cardio-vascular disease (heart attack, stroke), cancer and chronic respiratory disease. However, apart from natural ageing, these non-communicable diseases are also influenced by lifestyle choices. As a result, they affect all age groups. In 2013, 16 million people worldwide under the age of 70 years died from degenerative diseases; 82 per cent of these premature deaths were in low- and middle-income countries. Most of these premature deaths were linked to unhealthy diets, physical inactivity, smoking and excessive alcohol consumption.

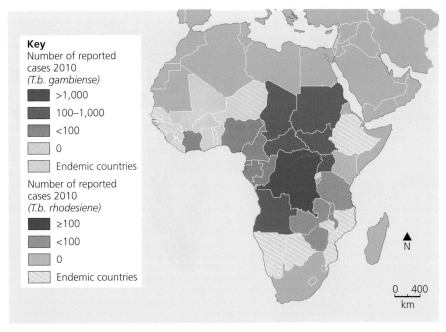

Figure 11.1 Distribution of the two forms of human African trypanosomiasis (sleeping sickness) (Source: WHO)

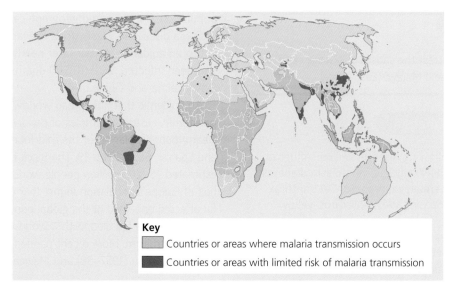

Key
　　Countries or areas where malaria transmission occurs
　　Countries or areas with limited risk of malaria transmission

Figure 11.2 Global distribution of malaria, 2010 (Source: WHO)

Global distribution of malaria, HIV/AIDS, tuberculosis, diabetes and cardiovascular disease

Malaria

Malaria is an infectious but non-contagious tropical disease. It is concentrated in Africa, Latin America, South Asia and Southeast Asia. (Figure 11.2). In total around 3.2 billion people are at risk in 97 countries. The malarial parasite is transmitted to humans by *Anopheles* mosquitoes which thrive in warm, humid environments. Where conditions are cooler and drier – deserts, high mountains and plateaux – malaria is absent. The disease is also absent from large urban areas in the tropics. In the US states bordering the Gulf of Mexico, and in northern Australia, effective public health measures have eliminated the disease.

HIV/AIDS

HIV/AIDS is an infectious and contagious disease. The HIV virus is spread by human body fluids such as blood and semen. In 2015, 35 million people were infected with HIV/AIDS worldwide. However, the global distribution of the disease is highly uneven (Figure 11.3).

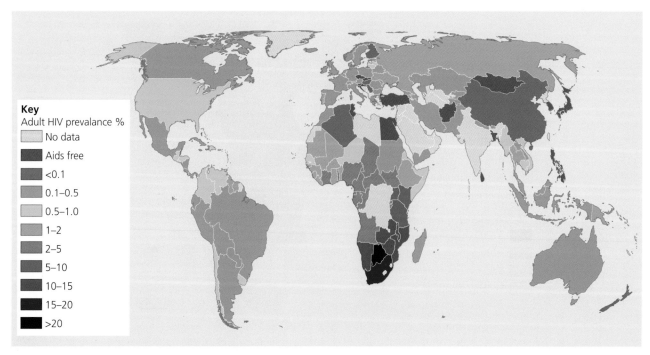

Key
Adult HIV prevalance %
　　No data
　　Aids free
　　<0.1
　　0.1–0.5
　　0.5–1.0
　　1–2
　　2–5
　　5–10
　　10–15
　　15–20
　　>20

Figure 11.3 Global distribution of HIV

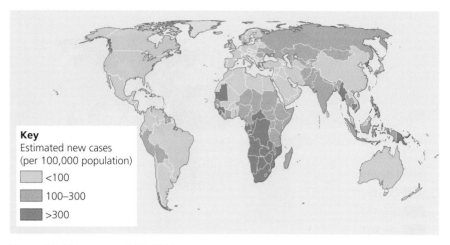

Figure 11.4 New cases of TB, 2010

The main concentration of infection is in sub-Saharan Africa. South Africa and Nigeria have the largest numbers of HIV/AIDS cases, while in Swaziland and Lesotho more than a quarter of the population carry the disease.

Tuberculosis (TB)

In 2013 there were nearly 9 million cases of TB worldwide and 1.5 million deaths. TB is an infectious and highly contagious disease associated with poverty and overcrowded living conditions. It is present in all regions (Figure 11.4), though 95 per cent of deaths occur in low- and middle-income countries. Africa has by far the highest number of TB deaths, with a large proportion among HIV/AIDS sufferers. Mortality rates from TB in 2013 were 94/100,000 in Nigeria and 69/100,000 in Mozambique. Outside Africa, TB mortality rates are high in many parts of Asia, and especially in Afghanistan, Myanmar and Cambodia.

Diabetes

Diabetes is a non-communicable disease caused by a deficiency of insulin, a hormone secreted by the pancreas. Globally the disease afflicts nearly 250 million people and is responsible for nearly 4 million deaths annually (Figure 11.5). Diabetes is widespread in both the developed and the less developed world but is most strongly concentrated in North America, East and South Asia. Type-1 diabetes develops in childhood and is genetic. Type-2 occurs in adulthood and is often linked to obesity, poor diet and physical inactivity.

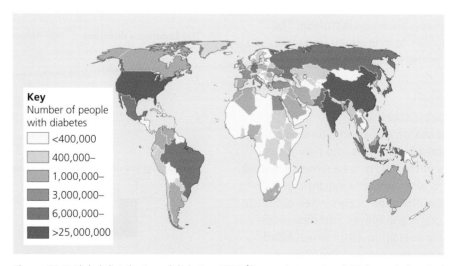

Figure 11.5 Global distribution of diabetes, 2010 (Source: International, Diabetes Federation)

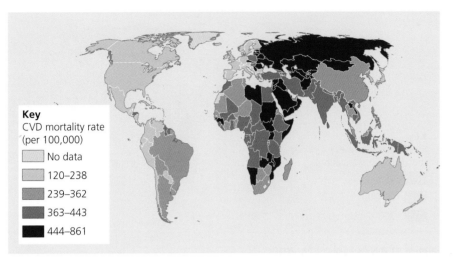

Figure 11.6 Global distribution of CVD mortality, 2011 (Source: WHO)

Cardiovascular disease (CVD)

CVD covers a range of illnesses including coronary heart disease, stroke, hypertension and angina. Because its incidence rises steeply with age, CVD is a major cause of mortality and morbidity in ageing populations in high-income countries. However, when standardised by age, the highest CVD mortality rates are found in Russia, sub-Saharan Africa and the Arabian Peninsula (Figure 11.6). CVD is responsible for 17 million deaths a year, with 80 per cent occurring in low- and middle-income countries. Premature death from CVD is linked to lifestyle, particularly to tobacco consumption, unhealthy diet and physical inactivity.

Disease diffusion

Diseases spread outwards from their origin and across space in a process known as **diffusion**. The diffusion concept applies to innovations, migration and settlement as well as the spread of disease.

Types of diffusion

The diffusion of disease takes several forms (Figure 11.7). In **expansion diffusion** a disease has a source and spreads outwards into new areas. Meanwhile, carriers in the source area remain infected. An outbreak of TB is an example of expansion diffusion. **Relocation diffusion** occurs when a disease leaves the area or origin and moves into new areas. The cholera epidemic in Haiti in 2010, which killed 7000 people, illustrates this process. The disease, which originated in Nepal, was brought to Haiti

by international aid workers flown in to tackle the earthquake disaster of that year.

Contagious diffusion describes the spread of disease through direct contact with a carrier. It is strongly influenced by distance. The Ebola epidemic in West Africa in 2014–15 is a classic example of contagious diffusion. In **hierarchical diffusion** a disease spreads through an ordered sequence of places, usually from the largest centres with the highest connectivity to smaller, more isolated centres. Diffusion is also channelled along road, rail and air transport networks which facilitate contact between carriers and a susceptible population. In 2009 the H1N1 flu virus became a pandemic via international flight routes and airports. In the USA it resulted in 61 million cases and 12,500 deaths.

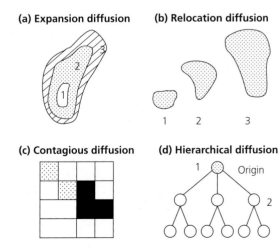

Figure 11.7 Types of diffusion

Barriers to diffusion

Barriers which limit the spread of disease are both physical and socio-economic. The most important physical barrier is distance. In general, the probability of a contagious disease spreading to an area is inversely proportional to distance from its source. Other physical barriers which slow or halt diffusion include mountain ranges, seas, oceans, deserts and climate. Climate is a major factor in the epidemiology and distribution of diseases such as malaria and sleeping sickness.

Political borders check the international movement of carriers of infectious disease. The spread of disease can also be controlled by imposing curfews to limit contact between people. This happened on several occasions in Sierra Leone in 2015 in an effort to contain the spread of the Ebola virus. Quarantining of western aid workers infected with Ebola on their return from West Africa also minimised the risks of the disease spreading to the UK and other countries. Other precautions often taken to check the spread of viruses include wearing face masks in public places and cancelling public events. Mass vaccination programmes to protect populations against diseases like flu and health education constitute additional barriers to the spread of disease.

Hägerstrand's diffusion model

Torsten Hägerstrand's original diffusion model was developed to simulate the spread of farm subsidies in the small area in southern Sweden. Later the model was applied to the contagious diffusion of diseases. The model is probabilistic rather than fixed or deterministic. As a result it produces slightly different outcomes each time it is run.

The model features several important concepts:

- A neighbourhood effect: the probability of contact between a carrier and non-carrier is determined by the number of people living in each 5 × 5 km grid square, and their distance apart. Thus people living in proximity to carriers have a greater probability of contracting a disease than those located further away. This distance-decay function is assumed to be geometric in character.
- The number of people infected by an epidemic approximates an S-shaped or logistic curve over time (Figure 11.8). After a slow beginning, the number infected accelerates rapidly until eventually levelling out, as most of the susceptible population have been infected.
- The progress and diffusion of a disease may be interrupted by physical barriers.

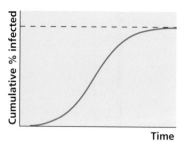

Figure 11.8 Logistic curve

Key idea

→ There is a relationship between physical factors and the prevalence of diseases which can change over time

Global patterns of climate and relief and their effect on disease

Temperature and precipitation are important drivers of vector-borne diseases and epidemics. Many diseases, including malaria, dengue fever, yellow fever and sleeping sickness, whose epidemiology depends on warm, humid conditions, are endemic to the tropics and sub-tropics. Diseases influenced by climate often show seasonal patterns. This is partly because temperature determines rates of vector development and behaviour as well as viral replication. Also, precipitation, which in the tropics is often seasonal, creates aquatic habitats such as ponds and stagnant pools, which allow insects and disease vectors to flourish and complete their life cycles.

Relief affects global patterns of disease because altitude causes abrupt changes in climate and disease habitats. Thus in Ethiopia, malaria is concentrated in the humid lowlands but is largely absent in the cooler highlands. Many diseases are water-borne. In the

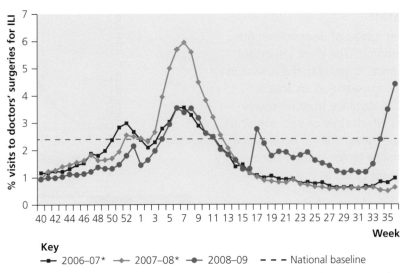

Figure 11.9 Reported cases of influenza-like illness (ILI), USA, 2009 (Source: ILINet)

developing world millions of people rely on water from wells and surface supplies contaminated by sewage. Bacteria responsible for cholera and other infectious diseases thrive in these conditions. Unprotected and stagnant drinking water supplies also provide habitats for disease vectors – for example, copepod vectors which transmit the parasitical disease Guinea worm to humans in West Africa.

Physical factors and disease vectors

Dengue fever is widespread in the tropics. Annually it infects around 400 million people and is responsible for 25,000 deaths. Climate controls dengue fever epidemiology and the life cycle of *Aede* mosquitoes that transmit the dengue virus to humans. Mosquitoes thrive in warm, humid conditions, which in turn favours the outbreak of dengue. In the South Pacific, sustained temperatures of more than 32°C and humidity levels above 95 per cent trigger waves of dengue epidemics.

These conditions occur in the summer months, but short-term weather changes and exceptional rainfall events can also lead to outbreaks of the disease.

Seasonal variations in disease outbreaks

The health risks associated with diseases often have a strong seasonal bias. In temperate regions in the northern hemisphere, epidemics of influenza, a contagious respiratory illness, peak in the winter months (Figure 11.9). Although the precise reason for this seasonality is not fully understood, it is known that transmission of the flu virus is most efficient at lower temperatures (e.g. 5°C) and when atmospheric humidity is low. These conditions occur most often in winter.

In the topics and sub-tropics, vector-borne diseases, transmitted by mosquitoes, flies, ticks, fleas and worms, often reach a peak during the rainy season (Figure 11.10). For example, diarrhoeal disease in

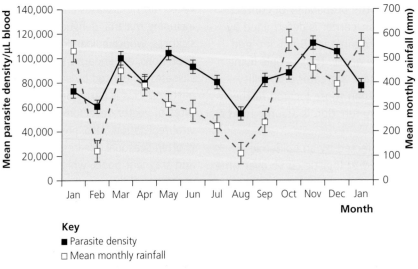

Key
■ Parasite density
□ Mean monthly rainfall

Figure 11.10 Rainfall and malarial parasite density, Entebbe, Uganda

South Asia surges in the pre-monsoon (March–April) and end of monsoon (September–October) periods when fly populations are highest. Sandflies, which transmit the protozoan-causing leishmaniasis to humans, are most abundant in the rainy season, when infection peaks.

Bilharzia or schistosomiasis, caused by a trematode flatworm hosted by freshwater snails, kills an estimated 200,000 people a year. Infection rates follow the life cycle of snails, which in turn is closely linked to seasonal precipitation and temperatures in the range 10–30°C.

➡ Activities

Study Figure 11.10, which shows the monthly pattern of rainfall (white squares) and malarial parasite infection (black squares) in Uganda, in equatorial Africa.

1 Describe the annual pattern of rainfall and malarial infection.
2 Describe and explain the relationship between rainfall and malarial infection.

Climate change provides conditions for emerging infectious diseases

Climate change – increases in temperature, rainfall and humidity – has stimulated transmission of vector-borne diseases and extended their geographic range. Warmer and wetter conditions have favoured the growth and spread of mosquitoes carrying tropical and sub-tropical diseases, such as West Nile virus (WNV), malaria and dengue fever.

WNV, first identified in Uganda in 1937, is transmitted by *Culex* mosquitoes. Birds are the main hosts for the virus. Today WNV has spread globally. It is prevalent throughout Africa; in the Americas its range extends from Venezuela to Canada; and it is also found in parts of Europe, West Asia and Australia. In 2012, 5500 cases of WNV were reported in the USA (Figure 11.11). High temperatures favour transmission – hence Texas is one of the US states most severely affected.

Climate change is also responsible for the spread of Lyme disease and trypanosomiasis. Ticks, which transmit Lyme disease, thrive in warmer conditions. In the USA, Lyme disease is expanding northwards and as temperatures rise the ticks will eventually colonise Canada.

In Africa, trypanosomiasis, or sleeping sickness, is endemic in 36 sub-Saharan countries and affects 70 million people. Trypanosomiasis is transmitted to humans by the tsetse fly. Outbreaks of the disease occur when average temperatures are in the range 20.7–26.1°C. In future climate change will affect the vector's growth rate and the geographical distribution of the disease. As temperatures rise, trypanosomiasis is likely to spread into southern Africa, and according to WHO will affect up to 77 million more people by 2090. However, the disease may disappear from East Africa where the climate may become too hot for tsetse larvae to survive.

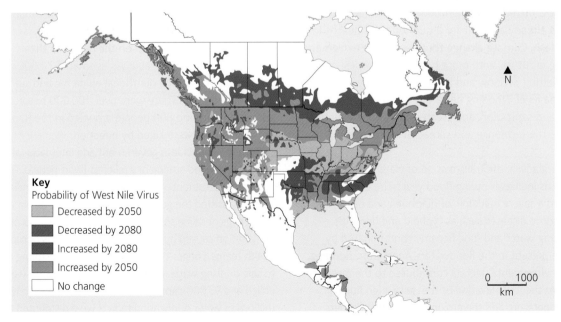

Figure 11.11 Spread of WNV in North America (Source: Global Change Biology)

The spread of zoonotic infectious diseases to humans

Zoonotic diseases, spread from animals to humans by viruses, bacteria, parasites and fungi, are very common. Major infectious diseases such as malaria, sleeping sickness and dengue fever are zoonotic. Infection may be transmitted by domestic as well as wild animals. For instance, dogs, in addition to bats, foxes, raccoons and other mammals, transmit rabies; poultry present a greater risk of transmitting Asian flu than wild birds. However, in most instances, transmission occurs only with close contact between people and animals.

The probability of zoonotic diseases being transmitted to humans increases where:

- the movement of infected wild animals is unrestricted by physical barriers (e.g. mountain ranges), or in the case of domestic animals, political boundaries
- controls on the movement of diseased domestic animals within countries are ineffective
- urbanisation creates suitable habitats for animals such as foxes, raccoons and skunks
- vaccination of pets and domestic livestock is sparse
- there is limited control within urban areas of feral dogs, cats, pigeons and other animals
- hygiene and sanitation are poor; drinking water is contaminated by animal faeces, blood and saliva; man-made habitats (e.g. surface pools, ponds) encourage insect vectors to breed
- there is prolonged contact between humans and animals, e.g. poultry farms and avian flu, cattle farming and anthrax.

> ### Key idea
> → Natural hazards can influence the outbreak and spread of disease

🔍 Case study: River flooding in Bangladesh, 2007

In August 2007, heavy monsoon rainfall caused the worst river floods for decades in Bangladesh – 60 per cent of the country was inundated and 9 million people living in the Ganges–Brahmaputra Delta were displaced from their homes. Exposure to flooding is a recurrent hazard in the delta, where three major rivers – the Ganges, Brahmaputra and Meghna – converge. These rivers, swollen by the annual monsoon rains and seasonal melting of glaciers in their Himalayan headwaters, pose a continual threat to the population of the low-lying delta.

In Bangladesh the transmission of water-borne pathogens increases during flood years. Water-borne diseases already account for a quarter of all deaths in Bangladesh. Climate change threatens more extreme flooding in future, with more powerful cyclones, more rapid melting of snow and ice in the Himalayas, and rising sea level. Flooding has a direct impact on water supply and sanitation, and creates conditions which increase the incidence and spread of infectious diseases.

In Bangladesh, the transmission of water-borne pathogens increases during flood years. The 2007 flood triggered a severe epidemic of diarrhoea and other water-borne diseases such as typhoid and hepatitis, as drinking water and food became contaminated by bacteria present in the floodwaters. Apart from flooding, other environmental factors contributed to the disease outbreak: the low elevation of the delta, high humidity and temperature and the proximity of large numbers of people close to aquatic environments.

Socio-economic factors also aided disease transmission. They include:

- exceptionally high rural population densities which in many districts exceed 1000 persons/km^2
- high levels of poverty: according to the World Bank over 40 per cent of Bangladeshis survive on less than US$1.25 a day. During the floods, the poorest groups suffered disproportionately
- a lack of access to adequate sanitation
- the displacement of nearly 14 million people caused by the floods.

One in eight wells were contaminated by bacterial pathogens derived from sewage.

The impact of the floods on the health of the population was considerable. An estimated 70,000 people were hit by the diarrhoeal epidemic and suffered acute dehydration. There were over 100,000 hospital admissions. Around 800 people drowned in the floods, but many more were killed by infection.

Both the Bangladesh government and international agencies provided emergency relief to flood victims. The government distributed food aid to the poorest families. UNICEF assisted the government by providing essential drugs, bags of saline solution (to combat the effects of dehydration caused by diarrhoea) and hundreds of mobile health teams. Longer-term relief focused on restoring access to safe drinking water. Hundreds of new tube wells were drilled and 93,000 damaged wells repaired. Meanwhile, millions of water purification tablets were distributed to reduce the post-flood risks of water-borne infection. ➔

Activity

Study Figures 11.12–11.14 and analyse:

a) the relationship between the areas flooded and the distribution of poverty

b) the link between flooding, poverty and diarrhoea disease.

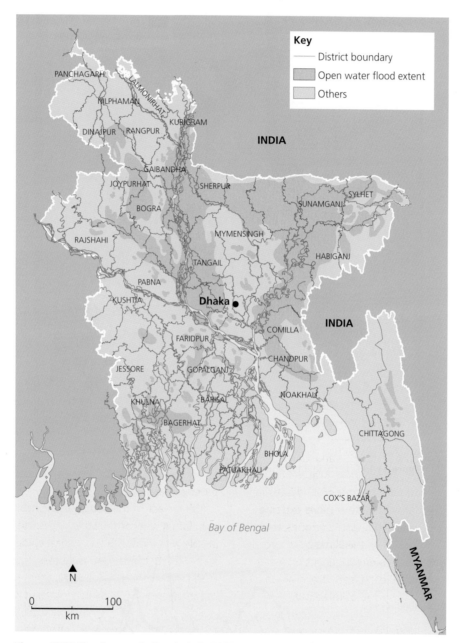

Figure 11.12 Flood extent in Bangladesh, 2007

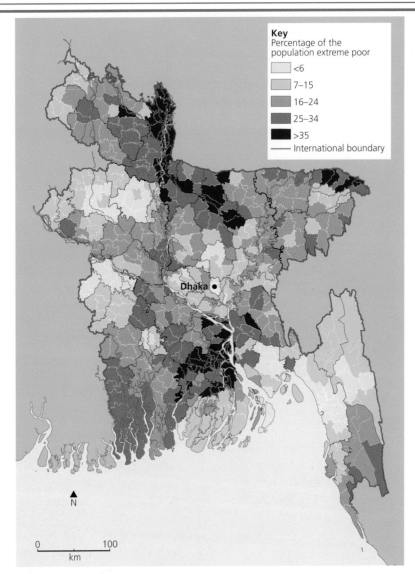

Figure 11.13 Poverty in Bangladesh, 2007

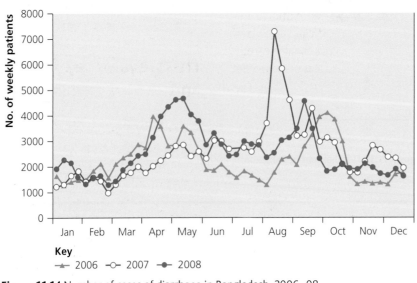

Figure 11.14 Number of cases of diarrhoea in Bangladesh, 2006–08

11.2 Is there a link between disease and levels of economic development?

Key idea
→ As countries develop economically, the frequency of communicable diseases decreases while the prevalence of non-communicable diseases increases

The epidemiological transition

Abdel Omran's model of the epidemiological transition describes the relationship between development and changing patterns of population age distribution, mortality, fertility, life expectancy and causes of death. Changes are driven by improvements in health care, standards of living and the quality of the environment.

According to Omran, societies undergo change in three epidemiological phases:

1 *The age of pestilence and famine.* In pre-industrial societies mortality is high and fluctuates from year to year. Life expectancy is low and variable, averaging around 30 years. Poor sanitation, contaminated drinking water and low standards of living make people more susceptible to infectious diseases which dominate mortality. Population growth is slow and intermittent. Today all countries have passed through this phase.
2 *The age of receding pandemics.* In industrial societies with advances in medical technology,

diet and hygiene, and improvements in living standards, epidemics causing large-scale mortality become rare. Life expectancy rises above 50 years and population growth is sustained. There is a shift in the main cause of death from infectious diseases to chronic and degenerative diseases. This phase includes many LIDCs and MIDCs today.

3 *In post-industrial societies the rate of mortality slackens.* Further improvements in medical technology, hygiene and living standards mean that mortality related to infectious disease is rare. Degenerative disease becomes the main cause of mortality. Man-made diseases associated with environmental change become more common. Many emerging economies such as Brazil and China are in this phase.

Some observers suggest a fourth phase – the age of delayed degenerative diseases – where medical advances delay the onset of degenerative CVD. The so-called cardiovascular revolution of the past 40 years has raised average life expectancy in ACs from the early seventies to the mid-eighties. Obesity and diabetes become increasingly common and problematic health factors.

→ Activity

Log into the World Bank databank website (data. worldbank.org/data/databases.aspx). Access data sets on infant mortality, maternal mortality and respiratory infection for high-, middle- and low-income countries.

a) Download the relevant data and represent the data in appropriate charts (see pages 513–17).
b) Describe the trends and changes evident in the charts.

The prevalence of communicable and non-communicable diseases

Non-communicable disease in ACs

Non-communicable diseases such as CVD and cancer dominate mortality and morbidity in ACs. Communicable diseases have largely been eliminated, thanks to advancements in medical diagnoses and treatments, high standards of living, proper sanitation, clean water supplies and appropriate food intake. The result is comparatively healthy populations that have long average life expectancies. However, prolonged life expectancy inevitably increases the proportion of deaths and illnesses connected to degenerative diseases and old age.

In ACs, **overnutrition** and excessive consumption of sugar, carbohydrates, fats and salt are increasing health risks and the prevalence of non-communicable diseases such as CVD, type-2 diabetes, hypertension and several types of cancer. Moreover, these heightened risks, exacerbated by obesity and physical inactivity, are increasingly apparent in younger age groups.

Meanwhile, overnutrition, once confined to the developed world, is becoming a significant health problem in the developing world. In 1974 in Brazil there were two cases of underweight adults for everyone who was obese. By 1997 the ratio had reversed, with obese adults outnumbering those who were underweight by two to one.

As standards of living rise, non-communicable diseases become more prevalent.

Cancer is often regarded as a disease of high-income countries. Yet 70 per cent of all cancer deaths today are in low- to middle-income countries. More worryingly, the incidence of cancer in poorer countries is rising rapidly and is expected to double by 2030. However, if we ignore absolute numbers, the incidence of cancer standardised by age structure remains much greater in ACs than in LIDCs and MIDCs (Table 11.2). Crude cancer rates for ACs average 316/100,000 for males and 253/100,000 for females, compared with 103 and 123 respectively in LIDCs.

Table 11.2 Incidence of cancer by continent in 2012 (Source: Cancer Research)

	Cases per year (millions)	Cases per 100,000 adults
Europe	3.44	255
Asia	6.76	152
Americas	2.88	242
Africa	0.85	123
Oceania	0.16	298

👟 Fieldwork idea

Obesity among young people is a major health problem in the UK. Poor diet is a key determinant of obesity. Select an urban area and investigate where young people are at most risk by:

a) mapping schools, youth facilities and leisure centres
b) mapping all fast-food outlets (fish and chips, take-away, kebabs, fried chicken, etc.) within 400m of each school, youth and leisure centre
c) analysing the pressure on each school, youth and leisure centre exerted by access to unhealthy food by constructing a 400 m radius catchment around each food outlet and counting the number of outlets that overlap each 400 m radius.

Communicable disease in LIDCs

Communicable diseases dominate mortality in the world's poorest countries. These diseases are classified into three groups: animal-borne, water-borne and food-borne. Water-borne diseases such as cholera, typhoid and polio, now eliminated in ACs, remain endemic in most LIDCs. Their prevalence in LIDCs is due to a number of factors, though most are related in some way to poverty. Failure to control communicable disease reflects inadequate health care services and a lack of resources to tackle the causes of disease. Other factors include inadequate nutrition, poor environmental and living conditions, and geography.

Inadequate food intake gives rise to **undernutrition** and **malnutrition**. Undernutrition results from too little food intake to maintain body weight. Malnutrition is the result of an unbalanced diet, in particular shortages of protein and essential vitamins. Both undernutrition and malnutrition are widespread in the poorest countries in the world.

Analysis by the UN Food and Agriculture Organization (FAO) shows that although globally, food intake per capita rose significantly between 1965 and 2015, the situation in the world's poorest regions showed little change. In sub-Saharan Africa, food intake rose by a mere 37 kcals/day/person, and in South Asia the increase was a modest 396 kcals/day/person.

Health and diet are closely linked. Undernutrition and malnutrition weaken the immune system and increase the risks of bacterial and viral infections. Malnutrition, caused by protein deficiency, is responsible for non-communicable diseases such as kwashiorkor and marasmus. Diseases caused by lack of vitamins include rickets (vitamin D), scurvy (vitamin C) and pellagra (vitamin B).

Poor environmental conditions are instrumental in the spread of communicable diseases. Water pollution is mainly caused by a lack of proper sanitation and hygiene. Polluted water from wells and surface streams provides a disease reservoir for cholera, typhoid and diarrhoea. Meanwhile, poor drainage provides breeding sites for disease vectors such as mosquitoes and water snails that transmit diseases like malaria and bilharzia. The threat of infectious disease is increased in LIDCs by the appalling conditions in which millions of people live. Slum housing and overcrowding are closely linked to TB and other respiratory diseases.

Geography also plays an important role in the prevalence of communicable diseases in LIDCs. Most of the world's poorest countries are in the tropics and sub-tropics. High temperatures and abundant rainfall create the epidemiology for a wide range of infectious diseases – malaria, dengue fever, sleeping sickness, filariasis, yellow fever, Ebola – which are absent in cooler climates of higher latitudes.

Case study: Air pollution and cancer in India

Causes and health impacts

Levels of air pollution in India are among the highest in the world. Research published in 2015 showed that air pollution reduces the average life expectancy of 660 million Indians by more than three years and that 99 per cent of India's 1.2 billion people breathe polluted air above safe levels, as defined by WHO (Figure 11.15).

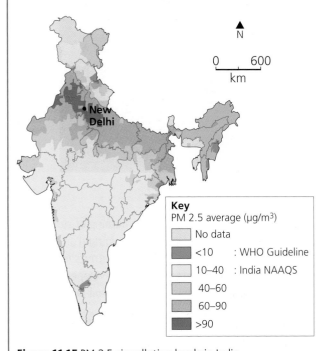

Key
PM 2.5 average (μg/m³)

	No data
	<10 : WHO Guideline
	10–40 : India NAAQS
	40–60
	60–90
	>90

Figure 11.15 PM 2.5 air pollution levels in India

Air pollution is due to emissions of particulates, nitrogen dioxide (NO_2), sulphur dioxide (SO_2) and ozone, principally by motor vehicles, coal-burning power stations and factories. However, indoor air pollution is also a problem. This is especially so in rural areas, where households often lack electricity and depend on biomass fuels such as animal dung for heating, and paraffin for cooking and lighting. Indoor air pollution from these sources is responsible for about 1 million premature deaths a year.

Particulate pollution is the biggest threat to human health. Tiny air-borne particles smaller than 2.5 micrometres (PM 2.5) are released by burning fossil fuels and penetrate deep into people's lungs. They cause serious respiratory problems (asthma, bronchitis) as well as lung and heart disease, and cancer (as a result of PM 2.5 pollution residents' risks from lung cancer in Delhi are increased by nearly 70 per cent). WHO's guidelines set safe PM (particulate matter) 2.5 pollution levels at 10 micrograms/m³. India, however, sets permissible levels at 40. Even so, PM 2.5 pollution

exceeds this threshold in most large cities (Figure 11.16) and in Delhi in the winter months it often exceeds 600, earning it the dubious title of the world's most polluted city (Figure 11.17).

Studies have shown a close relationship between air pollution in Delhi and elevated levels of mortality and morbidity. Compared with less-polluted rural areas in India, respiratory symptoms (e.g. breathlessness, chest discomfort) and diseases such as asthma are 1.7 times higher in Delhi. Lung function among Delhi's inhabitants is on average 40 per cent reduced, compared with 21 per cent in rural areas. And hypertension, linked to air pollution, is 40 per cent higher.

In recent years, as air pollution levels have increased, Delhi has seen a significant rise in pollution-related cancers such as lung and bladder cancer. While smoking remains the biggest cause of lung cancer, one in five cases now occurs among non-smokers. This represents a 20 per cent increase in the past 10 years. Between 2008 and 2010, lung cancer rates in Delhi for men increased from 14 to 15.5 per 100,000, and for women from 4.2 to 4.6 per 100,000.

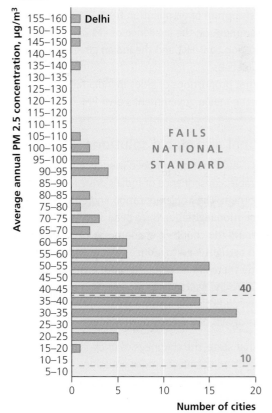

Figure 11.16 Indian cities PM 2.5 pollution (Source: WHO)

Figure 11.17 Air pollution in Delhi

Activities

Study Figures 11.15 and 11.16.

1 Refer to an atlas, and describe in detail the distribution of air pollution from PM 2.5 in India.
2 Comment on the incidence of PM 2.5 pollution in relation to WHO and the Indian government's safe guidelines.
3 What proportions of cities in India fail to conform to the Indian government's own PM 2.5 guidelines?

National and global solutions

Hitherto, air pollution has been seen as an inevitable, if undesirable, consequence of India's drive for economic development. The implementation and enforcement of anti-pollution legislation have been slow and have lagged well behind the country's economic growth. Standards need to be tightened to reduce vehicle emissions, yet currently there are no regulations for emissions of NO_2 and SO_2. As living standards increase it is expected that pollution controls will assume greater priority in future. Some progress has already been made:

- In Bihar state, the chimneys of brick-kilns have been retro-fitted to reduce smoke emissions.
- Fourteen Indian cities are currently building rapid-transit metro systems.
- Subsidies for petrol and diesel will be scrapped – one-third of all electricity is produced by noxious generators powered by petrol and diesel fuel.
- Restrictions will be placed on the burning of stubble in fields, a major cause of air pollution in rural areas.
- The world's first market for trading permits in emissions of particulates has been developed in the states of Gujarat, Maharashtra and Tamil Nadu.

Policies devised at the global scale to tackle climate change will have benefits for human health at national and local levels by cutting CO_2 and other GHG emissions. In 2012, 37 countries and the EU states agreed targets to cut GHG emissions by 18 per cent of 1990 levels by 2020. EU countries have set their own, even more ambitious targets: a reduction of 20 per cent of 2005 levels by 2020. The EU also has the world's largest carbon cap-and-trade scheme and each EU state has targets for expanding renewable energy and policies to comply with the European Climate Change Programme (see Chapter 10, page 308).

Countries which participate in these international initiatives should see reductions in air pollution, particularly in urban areas, and with this, improvements in life expectancy and the disease burden caused by emissions of PM 2.5, NO_2, SO_2 and ozone. Other global initiatives include the annual World Cancer Day, drawing attention to the current global cancer 'epidemic', and pressing governments to take more action to tackle the disease. Most recently, WHO published a 'Draft Road Map' to address the problem of cancer as a leading avoidable cause of death, and to confront the adverse health effects of air pollution.

Activity

Table 11.3 Human development and incidence of cancer in selected countries in 2013

Country	Human Development Index	Cancer rate per 100,000 people	Country	Human Development Index	Cancer rate per 100,000 people
Algeria	0.337	123.5	India	0.586	94
Argentina	0.808	216.7	Indonesia	0.684	133.5
Australia	0.933	323	Niger	0.337	63.4
Brazil	0.744	205.5	Norway	0.944	318.3
Canada	0.902	295.7	South Africa	0.658	187.7
China	0.719	174	Tanzania	0.488	123.7
France	0.884	324.6	Thailand	0.722	137.5
Germany	0.911	283.8	USA	0.914	318
Greece	0.853	163	UK	0.892	272.9
Guinea	0.392	90	Uruguay	0.790	251.1

Table 11.3 shows the Human Development Index (HDI) and the incidence of cancer in selected countries in 2013. The HDI is a UN measure of development based on economic, demographic and social criteria. The higher its value, the more developed a country is. Analyse the data to assess the link between economic and social development and the incidence of cancer, using Excel or an equivalent spreadsheet to:

a) Plot the data as a scattergraph where the HDI is *x* and cancer rates are *y* (see page 517).

b) Fit a trend line to the scatter of points and find the correlation coefficient between *x* and *y* (see pages 541–44).

c) Describe your results and suggest possible reasons for any relationship between development and the incidence of a non-communicable disease like cancer.

Skills focus

Real-time data on air quality for cities in Europe, Asia and the Americas are available online at www.aqicn.org/map. An air quality index used on this site describes levels of air pollution viz: good (0–50), moderate (51–100), unhealthy for sensitive groups (101–150), unhealthy (151–200), very unhealthy (201–300), hazardous (> 300). Values for specific pollutants such as PM 2.5 and NO_2 and the pattern of pollution over the previous 24 hours are also given.

a) Find the real-time air quality indexes for Delhi, Beijing, Shanghai, London and Paris.

b) Plot the different air pollutants for each city as a series of bar charts.

c) Describe and give possible explanations for differences between the cities in the incidence of specific pollutants.

d) Suggest how changes in diurnal air pollution could affect human health in cities.

1 How does an ageing population affect the incidence of non-communicable diseases?
2 Outline two environmental factors that are linked causally to communicable diseases.
3 What aspects of lifestyle are associated with the incidence of CVD and cancer?
4 What is the epidemiological transition?
5 What is the difference between undernutrition and malnutrition?
6 How does overnutrition contribute to disease?
7 Name three diseases caused by a lack of essential vitamins.
8 Name a disease caused by protein deficiency.

11.3 How effectively are communicable and non-communicable diseases dealt with?

Key idea

→ Communicable diseases have causes and impacts with mitigation and response strategies which have varying levels of success

🔍 Case study: Malaria in Ethiopia

Caused by a tiny plasmodium parasite, malaria is the world's most deadly disease. The parasite has two hosts: the *Anopheles* mosquitoes and humans. Mosquitoes act as vectors, transmitting the disease from person to person. Parasites enter people through the bite of an infected mosquito and exit by the same route. In 2013, malaria claimed 584,000 lives; three-quarters of deaths were children under the age of five years.

Incidence and patterns of malaria

Malaria is endemic in 75 per cent of Ethiopia's land area. Two-thirds of the country's population live in areas at risk from the disease, which kills around 70,000 people a year. However, malaria is not evenly distributed within the country. The areas of highest risk are the western lowlands, in Tigray, Amhara and Gambella provinces (Figure 11.18). There transmission rates peak after the rainy season, between June and November. In the midlands, where altitude ranges from 1000 m to 2200 m, transmission is also seasonal, with occasional epidemics. In Afar and Somali provinces in the eastern lowlands, the arid climate confines malaria to river valleys. The central highlands, comprising around one-quarter of the country, are malaria-free.

Environmental and human causes of malaria

Malaria thrives in warm, humid climates and where stagnant surface water provides ideal breeding habitats for mosquitoes. In Ethiopia these habitats are strongly influenced by altitude. The disease is endemic in the western lowlands where temperatures and humidity are high throughout the year. The absence of malaria in

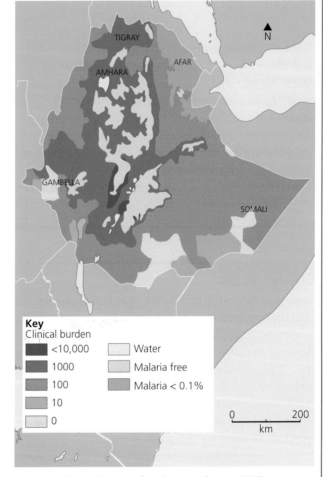

Figure 11.18 Distribution of malaria in Ethiopia, 2007

Key
Clinical burden
- <10,000
- 1000
- 100
- 10
- 0
- Water
- Malaria free
- Malaria < 0.1%

0   200
km

the highlands is explained by low average temperatures, which slow the development of mosquitoes and the plasmodium parasite (Figure 11.19).

Figure 11.19 Ethiopian highlands

Human factors also have an important role in malaria's epidemiology. Population movements, urbanisation, irrigation schemes and the misuse of malarial drugs have encouraged the spread of the disease.

Every year at harvest and planting time, large-scale, seasonal population movements take place between the malaria-free highlands and the agricultural lowlands. The timing of this migration coincides with the rainy season and the peak malarial transmission period (June–September) in the lowlands. Infection is also increased because harvesting often continues after sunset when mosquitoes are most active, and most migrant workers sleep in the fields overnight.

Irrigation projects in the Awash Valley and in Gambella province, with the construction of canals, micro-dams and ponds, and the cultivation of rice, have expanded the breeding habitats for mosquitoes. Urbanisation has had a similar effect: flooded excavations, garbage dumps, discarded containers and so on provide countless breeding sites. Meanwhile, malarial parasites are becoming increasingly drug-resistant.

Socio-economic impacts of malaria

The burden of malaria in Ethiopia, and throughout sub-Saharan Africa, has significant social and economic impacts. Hardest hit are the poor, often living in crudely built dwellings that offer few barriers to mosquitoes. Ethiopians suffer approximately 5 million episodes of malaria a year, which kill around 70,000 people. Malaria also has debilitating effects, causing absenteeism from work, slowing economic growth and reinforcing the cycle of poverty. Lost production in sub-Saharan Africa due to

malaria is estimated to be US$12 billion a year. The cost to health services is also considerable: in Ethiopia, malaria absorbs 40 per cent of national health expenditure and accounts for 10 per cent of hospital admissions and 12 per cent of health clinic visits. Dealing with malaria epidemics can overwhelm the country's health services as well as damaging tourism and curtailing inward investment.

Malaria also has implications for food security and the environment. The western lowlands, for example, are resource-rich, with considerable potential to raise food production. But malaria, which is endemic to the region, holds back development. This problem has a knock-on effect in the highlands. Because this region is malaria-free, it supports unusually high population densities. As a result, its meagre farming resources have been overexploited for generations, resulting in widespread land degradation. This situation contributed to devastating famines in the 1980s.

Strategies to control malaria

Since 2005 Ethiopia has benefited from the President's Malaria Initiative (PMI) and the Global Health Initiative (GHI) to scale-up malaria prevention and treatment throughout sub-Saharan Africa. Between 2008 and 2013 Ethiopia received grants of US$20–43 million a year for malaria control.

In 2011 the Ethiopian government also implemented a five-year plan (2011–15) for malaria prevention and control. The plan operates in partnership with a number of agencies, including UNICEF, the World Bank and the World Health Organization (WHO), non-governmental organisations (NGOs) and OECD donor countries. Its strategy is both direct and indirect. Direct action involves measures to eradicate mosquitoes, including periodic spraying of dwellings with insecticides and managing the environment to destroy breeding sites (e.g. stagnant water) for mosquitoes.

Indirect strategies focus on mass publicity campaigns to minimise potential mosquito breeding sites; providing early diagnosis and treatment of malaria (within 24 hours of the onset of fever); and distributing insecticide-treated bed nets to all households in infected areas.

Partly as a result of these efforts death rates from malaria halved between 2000 and 2010. In the past cycles of malaria have occurred every five to eight years. However, since 2003 there have been no malaria epidemics in the country. In Amhara, one of the provinces worst affected by malaria, the prevalence of the disease fell from 4.6 per cent of the population in 2006 to just 0.8 per cent in 2011.

Key idea
→ Non-communicable diseases have causes and
impacts with mitigation and response strategies
which have varying levels of success

 Case study: Cancer in the UK

Social, economic and cultural causes

Although some cancers are caused by occupational and
environmental hazards such as exposure to radiation,
toxic chemicals or asbestos, most are related to lifestyle.
Increased risks of cancer are associated with obesity, poor
diet, lack of exercise, smoking and alcohol abuse. Largely
as a result of changing lifestyles, since the 1970s cancer
rates in the UK have risen by 23 per cent for men and
43 per cent for women.

Cancers associated with lifestyle choices are
preventable. Sunbathing and the use of sunbeds, for
example, indicate a cultural preference for a tanned 'look',
despite the evident risks of skin cancer. Also opportunities
for sunbathing have increased in the past 50 years, with
growing wealth and the advent of affordable package
holidays to destinations such as the Mediterranean and
Florida. Along with wealth comes changes in diet and a
preference for meat and dairy products, fast food and
pre-packed 'ready' meals – changes which are linked to
a rise in the incidence of bowel cancer. And with higher
incomes, alcohol consumption invariably rises, increasing
the risks of oral, oesophageal and liver cancer. Finally, lack
of exercise and more sedentary lifestyles, together with
changes in diet, have driven an epidemic of obesity in the
UK and other ACs and increased risks of cancer and CVD.
Despite a decline in the popularity of smoking, it remains
the biggest single cause of cancer among both men and
women. Nearly one-fifth of all cases of cancer diagnosed
each year are smoking related.

Socio-economic impacts

Two million people are living with cancer in the UK today,
costing the UK economy £15 billion a year due to early
deaths, patients taking time off work, treatment on the
NHS and the cost of unpaid care. The Macmillan charity
estimates that the average cost to cancer patients is
around £570 a month. This figure includes loss of income,
the costs of medical appointments and prescriptions,
and extra heating costs. In addition, there are social and
psychological costs: cancer sufferers often experience
social isolation, anxiety resulting from loss of income, and
further physical as well as mental health problems.

The link between the incidence of cancer and socio-
economic deprivation is well known. Deprivation increases

the likelihood of smoking, alcohol consumption and
obesity. All are major causes of cancer. In the UK, cancer
rates in some of the poorest areas are three times greater
than in the most affluent (Table 11.4). Glasgow has the
highest cancer rate of any UK health authority and it is
no coincidence that in the wider central Scotland region
over half the population lives in wards which are among
the 20 per cent most deprived in the UK. This association
between deprivation and cancer is also strongly
entrenched in former industrial areas in northern England,
south Wales and London. Of the 50 most deprived small
census areas in England, 34 are in the northwest region
and 17 in the Merseyside conurbation alone. Significantly,
the two health authorities with the highest cancer
rates in England, Liverpool and Manchester, are in the
northwest region.

Cancer survival rates are also affected by socio-
economic status. For all types of cancer there is a
deprivation gap, with the more affluent having better
survival chances than the most deprived. For example,
14.2 per cent more women in the 'most affluent
group' survive bladder cancer compared with their
most deprived counterparts. This difference is largely
explained by pre-existing health status and speed of
diagnosis.

Table 11.4 Incidence of all cancers: % difference from the
UK and Ireland average

Region	Males	Females
North and Yorkshire	+1.6	−1.3
East Midlands	−3.3	−3.9
West Midlands	+0.4	−2.2
Northwest	+4.6	+1.8
East	−10.3	−5.3
London	−2.4	−3.2
Southeast	−4.2	−0.6
Southwest	−3.7	0
Wales	+7.9	+5.9
Scotland	+16.4	+13.2
Northern Ireland	+0.1	+2.6
Ireland	−2.5	8.7

→

Activities

1 Plot the regional and gender inequality in cancer incidence in Table 11.4 as a bar chart.
2 Summarise the geography of inequality in cancer incidence in the UK.

Government and international agency strategies to mitigate against cancer

The UK government's targets in its fight against cancer are to save 5000 lives a year, increase survival rates and reduce the gap in survival rates that currently exists between the UK and other European countries.

The strategies employed to achieve these targets are both direct and indirect. Direct strategies include investment in advanced medical technology, such as more precise forms of radiotherapy, and diagnostic methods such as endoscopy for early diagnosis and intervention. Mass screening for breast, cervical and bowel cancer is already well established and has proved highly effective. However, survival rates could be improved further by reducing waiting times between diagnosis and treatment and by giving more support to GPs in referrals to consultants. Meanwhile, cancer research focuses on improving understanding of the disease, developing new treatments, discovering new drugs and exploiting the potential of genetic engineering.

Indirect approaches emphasise changes in lifestyle and cancer prevention. Education and health campaigns informing the public of the dangers of smoking, excessive drinking and unbalanced diets can reduce the incidence of preventable cancers.

International agencies and charities are also involved in the fight against cancer. The International Agency for Research on Cancer is part of WHO. It conducts epidemiology and lab research into the causes of the disease. Cancer UK is a charity that researches the prevention, diagnosis and treatment of cancer. Funded by donations, legacies and charity events, it operates at hospitals and universities throughout the UK.

Skin cancer has increased significantly in the past three or four decades and current rates of skin cancer show a year-on-year rise of 3 per cent. The government has intervened directly by legislating to control the commercial use of sunbeds, with age limits for users, and standards of supervision and staff training. Direct clinical treatment involves surgery to remove malignant melanomas and chemotherapy. Publicity campaigns warn of the dangers of sunbathing and the unsupervised use of sunbeds, and advise on sunscreens, clothing and self-examination for cancerous lesions. During the summer months the Meteorological Office Advice regularly issues forecasts on UV intensities and safe limits of exposure. Skin cancer is a preventable disease, which can be controlled by modifications of behaviour and attitudes towards tanning.

Review questions

1 Outline the epidemiology of malaria.
2 Describe the geography of malaria in Ethiopia.
3 Why are irrigation schemes often accompanied by a rise in the incidence of malaria?
4 Describe three strategies that can be used to control malaria.
5 How can lifestyle choice affect the incidence of cancer?
6 Explain the relationship between socio-economic deprivation and cancer rates.
7 What can be done to improve cancer survival rates?
8 Suggest two reasons for the current epidemic of skin cancer in the UK.

11.4 How far can diseases be predicted and mitigated against?

Key idea
→ Increasing global mobility impacts the diffusion of disease and the ability to respond to it

The World Health Organization

Established in 1948 and headquartered in Geneva, WHO is the directing and co-ordinating authority on international health within the UN system. It works closely with other international organisations, including agencies such as UNICEF and the World Bank and NGOs such as the International Red Cross and Red Crescent Movement.

WHO has a wide-ranging brief, which includes:

- gathering health data
- providing leadership and identifying priority areas in matters critical to health
- researching health problems
- monitoring the international health situation
- supporting UN member states to devise health strategies
- providing technical support during health crises.

It collects data from the 194 member states and publishes them annually in its *World Health Statistics*. These data, available for each country, provide an insight into health risks, mortality from communicable and non-communicable diseases, government spending on health care and so on. However, the quality and completeness of these data are highly variable. For instance, WHO receives causes of mortality data from only 100 member states, while globally, two-thirds of all deaths are not even registered.

> ### ► Activity
>
> Access the health data from WHO's website for Brazil, Denmark and Liberia: www.who.int/gho/countries/en/
>
> a) For each country record the following statistics: malaria cases; death rates from TB; deaths from HIV/AIDS; per capita government spending on health.
> b) Present the information as a table.

WHO also researches health issues. Among its many research groups are those dedicated to influenza, tropical diseases, mental health and vaccines. Research projects are often partnerships with other international agencies. For example, it is currently collaborating with the multi-agency Stop TB Partnership, which aims to eradicate TB by 2050.

WHO takes a leading role in increasing awareness of epidemics and the outbreaks of new diseases, such as the Zika virus in 2016, and develops global strategies to combat diseases such as HIV/AIDS,

malaria and tuberculosis. It sets targets to improve prevention, diagnosis, treatment and care. Promoting research into new drugs and insecticides is also important. So too is its work in predicting the spread of diseases.

Support programmes for member states are an important part of WHO's brief. Following the 2015 Nepal earthquake disaster, WHO delivered emergency health services in the form of mobile medical units and supported foreign medical teams in areas worst hit by the quake. In Liberia, the 2014–15 Ebola epidemic caused the total collapse of the country's health care services, leaving it unable to cope with a serious outbreak of measles. WHO, together with UNICEF and the US CDC, stepped in and organised a country-wide measles vaccination programme to control the spread of the disease.

The 2009–10 H1N1 influenza pandemic

In April 2009 a new influenza virus, H1N1, was identified in Veracruz state in Mexico. The virus was similar to the infamous Spanish flu pandemic of 1918 which killed tens of millions of people. Also known as swine flu, H1N1 quickly spread to other parts of Mexico and North America. In April of that year WHO declared H1N1 an international public health emergency; by June it had upgraded H1N1 to pandemic status, with the disease recorded in 74 countries (Figure 11.20).

Influenza spreads easily through coughing and sneezing and is highly infectious. At its peak in the second half of June, the number of new H1N1 cases doubled every fifteen days. H1N1 swept the northern hemisphere in two waves. In the UK there were two peaks: in July and October (Figure 11.21). However, by late autumn the pandemic had begun to subside and by May 2010 was in steep decline.

Estimates of the number of deaths caused by the pandemic vary. Official figures quote 18,000 deaths, but this number, based only on lab-confirmed deaths, is likely to be a gross underestimate. Total mortality estimates by the end of 2010 range from 151,700 to 575,400, with a WHO average of 284,500. Half of all deaths were reported in Africa and Southeast Asia where the number of cases ranged from 43 million to 89 million.

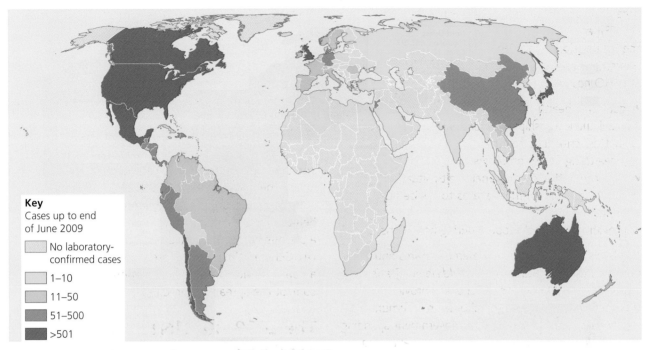

Figure 11.20 Global distribution of H1N1, June 2009 (Source: WHO)

Key
Cases up to end of June 2009

- No laboratory-confirmed cases
- 1–10
- 11–50
- 51–500
- >501

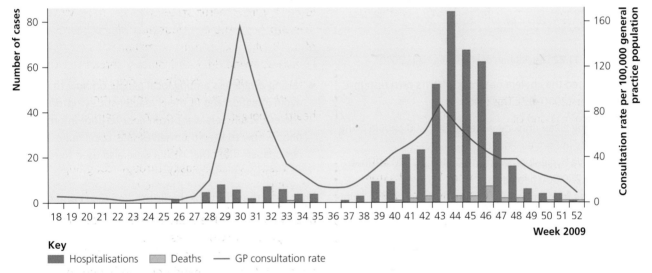

Key
- Hospitalisations
- Deaths
- GP consultation rate

Figure 11.21 H1N1 cases in the UK, 2009

🔍 **Case study:** The British Red Cross and the cholera epidemic following the Haiti earthquake of 2010

In January 2010, Haiti, on the Caribbean island of Hispaniola, was hit by a powerful magnitude 7 earthquake. The quake caused a huge natural disaster: 220,000 people were killed, 300,000 were injured and 1.3 million were made homeless. This is one of the poorest countries in the world, where 60 per cent of the population survives on less than US$2.5/day.

Following the quake, hundreds of thousands of homeless people were housed in makeshift camps. In the capital, Port-au-Prince, where prior to the quake 86 per cent of the inhabitants lived in slums, half the population had no access to toilets. Given the insanitary conditions, drinking water contaminated by sewage and overcrowding, an outbreak of cholera was inevitable. It finally occurred on 10 October 2010. The disease spread rapidly. Between the initial outbreak and November 2014, nearly 720,000 cases of cholera were recorded, with 8700 deaths (Figure 11.22).

The British Red Cross was one of several NGOs such as Oxfam and Red Crescent working in the disaster area.

➡

355

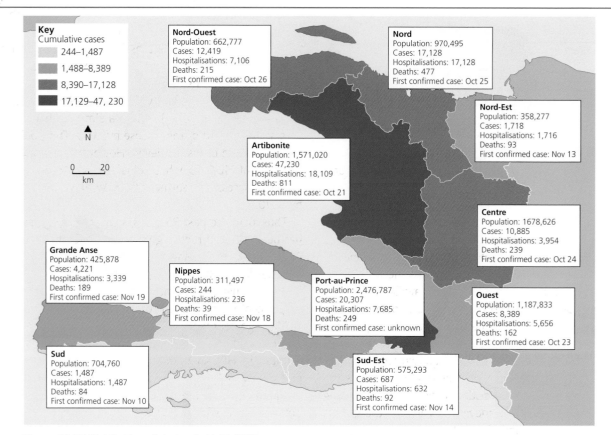

Figure 11.22 Distribution of cholera in Haiti, 2010

It targeted the cholera outbreak with its own response programme (2010–12). This involved:

- delivering clean drinking water to 300,000 people living in camps in Port-au-Prince
- a massive hygiene programme, building 1300 latrines serving 250,000 people
- providing medical supplies to the main hospital in Saint Marc in the affected area
- treating 18,700 cases of cholera in treatment units in La Piste camp in Port-au-Prince and in Port-a-Piment camp in southwest Haiti

- raising awareness among local people on how to avoid infection and of the characteristic symptoms of cholera. Volunteers went from door to door informing people how to keep themselves safe. Local radio, newspapers and other media were also used to reach as many people as possible.

Thanks to the efforts of the British Red Cross and other NGOs, the cholera epidemic was brought under control. In 2011 there were 35,000 new cases a month; by 2014 this figure had fallen to 2200. Even so, cholera remains a threat and a leading cause of infant mortality in Haiti.

> **Key idea**
> → Mitigation strategies to combat global pandemics and overcome physical barriers

Physical barriers and disease

Physical barriers often isolate communities and restrict population movements. In the face of a pandemic, remoteness can be an advantage, reducing the risk of infection and the spread of disease. However, it can also create problems. For instance, disease outbreaks within isolated communities may delay the arrival of medical assistance and emergency aid.

Until the mid-twentieth century the vast rainforest of the Amazon Basin isolated hundreds of indigenous Indian tribes from the outside world. For many tribes this isolation was a disaster. With little or no immunity to common 'western' diseases such as influenza, measles and chickenpox, contact with cattle ranchers, oil explorers and loggers often proved fatal. In Peru, half of the Nahua tribe, contacted for the first time in the early 1980s, was wiped out by disease following oil exploration on their land. A similar tragedy happened to the neighbouring Murunahua tribe, contacted for the first time in the mid-1990s by illegal mahogany loggers.

Outbreaks of water-borne disease, resulting from poor hygiene and contamination of water supplies, often accompany natural disasters such as earthquakes and floods. In remote regions that are difficult to access, disease can quickly get out of control and assume epidemic proportions. Pakistan and Nepal experienced major earthquakes in 2005 and 2015 respectively. Endemic, water-borne diseases such as typhoid, cholera and dysentery presented the greatest threat in isolated rural areas, which medical teams and emergency aid have difficulty reaching. In the remote Gorkha region of Nepal, many settlements are one or two days' walk from the main village. During the 2015 earthquake these settlements were cut off by landslides and were without clean water and medical supplies. Inaccessibility meant that it took weeks for medical help to arrive, giving time for epidemics to take hold. Cholera, for example, can be easily contained in its early stages if protective vaccines are available.

Yet remoteness can protect the wider population from disease risks. For instance, the Ebola virus first appeared in equatorial Africa in the 1980s. However, the communities affected were so isolated in the Congo rainforest that the disease was contained. In Maryland, USA, Chesapeake Bay divides the state into two distinct regions. The bay, acting as a barrier to population movements, limited the spread of measles in the period 1917–38.

Mitigation strategies to combat the global HIV/AIDS pandemic

Human immunodeficiency virus (HIV) was first identified in the USA in 1981. The virus weakens the human immune system, leaving those infected with little resistance to life-threatening diseases. This second stage of the disease is known as acquired immunodeficiency syndrome, or AIDS.

Although the HIV/AIDS pandemic has declined in intensity, it has not gone away (Figure 11.23).

Worldwide, 35 million people are currently living with HIV; and in 2013 there were 2.1 million new HIV cases recorded and 1.5 million deaths from AIDS. The disease affects all countries but 70 per cent of HIV infections are concentrated in sub-Saharan Africa.

HIV/AIDS was targeted by the UN as a Millennium Development Goal (MDG). MDG 6a aimed to halt and reverse the spread of the disease by 2015. This goal has been more or less achieved. However, the world fell short of MDG 6b, whose objective was to achieve universal access to treatment for HIV/AIDS sufferers by 2010.

Thus, over the past 35 years, significant progress has been made in reducing the prevalence of HIV/AIDS. Thanks to funding by governments, multilateral agencies like WHO and UNICEF, and the Global Fund and its partners, the number of new HIV infections was 33 per cent lower in 2013 compared with 2001 levels.

Many governments, supported by multilateral agencies and NGOs, have their own strategies to counter the HIV/AIDS epidemic. But the poorest, particularly in Africa, often struggle to fund comprehensive prevention and treatment programmes. Botswana, one of the most developed countries in sub-Saharan Africa, implemented two HIV/AIDS programmes between 2004 and 2016. While substantial progress was made, in 2013 nearly a quarter of the population aged 15–49 years was HIV positive, and there were nearly 5700 deaths from AIDS out of a total population of just 2 million. Even in the wealthiest countries with the most sophisticated health systems, HIV/AIDS remains a major health problem. In the USA, 1.1 million people are living with HIV and there are around 55,000 new cases every year.

Strategies used to fight the HIV/AIDS pandemic focus on three main areas: prevention, diagnosis and treatment.

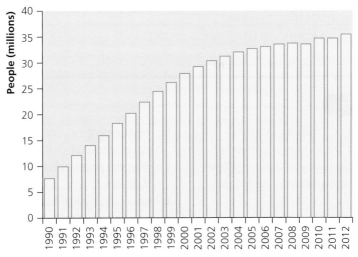

Figure 11.23 Number of people globally living with HIV (Source: AVERT)

Prevention

Prevention aims to modify high-risk behaviour associated with infection and relies on education and better public understanding of HIV. The fundamental message of education programmes is that the HIV virus is transmitted by body fluids such as blood and semen. Publicising the dangers of drug abuse via injection, and promoting safe sex through the use of condoms, helps people to protect themselves.

But prevention also addresses the issue of human rights. Promoting human rights and equality for groups most at risk from HIV/AIDS – women and girls, gay men, migrant workers and refugees – helps to reduce discrimination. A more tolerant attitude by society encourages these groups to access HIV/AIDS health programmes. Male circumcision has also reduced infection risks, though cultural obstacles often limit its acceptance. Meanwhile, better data gathering by governments and agencies improves our understanding of the epidemiology of the HIV virus, allowing funding for prevention to be better targeted.

Diagnosis and treatment

Early diagnosis of HIV can be achieved by screening for HIV antibodies in blood and saliva samples. This method is widely used in high-income countries where health services are well resourced. It is known that early diagnosis reduces the chances of HIV carriers developing AIDS. Huge reductions in mortality and morbidity have been achieved by the use of antiretroviral drugs (ARVs), which suppress the HIV virus and halt the progression of the disease. Today low-cost ARVs are widely available: 8 million people are currently receiving drugs to treat HIV/AIDS, of whom 6 million (mainly in the developing world) are supported by the Global Fund. However, in the poorest countries some of the newer and most effective drugs are not available at suitably low prices.

Educational programmes, aimed at raising individual awareness of the epidemiology and spread of diseases, are an important aspect of prevention. The media campaign against HIV/AIDS in ACs in the 1980s proved particularly effective in reducing risk. Progress in sub-Saharan Africa met greater cultural resistance, but gradually messages relating unsafe sex and drug abuse to HIV/AIDS are being accepted.

Activities

1 Complete Table 11.5 by calculating the projected number of deaths in London and some London boroughs given variables rates of attack (infection) and mortality.
2 Describe the possible impact on economy and society in London of a flu pandemic which infected half the population and where one person in 40 died.
3 Suggest possible mitigation strategies that could be used to control the spread of a flu pandemic in a large city like London.

Table 11.5 Possible rates of attack and mortality in a flu pandemic in London

	Population	Attack 25% Mortality 0.4%	Attack 35% Mortality 1%	Attack 50% Mortality 2.5%
London	7,541,638			
Camden	231,540			
Islington	187,470			
Newham	249,210			
Wandsworth	281,320			

Review questions

1 State two possible causes of cholera outbreaks.
2 How does climate affect the incidence of malaria?
3 Why are flu epidemics most likely to occur in winter?
4 Why has the threat of pandemics increased in the past 30 or 40 years?
5 Why do disease outbreaks often occur following natural disasters such as earthquakes and floods?
6 What emergency steps can be taken to prevent the global spread of disease?
7 How can physical barriers help to prevent the spread of disease?
8 How can physical barriers hinder the control of a disease outbreak?
9 What preventative measures can be taken to control the spread of HIV/AIDS?

11.5 Can diseases ever be fully eradicated?

> **Key idea**
> → Nature has provided medicines to treat disease for thousands of years

Medicines from nature: habitats and growing conditions

'Healing with medicinal plants is as old as humankind itself.'

(Petrovka, 2012)

Many modern medicines originate from natural compounds found in wild plants. Some medicinal plants and their healing properties have been known for thousands of years, long before their chemistry was understood. Hippocrates, the Greek physician (459–377 BC), recorded more than 300 medicinal plants and herbs which he classified according to their physiological action. Thus common centaury was used to treat fever, garlic to purge intestinal parasites, and opium and deadly nightshade as narcotics.

The importance of herbal medicines was underlined in London in 1673, when the Society of Apothecaries established the Chelsea Physic garden. Before the advent of scientific medicine, the healing qualities of many plants were so well known that their common names reflected this. Examples include woundwort, bladderwort, lungwort and liverwort. Now scientists have been able to identify the active constituents in these plants and synthesise them in the lab.

The first naturally derived medicine isolated from a plant in this way was morphine in the early nineteenth century. Morphine and other opioid-related drugs are extracted from the latex produced by unripe seed pods of several poppy species. They are used as analgesics to reduce pain. During the nineteenth century many more **alkaloids** (e.g. quinine) were extracted from wild plants, and towards the end of the century **glycosides** were discovered. The first semi-synthetic glycoside drug, aspirin, was based on salicin, isolated from the bark of the white willow (*Salix alba*) in 1899 (see Table 11.6).

Table 11.6 Some medicinal drugs derived from natural compounds

Drug	Source	Growing conditions	Medical usage
Salicin	Bark of white willow and other willow species	Widespread on river banks, floodplains and wetland throughout the temperate zone. Thrives on a range of soils, from light sands to heavy clay. Soil pH from 5.5 to 8.0.	Acts like aspirin. Used for pain relief, gout, osteoarthritis, etc.
Caffeine	Tea, coffee, coca and other plants	Tropical and sub-tropical conditions. Temperatures averaging 20–27°C. Abundant rainfall (1000–2000 mm/year). Soils which are well drained, with good organic content and nitrogen.	Stimulant for central nervous system, heart, muscles, etc. Migraine, epidural, anaesthesia, etc.
Quinine	Dried bark of cinchonas evergreen tree	Average temperatures above 20°C. Humid conditions with annual rainfall in excess of 2000 mm over at least eight months. No frost. Well-drained, fertile soils with abundant organic matter and good moisture-holding capacity.	Malaria. Kills malarial parasites in red blood cells.
Colchicine	Autumn crocus	Moist, temperate climate conditions. Deep, well-drained soils with slightly acidic pH (c6.5) and good moisture retention.	Cancer and gout.
Nicotine	Tobacco plant	Optimal mean daily temperatures 20–30°C. Rainfall 600–800 mm, with 20–30 mm every two weeks in the growing season and frost-free conditions. Light to medium textured soils with good drainage.	The main active ingredient in new drugs to treat wounds, Alzheimer's, depression, etc.
Morphine	Dried latex from seed pods of several species of opium poppy	Warm, humid conditions. Clear sunny days with temperatures 30–38°C. Susceptible to frost and wet weather. Deep, clay-loam, well-drained soils rich in humus. Soil pH 6–7.5.	Pain reliever.
Artemisinin	*Artemesia annua* plant leaves	Temperate climate. Optimal temperatures 13–29°C. Frost tolerant. At least 600–650 mm rainfall. Soils light to medium textured, well drained and fertile. Soil pH 6–8.	Anti-malarial agent.
Digitalis	Foxglove	Temperate climatic conditions. Tolerates high rainfall, cool summers and acidic soils.	Dropsy, heart failure.

The rosy periwinkle (*Catharanthus roseus*) is a small evergreen shrub which is native to Madagascar, although it is now common in many tropical and sub-tropical regions (Figure 11.24). The plant requires a warm tropical climate, without frost, and where soils are well-drained but moisture retaining, and slightly acidic.

The rosy periwinkle's use in traditional medicine is long established, from the treatment of wasp stings in India, to diabetes in China and the Philippines. It is also a popular ornamental garden plant. However, the plant came to

the attention of scientists only in the late 1960s when analysis revealed that it contained 70 known alkaloids, several of which have significant medicinal value. Two of these previously unknown alkaloids, vincristine and vinblastine, have been developed as powerful drugs in the treatment of various cancers. Vincristine is successfully used in chemotherapy in childhood leukaemia and has helped increase survival rates from around 10 per cent in 1970 to over 90 per cent today. Vinblastine has also proved highly effective in treating Hodgkin's lymphoma.

Currently scientists have been unable to synthesise these alkaloids and production of the drugs relies on commercial cultivation of the rosy periwinkle mainly in India, central Asia and Madagascar. Global sales of vincristine and vinblastine are worth hundreds of millions of dollars annually to Eli Lilly, the US pharmaceutical giant that developed them. However, few of these profits are channelled back to Madagascar and its indigenous rainforest people. This exploitation of biological resources is described as **biopiracy**. It deprives LIDCs like Madagascar of valuable international trade, potential exports and value added. In doing so, biopiracy hinders economic growth and progress in tackling poverty and inequality.

Figure 11.24 The rosy periwinkle

Conservation issues and medicinal plants

Supply and demand

Medicinal plants are mainly sourced from wild populations; only a small number of medicinal species like the rosy periwinkle and the foxglove are cultivated. They supply the raw materials for pharmaceutical medicines, though the majority of pharmaceutical drugs are made from synthesised products.

In contrast, virtually all products used in traditional medicine are harvested from wild plants – 80 per cent of the population in the developing world (nearly 5 billion people) rely on traditional medicines, so demand is huge. As a consequence, the most sought-after species collected and delivered to market by international and intra-national trade are under enormous pressure.

Survival of wild medicinal plant species

Increasingly, the sourcing of wild medicinal plants for traditional Chinese medicine (TCM) and other markets is unsustainable. Over-harvesting is widespread. It reduces plant populations and their genetic diversity, endangering their survival. In the worst cases it results in extinction. Slow-growing plants, and those occupying highly specialised niches, are particularly vulnerable. Current estimates suggest that at least 4000 medicinal

plants are threatened and 14 are listed as acutely endangered by the Convention on International Trade of Endangered Species of Wild Fauna and Flora (CITES). They include several species of yew (a source of the anti-cancer drug taxol), goldenseal, used as a herbal medicine and native to the USA and Canada, and *Nardostachys jatamansi*, a member of the honeysuckle family, found in the Himalayas in Nepal and India. Additional reasons for the decline of wild medicinal plants are the preference of TCM for wild plants rather than cultivates, and the preference for roots rather than leaves, flowers and seeds.

Protection of habitats and natural ecosystems

It is not just collecting that threatens medicinal plants. Habitat destruction, in particular deforestation in the tropics, is of even greater concern. Tropical rainforests are extraordinarily biodiverse, containing 70 per cent of terrestrial plant species, and yet no more than 1 per cent have been screened for potential medical use. Thus, on medical grounds alone, there are powerful reasons to conserve the rainforest and protect its genetic diversity. But with deforestation rates in recent decades averaging 325 km^2/day, many species have become extinct before scientists have had a chance to investigate or even discover them. Indeed, the Center for Biological Diversity estimates that at least one potential major drug is lost every two years due to tropical deforestation.

In the past, pharmaceutical companies exploited rainforest ecosystems, targeting medicinal plants for cultivation and for synthesising chemical compounds. But while the development of drugs such as vincristine (childhood leukaemia), turbocuarine (Parkinson's disease) and neostigmine (glaucoma) have proved highly profitable for the pharmaceutical industry, benefits accruing to indigenous rainforest people have been negligible. To many people this is seen as a type of theft, and has been dubbed **biopiracy**.

One response is to use medicinal plants as a means of conserving not only habitats but entire ecosystems. In return for indigenous people conserving forests, pharmaceutical companies divert part of their profits from drug royalties to help local communities. Such schemes already exist in Samoa and Costa Rica. In the

1980s scientists identified and extracted prostialin, a powerful new drug for treating HIV, from the bark of the mamala tree found in the Samoan rainforest. Part of revenues from sale of the drug are returned to Samoa as compensation for protecting the rainforest and to assist economic development in forest communities. The National Cancer Institute in the USA and the Swedish International Development Authority also provide funds for economic development and forest protection in Samoa. Thus local people benefit directly from forest conservation, a sustainable supply of valuable medicinal plants is assured, and ecosystems flourish and provide a range of free ecological services at both local and global scales.

> **Key idea**
> → Top-down and bottom-up strategies deal with disease risk and eradication

---

Case study: GlaxoSmithKline – a pharmaceutical transnational

GlaxoSmithKline (GSK) is a major pharmaceutical company headquartered in the UK. In 2013 the company's turnover was £23 billion. Its operations are global, with 84 manufacturing sites in 36 different countries and large research and development (R&D) centres in the UK, the USA, Spain, Belgium and China. Pharmaceuticals, including medicines for a range of acute and chronic disease, account for two-thirds of GSK's turnover. The rest is from consumer health care products and vaccines.

GSK's vaccines business is one of the largest in the world. In 2014 it distributed over 800 million doses of vaccine, of which 80 per cent were to countries in the developing world. Other important areas of operation include medications for type-2 diabetes, bacterial infection and oncology. It has produced a number of well-known medications, such as amoxicillin to fight bacterial infections, zidovudine for HIV infection, and bendazole to combat parasitic infections. These drugs and others are on WHO's list of essential medications.

A feature of international pharmaceutical companies is their huge investment in R&D. GSK employs 13,000 people in R&D and spends more than £3 billion a year researching new medicines. R&D is usually undertaken in partnership with other companies, universities and research charities. GSK is one of the few health care companies currently researching treatments for WHO's three priority diseases: HIV/AIDS, malaria and TB. Developing and testing new drugs is a long and costly

process and subject to a high failure rate. In part, this explains why many drugs are so expensive.

A problem that GSK and other pharmaceutical transnationals face is that demand for new drugs in LIDCs, whose economies are weak, is often too small to recoup development costs. Yet drugs are urgently needed in these countries and could bring huge benefits. Despite the problems, GSK devotes significant R&D resources to the needs of the developing world. For example, its research centre in Spain focuses primarily on TB, malaria and other tropical diseases and the company is currently close to launching the first effective vaccine against malaria. It is also developing a vaccine for the Ebola virus.

Although the profit motive is the driver of all transnationals, including GSK, the company has adopted an ethical policy towards the developing world. This policy includes:

- a commitment to a small return – 5 per cent – on each product sold
- providing three HIV/AIDS drugs to LIDCs at significant discount
- granting licences for the manufacture of cheap generic versions of its patented drugs
- capping the price of patented drugs to developing countries to 25 per cent of the UK price
- investing 20 per cent of its profits from sales in each developing country into that country's health infrastructure.

Strategies for disease eradication at global and national scales

Global campaigns and disease eradication

Global campaigns to eradicate major human diseases have had limited success. A notable exception was the worldwide eradication of smallpox in 1980. Currently WHO and other agencies are working to eradicate two diseases: poliomyelitis (polio) and dracunculiasis (Guinea worm). Both diseases are close to eradication, yet significant pockets of infection remain. In the past, similar attempts to eradicate other diseases, including malaria, hookworm and yaws, have failed.

Prior to the development of a polio vaccine in 1952, polio killed or paralysed 600,000 people a year. The Global Polio Eradication Initiative (GPEI), supported by WHO, UNICEF, the CDC and other agencies, began in 1988. A programme of vaccination had successfully eliminated the disease in the Americas and by 2011 the polio virus was endemic in just three countries: Afghanistan, Pakistan and Nigeria. Within these countries, the uptake of vaccination to provide immunity for children remains uneven. Political instability, and the murder of more than 80 health care workers in Pakistan and Nigeria in recent years by militant extremists, has interrupted vaccination programmes, presenting a major setback for the GPEI. In addition, since 2013, new outbreaks of polio have occurred in war-torn Syria and Iraq, where vaccination and basic hygiene have broken down.

In the past, global agencies and governments had assumed universal support for the eradication of a scourge like polio. However, this has not always been the case. Resistance to vaccination programmes, related to political and cultural factors as well as ignorance, provided unexpected challenges to its eradication, particularly in Pakistan and Nigeria.

National campaigns and disease eradication

Mauritius, a small island state in the Indian Ocean (Figure 11.25), illustrates how top-down national initiatives can successfully eliminate a disease like malaria and prevent its reintroduction. Malaria became endemic in Mauritius in the mid-nineteenth century and in 1867 an epidemic killed one in eight of the island's population.

A major government-backed campaign to eliminate malaria was launched between 1948 and 1951. Spraying buildings and the breeding sites of mosquitoes with DDT reduced mortality rates from malaria from 6 per 1000 in 1943 to 0.6 per 1000 by 1951, allowing WHO in 1973 to announce that malaria had been eliminated from the island.

Then, two years later, Cyclone Gervaise hit Mauritius. In the aftermath of the disaster, migrant workers employed in reconstruction reintroduced malarial parasites to the island. At the same time the destruction caused by the cyclone provided new breeding opportunities for malarial mosquitoes. In 1982 a malaria epidemic forced the government to embark on a second elimination campaign. Spraying of mosquito breeding sites and indoors resumed, predatory fish that feed on mosquito larvae were introduced, and there was mass administration of the anti-malarial drug, chloroquine, to the population.

Since 1998 the government has taken steps to prevent the reintroduction of malaria. Rigorous passenger screening has been implemented at the international airport. Currently, 175,000 passengers per year are screened. They comprise potential carriers of disease, such as arrivals from malaria-endemic countries and those showing symptoms of fever. These people are kept under surveillance by health workers for up to four months, during which time blood samples are taken and analysed. At the same time insecticide spraying, both indoors and outdoors, continues throughout the island, with particular attention to mosquito breeding sites and the residences of migrant workers.

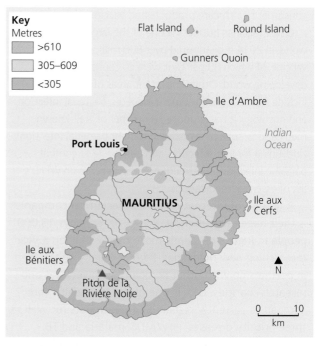

Figure 11.25 The island of Mauritius

So far, government efforts to prevent reintroduction have proved highly effective: only one imported case of malaria has been seen since 1997. The example of Mauritius shows how 'top-down' schemes can successfully eliminate a disease like malaria and prevent its reintroduction.

Activities

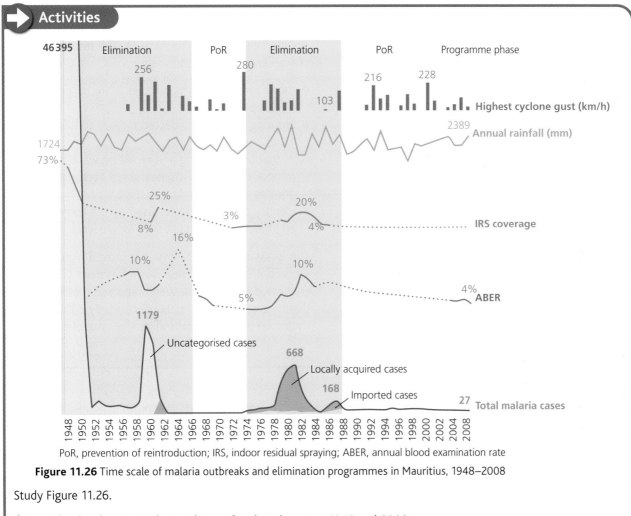

PoR, prevention of reintroduction; IRS, indoor residual spraying; ABER, annual blood examination rate

Figure 11.26 Time scale of malaria outbreaks and elimination programmes in Mauritius, 1948–2008

Study Figure 11.26.

1 Describe the changes in the incidence of malaria between 1948 and 2008.
2 Analyse the relationship between tropical cyclones and outbreaks of malaria before and after 1988.
3 Outline two responses to malaria epidemics suggested by Figure 11.26.

Grass-roots strategies for disease eradication

Top-down strategies for disease eradication, controlled by governments and international agencies often encounter resistance at local levels, and may exclude some groups, especially women. An alternative approach is a 'grass roots' strategy which involves and empowers local communities.

Guinea worm eradication programme

In Ghana in West Africa the Guinea worm eradication programme has partnered the Ghana Red Cross women's clubs to reduce the transmission of Guinea worm (Figure 11.27). This innovative programme involves teaching women volunteers how the Guinea worm is transmitted and how transmission can be prevented. The volunteers then visit villages and educate local communities.

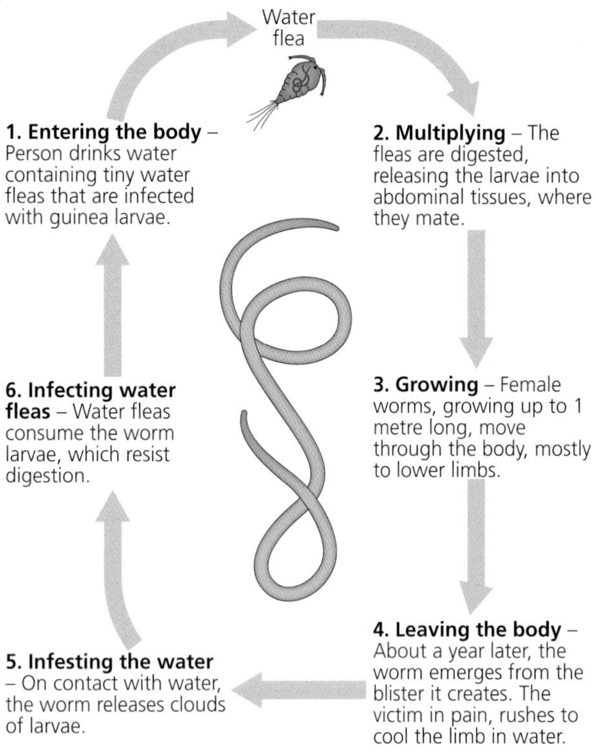
Water flea
1. Entering the body – Person drinks water containing tiny water fleas that are infected with guinea larvae.
2. Multiplying – The fleas are digested, releasing the larvae into abdominal tissues, where they mate.
3. Growing – Female worms, growing up to 1 metre long, move through the body, mostly to lower limbs.
4. Leaving the body – About a year later, the worm emerges from the blister it creates. The victim in pain, rushes to cool the limb in water.
5. Infesting the water – On contact with water, the worm releases clouds of larvae.
6. Infecting water fleas – Water fleas consume the worm larvae, which resist digestion.

Figure 11.27 The life cycle of a Guinea worm

In the past this work was invested in male volunteers and met with limited success. This was because men frequently work outside villages and it is mainly women who are responsible for sourcing water and its use for household consumption. Women were able to appreciate the value of filtering drinking water and avoiding contamination of water sources by people already infected with the parasite. The responsibilities of women volunteers included:

- monitoring, identifying and reporting all new cases of Guinea worm
- ensuring that those infected did not contaminate water sources
- distributing, checking and replacing water filters that remove water fleas (Guinea worm vectors) from drinking water
- identifying water sources used by the community and requiring treatment with larvicides.

This 'grass roots' programme has proved highly successful, and the Guinea worm has been effectively eradicated from Ghana. WHO reported that in 1989 there were more than 179,000 cases. By 2010 the country reported its last indigenous case of the disease.

Activity

Table 11.7 Average life expectancy: more developed and least developed countries, 1990–2015 (Source: IDB US Census)

	1990	1995	2000	2005	2010	2015
More developed	73	74	76	76	77	79
Least developed	50	52	54	56	58	60

With reference to Table 11.7:

a) Plot the data as a line graph.
b) Compare the differences in trends in life expectancy between more developed (i.e. high-income) and least developed (low-income) countries between 1990 and 2015.
c) Suggest possible reasons for the differences.

Skills focus

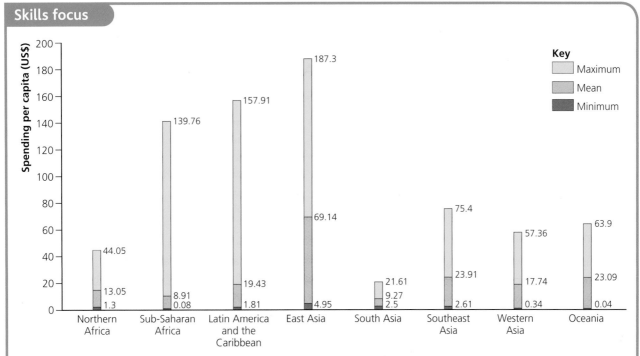

Figure 11.28 Variations in expenditure on medicines by countries within world regions

Study Figure 11.28.

a) Describe the regional variations in average spending on medicines.
b) Suggest and justify an alternative to:
   i) the mean value, as a measure of differences in spending between the regions
   ii) the range, as a measure of dispersion (see pages 533–34)

✓ Review questions

1 What is meant by disease eradication?
2 Name the three diseases targeted by the UN in 2000 for eradication.
3 What are the advantages of grass-roots disease eradication campaigns compared with top-down campaigns?

4 Why are many new drugs for treatment of major diseases so expensive?
5 Explain why the availability of generic drugs is so important in LIDCs.
6 What steps can transnational pharmaceutical companies take to make patented drugs more affordable?

Practice questions

A Level

Section A

1 a) Describe **one** alternative graphical method of presenting the data on rainfall and parasite density in Entebbe, Uganda, shown in Figure 11.10 on page 340. [3 marks]
  b) Explain how seasonal patterns of rainfall can influence outbreaks of malaria. [6 marks]

Section B Synoptic

2 Examine how climate change might affect the distribution of infectious diseases. [12 marks]

Section C

3 To what extent is there a link between the prevalence of non-communicable diseases and levels of economic development? [33 marks]

Table 1

	Life expectancy	GNP per capita
Australia	73	44,700
Brazil	65	15,570
China	68	13,170
France	72	40,100
Guinea	50	1,130
India	58	5,630
Niger	51	910
Norway	71	67,100
South Africa	52	12,700
UK	71	39,500

AS Level

Section A

1 a) Explain the relationship between economic development and a country's epidemiological transition. [4 marks]
  b) Show how seasonal variations in climate influence disease outbreaks. [6 marks]
  c) Study **Table 1** which shows average healthy life expectancy at birth (years) and GNP per capita (US$).
    i) Outline and justify a statistical method you could use to analyse the data in Table 1. [4 marks]
    ii) Using the evidence in Table 1 analyse the variations in life expectancy. [6 marks]

Section B Synoptic

2 a) With reference to **Figure A** suggest how the diffusion of communicable diseases can be influenced by physical landscape systems. [8 marks]
  b) Examine how some human factors influence disease outbreaks in different landscape systems. [8 marks]

Section C

3 Discuss the view that environmental factors are the main cause of ill-health in LIDCs. [20 marks]

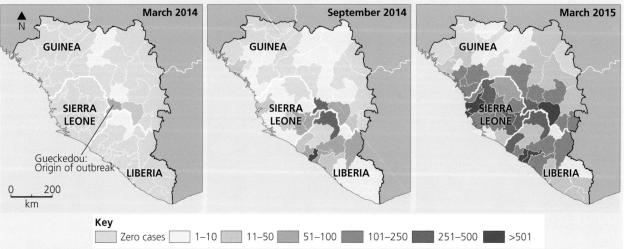

Key
Zero cases   1–10   11–50   51–100   101–250   251–500   >501

Figure A Diffusion of Ebola

Chapter 12

Exploring oceans

The oceans remain largely unexplored and yet they are vital to life on Earth. About half of the oceans are more than 3 km deep, with the deepest parts of the ocean basins making up about half of the entire Earth's surface. We have explored only about 1 per cent of these regions and know more about the Moon's surface than we do about the ocean deeps.

12.1 What are the main characteristics of oceans?

> **Key idea**
> → The world's oceans are a distinctive feature of the Earth

The global distribution of the world's oceans

Viewed from space the Earth appears as a 'blue planet', with oceans and seas making up about 71 per cent of the surface (Figure 12.1). The oceans and seas interconnect but five major individual oceans are recognised as well as many seas (Figure 12.2, Table 12.1).

Table 12.1 Fact file of the world's oceans and major seas

Name of ocean/sea	Area (million km²)	Average depth (km)	Greatest known depth (km)
Pacific Ocean	155.6	4.0	11.0 Mariana Trench
Atlantic Ocean	76.8	3.9	9.3 Puerto Rico Trench
Indian Ocean	68.6	3.9	7.5 Sunda Trench
Southern Ocean	20.3	4.0–5.0	7.2 South Sandwich Trench
Arctic Ocean	14.1	1.2	5.6 un-named
Mediterranean Sea	3.0	1.4	4.6 off coast of Greece
Caribbean Sea	2.7	2.6	6.9 Cayman Trench
South China Sea	2.3	1.6	5.0 west of Luzon, Philippines
Bering Sea	2.3	1.5	4.8 Aleutian Trench
Gulf of Mexico	1.6	1.5	3.8 Sigsbee Deep

Ocean basin relief

With mountains that dwarf the Himalayas, waterfalls bigger than Niagara and more active volcanoes than on the land, the ocean basins contain a dramatic landscape. All the oceans have a similar basic structure. Starting at the edges of the continents the water gradually deepens across the relatively flat, wide margin

Figure 12.1 Satellite image of Earth

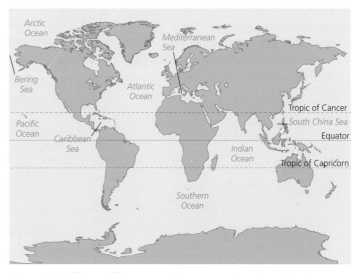

Figure 12.2 The world's oceans and seas

367

of the **continental shelf**. Its average width is 70 km, with a slope angle of just under 2 metres per kilometre.

The continental shelf ends at the **continental slope** where the slope angle increases to a much faster rate, about 70 metres per kilometre. It is not continuous as submarine canyons and gullies are cut into it. On average the continental slope is quite narrow, about 16 km. The steep slope then gives way to the **continental rise**, a wide and very gently sloping zone.

Eventually the deepest part of the oceans, the **abyssal plain**, is reached (Figure 12.3). Across these vast areas, which make up about half of the entire Earth's surface, whole chains of mountains, areas of high relief and isolated peaks rise up. Collectively these are known as **seamounts** and can climb to over 3000 m above the abyssal plain. The New England seamount contains more than 30 peaks along its 1600 km length. **Guyots** are peaks that once rose above the surface of the ocean. Erosion then reduced their height below sea level, leaving the peak with a flat top. Through geological time, many guyots have sunk well below sea level as the weight of the guyot on the oceanic crust makes it subside into the upper mantle.

In the Atlantic Ocean, a thick accumulation of sediments transported off the surrounding land areas has largely buried these peaks. However, the Pacific floor has thousands of guyots (former volcanoes) rising up above its abyssal plain.

Stretch and challenge

Suggest why the abyssal plain in the Pacific is not covered in a thick accumulation of sediment. Look around the edge of the Pacific for clues (see pages 466–71).

Crossing the abyssal plains are very long chains of mountains marking **mid-oceanic ridges** – cracks (transverse faults) up to 1600 km long run at right angles to these ridges. Along the centre of the ridge is a rift valley, a deep notch with steep sides. Magma rises up from the upper mantle along the **rift valley**, causing the crust to be pushed apart. At the mid-Atlantic ridge, one

side of the abyssal plain flows east, the other west, each moving at about 1 cm a year. This process of **sea-floor spreading** is of fundamental importance to the shape and size of the ocean basins and to the position of the land masses across the globe. The Atlantic Ocean, for example, has opened and closed on several occasions during geological time as the convection currents powering the magma movements change.

Stretch and challenge

Research the evolution of the Atlantic Ocean through geological time to produce a presentation outlining the Atlantic's geological history.

As the magma cools, the magnetic orientation of the poles is 'locked' in the iron particles in the new rock. It is now known that the polarity of the Earth's magnetic field flips so that magnetic north becomes magnetic south and vice versa. Evidence from the past few million years suggests that this happens every 200,000–250,000 years. Research has identified symmetrical patterns of magnetic 'stripes' either side of the ridges. These indicate the rate of sea-floor spreading and **palaeomagnetism** has provided important evidence to support the evolving theory of plate tectonics. Beginning with the work of Alfred Wegener in the early 1900s, followed by increases in data on seismicity and paleomagnetism from the 1950s onwards, the radical theory of plate tectonics evolved. Our knowledge and understanding of the key features of the oceans is a vital element of this theory (see pages 466–69).

Some ocean margins are characterised by **subduction zones** (Figures 12.4, 14.5 and 14.6). Here, crustal plates are converging, with the result that one is forced down into the mantle. A feature of the ocean basin landscape at this point is a **trench**. These are the deepest places in the ocean, varying from 7 km to 11 km. Only two people have ever ventured to the bottom of the deepest place in the oceans. In 1960, Jacques Piccard and Lieutenant Don Walsh of the United States Navy travelled to the Challenger Deep in the Mariana Trench in the northwest Pacific.

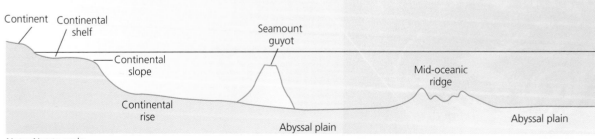

Note: Not to scale
(vertical nor horizontal)

Figure 12.3 Simplified cross-section of an ocean basin

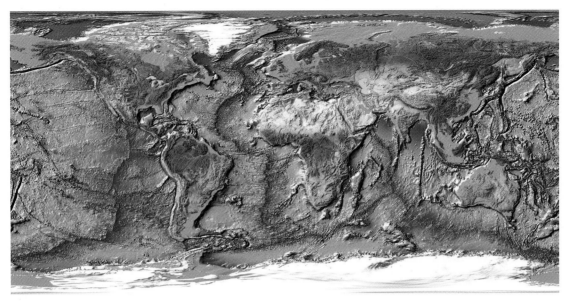

Figure 12.4 The Earth's surface without the oceans

 Activity

a) Draw a map showing the location of the Mariana Trench.
b) Describe and explain the formation of the trench.

Stretch and challenge

Research the term 'hypsographic curve'. Describe the distribution of the Earth's surface above and below sea level, in particular the areas of the various components making up the ocean basins.

Key idea

→ Water in the world's oceans varies horizontally and vertically

In 1872, HMS *Challenger* sailed from Portsmouth and for the following three and a half years its crew undertook the first scientific investigation of the world's oceans. For the time, this pioneering mission was as adventurous as a space probe mission to another planet is today. Technological advances are allowing much progress to be made in exploring the oceans. Since 2011, NASA has been using its Aquarius satellite to observe salinity levels around the world (Figure 12.5).

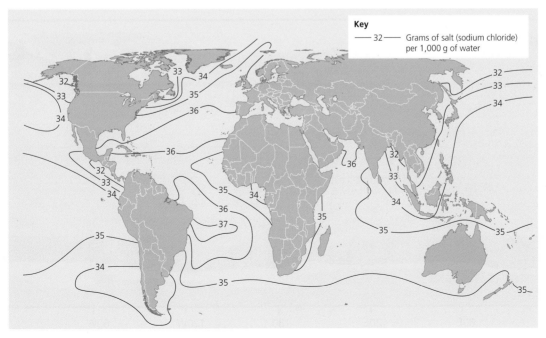

Figure 12.5 Global variations in surface salinity

More and more is being discovered about how and why sea water varies in three dimensions.

Salinity

Salinity is a measure of the concentration of sodium chloride (salt). It is expressed as grams (parts) of sodium chloride per 1000 g of water. Fresh water salinity is usually less than 0.5 ppt (parts per thousand) whereas for sea water the average is about 35 ppt.

Activities

1 Describe the global pattern of surface salinity of the oceans.
2 Suggest reasons for variations in surface salinity of the oceans. Include the following factors in your discussion: precipitation over oceans; evaporation of sea water; freezing of sea water; melting of glaciers and icebergs; groundwater and river flows into the oceans.

Salinity also varies with depth (Figure 12.6). The rapid change in salinity close to the surface is known as the **halocline**.

Variations in salinity influence water density. Differences in density affect water movements, such as the flow of ocean currents that move heat from the tropics to the poles and affect global climate. Just as significant are density changes in different layers of water. These affect how, where and when water moves vertically in the oceans. As records of ocean salinity grow, they add to our understanding of how the ocean system operates and how oceans link with climate change.

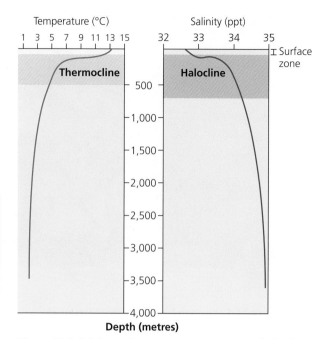

Figure 12.6 Salinity and temperature variations with depth based on the northeast Pacific

Temperature

Because oceans absorb, store, transfer and release heat in vast quantities they are very important in controlling climate and influencing weather from global to local scales (Figure 12.7). Water has a very high specific heat capacity. Much energy is needed to raise water temperature, but once heated, water retains its heat better than almost any other substance. By contrast, land heats up and cools down relatively quickly.

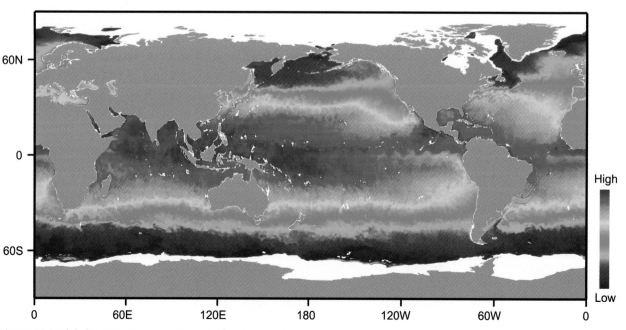

Figure 12.7 Global variations in average sea surface temperatures

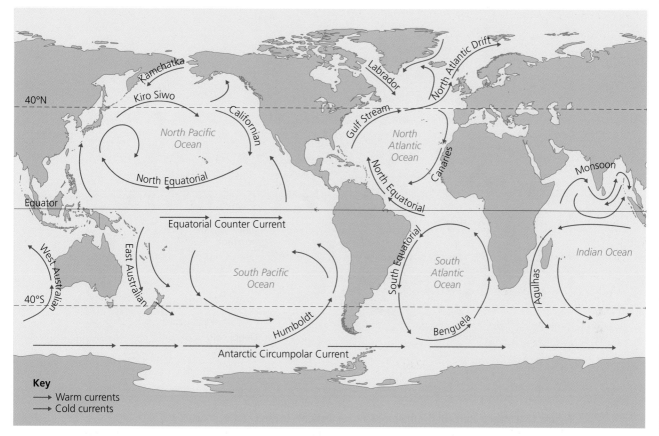

Figure 12.8 Surface ocean currents

Water temperature decreases rapidly as depth increases close to the surface. This is known as the thermocline. The depth of the warm surface layer varies both with season and location. Generally, below about 1 km, water temperature hardly changes with increasing depth (Figure 12.5).

> ### Activity
>
> Log on to the website www.argo.ucsd.edu. This is the home site of the Argo project.
> a) Describe the aim of the Argo project.
> b) Outline how the data are being collected.
> c) Follow the links that allow the Argo data to be linked with Google Earth. Generate a map showing the data. Describe the pattern shown.

Warm and cold ocean currents

Most of the oceans hold warm water at the surface and cold water in their depths. These two zones tend not to mix. There is a basic movement of surface water from the low latitudes to the middle and high latitudes.

Towards polar regions, ocean surface currents become cooler, more saline and therefore denser (Figure 12.8). Water sinks into the ocean's depths and starts to disperse horizontally. Deep currents flow back to equatorial regions where water rises, resulting in the **thermohaline circulation** or **ocean conveyor belt**. This operates as a continuous, very slow-moving flow around the globe (Figure 12.9).

Gyres are distinctive features of the surface circulation which are generated by winds. There are north and south gyres in the Atlantic and Pacific Oceans and a southern gyre in the Indian Ocean (Figure 12.18).

> ### Stretch and challenge
>
> Investigate what the Coriolis effect is. Explain the influence of the Coriolis effect on the pattern of ocean surface currents.

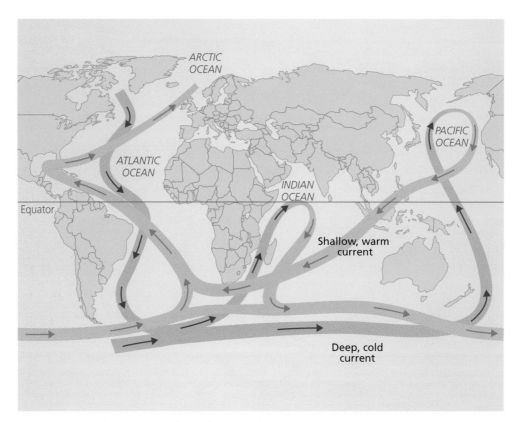

Figure 12.9 Main features of the thermohaline circulation – oceanic conveyor belt

The circulation in the North Atlantic

Relatively warm saline water flows from off the coast of Florida to the northeast as the Gulf Stream. Vast quantities of heat energy are thereby transferred to the mid-latitudes where it has a significant impact on the weather and climate of western Europe. The current, called the North Atlantic Drift in its northern course, tracks all the way to the Norwegian coast and on into the Arctic. As the water cools it increases in density and so starts to sink. This is one of the two most important locations in the oceans where deep-water currents form. Water that sinks here will surface at the Equator some 150–250 years later.

The far north of the northwest Atlantic is where the cold Labrador Current flows southwards from the Arctic. It is responsible for the harsh cold winters of northeast Canada and the USA. A northward branch of the warmer Gulf Stream flows along the western coast of Greenland. This has the effect of helping to keep this coastline relatively ice-free.

The North Atlantic also receives a major inflow of water from the Mediterranean. This water is warm (13°C) and salty (37.3 ppt) as it passes through the Straits of Gibraltar and moves out into the Atlantic as a recognisable tongue of water at about 1000 m deep. Eventually its density matches that of the surrounding water and mixing occurs.

Much monitoring and research are being carried out on ocean circulation in the North Atlantic in the context of climate change. It has been suggested that if vast quantities of fresh water were released from melting ice in and around the Arctic, salinity in the region would be significantly reduced. Some models predict that the resulting reduction in sea water density could be sufficient enough to prevent water sinking in the North Atlantic, thus disrupting the ocean conveyor. In turn, weather and climate would be altered, which could have serious impacts on millions of people. For example, the climate of northwest Europe would cool appreciably, making it similar to present-day northeast Canada (Labrador).

Key idea
→ Changes in light, temperature and nutrient supply influence the biodiversity of oceans

All organisms need a source of energy to live. All but a minute fraction of this energy comes from the Sun. Essential nutrients such as minerals (for example, potassium, phosphate and calcium) are also required.

Changes in light levels and temperature

Incoming long-wave light energy from the Sun is not distributed evenly across the oceans nor with depth. Light is at its most intense in the equatorial regions as the Sun's rays strike the surface at angles close to 90°. With increasing distance from the Equator, these angles progressively reduce so an equivalent amount of light energy is spread over a greater surface area.

Light is able to penetrate water but with increasing depth of water, light levels decrease. The layer of water where there is enough light for photosynthesis to take place is known as the **photic zone**.

At depths below where light reaches, some organisms use **bioluminescence**. Their biochemistry generates small amounts of light from specialised parts of their body, either to lure their prey or to evade a predator.

Table 12.2 Degree of light penetration with depth in clear and cloudy water

Water depth (metres)	% light penetration	
	Clear ocean water	Cloudy coastal water
0	60	60
10	45	20
20	35	10
30	25	7
40	18	3
50	15	2
60	12	0
70	10	0
80	8	0
90	6	0
100	5	0
120	4	0
140	4	0
160	3	0
180	2	0
200	1	0

Ocean temperatures are largely the result of the energy transfer from sunlight to water molecules. Global sea surface temperatures are therefore closely related to variations in sunlight. As a result, with increasing depth, temperatures decrease (Figure 12.6).

Occasionally, regional sea temperatures are influenced by strong winds blowing away surface water, allowing deeper cold water to rise to the surface, for example along the Pacific coast of South America. This occurs for periods of usually just a few days.

Changes in nutrient levels

A variety of nutrients such as nitrogen, iron and zinc is needed by all organisms. Dissolved nutrients from weathered rocks are transported by rivers into the oceans and from dust carried by winds. Plankton incorporate nutrients into their tissues and these are then passed through ecosystems as consumer organisms feed on plankton and each other. Nutrients are returned to the water when organisms die and via their waste products.

Unlike light and temperature, nutrient levels are relatively low at the surface, especially away from the continents where rivers flow into the ocean, bringing dissolved minerals. Warm waters in equatorial regions are also low in nutrients. Any nutrients that are deposited at the surface, from rivers or airborne dust, are soon used up or sink to the ocean floor. Because there is a strong decrease in temperature with increasing depth, deeper water which is colder and denser cannot rise to the surface through the thermocline. Nutrients from deeper down are therefore not generally brought to the surface.

However, surface nutrient levels are higher at locations where there is an upwelling of water from beneath the thermocline. Locations such as the Southern Ocean around Antarctica have strong upward movements of deep water at certain times of the year. The nutrients carried by the rising water support much biological activity (Figure 4.10).

Stretch and challenge

Investigate the causes of the seasonal upwelling of deep water around Antarctica. Websites such as the British Antarctic Survey site (www.bas.ac.uk) offer valuable information.

Although the deepest marine locations (abyssal plains) have very low nutrient levels, they are not without life. The Sun's energy is transferred down from the surface by marine 'snow' falling. This is made up of small particles, the remains of organisms living near the surface. Various organisms capture the 'snow' and this provides the basis of food chains and webs for deep-water ecosystems.

The discovery in the late 1970s of **hydrothermal vents** at mid-oceanic ridges added revolutionary knowledge about deep ecosystems. The hot water vents or springs, sometimes reaching temperatures as high as 380°C, are rich in silica, manganese, hydrogen, sulphur and methane. Specialised bacteria fix the energy from these erupting 'smokers' and communities develop that are unlike any other on Earth. Because they are not dependent on organic matter from the surface they are not powered by the Sun's energy. Instead organisms such as tube worms, shrimp and small crabs obtain their energy from the chemical energy in the hot water.

In the last decade, **cold seeps** have been discovered, for example in the Gulf of Mexico. These are low-temperature flows of water found in shallower water than hydrothermal vents but are similar in that their communities are dependent on chemical rather than solar energy. Hydrocarbons such as methane or hydrogen sulphide supply the energy.

Activity

Describe the communities that develop around hydrothermal vents and cold seeps and explain how they function. Organisations such as Scripps Institution of Oceanography, Woods Hole Oceanography Institution and National Geographic have websites with relevant materials.

Biodiversity in the oceans

About 80 per cent of all life on Earth is found in the oceans. To date, some 250,000 different marine species have been identified, but no reliable estimate of the total number of species can be made. This is due in part to the very limited knowledge we have of the oceans. But advances in genetic analysis are also revealing more of the ocean's biodiversity.

As with life on the land, ocean ecosystems start with organisms capable of trapping sunlight. **Producer organisms** convert sunlight to chemical energy and organic matter by photosynthesis. Microscopic plants called **phytoplankton** are the key oceanic producers. Nitrogen, phosphorus and silicon are particularly important for these producers. Food webs and chains built around producers make up ecosystems (Figure 4.33).

It is tempting to assume that all parts of the oceans are teeming with life. Living organisms have indeed been found throughout the oceans and in some very extreme environments. However, life is not abundant everywhere in the oceans and there are significant variations by both latitude and depth.

As with ecosystems on land, it is useful to start by considering **net primary productivity** (NPP). This is a measure of how much of the Sun's energy is captured and is usually measured as grams of carbon per unit of area per year. Total annual NPP is approximately 100 billion tonnes of carbon. Just under half of this occurs in the oceans and is a key part of the carbon cycle. (This is discussed in greater detail in Chapter 4 on the carbon cycle.) Remote sensing from satellites has allowed much more accurate information to be obtained about oceanic NPP (Table 12.3).

Table 12.3 Marine NPP by region

Region	% marine NPP	% global NPP	Estimated NPP (g/cm²/year)
Coastal	20	10	250
Deep ocean	80	40	130

The highest productivity per unit of volume occurs in locations where the supply of dissolved nutrients is greatest. Although the percentage of total marine NPP from coastal regions appears limited, in relation to the area and depth of water, productivity is high. Dissolved nutrients are brought down from the land into the coastal zone by rivers making places such as estuaries areas of high productivity. The coastal shelves sustain high productivity by run-off from the land and the mixing of the relatively shallow water.

By contrast, large areas of the deep oceans are comparatively unproductive given their vast volume of water. The sub-tropical gyres for example receive virtually no direct supply of dissolved nutrients and have been described as 'biological deserts'. Locations such as the eastern central and northern Pacific and much of the Southern Ocean are so far away from land that they receive neither dissolved nutrients from run-off nor mineral dust blown from the land. As on land, much marine NPP comes from relatively confined regions.

Nutrient inputs, in particular nitrogen, influence NPP. In temperate coastal areas, kelp forests can have levels of NPP approaching those found in tropical rain forests. 'Meadows' of sea grasses such as eel grass play important roles in near-shore ecosystems. Floating seaweeds such as sargassum are key producers in some locations.

> **Activity**
>
> Research the distribution of sargassum in the oceans using websites such as www.noaa.gov. Describe the ecosystem based on this seaweed and assess its importance as a marine habitat.

Pelagic ecosystems exist away from the coastal zone in the deep waters of the open seas.

The Antarctic marine ecosystem – a deep-water ecosystem

Despite the low temperatures, the waters around Antarctica are one of the more productive ocean regions. The cold water temperature allows more oxygen to dissolve in the ocean, which is an advantage for marine life. Phytoplankton productivity is high during the summer (November to March) when there are never fewer than 12 hours of sunlight and more than 20 hours for several weeks. The surface waters in the ocean around Antarctica are rich in nutrients. This is due to the following sequence:

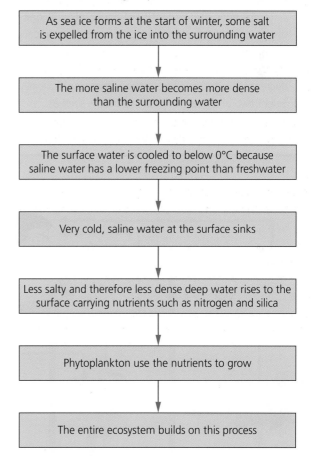

As sea ice forms at the start of winter, some salt is expelled from the ice into the surrounding water

↓

The more saline water becomes more dense than the surrounding water

↓

The surface water is cooled to below 0°C because saline water has a lower freezing point than freshwater

↓

Very cold, saline water at the surface sinks

↓

Less salty and therefore less dense deep water rises to the surface carrying nutrients such as nitrogen and silica

↓

Phytoplankton use the nutrients to grow

↓

The entire ecosystem builds on this process

For example, vast numbers of squid live in the Southern Ocean, feeding on primary consumers such as zooplankton and krill. Squid are very important food for a wide variety of organisms such as sperm whales, seals, penguins, fish and seabirds. They are also harvested by humans.

At the end of the Antarctic winter, some 100 million seabirds breed along the coast and offshore islands of Antarctica. Their chicks develop quickly to take advantage of the brief summer and then the birds migrate north, some even to the Arctic.

The physical environment greatly influences the ecosystem, for example seasonal changes in the extent of sea ice. In February, sea ice usually extends to about 3 million km² and by late September this has grown to some 20 million km².

Compared with many other ecosystems, marine **food webs** and **food chains** in the Antarctic are relatively simple and biodiversity is comparatively low. For example, one food chain starts with phytoplankton; krill feed on these and are in turn eaten by filter-feeding whales (baleen whales).

Activities

1 Describe one food chain and one food web in the Antarctic ecosystem. Identify each organism as one of primary producer, primary consumer or secondary consumer.
2 Using these two examples, explain why biomass decreases from the lower to the higher trophic levels.
3 Select two or three groups of organisms such as penguins, whales, seals or seabirds. For each group, research their habitats, sources of food and life cycles.

Salt marsh ecosystem – an inter-tidal ecosystem

Salt marshes are coastal wetlands where the ocean meets the land and are particularly common in the mid to high latitudes. The zone between low and high tide is a very dynamic one. Large volumes of salt water move backwards and forwards across it, leaving the zone alternately flooded or exposed. Organisms living in this zone have evolved to cope with the sometimes rapid and frequent changes in environmental conditions.

Accumulation of sediment is a crucial factor in salt marsh development. Weathered and eroded material from inland is carried into the coastal zone by rivers. Sands and in particular clays are deposited and may develop into salt marsh.

Stretch and challenge

Describe and explain the process of **flocculation**. Why is this process considered such a vital one in salt marsh development? Clue – consider water movements and size of clay particles.

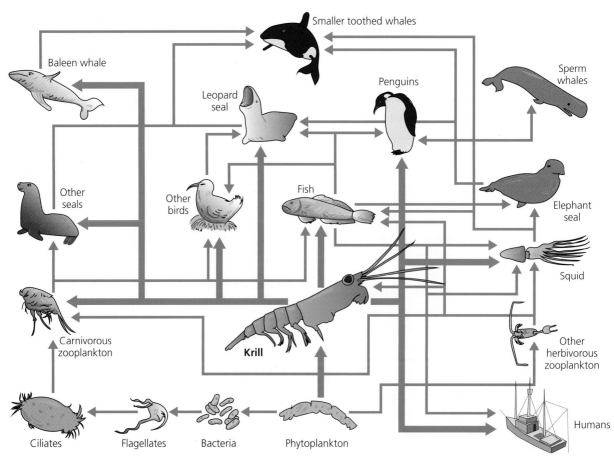

Figure 12.10 Food chains and webs in the Antarctic ecosystem

Activities

Study the following data about the different trophic levels in a typical salt marsh.

- Primary producers: blue-green algae; grasses (e.g. eel grass/cord grass); sea lavender
- Primary consumers: invertebrates (e.g. shellfish/crabs/flies/moths/butterflies)
- Secondary consumers: bird species (e.g. Oyster-catcher/Redshank/curlew); geese (e.g. Barnacle/Brent); ducks (wigeon); fish (e.g. flounder/sea bass/plaice); amphibians (e.g. natterjack toad); mammals (e.g. rabbits/voles/shrews)
- Tertiary consumers: birds (e.g. kestrel/owl); mammals (e.g. fox)

1 Using the data above, construct a diagram showing the food chains and webs found in a typical salt marsh ecosystem. Use the Antarctic example as a guide for the layout.
2 Describe one food chain and one food web in the salt marsh ecosystem.
3 Assess the extent to which salt marsh ecosystems have a 'value' to human society.

Looking across a salt marsh may give the impression that little biological activity takes place there. However, salt marsh is one of the more productive ecosystems on the Earth, rivalling arable land in terms of productivity. Dissolved and solid nutrients are transported by rivers down to the coast while regular tidal movements of water mix the water and nutrients. In this way, there is always plenty of food for the ecosystem. Light levels vary seasonally but are a limiting factor only during the winter period.

The rhythm of the tides has a strong influence on the marsh. A clear arrangement of zones usually evolves, extending from the low tide area which is covered by salt water most of the time to the part of the marsh where only occasionally the very highest tides reach. This zonation can be seen in vegetation change and is called **plant succession**. As vegetation changes, so do other organisms such as insects, birds and animals.

Fieldwork idea

Fieldwork aims:

- Measure changes in vegetation across a salt marsh system.
- Measure changes in soil conditions across a salt marsh system.
- Assess changes in non-plant organisms across a salt marsh system.

Fieldwork methods:

- Sample using transects stretching from low tide to where terrestrial ecosystems are established.
- Quadrat sampling to record vegetation type and percentage cover.
- Test soil in each quadrat for pH, moisture and organic content.
- Note evidence for non-plant organisms, e.g. invertebrates for each quadrat.

Presentation and analysis:

- Graph changes in conditions along transect; kite diagram for vegetation change.
- Calculate correlation coefficients for changes in conditions with distance from low tide mark.

(For detailed guidance on techniques, see Chapter 15, Geographical Skills, pages 506–44.)

Activity

Choose examples of organisms living in your chosen ecosystems. Describe and account for any adaptations that enable the organisms to survive in their respective environments.

Review questions

1 Using a labelled sketch cross-section, describe the relief of an ocean basin.
2 What is the significance of the process of sea-floor spreading to the formation of ocean basins?
3 How does salinity vary with increasing depth?
4 What is the thermocline?
5 Describe and explain the flow of water along the ocean conveyor belt.
6 What is the significance of the 'photic zone' for ocean biodiversity?
7 Describe the main sources of nutrients in oceans.
8 What is the significance of phytoplankton to ocean ecosystems?
9 Outline why Antarctic waters are so rich in nutrients.
10 Using named examples of organisms, explain the differences in trophic levels found in coastal salt-marsh ecosystems.

12.2 What are the opportunities and threats arising from the use of ocean resources?

> **Key idea**
> → Biological resources within oceans can be used in sustainable or unsustainable ways

Advances in technology are allowing humans to explore more of the oceans. In turn this is making possible more intensive use of biological resources such as fish. In addition, previously unused resources such as krill are now exploited. Not only does the use of biological resources in the oceans offer much potential to humans (e.g. for food), it also poses significant issues regarding the management of the oceans.

The growing human population increases demand for resources of all types. Existing and new sources of food are eagerly investigated to assess their potential for additional food supply.

The 'value' of biological resources

The term **natural capital** can be used to describe goods or services that are not manufactured but have a value to humans. This includes resources such as fish but also phytoplankton, the organisms responsible for photosynthesis that provides oxygen. Natural capital therefore yields **natural income**, such as harvests of shellfish, tuna or krill.

Table 12.4 Categories of ecosystem services

Ecosystem service categories	Definition
Provisioning services	Direct products of ecosystems such as food
Regulating services	Benefits from natural regulation of, for example, CO_2
Cultural services	Non-material benefits obtained from natural systems, such as swimming in the sea or aesthetic pleasure from looking at scenery
Supporting services	Ecosystem processes which support other services such as nutrient cycling

Today management of ecosystems extends beyond simply considering, for example, the sustainability of fish catches. Ecosystems also provide free services such as provisioning or regulating services that are both essential and beneficial to humankind (Table 12.4).

If it is not possible to use an **ecosystem service** (e.g. a contaminated beach due to an oil spill), then there is a loss and a cost to humans. However, it should be recognised that dissimilar human groups may regard the same component of an ecosystem in different ways and assess its benefits differently. In the Arctic, the hunting of whales and seals is part of Inuit culture, but this is seen as an unacceptable practice by people living in very different environments. Some urban dwellers in ACs campaign against such hunting traditions.

Case study: Krill – a new marine resource

Krill are the 'engine' that powers the Antarctic ecosystem. They are small shrimp-like crustacea that can live for about five years. They inhabit the upper parts of the water column and are able to swim up or down but not against currents.

Krill occupy a very special place in the ecosystem as they are so abundant; their biomass is estimated at more than that of the human race. They exist in swarms, with the larger ones estimated to contain 2 million tonnes extending up to 400 km². The food chains and webs involving krill include most of the Antarctic ecosystem. They are primary consumers feeding on plankton and in turn a wide range of predators feed on them. Seals, fish, penguins, squid, whales and many bird species such as albatross rely on krill as a staple part of their diet.

Human impacts on krill populations

Commercial harvesting of krill began in the early 1970s and has become a significant industry in the Southern Ocean (Figure 12.11). Krill is mainly processed into a variety of products for direct human consumption (paste, frozen tails, oil), for animal feeds, or used as bait for sports fishermen.

The five countries harvesting krill currently are Chile, China, Republic of Korea (South Korea), Norway and Ukraine. Of these, Norway accounts for just over half of the annual catch with China and Republic of Korea each taking about 18 per cent.

During the 1970s and 1980s concerns grew that krill fishing would follow a 'boom and bust' pattern. Large-scale exploitation would be followed by a collapse in krill stocks as the threshold for sustainable fishing was exceeded. This was a pattern that had been seen

→

before around Antarctica with the hunting of fur seals and various whale species.

The Commission for the Conservation of Antarctic Marine Living Resources (CCAMLR) was set up in 1982. Its 25 member states are made up of both ACs and EDCs such as Australia, Japan, Norway, Republic of Korea, UK, Brazil, Chile and China. No LIDCs currently fish for krill nor are members of CCAMLR. One of its aims is to monitor and regulate commercial interests in krill. It adopts a holistic approach which looks at not just the exploited species but also all the other dependent ecosystem components. It is trying to establish catch limits but the interests of the commercial krill fishing industry make this a contested issue.

As more data are collected, much of it from hydro-acoustic (sonar) surveys of krill shoals, there is growing evidence that krill stocks may have dropped severely since the 1970s; some research suggests a decline of as much as 80 per cent. The decline is closely linked to warming seas, a trend that is likely to continue for the foreseeable future. It is now known that krill feed on algae on the under-side of sea ice and that this location is an important 'nursery' for young krill during the Antarctic winter. A key breeding location for krill is the Antarctic Peninsula where the area of sea ice is melting rapidly.

The quantity of krill removed from the ecosystem is beginning to cause concern. CCAMLR now sets a total allowable catch (TAC) for each fishing season aimed at maintaining enough krill for a healthy breeding population and enough for predators. This is based on mathematical modelling which simulates variations in krill populations over 30-year periods. The models are run thousands of times, each time with slight changes to inputs and processes within the ecosystem such as number of breeding krill and different quantities of catch. The current TAC threshold or 'trigger' level for krill catch is set at 620,000 tonnes per year. This is well below the catch limit for the whole of the area around Antarctica of some 5.6 million tonnes.

The concern is that if the TAC were to be increased, krill harvesting would need to be spread out across the whole region to avoid over-fishing in specific locations, such as around the Antarctic Peninsula. The krill trawlers tend to operate in sheltered locations around islands where dense concentrations of krill are predictable. These areas are also the main foraging areas for many predators such as penguins and baleen

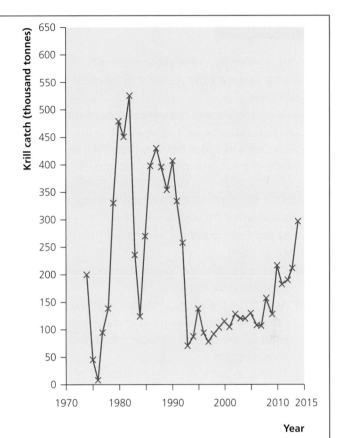

Figure 12.11 Krill catch in the Southern Ocean, 1974–2014

whales. Away from these locations, krill exist at densities that make commercial fishing economically unviable.

The stakeholders in CCAMLR are dominated by ACs. Some are active in fishing for krill such as Norway but the majority do not have boats in the Southern Ocean but do have long-term scientific interest in the Antarctic such as Australia and the UK. China's significant role in the krill industry is in part a reflection of its developing maritime power but also the need to feed its large population. To date, the member countries of CCAMLR are pursuing a broadly sustainable approach to krill harvesting. Conservation is balanced with fishing so that existing ecological relationships within the Antarctic ecosystem are maintained. Challenges such as trying to monitor and regulate boats fishing illegally, albeit very few in number and tending to be after fish not krill, as well as trying to predict the impacts of climate change, are likely to grow in the future. Reliable and accurate data are increasingly available but as with many marine areas, international management is contested.

Activities

1 What is meant by a **renewable** resource?
2 To what extent are commercial krill fisheries sustainable?
3 Research the use and management of another biological resource such as whales or a fish species such as cod or herring. (Possible sources include International Whaling Commission, https://iwc.int/home; World Wildlife Fund, www.wwf.org.uk; National Oceanic and Atmospheric Administration (NOAA), www.fishwatch.gov; European Commission, ec.europa.eu/fisheries/index_en.htm).
   a) Outline the history of exploitation including locations and catch sizes.
   b) Assess how the values, attitudes, socio-economic status and political situations of stakeholders influence the use and management of your chosen resource.
   c) To what extent is your chosen resource being well managed? Use concepts such as 'resilience' and 'threshold' in making your judgement. Look at 'How is the ocean owned?', pages 384–85.

Key idea
→ The use of ocean energy and mineral resources is a contested issue

Unlike biological resources, ocean energy and mineral resources have only recently been exploited by humans. Drilling for oil from piers extending out from the shore started at the end of the nineteenth century while the first free-standing rig only began operating in the late 1930s. As demand has grown and technology advanced, the offshore oil and gas industry has moved into deeper and more hostile waters.

Non-renewable ocean resources: oil and gas

Oil and gas are **non-renewable**, finite resources. Since the mid-twentieth century increasing demand for oil and gas has stimulated the extension of exploration and production to the edge of the continental shelf. Today significant production comes from locations such as the Gulf of Mexico, the Persian Gulf and the North Sea. Technological advances have made possible drilling in deeper and stormier environments. Commercial drilling rigs now regularly operate in water depths of up to 2000 m. The next generation of rigs will be capable of operating at depths of up to 3500 m.

Oil and gas prices are highly volatile, being very sensitive to changing demand in the global economy. Variations in margins (the difference between cost of production and market price) influence investment decisions. This is particularly the case in deep-water exploration where production costs and risks are high.

Investment in infrastructure, both offshore and onshore, is needed to service the oil and gas industry. This includes terminals where oil and gas are transferred, storage tanks, pipelines and refineries. Supertankers, weighing more than 250,000 tonnes, because of their great size require a water depth of 30 m or more, and not less than a 2 km stretch of open water in which to turn.

Oil and gas exploitation has both positive and negative economic and environmental impacts (Table 12.5).

Table 12.5 Positive and negative impacts of oil and gas exploitation

Positive impacts	Negative impacts
• Employment opportunities • Wealth creation • Raw materials for a wide range of products such as fertilisers, paints, plastics, medicines, fibres, which enable a higher standard of living • The rigs act as artificial reefs under water, increasing local populations of some organisms (e.g. red snapper in Gulf of Mexico, corals and shellfish such as barnacles)	• Local communities becoming overdependent on this one industry for jobs • Ecosystem disturbance such as noise and stirring up the sea bed • Visual impact • Pollution from oil spills

The Gulf of Mexico

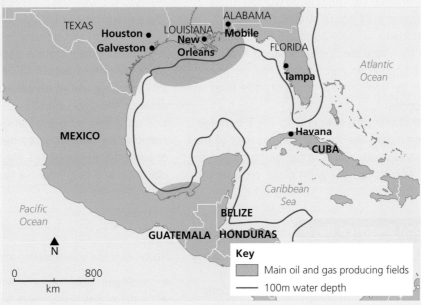

Figure 12.12 Gulf of Mexico oil and gas fields

The Gulf of Mexico region is one of the most important regions in the world for oil and gas production and processing. The undersea geology of the Gulf of Mexico is rich in **hydrocarbons**. Offshore drilling has operated for the past 70 years, first in shallow water using rigs resting on the sea bed, then into deeper waters as floating rigs were developed. Today, a range of high-tech equipment is used, including drill ships, semi-submersible rigs and wells located on the sea bed.

Both the USA and Mexico have substantial offshore oil fields in current production and have extended production into ever deeper water. About 75 per cent of the US's Gulf production of oil comes from wells at depths of more than 300 m. Recent discoveries tend to be in water that is deeper than 1500 m. Supporting oil and gas production is more than 40,000 km of active pipelines criss-crossing the

sea floor, and more than 29,000 km of defunct pipeline. In addition, 45 per cent of the US's oil refining capacity and half of its natural gas processing capacity are located along the Gulf coast.

As the energy industry's economic fortunes rise and fall with the price of energy, so does its socio-economic impact. As a result of the **multiplier effect** jobs in the energy industries create additional employment, which in turn attracts more employment, services and wealth. This is a case of 'success breeding success'. In total around 240,000 jobs are linked directly with the US Gulf energy industries and New Orleans and Houston are among the top ten ports in the world in terms of weight of cargo handled. A large part of this trade is linked to the energy industries.

Renewable ocean resources: waves and tides

The world is overwhelmingly dependent on fossil fuels as its primary source of energy. Currently, coal, oil and gas account for between 75 and 80 per cent of global energy consumption. However, in the longer term, alternative energy sources will be required. This is partly because demand for energy will increase with rising standards of living as well as population increases. But more important is the need to address the problem of climate change linked to the burning of fossil fuels and emissions of CO_2. It is this that is helping drive the exploration of wave and tidal energy.

Wave and tidal energy are examples of **flow resources** which are naturally regenerated by the Sun's energy and the gravitational pulls of the Moon and Sun.

Tidal energy

Tidal energy makes use of the flow of water with the rise and fall of tides. This is not new as mills driven by water wheels turned by tides have existed for centuries. Although large-scale schemes have been investigated for several decades, few have been built.

As with hydroelectric power, tidal power potential is dependent on local physical geography, for example the shape of the coastline and **tidal range**.

Tides are reliable, regular and predictable. Most coastlines experience two high and two low tides approximately every 24 hours. Although many suitable locations exist, only a limited number are close enough to where electricity demand is high enough (there is an economic limit to the distance electricity can be transmitted) to justify the huge cost of development.

One way to obtain tidal energy is to use a **barrage** by constructing a dam-like structure across part of the coast, usually an estuary. Gates in the barrage open as the tide rises. At high tide the gates close, creating a tidal lagoon. As the tide falls, this stored water is then released through the barrage's turbines and back out to sea, generating electricity.

The Shiwa Lake scheme in South Korea is the world's largest tidal power station, generating 254 MW of power at its maximum output. It became operational in 2011, making use of an earlier sea wall built for flood defence and agricultural purposes. It uses the incoming tide only. This is because after the sea wall was constructed, water quality declined. Water is released at low tides in controlled ways so as to flush out pollution and help improve water quality.

Tidal Lagoon Swansea Bay

A 320 MW tidal power station is planned for Swansea Bay in south Wales. A 9.5 km long breakwater will enclose an area of 11.5 km^2 and 16 turbines will spin using both the incoming and outgoing tides to generate 320 MW of power. This is enough to power about 155,000 homes.

The Swansea Bay scheme (Figure 12.13) is based on large-scale, advanced engineering and is intended to pave the way for other schemes both in the UK and elsewhere. It will cost an estimated £1 billion (2015 prices).

There are a number of advantages to the Swansea scheme:

- It uses a renewable flow energy source.
- It will generate electricity for 16 hours in every 24 hours.
- It has low CO_2 emissions – construction and materials used produce CO_2 but it is estimated that the first four years of full operation will save the equivalent quantity of CO_2.
- Its lifetime carbon footprint will be very small.
- It has an estimated working life of 120 years.
- It uses existing technologies, e.g. breakwater and turbine design.
- It will create local employment – short-term during construction, longer-term in maintenance and operation, possible orders for machinery from other similar power stations, a visitor centre.
- It will create recreation and tourism opportunities, e.g. a cycle path across the breakwater, recreational fishing, sailing and rowing on the lagoon, a visitor centre.

However, there are concerns, in particular the impact on:

- sediment movements within Swansea Bay
- water quality due to more limited flows within the bay
- the marine ecosystem – physical disturbance of habitats, noise and vibration, the movement of marine creatures and larger marine organisms passing through turbines
- local inshore fisheries
- the environment in Cornwall (a conservation zone) where stone for the breakwater will be quarried.

Controversy also surrounds the cost of the electricity generated. Subsidies will be needed, similar to those provided to other renewables such as wind and solar power and these are subject to changing political decisions. This affects the economic viability of these schemes. However, the developers claim that future tidal schemes will benefit from the lessons learned at Swansea. There is also the possibility of earnings from building tidal schemes overseas.

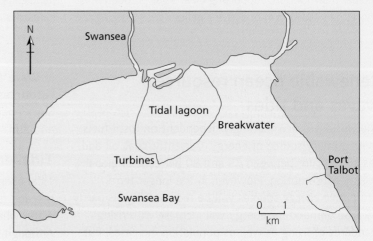

Figure 12.13 Tidal Lagoon Swansea Bay power scheme

Wave energy

Converting the rise and fall of water associated with the passing of a wave into energy has been an aim for many decades. It is an attractive proposition as the potential energy is enormous and much greater than tidal power.

Some wave technology systems are submerged, such as the Waveroller. This Finnish technology consists of a series of large flaps sticking up vertically from the sea bed in relatively shallow water. As waves pass, the flap moves back and forth developing pressure in a hydraulic system which in turn powers an electricity generator.

Other devices are installed at the surface. Wave energy is used to compress air or fluids to drive a generator.

One significant obstacle to the development of wave technology has been wave energy itself. For a device to be effective, high wave energy is required. But in locations where this occurs consistently, such as the west coast of the British Isles, many devices cannot survive the rough seas. Making them more robust means a substantial increase in size and weight. This reduces the efficiency of energy conversion to electricity, increasing the cost of the electricity generated. In addition, servicing equipment in such harsh wave environments adds to costs.

One of the wave machines is Pelamis, an elongated, articulated steel tube. As waves pass, the structure rises and falls, driving hydraulic systems which turn generators to produce electricity. However, it is some way off being a viable source of wave energy.

Even so, wave energy technology is making progress. CETO is a technology that uses several buoys linked to hydraulic cylinders. The buoys float on the surface and as they move up and down with the waves, they create hydraulic pressure. This either generates electricity offshore or is transmitted onshore to drive a generator. Testing has gone on near Perth, Western Australia with increasingly large-scale machines. The hope is to develop a commercially viable system in the next few years.

Sea-floor mining: the final frontier?

Discoveries of mineral deposits, **ferrous** and **non-ferrous**, on the sea bed opened up the possibilities of underwater mining. In relatively shallow waters (less than 300 m) off the coast of southern Africa, diamonds are recovered.

About 40 years ago there was much interest in mining minerals from the sea bed. Potato-sized manganese nodules had been found covering large areas of the central Pacific. However, operations to recover them were unprofitable. Increases in the prices of minerals such as gold, silver, copper, lead, iron and zinc during the late twentieth and early twenty-first centuries have re-ignited commercial interest in sea-floor mining although, as with oil and gas, prices are volatile.

The growth in the use of Rare Earth Elements (REEs) in electronic technology has also given a boost to sea-floor mining. Rare Earths are crucial to modern high technology. Their unique properties are essential to miniaturisation in electronic products such as smart phones.

There is also a geopolitical dimension relevant to mining REEs as they have a strategic importance, for example in telecommunications and military hardware such as missiles. Currently, production of REEs is dominated by China. The reliance on one source for the supply of these vital commodities is a significant concern for the USA, Japan and the EU. This is likely to influence decisions about granting permission for ocean mining to proceed within the Exclusive Economic Zones (EEZs) of some countries (Figure 12.14).

Deep sea remotely operated vehicles (ROVs) are being used to survey and sample the sea bed. A number of different ways of extracting minerals can be used; most mining companies choose a hydraulic suction system which is similar to a giant vacuum cleaner.

Sea-bed mining is not without concerns. Mining is focused in and around the mid-ocean ridges and hydrothermal vents which are sources of some minerals. Given the absence of knowledge about the environment and ecosystems in the ocean deeps, it is difficult to assess potential damage from mining. Different vent sites have ecosystems unique to them. So little of the deep ocean has been systematically researched that mining could severely damage these sensitive locations before they have been fully investigated.

However, many oceanic minerals exist in higher concentrations than land-based ones. For example, land-based copper ore is about 4 per cent metal content while some sea-bed copper ores are ten times more concentrated. The result is that less material needs to be mined to obtain the same quantity of mineral.

Another concern is how to dispose of mining waste. **Tailings**, as such waste is called, are released back into the ocean. This can cause cloudy or turbid water. Where the suspended material settles back to the sea floor it can smother sea-bed ecosystems.

It is true that mining companies add to our knowledge and understanding of the deep oceans. But sea-bed mining is a highly contested activity and it is likely that as the demand for minerals increases and technology advances, pressure on the sea-bed environment will grow in the next few decades.

> **Key idea**
> → Governing the oceans poses issues for the management of resources

How is the ocean 'owned'?

In 1968 an American ecologist, Garrett Hardin, published an article entitled 'The Tragedy of the Commons'. He argued that tension exists between the interests of everyone (the common good) and self-interest. This is particularly so when a resource is seen as belonging to all, such as the oceans or the atmosphere. Such resources are known as the **global commons**. People tend to exploit the resource without considering their impact on it. The advantage to the individual is greater than its cost because the cost is shared among very many. In the short term, it is worth an individual taking all they can, for example of a fish stock, because if they do not, then someone else will. This view assumes that humans act selfishly and do not consider the needs of others. Exploitation of ocean resources is an interesting example of the **tragedy of the commons**.

Various frameworks for managing oceans as a common resource have been developed. The United Nations Convention on the Law of the Sea (UNCLOS) is an international agreement that attempts to define the rights and responsibilities of countries with regard to the coastal zone and beyond (Figure 12.14). Most countries have signed the convention. Landlocked countries are given free right of access to and from the sea under UNCLOS rules.

The system of coastal zones can be disputed by countries and is difficult to implement. Exact boundaries are often disputed as countries try to maximise their rights, especially if there is the prospect of access to resources. For instance, the UK claims the tiny island of Rockall, some 460 km west of the Outer Hebrides. Part of the UK's EEZ gives legal access to the sea bed around it. Surveys indicate substantial reserves of oil and gas in the area.

Concerns also focus on the existence of separate treaties governing other uses of the ocean. Laying sea-floor cables, dumping waste and fishing are governed in different ways. There are, for example, more than 30 fisheries management organisations worldwide, including the European Common Fisheries Policy.

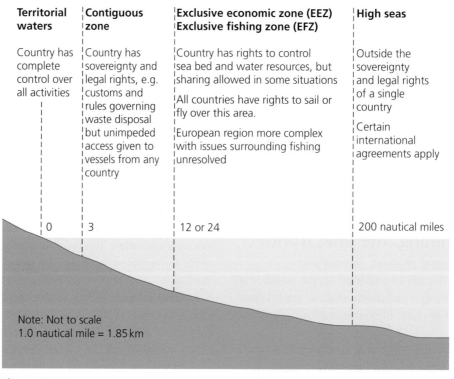

Territorial waters	Contiguous zone	Exclusive economic zone (EEZ) Exclusive fishing zone (EFZ)	High seas
Country has complete control over all activities	Country has sovereignty and legal rights, e.g. customs and rules governing waste disposal but unimpeded access given to vessels from any country	Country has rights to control sea bed and water resources, but sharing allowed in some situations. All countries have rights to sail or fly over this area. European region more complex with issues surrounding fishing unresolved	Outside the sovereignty and legal rights of a single country. Certain international agreements apply
0	3	12 or 24	200 nautical miles

Note: Not to scale
1.0 nautical mile = 1.85 km

Figure 12.14 Ocean management zones according to UNCLOS

Various management issues have emerged since the UNCLOS came into force in 1994 and so are not covered by the agreement. These include:

- ocean acidification
- the patenting of genetic resources through **bio-prospecting**
- fishing in the deep oceans
- the lack of any agreement for the establishment of marine reserves in the high seas
- the absence of regulation of under-water noise and its potential impacts on marine life.

The International Seabed Authority was created by the UN as part of the UNCLOS. It is intended to oversee the exploitation of sea-bed resources in the oceans. With increasing attention given to sea-bed mineral mining, the Authority's management role is gaining in importance. However, hydrothermal vents, their minerals and unique ecosystems had not been discovered when the UNCLOS was negotiated so are not included.

The International Whaling Commission (IWC) is the organisation responsible for the management of whale species. Founded in 1946, it has 88 member countries. In 1986 a moratorium banned commercial whaling and remains in force. Limited catches are allowed for communities in which whaling is important to their culture and economy such as some Inuit communities in the Arctic. However, commercial hunting of whales has continued and is a highly contested topic. It is important to appreciate that whales also face non-whaling risks. Becoming tangled in fishing nets and drowning, being hit by ships, ingesting debris and the impacts of climate change on prey species (e.g. krill) are all issues the IWC researches in order to conserve

whale stocks (see also pages 378–79, Case study: Krill – a new marine resource).

Marine reserves

Like national parks on land, marine reserves are intended to protect marine habitats and ecosystems. It is recognised that some marine locations have unique biological, geological, historic and cultural features, which should be protected in their entirety. Protecting these areas is important not just for present generations but for those in the future. Marine reserves are also seen as an important means of increasing the oceans' resilience to the challenges of climate change, including water warming and acidification.

In 2010, members of the International Convention on Biological Diversity committed to establishing 10 per cent of the oceans as Marine Protected Areas (MPAs) by 2020. Some members, such as the EU, pushed for a higher target (30 per cent), but currently only around 3 per cent of the world's oceans are designated as marine reserves.

Around the UK there are 207 MPAs of different types. Some are for conservation, for example the No Take Zone (NTZ) around Lundy Island in the Bristol Channel. No fishing, dredging or any other type of human disturbance, such as dumping of waste material, is allowed. Others protect underwater reefs such as those around St Kilda in the Western Isles of Scotland or estuaries such as the Exe in Devon.

The UK also has responsibilities around the world because of its fourteen Overseas Territories, legacies of a colonial past. Together they add up to 6.8 million km², nearly 30 times the size of the UK and more than twice that of India.

The Chagos Marine Reserve

The Chagos **Archipelago** is located in the middle of the Indian Ocean (Figure 2.15) and is administered by the UK as one of its Overseas Territories. In 2010 it was designated as a complete 'no-take' (no fishing) marine reserve covering more than 640,000 km², an area two and a half times the size of the UK (www.chagos-trust.org). The land area within this is only 55 km². Within the reserve are tropical islands, coral reefs, ocean trenches, abyssal plains and seamounts. The archipelago is comparable in its combination of marine environments and biodiversity to the Great Barrier Reef or the Galapagos Islands. All extractive activities, such

as industrial-scale fishing and ocean-bed mining, are prohibited.

Although the Chagos coral reefs have been damaged by the effects of global warming (e.g. coral bleaching), because the human impact is small, they have recovered strongly. Most of the reefs are healthy and have the advantage of some of the cleanest sea water in the world.

Marine species benefiting from protection are fish including migratory species such as yellowfin and bigeye tuna and silky sharks. Turtles (green and hawksbill) and

→

many seabirds (sooty terns and red-footed boobies) use the islands for breeding. On some atolls, rats introduced accidentally by humans have been eradicated, which removes one threat to nesting birds. Habitat recovery includes the removal of coconut trees introduced for commercial plantations.

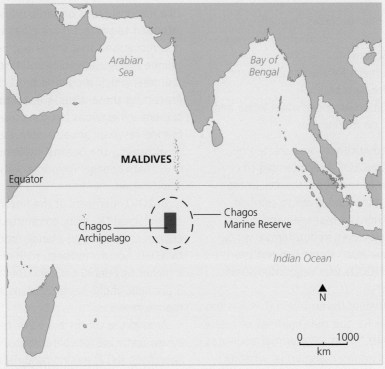

Figure 12.15 The location of the Chagos Archipelago and Marine Reserve

Review questions

1 What is meant by the terms 'natural capital' and 'natural income'?
2 Outline how an ecosystem service can be a cost and a benefit to humans.
3 Describe the significance of krill within the Antarctic ecosystem.
4 Draw up a table outlining the advantages and disadvantages of the exploitation of oil and gas reserves from beneath the ocean bed.
5 What is meant by a 'flow resource'?
6 Outline the ways wave and tidal energy might be exploited.
7 Draw up a table outlining the advantages and disadvantages of the exploitation of ocean bed minerals.
8 Summarise what is meant by the 'tragedy of the commons'.
9 What are the threats that whales face?
10 Outline the characteristics that make the Chagos Archipelago such a good location for a marine reserve.

12.3 How and in what ways do human activities pollute oceans?

Key idea
→ There are a variety of pollutants that affect the ocean system

Pollution occurs when human activity adds a substance to the environment that affects organisms adversely and at a rate greater than that at which it can be rendered harmless. **Point source** pollution is the release of pollutants from a single, clearly identifiable location such as a leaking sea-bed pipeline. **Non-point source** pollution is the release of pollutants from several, widely dispersed origins such as exhaust fumes from a ship's diesel engines.

Many human activities, for example industrial production, burning of fossil fuels, and agriculture, generate pollutants that can find their way into the oceans. At one time, it was thought that the vastness of the ocean could dilute pollutants sufficiently to render them harmless. However, some pollutants are so toxic and persistent that they can seriously damage marine ecosystems and food chains. For example, the fishing grounds off Minamata, Japan, remain dangerously contaminated with mercury, even though the release of the pollution (industrial waste water) ceased in the late 1960s.

Many pollutants originate far inland and are transported to the ocean via rivers or through fallout and absorption from the atmosphere. Pollutants of particular concern include petroleum, nitrate and phosphate fertilisers, pesticides such as DDT, and industrial contaminants such as heavy metals. Even noise pollution, from activities such as shipping, seismic exploration and sonar, can affect the behaviour and health of marine mammals like whales and dolphins.

Pollution from fossil fuel burning at sea

The combustion of fossil fuels produces several pollutants, such as carbon dioxide, sulphur dioxide, nitrogen oxides and particulates. Shipping of all types relies on fossil fuels for power. There is, however, a perception that sea transport produces low levels of pollution. But because ocean routes account for a very great tonnage of goods and distances can be long, total emissions by shipping (in contrast to emissions per tonne/km) are significant at the global scale.

There are 90,000–100,000 ocean-going cargo vessels plying the world's oceans. They operate 24 hours a day and for 280 days a year on average. The engines consume huge amounts of low-grade oil known as bunker fuel, which is high in sulphur and other pollutants. The largest vessels emit about 5000 tonnes of sulphur a year. Overall sea transport is responsible for 9 per cent of annual SO_2 emissions and 15–30 per cent of NO_2 pollution. Moreover, shipping accounts for 3.5–4 per cent of all greenhouse gas emissions from human activities. Most air pollution from ships is released in the northern hemisphere where the busiest international shipping routes are concentrated (Figure 12.25).

Locally in and around major ports, air pollution from ships can be a serious hazard. Long Beach in Los Angeles, San Francisco, Galveston and Pittsburgh are examples of major US ports where air pollution from ships is a serious issue. Long Beach port is ranked second in the USA and tenth in the world in terms of tonnage handled, around 78 million tonnes a year. It is a major transport node on the Pacific Rim. Between 6 million and 7 million containers move through the port annually. With about 30,000 jobs directly linked with the port and some 300,000 connected in one way or another to southern California, port operations are vital to the region's economy. Given the scale of port activities, it is not surprising that air pollution is 2–3 times higher in the immediate vicinity of the port than further away.

With growth in the number and size of cruise ships, localised air pollution has become a serious problem in popular cruise destinations such as the Caribbean and Alaska. The concentration of these often very large vessels into restricted coastal inlets, such as the steep-sided Alaskan fjords, can lead to the heavy concentration of diesel fumes and the formation of haze. Although it is very difficult to regulate marine shipping emissions, many of the world's shipping lines have taken measures to address the issue of air pollution. Increased fuel-efficient engines save costs as well as reduce emissions. Using fuel lower in sulphur content is another option increasingly taken.

There has also been a trend towards reducing ship speeds. 'Slow steaming', as it is called, involves cutting the speeds from 27 knots to less than 20 knots. Significant savings in fuel consumed are associated with much reduced emissions. Typically, the larger container ships would burn 300 tonnes of fuel and emit 1000 tonnes of CO_2 per day.

Better ship design, with more efficient hulls and propeller shapes, is improving fuel efficiency. Wind power is nothing new to shipping, but some modern vessels have been fitted with sails and there is even a design that uses kites.

Pollution from domestic and industrial sources

The oceans are where much pollution generated on land finally ends up. Rivers discharge dissolved chemicals or solids into the sea. Pollutants discharged to the atmosphere can be washed out by precipitation. The sea can be used as a convenient place to dump solid and liquid waste such as radioactive water from nuclear plants, for example from Sellafield into the Irish Sea.

Until recently, most coastal communities and those inland used either rivers or the sea to remove raw sewage and industrial effluent such as the waste from the tanning of leather. Today few ACs routinely discharge untreated effluent into seas and rivers. Progress is being made in EDCs but environmental improvements are not yet a top priority.

The situation in most LIDCs is less favourable. The list of pollutants is long and some are extremely toxic and harmful to human health and the environment. Suspended solids, organic waste, heavy metals (e.g. mercury), nitrates and phosphates (e.g. dissolved from detergents) and pesticides, fungicides and growth hormones from agriculture pollute waterways, lakes and inshore waters.

Algal blooms

Pollutants such as nitrates and phosphates in lakes and seas create a rich 'soup' of nutrients which stimulate blooms of algae. This can lead to the reduction in the level of dissolved oxygen in the water and the consequent death of fish.

Radioactive waste

The nuclear industry expanded rapidly after the Second World War in Europe (including Russia), China and the USA, with its increased application to energy production, medical and industrial purposes and the manufacture of nuclear weapons. This created the problem of disposing of large quantities of nuclear waste. At first, the oceans were seen as the ideal dumping ground as radioactivity would be dispersed throughout the vast volume of water.

From 1946 to 1993, 13 countries with nuclear industries disposed of radioactive liquids and solids including reactor vessels that still contained nuclear fuel in the oceans. Tens of thousands of steel drums containing radioactive waste were dumped in the ocean. In addition, eight nuclear submarines (six Soviet Union/Russian and two American) have sunk or were scuttled. The Russian K-27 was sunk in 1981 and is lying on the floor of the Kara Sea (part of the Arctic) in 33 m of water. Its nuclear reactor with its enriched uranium fuel remains on board. From underwater surveys it appears that the reactor is still enclosed in its

Activities

1 The following statements are in the wrong order. Write them out so that they appear in the correct order to show the cycle of algal bloom:
 ● Rapid growth of algae boosted above normal to create an algal bloom.
 ● Large numbers of dead algae form a biomass on which bacteria and fungi feed.
 ● Leaching of nutrients such as phosphates and nitrates into sea.
 ● Algae cover sea surface, preventing light reaching majority of algae lower in the water column which then die.
 ● Loss of dissolved O_2 impacts on organisms requiring it such as fish which die, adding to the biomass available to the decomposers.
 ● Aerobic respiration of biomass by decomposers removes much of the dissolved O_2 in the water.
2 Describe the possible sources of minerals which act as nutrients for algae.
3 Investigate the impacts an algal bloom might have on:
 a) local fishing industry
 b) local tourist industry
 c) the health of humans.
4 Discuss why the management of algal blooms (also known as 'red tides') is not straightforward.

protective shield. However, sea water is very corrosive and in time, the submarine's structure will weaken and eventually release high-level radioactive material into the sea.

The Fukushima nuclear power plant in Japan leaked substantial quantities of radioactive material following damage caused by the effects of the tsunami in 2011. Radiation is known to accumulate in the marine food chain. Strict bans have been imposed by the Japanese government on the catching and consumption of shellfish and fish in the immediate vicinity of the Fukushima plant. Given that leaks of radioactive water continue to seep into the sea, these bans are likely to be in place for a long time. Air-borne radiation has been carried much further afield. Low-level radiation originating from Fukushima has been identified off the coast of the northwest USA. Oceanographers generally believe that there is little threat to the Pacific as a whole; the serious risks are local in scale.

Case study: The Deepwater Horizon disaster

In April 2010, a BP oil rig, located 40 miles off the Louisiana coast and in 1500 m water depth, exploded, killing 11 workers and injuring 17. This followed the failure of the 'blow-out' device designed to prevent high-pressure oil and gas from blasting up the drill pipe.

For 87 days oil gushed out of the well-head on the sea bed (Figure 12.16). The amount of oil released is disputed; upper estimates put the volume at roughly 4.9 million barrels but experts and BP have come up with figures considerably lower than this. However assessed, the Deepwater Horizon disaster was the largest oil spill in history. At its maximum, 180,000 km^2 of the Gulf was affected and just over 1,600 km of shoreline was polluted.

Various measures were tried to seal the well. Attempts to close the blow-out valves were unsuccessful as too much damage had taken place to the mechanisms. A large containment box was lowered to cover the well-head but this failed to stop the oil. Then a strategy known as 'top kill' was tried. This is when materials such as heavy and dense mud and concrete are poured into the well to seal it. It took several weeks to force sufficient material into the well to plug it.

Meanwhile various strategies attempted to deal with the escaped oil:

- skimming surface oil – booms towed by small manoeuvrable boats collect oil; the oil is then scooped up and taken away
- burning surface oil – surface oil is collected in a fire-proof boom; the oil is then burned
- dispersants – chemicals break the oil into smaller particles to prevent oil slicks forming; these were sprayed from the air and by boats; 1.84 million gallons of dispersants were used; evaporation and biological degradation by oil-consuming bacteria then removes the oil
- artificial barrier islands were constructed just off-shore in some places; these were to act as physical barriers to stop the oil reaching the shore; they tended, however, to be soon washed away by the waves and currents in the Gulf
- beach cleaning – oil washed up on beaches mixed with sand; the contaminated sand was then scooped up mechanically or shovelled by hand into piles for collection

Figure 12.16 Deepwater Horizon rig after the 2010 explosion

- beach cleaning – sand mixed with oil was collected and 'washed' by equipment using very hot water; the sand could then be returned to the beach while the separated oil was taken away for processing.

The entire marine ecosystem along the Gulf coast was affected, with devastating (though short-term) impacts. The media published images of oil-soaked birds, fish, turtles and mammals such as dolphins. Mortality among such heavily affected creatures was high.

The fishing industry along the US Gulf coast is one of the most productive in the world (shrimp, oysters and fish) and important for the regional economy and employment. In the short term, fishing stopped and there was a loss of income for those involved in the industry.

The tourism industry along the Gulf coast was also badly hit. Media images of contaminated beaches and oil-covered wildlife meant that few people wanted to vacation in the immediate aftermath of the spill. There were economic and social hardships for those directly employed in tourism as loss of income and unemployment followed. This then rippled through the local economies of the coastal communities in a general downward spiral.

Much research and monitoring continues in the areas affected. The rate at which the ecosystem has recovered has surprised many. Five years after the spillage the Gulf's fish and oyster industries are allowed to sell

→

their catch and virtually all the beaches are open. Some species such as brown pelicans and seaside sparrows show no long-lasting effects on their numbers. However, oil does still wash up and in places deposits remain on the sea bed and covers now dead coral reefs. Perhaps the location most adversely affected was the coastal salt marshes. Oil trapped here tended to accumulate in the mud which meant that in the **anaerobic** conditions, the oil was not broken down by bacteria.

One question still to be answered is where did all the oil go? (Figure 12.17) A key factor which meant that the disaster was not as severe as it might have been was the nature of the oil. Not all crude oil has the same chemistry. Oil from the Deepwater Horizon well is a light crude and so dissolves in water more readily than heavy crude. Some floated to the ocean surface where it evaporated. Surface oil was driven by currents and winds to the shore. Some oil became trapped in the deep ocean layers below 1000 m while yet more sank to the ocean floor.

One difficulty that researchers into the disaster face is that the Gulf is not a clean ocean. It has a long history of oil spills as well as natural seeps of oil from the sea bed. Some 2,000,000 gallons of oil per day leak naturally, something that has being occurring for thousands of years. Identifying

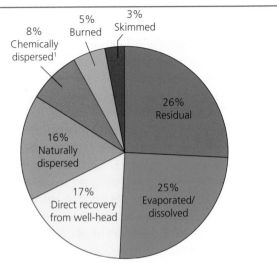

Residual = oil in ocean, washed ashore and collected or in sand and sediments
Naturally dispersed = oil currently being degraded naturally, e.g. by bacteria

Figure 12.17 Where oil from Deepwater Horizon has gone

the oil from the Deepwater Horizon spill is very difficult. The Gulf also receives large quantities of pollutants from the land via the rivers. Run-off from farmland carries pesticides, fungicides and insecticides as well as excess nutrients, all of which end up in the Gulf waters.

See also pages 380–81.

See also pages 380–81.

Activities

1 Research the ecosystems found in the Gulf region. Using a simple diagram of the food chains and webs, describe and explain how an oil spill impacts the ecosystem.
2 Research the socio-economic impacts the oil spill had on various groups living along the Gulf coast. Construct a mind map summarising the impacts by stakeholder groups such as fishermen, those employed in the tourism industry, oil workers and those in retail in coastal settlements. Highlight both direct and indirect impacts of the spill.

Key idea
→ The pattern of ocean currents can disperse and concentrate pollution

Plastics, metals, rubber, paper, timber, textiles, fishing gear and countless other items enter the marine environment every day, making marine debris a widespread pollution problem. Debris finds its way into oceans in a variety of ways: from rivers and beaches to dumping off the sides of ships at sea. Accidental discard can happen when ships are caught in storms. For example, in 1992, a container holding thousands of plastic bath toys was swept off a ship in the Pacific, northwest of the Hawaiian Islands. The toys, ducks, turtles and frogs have spread around the oceans and have inadvertently advanced understanding of the operation of ocean currents.

But marine debris can be persistent, potentially over many decades. It can pose physical dangers to ships as well as damage ecosystems. Larger marine creatures such as seals and sharks can become entangled in discarded fishing nets and drown. Perhaps the most serious impact comes from plastics. Most do not **biodegrade** but break down into small pieces called micro-plastics as a result of photodegradation (the action of sunlight on the polymer chains that make up plastic.) The plastic items in our lives tend to start as tiny manufactured pellets of plastic, known as 'nurdles'. These are shipped round the world in vast quantities and being so small and light can be easily spilled and dispersed. They are found in every ocean and along the vast majority of coastlines.

Case study: Great Pacific Garbage Patch – the concentration of pollution in gyres

An ocean **gyre** is a system of circular currents formed by wind patterns and the rotation of the planet (see page 371). These principal gyres have their centres approximately 30° north or south of the Equator (see also page 371). The area in the centre of a gyre is usually calm and once debris has been carried to its centre it tends to stay there. The accumulation of debris in the North Pacific gyre has led to the term 'Great Pacific Garbage Patch'.

The use of this name has led to the assumption that this 'patch' is a large and continuous area of debris rather like a floating landfill made up of material such as bottles, nets, yoghurt pots and cotton bud shafts. Much is plastic and it not only exists at the surface but is also suspended below the surface for several metres. While there are locations with high concentrations of relatively large debris, most of the garbage is actually micro-plastics, barely visible to the naked eye.

Neither is the Great Pacific Garbage Patch one enormous mass stretching across most of the Pacific. Rather, there are two principal areas of accumulation, one east and one west, with a lower-density zone just to their north (Figure 12.18).

Large pieces of plastic can become wrapped around animals such as seals and strangle or drown them. Smaller pieces are ingested by fish and birds. The plastic interferes with digestion both physically and chemically. It is not just the chemicals contained in the plastics but the pollutants that plastics attract once they are in the oceans that also harm organisms. Once this pollution is in the food chain, there is the risk of it being transferred to humans.

The key point is that regardless of actual volume of debris and where it is, marine debris is pollution which potentially has very harmful effects on the marine ecosystem. And the Pacific Ocean is not the only ocean so polluted. Debris, and in particular plastic debris, is being found in all oceans and marine ecosystems.

Although research on marine plastic pollution is still in its infancy, we know that this problem is likely to become more serious in the future. Plastics are long-lived in the environment – items from the 1940s are being discovered in the stomachs of seabirds like albatross. But plastic in itself is not the problem – the problem is how we dispose of it.

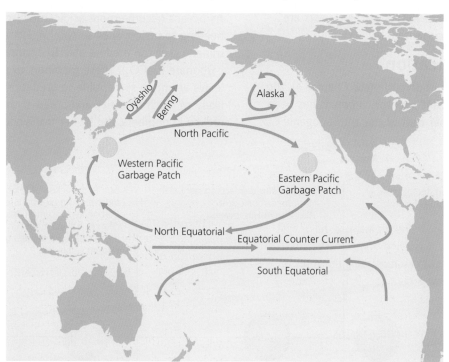

Figure 12.18 Pacific garbage patches

Review questions

1 Distinguish between point and non-point source pollution.
2 Describe the ways ocean shipping contributes to atmospheric pollution.
3 What measures can be taken to reduce atmospheric pollution from ocean shipping?
4 What is an algal bloom?
5 Describe three sources of radioactive pollution affecting the oceans.
6 What is meant by an ocean gyre?
7 Explain why plastic pollution in the oceans is so serious an issue.

12.4 How is climate change impacting the ocean system?

> **Key idea**
> → Climate change is altering the nature of the ocean's water

The Earth's climate has changed on many occasions over geological time. There is also evidence of shorter-term variability change. Today there is overwhelming evidence that the Earth's climate is currently experiencing significant change. These changes have a direct effect on the oceans, which are closely connected to the atmosphere.

There is clear evidence that greenhouse gas (GHG) concentrations have been rising since the mid-eighteenth century. There is debate as to how sensitive the global climate is to increased levels of gases such as carbon dioxide (CO_2) and methane (CH_4). However, the current widely held scientific view is that increases in GHG concentrations in the past 200 years are responsible for the current global warming.

Ocean acidification

Within the Earth's carbon cycle (see Chapter 4), the oceans are a significant **sink** for carbon. It has been estimated that 30 per cent of the anthropogenic CO_2 produced over the past 250 years has been absorbed by the oceans. Without this absorption in the oceans, the level of atmospheric CO_2 would probably have already gone past a critical **tipping point**, resulting in rapid rises in global temperatures and sea levels. Of concern is that as surface water temperature rises, the ability of the water to absorb CO_2 reduces.

However, present-day ocean chemistry has been changing due to the addition of CO_2. The fundamental and rapid change in the ocean's pH has far-reaching consequences for marine organisms, ecosystems and potentially for human societies. Average global surface ocean pH has already fallen from a pre-industrial level of 8.2 to 8.1, still in the alkaline range of pH, but the process is known as **acidification** (Figure 12.19). Because the pH scale is a logarithmic one this represents a 30 per cent increase in acidity.

The forecast is for ocean pH to reach 7.8–7.9 by 2100, a doubling of acidity.

The impact of ocean acidification on marine ecosystems

As oceans become more acidic, molluscs and crustacea needing to build a shell or coral needing to build its mineral skeleton are less able to accumulate calcium carbonate ($CaCO_3$). These members of the lower trophic levels are more susceptible to predation and less likely to reach maturity and breed. If these organisms reduce in numbers, the higher trophic levels in marine

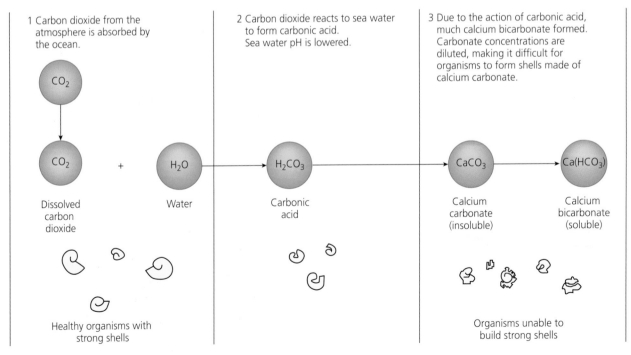

1 Carbon dioxide from the atmosphere is absorbed by the ocean.

2 Carbon dioxide reacts to sea water to form carbonic acid. Sea water pH is lowered.

3 Due to the action of carbonic acid, much calcium bicarbonate formed. Carbonate concentrations are diluted, making it difficult for organisms to form shells made of calcium carbonate.

CO_2

CO_2 + H_2O → H_2CO_3 → $CaCO_3$ → $Ca(HCO_3)$

Dissolved carbon dioxide

Water

Carbonic acid

Calcium carbonate (insoluble)

Calcium bicarbonate (soluble)

Healthy organisms with strong shells

Organisms unable to build strong shells

Figure 12.19 The process of ocean acidification

ecosystems are put at risk. For example, the sea butterfly is a small snail-like creature about the size of a pea. They are part of the zooplankton community found in the upper zone of the oceans ranging from the poles to the tropics. In good conditions they exist at about 2000 individuals/m³. Their role in ecosystems is vital as they are a significant food source for krill, many species of fish such as herring and mackerel and some whales. As with all components of an ecosystem, if populations of organisms at the base of the food chain collapse, the entire system is put at risk. Beyond a loss in biodiversity, acidification threatens fisheries and aquaculture (e.g. fish farms). Food security for millions of people is thereby jeopardised, as is eco-tourism, for example to coral reefs.

Declining populations of marine organisms with $CaCO_3$ shells and skeletons affect the carbon cycle. When these creatures die, they sink to the sea bed. The shells become sediment and eventually rock, storing carbon for millions of years.

Some shellfish have been shown to have difficulty making their shells in acidic water. Mussels face additional risks from the acidification of sea water. They cling to rocks by a thread-like material which is less effective in more acidic water. Massive die-offs of oysters have been recorded in parts of the Pacific northwest of the USA, putting valuable commercial and subsistence fisheries at risk.

A significant unknown is the impact of acidification on jellyfish. One suggestion is that they will thrive in warmer and more acidic conditions. Jellyfish compete with various predators for small fish and are already something of a problem. Their numbers can increase rapidly until they dominate and disturb the equilibrium of marine ecosystems.

There are areas of the oceans where CO_2 naturally seeps up from the sea bed, such as around underwater volcanoes. These more acidic locations do have abundant life but only a few species have adapted to this particular environment. They include diatoms (a type of algae) with silica shells, which are not affected by acidification. Sea grasses that absorb and store carbon flourish. Some of the world's most productive fisheries, such as those off the coast of Peru, are found where water upwells from the depths, bringing CO_2 and nutrients to the surface.

The impacts on people of depleting fish stocks due to ocean acidification

Changes in marine ecosystems could significantly alter marine harvests of shellfish, crustacea and fish which are examples of provisioning ecosystem services (Table 12.4). Research is also indicating that early life stages of fish

species (eggs, larvae) are vulnerable to increasingly acid sea water. Globally, some 200 million tonnes of seafood are produced annually including from aquaculture such as fish farms. Just under half of the catch comes from the Pacific and a quarter from the Atlantic.

Some of the countries most dependent on seafood for protein in people's diets include LIDCs and EDCs such as The Gambia, Bangladesh and island states in the Pacific such as Fiji. In some countries seafood can account for about 50 per cent of dietary protein and many of these countries have limited agricultural alternatives for protein production. By 2050, both population increases and the impacts of ocean acidification on marine ecosystems are likely to cause greatest stress on fishing in the tropics. It is also the case that many ACs gain considerable provisioning services from marine harvests such as Japan, Norway and Canada.

Although changes to some fish stocks are being strongly linked to ocean acidification, the overall picture is not clear. Some predatory species switch prey and seem to show few negative effects but then such behaviour puts them in conflict with other predatory species which decline. Research into marine ecosystems is particularly challenging given the three-dimensional and often hostile nature of the marine environment. What is clear is that marine ecosystems are coming under increasing pressures from human activities which are adding to the stress coming from ocean acidification.

Warming oceans and coral reefs

Corals are marine polyps, a type of invertebrate. Each polyp is a sac-like animal typically only a few millimetres in diameter and a few centimetres in length. They secrete a protective skeleton around themselves made up of $CaCO_3$. Some corals catch their food by using small stinging tentacles, but most live in symbiosis with algae known as zooxanthellae. These algae release nutrients via photosynthesis, which the polyps feed on. In return, the algae are sheltered within the hard coral skeleton and obtain some minerals from the coral. The algae contain pigments which give coral its colours.

Coral requires particular environmental conditions if it is to grow well.

- Temperature: mean annual water temperature not less than about 18°C, ideally around 26°C.
- Salinity: corals require salinity levels greater than 30,000–32,000 ppm (see pages 369–70).
- Water depth: 25 m or less. But corals die if exposed to the air for too long so grow only as high as the lowest tides.
- Light: algae require light for photosynthesis.

- Clear water: sediment reduces the light available for photosynthesis. It can also cover the coral and clog up the feeding tubes.
- Wave action: the water needs to be well oxygenated so some wave action is needed, but not so much that it might physically damage the coral.

As oceans warm, corals are pushed to their thermal limits. If the temperature exceeds their tolerance, the relationship between coral polyp and algae is disturbed. At its most extreme, many of the algae are expelled, depriving the coral of colour. These **coral bleaching** episodes have been noticed all around the tropics. Intensive studies are indicating that bleaching is a highly complex event. It is also associated with changes in salinity, for example a sudden flood of freshwater run-off from the land. Such jets of water often carry high levels of suspended sediment, which also stress the coral ecosystem.

Mass bleaching episodes

Observations of entire coral communities bleaching have been increasing since the 1980s. Since then mass bleaching has grown in frequency, intensity and geographical extent. In the late 1990s, coral reefs in most parts of the world were experiencing mass bleaching. In regions such as the Seychelles, the Maldives and parts of the eastern Pacific, coral mortality reached 80 per cent. Some reefs have since recovered, although there may be less coral overall.

Satellite measurements of sea surface temperatures are used to predict bleaching events several weeks in advance. The rate of temperature increase is charted and then correlated with coral mortality. Thanks to such research, more is being understood about bleaching and whether anything can be done to improve coral's resilience.

The Great Barrier Reef, Australia

Coral reefs are known as the 'rainforests of the oceans' due to the similarities between the two ecosystems. Their features include:

- very high biodiversity – more than 25 per cent of all marine life live on coral reefs. The Great Barrier Reef, Australia, for example is home to approximately 1500 fish species; 350 coral species; 4000 mollusc species; 500 algae species; 6 of the 7 turtle species in the world; 24 seabird species; >30 species of whale, dolphin and the dugong, a seal-like mammal, unique to the Reef
- efficient recycling of energy and matter, with the vast majority of nutrients tied up in reef organisms.

Many fish and molluscs such as limpets eat algae growing on the coral thereby preventing the coral being smothered. Predatory fish such as sharks and groupers eat smaller fish keeping populations in balance. Many animals such as sea squirts and marine worms filter sea water, removing organic particles that would soon turn the water cloudy. Throughout any coral reef, complex food chains and webs sustain what is probably the most diverse ecosystem on the planet.

Threats to the biodiversity of coral reefs

The danger posed to the biodiversity of coral reefs from climate change is four-fold:

- increased sea water temperatures lead to coral bleaching events
- sea level rise increasing the depth of water over corals thereby reducing light levels

- increased wave energy from greater number of more intense tropical storms (hurricanes/cyclones/typhoons)
- ocean acidification reduces coral's ability to build carbonate structures.

All of these, either singly or in combination, can reduce a coral reef's ability to shelter or offer niches to occupy for so many marine species.

Threats to local communities from disruption to coral ecosystems

Threats to coral ecosystems represent both direct and indirect impacts for human communities. Coral reefs provide provisioning, regulating, cultural and supporting ecosystem services (Table 12.4) They directly act as a buffer for coastlines from high energy waves such as during a tropical storm or from a tsunami. Without them, losses of life, property and amenities such as beaches would be greater. Coastal development would cost much more as investment would be required in expensive engineering such as sea walls. The global economic value of shoreline protection from coral reefs has been estimated at about US$10 billion per year.

Coral reefs play a significant role in acting as 'nursery locations' for many fish species which are then harvested by humans. Loss of this local supply of fresh protein would be extremely detrimental to communities in LIDCs and some EDCs already suffering from widespread under-nourishment. These are communities which are vulnerable to small changes in their lifestyles as they have limited resilience because of their lack of resources. →

Recreational fishing also generates significant employment and income for some communities which spreads through the area due to the multiplier effect. Other recreational activities such as scuba diving and snorkelling also attract visitors. The white sand beaches often associated with coral reefs are made up of broken-down fragments of the coral. It is not just tourists who value the cultural services of coral reefs. Coral reefs have long played a significant role in the songs, poetry, art and folklore of local communities.

> ### ➡ Activities
>
> 1 Research the management strategies currently employed along the Great Barrier Reef (www.gbrmpa.gov.au/managing-the-reef/).
> 2 To what extent might the management of the Great Barrier Reef act as a model for reefs elsewhere?

> ### Key idea
> → Climate change is altering sea levels

Just like the Earth's climate, sea level has also changed considerably over geological time. In the past 2 million years, coastlines worldwide have experienced considerable variation in sea level. In some places, the variation has been as much as 120 m.

Changes in absolute sea levels are called **eustatic** changes. They are global because all the oceans and seas are interconnected. Changes in the absolute level of the land are called **isostatic** changes. They are localised and often the result of tectonic movements such as earthquakes.

Recent sea level change

Sea level rise (Figure 12.20) is one of the key areas studied by the **Intergovernmental Panel on Climate Change (IPCC)**. This is a large international and interdisciplinary group of leading scientists. They research into the causes and effects of climate change to the environment and human societies.

It has become clear over the past few decades that a eustatic rise in sea level is taking place and at an accelerating rate. Measuring this change is increasingly accurate thanks to the use of sophisticated satellite technology. During the early years of the twenty-first century, the rise has averaged about 3.0 mm per year.

Causes of recent sea level change

Several factors are responsible for the present-day rise in the level of the oceans.

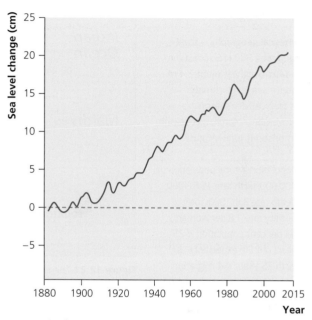

The figures are from tidal gauges. The blue line represents the three-year running mean

Figure 12.20 Sea level change, 1880–2015

- **Thermal expansion of water**: global warming is resulting in an increase in surface water temperatures. As its temperature rises, the density of sea water decreases, which brings about an increase in the volume of water in the ocean basins. This thermal expansion of water causes a rise in sea level.
- The melting of glaciers and small ice caps: global warming is resulting in an increase in air temperatures. As a result, glaciers and small ice caps such as in the Andes and Himalayas are melting. The water released flows via rivers into the oceans, thereby contributing to rising sea level.
- The melting of the Greenland and Antarctic ice sheets: a complex picture exists for these vast areas. In general, the central portions of the ice sheets are thickening slightly as ice accumulates. The margins of the ice sheets are, however, ablating and thinning rapidly, and flows of ice towards the sea are accelerating.

The Maldives are a group of coral atolls in the Indian Ocean to the southwest of southern India. They face a number of issues similar to other low-lying islands around the world. Sea level rise can be directly linked to some issues; other issues are increasingly affected by the rising oceans.

Physical geography – total land area 300 km²; coastline 644 km; highest point 2.4 m above sea level; climate is tropical, dry season November–March, wet season (monsoon) June–August.

Population – total population of 400,000; birth rate 15.6/1000; death rate 3.8/1000; total fertility rate 1.8 per woman; 44 per cent population < 25 years old; life expectancy at birth 75 years; net migration −12.7/1000; strong population growth puts pressure on all the other factors; substantial flows of emigrants tend to be of young adults, depriving the Maldives of a work force and splitting families and communities; see point about youth unemployment in Employment box.

Agriculture – 10 per cent land area arable, 10 per cent permanent crops, e.g. coconut plantations; land lost to the sea threatens agricultural production; salt water is toxic to plants.

Industry – fish processing, mostly tuna; building of small boats; coconut processing; handicrafts, e.g. products made from coconut fibres; long-term threats to fish stocks from ocean acidification and loss of coral reefs as nursery areas; coconut trees threatened by sea water contaminating the soil they grow in with salt.

Employment – primary (agriculture and fishing) 15 per cent; manufacturing industry 15 per cent; services 70 per cent; unemployment rate 28 per cent (15–65 year olds); agricultural production threatened, as is tourism.

Fresh water – total renewable water (freshwater aquifers) 0.03 cubic km per year; water consumption 0.01 cubic km per year; rising sea levels threaten to contaminate aquifers with salt water.

Economy – GDP (ppp) US$9100; origin of GDP agriculture 3 per cent, industry 17 per cent, services 80 per cent, of which tourism makes up about 30 per cent; tourist numbers fall as some locations lost to the sea; threat of rising unemployment and increased emigration.

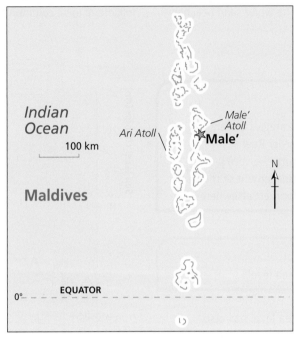

Figure 12.21 The threats facing the Maldives

Activities

1 Write an analysis of the impacts rising sea levels are having on the Maldives, or on any other island community facing similar concerns, for example in the Pacific or the Caribbean.
2 What strategies can island societies adopt to combat sea level rise in both the short and the long term?

Key idea

→ Climate change is altering high-latitude oceans

The effects of climate change vary from one geographical region to another. The high-latitude oceans, the Arctic and the Southern Ocean around Antarctica, are recognised as locations where climate change is having serious impacts.

Why does sea water freeze in the high latitudes?

Sea water freezes at −2.0°C rather than zero because of its salt content. When sea water freezes, the ice contains very little salt because only the water part freezes.

In the high latitudes, the annual heat budget has a net deficit (Figure 12.22); more heat leaves the system than is input. This is due to the low angles at which the Sun's rays hit the surface. Because the angles are low, the Sun's energy is less intense, being spread over a greater surface area than in the mid and low latitudes.

Additionally, ice that covers nearly all of Antarctica and much of the Arctic has a high **albedo**. Thus a large proportion of incoming solar radiation is reflected back into space. In Antarctica, monthly mean temperatures around the coast are between −10°C and −30°C during the southern hemisphere winter. The lowest

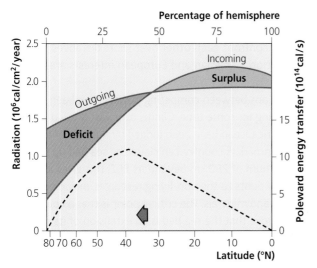

Figure 12.22 Energy balance variations by latitude

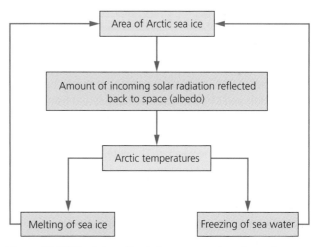

Figure 12.23 The interaction between area of sea ice and warming of the Arctic

temperature ever recorded on Earth was −89.2°C at Vostok, a Russian research station in Antarctica. During the Arctic winter, temperatures average −30°C to −35°C, dropping to −50°C at their coldest. The summer months are close to or just below 0°C.

The relationship between global warming and sea ice

Since its formation, the IPCC has been concerned about the effect of global warming on sea ice. This primarily concerns the Arctic, which is a frozen ocean.

Concern regarding Arctic sea ice focuses on two key aspects of the ice: area and thickness. Satellites, such as the European Space Agency's CryoSat-2, allow year-round accurate monitoring of the Arctic sea ice. Figure 12.23 shows the interaction between the area of sea ice and warming of the Arctic.

Between 1978 and 1996 the area of Arctic sea ice decreased by 2.9 per cent per decade. April marks the start of the summer melting season, with the area of sea ice at its maximum. April 2015's figure was the second lowest since satellite observations began. The ice covered 4.0 million km² compared with the long-term average (1981–2010) of 15.0 million km² (Figure 4.30).

In addition to sea ice area, sea ice depth is investigated. Several navies deploy their submarines beneath the ice in the Arctic Ocean to monitor each other and to try to 'hide' from surveillance from satellites. One benefit from this military activity has been the measuring of ice thickness. At the height of the Cold War in the 1960s, Arctic sea ice was typically 4 m thick at the centre of the Arctic Ocean; that figure is now about 1.25 m. Given the data now available, there is much confidence in predictions that Arctic sea ice will continue to shrink in terms of both area and thickness, albeit with occasional years of increase.

A major concern is that as the area of the Arctic covered by ice declines and the ice is thinning, a point will come when a threshold is crossed. After that, so much of the Arctic will be free of ice that increasing amounts of solar radiation will not be reflected but will be absorbed by the sea and lead to the warming of the Arctic Ocean. In turn this will encourage further melting of sea ice and so an irreversible cycle operates. Perhaps most worrying is that the observed reductions in Arctic sea ice extent and thickness exceeds those predicted by the more recent computer modelling.

Indigenous peoples

There are 4 million people living in regions bordering the Arctic Ocean. The **indigenous peoples** of the Arctic such as the Inuit and Chukchi continue to hunt and fish in the Arctic Ocean. Their sustainable harvests of animals such as walrus and seals are greatly affected by patterns of sea ice as well as the weather. The potential disruptions to the Arctic marine ecosystem caused by thinning ice, decreasing ice cover and increasingly severe weather make hunting less reliable and more dangerous. Such activities are core elements to the indigenous societies, with their loss having profound impacts on culture and the ways societies operate. It may also impact health if there is a reduction in food diversity and availability.

Geopolitics

Viewing a polar-centred map highlights the geopolitical importance of the Arctic region (Figure 12.24). Two of the world's superpowers, the USA and Russia, confront each other over short distances. In addition, Canadian and European interests are prominent in the Arctic.

Tensions between competing powers have been increasing for some time. Claims over vast areas of the Arctic Ocean have been lodged by Russia, Canada and Denmark. A country can claim the sea bed up to a maximum of 280 km beyond its EEZ. This gives a country rights to the non-living resources, including oil, gas and minerals. The critical point is that a country must prove that the sea bed is an extension of its continental shelf. Recently, Russia ceremoniously planted its flag on the submarine mountain chain, the Lomonsov ridge. This attempt to claim rights now lies with the UNCLOS, which will adjudicate.

Currently the **militarisation** of the Arctic is accelerating. All the nations bordering the Arctic Ocean have been investing in military infrastructure designed for Arctic operations.

Figure 12.24 The geopolitics of the Arctic region

Minerals

A significant incentive for countries to focus on the Arctic is its mineral wealth. As warming melts both sea ice and the tundra, exploration is revealing vast reserves. One estimate suggests as much as 90 billion barrels of oil and 47 billion m^3 of natural gas are in the Arctic region. Global transnational corporations in the energy and mining industries are taking seriously the prospect of recovering these vast reserves. This a long-term project dependent on further technological advances and relatively high commodity prices, but it is unlikely to go away.

Transport routes

Connected to mineral exploration is shipping across the Arctic. Until recently the sea ice has made the Arctic Ocean impassable to shipping. Explorers had long tried to find and sail through the North-West Passage (NWP) from the northwest Atlantic to the Pacific Ocean. In 2014 the first cargo ship, unescorted by an ice-breaking vessel, sailed through the NWP. Interest in the Northern Sea Route (NSR) across the Siberian coast is also growing. Shipping companies could reduce their costs significantly by using either of these routes. However, some costs rise, such

as the high fuel consumption of vessels sailing through sea ice. Shipping companies would have to pay more for crew experienced and trained for Arctic conditions, higher insurance and for an escorting ice-breaker.

At the outset, most shipping is likely to transport resources out of the region. The Russian Yamal liquid natural gas plant will be a shipping point for the next few years along the NSR. It will be several years before ships are regularly crossing the Arctic, but if sea ice continues to melt at its current rate, then by 2025, 200–300 ships are projected to operate in the Arctic Ocean.

Managing the Arctic Ocean

Unlike the Antarctic, the Arctic does not have a comprehensive treaty protecting it from economic activities such as mineral and energy extraction. However, the Arctic Council, established in 1996, is acquiring a substantial role in governing the region. The original members are Canada, the USA (Alaska), Russia, Norway, Denmark (Greenland), Sweden, Finland and Iceland. The diversity of indigenous peoples is represented on the Council.

An indication of the growing international interest in the Arctic is the twelve observer states, which include the UK, China and Singapore.

Review questions

1 What is meant by ocean acidification?
2 What is the link between ocean acidification and food security?
3 Describe the environmental conditions coral needs to grow.
4 Outline the process of coral bleaching.
5 What is meant by eustatic change?
6 Describe three causes of rising sea level.
7 Outline strategies that communities living on low-lying islands can adopt to cope with rising sea levels.
8 Explain why sea water freezes in the high latitudes.
9 Why are geopolitical tensions rising in the Arctic Ocean?
10 What might be the impacts of more commercial shipping sailing through the Arctic Ocean?

12.5 How have socio-economic and political factors influenced the use of oceans?

Key idea
→ Oceans have been and continue to be vital elements in the process of globalisation

The importance of oceans in the process of globalisation

Over many centuries oceans have been important as a means of transportation. Both goods and people have travelled over increasing distances. Since the growth of intercontinental air travel, people now tend to travel mainly short distances by sea, for example ferries between the UK and Europe. Freight dominates maritime transport.

Globalisation in the twenty-first century differs significantly from earlier activities such as gold and silver shipped from South America to Europe in the sixteenth century, or the slave trade. Longer and more extensive connections have led to increasing interdependence of people and places. Trading of goods of all sorts criss-crosses the oceans: Australian coal is shipped to Japan, furniture made in China makes its way to Europe, grain produced in the USA and Canada arrives in the UK (Table 12.6). Total world trade has more than trebled to 45 per cent of global GDP since the 1950s.

Global connections affect the everyday lives of billions of people; just a century ago, few people had links beyond their own country. Today there are global brands recognised in most countries and billions of people purchase products made outside their own country.

Oceans have a role in a key aspect of globalisation, known as **time-space compression**. This refers to the world being considered a 'smaller place' as interconnections grow. New technologies have revolutionised connectivity. Developments in ocean transport have increased the speed and reliability of delivery. Globalisation means that oil tankers from the Gulf queue up in the Channel waiting for the Hong Kong-based owners of their cargoes to give permission to head for Rotterdam, when the global price per barrel of oil rises to its most profitable level.

Table 12.6 Growth of global seaborne trade

Year	Oil	Bulk cargoes[1]	Dry cargo[2]	Total cargoes
1970	1,442	448	676	2,566
1980	1,871	796	1,037	3,704
1990	1,755	968	1,285	4,008
2000	2,163	1,288	2,533	5,984
2010	2,752	2,333	3,323	8,408

[1] Iron ore, grain, coal, bauxite, phosphate

[2] Wide range of products e.g. textiles

The pattern of global shipping routes

The principal shipping routes follow a relatively simple pattern (Figure 12.25). An east–west corridor links North America, Europe and Pacific Asia through the Suez Canal, the Strait of Malacca and the Panama Canal. A major route also extends from Europe to eastern South America and then various secondary routes, such as between Brazil and South Africa, add to the pattern.

Factors influencing global shipping routes

Physical geography has a key role to play in the pattern of ocean routes. The shape of coastlines, winds, water currents, water depth, reefs, sea ice and icebergs influence shipping routes. Two major influences on east–west routes was the long detour to the south round either Cape Horn or the Cape of Good Hope.

Impacts of the Suez and Panama Canals

Planned by the French but constructed by the British, the Suez Canal opened in 1869. The Panama Canal, also initiated by the French, was completed and opened in 1914 under USA control (Table 12.7). These engineering projects are two of the most significant maritime 'shortcuts' ever built and have had far reaching consequences for trade and geopolitics.

Table 12.7 Fact file: the Suez and Panama Canals

	Suez Canal	**Panama Canal**
Location	Red Sea to Mediterranean	Atlantic to Pacific
Length	192 km	82 km
Maximum size of ship	150,000 tonnes	65,000 tonnes
Average time to sail through	About 14 hours	About 17 hours
Distance saved	About 8,900 km	About 13,000 km
Sailing time saved	About ten days	About twenty days

Both canals are currently being upgraded to allow more and larger ships to pass through them. The Suez Canal can only accommodate ships travelling in one direction at a time while the Panama Canal has locks, which slow the travelling time.

Stretch and challenge

Investigate what is meant by a 'great circle route'. Explain why they are significant to intercontinental shipping trade. Use examples in your explanation.

The physical geography of the coast is an important influence on port location. Depth of water, tidal range and shelter are key factors. Natural harbours such as Sydney, San Francisco and Singapore are long-established ports. But with increasingly ambitious engineering, harbours can be developed in previously little used locations. For example, extensive engineering such as dredging and dock construction have allowed Europoort to develop at the mouth of the River Rhine in the Netherlands.

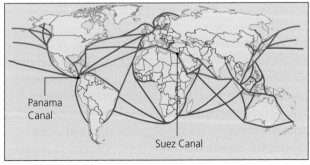

Key
— Core route   — Secondary route

Figure 12.25 Early twenty-first century shipping routes

The direction and type of trade across oceans

Ocean trade connects producers and consumers of raw materials and manufactured goods. Market size exerts a strong influence on the volume of shipping visiting a port. For example, Europoort serves a large part of western and central Europe. Total population is an indication of market size but an imperfect one.

Income levels are also significant as they affect the ability to purchase goods. In addition the type of goods traded influence the volume and direction of ocean trade. For example, there are clear patterns in the trade in unprocessed primary products such as crude oil, mineral ores and agricultural products compared to manufactured goods such as vehicles, electronic equipment and clothes.

 Activities

Study Tables 12.8 to 12.13, which show the percentage of world trade in various categories of goods.

Table 12.8 Leading exporters of agricultural products (2014)

Country	Value (million US$)	% share of world trade
EU	613,000	37
USA	172,000	10
Brazil	86,400	5
China	66,175	4
Canada	62,800	4
Indonesia	45,000	3
Argentina	43,150	3
India	42,400	3
Thailand	42,000	3
Australia	38,400	2
Malaysia	34,000	2

Table 12.9 Leading importers of agricultural products (2014)

Country	Value (million US$)	% share of world trade
EU	623,000	36
China	157,000	9
USA	142,000	8
Japan	94,000	5
Russian Federation	42,000	2
Canada	38,000	2
South Korea	33,700	2
Saudi Arabia	29,300	2
Mexico	27,000	2
India	25,700	1
Hong Kong, China	25,000	<1

Table 12.10 Leading exporters of fuels and mining products (2014)

Country	Value (million US$)	% share of world trade
EU	682,400	16
Russian Federation	377,300	9
Saudi Arabia	325,600	8
USA	187,200	5
Australia	160,300	4
Canada	149,000	4
United Arab Emirates	129,200	3
Norway	121,000	3
Qatar	116,000	2
Kuwait	112,300	2
Nigeria	104,000	2

Table 12.11 Leading importers of fuels and mining products (2014)

Country	Value (million US$)	% share of world trade
EU	1,263,000	30
China	534,000	13
USA	486,000	12
Japan	361,000	8
South Korea	227,300	5
India	210,300	4
Singapore	130,000	3
Taiwan	89,000	2
Turkey	68,100	2
Canada	64,800	2
Germany	60,100	2

Table 12.12 Leading exporters of manufactured goods (2014)

Country	Value (million US$)	% share of world trade
EU	4,385,000	38
China	1,925,000	17
USA	1,102,000	10
Japan	709,600	6
South Korea	462,600	4
Hong Kong, China	423,200	4
Singapore	283,000	2

Table 12.13 Leading importers of manufactured goods (2014)

Country	Value (million US$)	% share of world trade
EU	3,905,000	33
USA	1,618,000	14
China	1,060,000	9
Hong Kong, China	489,000	4
Japan	418,000	3
Canada	340,000	3
Mexico	290,000	2
South Korea	257,400	2
Russian Federation	253,100	2

1 On an outline world map, represent the percentage share figures for each category of trade. Use a located graphical method such as a bar graph.
2 Describe and suggest reasons for the patterns.
3 The figures are for *total trade*, not just ocean trade. Suggest the role that ocean transport is likely to play in the trade for each category of product.

Marine technology: a revolution in transport

Technological changes in ocean shipping, including ports, have revolutionised movement of all types of products by sea. The huge increase in the sizes of ships over the past few decades is unprecedented (Figure 12.26).

Containerisation is fundamental to the process of globalisation. By using standard-sized metal boxes to move a wide variety of goods, costs have been reduced dramatically. The elimination of item by item, or 'loose cargo' handling reduces costs at every stage of a journey from factory to final distribution centre. That so many containers can now be moved by one ship adds to the **economies of scale**. Ocean freight rates have reduced as a consequence, and it is not just costs which are saved; the time taken for goods to travel around the globe has greatly reduced. Loading and unloading containers is highly mechanised and benefits from every single container having its own unique code. Computers track each container allowing the logistics of distribution to be very efficient.

Bulk carriers of goods such as oil, mineral ores and grains have also increased in size and achieved similar scale economies. The largest oil tankers carry some 3 million barrels of oil, equivalent to some 440,000 tonnes. Iron ore carriers can be nearly as big at 400,000 tonnes (see also pages 199–203).

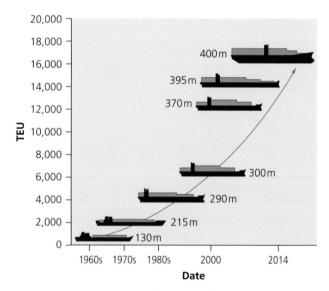

Figure 12.26 Increasing size of container ships in metres. TEU: 20-foot equivalent unit – the measure of the number of containers carried by a ship

Although oceans are primarily highways for goods, cruise ships have also been a growing presence. The largest of these are up to 225,000 tonnes, 360 metres long and can carry around 6,000 passengers and 2,300 crew. Unsurprisingly, this class of vessel is the most expensive ever constructed.

When large vessels come to port, the scale of the handling facilities matches that of the ships.

Table 12.14 Port of Singapore – vital statistics per year

Factor	Statistic
% world's containers handled	20
% world's crude oil handled	50
Number of ships docked	130,000
Number of containers handled	About 33 million

Until 2005, Singapore was the world's busiest port in terms of total tonnage handled. It is now second after Shanghai. However, Singapore remains the world's largest trans-shipment port in terms of goods in and out (Table 12.14).

The importance of ocean-going vessels in increasing the interconnectivity of the world cannot be underestimated. Although much is rightly made of the impact of the internet, a strong argument can be made for the revolution in ocean transport as the driver of globalisation.

Submarine cables: unseen connections

Oceans can also be crossed below the surface (Figure 12.27). The first submarine cables were laid in the second half of the nineteenth century, notably the first trans-Atlantic telegraph cable in 1866. By the early twentieth century a global telegraphic network had been laid. Telephone cables were laid during the 1950s. Today, fibre optic cables criss-cross the oceans forming an essential part of the globalised telecommunications network. The internet could not exist without these cables.

When laying a new cable route, several cables are laid at the same time so that if one were to fail, traffic can be swiftly re-routed. In the past two decades, the greatest amount of cable laying has been in Pacific Asia. A submarine link across the Arctic is also being developed between locations such as London and Tokyo for example.

> **Activities**
>
> 1 Describe the pattern of submarine cable networks.
> 2 Suggest reasons for this pattern.

> **Key idea**
>
> → Oceans are important spaces where countries challenge each other

Rival countries have long argued about the 'right' of their ships to sail freely across the seas and to have access to ocean resources. The establishment of sea areas in international law such as the 'exclusive economic zone' has not been straightforward. In some locations marine boundaries continue to be disputed. Geopolitics can be especially contentious in ocean settings. The Arctic is likely to witness increasing tensions as the region warms and the Arctic Ocean becomes more accessible.

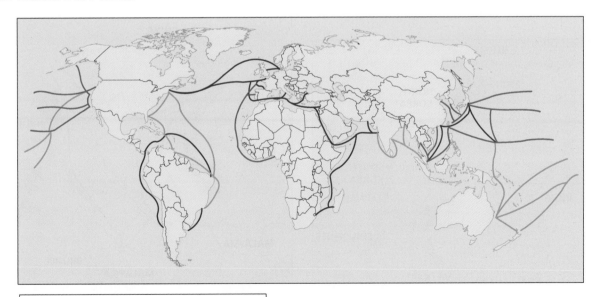

Key

— Higher speeds (> 3,000 gigabytes/second)
— Medium speeds (2,000–3,000 gigabytes/second)
— Lower speeds (< 2,000 gigabytes/second)

Figure 12.27 Main global submarine cable network

Over the centuries, gaining naval supremacy has been important to the political and economic ambitions of countries. Sometimes a clash at sea has been a pivotal moment in a nation's history. The defeat of the Spanish Armada in 1588, the Battle of Trafalgar in 1805 and the Battle of the Atlantic during the Second World War, can all be seen as turning points in the course of British history. Today naval power remains a significant element for many countries wishing to preserve or expand their spheres of influence. A **superpower** such as the USA and emerging powers such as India and China either maintain or are growing their ability to exert power via the oceans.

China's growing naval power

China has for centuries been a naval power within Asia. In recent years, significant investment has been made into its naval capabilities. This is part of China's emerging role both regionally within Asia and the Pacific but also globally. Economic strength and increasing technological expertise are allowing the Chinese navy to expand and modernise at a rapid rate.

China's coastline extends for 14,500 km, which ranks it tenth in the world simply in terms of length. There are four primary domestic naval ports covering the north, central and southern stretches of coastline. It is significant that China has been establishing bases in other countries, which allow its navy to operate further away from home. The term **blue water navy** describes one that is able to operate away from

its home bases. That China is extending its naval military capacity is a strong indication of its emerging superpower status (Figure 12.28).

The spread of Chinese naval power into the Indian Ocean is of increasing concern for India. These concerns are made more serious for India because potential Chinese naval bases are in countries such as Pakistan, with which India does not necessarily have good relations. The level of infrastructure in each of these overseas bases, however, is not developed enough to support substantial Chinese military power. But China's developing commercial interests in the trade across the Indian Ocean, such as that originating from the Persian Gulf and the Suez Canal, and its involvement in Africa, mean that its is likely to retain a strong strategic interest in the Indian Ocean.

Marine conflict: the South China Sea

For centuries, various countries have argued, and on occasions fought, over territory in the South China Sea. Two island chains, the Paracels and the Spratlys, as well as areas of adjacent sea are claimed either in whole or in part by various countries (Figure 12.29). The island chains include many rocky outcrops, coral atolls and reefs such as Scarborough Shoal, and sandbanks. China claims the most ocean as defined by the 'nine-dash line'. This claim is based on historical factors which Vietnam disputes. Taiwan mirrors China's

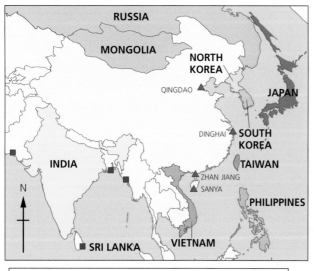

Key

▲ Chinese naval base
■ Ports receiving substantial Chinese investment in its facilities

Figure 12.28 China's naval bases

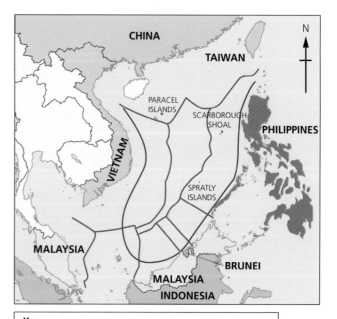

Key

—— China's claimed territorial borders
—— UNCLOS 200 nautical mile exclusive economic zone

Figure 12.29 Marine conflict in the South China Sea

claim while the proximity to the Spratlys of the Philippines archipelago is the basis of its claim. Other countries also involved in claiming some of the South China Sea are Malaysia and Brunei. The attractions of the islands include reserves of oil and gas under the sea bed while trade routes passing through the region are important to China's economy.

Recent incidents in the South China Sea

- 1974 and 1988 – armed clashes between China and Vietnam over the Paracels and Spratlys; some 130 Vietnamese military personnel killed
- 2012 – China and Philippines accuse each other of incursion in the Scarborough Shoal
- 2012 – China formally creates Sansha City in the Paracels to administer Chinese territory in the region; Vietnam and the Philippines protest
- 2013 – Philippines challenges legality of China's actions under the UN Convention on the Law of the Sea
- 2014 – China sets up a drilling rig near to Paracels; multiple collisions occur between Chinese and Vietnamese vessels
- 2015 – US satellite and spy plane reconnaissance shows China building infrastructure on some of the Spratly Islands, e.g. an airstrip.

The USA has significant strategic interests in the region. The region, which is worth US$1.2 trillion to the US economy, is a very important trading location for the USA. The USA also has several long-standing allies such as South Korea and Taiwan whose security it needs to safeguard.

Militarily, the US Seventh Fleet based in Japan is by far the strongest in the region. However, the chance of direct confrontation between the USA and China is low if past tensions are anything to go by, even though the USA has not signed up to the UNCLOS agreement. It only voluntarily agrees to its principles and so this weakens the argument for a legal and rules-based solution to disputes in the South China Sea.

> Key idea
> → Oceans present hazardous obstacles to human activities

Over the centuries oceans have been used as escape routes for some despite the dangers of sea crossings. Numerous individuals and groups have escaped persecution by taking to the sea. The Pilgrim Fathers sailed across the North Atlantic to escape religious persecution in the early seventeenth century. Today,

significant numbers of refugees use sea crossings as part of their attempt to find security. Large numbers of refugees use the Mediterranean to cross from Libya into Italy and Turkey into Greece.

But oceans have also been used by some to attack and threaten others. Novels and films often portray pirates in rather romantic ways, but both in the past and today, piracy can be a savage and potent threat.

The threats to shipping from piracy

Piracy has various legal definitions but essentially it is 'the act of boarding any vessel with intent to commit theft or any other crime, and with an intent or capacity to use force in furtherance of that act'. Piracy takes place not just out at sea but also close to shore and in harbours.

With the rise in transcontinental shipping associated with globalisation has come a rise in piracy. This has occurred recently and more than a century after the virtual disappearance of marine pirates. And as with piracy in the past, the geography of trading routes helps describe where most piracy takes place (Figure 12.30).

There are two areas where significant and recent piracy activities have occurred: the western Indian Ocean and Southeast Asia. The former region includes the strategically important route through the Gulf of Aden and Red Sea, which leads to the Suez Canal. In Southeast Asia, the Malacca and Singapore Straits are maritime 'choke points' adding to the vulnerability of shipping. The Gulf of Guinea in West Africa is the other main location where piracy has been taking place.

Pirates have tended to attack large vessels such as bulk carriers, container and cargo ships. The ship and its crew have then been held for ransom.

There is a clear seasonal pattern to attacks in the western Indian ocean (Figure 12.31). The monsoon seasons, and in particular the summer one, register fewer incidents. This is due to the stronger winds of the monsoons creating rough sea conditions. The smaller boats from which pirates tend to operate are difficult to control in stormy conditions when coming alongside a large vessel and boarding is extremely hazardous.

The threat posed by the sharp increase in attacks since the end of the twentieth century resulted in international action. A substantial maritime coalition involving the EU, NATO, the USA and a host of countries such as Russia, India, China, Japan and South Korea patrolled the most vulnerable areas. As a result of intense military surveillance, attacks in the Indian Ocean have reduced significantly in the past few years.

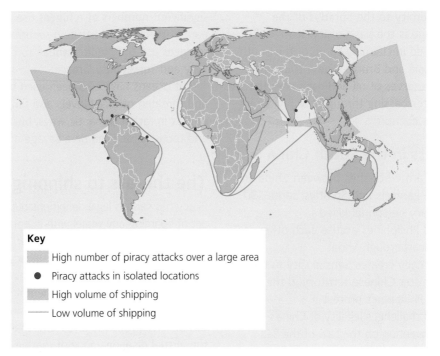

Figure 12.30 Main trading routes and locations of acts of piracy

Military deterrence has also been stepped up in the Gulf of Guinea as 'petro-piracy' has increased. Tankers, drilling platforms and supply vessels have become targets.

The reasons behind the rise in piracy are complex. It is tempting to relate it solely to the relative poverty of the countries like Somalia where the pirates have their bases. Certainly this has played a part, but other factors operate as well, such as

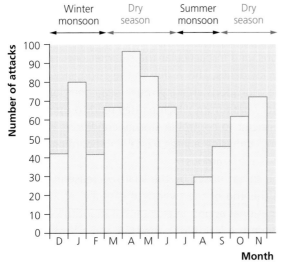

Figure 12.31 Monthly distribution of pirate attacks, western Indian Ocean

dysfunctional government. Piracy often has strong links to organised crime networks. The 'profits' from piracy are channelled (laundered) through various economic systems, such as land and house purchases. The fishermen of Somalia link the growth of piracy to the loss of their traditional fisheries due to industrial-scale fishing in the area by boats from outside the region. As well as sea-based policing, efforts are being made to help governments function more effectively and to offer training opportunities to younger men to divert them from criminal activities.

Ocean routes for refugees

There are many reasons why groups and individuals wish to leave one location and move permanently to another. It is often a balance of positive and negative feelings about both the source and destination locations, which potential migrants take into account. When the balance shifts in favour of a potential destination, migration occurs.

Economic reasons figure prominently in many moves, with migrants driven by a desire to achieve a higher standard of living. There are, however, a range of reasons forced upon people. Political and religious persecution may pose direct threats to safety and cause people to take great risks to escape.

A **refugee** is 'a person who owing to a well-founded fear of being persecuted for reasons of race, religion, nationality, membership of a particular social group, or political opinion, is outside the country of his or her nationality' (UN Refugee Convention). In addition, the term is given to those who flee a location due to environmental disasters such as droughts and floods.

It is not just countries with a coastline that supply refugees who use the oceans as escape routes; often sea-borne refugees come from landlocked countries, hundreds of kilometres from the ocean.

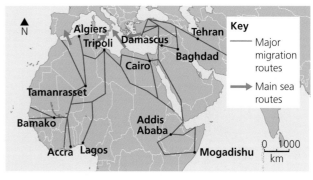

Figure 12.32 Refugee routes using the Mediterranean

Refugees by boat in the Mediterranean and Asia

Migrants crossing the Mediterranean from Africa to Europe are not new. The relatively wealthy countries of western Europe have long attracted migrants from North Africa, especially if strong historic links exist between countries such as Algeria and France and Morocco and Spain. What is a more recent phenomenon is the smuggling of people on a large scale. Continuing political instability and a lack of progress in raising standards of living mean that in several countries, traffickers find plenty of desperate people willing to pay substantial sums of money to be smuggled.

Refugees head towards the Mediterranean from a number of different countries in sub-Saharan Africa and the Middle East (Figure 12.32), many as the result of a surge in armed conflict, for example Syrians fleeing civil war or Eritreans escaping from a country with one of the worst human rights records in the world. By the time some migrants have reached the Mediterranean they have already survived the crossing of the Sahara.

Refugees travelling by boat target various destinations. Australia is often the intended destination for Vietnamese boat people or Tamils fleeing from Sri Lanka. People leaving Bangladesh and Myanmar tend to travel south, ending up in Thailand, Indonesia and Malaysia.

Taking advantage of the desperation of these refugees, people traffickers control the smuggling routes to the coast and on to boats. The perils of the sea crossings and the poor physical condition of the boats have resulted in many migrant deaths.

As with piracy, the response to the marine-based issue of 'boat people' is international. Global governance is playing an increasing role in dealing with both the causes and the consequences of the flows of refugees using the oceans as escape routes. The United Nations High Commissioner for Refugees (UNHCR) and the EU, for example, try to manage the flows of forced migrants. Maritime patrols in the Mediterranean rescue refugees adrift in un-seaworthy boats and resettlement programmes try to avoid refugee camps becoming permanent homes for displaced people (see also pages 218–19).

Review questions

1 What is meant by the term 'time-space compression'? Use examples from this chapter to illustrate your answer.
2 Describe the global pattern of principal ocean trading routes.
3 For either the Suez or Panama Canal, describe the impacts the canal has on ocean shipping.
4 How has technology changed the way goods are transported by ocean shipping?
5 What is the significance of possessing a 'blue water navy' to a superpower?
6 Why is there such geopolitical tension in the South China Sea?
7 Describe and explain the location of twentieth-century piracy hot spots.
8 Why has the Mediterranean become a focus for refugees travelling by boat?

Practice questions

A Level

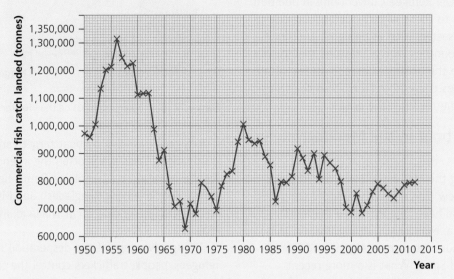

Figure A Commercial fish catch landed in USA ports along the Atlantic coast, 1950–2012

Section A

1 Study **Figure A**. Identify **three** limitations with the data evidence in Figure A. [3 marks]

2 Explain the level of biodiversity in a salt marsh (inter-tidal) ecosystem. [6 marks]

Section B Synoptic questions

3 Examine how changes in the carbon cycle may be impacting on coral reefs. [12 marks]

Section C

4 Assess the extent to which countries use oceans to exert their influence regionally and globally. [33 marks]

AS Level

Section A

1 a) Explain the process of the acidification of ocean water. [4 marks]

b) With reference to **one** biological resource, suggest why its management can be challenging. [6 marks]

c) Study **Table 1**, which shows changes in the level of dissolved oxygen in the Pacific Ocean, south of California.

Table 1

Depth below ocean surface (metres)	0	500	1000	1500	2000	2500	3000	3500	4000
Dissolved O_2 (ml/l)	6.0	0.9	0.5	0.9	1.3	2.0	2.3	2.5	2.2

  i) Using the data in Table 1, calculate the mean and the median level of dissolved oxygen. You must show your working. [4 marks]

  ii) Using evidence from Table 1, analyse why oceanic biological productivity is concentrated close to the surface. [6 marks]

d) Assess the extent to which it is possible to manage pollution from human activities in marine environments. [12 marks]

Section B Synoptic questions

2 a) With reference to Figure 12.10 on page 376, suggest how the Antarctic ecosystem influences the region's sense of place. [8 marks]

b) Examine how the exploitation of wave and tidal energy affects place profiles. [8 marks]

Section C

3 To what extent do countries use oceans to exert their influence? [20 marks]

Chapter 13

Future of food

The question of how humankind will feed itself during the twenty-first century is one that affects us all. The global mismatch between demand and supply of food has led to startling inequalities among the world's populations. Hunger and obesity exist across the development spectrum. While there is a consensus that we can produce enough food, this on its own does not mean that everyone will enjoy a healthy and balanced diet. The future of food will be determined by the efficient functioning of the global food system and by the physical, social, economic and political factors that affect it.

13.1 What is food security and why is it of global significance?

> **Key idea**
> → The concept of food security is complex and patterns of food security vary spatially

Defining what it means to be food secure

A widely used definition of food security comes from the United Nations' Food and Agriculture Organization (FAO):

'Food security exists when all people, at all times, have physical and economic access to sufficient, safe and nutritious food that meets their dietary needs and food preferences for an active and healthy life.'

From this definition the World Food Programme (WFP) identifies the 'three pillars' of availability, access and utilisation. The FAO identifies, in addition, stability (Table 13.1).

Food security analysts have defined two general types of food security: chronic and transitory (Figure 13.1).

Food security and the Millennium Development Goals

At the beginning of the new millennium world leaders at the UN agreed on eight goals – the Millennium Development Goals (MDGs) – to fight poverty between 2000 and 2015. MDG 1 aimed to 'eradicate extreme poverty and hunger'. Progress in shown in Figure 13.2.

While the proportion of undernourished people in developing regions fell between 1991 and 2015, from 23.3 per cent to 12.9 per cent, progress was uneven across regions and countries.

Latin America, the east and southeastern regions of Asia, the Caucasus, Central Asia and the northern and western regions of Africa have all made fast progress. Progress was recorded in southern Asia, Oceania, the Caribbean and southern and eastern Africa, but at too slow a pace to reach the MDG 1 target of halving the proportion of chronically undernourished.

Table 13.1 FAO Dimensions of food security (based on FAO Dimensions of food security)

Physical **AVAILABILITY** of food	This addresses the 'supply side' of food security and is determined by the level of food production, stock levels and net trade.
Economic and physical **ACCESS** to food	An adequate supply of food at the national and international level does not in itself guarantee household level security. Concerns about insufficient food access have resulted in a greater policy focus on incomes, expenditure, markets and prices in achieving food security.
Food **UTILISATION**	This is the way the body makes the most of various nutrients in food. Sufficient energy and nutrient intake is the result of good care and feeding practices, food preparation, diversity of diet and intra-household distribution of food.
STABILITY of the three dimensions over time	Even if food intake is adequate today, food insecurity remains if access to food is inadequate on a periodic basis. Adverse weather, political instability or economic factors (unemployment or rising food prices) may impact food security status.

	CHRONIC FOOD INSECURITY	TRANSITORY FOOD INSECURITY
is...	long term or persistent.	short-term and temporary.
occurs when...	people are unable to meet their minimum food requirements over a sustained period of time.	there is a sudden drop in the ability to produce or access enough food to maintain a good nutritional status.
results from...	extended periods of poverty, lack of assets and inadequate access to productive or financial resources.	short-term shocks and fluctuations in food availability and food access, including the year-to-year variations in domestic food production, food prices and household incomes.
can be overcome with...	typical long-term development measures also used to address poverty, such as education or access to productive resources, such as credit. They may also need more direct access to food to enable them to raise their productive capacity.	different capacities and types of intervention, including early warning capacity and safety net programmes. Transitory food insecurity is relatively unpredictable and can emerge suddenly. This makes planning and programming more difficult.

Figure 13.1 The duration of food security (Source: FAO)

In West Africa, Southeast Asia and South America, undernourishment declined faster than the rate of children underweight, suggesting room for improvement in hygiene, access to clean water and the 'quality' of diets, especially in the poorest populations.

Current trends in global food security

A variety of measures and data sources exists to measure food security.

The Global Hunger Index

The Global Hunger Index (GHI) measures hunger at global, regional and national scales. Calculated each year by the International Food Policy Research Institute (IFPRI), the GHI is designed to raise awareness of geographical differences in hunger and how these change over time. It combines four equally weighted indicators:

1 Undernourishment – as a percentage of the population.

2 Child wasting – the proportion of children below the age of five with low weight for height, reflecting acute undernutrition.
3 Child stunting – the proportion of children below the age of five with low height for their age, reflecting chronic undernutrition.
4 Child mortality – the mortality rate of children under the age of five.

The GHI ranks countries on a 100-point scale: zero is the best score – no hunger – and 100 is the worst. The most recent data for 2015 show that hunger has fallen (Figure 13.3). However, there remain individual countries where hunger is still serious or alarming.

National pattern
Between 1990 and 2013, 26 countries reduced their GHI score by more than 50 per cent, but 17 achieved reductions of less than 25 per cent (Table 13.2).

	Africa		Asia					Latin America and the	Caucasus and
	Northern	sub-Saharan	Eastern	Southeastern	Southern	Western	Oceania	Caribbean	Central Asia
Reduce extreme poverty by half	low poverty	very high poverty	low poverty	moderate poverty	high poverty	low poverty	–	low poverty	low poverty
Productive and decent employment	large deficit	very large deficit	moderate deficit	large deficit	low deficit	large deficit	very large deficit	moderate deficit	large deficit
Reduce hunger by half	low hunger	high hunger	moderate hunger	moderate hunger	high hunger	moderate hunger	moderate hunger	moderate hunger	moderate hunger

The text in each box indicates the present level of development.
The colours show progress made towards the target according to the key below:

■ Target met or excellent progress   ■ Good progress   ■ Poor progress or deterioration   ■ Fair progress   ■ Missing or insufficient data

Figure 13.2 Millennium Development Goal 1: Eradicate extreme poverty and hunger, 2015 progress chart (Source: UN)

Table 13.2 Global Hunger Index (GHI) scores

Best performers	Worst performers
• Kuwait – invasion by Iraq in 1991 was responsible for unusually high scores in the past. • Thailand – high levels of economic growth. • Vietnam – high levels of economic growth. • Ghana – political stability and government investment. • Mexico – high levels of economic growth.	• Swaziland – HIV and AIDS epidemic. • Iraq – conflict and the burden of Syrian refugees. • Burundi – decades of economic decline. • Eritrea – conflict. • Sudan – drought and conflict.

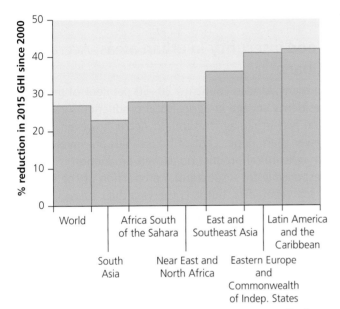

Figure 13.3 Progress in reducing Global Hunger Index (GHI) scores

Poverty is clearly one of the most important constraints to reaching national food security goals, while political stability and economic growth promote food security. Countries that experience chronic food insecurity are generally those suffering problems such as prolonged military conflict, economic stagnation and decline, and environmental degradation.

The Global Food Security Index

The Global Food Security Index provides a worldwide overview of countries most and least vulnerable to food insecurity. It combines 28 indicators which cover three key areas: the affordability, the availability and the quality and safety of food.

> ### Activities
>
> 1 Describe the pattern shown in Figure 13.4.
> 2 For the category 'needs improvement', what factors could explain the pattern?
> 3 Identify one country within the 'needs improvement' category and research the obstacles to food security.

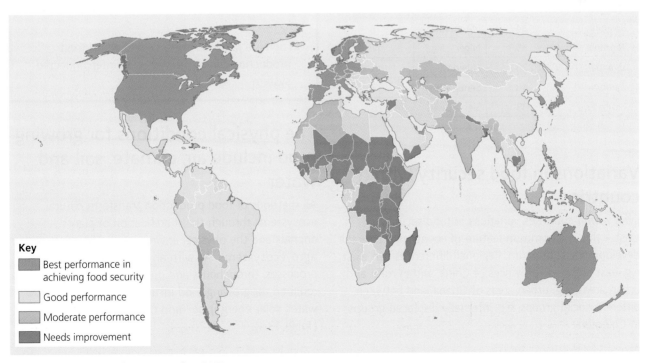

Key
- Best performance in achieving food security
- Good performance
- Moderate performance
- Needs improvement

Figure 13.4 Global Food Security Index, 2015

Variations in food security within countries

Even *within* countries variations in food security exist – this is a common feature of countries across the development continuum. Regional differences can exist, e.g. east and west provinces of China, within rural and urban areas, e.g. urban areas of Ghana, and between different social groups, e.g. internally displaced groups in Colombia.

The east and west provinces of China

Despite its economic growth China has the second largest number of poor people in the world, after India. There are significant differences in food production across China leading to food deficit and food surplus regions. Out of 31 provinces, 9 have been classified as food insecure (including Tibet and Yunnan). In these areas 60 per cent of people consume less than the recommended target of grain products which form the staple diet.

Food insecurity in urban areas, Accra, Ghana

In many African countries 20–50 per cent of urban residents engage in agricultural production (urban and peri-urban agriculture). In Accra, however, fewer than 15 per cent of households are involved in agricultural production. Urban production is responsible for a substantial proportion of the city's fresh vegetables, but it is the wealthier classes that can afford them.

Vulnerable households often rely on cheaper, less-preferred foods or ration money for the purchase of street foods.

Food insecurity among the internally displaced people of Colombia

Colombia has been plagued by 50 years of armed internal conflict, 6.2 million people having been internally displaced as a result – 95 per cent of internally displaced people are food insecure.

> **Key idea**
> → Food is a precious resource and global food production can be viewed as an interconnected system

The physical conditions for growing food include air, climate, soil and water

All methods of food production transform natural ecosystems through the modification of plants, animals and the physical environment. The result is **agro-ecosystems**. As with any system there are inputs, processes, throughputs and outputs. The physical inputs required for growing food include air, climate (heat, water, solar energy), soil and mineral nutrients (Table 13.4).

Table 13.4 The physical conditions required to grow food

Temperature	Light	Water	Air	Soil
Crops can grow at below optimum temperatures; however, temperatures that are too high or too low result in reduced yields. Tropical crops such as rice require temperatures between 16°C and 27°C, temperate crops such as wheat grow at optimum mean daily temperatures of between 15°C and 20°C.	Photosynthesis uses sunlight. Plants differ in light requirements – light intensity and duration are important for crop growth.	Water comprises 80 per cent of living plants and is a major determinant of crop productivity and quality. Water is essential for the germination of seeds and crop growth. In terms of biological functions, water is used in the photosynthesis to produce sugars from light energy and water acts as a solvent and means of transport for minerals and sugars throughout the plant.	Photosynthesis involves the absorption of CO_2 from the atmosphere and release of O_2. Plants also require some oxygen for respiration to carry out their functions of water and nutrient uptake. Some plants fix nitrogen from the atmosphere.	Soils are the mixture of mineral and organic matter in which plants grow. They supply water, nutrients and material in which root systems can develop. Plants absorb essential minerals mostly through their roots: the main ones are nitrogen, phosphorus, potassium and calcium.

Feeding the world is a complex system of growing, processing, transporting and disposing of consumer waste

The global food system is made up of a complex sequence of chains through which societies gain their food supply. Farming dominates the production side and starts the chain. Beyond this there are a number of **downstream functions** through which food passes before reaching the consumer (Figure 13.5).

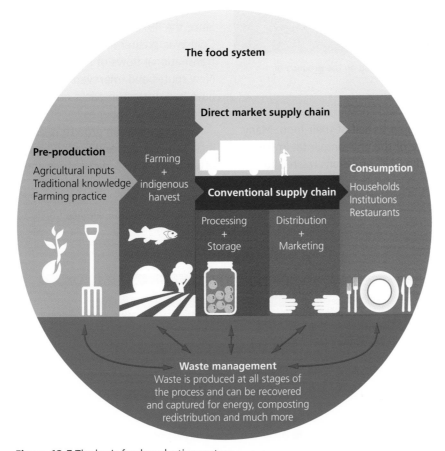

Figure 13.5 The basic food production system

Waste food

Waste food is an output of the system and can occur upstream at the production end when crops fail or do not pass stringent quality control measures, as well as at the consumer end (Figure 13.6).

Figure 13.6 Waste vegetables rejected on the grounds of quality control

Farmers, supermarkets and consumers in the developed world discard up to half of their food. Many consumers are unaware of the environmental impact of deforestation and the GHG emissions related to producing food that will never be eaten. Nor is the issue of waste food confined to the developed world (Figure 13.7); in developing countries crops often rot because farmers lack efficient means to transport, process and store food or because they do not pass quality control measures.

Food production methods vary from intensive to extensive and subsistence to commercial

Food is produced through a wide range of farming systems (Figure 13.8). Any one farming system will incorporate several different features. For example, market gardening in the Vale of Evesham in the UK could be described as temperate, sedentary, owner-occupied, capital-intensive, labour-intensive and commercial. Table 13.5 describes the main types of farming systems.

> **Key idea**
> → Globalisation is changing the food industry

The influence of globalisation on the food industry

Since the 1970s, and especially since the end of the Cold War, greater interconnectedness has increased transnational flows of people, goods and information. New routes and improved access to global food sources have had distinct impacts on international trade in food, diets and societies.

World population growth is one factor that has increased the demand for food and this growth has often been fastest in countries and regions where it is

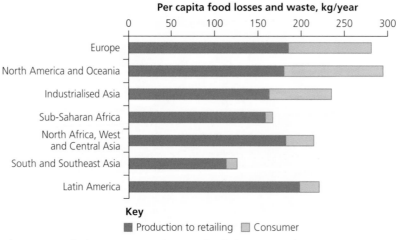

Per capita food losses and waste, kg/year

Key
■ Production to retailing ■ Consumer

Figure 13.7 Which regions waste the most food? (Source: FAO)

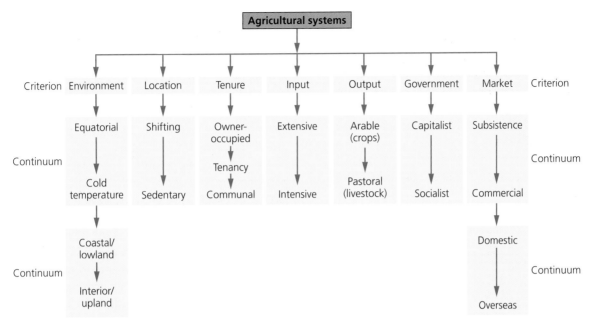

Figure 13.8 Criteria by which farming systems are classified

Table 13.5 Food production methods

Arable and pastoral	Subsistence and commercial	Shifting and sedentary	Extensive and intensive
Arable – the growing of food crops, often on fairly level, well-drained soils, are of good quality. **Pastoral** – the raising of livestock, often in areas unsuitable for arable farming (too cool, wet or dry or too steep). Soils often have limited fertility. Livestock farming is sustainable only when the **carrying capacity** of the area is not exceeded.	**Subsistence** – provision of food by farmers for their own consumption and for the local community. Subsistence farmers are vulnerable to food shortages due to the lack of capital and other entitlements. **Commercial** – farming for profit, often on a large scale with high capital inputs.	**Shifting cultivation** – confined to a few isolated places with low population density, large areas of land and limited food demands (e.g. indigenous groups in tropical rainforests). The system, which is sustainable at low population densities, is essentially a rotation of fields rather than a rotation of crops. **Sedentary** – farmers remain in one place and cultivate the same land year after year.	**Extensive** – large-scale commercial farming. Inputs of labour and capital are small in relation to the area farmed. Yields per hectare are low but yields per capita are high. **Intensive** – small-scale with high labour and/or capital inputs and high yields per hectare.
Examples: Arable farming – the Nile Valley, the Great Plains. Pastoral farming – hill sheep farming in Wales, nomadic herding in the Sahel.	Examples: Subsistence – wet-rice farming in India. Commercial – cattle ranching in South America; oil palm plantations in Malaysia.	Examples: Shifting cultivation – the Amazon Basin and Indo-Malayan rainforest. Sedentary – dairy and arable farming in the UK.	Examples: Extensive farming – Canadian Prairies cereal farming. Intensive – horticulture in the Netherlands.

most difficult to grow food, e.g. Sudan, Burkina Faso, Ethiopia and western Sahara.

Changing global tastes mean that in many regions consumers do not have to wait for seasonality of foodstuffs because they can be sourced from across the globe, for example year-round fruit and an abundance of choice for 'exotic' foods which have entered the diet in ACs.

Globalisation of the food industry causes a number of issues

The globalisation of the food industry raises several issues, including food miles, inequalities between TNCs and small food producers, obesity and food prices.

Food miles

Food miles indicate how far food has travelled from producer to consumer. Owing to globalisation, an increasing proportion of food products is being transported over long distances. It is estimated that supermarket food travels an average 2400 km before it arrives on supermarket shelves. The broadening of food tastes and a desire by consumers for year-round supplies of fresh fruit and vegetables have meant that food is transported vast distances, and most often by air (Table 13.6).

Table 13.6 UK average food miles of selected food products

Food	Source	Distance (miles)
Apples	USA	10,133
Asparagus	Peru	6,312
Pears	Argentina	6,886
Grapes	Chile	7,247
Lettuce	Spain	958
Strawberries	Spain	958
Broccoli	Spain	958
Spinach	Spain	958
Red peppers	Holland	62
Chicken	Thailand	6,643
Brussels sprouts	Australia	10,562

Sourcing food products on an intercontinental scale has implications for greenhouse gas (GHG) emissions and the environment. However, there are instances where imports are less environmentally damaging than home-grown products. Lettuces grown in winter in greenhouses in the UK create higher GHG emissions than those grown outdoors in Spain in winter and transported to the UK.

Inequality between TNCs and small suppliers

The growth in agri-food systems across the world is reflected in a shift from traditional food systems and small-scale farming towards agribusiness, TNCs and major food retailers. National food systems are increasingly linked through patterns of trade and investment, referred to as **global value chains**.

TNCs control the terms by which farmers can participate in the food system. They often favour large, capital-intensive growers, leaving small producers disadvantaged and marginalised.

In Brazil, India, Mexico and South Africa, **Foreign Direct Investment** (FDI) by large TNCs has reduced the power of national governments to regulate their own food systems. In today's interconnected world, decisions are often made globally, with individual governments in EDCs and LIDCs limited in their ability to promote the interests of their small farmers. In Brazil, large agribusiness firms account for 62 per cent of the value of agricultural production and the majority of food exports. There the government has adopted specific policies to help small farmers, such as credit programmes and the purchase of food to be used in food assistance.

Obesity

Globalisation has had a major influence on dietary patterns. As countries develop and affluence increases, consumption shifts from cereals towards foods that are more expensive, such as meat and dairy products, and more processed. In addition, the number of fast-food outlets promoted by TNCs increases. One of the leading examples of dietary change is China. In the past 30 years the consumption of meat products in China has increased six-fold (Figure 13.9). Along with

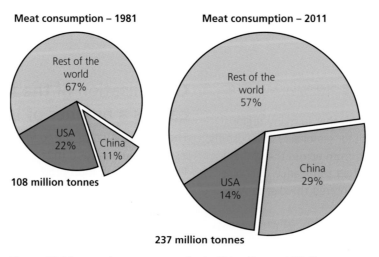

Figure 13.9 Increased meat consumption in China (Source: USDA)

Figure 13.10 International fast food (Source: www.worldmapper.org) This cartogram – part map, part pie chart – keeps the continents in the same place while changing their size to reflect the value of the variable – in this case, fast-food giants. For continental base map visit www.worldmapper.org

many parts of the developing world, China is a major investment focus for fast-food retailers (Figure 13.10): it accounts for half the total revenue of Yum brands (6400 outlets) and has the largest number of KFC outlets in any country (4500). McDonald's has 2000 outlets and plans a further 300 by the end of 2015. Supporting the fast-food retail industry is a vast food-processing enterprise delivering low-cost, consistent products.

McDonald's

Cold-storage, logistics and food-processing companies supporting the fast-food industry have followed McDonald's to China. They mainly employ ex-pat labour, while the retail outlets depend on local labour.

KFC

KFC (Figure 13.11) has developed local supply chains and has adapted its products to local tastes with Peking Duck rolls, rice porridge (congee) and fried dough sticks.

Figure 13.11 KFC in China

Such changes in diet have led to increased levels of obesity, particularly among the rising middle classes of EDCs. In Brazil, the number of McDonald's restaurants grew 380 per cent between 1993 and 2002.

In the Philippines, cultural influences from the developed world, transmitted through television, advertising, tourism and education, have led to changes in lifestyle and food consumption patterns. Dietary changes include an increase of more than 50 per cent in imported bread and bakery products, a decrease in vegetable consumption, and an increase in fried street foods. While there is still malnutrition in the poorest areas, the number of overweight children in the Philippines has doubled.

Price crises

The FAO produces a Food Price Index, which is a measure of the monthly change in international prices of a basket of food commodities. Compared with 2011, the index shows a declining trend, although there have been some small fluctuations, e.g. in 2012.

However, global food prices are extremely volatile and vulnerable to a range of **price shocks**. Each month the FAO publishes a report which gives an insight into the factors affecting food prices. The report includes a map highlighting price warnings (Figure 13.12).

Out of the eleven warnings in October 2015, six were due to prolonged periods of exceptionally dry weather, three to increased transport costs and one each to flooding and unusually high demand due to a supply shock in a neighbouring country.

The last major international food crisis was in 2008 when rising food prices sparked worldwide unrest and threatened political stability in the Middle East, Haiti, Indonesia, Cameroon and the Ivory Coast.

Globalisation of the food industry creates a number of opportunities

Technological innovation

Global sharing of technological advances in farming has increased food production. However, it is important that new technology is shared carefully and is not simply a tool for wealthy farmers and businesses wanting to maximise income rather than improve food security. The Asian Green Revolution, for example, was criticised for its dependency on high inputs of fertilisers and pesticides which benefited wealthy farmers but led to increased debt among a large proportion of the rural population.

Equally, advances in technology may have a damaging effect in countries that cannot keep pace with change. For example, farmers in some African countries, e.g. Ethiopia, have suffered from the increasingly efficient production of oil palm, cocoa and coffee grown in Asian countries.

Modern biotechnology is capable of altering the DNA of crops. Recent research has focused on the production of plants that will be better able to withstand the harsh environmental conditions associated with global warming, particularly regarding tolerance to drought and soil salinity. This focus is significant as many of the world's food insecure nations experience the negative impacts of desertification and salinity. A further significant area of development in these **genetically modified (GM) crops** is that of nitrogen-use efficiency. By improving the efficiency with which plants use the available nitrogen, fertiliser applications can be reduced. The manufacture of fertiliser emits large amounts of CO_2.

Figure 13.12 Food price warnings, 2015 (Source: FAO)

An alternative to large-scale, high-cost initiatives is small-scale, low-cost, **appropriate technology** with low inputs. One example is fertiliser deep displacement (FDP), which works by using a specialised fertiliser called 'briquette'. Planted 7–10 cm below the soil, the fertiliser slowly releases nitrogen. The deep placement means that less nitrogen is lost through run-off, crop yields are improved and fertiliser use is reduced. FDP has been successful in Niger, Nigeria and Burkina Faso.

The future

A mobile app called Vet Africa allows animal health workers to diagnose and treat livestock illness. Another innovative app, Farming Instruction, provides agricultural information to rural farmers. Access to mobile phone applications is widespread in many developing countries and can help farmers share information on issues such as cropping methods, markets and prices (Figure 13.13).

Figure 13.13 Agricultural mobile applications

Short-term food relief

The co-operation and co-ordination required to assist countries in need of emergency food aid have been enhanced by globalisation. International aid comes through different donor routes – bilateral aid (from one country to another), multilateral aid (provided by a number of countries and agencies, e.g. the UN) and non-government aid (provided through a voluntary organisation).

One example of a global effort has been the response to the civil war in Syria since 2011. The scale and cost of the WFP relief project is summarised in Figure 13.14. The Syria operation is WFP's largest and most complex. 6.5 million people have been displaced from their homes in Syria and there are 4.7 million refugees in the region.

- US$25 million are needed each week to meet the basic needs of people affected by the conflict.
- Food supplied includes bulgur, canned foods, lentils, oil, rice, salt, sugar and wheat flour.

For up-to-date information go to www.wfp.org/sites/default/files/WFP%20Syria%20response_info-01.png.

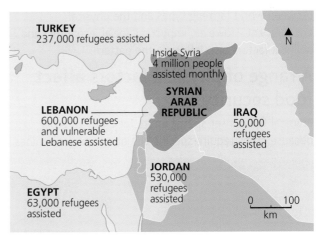

Figure 13.14 WFP's Syria response, 2016

Consumer choice

The increase in global trade has created an abundance of food products available to consumers through retail outlets and online. The types of food with global appeal include brands such as Pepsi and Coca-Cola and fast-food outlets such as KFC, McDonald's and Subway. There now exists a consumer-driven global food industry serviced by retail giants such as Tesco and Carrefour.

13.2 What are the causes of inequality in global food security?

Key idea
→ A number of interrelated factors can influence food security

Figure 13.15 shows a classification of world farming types. On a continental scale there is a close relationship between world farming types and the physical environment. However, important social, economic and political factors also influence the pattern.

A range of physical factors affect food security

Physical factors set spatial limits for crops to grow because crops require specific climatic (also determines length of growing season) and soil conditions. Despite technical advances the ability of countries to produce sufficient food remains heavily influenced by the physical environment.

Soil

Soil is a mixture of mineral and organic matter in which crops grow. Farming is influenced by the depth, drainage, texture, structure, pH and mineral content of soils (Table 13.7). A range of factors influences the formation of soil – these are illustrated in Figure 13.16.

From an agricultural point of view, the most important soil characteristics are:

- texture – the size of mineral particles in the soil. It determines the soil's ability to store nutrients and hold moisture. Coarse-textured soils are often leached and acidic
- structure – this is the way the soil particles are bound together to form soil aggregates, which allow air, water and plant roots to penetrate the soil
- nutrient supply – soils supply plants with the chemical elements for their growth, the most important being nitrogen, phosphorus and potassium. Partly decomposed organic matter or humus increases nutrient supply. In natural ecosystems nutrients are recycled; in agro-ecosystems harvesting removes nutrients.

Temperature and growing season

Each crop requires a minimum threshold temperature for growth and a growing season of a specific length. For example, in the UK most cereals require a minimum

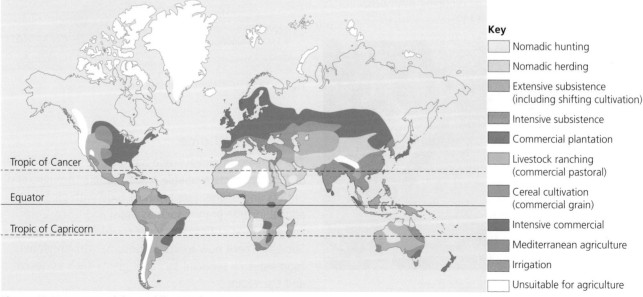

Key
- Nomadic hunting
- Nomadic herding
- Extensive subsistence (including shifting cultivation)
- Intensive subsistence
- Commercial plantation
- Livestock ranching (commercial pastoral)
- Cereal cultivation (commercial grain)
- Intensive commercial
- Mediterranean agriculture
- Irrigation
- Unsuitable for agriculture

Figure 13.15 Location of the world's main farming types

Table 13.7 The influence of soils on farming and their modifications

Soil	Description	Possible problem	Response
Clay soils	Heavy, acidic. High levels of nutrient and organic matter. Small particle size, high water retention can lead to waterlogging.	Poorly drained. Acidity.	Dykes and drainage ditches, underground drainage pipes. Add lime.
Sandy soils	Well drained, lighter, less acidic	Leaching, sometimes too well drained; can be prone to drought and nutrients are leached out.	Irrigation Add clay Fertiliser added to replace leached nutrients.
Silty soils	Lack of minerals and organic matter. Small pore size means that moisture is retained.	Moisture deficiency.	Irrigation.
Loam soil	Mixture of silt and sand and some clay. Well aerated and light. Little erosion.	Best for agriculture due to mix, 20 per cent clay, 40 per cent each sand and silt.	

temperature of 6°C. Cotton requires a growing season of 200 days and some varieties of wheat 90 days. Within the tropics temperatures are high enough for a year-long growing season providing there is adequate precipitation (the hydrological growing season). Increases in latitude and altitude reduce the length of the growing season.

Precipitation and water supply

The average annual precipitation determines the growth of cereals, grass, root crops or fruit/tree crops. However, in addition to the amount of precipitation, its effectiveness (taking account of **evapotranspiration**) and its seasonal distribution are important. Some crops have very specific requirements: coffee needs a dry period before harvesting, while maize requires high levels of precipitation to ripen. In some parts of the world a precarious balance exists between precipitation and food production. India, for example, depends heavily on the annual monsoon rains: if they are late or fail, the impact on food security is severe.

The intensity and duration of precipitation are also important. Prolonged periods of moderate intensity can infiltrate the soil, whereas heavy downpours promote rapid run-off, and moisture often fails to reach the root zone of crops.

Altitude

Altitude is a good example of how climate, soils and growing season are all interlinked. As height increases, temperature decreases, snow and precipitation increase, and the growing season decreases. At the same time soils take longer to develop, nutrient recycling is slower and leaching becomes prevalent.

Aspect

Mountainous areas are characterised by steep slopes with different directions or **aspects**. Aspect is an important factor in determining **microclimate**. In the northern hemisphere, south-facing slopes receive much more sunlight than north-facing slopes. South-facing slopes are therefore warmer and have drier soils. Crops

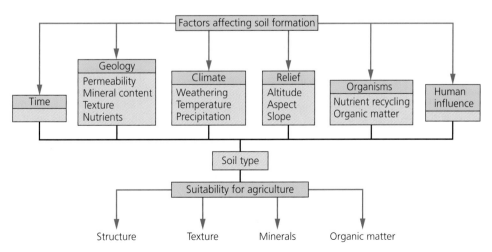

Figure 13.16 Soils and agriculture

on these slopes can grow at higher altitudes than on north-facing slopes. Evapotranspiration rates as well as temperatures are lower on north-facing slopes.

Slope

Slope angle affects rates of erosion and use of machinery. Slope also has a significant impact on soil depth and drainage and thus the crops that can be grown. On steep slopes, soils are often thin, poorly developed and excessively drained. Soils at the base of a slope can become waterlogged. On gentler slopes there is less movement of water through the soil and as a result less erosion and leaching.

➡ Activities

1 Various theories and models have been put forward to explain the location of farming types. Research one such theory, such as the optima and limits model (McCarty and Lindberg), locational rent theories or the theory of three agricultural revolutions in the UK. Summarise your chosen theory.

2 Modern technology can be used to combat problems with the physical environment such as infrared technology in Kenya and famine early-warning systems (FEWS) in the Sahel. Research an example of the use of modern technology to combat problems with the physical environment in an AC and an LIDC.

The social, economic and political factors affecting food security

Land ownership

Farmers may be owner-occupiers, tenants or landless labourers/employees on state-owned farms or on commercial enterprises. There are many different types of land ownership (Table 13.8).

 **Fieldwork idea**

An investigation of the factors affecting food production on a located farm

The success of farm studies is highly dependent on the help and co-operation of the farm owner or manager. A local contact can be used or the organisations listed below offer support to schools in finding a suitable farm:

● Farms for schools – www.farmsforschools.org.uk
● Farming and countryside education – www.face-online.org.uk

a) Fieldwork focus: the range of factors – physical, human and political – which explain production, i.e. what is grown, where, in what quantity and why.

b) Primary data can include:
   i)   mapping of field use and layout – 1:10,000 map, classification agreed in class or researched (AC – arable cereals, SA set aside, etc.)
   ii) soil sampling – texture, organic content, soil profiles
   iii) slope study – using clinometers and ranging poles
   iv) soil infiltration – inserting a section of plastic piping into the ground, filling it with water and measuring the fall in water level in 1 minute intervals.

c) Supporting secondary data on government policy, soil and climate can be researched using, for example:
   ● National Farmers Union – www.nfu.org.uk
   ● Natural England – www.gov.uk/government/organisations/natural-england
   ● Countryside Grant and Stewardship Schemes – www.defra.gov.uk
   ● Soil Association (for maps and theory) – www.soilassociation.org

d) Qualitative data – written transcript of interview with the farm manager.

Table 13.8 Global examples of land ownership

Russia	China	Bangladesh
In 2001 President Vladimir Putin sought to address the issue of land ownership in Russia. Regarding agricultural land, overall, the demand for private ownership has been low. In some regions where farmers have purchased their own land, such as Saratov and Tatarstan, farming has become more productive. Private farmers now total 26,000 in Russia, but the vast majority of land remains under state ownership.	China has gone through several types of land ownership: ● Pre-1949 – small intensive farms, owned by absentee landlords, were worked by tenant farmers. They often paid half of their produce to the landlord. ● Post-1949 – the Communist Party appropriated farmland and redistributed it among peasant farmers. Output was low and so the farms were joined together into state-run communes. ● 1982 – the commune system was abolished and farmers took out contracts with the government to farm land rent-free. After producing a quota they could sell any surplus and this improved food security. Farmers are now able to sub-let land.	Share-cropping is when the farmers have to pay a 'rent in kind' to the landowner in order to occupy the land. In Bangladesh this usually involves an arrangement where the landlord supplies fertilisers, seeds and machinery in exchange for a share of the harvest, which is usually 50 per cent plus. As a result, many farmers remain poor and are food insecure.

Capital

In ACs, farming is **capital intensive**, with investment supplied by banks, private investments and governments. In LIDCs there is often a shortage of capital leading to **labour-intensive** methods of farming. This situation can limit output and lead to food insecurity.

Competition

Competition is a crucial dimension of food security. There are two main issues:

Competition in food markets

The growing dominance of retail chains, agribusinesses and TNCs means a lack of competition in food markets. This leads to concerns over the prices paid to farmers for their produce and the prices consumers must pay for food, particularly the poorest income groups in both ACs and LIDCs who spend a high proportion of their income on food.

In addition, where increased competition in agricultural trade does exist between ACs and LIDCs, any comparative advantage of producers in LIDCs is often off set by the ability of ACs to offer subsidies, for example in the European market where this resulted in food surpluses and low global prices.

Competition for scarce resources

Food producers are experiencing greater competition for land, water and energy resources. Increasingly agricultural land has been lost to urbanisation and government decisions to grow biofuels on good quality agricultural land.

Technology

Technological developments such as new strains of seeds and fertilisers, advances in mechanisation and land management such as new methods of irrigation can improve production. ACs are more able to take advantage of technological innovations than LIDCs, thanks to their greater capital reserves and expertise. In LIDCs, appropriate technology is a more effective way to reduce food insecurity – for example, small-scale drip irrigation schemes based on bore holes rather than huge multi-purpose dams, and simple tools manufactured locally rather than tractors, grain dryers and combine harvesters.

Land grabbing

Land grabbing refers to the acquisition of farmland in developing countries by other countries seeking to ensure their own food security. A number of 'push' factors, such as water scarcity, export restrictions on major producers and price fluctuations in global markets, have forced countries short of productive land and water, such as China and India, to find alternative ways of sourcing food (Table 13.9). As a result, poor people in the target country risk losing access to the land and food supplies on which they themselves depend.

The main investors in Table 13.9 form two groups:

- countries with land and water constraints but rich in capital, e.g. the Gulf states
- countries with large populations and food security concerns, e.g. China and India.

Benefits to target countries include the creation of local employment, the development of rural infrastructure, the resourcing and introduction of new agricultural technologies, the creation of local food surpluses and enhanced food security.

But there are disadvantages for target countries – for instance, local farmers may be displaced from their land with no prospect of alternative employment, and the creation of unequal power relations between

Table 13.9 Overseas land investments to secure food supply in Africa (Source: IFPRI)

Target country	Investor country	Nature of the deal
Democratic Republic of Congo	China	2.8 million ha for biofuel
Ethiopia	India	$4 billion invested for growing of flowers and sugar
Kenya	Qatar	40,000 ha for fruit and vegetable cultivation in exchange for funding a $2.3 billion port
Malawi	Djibouti	Unknown area of farmland
Mali	Libya	100,000 ha for rice
Mozambique	China	$800 million investment for rice
Sudan	Egypt	Land for 2 million tonnes of wheat annually
Sudan	Jordan	25,000 ha for crops and livestock
Sudan	Kuwait	Unknown
Sudan	Qatar	Unknown
Sudan	Saudi Arabia	10,000 ha for wheat and vegetables

foreign national governments and local farmers who face growing food insecurity.

Activities

1 Referring back to the definition of food security on page 409, identify further examples of the physical and socio-economic factors that affect food security. Include a brief summary of the nature of the impact (is it positive or negative?).
2 What are the advantages and disadvantages of land acquisitions for target and investor countries? Relate your points to specific country examples and present your findings to the class.

Theoretical positions on food security

Thomas Malthus

Thomas Malthus based his 'Essays on the Principle of Population Growth' (first published in 1798) on the theory that an optimum population exists in relation to food supply and that any increase in population beyond this threshold will lead to 'war, famine and disease'. His two principles were:

● In the absence of checks, human population will grow at a geometric rate (i.e. 1, 2, 4, 8, 16, etc.). On such a basis population will double every 25 years.

● Yet food supply increases only at an arithmetic rate (i.e. 1, 2, 3, 4, etc.). Thus population growth would inevitably outstrip the growth in food supplies.

Malthus argued the 'natural checks' to population growth of famine, war and disease could be avoided only if people adopted 'preventative checks' such as abstinence and later marriage to control fertility.

However, Malthus's doom and gloom prophesy has proved inaccurate. In the past two centuries food production has increased massively thanks to new high-yielding crops (HYVs), new foods such as soya, the use of agrochemicals, greenhouses and polytunnels, and land reclamation (e.g. drainage of wetlands).

Ester Boserup

In 1965 Ester Boserup presented an alternative to Malthus's theory. She believed that although population growth would increase the demand for food, it would push up prices and incentivise farmers to raise production. This would be done by cultivating more land, using more advanced technology and intensifying production (e.g. by irrigation or multi-cropping).

Activity

Investigate recent events in relation to population and food supply, e.g. famines in East Africa, to assess the relevance of Malthus's theory.

Case study: Food security in India

Despite rapid economic growth in India in the past three decades, millions of Indians suffer poverty and hunger. Previously self-sufficient in wheat, India now imports vast quantities of grain because rapid population growth has led to 17 million extra people to feed each year.

India has a range of natural environments – mountains, deserts, grasslands, tropical and temperate forests. These give rise to a distinctive geography of agriculture (Figure 13.17). There are three major ecological zones: the Himalayan mountains to the north, the Indo–Gangetic Plain in the centre of the country and the Peninsular Plateau to the south.

The recent decline in food production in India is largely due to poor wheat harvests. The Punjab region in the northwest (Figure 13.18) occupies less than 2 per cent of the land area of the country yet produces two-thirds of its food grains (wheat and rice are the most common crops). Once called 'the bread basket of India', the Punjab

now faces a number of physical and human threats to food production.

Physical challenges

Water shortages

The Punjab's climate is semi-arid. Mean annual precipitation is around 630 mm and the annual temperature averages 21°C. The climate is divided into three seasons:

● Hot season (April–June).
● Rainy season (July–September, 70 per cent of the annual rainfall)
● Cold season (October–March)

Unreliable monsoon rains in the past sixteen years have caused frequent droughts. As a result, farmers have had to draw more **groundwater** to irrigate crops, accelerating decline in the **water table**. Eighty per cent of the groundwater sources are overexploited, with farmers having to drill ever-deeper wells. Expensive equipment

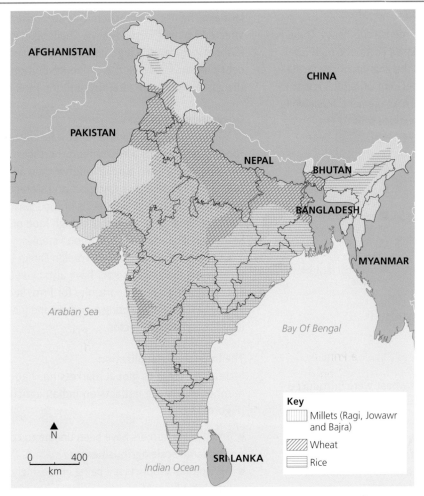

Figure 13.17 Map of food crops in India. Note overlap of crops in many areas (Source: www.mapsofindia.com)

has to be purchased to drill to such depths and the financial stress has led to high suicide rates among farmers and many leaving agriculture.

Increased temperatures due to climate change

Rising temperatures and more frequent heatwaves have meant that wheat crops have reached their maximum heat tolerance. Vulnerability to short-term heat greatly reduces crop yields.

Extreme events such as floods and droughts are also occurring, exacerbating the fall in the water table across the region (Figure 13.19).

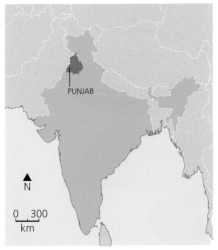

Figure 13.18 The Punjab region in northwest India

Figure 13.19 An Indian farmer in the Punjab carries a pitcher of water

Soil erosion

When the rain does come it is intense and heavy. There is little **infiltration** of the hard baked earth and high levels of **overland flow**. The result is the erosion of the fertile upper layers of soil. Up to 40 per cent of rainfall is lost to run-off, with inadequate water management. Meanwhile, deforestation and overgrazing expose soils to erosion by water and wind (Figure 13.20).

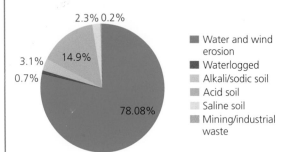

Figure 13.20 Reasons for land degradation

Human challenges

The Green Revolution

In the 1960s HYVs of rice and wheat were introduced as part of the Green Revolution. The aim was to make India self-sufficient in food grains and reduce dependence on imports. Apart from increases in yields, the new crops had more resistance to heavy rain and wind damage than traditional varieties, and as a spin-off created employment in the manufacture of agrochemicals. Nonetheless, there were socio-economic and ecological disadvantages: HYVs were dependent on irrigation and the intensive use of chemical fertilisers and pesticides. This favoured larger farms and wealthier farmers. The use of agrochemicals also adversely affected soil and water quality. Challenges to small farmers included the high costs of inputs and resulting debt problems. Overall the Green Revolution widened the gap between the rural rich and the rural poor.

Government policy

In 2013 the Indian government introduced the National Food Security Bill to alleviate food shortages. Its focus was subsidising grain purchase rather than addressing supply issues. Inefficient transport and storage infrastructure created high levels of food wastage. In addition, there were problems of high food prices and limited quality and quantity of food staples.

Government limitations on Foreign Direct Investment (FDI) in the food retail sector have also been unpopular. FDI is regarded as an opportunity for firms such as Wal-Mart and Carrefour to bring expertise and innovation into the agricultural sector.

The impact of globalisation

Increased access to global markets has had a number of adverse impacts on Indian agriculture (Figure 13.21):

- Small-scale farmers have been unable to compete with large-scale **agribusinesses**.
- Agricultural products are being increasingly imported into India from ACs.
- Small-scale farmers have been forced into high-value crops, which has made them food insecure.
- **Genetically modified crops** are being sold by multinational companies charging exorbitant prices.

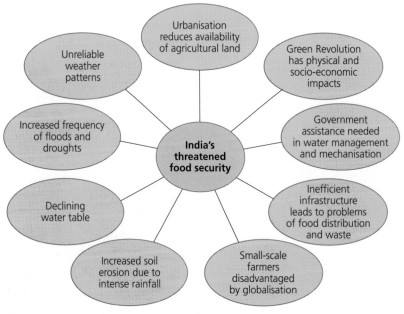

Figure 13.21 India – food security under threat

The future

Key requirements for the future include the following:

- Increased government investment in agriculture – the focus has been on price support for wheat and rice, but farmers need direct investment in machinery, fuel to draw water, seeds, fertilisers and pesticides.
- Better prices for farmers – evidence suggests that direct selling through farmers' markets guarantees better prices.
- Reduced input costs – co-operative farming where machinery and labour is shared, and organic farming with less dependence on chemical fertilisers and pesticides, would reduce the cost of inputs.
- Move to local food security – despite being the second highest producer of wheat and rice in the world, India accounts for one-third of the world's hungry. Inefficient storage and distribution systems mean that food often does not reach those most in need, particularly in rural areas. Local food distribution centres and infrastructure improvements are seen as a solution.
- Appropriate technology – to ensure better management of scarce water supplies.

Stretch and challenge

Study Table 13.10. What are the challenges for future food production in India?

a) Select appropriate techniques to present the data, e.g. line graph of population growth, pie charts, divided bar charts or histograms to show the percentage of land under different uses (see Chapter 15, Geographical Skills).

b) Calculate the percentage change in different land uses between 1900 and 1991, and 2009 and 2010 .

c) What proportion of the total land under crops is irrigated?

Table 13.10 Cropping pattern in India (million hectares) (Source: State of Indian Agriculture Report)

Years	1990–91	2003–04	2009–10
Total area under crops	186	190	192
Area under food crops	141	142	141
Area under non-food crops	45	48	51
Total irrigated area	63	78	86
Population (by census year)	846,387,888 (1991)	1,028,737,436 (2001)	1,210,726,932 (2011)

✔ Review questions

1. How do the following affect soil formation?
   i) geology
   ii) climate.
2. What is leaching?
3. Explain nutrient recycling in a natural ecosystem.
4. What is meant by appropriate technology?
5. List the main advantages and disadvantages of land grabbing for target countries.
6. State two ways in which globalisation has had negative impacts on small-scale farmers in countries such as India.
7. How can the following help farmers in food insecure regions?
   i) the provision of food at a local level
   ii) organic farming.

13.3 What are the threats to global food security?

Key idea
→ Risks to food security can be identified to highlight the most vulnerable societies

Regions, countries and people whose food security is most at risk across the development spectrum

According to FAO statistics, in 2015, 795 million people were hungry and did not gain sufficient energy from their diet; 98 per cent of these people lived in LIDCs. They included several widely dispersed, vulnerable groups:

- Rural dwellers: 75 per cent of hungry people live in rural areas, mainly villages in Asia and Africa. They are particularly vulnerable as they are overwhelmingly dependent on agriculture and have no alternative income.
- Farmers: the FAO states that half of the world's hungry people are from small-scale farming communities dependent upon crop yields from marginal land prone to drought and flooding.
- Children: UNICEF estimates that 146 million children in LIDCs are underweight as a result of acute or chronic hunger.
- Women: they are primary food producers in many LIDCs. Evidence shows that women are more affected by hunger and poverty than men. This has impacts on mortality rates during childbirth and low birth-weight children.

The global pattern of food insecurity

The global pattern of food insecurity is complex and ever changing. While progress has been made in many areas, there are countries, e.g. Sudan and regions, e.g. sub-Saharan Africa (SSA), where food insecurity remains a significant issue. In addition, even where progress has been made in regions such as Latin America, many areas remain fragile and vulnerable. Examples show how the impacts of political unrest, conflict or natural hazards can easily force these areas back into a position of food insecurity. The good news is that at the global scale undernourishment decreased from 18.6 per cent in 1990–92 to 10.9 per cent in 2014–16. Thus, despite a rapidly growing world population, the absolute number of undernourished people has declined. However, there are still widespread differences between regions (Table 13.11).

Table 13.11 Number of undernourished (millions) and prevalence (%) of undernourishment around the world (Source: FAO)

Region	1990–92		2014–16*	
	No.	%	No.	%
World	1,010.6	18.6	794.6	10.9
Developed regions	20.0	< 5.0	14.7	< 5.0
Developing regions	990.7	23.3	779.9	12.9
Africa	181.7	27.6	232.5	20.0
Northern Africa	6.0	< 5.0	4.3	< 5.0
Sub-Saharan Africa	175.7	33.2	220.0	23.2
Eastern Africa	103.9	47.2	124.2	31.5
Middle Africa	24.2	33.5	58.9	41.3
Southern Africa	3.1	7.2	3.2	5.2
Western Africa	44.6	24.2	33.7	9.6
Asia	741.9	23.6	511.7	12.1
Caucasus and Central Asia	9.6	14.1	5.8	7.0
Eastern Asia	295.4	23.2	145.1	9.6
Southeastern Asia	137.5	30.6	60.5	9.6
Southern Asia	291.2	23.9	281.4	15.7
Western Asia	8.2	6.4	18.9	8.4
Latin America and the Caribbean	66.1	14.7	34.3	5.5
Caribbean	8.1	27.0	7.5	19.8
Latin America	58.0	13.9	26.8	< 5.0
Central America	12.6	10.7	11.4	6.6
South America	45.4	15.1	Ns	< 5.0
Oceania	1.0	15.7	1.4	14.2

*2016 estimate

Thus, in spite of the progress in the Caucasus and Central Asia, eastern Asia, Latin America and northern Africa, there remain significant pockets of food insecurity within these regions. Overall improvements in food security have been especially slow in southern Asia and sub-Saharan Africa.

In northern Africa the position is very different. In countries like Algeria, Egypt, Morocco and Tunisia, food insecurity is close to being eradicated. However, the situation is fragile as food prices have remained low due to government subsidies. In the long term these subsidies are unsustainable. Moreover, there are dietary concerns in relation to malnutrition, with increasing overweight and obesity issues. A further concern is that the region is vulnerable because of ongoing political and economic instability.

In many food insecure areas, natural and human-induced disasters and political instability have contributed to crisis situations:

- Since March 2011, violence in **Syria** has claimed hundreds of thousands of lives. Over 50% of the population have fled their homes. Approximately 9.8 million people are in need of food and agriculture assistance, 6.8 million in 'critical' need. In addition, food price increases are dramatic: flour up 197%, rice up 403% and bread up 180%. Coping strategies include eating fewer meals a day, eating less nutritious food and buying food on credit.
- 6 million people in **Yemen** are classified as facing food insecurity at emergency level according to the WFP. Conflict continues to disrupt markets and livelihoods; this has resulted in a serious fall in income and access to basic staple foods. 29% of the cropped area is irrigated by wells and water transported by trucks, fuel shortages have seriously disrupted water supplies with impacts in the long term on agriculture and food prices.
- The **Nepal** earthquake of 2015 has severely impacted food security. 1.4 million people are in need of food assistance. Most of these people live in remote mountain areas along the seismic belt where 70% of households are battling borderline food consumption. In the worst-affected areas 80% of households have lost their entire food stock. Seed losses are a major concern as many households are dependent on trade for income. Although food assistance is helping, this will be a long-term problem.
- 4.4 million people in **Iraq** are food insecure and require urgent food assistance, 57% more than

2014 estimates. Conflict and displacement are eroding food security. Conflict has also affected food prices, which remain high and volatile – wheat, flour, sugar and lentils are particularly affected. In addition, the conflict has meant that government-run distribution schemes have been hampered.

Activities

1 Comment on the data shown in Table 13.11. The data can be summarised in a choropleth world map (refer to Chapter 15, Geographical Skills). Apply some descriptive statistical techniques such as percentage change in number of undernourished and mean values by region.
2 Visit the WFP website at www.wfp.org. A global food security update is published quarterly on this site. Use back copies of this update to track food security trends over the past ten years.

Food storage and distribution issues

The process by which food is transferred from the farm to our plates is referred to as the **food supply chain**. Processes include production, harvesting, processing, distribution, consumption and disposal.

Problems can occur at any of the stages of the supply chain and this in turn puts food security at risk.

Pinch points are places in the chain where disruption often occurs. These places are found in every stage of the food supply chain. In an increasingly globalised world, food supply chains are subject to political, economic, social, environmental and technological pressures, creating pinch points at the local, regional and global scales. In addition, retailers cut costs and wastage by using 'just-in-time' (JIT) delivery systems. With such systems any disruption to the supply chain can cause widespread disruption (Table 13.12).

Activities

1 Explain how the food supply chain works.
2 What is a pinch point?
3 How can pinch points put the supply chain and food security at risk?

Table 13.12 Examples of food supply chain disruption

Source of disruption to the food supply chain	Example
Energy	UK fuel strikes, 2012
Natural disaster	Icelandic volcanic ash cloud, 2010
Disease	H1N1 pandemic, 2009
Conflict	Syrian civil war, which has destroyed regional infrastructure, 2014
Transportation networks	Vulnerability of the Suez Canal to political unrest – the canal is a major supply route between Europe and southern Asia

The physical and human causes of desertification and how this changes ecosystems to increase the risk of food insecurity

What is desertification?

Desertification leads to a reduction in agricultural capacity as a result of human or natural processes. Full desert-like conditions are rare, but productivity is greatly reduced. Desertification is the outcome of persistent **land degradation**, most often in dryland environments, frequently caused by deforestation, overgrazing and overcultivation.

Desertification now ranks among the greatest environmental challenges of our time, with 168 countries affected (Figure 13.22). Every year an area three times the size of Switzerland becomes desertified. 74 per cent of the poorest people in the world are directly affected, including large areas in LIDCs where people depend on farming for survival. Desertification in these areas adds to already high levels of chronic hunger. The situation is particularly acute in parts of the Sahel in Africa. In Somalia, Ethiopia, Djibouti and Kenya, a combination of weak governments and drought (possibly linked to climate change) are driving desertification.

What causes desertification?

Desertification is not just confined to drylands. It occurs in a number of different environments, highlighting its complexity. The causes of desertification are economic, social, political and environmental.

Changing farming practices

Historically, traditional dryland economies have been based on a mixture of small-scale hunting, gathering, cropping and livestock herding by nomadic pastoralists, e.g. in the Sahel. For centuries the nomadic pastoralist way of life was sustainable and suited the limited **carrying capacity** of some dryland ecosystems. Recently population pressures, e.g. in Niger and Chad, urbanisation and an increase in agricultural exports, e.g. in Kenya, have led to more intensive methods of farming. These new systems

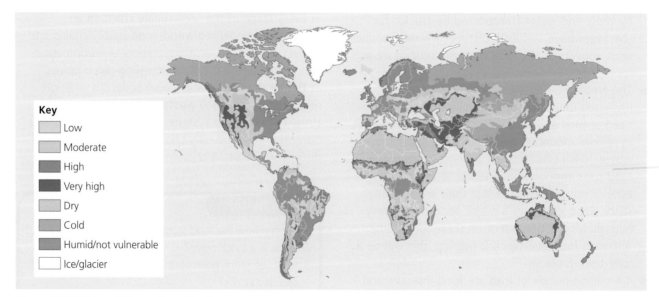

Key
- Low
- Moderate
- High
- Very high
- Dry
- Cold
- Humid/not vulnerable
- Ice/glacier

Figure 13.22 Vulnerability to desertification

deplete the soil nutrients as traditional fallow periods and crop rotation are abandoned. In addition, clearance of surface vegetation allows wind and water erosion to aggravate the damage; top soil is carried away, leaving behind a mix of dust and sand. Farmers are often forced onto areas of marginal land and the cultivation of already fragile soils.

Water scarcity

Increased intensity of cultivation in many drylands has been based on irrigation projects that are often unsuitable, poorly managed and underfunded. Lake Chad on the border of Chad, Nigeria and Cameroon has shrunk significantly in the past 30 years, from 25,000 km² to 2000 km². The causes are complex but include the high demand for irrigation water, damming of rivers feeding the lake and diversion of water away from the lake for large-scale schemes such as the South Chad irrigation project.

Demand for fuelwood

Demand for firewood for cooking and lighting is a major cause of deforestation and desertification in developing countries. The clearing of trees and shrubs increases the vulnerability of soils to wind and water erosion. Many African countries still have a high dependency on fuelwood, e.g. Burkina Faso, where 90 per cent of energy is provided by wood, Chad, where the figure is 89 per cent, and Mali, where it is 80 per cent.

Fires

In the past dryland pastoralists used controlled fire to release the nutrients stored in vegetation and stimulate the growth of new pasture. However, more intensive farming has led to the use of fire to clear large areas of new farmland. This has led to widespread land degradation, for example in the large-scale coffee plantations in Ethiopia and Kenya.

Increased periodic drought and changing rainfall patterns

Desertification is exacerbated by climate change. As extreme weather events such as drought become more frequent, land degradation and desertification increase. There has been a significant decline in precipitation in the Sahel region of Africa (Figure 13.23), which has resulted in an increase in periodic drought. In addition, dryland areas are particularly vulnerable to changes in rainfall patterns – the amount, distribution and seasonal timing of rainfall. Even a slight change can lead to over-exploitation of scarce resources and contribute to further degradation.

Impacts of desertification

Desertification has serious implications for food security worldwide (Table 13.13).

What can be done to prevent desertification and improve food security?

The 2005 Millennium Ecosystem Assessment, co-ordinated by the UN Environment Programme (UNEP), proposed a series of measures to help prevent desertification and restore degraded land. The measures involve policy intervention at the local and global levels

Table 13.13 Impacts of desertification

Environmental	Economic	Social and cultural
• Continual cropping and fewer fallow periods exhaust soil nutrients and decrease fertility, leading to food shortages. • Soil nutrients are lost through wind and water erosion, reducing yields. • Loss of biodiversity as vegetation is removed – this impacts the functioning of the ecosystem. • Increased dust formation, which can affect cloud formation and rainfall patterns, which increases vulnerability to fluctuating harvests.	• Reduced availability of fuelwood leads to increased purchase of kerosene, with health issues such as sore eyes and respiratory problems. • Food shortages, with a risk of growing dependency on food aid. • Reduced income from traditional economies – pastoralism and cultivation of crops. • Widespread rural poverty, leading to food security issues as people cannot afford food.	• Dryland populations are often socially and politically marginalised due to poverty and remoteness. • Forced migration, due to scarcity of productive farmland, and hunger. • Increased male out-migration, leaving women to undertake responsibility for farming and water management. • Loss of traditional knowledge and skills, leading to falling yields.

Downward spiral leading to desertification　　　　　　　　　　　　　**Approach to avoid desertification**

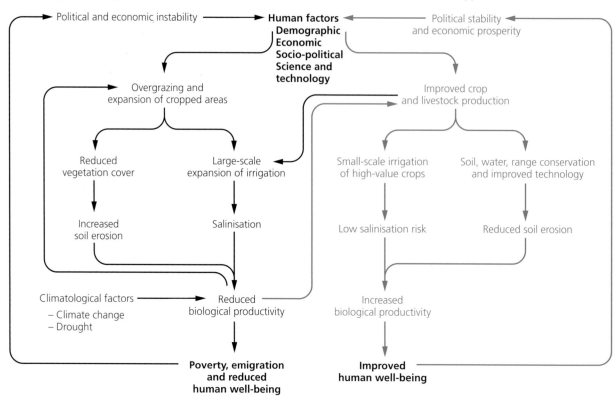

Figure 13.23 The downward spiral leading to desertification (Source: Millennium Ecosystem Assessment)

and are closely linked to efforts to reduce poverty. Specific proposals include:

- investing in integrated land- and water-management techniques to protect soils and reduce overgrazing, e.g. rainwater harvesting and water storage
- the development and introduction of drought-resistant crops
- maintaining vegetation cover to protect soils and restore soil fertility through the use of organic fertilisers

- reducing clearance of shrubs and trees by developing non-wood energy supplies which are naturally available in dryland ecosystems, e.g. solar, biogas and wind power
- encouraging mixed farming practices where livestock rearing and cropping are combined to allow more efficient recycling of nutrients
- diversifying land use to generate income through alternative livelihoods, e.g. dryland aquaculture, greenhouse agriculture and tourism-related activities.

Activities

1 Describe the relationship between poverty and desertification as a flow diagram. Include the following terms in your diagram: sedentarisation of nomads, soil erosion, drought, concentration of human and animal impact, expanding population, tree and shrub destruction, soil compaction, increased demand for building and fuel materials, increased run-off, less water penetration.

2 Research and outline one of the following water-management techniques used in LIDCs: rainwater harvesting, gravity-fed schemes, tube wells, boreholes, recycling projects.

Case study: The Sahel, a dryland area

Many dryland regions in Africa suffer desertification. The Sahel, which extends from West to East Africa across ten different nations, is of particular concern (Figure 13.24).

In 2006 a report by UNEP and the International Centre for Research in Agroforestry (ICRAF) on climate change and variability in the Sahel stated that 'without urgent investment, feeding the Sahel is "mission impossible"'.

There is a range of physical and human threats to food security in the Sahel (Tables 13.14 and 13.15).

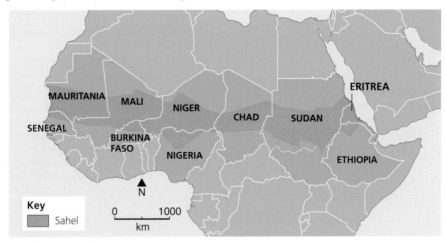

Figure 13.24 The Sahel region of Africa

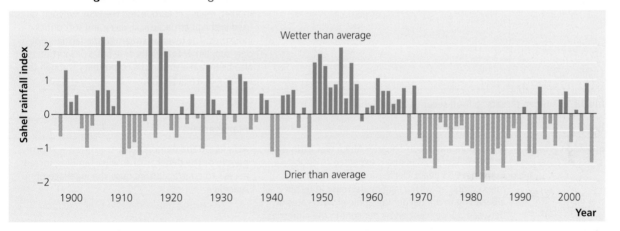

Figure 13.25 Rainfall in the Sahel, 1900–2004

Table 13.14 Physical threats to food security in the Sahel

Threat	Explanation	Management
Increased periodic drought	There is an increased frequency of periodic drought (Figure 13.25). The limited amount of moisture leads to low productivity in the ecosystem and low biodiversity. Soils cannot support agriculture and only specially adapted **xerophytes** can survive.	Irrigation is needed and must be managed to ensure unintended side effects such as salinisation. Contour stone barriers can also be used to slow run-off, increase infiltration and capture sediment leading to groundwater recharge.
Changes in rainfall pattern – amount and seasonal timing, linked to climate change	The variable amount, type and timing of rainfall can impact on water availability. Winter rain can penetrate the soil but in the summer light showers fall on hot surfaces and are quickly lost to evapotranspiration. Heavy rain can quickly erode soils.	Storage of water during wet periods in 'planting pits'. Planting of more drought-tolerant GM crops.
Exposure to high winds	The scarcity of vegetation in the Sahel means that high winds in winter can remove moist air around plants and soil and increase evapotranspiration. Soil erosion increases and silt; clay and organic matter are removed from the surface layers, leaving a sandy, infertile soil.	Vegetation barriers – suitable grass species form a barrier to wind and water erosion.

Threat	Explanation	Management
Infertile soils	Due to the lack of vegetation there is little deposition and accumulation of organic litter in the soil. When land is cultivated this limited organic matter is quickly lost. Erosion leaves shallow, sandy soils with poor structure and water retention.	'Planting pits' are dug in existing fields before the onset of the rains. The pits collect and store water and run-off. Organic matter is also placed in them to improve soil fertility and attract termites which dig channels and increase the soil's water-holding capacity. Planting of nitrogen-fixing trees. The farm management of regenerating vegetation and the integration of trees and crops is referred to as 'the re-greening of the Sahel'.
Changes in the albedo of the surface	The surface is light-coloured rock which reflects the insolation. The atmosphere absorbs the reflected heat and air temperatures rise. The hot air can hold more vapour so there are few clouds and little likelihood of rain. Desertification has occurred.	Re-greening projects. Projects such as the FAO Acacia Project have helped to regenerate the Sahel region in Africa. From 2004–07 the FAO in partnership with the Senegalese Forestry Service provided acacia seeds and seedlings to Senegal and five other partnership countries (Burkina Faso, Chad, Kenya, Niger and Sudan). Farmers were taught how to sow and plant the acacia seedlings using donated tractors specially adapted to ploughing in dry conditions. The trees capture and restore nitrogen to the soil – enhancing fertility, they provide shelter and fodder for livestock and allow the cultivation of low level crops (peanuts and millet). In addition, the 'gum' they produce has an international market, mainly in Europe and the USA, where it is used in the pharmaceutical, dairy and soft drinks industries. The revenue stream to the farmers not only supplements their income but a proportion has also been reinvested in the community through the building of a mill facility.

Table 13.15 Human threats to food security in the Sahel

Threat	Explanation	Management
Population growth	The population of sub-Saharan Africa was estimated at 920 million in 2014 and is expected to double in the next 35 years. Decreasing infant mortality rates combined with a lagging decrease in fertility are leading to annual demographic growth rates of between 2.5 per cent and 4 per cent in the countries of the Sahel region. Population will continue to grow beyond the region's capacity to feed itself. This growth, combined with a suggestion of a 3–5°C temperature change by 2050, could mean a fall in agricultural production by 50 per cent in Sudan alone.	Accelerate the demographic transition – inform people of the benefits of smaller families, increase access to contraception, raise the legal age of marriage.
Increasing deforestation	Increased deforestation for fuel wood is a direct consequence of population and settlement growth. It leads to increased wind and water erosion of soil and loss of organic matter.	Agro forestry and afforestation projects.
Overgrazing and overcultivation	Population growth also results in overuse of land to grow crops. The soil does not have time to recover and infertility leads to crop failure.	Educate farmers in methods to increase productivity and reduce land degradation. Focus on increasing investment in irrigation systems. Wider adoption of drought-resistant crop varieties. Diversify crops by rotating millet with cassava and sorghum, and use natural mulch as fertiliser. Identify trees and bushes that act as natural fertilisers. Land regeneration through simple irrigation systems – planting pits and half-moon water catchments. →

Threat	Explanation	Management
Land grabbing (Figure 13.26)	Agribusinesses are increasingly looking to acquire land in Sudan for agricultural production. Recent acquisitions have been allowed for the UAE, Saudi Arabia, Kuwait, China and South Korea. This has displaced indigenous farmers with no legal land tenure or negotiating rights. Many migrate to towns as drought and famine intensify on marginal land.	Improve the legal rights of farmers to land ownership.

Figure 13.26 Land grabbing

Key idea

→ The food system is vulnerable to shocks that can impact food security

How extreme weather events can affect food production

Extreme weather events such as heatwaves, droughts, wildfires, periods of intense rainfall, floods, hurricanes and tornadoes, are becoming more frequent and more severe around the world. Warming of the atmosphere increases the number of times that temperatures reach extreme levels. In addition, more water evaporates from the oceans. A resulting increase in the water vapour in the atmosphere leads to intense rainfall and because water vapour helps the Earth hold on to more heat energy from the Sun, further global warming occurs. Longer events such as heatwaves and prolonged rainy periods also occur more often.

Figure 13.27 outlines how extreme weather events such as heatwaves can affect food production.

Although the impact of climate change on food production will vary across the world, there are the following key global links:

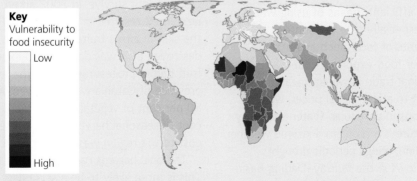

Heatwave – a heatwave is a period of several days or weeks of abnormally hot weather, and can be critical if it coincides with key stages of crop development. In the 2003 heatwave in Europe there were 3000 deaths and Portugal lost 3500 ha of forest and farmland to fires. Some food-exporting countries in Europe had to import food for the first time in decades. Research suggests that world-wide heatwaves that previously occurred once every 3 years are now happening every 200 days

Drought results in crop failure and reduced water quality and quantity. In 2003 southern Ethiopia experienced the longest drought on record – 20 million people needed food aid. Drought leads to further environmental impacts such as soil erosion, gullying, subsidence, rockfalls and weathering. Meteorological drought is projected to increase in frequency and duration

Key
Vulnerability to food insecurity
Low
High

Floods – as climate warms there will be more heavy rainfall events leading to flooding, which can destroy crops, disrupt food distribution, erode soil and damage infrastructure. In the UK Defra estimates that 35,000 ha of high-quality arable land will be flooded at least once every 3 years by the 2020s. 58% of the UK's most productive farmland lies within a floodplain

Tropical storms – within the Tropics many dry regions receive annual rainfall from tropical cyclones. Research suggests that climate change will cause hurricanes and tropical storms to become more intense, unleashing stronger winds and causing more destruction of farmland and infrastructure. Ocean temperatures will rise and this will lead to higher energy storms

Figure 13.27 Vulnerability to food insecurity as a result of climate change, 2016 (Source: based on WFP map and Met Office map)

- Food production is a driver of climate change, accounting for almost one-third of greenhouse gas emissions, e.g. CO_2 from deforestation, land clearance, and methane from intensive crop and livestock production.
- The global food system will face significant modification from climate change through extending growing seasons in some areas and reducing them in others. Climate change also affects food production through increased frequencies of droughts, wildfires and river and coastal floods.
- Farming can be a sink for carbon. Soils do have a limit for storage of carbon in the form of organic matter from crop residues and manure. In developed countries permanent set-aside land can sequester large amounts of carbon if left unmanaged or reforested.
- The increased frequency of extreme weather events has the potential to destroy not only crops but also key infrastructure, thereby disrupting distribution, exacerbating poverty and threatening food security.

How water scarcity can affect food production

Less than 1 per cent of all water is available for human use; of this, agriculture is the largest water user, consuming 68 per cent of water drawn from rivers, lakes and aquifers.

Water scarcity is one of the most urgent food security issues facing the world. Without water people do not have a means of watering their crops to provide food. Of the water available for agriculture, up to 60 per cent is lost due to poor irrigation systems and high levels of evapotranspiration. In areas such as Southeast Asia, a boom in groundwater irrigation in the 1990s has depleted aquifer supplies.

In attempts to resolve issues of water scarcity many countries across the development spectrum are trying to find more efficient means of using water in agriculture. This is especially true in countries that produce large amounts of food, such as Australia.

Australia is a major food-exporting country. Its volume of international trade in wheat, meat and dairy products is sufficient to affect global prices, which impacts poorer nations in particular. Water insecurity in Australian agriculture is a problem. Extreme weather, often driven by El Niño, brings periodic droughts which disrupt river flows in the Murray–Darling Basin, Australia's largest river system. Australia has put in place water-efficiency measures, including a cap on the amount of water extracted from major rivers, withdrawal of subsidies for irrigation and trade in water between farmers.

In Kenya, appropriate technology is required to address water scarcity. Initiatives include the following:

- mulching – laying plant leaves in between rows of cultivated crops to reduce soil erosion and retain water
- drip irrigation, which reduces loss by evaporation by directing a slow-moving supply of water to the base of the crops
- training farmers in water-harvesting technology.

Virtual water

The growing trade in food products requiring a high input of water is one factor influencing water scarcity. In water-scarce countries, importing water-intensive food products relieves pressure on domestic water resources. Pakistan has offered farmland to water-scarce Gulf states. However, this arrangement comes with a risk to future food security in Pakistan. The major virtual water exporters are Australia, the USA and Brazil; the largest importers are Japan and South Korea.

How tectonic hazards can affect food production and distribution

The negative impact of volcanic activity on livestock and food production

Volcanic ash fall can adversely affect food production in a number of ways, with economic, social and environmental consequences.

Ash falls destroy pasture land and as a result livestock need to be supplied with all of their feed in order to survive in the short term. If the ash contains fluorine, it can cause a condition known as flurosis when consumed by livestock. In 1996, 2000 animals died on the pasture land near Mount Ruapehu in New Zealand following a volcanic eruption.

Ash falls also affect crops – increased sulphur levels and lowered pH can alter a soil to such an extent that crops cannot survive. In 2002 an eruption of Mount Etna, Italy, resulted in a light ash fall which adhered to the skin of citrus fruits grown in the area. The crop was destroyed as it was deemed too expensive to clean the fruit before processing. Half of the orange crop in the province of Catania was destroyed. At the same time 80 per cent of vegetable crops were lost and 75 per cent of seasonal harvesting jobs. The total estimated cost to the region was €140 million.

Tectonic hazards can also impact food distribution. This is more likely to be a secondary impact and will affect remote communities where transport links are poor. After the earthquake of 1998 in Afghanistan, the mountainous environment and isolation of many rural villages meant that once transport links and bridges were damaged or destroyed, food distribution became a serious concern. Helicopters were provided as part of the foreign aid operation to distribute emergency food aid. More recently in Nepal in 2015, similar difficulties in food distribution became an issue for an extended period of time as governments of poor countries do not have the resources to rebuild communication links.

The impact of the 2015 earthquake in Nepal on food production and distribution

In April 2015 a 7.8 Mw earthquake struck Nepal. It killed more than 8000 people and left 35 million in need of emergency food aid (Figure 13.28). In the immediate aftermath of the quake the FAO appealed for disaster aid to help Nepalese farmers resume rice planting ahead of the growing season.

Other impacts of the earthquake on Nepal's agriculture, apart from the destruction of crops, were:

- Many farmers missed the planting season from May onwards and were unable to harvest rice, the country's staple food.
- Stocks of wheat and maize were destroyed.
- Livestock were killed and machinery damaged or destroyed.
- Widespread damage to roads, bridges, etc. meant that markets could not function.

- Farmers suffered a loss of income – in a poor country like Nepal, the government cannot reimburse farmers.
- There was damage to vital irrigation and drainage channels.

Figure 13.28 The earthquake of 2015 had a major impact on vulnerable agricultural areas of Nepal

Case study: Food production techniques in an extreme environment, the Arctic

The physical conditions of the environment

The Arctic covers 14.5 million km² consisting of the ice-covered Arctic Ocean and the surrounding land of northern Alaska, Canada, Scandinavia and Russia (Figure 13.29). Some parts of the Arctic are covered with ice sheets, e.g. the majority of Greenland, while others, such as northern Alaska, are tundra.

The thick ice and snow cover, bitter cold and frequent storms make the Arctic one of the world's most extreme environments. Summers may have periods of continuous daylight, but because the angle of the Sun is so low in the sky, temperatures do not rise much above freezing point and as a result the growing season is exceptionally short.

Arctic tundra

The tundra ecosystem has a very low organic productivity – the **net primary productivity** of 140 g/m²/year is the second lowest of land biomes. Due to the low moisture availability for much of the year (as it is stored as ice) plants must be adapted to moisture-deficient conditions. In addition, they are compact, low and slow growing in order to adapt to the high winds and limited growing season. Short roots avoid the **permafrost** and make use of the summer surface thaw. Low shrubs, lichens and mosses are the dominant plant species. The lack of nitrogen-fixing plants limits fertility and the cold, wet conditions inhibit the breakdown of plant material.

The tundra is an extremely fragile ecosystem in a delicate balance. The indigenous people of the Arctic have adapted to mainly traditional livelihoods, which make full use of the environment in which they live.

Food production methods used by indigenous people in the Arctic

Due to the harsh environmental conditions of the Arctic, farming techniques centre on a 'hunting, herding, fishing and gathering' culture of food production.

Wakeham Bay, Quebec, northern Canada

Fresh food sources come from the ocean where the coastal sea tide allows mussel farming. In spring the tide changes twice a day – as the tide goes out, the level of the sea falls and the sea ice covering it drops down. Where the ice touches the sea bed, caves are created where blocks of ice lean together. Mussels are harvested here and provide a rich source of minerals and vitamins to the indigenous populations.

Qaanaaq, northern Greenland

The Inuit people are traditional hunters of marine mammals such as the narwhal (type of whale), which provides a rich source of vitamin C. In the summer months the hunters camp on the edge of the sea ice and travel in kayaks among the fragmenting blocks of ice to catch their prey using handmade harpoons. The maktaaq, or skin of the narwhal, contains more vitamin C than oranges.

Barrow, Alaska

Whale hunting is an important aspect of life for the Inupiat people in Alaska despite the fact that many now have access to food shops. They hunt bowhead whales, which pass the coast twice a year. Whale meat contributes an essential part of their diet, with most ➔

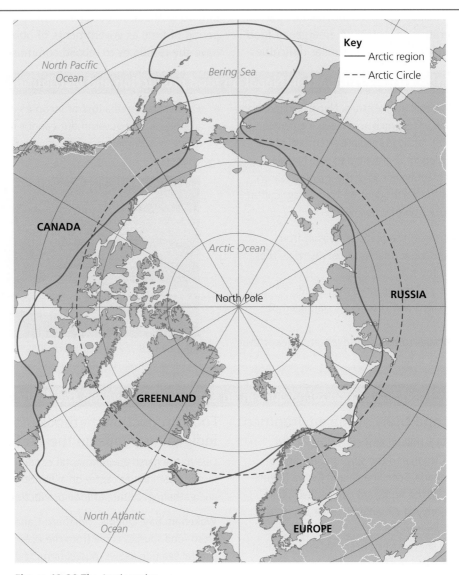

Figure 13.29 The Arctic region

people eating it twice a week on average. Whale oil is also added to other food sources such as caribou. Traditionally the meat prepared and distributed when the boats come ashore is kept in caves dug in the permafrost.

Threats to food security

The extreme environmental conditions of the Arctic support low species diversity and delicate food webs which make the ecosystem vulnerable to change. As the sea ice, snowfall and permafrost diminish due to the effects of climate change, the future food security of indigenous populations still largely dependent on foods harvested from the environment is severely threatened.

Impact of climate change on traditional food sources

The hunting terrain, migration patterns, quantity and quality of traditional food sources have all been affected by climate change.

Slushy ice does not give a firm base from which to hunt and retrieve seals and whales. In addition, travel routes are made longer, more dangerous and unpredictable. The decreasing ice is forcing species such as polar bears and walrus onto the land. The polar bears have been forced to scavenge berries, mosses, lichens and bird eggs, leading to a lowering of their body mass and also disrupting the food supply of other species. Fish stocks dependent on ice cover have declined and their feeding territory has been invaded by warm-water species. Herding activities are delayed as the lakes the herders cross are not freezing over.

Grazing animals such as reindeer and Peary caribou have dramatically declined in numbers. An earlier spring means that calves are born past the peak of prime foliage availability, meaning that many do not survive. As the Arctic warms, woody shrubs are expanding at the expense of lichens and other caribou food sources. ➡

More 'southerly' animals such as deer are moving north and bringing diseases with them, such as meningeal brain worm, which is lethal to caribou.

Impact on diet of indigenous people

In addition to problems with the availability of food, increasing pollution and contamination are leading to a belief among indigenous people that the safety and quality of 'wild resources' are also in decline. Contaminants are showing up in the animals, fish and waters of the Arctic. As a consequence, people are turning to alternative store-bought foods, which brings further social problems, such as obesity, malnutrition, health issues and economic problems, e.g. high food prices due to remoteness but low budgets due to low income and restricted job opportunities.

Foods high in salt, sugar and unhealthy fats are leading to a decline in general health. Some research suggests that 30 per cent of Inuit children suffer malnutrition and incidences of obesity, diabetes and heart disease in adults are increasing (Figure 13.30).

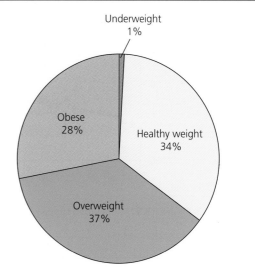

Figure 13.30 Weight status, adults in Alaska, 2012 (Source: AK BRFSS)

✓ Review questions

1 Choose **one** of the regions from Table 13.11 and give two reasons for the change in the number of undernourished people between 1990–92 and 2014–16.
2 What is JIT delivery? List some advantages and disadvantages of this method of food delivery.
3 Define the term 'desertification'.
4 How do farming practices exacerbate desertification?
5 How can agriculture help mitigate climate change?
6 Explain the concept of virtual water.
7 Why are some countries importers of virtual water?
8 Compare the different ways ACs and LIDCs are tackling water-use efficiency.

13.4 How do food production and security issues impact people and the physical environment?

> **Key idea**
> → Imbalance in the global food system has physical and human impacts

Global food supply has increased steadily over recent times (see Figure 13.31) as a result of higher yields per unit of land, crop intensification and an increase in the amount of land being farmed globally. However, there has been a range of impacts on both humans and the environment as a result of changes in the global food system.

How attempts to increase food production and security can impact the physical environment

Agriculture is now one of the world's largest users of land, but this has come at a cost to the environment (Figure 13.32).

Irrigation and salinisation

Salinisation is a form of land degradation in arid and semi-arid climates (Figure 13.33). It is an increase in the amount of salts in the soil, which are brought to the surface when high rates of evaporation and transpiration combine with low precipitation and poor soil drainage (Figure 13.34). Salts appear naturally in soil, streams, rivers and groundwater; concentrations increase as plants intake water but leave the salt behind. The salt layer is poisonous to plants and can inhibit water absorption.

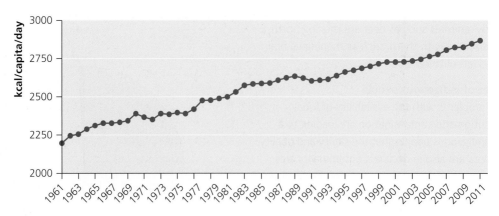

Figure 13.31 Global food supply, 1961–2011

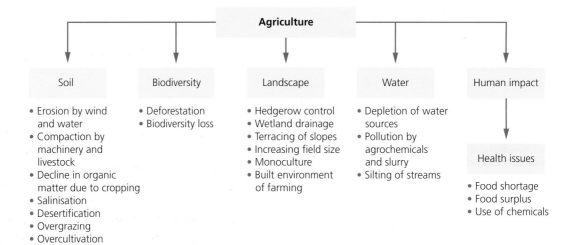

Figure 13.32 Summary of the environmental impacts of food production

Figure 13.33 Salinated soil in Colorado

Salinisation is a process that occurs naturally and is often made worse by human activity, for example the risk of salinisation is high where gravity flow methods of irrigation supply more water than crops can use.

There are two main causes of salinisation linked to irrigation:

1 Water used to irrigate crops evaporates under dry conditions; capillary force brings water to the surface and deposits salts (chloride, sulphate and carbonate) within the upper layer of the soil, forming a pan.
2 In coastal areas the excessive withdrawal of underground water for irrigation leads to the infiltration of marine, saline water into fresh groundwater supplies.

One of the most dramatic examples of irrigation-induced salinisation is the Aral Sea in Asia. Land irrigated with the water from the lake has lost fertility due to salt build-up and the volume of the Aral Sea itself has fallen by 90 per cent. FAO statistics suggest that as much as 20 per cent of irrigated land globally has reduced productivity due to salinisation. Countries with the largest areas salinised are Pakistan, China, the USA and India.

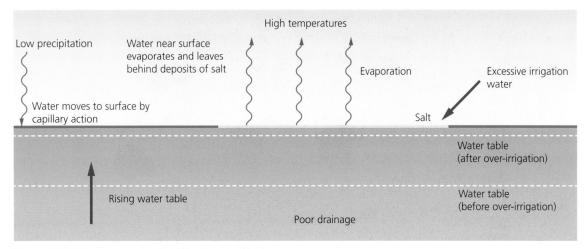

Figure 13.34 The process of salinisation in irrigated areas

Labels in figure:
- Low precipitation
- Water near surface evaporates and leaves behind deposits of salt
- High temperatures
- Evaporation
- Excessive irrigation water
- Water moves to surface by capillary action
- Salt
- Water table (after over-irrigation)
- Rising water table
- Water table (before over-irrigation)
- Poor drainage

Land-management solutions to salinisation include the following:

- Avoiding over-irrigation of crops by using techniques such as drip irrigation, soil moisture monitoring and accurate determination of water requirements.
- Good crop selection – use deep-rooted plants to maximise water extraction.
- Minimising fallow periods, since bare soil aids water infiltration.
- Adopting crop rotation to help minimise fallow periods. 'Break crops' such as cannula, lupins and peas improve soil structure and fertility.
- Avoiding deep-ripping and over-tillage to minimise infiltration of water.
- Good soil management – maintain satisfactory fertility levels, pH and structure of soils to encourage growth of high-yielding crops.
- Maximising soil surface cover using multiple crop species with different growth habits.
- Mulching exposed ground to help retain soil moisture and reduce erosion.
- Establishing and maintaining trees and shrubs on the property and public lands to maintain the water table.
- Introducing salt-tolerant crops such as sugar beet and barley.
- Drip feed irrigation, which efficiently directs water to plant roots.

Activities

1 Draw a diagram to summarise the causes and effects of salinisation.
2 Research two contrasting management schemes to help prevent salinisation, the World Bank (www.worldbank.org) and the FAO (www.fao.org).

Deforestation and the impacts on biodiversity

Increasing food production remains a reason for conversion of natural forested land to agricultural land. In 2012 the Organization for Economic Co-operation and Development (OECD) published an Environmental Outlook Report in which it stated: 'Globally, the area of agricultural land is projected to expand in the next decade to match the increase in food demand from a growing population.' The clearance of forest in order to gain agricultural land remains at the forefront of the expansion referred to by the OECD. Each year an area roughly the size of Costa Rica is destroyed globally, according to the FAO. Between 2000 and 2010, 13 million ha of forest were converted annually to other uses, predominantly food production.

Figure 13.35 summarises the pattern of net change in forests by country. The reasons for continuing high rates of deforestation in two 'hot spots', Brazil and Indonesia, are outlined in Table 13.16.

Biodiversity is defined as the variability among living organisms from all sources – air, land and water. According to the FAO World Agricultural Report: 'Loss of biodiversity owing to agriculture continues unabated, even in countries where nature is highly valued and protected.' Species richness is closely related to the occurrence of wild habitats. Deforestation, acquisition of agricultural land, hedgerow removal, grazing and drainage of wetlands all reduce the natural habitats available to wildlife. The irony is that decline in biodiversity impacts not only on people and ecosystems but also on agriculture itself as some affected species are important to pollination, pest control and the recycling of soil nutrients.

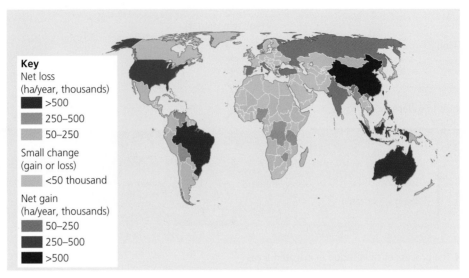

Figure 13.35 Net change in forest area by country, 2005–10

Table 13.16 Reasons for high rates of change in forest area in Brazil and Indonesia

Brazil	Indonesia
Increased demand for soybeans has led to critical habitat loss in Brazil, particularly the centre west states of the Cerrardo savanna ecoregion. Deforestation is projected to increase threefold over the next 35 years to reach 54 million ha by 2050.	Oil palm is now planted in large-scale plantations throughout the tropics. The top two producing countries are Malaysia and Indonesia. Figures suggest that oil palm expansion can be linked to the loss of 700,000 ha^2 of tropical forest in Malaysia alone.
Soybean is a leguminous crop used for a variety of commercial food products – soy sauce, cooking oil, miso and soy milk. It is also the basis of a number of tofu products. All of these products are becoming increasingly popular in Western diets where soybean products provide a low-calorie protein source.	Sumatra is a particular focus for cultivation as it is considered to have the best climate and soils together with a processing infrastructure. Forests are cleared through controlled forest fires, with a loss of 80% of plant species. Of particular concern is the loss of lowland forest in Riau and Jambi, which harbour endangered species such as the Sumatran tiger and elephant and orangutans, which are on the edge of extinction.
Soybean expansion poses a threat to areas of tropical forest rich in biodiversity, i.e. they contain an abundance of rare and threatened species. The Cerrardo ecoregion is the most extensive woodland/savanna region in South America. The region is of enormous biological significance, containing 10,000 plant species, many of which are endemic to central Brazil. The Cerrardo is also home to a number of 'focal species', such as the maned wolf, the giant armadillo and the giant anteater. It is estimated that only 35% of the Cerrardo remains in its natural state. Biodiversity loss is huge in large-scale monoculture plantations where native vegetation is cleared using machinery and large quantities of pesticides and herbicides. Not only do these have a long-term effect on the areas planted with soybean but the effects have been carried downstream to the Pantanal wetlands where hundreds of bird species such as kites and hawks, as well as jaguars, monkeys and river otters, are being threatened.	While much of the land cultivating palm oil in Indonesia has been designated for this purpose, there is growing evidence that it has expanded into a number of national park buffer zones and forest areas of high conservation value. The areas of Sumatra and Kalimantan are some of the most species-rich forests on Earth.

Activities

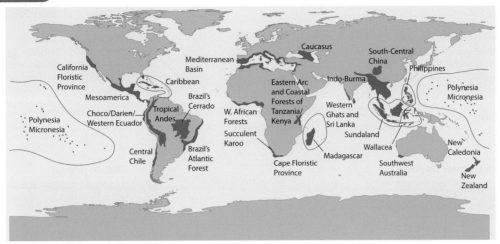

Figure 13.36 Global biodiversity hotspots

1 Figure 13.36 shows a world map of **biodiversity hotspots**. Compare this with the map in Figure 13.35 of net change in forest area by country. Which areas seem to present the greatest challenges?

2 Draw a diagram to summarise the economic and environmental pressures of plantation agriculture in countries like Brazil and Indonesia.

3 The World Wide Fund for Nature (WWF) was initially concerned with the protection of endangered species. Now it works with governments, NGOs, local people and businesses on a range of issues including agriculture and species diversity. Visit the WWF at www.worldwildlife.org/threats/deforestation and summarise the ways in which it is working to reduce deforestation, e.g. creating protected areas and stopping illegal logging.

Skills focus

Spearman rank exercise

Table 13.17 Deforestation and agricultural pressure in selected countries

Country	Annual rate of deforestation %	Population density per km² of arable land	% arable land
Brazil	−0.42	318	7
Indonesia	−0.71	1,136	11
Australia	−0.61	43	8
Venezuela	−0.61	1,005	3
Bolivia	−0.53	294	3
Mali	−0.62	248	4
Tanzania	−1.16	1,008	4
Zimbabwe	−1.97	382	8
Kenya	−0.31	766	8
Nigeria	−4.0	428	33
Argentina	−0.80	144	10
Peru	−0.22	761	3

Use the data in Table 13.17 for a Spearman rank test.

a) Plot the annual rate of deforestation (y) against one of the other two criteria (x) as a scatter graph.

b) Describe the relationship between the two variables.

c) Test the significance of the relationship by calculating the Spearman rank correlation coefficient.

d) Comment on the relationship.

e) Research one other variable which could affect deforestation rates, e.g. percentage of agricultural exports.

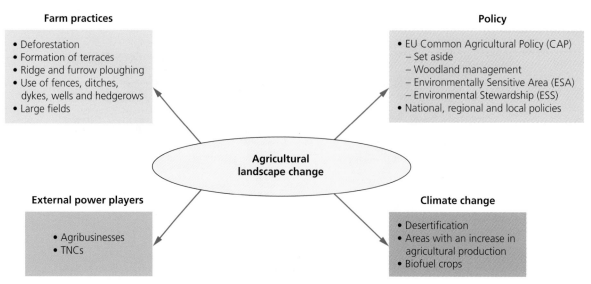

Figure 13.37 The driving forces in agricultural landscape change

Changing landscapes

In many parts of the world agricultural landscape is an integral part of a country's development, but the degree of change on landscapes has been greatest in the last 100 years. Due to the rise of modern industrial farming methods in many parts of the world and to population pressures in the developing world, the impact of farming on natural landscapes has evolved from a fairly low level to a dramatic and concerning one. Figure 13.37 summarises the various driving forces behind agricultural landscape change.

The growth of agribusiness and the industrialisation of farming have had a range of impacts on the rural landscapes of the UK. Changes include:

- Field sizes have increased to accommodate the mechanisation of agriculture and include the removal of hedgerows and ditches as boundaries.
- Natural habitats have been converted to food production. These include wetlands that have been drained, small woodland copses that have been felled, moorlands that have been ploughed up and natural grasslands that have been converted to intensive grassland.
- Mono culture has changed the farming landscape and gives rise to a less varied and diverse landscape pattern.
- The 'built environment' has become more intrusive, with glass houses, polytunnels, livestock sheds and silos.

Examples of landscape management in Europe

In the UK, Defra has introduced the Entry Level Stewardship Scheme (ELS). Farmers receive £30 per hectare for small-scale environmental management projects such as hedgerow management, stone wall maintenance, creating buffer strips and pond management.

Belgium also has packages available for the management of landscape. Each has a focus on a particular feature – for example, hedgerows, trees, ponds. There are strict guidelines on length and width of hedgerows, and they must contain native species.

Water quality from agrochemicals

The increase in agricultural productivity of many areas has been enhanced by the widening use of agrochemicals – herbicides, pesticides, insecticides and fertilisers. However, this has also come at a cost to the environment, in particular with regard to the quality of surface and underlying groundwater supplies (Table 13.18). Developed countries have had major problems with water pollution and it is now evident that the problems are spreading to emerging economies where there is less monitoring and control of the use of agrochemicals.

Table 13.18 Agricultural impacts on water quality (Source: FAO)

Agricultural activity	Impacts	
	Surface water	Groundwater
Fertilising	Run-off of nutrients, especially phosphorus, leading to **eutrophication**, causing taste and odour in public water supply, excess algae growth (**algal bloom**), leading to deoxygenation of water and fish death.	Leaching of nitrate to groundwater; excessive levels are a threat to public health.
Manure spreading	Carried out as a fertiliser activity; spreading on frozen ground results in high levels of contamination of receiving waters by pathogens, metals, phosphorus and nitrogen, leading to eutrophication.	Contamination of groundwater, especially by nitrogen, which is a soluble compound that can easily **leach** from soil into deep **aquifers** by **percolation**.
Pesticides	Run-off of pesticides leads to contamination of surface water and biota; dysfunction of ecological system in surface waters by loss of top predators due to growth inhibition and reproductive failure; public health impacts from eating contaminated fish. Pesticides are carried as dust by wind over very long distances and contaminate aquatic systems thousands of miles away (e.g. tropical/sub-tropical pesticides found in Arctic mammals). The widespread impact of pesticides is shown in Figure 13.38.	Some pesticides may leach into groundwater, causing human health problems from contaminated wells.

🥾 Fieldwork idea

An investigation of environmental management on a located farm

The success of farm studies is highly dependent on the help and co-operation of the farm owner or manager. The organisations listed on page 422 offer support to schools in finding a suitable farm.

a) Fieldwork focus – how farm owners and managers seek to farm within a context of environmental protection; a good located example would be an organic farm.

b) Primary data collection – mapping of land use (including any crop rotation patterns), drainage, woodlands, hedgerows and patterns of change.

  i) Field sketches and photographs of the visual impact of environmental projects such as solar panels, wind turbines, and of structures such as polytunnels.

  ii) Evidence collected through mapping, photographs, farm manager interview on the implementation and impact of government schemes and organic farming methods, e.g. waste disposal, use of water resources, energy supply. Environmental quality assessments carried out.

  iii) Statistical data on the use of herbicides and pesticides – trends can be graphed over time depending on data availability.

  iv) Water samples can be collected if the location is suitable, upstream and downstream of the farm.

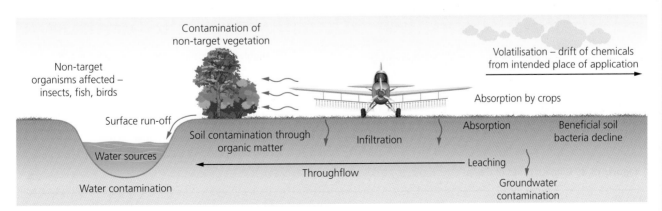

Figure 13.38 The widespread impact of pesticide drift

Case study: Land degradation of the steppe, China

Steppe is a dry, grassy plain which occurs in temperate climates between the tropics and polar regions. It is mostly found in the USA, Mongolia, Siberia, Tibet and China. Characterised by long cold winters (temperatures of −40°F) and short hot summers (temperatures of 100°F), these are semi-arid environments (25–50 cm of rain per year) prone to drought, strong winds and intense thunderstorms.

Previously the steppe area of China was occupied by nomadic pastoralists who moved around the grasslands with herds of domesticated animals (horses, cattle, sheep). These herds easily adapted to the natural environment and fulfilled an ecological niche as they grazed. The size of the herd was key to keeping the natural system in balance. However, with a diet based on meat and milk products, the growing steppe communities required cereals and vegetables to supplement their diet. With trading options limited the nomads organised themselves into farming units.

With continued population growth, both in rural areas and in the emerging cities of China, the state offered incentives such as tax cuts to farmers in order to expand the agricultural sector and meet the growing food demands in urban areas. Further investment by the state in agricultural technology and trade in agricultural products led to more growth

in food production. New crops were introduced such as corn from North America. This could be grown in areas where traditional Chinese crops could not, bringing marginal land into production. Driven by a state objective to ensure national food security and rural stability, agriculture was further intensified by the use of fertilisers and chemicals and infrastructure investment.

However, the growth in cultivated crops gradually resulted in land degradation as the fragile steppe was overexploited by:

- overgrazing – more animals than the land can support and less mobility due to changes in land tenure
- overcultivation – grasslands removed, land degraded with fertilisers and chemicals, ploughing destroying soil structure and leading to erosion of the nutrient rich top soil.

Table 13.19 outlines the impacts of land degradation and Table 13.20 offers possible solutions.

Despite the success of some land-management practices, the rate of land degradation exceeds the rate of environmental restoration on the steppe. This threatens China's overall, long-term development and standard of living.

Key

- **Sub-arctic** – very short, cool summers; long, severe winters; brief springs and autumns; modest precipitation
- **Humid continental, cool summer** – short, mild summers; long, cold winters; rain greatest in summer; heavy snowfall
- **Semi-arid (steppe)** – short, hot summers; long, cold winters; little rain or snow
- **Humid continental, warm summer** – long, warm summers; short, cold winters; rain greatest in summer; snow in winter
- **Arid (desert)** – long, very hot summers; short, cool winters; sparse rain
- **Highlands** – local climates vary with altitude and exposure
- **Humid sub-tropical** – long, warm summers with heavy rain; short, mild winters with lighter rainfall
- **Tropical wet, rainforest** – uniformly high temperatures with no winters; heavy rainfall distributed throughout year with no distinct dry season

Figure 13.39 Climate zones of China

Table 13.19 The impacts of land degradation

Environmental	Economic	Socio-cultural
• The physical processes of water and wind erosion (erosion of 5 billion tonnes of top soil per year) and the chemical degradation of land due to the process of salinisation both lead to long-term environmental impacts (desertification of 2500 km² of land annually). • Biological degradation from pollution and infestations also results in a long-term impact on biodiversity. • The physical degradation caused by the compaction and hard setting of surfaces by grazing animals can be short term once there is improved management of herd sizes. • Exposure of soil due to cultivation and overgrazing leads to loss of soil and nutrients through wind (20 times increase in the frequency of strong winds in China in the 1990s) and water (nutrients washed away equivalent of 54 million tonnes of fertiliser) erosion. • Increased sedimentation of rivers. • Overgrazing reduces vegetation cover and results in an environment that favours rodent and weed infestations. • Addition of fertilisers and chemicals pollutes water sources and further degrades soils (2 million ha of pasture land degraded each year). • Drilling of deep wells for irrigation reduces the water table, leads to soil degradation and salinisation. • Compaction of soil by grazing animals leads to less infiltration and more run-off, lowering the water table and leading to salinisation.	• Reduced income from traditional economies (direct economic loss valued at US$6.5 billion). • Increased dependence on food aid. • Increased rural poverty. • Expansion of agricultural land due to diminishing returns season by season.	• Loss of traditional skills. • Forced migration due to food security issues. • Tension in receipt areas of migrants.

Table 13.20 Solutions to land degradation

Cause of degradation	Strategies for prevention	Problems and drawbacks
Overgrazing	Improved stock quality: through vaccination programmes and the introduction of better breeds, yields of meat, wool and milk can be increased without increasing the herd size. Better management: reducing herd sizes and grazing over wider areas would reduce soil damage.	Vaccination programmes improve survival rates, leading to bigger herds. Population pressure often prevents these measures.
Overcultivation	Use of fertilisers: these can double yields of grain crops, reducing the need to open up new land for farming. New or improved crops: many new crops or new varieties of traditional crops with high-yielding and drought-resistant qualities could be introduced. Improved farming methods: use of crop rotation, irrigation and grain storage can all increase and reduce pressure on land.	Cost to farmers. Artificial fertilisers may damage the soil. Some crops need expensive fertiliser. Risk of crop failure. Some methods require expensive technology and special skills.

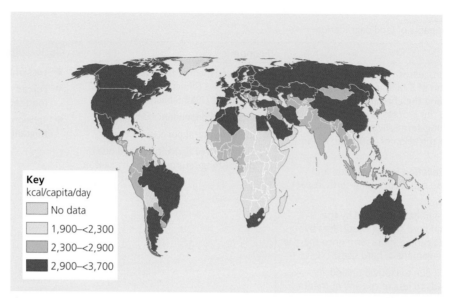

Key
kcal/capita/day

- No data
- 1,900–<2,300
- 2,300–<2,900
- 2,900–<3,700

Figure 13.40 The global distribution of daily calorie intake per person, 2009–11 (Source: FAO)

> **Activities**

1 Divide the impacts of land degradation in the steppe of China into short term and long term.
2 What human and physical obstacles could prevent environmental restoration?

How food security issues impact on people

Health issues associated with food shortages

Ensuring food security inevitably involves the consumption of adequate quantities of safe and good-quality foods that make up a healthy diet. However, nearly 30 per cent of the world's population suffer from some form of **malnutrition**. When the daily calorie intake of individuals is low, they cannot sustain healthy, active lives. The result is a spectrum of decreasing health issues. In addition, hundreds of millions of people suffer ill-health as a result of excessive calorie intake or unhealthy diets. Figure 13.40 shows high levels of consumption in North America, Europe, parts of North Africa, Australia and New Zealand, and low levels of consumption in most of Africa, central and northern Asia. These extremes of high and low consumption can be explored more fully through the **nutrition spectrum** (Figure 13.41), which recognises critical conditions between the two extremes.

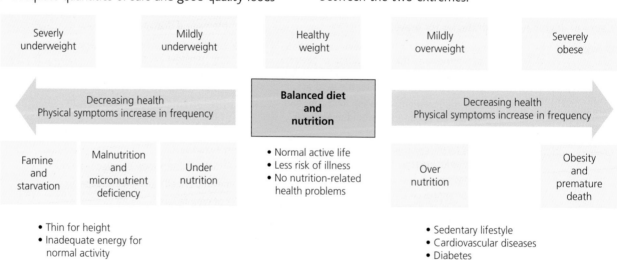

Figure 13.41 The nutrition spectrum

Table 13.21 Nutrient-deficiency diseases

Nutrient	Function	Source	Deficiency disease	Symptoms	Scale of disease
Vitamin A	Vision; body growth and healing	Milk, cheese, liver and fish	Xerophthalmia	Poor sight, blindness, reduced resistance to infection	Affects 50% of children in the developing world
Vitamin B	Release of energy	Liver, some grains and pulses	Pellagra	Weight loss, diarrhoea, mental disorder	Prevalent where maize diets predominate
Vitamin B1	Release of energy; nerves	Grains, milk, eggs, dried peas and beans	Beriberi	Loss of appetite, swelling, heart failure	Prevalent where overcooking predominates
Vitamin C	Wound healing, iron absorption	Citrus and other fruits; potatoes and green vegetables	Scurvy	Slow healing of wounds, bone weakening	Not known
Vitamin D	Calcium absorption	Sunlight, dairy produce, oily fish	Rickets and osteomalacia	Bone deformities	Prevalent where insufficient exposure to sunlight
Protein	Growth and repair of body tissues	Meat, cheese, eggs, nuts and pulses	Malnutrition – kwashiorkor and osteomalacia	Muscle wasting and weight loss	Affects about a quarter of the population of the developing world
Iron	Formation of red blood cells	Liver, meat, vegetables with green leaves	Anaemia	Blood disorders causing fatigue, loss of appetite and low blood pressure	Affects 917 million people, especially women in the developing world
Iodine	Vital to brain activity	Fish, seafood, eggs, milk and cheese	Stillbirths, endemic cretinism, goitre	Brain damage and mental retardation	600 million people affected

The nutrition spectrum and key definitions

The specific health consequences of food shortages are many and varied, but high levels of sickness and disability, shortened life spans and diminished productivity are common outcomes. Chronic hunger is therefore reflected in economies and contributes to a long-term cycle of household hunger and poverty. Malnutrition is associated with a number of specific diseases (Table 13.21).

Activities

1 Construct a cause-and-effect diagram to show the cycle of household hunger and poverty.
2 How can the cycle be broken?

Health issues associated with food surpluses and poor diet

A growing concern is that obesity and overweight are no longer confined to developed nations. For the first time in history the number of overweight people exceeds those underweight (Figure 13.42). Developing nations with high numbers of people overweight include Colombia (41 per cent), Brazil (36 per cent) and Zimbabwe (23 per cent). Such is the spread of obesity that it is now referred to as **globesity**.

The cause of obesity is an energy imbalance between calories consumed and calories expended. Globally there has been an increased intake of energy-dense foods that are high in fat and sugars but low in vitamins and minerals and a decrease in physical activity. Poverty in ACs is also leading to a dependence on low-value foods, ready meals and takeaway foods with a high content of fat, sugar and salt and low levels of vitamins and minerals.

The health consequences of obesity include a risk of non-communicable diseases such as cardiovascular diseases, diabetes, musculoskeletal disorders and some cancers. Obesity in children can lead to breathing difficulties, increased risk of fractures, hypertension, early markers of cardiovascular disease and possible psychological effects.

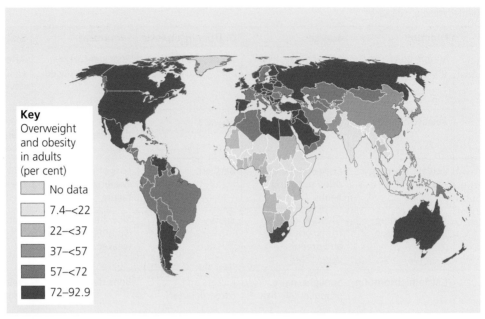

Figure 13.42 Prevalence of overweight and obesity in adults, %, 2010

Obesity, being overweight and the related health consequences are preventable by simple lifestyle changes such as exercising more, reducing calorie intake and consuming more fruit and vegetables. Governments also have a responsibility to ensure that a healthy diet is affordable and accessible, particularly for those suffering poverty and deprivation. The food industry has responsibilities to reduce unhealthy components of processed foods and review pricing and marketing practices to ensure that healthy choices are a real option.

Facing a double burden

Many developing nations, with resources already stretched, must now cope with serious health issues at each end of the nutritional spectrum. Obesity is often a sign of poor nutrition. As urbanisation increases, diets change and lifestyles become more sedentary. Diets of vegetables and grains give way to those of more fat and sugar content. The result is that at one end of the spectrum developing nations are still struggling to feed many of their people and at the other they are facing increasing costs of treating illnesses related to obesity. It is not uncommon to find undernutrition and obesity existing side by side within the same country, region and community.

Skills focus

Research current statistics produced by organisations such as WHO and the FAO to answer the question:

- What is the extent of the relationship between obesity and affluence?

You may include a range of statistical techniques to analyse the relationship: descriptive techniques such as inter-quartile range and frequency distribution or more analytical tests such as the Mann–Whitney U test or the Spearman rank correlation test (see Chapter 15).

Harmful impacts on human health as a result of the increased use of chemicals and pesticides

The use of chemicals and pesticides has increased over the last 50 years due to the Green Revolution and the benefits to increase crop yields. The fast-growing developing countries are becoming the greatest users of these agricultural inputs. Data from the FAO show that countries such as Colombia, Japan, the Republic

India withdraws noodles from shops in mounting food safety scare

Guardian. June 2015

A major multinational food company was forced to pull its instant noodles from supermarkets across India after it was found that some packets contained excess lead and an unlisted chemical flavour enhancer, monosodium glutamate. The Indian Food Safety Regulator said that it had recalled the noodles as they were 'unsafe and hazardous to human health'. However, food specialists have pointed out that multinationals are a 'soft target' and that 'street food' would contain much higher levels of contamination.

Wind-blown agrochemicals prompt local action in Argentina.

BBC News, May 2014

In Chaco, Argentina, the minister of public health has commissioned a report to investigate the link between pesticide use and increased levels of cancer and birth defects. Local people claim that the wind blows agrochemicals from small crop-spraying aeroplanes into their homes.

Soya growing dominates the landscape in an area once famous for its cotton.

Argentina has become one of the world's largest producers of GM soya, which is now a leading export. However, the use of agrochemicals to support the crops has become a concern to local populations.

Japanese ham, sausage and other meats follow noodles off the shelves in contamination scare

Telegraph. October 2008

Meat products were recalled after a factory near to Tokyo was found to have a toxic water supply. This followed just days after packs of instant noodles in Japan were found to have insecticide contamination and frozen green beans from China contained thousands of times the permissible level of pesticide.

Figure 13.43 Recent food health scares making the headlines

of Korea and Chile are among the highest users of insecticides and herbicides, but all continents use chemicals to control disease and enhance soil nutrients.

Monitoring and management of the impacts of agrochemicals on human health have led to a range of health scares, which are a frequent occurrence in the world's media (Figure 13.43).

The Food Standard Agency (FSA) is responsible in the UK for protecting public health regarding food safety issues. The number of incidents requiring attention has continued to increase, while the vast majority of contamination is in meat products. The surprising entry of the USA as the second highest source of contamination is due to unauthorised ingredients in soft drinks and body building products (Table 13.22).

Table 13.22 Origin of non-UK food contamination, 2014 (Source: FSA)

Country	Number of contamination incidents
China	66
USA	65
India	42
Netherlands	24
France	22
Nigeria	22
Poland	22
Germany	18
Bangladesh	17
Ireland	17

Case study: The impact of poor food security on the lives of people in Kenya

Kenya in East Africa has a population of 44 million people (World Bank, 2013). It has a growing economy based on tourism, telecommunications, construction and agriculture; commercial crops include tea, coffee, vegetables (French beans, peas, corn) and also flowers. The physical geography of Kenya encompasses savanna, lakelands, mountain highlands, the Great Rift Valley and in the north the deserts of Chalbi and Kaisut Losai (Figure 13.44).

Despite a growing economy, half of the population live below the poverty line and there is severe food insecurity in many regions – 10 million people suffer chronic food insecurity. The main factors leading to this situation are

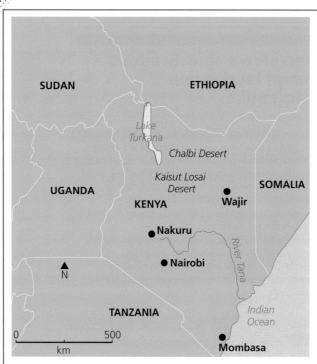

Figure 13.44 Map of Kenya

issues with access and distribution, tribal divisions, and displacement of farm workers to find jobs in urban areas. Two million people rely on food assistance and 31 per cent of the population are undernourished. However, in a country where only 20 per cent of the land is considered suitable for agriculture, export of vegetables (and flowers) is big business and accounts for 30 per cent of GDP and an estimated 5 million jobs. In addition, some of the most productive farmland around the Tana river delta is leased to foreign investors and affluent countries, e.g. Qatar.

Urban food insecurity

Kenya is a rapidly urbanising country – by 2050, 50 per cent of the population will live in urban areas. Rising food prices have put the urban poor in a position of food insecurity. In many parts of the world urban agriculture is a growing solution, but in the slum areas there is often a lack of space. In the Kibera slums of Nairobi, people have begun an innovative form of urban agriculture called 'sack gardening'. A series of sacks is filled with manure, soil and small stones to enable water to drain. They are then used to cultivate crops such as spinach, kale, onions, tomatoes and arrowroot.

Advantages of the project

- Young people are often employed as sack farmers and receive a small wage, a guaranteed healthy meal a day and training in materials and management.
- There is increased access to food for the urban poor.
- A proportion of the money from selling the crops goes into a credit society to help young people set up their own business.
- In an environment where many are living with HIV, vegetables boost nutrients.

Difficulties with the project

- There are challenges in finding contamination-free sites.
- Kibera still has a high dependency on food aid; WFP initiatives provide more than 1 million schoolchildren with their only meal of the day.
- Kenya remains the fourth largest recipient of food aid, with only Ethiopia, Sudan and Zimbabwe above it.

Case study: The impact of poor food security on the lives of people in the USA

In 2013, 14 per cent of households and 49 million people suffered food insecurity in the USA (Figure 13.45). Food insecurity in ACs such as the USA is associated with economic, social and political factors driven predominantly by poverty.

Causes of food insecurity in the USA due to the lingering effects of economic recession

- High unemployment and under-employment
- Stagnant and falling wages
- Failure of government initiatives to boost jobs and wages
- Support systems such as Food Stamps and Nutrition Assistance Programmes have been cut in recent years due to the recession

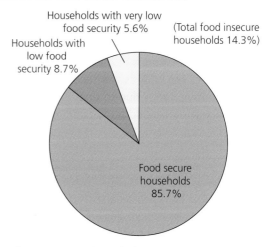

Figure 13.45 US households by food security status, %, 2013 (Source: US economic research survey)

Dimensions of hunger in the USA

Hunger and poverty

Unemployment, stagnant wages and a rising cost of living has meant that an increasing number of families are living below the poverty line in America. The Feeding America organisation reports 54% of the households receiving support have at least one family member unemployed.

Child hunger

In the USA 15 million children face hunger. In addition many poorer households are forced to purchase cheap, unhealthy food. Not having enough nutritious food can lead to physical and mental health issues in children in addition to academic underachievement which affects future life opportunities.

Senior citizens

Millions of senior citizens are faced with daily choices between purchase of food or medicines. Diabetes and heart conditions are the main medical issues. Older people also have mobility issues which put them at risk of hunger.

Ethnic groups

In the USA both African-Americans and Latino populations are disproportionally affected by unemployment, poverty and food insecurity.

Urban food insecurity

In the Bronx district of New York, urban neighbourhoods have high levels of unemployment and poverty. 37 per cent of people in the Bronx are food insecure (one in four children and one in ten senior citizens) and yet the area is crammed with fast-food outlets and a lack of grocery stores, earning it a reputation as a 'food desert'. Many people depend on food banks for staples; they also rely on stores offering food credit or donations.

Attempts to help people include national projects such as food stamps, the Special Supplemental Nutrition Program for Women, Infants and Children (WIC) which provides food vouchers, education and health referrals, and Temporary Assistance for Needy Families (TANF) – a direct block of funding to states to help families living below the poverty line.

The link between poverty and obesity is one to plague many of the most developed nations in the world.

Review questions

1 What is salinisation and how is the process intensified by inappropriate irrigation methods?
2 Why is there deforestation in many developing countries?
3 Explain the negative environmental impacts of a reduction in biodiversity. Make reference to the functioning of ecosystems in your explanation.
4 Summarise the landscape changes that have arisen from modern farming techniques in the developed world.
5 How are surface water and groundwater supplies affected by agrochemicals?
6 What is micronutrient deficiency?
7 Explain the 'double burden' resulting from food insecurity issues in advanced countries.

13.5 Is there hope for the future of food?

Key idea
→ Food is a geopolitical commodity; a number of key players will continue to influence the global food system

The geopolitics of food

The term 'geopolitics' refers to the concept of space, nations and the relations between them. In recent years, food has become a geopolitical commodity. In the early twenty-first century, global economic recession, food supply shocks, civil unrest and food riots in countries such as Egypt, Somalia and Tunisia have raised Malthusian fears that by mid-century there may not be enough food to feed the world's rapidly growing population (Figures 13.46 and 13.47).

While world leaders attending the G8 summit of 2009 in Italy pledged more and better food to those in need, some nations were already taking defensive measures such as food export bans, e.g. India and also land acquisitions by Gulf states.

However, there is evidence to show that the world 'food problem' is not caused by a scarcity of food. It is more a matter of the *distribution* of food resources and *access* to markets, technology, commercial opportunities, land and water. Currently the divergence between where food is grown and where it is most

Figure 13.46 Violence erupts when staple foods are scarce

- international organisations such as the WTO
- profit-making organisations – agribusinesses, TNCs and food retailers
- NGOs such as the World Fair Trade Organization.

Opportunities exist between countries to ensure food security

Trade

Trade is a critical component of food security, requiring governments to work together for the benefit of all. The scale of food trade today is unprecedented: trade has increased five-fold over the past five decades. Agriculture still accounts for more than one-third of export earnings in 50 developing countries. But export subsidies and import tariffs by developed countries mean that some poorer countries are unable to compete fairly in international markets.

In the long term, expenditure on staples such as rice and wheat has declined while spending on higher-value products such as meat, dairy, fruits, vegetables and processed foods has increased. These high-value products are subject to more safety and quality checks in developed countries, creating barriers to food exports from developing countries.

The global geography of food trade is dominated by Europe, the USA and Asia. Changing diets in emerging economies such as Russia, China, India and Brazil are

needed is putting increasing pressure on global political relations.

A number of key players influence the global food system (Figure 13.48). These include:

- national governments, whether acting independently or together

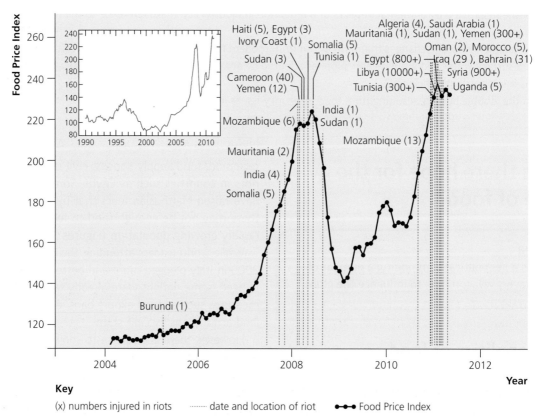

Key

(x) numbers injured in riots   ······· date and location of riot   ●━●━● Food Price Index

Figure 13.47 The relationship between civil unrest and global food prices in 2008 and 2011

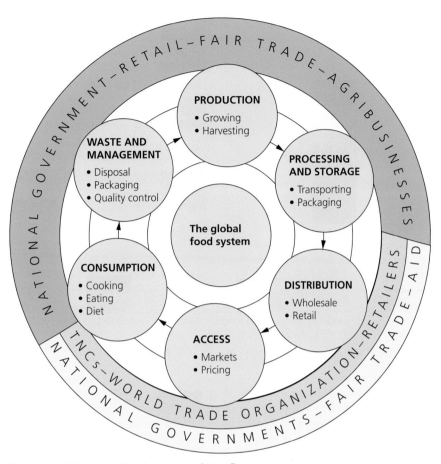

Figure 13.48 The global food system and its influences

creating new trade flows while these countries are gaining an increased share of food exports (Figures 13.49 and 13.50).

Trading blocs

A trading bloc is an agreement between a number of countries to promote free trade among its members but to impose tariffs on imports from non-member states. Examples include the European Union (EU) and the Association of Southeast Asian Nations (ASEAN). The EU is the largest and most established trading bloc.

In 1963 the EU introduced its Common Agricultural Policy (CAP), which aimed to provide affordable food for EU citizens and at the same time a fair standard of living for farmers. Being one of the oldest EU policies the CAP has been reformed on many occasions. A major criticism of the policy in its first 25 years was its focus on guaranteed prices paid to farmers for food production, regardless of demand. This resulted in massive food surpluses and depressed world prices, making it impossible for farmers outside the EU to compete. Subsidies have now been decoupled from production.

Multilateral agreements

These exist where several countries engage in a trading relationship with a third party. An example is the Lomé Agreement (1975) where the EU gave Africa, Caribbean and Pacific (ACP) nations almost completely free trade access to EU markets.

However, the EU's CAP has had an ongoing negative impact on ACP nations as unfair terms of trade with the EU continue to suppress the development of commercial farming in developing nations (Figure 13.51).

Bilateral agreements

A bilateral agreement is one made between two political entities that has mutual benefits and is legally binding to the two parties. For example, both Sainsbury's and Waitrose supermarkets have trade agreements with St Lucia for fair trade bananas. Café Direct has a number

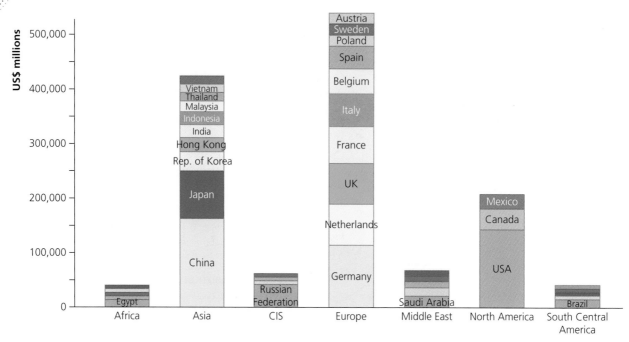

Figure 13.49 Leading importers of agricultural products, 2013 (Source: WTO)

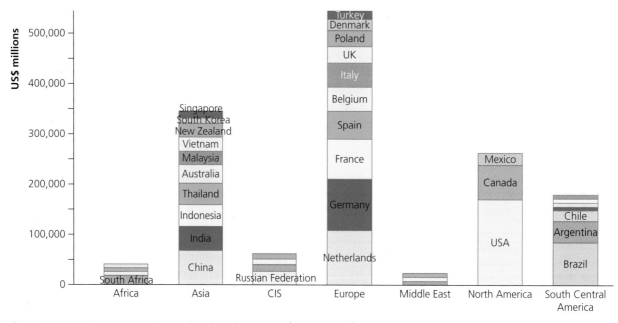

Figure 13.50 Major exporters of agricultural products, 2013 (Source: WTO)

of fair trade bilateral agreements for coffee, tea and cocoa products in Colombia and Peru.

The role of the World Trade Organization

Developing countries continue to push for improved access to the markets of ACs. Agriculture is the mainstay of many LIDCs' economies, underpinning food security, export earnings and rural development. Such countries often find themselves marginalised and unable to compete in a competitive global trade environment. Due to the growing complexity of

global food trade and the protectionist measures such as tariffs, quotas and subsidies, there is a need for regulation.

The WTO was set up in 1995 as a continuation of the General Agreement on Tariffs and Trade (GATT), which formed in 1947 when political trading ties were used to further economic development. The WTO, based in Geneva, Switzerland, has 161 members (2015) representing 97 per cent of total world trade. Its main role is to provide a forum for governments to negotiate trade agreements and as an organisation supporting

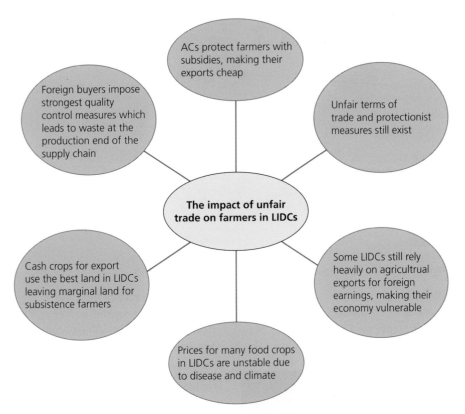

Figure 13.51 The impact of unfair trade on farmers in LIDCs

free trade, to persuade countries to abolish import tariffs and other barriers and to settle trade disputes (Figure 13.52).

> **Activities**

The Food and Agriculture Organization of the United Nations (FAO) is another key player in the quest to create a food secure world.

1 Visit the WTO (www. wto.org) and the FAO (www.fao.org). Compare and contrast the roles of the two organisations.
2 Which organisation appears to have achieved more?
3 What are the obstacles to progress for a fairer global transfer of food?

Food aid

In many LIDCs where trade agreements have failed to achieve food security, various food aid and food provision schemes can improve people's access to food. Examples include bilateral arrangements between governments, the work of international agencies such as the WFP, and the storing and distribution of major food staples such as rice and wheat.

There are different types of food aid: project food aid, programme food aid and emergency or relief food aid. In 2011 there was a major review of food aid through the Food Aid Convention (FAC). Many experts advised that reform of food aid was well overdue and that the effectiveness of food aid was questionable, often reflecting the interests of the donor countries rather than the needs of recipient countries.

- Donor-driven food aid centres on the use of food aid as a vehicle to 'dump' surpluses from ACs. Countries involved in such practices include the United States, Canada, EU states, Australia, China and Japan. Legislation has permitted the practice of sending surplus grains to LIDCs in the name of aid, but in effect this is seen as a branch of foreign policy aimed at promoting geopolitical interests.
- Creating a cycle of food aid dependency could be the long-term outcome of food aid for recipient countries. In many LIDCs food security for rural populations depends on the sale of agricultural products to local urban markets. Large quantities of food aid swamp these markets, driving down prices and reducing incomes of indigenous farmers.

On the positive side, food aid has made a crucial contribution to saving lives in emergency situations, for example in the civil conflict in Syria and the devastating earthquake in Nepal in 2015. If managed appropriately it can save lives, protect livelihoods and promote recovery.

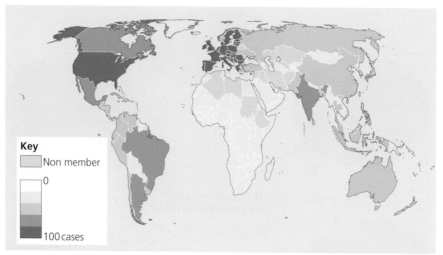

Figure 13.52 Disputes between member nations of the WTO, 2016

Key
- Non member
- 0
- 100 cases

Emergency food aid in Nepal

Emergency food aid was sent to the survivors of the two earthquakes that hit Nepal on 25 April and 12 May 2015 (Figure 13.53).

Figure 13.53 Emergency food aid for survivors of the Nepal earthquakes

Objectives:

- To deliver food to the survivors, focusing on some of the most remote mountain communities. Food was distributed to nearly 2 million people in seven districts. Cash was also given to survivors so that they could buy food locally, which revitalises markets.
- To run logistical services such as a relief despatch hub at Kathmandu's airport.
- An offer of cash for work so that critical community infrastructure was cleared of rubble and people could access food markets.

Alternative to direct food aid

'It was like a man without clothes. The mountains were bare from deforestation and every time it rained, the water washed away all the top soil.'
Tonkollu Letu, a farmer in Ethiopia

The MERET programme (Managing Environmental Resources to Enable Transition) was a joint venture between the WFP and the Ethiopian government. It helped to feed people while they work on the reclamation of **degraded** land (Figure 13.54). The project was started in 2003.

Inputs:

- Terracing hillsides to prevent erosion.
- Digging ditches and irrigation channels to capture rainwater.
- Building dams to create a sustainable water supply.
- Installing new 'cooking stoves' which burn three times less wood, so decreasing deforestation.

Outcomes:

- Enough corn, soy, coffee and fruits grown to feed families and raise a profit at local markets.

Figure 13.54 Working the land as part of the MERET programme

- Profit used to buy more land and livestock.
- Meat also sold at markets.
- New cooking stoves are safer as open fires are a danger to people and create toxic smoke.
- Time freed up from collecting wood so there is more time for farm management.

The role and responsibilities of profit-making organisations

Agribusinesses

Agribusiness refers to large-scale, capital-intensive, corporate farming by business enterprises. The activities of agribusinesses sometimes extend throughout the food chain, with involvement in the production, processing and distribution of food as well as the manufacture of farm supplies. DuPont is a massive agribusiness involved in several aspects of the food chain. The range of its activities is illustrated in Figure 13.55.

A criticism of agribusinesses is that in their pursuit of profit environmental issues have been compromised. Many agribusinesses are highly scientific, specialising in high-yielding, genetically modified crops, using large amounts of agrochemicals, and hormone and growth promoters. All can have damaging environmental side effects.

➡ Activities

1 Agribusiness activities include downstream links to processing, storing, packaging and delivery, as well as upstream links to seed companies and livestock research. What are the advantages of this spread of involvement in the food production chain?
2 Research one other agribusiness (not DuPont) and evaluate its impact on the food production system.

TNCs

TNCs are usually organised exclusively on an international scale. They often specialise in 'downstream activities' such as the processing and distribution of food. Examples of such companies include Nestlé, Unilever, Kraft and General Foods.

➡ Activity

Construct a table to explain the economic, social, political and environmental impacts of a food-related TNC on a host country, e.g. Unilever. The impacts could be further divided into advantages and disadvantages.

Sources include: www.unilever.com and www.oxfam.org.

Food retailers

The dominance of supermarket chains in the distribution and retailing of food today is not confined to the developed world. Research has shown that transnational retail firms are extending their operations in the developing world with wide-ranging effects on local markets.

Driving this development is population growth in rapidly expanding urban agglomerations in emerging and developing countries (EDCs). Supermarkets now control 60 per cent of food retailing in Latin America. With a combined population of 2.3 billion, China and India are targets for investment by global retailers. This process impacts adversely on local traders, who are unable to compete with the power of transnational retailers. In Brazil, the supermarkets have eliminated large numbers of small traders in the red meat sector. The major power players in food retail are household names: Tesco, Sainsbury's, Wal-Mart and Carrefour. The geographical spread of the fast-food retail giants such as KFC, Pizza Hut, Subway and in particular McDonald's into developing countries is equally remarkable (page 417).

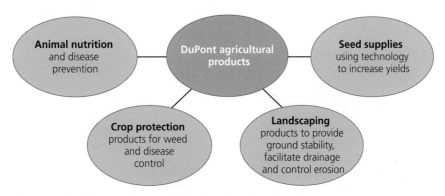

Figure 13.55 Agricultural product lines of DuPont

Fair trade organisations

The World Fair Trade Organization defines fair trade (Figure 13.56) as:

'a trading partnership, based on dialogue, transparency and respect, that seeks greater equality in international trade. It contributes to sustainable development by offering better trading conditions to, and securing the rights of, marginalised producers and workers – especially in the South.'

The worldwide fair trade network includes 586 producer organisations from 58 countries and more than 650 traders. Large TNCs such as Nestlé and Tate and Lyle and major food retailers including Asda, Waitrose and Sainsbury's support fair trade products. The UK has the largest fair trade market among EU nations.

> ### ➡ Activities
>
> Produce a report on the role and responsibilities of the World Fair Trade Organization. Address the following:
>
> 1 Who decides what is fair?
> 2 What are some of the myths surrounding fair trade?
> 3 Evaluate the success of two contrasting examples of fair trade practice.
>
> Useful websites for information are:
> www.fairtrade.org.uk
> www.wfto.com

> ### Key idea
> → There is a spectrum of strategies that exists to ensure and improve food security

Approaches to increasing food security

Food security is threatened by a wide range of physical and human factors, from climate change and land degradation to financial crises and unfair global food markets. However, a variety of approaches have led to some success.

Short-term relief

The WFP continues to supply food aid to the most desperate groups such as refugees and those suffering from political conflict, for example Syria, and natural disasters, for example Nepal. Food aid donations can alleviate emergency situations but it is not a long-term solution.

Long-term system redesign and capacity building

Capacity building refers to the ability of communities, countries and global institutions to build a resilient food system.

Despite progress towards food security at the global level, financial crises, high and fluctuating food prices, conflict and natural events such as droughts, floods

Figure 13.56 The principles of fair trade (Source: WFTO)

and tectonic hazards pose a considerable threat to future food security. The FAO has reported an increase of food emergencies from an average of fifteen per year in the 1980s to thirty per year in the new millennium.

Capacity building in the food supply system can be achieved by:

- economic development
- government monitoring of food supply and distribution
- efficiency of pricing and distribution within domestic markets
- access to fair trade agreements
- food safety
- investment in research and innovation
- investment in transport infrastructure
- efficient storage and distribution which minimises waste
- educating people in healthy and nutritious diets.

Techniques for improving food security

A range of techniques exist to improve food security from large-scale technological techniques to small-scale bottom up and appropriate approaches.

Large-scale technological techniques

These include GM crops, with higher yields and greater resistance to pests and also the development of crops more able to flourish in the harsh conditions of global warming. In addition the future development of water conservation and new irrigation schemes will be crucial to the improvement of food security.

Small-scale bottom-up and appropriate approaches

These seek the involvement of local farmers. Self-help schemes such as the use of simple tools manufactured locally, rainwater harvesting and sack gardening, for example in Kenya, the FAO Acacia Project in the Sahel, and fertiliser deep displacement (FDP) in Niger and Nigeria.

 Activity

The table below lists a range of strategies that can be used to improve food security.

Drawing on the case studies below of Cuba and the UK, and on other examples in this chapter, provide an example for each.

Strategy	Located example
Short-term aid relief	
Capacity building (developing and strengthening the skills and resources to achieve food security)	
System redesign (a country's strategic plan to achieve future food security)	
Large-scale technological techniques	
Small-scale 'bottom-up' approaches which focus on the participation of farmers	
Appropriate technology	

Case study: Achieving food security

Cuba

Background

In the 1960s Cuban agriculture was highly industrialised and relied on the former USSR for fuel, fertilisers and pesticides. When the USSR collapsed in 1991 Cuba lost more than 50 per cent of its oil imports, much of its food imports and 85 per cent of its external trade (primarily sugar exports to the USSR). Between 1989 and 1993

average daily calorie intake fell from 3012 to 2325. Yet by 2011 it had increased to 3277. With no access to chemicals and machinery this was largely accomplished through low-input organic farming. Despite now being 90 per cent self-sufficient in fruit and vegetables, large quantities of food products are imported, in particular cereals, meat and milk protein (Table 13.23). In 2008 Cuba spent $2.2 billion on food imports including $700 million on rice and $250 million on powdered milk.

Table 13.23 Food balance sheet for Cuba (Source: adapted from FAO statistics, 2011)

Food group	Domestic supply (1,000 metric tons)			Per capita supply		
	Produced	Imported	Total	Total kcal/day	Protein g/day	Fat g/day
Total[a]				3,277 (2,121)	84 (50)	66 (50)
Total vegetable products[a]				2,799 (1,727)		
Total animal products[a]				478 (394)		
Cereals	732	2,168	2,900	1,228	28	5
Wheat	0	828	828	374	10	1
Rice	378	552	930	630	12	1
Maize	354	713	1,067	225	6	3
Starchy roots	1,446	25	1,471	238	3	1
Cassava	486	0	486	69	0	0
Potatoes	166	26	192	27	0	0
Vegetables	2,249	11	2,260	84	4	1
Fruit	1,967	8	1,975	210	2	1
Coffee	10	8	18	2	0.5	0
Alcohol	393	13	406	63	0	0
Meat	292	268	560	264	17	21
Milk	603	579	1,182	145	8	7
Fish	55	18	73	10	2	1

[a]Figures in red are from 1961 for comparison

Challenges facing Cuban agriculture

- Increased frequency of severe weather events due to global climate change.
- Sugar plantations with degraded soils.
- Lack of mechanisation.
- Shortage of foreign exchange needed to buy imported food.
- Inefficient state-owned farms (most of the tomato crop of 2012 rotted as state trucks failed to collect on time).
- Insufficient food shops. Where they do exist they are expensive, carry little fresh produce and vary in what they stock.
- The existence of a 'black market' for products such as eggs and alcohol.

Towards improved food security

Urban agriculture

Urban agriculture is based on urban farms known as *organiponicos*: small scale co-operatives producing fruit and vegetables on urban waste ground (Figure 13.57).

Food is supplied directly to local people through market stalls. Around a half of Havana's demand for fruit and vegetables is supplied by 383,000 urban farms, which employ 140,000 workers. Shortages of fertilisers and pesticides mean that the produce is organic, while oxen are often used instead of fuelled farm vehicles. This method of farming supports the ethos of **permaculture** – food production in a closed loop that is sustainable.

One of the flagship farms is the Vivero Organiponico Alamar, an 11- ha farm on the outskirts of Havana with a work force of 164 people. It produces a diverse range of vegetables, herbs, spices and fruit to the local community, hotels and restaurants.

Changes in land tenure

Large, inefficient state-run farms have given way to small scale co-operatives. They comprise 10,000 acres of rent-free land leased from the state. Fifteen per cent of Cuba's arable land is now privately owned, allowing farmers the freedom to buy or rent basic equipment and to purchase their own fertilisers without depending on government distribution. Cuba has 6 million ha of flat

Figure 13.57 A Cuban *organiponico*

land, half of which is uncultivated. This allows scope for increasing production in the future.

New crops

Increasingly farmers are adopting drought-resistant crops such as avocado, oranges and guava. Crops resistant to wind and rain in the hurricane season, including sweet potato, squash and yams, are widely grown. The state has also supported the development of GM crops, with GM corn now planted across fourteen provinces.

Support for farmers

The government has raised guaranteed prices in support of the farmers and there is also assistance from programmes such as the Programme of Local Support for the Modernisation of Agriculture (PALMA), which provides irrigation systems and rents equipment to drill bore holes. In addition, Cuba has developed a specialism in research and training in the field of agro-ecology.

UK

Background

All countries, regardless of economic status, must address future food security. In 2009 the UK government published a consultation document, 'Food 2030'. It set out a strategy for food security in the UK in light of population growth, climate change and rising levels of obesity (Figure 13.58).

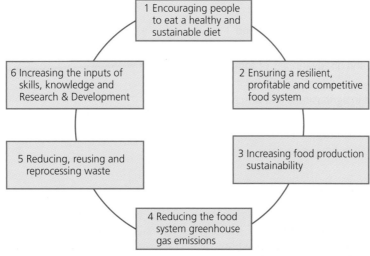

Figure 13.58 The core issues in the UK's food system (Source: Food 2030)

The UK currently imports 40 per cent of its food. Three times more food is imported than exported. Even with industrial-scale farming, to be self-sufficient the UK would need to produce two-thirds more food or eat 40 per cent less. As a result, international trade is crucial and a degree of dependency on external sources of food inevitable. 'Food 2030' recommends a range of actions to ensure food security in the UK in the future (Table 13.24).

Table 13.24 Recommendations to ensure future food security in the UK

For the producer	For the consumer
Grow more food per hectare but in a sustainable way that uses less energy and water.	Eliminate vast quantities of food waste.
Introduce more GM crops – increased spending on new crop varieties, livestock breeds and food innovation.	Tackle diet-related health issues by: • increasing the amount of fish in diets and developing aquaculture industries, e.g. aquaponics, where fish farming is integrated with hydroponics (growing plants such as lettuce in water) • improving information on the nutritional content of food, particularly with regard to salt, fat and sugar.
Consider the impacts of economic activities such as farming on our natural ecosystems and ensure that ecosystems that provide the inputs of water, soil and nutrients are protected and enhanced.	Increase availability of organic produce as this has the advantages of better soil fertility, natural fertilisers, no herbicides and restricted use of pesticides.
Ensure the food chain has efficient infrastructural support through seaports, transport systems, energy, water and sewage facilities.	Cut supply chains by encouraging local and regional sourcing of food products. Small independent food stores offer fresh produce and local diversity, e.g. Booths supermarkets in northwest England. Developments such as **farmers' markets** and direct marketing (door-to-door delivery of organic food boxes) offer the advantages of short supply chains. Eat more food that is in season to cut production and distribution costs.

✔ Review questions

1 Explain why food is described as a 'geopolitical commodity'.
2 What are trading blocs? How do trading blocs like the EU affect LIDCs?
3 What are the differences between bilateral and multilateral trading agreements?
4 State the advantages and disadvantages of food aid.
5 What is agribusiness?
6 Name a TNC involved in agribusiness and describe its operations in the food chain.
7 What is meant by food security?

Practice questions

A Level

Section A

1 a) Identify **three** limitations with the data evidence in Table 13.3 on page 412. [3 marks]

b) Explain how natural disasters can contribute to food insecurity. [6 marks]

Section B Synoptic

2 Assess how place profiles can be influenced by food retailers. [12 marks]

Section C

3 'Threats to global food security are physical.' How far do you agree with this statement? [33 marks]

AS Level

Section A

1 a) Explain the process of salinisation in soils. [4 marks]

b) Suggest why obesity is now a global issue. [6 marks]

c) Study **Table 1**, which shows agricultural machinery – tractors (wheel and crawler tractors) per 100 km² of arable land by country.

i) Using these data on agricultural machinery, calculate the standard deviation. You must show your workings. [4 marks]

Formula:

$$\sigma = \sqrt{\frac{\Sigma\left(x - \bar{x}\right)^2}{n}}$$

ii) Using evidence from Table 1, analyse the contrasts in usage of agricultural machinery. [6 marks]

d) Assess the extent to which government policy can impact food security. [12 marks]

Section B Synoptic

2 a) Examine the impact of modern farming methods on landscape systems. [8 marks]

b) With reference to **Figure A**, suggest how changing agricultural landscapes can affect people's perception of place. [8 marks]

Section C

3 Examine the extent to which technology is impacting food production. [20 marks]

Table 1

Country	Botswana	Brazil	Canada	Chile	Egypt	Spain	Iceland	Somalia	Sudan
Number of tractors	140	116	162	399	385	786	863	12	10

Figure A

Chapter 14

Hazardous Earth

Hazards occur when and where physical and human systems meet, sometimes in dramatic and violent ways. Extreme geophysical events such as volcanic eruptions and earthquakes present particular challenges for human activities. The concentrated release of energy poses huge and often unpredictable threats to human life and the built environment in LIDCs, EDCs and ACs.

14.1 What is the evidence for continental drift and plate tectonics?

Key idea
→ There is a variety of evidence for the theories of continental drift and plate tectonics

What is the basic structure of the Earth?

Studies of the way in which earthquake waves are transmitted through the Earth show that there is a concentric structure to the Earth's interior. Three primary concentric layers can be identified: the core, the mantle and the crust (Figure 14.1).

The core and mantle are separated by a sharp boundary at a depth of about 2900 km. The mantle–crust boundary is marked by the **Mohorovičić discontinuity**, or **Moho** as it usually abbreviated. Its depth below the continents averages around 35 km while under the oceans it is just 10–15 km below the surface. Table 14.1 lists the properties of the mantle and the crust.

As more research with improved techniques has investigated the Earth's structure, it has become clear that the upper mantle consists of two layers. The layer that extends from 100 km down to 300 km is known as the **asthenosphere**. This layer is semi-molten or viscous and is capable of flowing slowly. Lying immediately above this is the **lithosphere**, a rigid layer sandwiched between the crust and the asthenosphere.

Table 14.1 Summary of the properties of the mantle and the crust

	Thickness	Density (kg/m³)	Mineral composition
Crust	*Continental* Mean: 35 km Min: < 30 km Max: 70 km *Oceanic* 5–10 km	*Continental* 2.6–2.7 *Oceanic* 3.0	*Continental* Mainly granitic, silicon, aluminium (Sial) *Oceanic* Mainly basaltic, silicon and magnesium (Sima)
Mantle	To a depth of 2,900 km	3.3 at Moho 5.6 at core	Rich in magnesium and iron

The lithosphere varies in thickness and its boundary with the asthenosphere is difficult to define precisely as it starts to melt and becomes incorporated into the asthenosphere. Together, the lithosphere and crust make up oceanic and continental plates (Table 14.1).

It is thought that within the asthenosphere, convection currents exist, caused by vast amounts of heat generated deep in the mantle. As a result, the semi-molten asthenosphere flows carrying with it the solid lithosphere and crust.

Research into how and why the asthenosphere moves continues, with several possible mechanisms being investigated.

Stretch and challenge

1 Research the following ideas of how the crust and the lithosphere move:
   a) shallow and deep convection
   b) ridge-push
   c) slab pull.
2 Make an assessment of the pros and cons of each idea.

Continental drift and the theory of plate tectonics

During the eighteenth century maps of the world multiplied and became more accurate, and the shapes of the continents and their relationships to each other attracted interest. In 1801 the German geographer

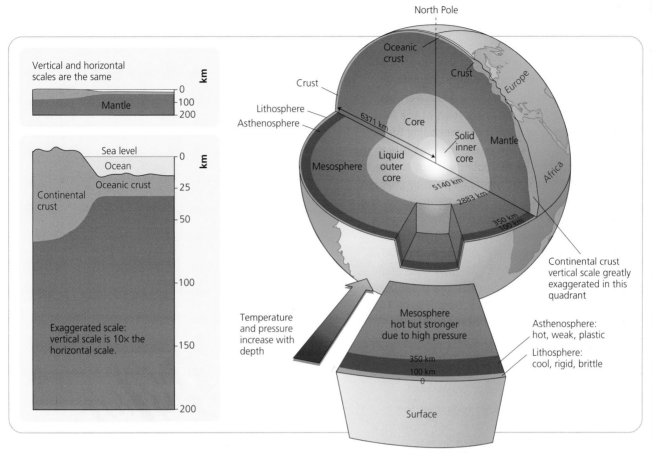

Figure 14.1 The Earth's internal structure (Source: *Advanced Geography; Concepts & Cases* by P. Guinness & G. Nagle (Hodder Education, 1999), p.332)

and explorer Alexander von Humboldt wrote about the apparent fit between northeast South America and the west coast of Africa. However, it was not until the early twentieth century that ideas concerning the geographical positions of the continents started to receive serious attention.

Continental drift – the idea of Alfred Wegener

In 1912 Alfred Wegener made a comprehensive case for **continental drift** (Table 14.2) and worked on his ideas until his death in 1930. He proposed that in the Carboniferous period, 250 million years ago, a large single continent, Pangaea, existed. This slowly broke apart into two large land masses: Laurasia to the north and Gondwanaland to the south. This movement continued to the present day as the continents separated and spread across the globe.

Although scientists could recognise the value of Wegener's theory it failed to provide a mechanism to explain the movement of continents. Without this the scientific establishment remained sceptical. Thus until

the mid-twentieth century Wegener's ideas existed as little more than footnotes.

Palaeomagnetism
Technology designed to track submarines measured very small variations in the Earth's magnetic field.

Table 14.2 Wegener's evidence for continental drift

Geological evidence	Biological evidence
• The fit of continents such as South America and Africa on either side of the Atlantic. • Evidence from about 290 million years ago of the effects of contemporaneous glaciation in southern Africa, Australia, South America, India and Antarctica, suggesting that these land masses were joined at this time, located close to the South Pole. • Mountain chains and some rock sequences on either side of oceans show great similarity, e.g. northeast Canada and northern Scotland.	• Similar fossil brachiopods (marine shellfish) found in Australian and Indian limestones. • Similar fossil reptiles found in South America and South Africa. • Fossils from rocks younger than the Carboniferous period, in places such as Australia and India, showing fewer similarities, suggesting that they followed different evolutionary paths.

When used by geologists the magnetic field showed up as a striped pattern across the ocean floor. Explaining this pattern begins with the igneous rocks, which form the oceanic crust and the ocean floor. These rocks, which originate as lava flows, contain iron particles. As lava erupts, it cools and the magnetic orientation of the iron particles is locked into the rock, depending on the Earth's polarity at the time. However, it was discovered that the Earth's polarity is not constant. Every 400,000–500,000 years, its polarity changes orientation and this is recorded in the rocks on the ocean floor. This ancient record of changes in the Earth's polarity is known as palaeomagnetism.

When the palaeomagnetic data from the mid-Atlantic and off the coast of California were made available to non-military scientists in the 1960s, the idea of **sea-floor spreading** was born (Figure 14.2).

The width of each strip of ocean bed with the same magnetic orientation was found to correspond with the time scale of each magnetic reversal. The symmetrical pattern of geomagnetic reversals on either side of mid-ocean ridges indicated that as fresh molten rock from the asthenosphere reached the ocean bed, older rock was 'pushed' away from the ridge. In this way, sea-floor spreading moves material across the ocean floors as a 'conveyor belt' operating on each side of the mid-ocean ridge. Eventually, the sea floor reaches an ocean trench where material is subducted into the asthenosphere and becomes semi-molten.

The ideas of sea-floor spreading and the movement of the continents were linked in the mid-1960s. With increased knowledge about the shape and size of crustal plates and their boundaries, it became clear that the plates were moved by sea-floor spreading from the mid-ocean ridges to subduction at ocean trenches. The continents were carried by the plates and this supported Wegener's idea of the break-up of Pangaea and the subsequent drift of the continents.

The age of sea-floor rocks

During the 1960s an ocean-drilling programme was established which investigated ocean sediments and crustal rocks on the deep ocean floor. Drilling recovered cores in water up to 7000 m deep and the cores revealed a spatial pattern of sediments that supported the theory of sea-floor spreading. The thickest and oldest sediments were found nearest to the continents. However, the cores also showed that

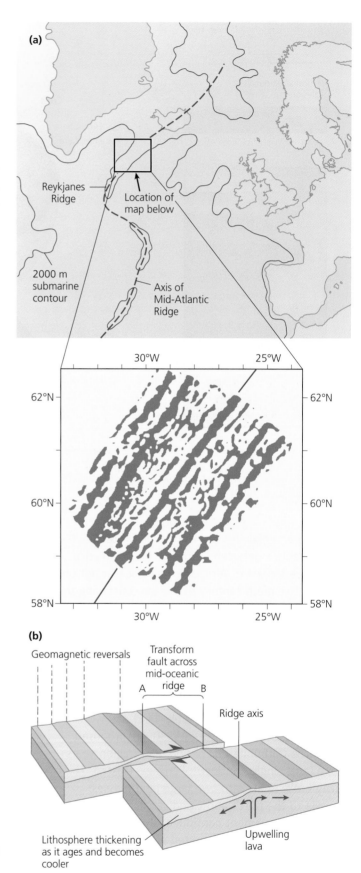

Figure 14.2 Sea-floor spreading and palaeomagnetism

nowhere in the oceans was rock older than 200 million years. This confirmed that the ocean crust was constantly recycled over this period (see also pages 367–69).

> **Key idea**
> → There are distinctive features and processes at plate boundaries

The global pattern of plates and plate boundaries

Although ideas about continental drift and plate tectonics were gaining acceptance in the 1950s, very little was known about tectonic plates. Again science being used for military purposes revealed invaluable data. From the mid-1960s nuclear tests were conducted underground. These released vast amounts of energy which were picked up on seismometers.

The detailed maps produced from seismic data worldwide showed that most earthquakes, and in particular the high-magnitude ones, were spatially concentrated in narrow bands. In between were relatively large areas that generated few earthquakes. This suggested that the rigid lithosphere and crust were broken up into **tectonic plates**. Moreover, these giant slabs of lithosphere and crust were not static: in some places they were moving apart, in others they were converging.

Today, the pattern of seven major plates and many more minor ones is well known (Figure 14.3), although the details of plate boundaries is still the subject of research.

The features and processes associated with tectonic plate boundaries

There are three types of plate boundary: divergent (constructive), convergent (destructive) and conservative. The relative movements of plates at these boundaries are critical in determining the different processes operating there. In turn, distinctive landforms and landscapes develop along the plate boundaries.

Divergent (constructive) plate boundaries

Locations where plates are diverging (moving apart) are where magma is rising through the asthenosphere and forcing its way to the surface. Mostly this takes place at the mid-ocean ridges (Figure 14.7a). Mid-ocean ridges are probably the most spectacular relief feature on the planet but remain hidden at an average depth of 2.5 km below the ocean surface. They consist of very long chains of mountains, in places rising 3000 m above the sea bed. Added together, these submarine mountain chains have a combined length of 60,000 km (see Figure 12.4).

Mid-ocean ridges are not continuous; at frequent intervals they are broken into segments by **transform faults**. These faults displace the ridge sideways by tens or in places hundreds of kilometres. Volcanic activity is absent along transform faults, but as they slip, energy is released in the form of earthquakes (Figure 14.2).

Mid-ocean ridges vary in shape depending on the rate of spreading and this is determined by the amount and rate of magma brought to the surface by convection currents (Table 14.3).

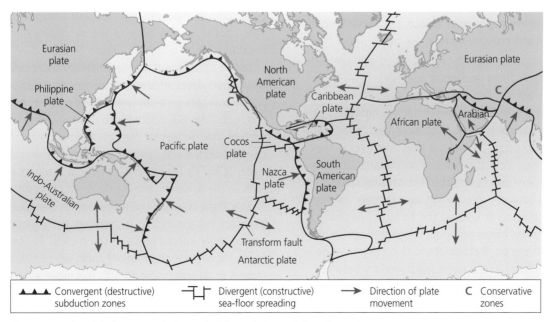

Figure 14.3 The global pattern of plates, their boundaries and directions of movement

Table 14.3 Types of spreading mid-oceanic ridges

Boundary type	Rate of spread	Type of landforms	Example
Fast spreading	Up to 16.5 cm per year	Broad and smooth mountain chains; no central rift valleys	East Pacific Rise (Pacific and Nazca plates)
Medium spreading	5–10 cm per year	Poorly defined central rift valleys; relatively smooth mountain chains	Galapagos Ridge (just south of the Nazca and Cocos plate boundary)
Slow spreading	2–3 cm per year	Steep slopes to mountain chains; clearly defined central rift valleys	Mid-Atlantic Ridge (North America and Eurasia and South America and African plates)

The eruption of magma along divergent boundaries occurs mostly underwater. Magma erupting directly on to the sea bed is cooled rapidly, forming rounded mounds called **pillow lavas**.

As magma rises to the surface, the overlying rocks can be forced up into a dome. The rigid lithosphere is placed under great stress and eventually fractures along parallel faults. This produces the underwater rift valleys found along mid-ocean ridges. In the North Atlantic, the extrusion of magma has been so great it has created the world's largest volcanic island – Iceland. There the central rift valley is a prominent landscape feature (Figure 14.4).

At mid-ocean ridges, sea water seeps into rifts and is superheated. As it rises towards the surface it causes chemical changes in the basaltic rocks. Superheated jets of water sometimes re-emerge on the ocean floor containing metal sulphides. These features are commonly known as **black smokers** and support unique and highly specialised organisms and ecosystems (see also pages 374–77).

Rifting away from mid-oceanic ridges

Rift zones are not confined to the ocean floor; they also occur on land and in part explain how continents can break up. The continental crust must thin considerably for rifting to occur. One of the best examples is the rift stretching from the Red Sea northwards to Turkey. Here the crust has been stretched, causing faulting and forming a sunken valley known as a **graben**. As the rift widened, magma erupted at the surface. Eventually the rift valley sank below sea level, forming the present-day Red Sea. If the crust continues to thin here, magma will well up to form a new spreading boundary between Africa and the Arabian Peninsula. Further north in Israel, the rift dips below sea level, forming the Dead Sea. At −400 m it is the lowest point on the continental surface (Figure 14.3).

Figure 14.4 Iceland's rift valley

Convergent (destructive) boundaries

Oceanic and continental tectonic plates may converge in one of three different combinations:

- oceanic–continental
- oceanic–oceanic
- continental–continental.

Oceanic–continental plate margins

Because of their different densities, when oceanic and continental plates converge, the denser oceanic plate is forced under the continental plate. This process is known as **subduction**.

Subduction causes a deepening of the ocean at the plate boundary and forms an ocean **trench**. Ocean trenches are long, narrow depressions with depths of 6000–11,000 m. They are usually asymmetric, with their steepest side towards the continent. Ocean trenches mark the zone of subduction where the oceanic crust descends into the asthenosphere. As it does so, the leading edge of the overriding plate is buckled to form a trench (Figure 14.5).

Layers of sediment and sedimentary rocks develop on oceanic plates adjacent to continents. As an oceanic plate converges on a continental plate, these sediments and rocks crumple, fold and are uplifted along the leading edge of the continental plate. In addition, the continental crust is buckled and uplifted, and vast amounts of molten material are injected into it. The result is mountain chains such as the Andes on the Pacific coast of South America.

The angle at which the oceanic plate is subducted is between 30° and 70°. As it descends, the oceanic plate comes under intense pressure and friction. Faulting and fracturing occur in the **Benioff zone**, where the descending plate is at an angle close to 45°. This process releases vast amounts of energy in the form of earthquakes.

Subduction also causes the oceanic plate to melt. Because the melted material is less dense than its surroundings, it rises towards the surface as **plutons** of

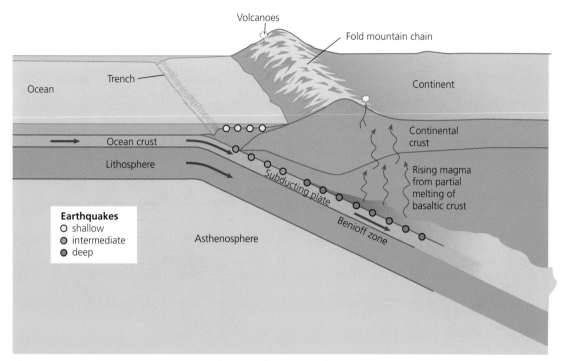

Figure 14.5 Ocean–continent convergent plate boundary

magma. Huge intrusions of magma create further uplift of fold mountains. Where rising magma reaches the surface it forms volcanoes.

Oceanic–oceanic plate margins

When two oceanic plates meet, the slightly denser one will subduct under the other, creating a trench (Figure 14.6). As the descending plate melts, magma rises to the surface and forms chains of volcanic islands known as **island arcs**. In the Central Atlantic, the North American plate is subducted beneath the smaller Caribbean plate, forming island arcs such as the Antilles. The northwest zone of the Pacific 'Rim of Fire' contains island arcs such as the Aleutian Islands, which extend westwards from Alaska. Relatively close by is the Challenger Deep, the deepest part of the world's oceans. Part of the Mariana Trench, it lies 11,000 m below the ocean surface and is located where the Pacific plate is subducted beneath the Philippine plate.

Stretch and challenge

1 Research the eruption of Santorini in 1628 BC.
2 From the information you have gathered, write an eye-witness account of someone living on the island of Crete at that time. Include details of the events of the eruption and describe what features you would have seen.
3 Conclude your account with comments following a visit to the island of Thira (Santorini) a few weeks after the eruption. Describe what you found.
4 Write a short account written several years after the eruption, describing what had happened to your way of life on Crete in the intervening years.
5 Write an analysis of the eruption from the perspective of a twenty-first-century scientist. Offer explanations as to the causes and consequences of the eruption.

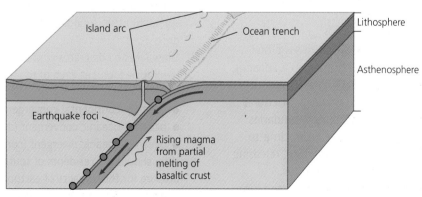

Figure 14.6 Oceanic–oceanic convergent plate boundary

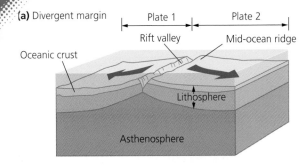

(a) Divergent margin

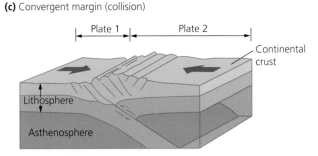

(c) Convergent margin (collision)

(b) Convergent margin (subduction)

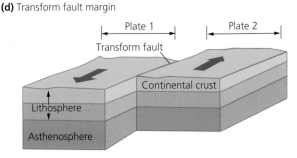

(d) Transform fault margin

Figure 14.7 Continental–continental convergent (collision) plate boundary (Source: *Advanced Geography: Concepts & Cases* by P. Guinness & G. Nagle (Hodder Education, 1999), p. 334)

Continental–continental plate margins

When two continental plates converge, little if any subduction takes place (Figure 14.7c). This is because the two plates have similar densities. In Europe, the collision of the African and Eurasian plates over the past 40 million years has created the Alps.

> ### Activities
>
> 1 Research the formation of the Himalayas.
> 2 Working in small groups, make a presentation that describes and explains how and why the Himalayas have been formed. Make sure you have the sequence of events in their correct order. Start with the break-up of Gondwanaland and the drift of India towards Eurasia.

Conservative plate margins

At some places, tectonic plates neither diverge nor converge but rather slide past each other in a shearing motion (Figure 14.7d). Volcanic activity is absent at these conservative plate margins. However, frictional resistance to movement along the plate boundaries often causes the build-up of pressure. From time to time, these pressures cause rocks to fracture, releasing enormous amounts of energy as earthquakes.

Conservative plate margins are not associated with spectacular landforms. However, in some places it is possible to discern active plate boundaries extending through the landscape like a giant tear. And where rocks are exposed at the surface, the extent of movement between strata may be visible. Drainage is also modified as river courses are deflected by movements along the faults.

The most famous example of a conservative plate boundary is in California, where the North American and Pacific plates slide past each other along the San Andreas fault system (Figure 14.8).

✓ Review questions

1 Describe the internal structure of the Earth.
2 What are the main differences between continental and oceanic crust?
3 What evidence is there for sea-floor spreading?
4 Outline the evidence (apart from sea-floor spreading) that supports the theory of continental drift.
5 Draw labelled diagrams to show the processes and landforms found at each of the following plate boundaries:
   ● oceanic–continental convergent (destructive)
   ● oceanic–oceanic convergent (destructive
   ● oceanic–oceanic divergent (constructive).
6 Why should the residents of southern California prepare for the impacts of earthquakes but not the impacts of volcanic eruptions?

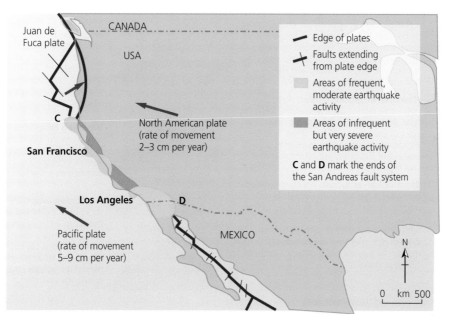

Figure 14.8 A conservative plate boundary: the San Andreas fault system, California

14.2 What are the main hazards generated by volcanic activity?

> **Key idea**
> → There is a variety of volcanic activity and resultant landforms and landscapes

The nature of vulcanicity

Volcanic eruptions produce a varied range of landforms. This diversity is related to where the eruption occurs, the type of lava, the variety of materials such as ash, pumice, gases produced, and how the eruption takes place. For example, differences in the chemistry of the magma (molten rock below the surface) influences the type of eruption and the shape of the volcano. Once magma is ejected at the surface, its behaviour is affected by its **viscosity**, which in turn is determined by its chemical composition and temperature.

Eruptions can be divided into two groups: **explosive** and **effusive** (Table 14.4).

Volcanoes can be classified according to the type of eruption (Figure 14.9).

The products of explosive eruptions

Strato-volcanoes, sometimes known as **composite cone volcanoes**, made up of layers of ash and acid lava, have concave symmetrical profiles (Figure 14.10). Most contain complex internal networks of lava flows which form minor igneous features such as **sills** and **dykes**.

Table 14.4 Characteristics of explosive and effusive eruptions

	Explosive eruptions	**Effusive eruptions**
Location	Convergent plate boundaries	Divergent plate boundaries
Type of lava	Rhyolite (more acid); andesite (less acid)	Basalt
Lava characteristics	Acid (high % silica), high viscosity, lower temperature at eruption	Basic (low % silica), low viscosity, higher temperature at eruption
Style of eruption	Violent bursting of gas bubbles when magma reaches surface; highly explosive; vent and top of cone often shattered	Gas bubbles expand freely; limited explosive force
Materials erupted	Gas, dust, ash, lava bombs, tephra	Gas, lava flows
Frequency of eruption	Tend to have long periods with no activity	Tend to be more frequent; an eruption can continue for many months
Shape of volcano	Steep-sided strato-volcanoes; caldera	Gently sloping sides, shield volcanoes; lava plateaux when eruption from multiple fissures

Because acid magma does not flow easily, the vents of strato-volcanoes are often filled with a mass of solidified magma. This plug prevents magma from rising freely from depth. As a result, enormous pressures can build up inside a volcano until eventually it erupts explosively, sometimes literally blowing its top off.

Calderas are volcanic craters, usually more than 2 km in diameter. They develop when an explosive

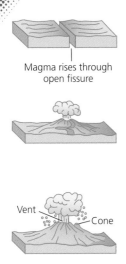

Magma rises through
open fissure

Icelandic lava eruptions are characterised
by persistent fissure eruption. Large
quantities of basaltic lava build up vast
horizontal plains. On a large scale they have
formed the Deccan Plateau and the
Columbia Plateau.

Hawaiian eruptions involve more
noticeable central activity than the Icelandic
type. Runny, basaltic lava travels down the
sides of the volcano in lava flows. Gases
escape easily. Occasional pyroclastic activity
occurs but this is less important than the
lava eruption.

Vent

Cone

Strombolian eruptions are characterised
by frequent gas explosions which blast
fragments of runny lava into the air to form
cones. They are very explosive eruptions
with large quantities of pyroclastic rock
thrown out. Eruptions are commonly
marked by a white cloud of steam emitted
from the crater.

In **Vulcanian eruptions**, violent gas
explosions blast out plugs of sticky or cooled
lava. Fragments build up into cones of ash and
pumice. Vulcanian eruptions occur when there
is very viscous lava which solidifies rapidly
after an explosion. Often the eruption clears a
blocked vent and spews large quantities of
volcanic ash into the atmosphere.

Vesuvian eruptions are characterised by
very powerful blasts of gas pushing ash clouds
high into the sky. They are more violent than
vulcanian eruptions. Lava flows also occur.
Ash falls to cover surrounding areas.

In a **Plinian eruption**, gas rushes up through
sticky lava and blasts ash and fragments into
the sky in a huge explosion. The violent
eruptions create immense clouds of gas and
volcanic debris several kilometres thick. Gas
clouds and lava can also rush down the slopes.
Part of the volcano may be blasted away
during the eruption.

Figure 14.9 Types of eruptions and their volcanoes

eruption destroys much of the cone and the underlying
magma chamber is largely emptied. Without the
support of the underground magma, the sides of the
volcano collapse to form a caldera. The 1883 eruption
of Krakatoa in Indonesia left a caldera 7 km wide.

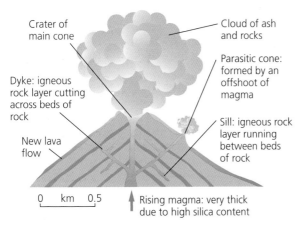

Crater of
main cone

Cloud of ash
and rocks

Dyke: igneous
rock layer cutting
across beds of
rock

Parasitic cone:
formed by an
offshoot of
magma

Sill: igneous rock
layer running
between beds
of rock

New lava
flow

0   km   0.5

Rising magma: very thick
due to high silica content

Figure 14.10 A generalised composite volcano

Convergent plate boundaries and explosive eruptions
often give rise to chains of strato-volcanoes. Indonesia,
for example, has 130 active strato-volcanoes along its
archipelago, while in the Andes, 66 strato-volcanoes
have erupted in recorded history.

The products of effusive eruptions

Lava plateaux
When basic magma erupts from multiple fissures,
vast areas can be covered by free-flowing lava. These
events are known as **flood basalts**. Examples include
the Deccan Plateau in central India, which covers more
than 500,000 km², while the Columbia Plateau in the
northwest USA covers 130,000 km². When first formed
these plateaux have a uniform slope of around 1°.

However, millions of years of **denudation** have created
more varied relief. No large-scale flood basalt events
have taken place in the past 50 million years.

Shield volcanoes
Effusive eruptions are found at divergent plate
boundaries. Because these boundaries coincide with
mid-ocean ridges, most effusive eruptions occur unseen
on the ocean floor. Iceland, however, is an exception.
It owes its formation to effusive volcanic activity and
is one of the most active volcanic regions in the world.
Eruptions of basic lava result in volcanoes with gently
sloping sides. If successive flows accumulate for long
enough, huge volcanoes, extending horizontally for
tens of kilometres, such as Skjäldbreidur on Iceland,
can develop.

Eruptions at hot spots
The Hawaiian chain of islands lies at the centre of the
Pacific plate, thousands of kilometres from the nearest
plate boundary (Figure 14.11). Its formation is due to
the existence of a **hot spot**. A hot spot is a fixed area
of intense volcanic activity where magma from a rising
mantle plume reaches the Earth's surface.

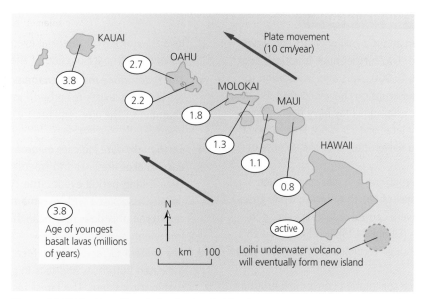

Figure 14.11 The Hawaiian chain of volcanic islands

As the Pacific plate has slowly moved northwest over the Hawaiian hot spot (at an average rate of 10 cm/year), vast amounts of basalt have accumulated on the ocean floor to produce the Hawaiian islands. The active volcanic peaks of Mauna Loa and Mauna Kea on the Big Island (Hawaii) reach over 4000 m above sea level and rise more than 9000 m from the ocean floor. Over millions of years, as the Pacific plate moves northwest and away from the hot spot, the volcanoes lose their source of magma and become extinct.

Thus on the older islands in the Hawaiian chain, such as Kauai, volcanism is no longer active. There, weathering and erosion have broken down the volcanic rocks into deep and fertile soils and spectacular valleys have been carved by rivers. Further along the chain to the northwest, the volcanic islands have sunk below the surface of the Pacific to form underwater mountains or seamounts.

Some 30 km to the southeast of the Big Island, the next volcano in the chain, Loihi, is slowly rising up from the ocean floor. Currently the summit of this seamount is 970 m below sea level. However, its position directly above the hot spot means that it will continue to grow. Eventually it will emerge above sea level to form a new island in the Hawaiian chain.

Running through East Africa is a 4000 km long rift valley containing several active volcanoes. Over the past 30 million years, the crust has been stretched, causing tension within the local rocks. The result is rifting, with magma forcing its way to the surface and creating a line of active volcanoes. Mount Kilimanjaro, the highest peak in Africa, was formed in this way. Erta Ale is another active volcano: a basaltic shield volcano, it is located in the Afar region of northeastern Ethiopia.

Not all hot spot volcanism leads to shield volcanoes. El Teide on Tenerife in the Canary Islands is a strato-volcano which has erupted on several occasions since 1700. The most recent was in 1909.

Super-volcanoes

A volcano that erupts more than 1000 km³ of material in a single eruption event is known as a **super-volcano**. Super-volcanoes exist as giant calderas. The Yellowstone super-volcano in Wyoming has a caldera measuring 75 km in diameter. Apart from Yellowstone, other super-volcanoes that have been active in the past 2 million years include Long Valley in eastern California and Toba in Indonesia. The most recent super-volcanic eruption occurred 27,000 years ago at Taupo, North Island, New Zealand (Figure 14.12).

The impact of these very high-magnitude events is deduced from the extent and depth of ash layers and their impact on plant, insect and animal species (palaeobiology).

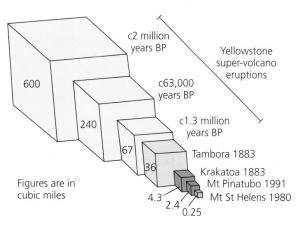

Figure 14.12 Volume of erupted material of selected volcanic eruptions

The Yellowstone super-volcano

Volcanic activity began about 2 million years ago when a giant caldera formed. Two further huge eruptions took place 1.3 million and 640,000 years ago. In addition to these, dozens of smaller eruptions have occurred. Because the evidence of earlier eruptions may have been either buried or destroyed by later ones, it is not known exactly how often the volcano has erupted, nor the volume of material extruded.

During the largest eruptions, ash was spread over much of the western half of North America, while gases and tiny particulates were carried around the globe in the upper atmosphere (Figure 14.13). It is likely that this caused a noticeable, though temporary, decrease in global temperatures.

There has been much speculation concerning eruption events by super-volcanoes and their impacts. Some key misconceptions are:

- *Yellowstone is overdue a very high-magnitude event.* Tectonic events do not follow a regular pattern, making accurate and reliable prediction impossible.
- *When Yellowstone erupts it will have severe global consequences.* The most probable eruption will be a lava flow or a hydrothermal event. These would have little if any impact outside of the Yellowstone region.
- *The Yellowstone magma chamber is growing.* No evidence exists for growth in a near-surface chamber of the size required for a super-volcanic eruption.
- *Earthquake data indicate magma is moving.* Recent earthquakes are linked to fractures in brittle rocks. Even if long-period events (the type of earthquake usually associated with magma movements) are registered, this does not mean an eruption is imminent.
- *The caldera surface is rising.* Measurements have indicated both rises and falls in ground surfaces within the caldera. The latest research suggests that volcanoes can display quite rapid ground movements without necessarily erupting.

Yellowstone last erupted about 70,000 years ago. Given that the hot spot under Yellowstone heats, stretches and weakens the crustal rocks, volcanic activity could occur at any time. However, the probability of a super-volcano style eruption is exceedingly low.

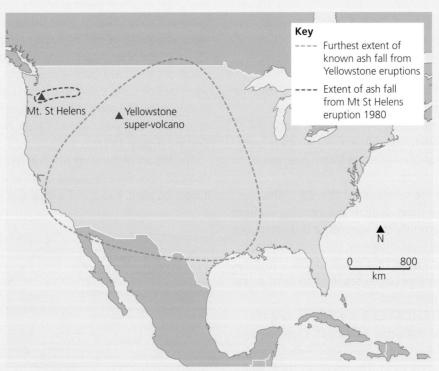

Figure 14.13 Known ash fall boundary for major Yellowstone eruptions

Measuring and assessing volcanic activity

Eruptions vary in the amount of material released, their type (effusive/explosive) and duration. It is, therefore, difficult to compare one eruption with another. However, two key factors need to be considered:

- Magnitude – the amount of material erupted.
- Intensity – the speed at which material is erupted.

Imagine two volcanic events. One is an eruption that produces small quantities of material day to day or week to week but which erupts more or less continuously over many months or even years. The other erupts the same quantity of material but in just a few hours. Although in terms of magnitude they are the same, they are very different types of volcanic event.

The most widely used measure of eruption is the **Volcanic Explosivity Index (VEI)**. It combines magnitude and intensity into a single number on a scale from 0, the least explosive, to 8, the most explosive. Each increase in number represents nearly a ten-fold increase in explosivity (Table 14.5).

In assessing the explosivity of an eruption, several factors are taken into account: the volume of erupted material, the height ejected material reaches, the duration in hours and various qualitative descriptions.

The VEI is not that useful for effusive eruptions such as those on Hawaii. However, it is valuable in suggesting the relative impacts volcanoes might have at different geographical scales.

Table 14.5 VEI comparisons for some selected eruptions

Name	VEI
Kilauea, Hawaii, present day	0
Montserrat, 1995	3
Eyjafjallajökull, 2010	4
Mount St Helens, 1980; Vesuvius, 79	5
Mount Pinatubo, 1991	5–6
Krakatoa, 1883	6
Tambora, 1815	7
Yellowstone, 640,000 Before Present	8

Key idea
→ Volcanic eruptions generate distinctive hazards

Hazards produced by volcanic activity

As with all tectonic activity, volcanoes become hazardous when they interact with human communities and activities (Figure 14.14).

- Lava flows. The impacts are dependent on type of lava. Basic (basaltic) lava is free-flowing and can run for considerable distances. On Hawaii in July 2015 a lava flow extended for 20 km before stopping and in August the same year a flow was reported to have covered about 800 m in a day. Acidic lavas

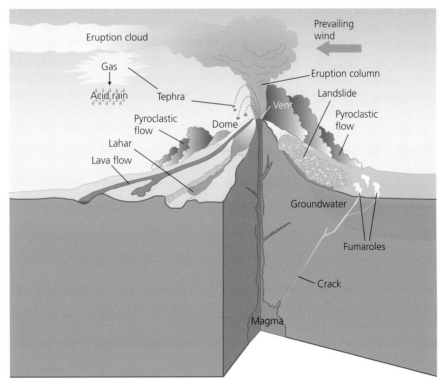

Figure 14.14 Types of volcanic hazard

477

such as rhyolite are thick and pasty so do not flow easily. Everything in the path of lava will be either burned, bulldozed or buried. Although they destroy infrastructure, property and crops, lava flows rarely cause injuries or fatalities.

- Pyroclastic flows. These are a combination of very hot gases (500°C+), ash and rock fragments travelling at high speed (100 km/h). They follow the contours of the ground and destroy everything in their path. The inhalation of such hot and poisonous gas and ash causes almost instant death. The Roman town of Pompeii was overwhelmed by a pyroclastic flow from Mount Vesuvius in AD 79.
- Tephra. This describes any material ejected from a volcano into the air. It ranges in size from very fine ash to large volcanic bombs (>6 cm across). It also includes lighter debris such as pumice. Tephra is potentially very hazardous, burying farmland in layers of ash and destroying crops. Transport can be disrupted both on the ground and in the air. The eruption of Iceland's Eyjafjallajökull volcano in April 2010 led to the cancellation of 100,000 flights. Buildings can collapse due to the weight of accumulated ash and people with respiratory diseases may have difficulty breathing.
- Eruptions emit a wide range of toxic gases, including CO, CO_2 and SO_2. These gases can pose a deadly (and silent) threat to human populations. When SO_2 combines with atmospheric water, acid rain is produced. This enhances weathering and can damage crops and pollute surface water and soils.
- Lahars. These are a type of mud flow with the consistency of wet concrete. Snow and ice on a volcano summit melt during an eruption and flow rapidly down the cone. Rock fragments large and small, as well as ash and soil, are mixed together. Lahars can travel at speeds up to about 50 km/h. Everything in their path is either destroyed or buried

under thick layers of debris. In 1984, following the eruption of Nevado del Ruiz, the Colombian town of Armero was overwhelmed by lahars, resulting in the deaths of 23,000 people. In places such as Southeast Asia, ash-covered slopes of volcanoes continue to generate lahar hazards after periods of heavy rain.
- Floods. Volcanic eruptions beneath an ice field or glacier cause rapid melting. In Iceland, several active volcanoes lie under the Vatnajökull ice field. During an eruption, vast quantities of water accumulate until they find an exit from under the ice. The resulting torrent of water, which can cause devastating floods, is known locally as jökulhlaup.
- Tsunami. The violent eruption of some island volcanoes can cause massive displacement of ocean water and tsunami waves capable of travelling at speeds of up to 600 km/h. In deep water they have a height that is usually less than 1 m and a very long wavelength of up to 200 km. Approaching the shore, tsunami waves increase rapidly in height and when they break, transfer vast amounts of energy and water along the shore and inland. The tsunamis created by the eruption of Krakatoa in 1883 are believed to have drowned 36,000 people.

Such is the nature of volcanic eruptions that most of the hazards they pose are relatively short term. However, eruptions emitting very large quantities of ash into the upper atmosphere can have implications for longer-term climatic change. Past eruptions of super-volcanoes such as Toba, Indonesia (some time between 69,000 and 77,000 years ago) led to reductions in global temperatures as ash blocked sunlight from reaching the Earth's surface. The release of SO_2 also added to cooling. Mixed with water in the atmosphere SO_2 forms sulphuric acid, which reflects insolation.

Lake Nyos, Cameroon

Lake Nyos is one of a number of deep lakes that occupy volcanic craters in Cameroon in West Africa (Figure 14.15). It is 2 km wide and 200 m deep. In August 1986, 1700 people and all animal life in the area around the volcano were asphyxiated. The cause was a leak of CO_2 from a volcanic crater lake. The gas had built up at the bottom of the lake after being emitted from the underlying magma chamber. CO_2 is a dense gas and

when it escaped it flowed down the volcano slopes as a 50 m thick ground-hugging layer travelling at about 70 km/h.

It is not known for certain what caused the CO_2 to escape the lake. Possible explanations are a deep movement of magma, an earthquake, a change in water temperature in the lake, or strong winds stirring up the lake waters. →

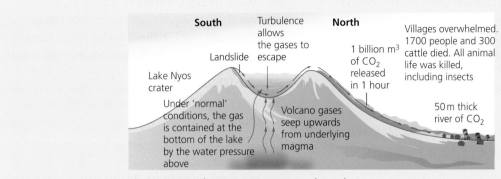

Figure 14.15 Lake Nyos: poisonous gas hazard

South
Turbulence allows the gases to escape
North
Landslide
1 billion m³ of CO_2 released in 1 hour
Villages overwhelmed. 1700 people and 300 cattle died. All animal life was killed, including insects
Lake Nyos crater
Under 'normal' conditions, the gas is contained at the bottom of the lake by the water pressure above
Volcano gases seep upwards from underlying magma
50 m thick river of CO_2

Activities

The hazard event profile is a diagram that represents the characteristics of hazards (Figure 14.16).

1 Justify the profile shown in Figure 14.16, which is for a volcanic eruption on Hawaii.
2 On a copy of Figure 14.16, draw the hazard event profile for an eruption from an Indonesian volcano.
3 Justify the profile you have drawn.
4 Add the event profile of a super-volcano.
5 Evaluate the usefulness of the hazard event profile in understanding differences in the hazards produced by volcanoes.

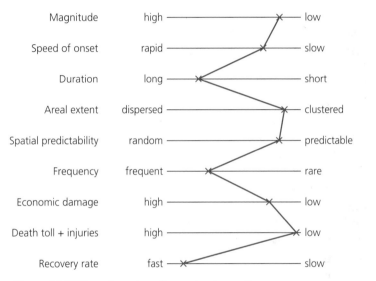

Magnitude	high — low
Speed of onset	rapid — slow
Duration	long — short
Areal extent	dispersed — clustered
Spatial predictability	random — predictable
Frequency	frequent — rare
Economic damage	high — low
Death toll + injuries	high — low
Recovery rate	fast — slow

Figure 14.16 Hazard event profile – volcanic eruptions

Review questions

1 What is meant by the viscosity of lava?
2 Draw labelled cross-sections of the following types of volcanoes to describe and explain their shape:
   ● strato-volcano
   ● shield volcano.
3 Draw a sequence of labelled diagrams to illustrate the formation of a caldera.
4 How have the Hawaiian chain of islands been formed?
5 What are 'super-volcanoes'?
6 What is the VEI and how is it calculated?
7 What is the influence of the type of lava on volcanic hazards?
8 What role can meltwater play in volcanic hazards?

14.3 What are the main hazards generated by seismic activity?

Key idea
→ There is a variety of earthquake activity and resultant landforms and landscapes

What is an earthquake?

Earthquakes represent the release of stress that has built up within the Earth's crust caused by tension, compression or the shearing of rocks. A series of seismic shock waves originates from the earthquake **focus**, the location where the stress is suddenly released. Immediately above this point, at the Earth's surface, is the **epicentre**.

Many earthquake events are preceded by a number of fore-shocks. Later a series of after-shocks follows the main quake, which gradually reduces in intensity.

Earthquake activity tends to be concentrated in one of four locations:

- Mid-ocean ridges – tensional forces associated with spreading processes and subsequent faulting and rifting.
- Ocean trenches and island arcs – compressive forces associated with the subduction of one plate below another.
- Collision zones – compressive forces associated with the grinding together of plates carrying continental crust.
- Conservative plate margins – shearing forces associated with the intermittent movement of one plate past another.

Although the great majority of earthquakes occur on or close to tectonic plate margins, stress within the crust is widespread. As a result, earthquakes can occur anywhere. Locations such as the UK, which were tectonically active in the past, still preserve old fault lines along which movement can take place. Smaller earth tremors are also associated with the collapse of old mine workings or the extraction of gas and oil from rocks deep underground.

Different types of earthquakes

Seismic waves

Seismic waves can travel both along the surface and through the layers of the Earth. There are three types of waves:

- Primary (P) waves – fast-travelling, low-frequency compressional waves. They vibrate in the direction in which they travel.
- Secondary (S) waves – half the speed of P waves, high-frequency waves. They vibrate at right angles to the direction in which they travel.
- Surface (L) waves – slowest of the three types, low-frequency waves. Some L waves have a rolling movement that moves the surface vertically while others move the ground at right angles to the direction of movement.

L waves travel through the outer crust only, S waves cannot pass through liquids (and so cannot travel through the outer core), while P waves travel through the Earth's interior (through both solids and liquids).

Depth of focus

Earthquakes are often categorised according to their depth of focus.

- Shallow focus – surface down to about 70 km. Shallow quakes occur in cold, brittle rocks resulting from the fracturing of rock due to stress within the crust. They are very common, with many releasing only low levels of energy, although other high-energy shallow quakes are capable of causing severe impacts.
- Deep focus – 70 to 700 km. Deep quakes are poorly understood. With increasing depth, pressure and temperatures increase to very high levels. Minerals change type and volume, which may contribute to a release of energy. It may be that water has a role in releasing energy but scientists continue to evolve their ideas about these less frequent but often powerful quakes.

Assessing earthquake energy

The magnitude and destructive power of earthquakes can be assessed in different ways. The **Richter scale**, developed in 1935, uses the amplitude of seismic waves to determine earthquake magnitude (Table 14.6). The scale is logarithmic so each whole-number increase in magnitude represents a ten-fold increase in the amplitude of the seismic wave. This represents a 30-fold increase in the release of energy.

The Richter scale has no upper limit, though the largest earthquakes record a magnitude of around 9 (e.g. Sendai quake off the Pacific coast of Honshu, Japan in 2011). It is important to understand that the Richter scale is not used to express damage. An earthquake's damage is determined partly by magnitude but also by other factors such as population density and levels of preparedness.

Table 14.6 Richter scale and energy released

Richter scale	Ground motion	Energy released
1	1	1
2	10	30
3	100	90
4	1,000	27,000
5	10,000	810,000
6	100,000	24,300,000
7	1,000,000	729,000,000
8	10,000,000	21,870,000,000
9	100,000,000	656,100,000,000

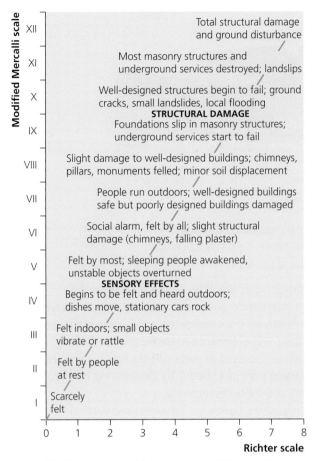

Figure 14.17 The relationship between the Richter and the Modified Mercalli scales

Table 14.7 Moment Magnitude scale

Moment Magnitude scale	Example
1.0	–
2.0	–
3.0	–
4.0	–
5.0	Lincolnshire, UK, 2008
6.0	Chile, 2015
7.0	Haiti, 2010; Nepal, 2015
8.0	Sumatra, 2012
9.0	Indian Ocean, 2004; Japan, 2011

Earthquakes with a magnitude of about 2.0 or less are usually called micro-earthquakes; they are not commonly felt by people and are generally recorded on local seismographs only. Events with magnitudes of about 4.5 or greater (there are several thousand such shocks annually) are strong enough to be recorded by sensitive seismographs worldwide.

The effects of earthquakes on landforms and landscapes

On geological timescales and across areas covering many thousands of square kilometres, earthquakes are associated with the formation of entire mountain chains such as the Himalaya–Karakoram Range in Asia. The northward drift of India into Eurasia and the subsequent continental collision led to a complex pattern of folding and faulting of rocks. It also created the world's highest and most impressive fold mountains: 96 of the world's 109 peaks of over 7000 m are located here.

To the north of the Himalaya–Karakoram Range lies the Tibetan Plateau. Averaging 4500 m above sea level it covers an area of 2.5 million km^2 (ten times the size of the UK). Major fault systems are evident in the rocks and these indicate considerable movement.

The entire region is tectonically active, as the 2008 (8.0 Mw) and 2013 (7.0 Mw) earthquakes in Sichuan province and the 2015 (7.0 Mw) event in Nepal demonstrated.

The **rift valleys** along mid-ocean spreading ridges, in East Africa and elsewhere, are also evidence of the effects of earthquakes on the morphology of the Earth's surface (Figure 14.18).

The inward-facing **fault scarps** or **escarpments** of rift valleys mark the location of faults caused by tension and compression within the crust. Rift valleys on the continents are altered by weathering and erosion. Over time fault scarps are worn away, blending into the landscape. They can even disappear under accumulated sediments.

The **Modified Mercalli scale** measures earthquake intensity and its impact. It relates ground movement to impacts that can be felt and seen by anyone in the affected location. In this respect – and unlike the Richter scale – it is a qualitative assessment based upon observation and description (Figure 14.17).

Increasingly, scientists are using the **Moment Magnitude scale (Mw)** (Table 14.7). This measures the energy released by the earthquake more accurately than the Richter scale. The amount of energy released is related to geological properties such as the rock rigidity, area of the fault surface and amount of movement on the fault. The Moment Magnitude scale provides the most accurate measurement of large earthquakes as it uses the amount of physical movement caused by an earthquake, which is a direct function of energy. However, it is not used for small earthquakes.

For comparison, Mw 2.0 is equal to a large quarry blast, Mw 5.0 is equivalent to the atomic bomb detonated over Nagasaki in 1945, while the largest thermo-nuclear devices today release about Mw 7.0.

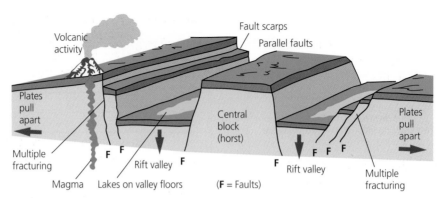

Figure 14.18 Faulting and the formation of rift valleys

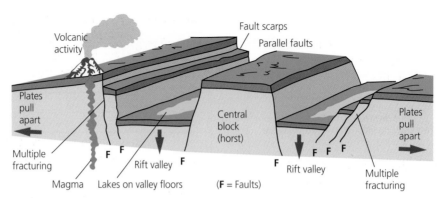

Stretch and challenge

1 There are different opinions concerning the origins of rifting. One involves the pulling apart of the rocks due to tensional forces in the lithosphere. This produces faults along which the rocks fracture and slip. Another possibility is that as magma rises from the asthenosphere, the rocks above are pushed up in a dome. The rocks forming the dome are stretched and fracture.

Research these two ideas, focusing on the roles of (a) tensional forces and (b) the upward movement of rocks into a dome. Specialist physical geography textbooks usually have a section on rifting in their plate tectonics chapter. Some university Geology/Earth Science departments have presentations accessible on the web, for example Leicester University Geology department. There are diagrams illustrating rift formation accessible via the web.

2 Draw annotated diagrams illustrating the two different ideas. Make sure you include information about volcanic activity.

Key idea
→ Earthquakes generate distinctive hazards

Types of hazards posed by earthquakes

As with all tectonic activity, earthquakes become hazardous when they interact with people. Every year there are around 100 large earthquakes with the potential to impact significantly on human societies. Between 2000 and 2015, an estimated 800,000–900,000 people were killed in earthquakes, with many more suffering injuries.

Ground shaking and ground displacement

This is the vertical and horizontal movement of the ground (Figure 14.19). The severity of ground shaking depends on:

- earthquake magnitude
- distance from the epicentre
- local geology.

Locations that are close to the epicentre of a high-magnitude quake and where the surface layers are relatively unconsolidated and have a high water content will experience extreme ground shaking. These conditions existed in parts of Mexico City and the Japanese port of Kobe hit by devastating earthquakes in 1985 and 1995 respectively.

In general, buildings can withstand vertical movements better than horizontal ones: it is the swaying of structures that is so dangerous to their stability. Ground movements that cause displacement of rocks along fault lines can rip apart pipelines and sewers, sever rigid structures such as railway tracks and roads, and cause buildings to collapse.

The displacement of the surface can also disrupt natural drainage, diverting streams and rivers and affecting the movement of groundwater in aquifers. This can have serious implications for public water supplies and irrigation for agriculture.

Figure 14.19 Movement across a fault

Liquefaction

When an earthquake strikes an area with surface materials of fine-grained sands, alluvium and landfill with a high water content, the vibrations can cause these materials to behave like liquids. As a result, these materials lose their strength; slopes such as river banks collapse and structures tilt and sink as their foundations give way. Liquefaction was a major issue during the Kobe earthquake as much of the port had been built on reclaimed land in Osaka Bay. Just under 200 berths in the port were destroyed, affecting not just the Japanese economy but trade throughout the world (Figure 14.20).

Figure 14.20 Dockside damage, Kobe 1995

Landslides and avalanches

Both ground shaking and liquefaction can cause slope failure. Some earthquake-prone regions are especially vulnerable to slope failure. Steep slopes in mountainous regions like the Himalaya–Karakoram range are notoriously unstable and vulnerable to landslides. Their vulnerability is increased by deforestation and heavy monsoon rains, so that even small tremors can trigger landslides. The Nepalese earthquake of 2015 triggered large numbers of landslides and avalanches caused by ground shaking.

Landslides block transport routes in mountainous regions where accessibility is already difficult. Movements of soil and rock on slopes can also block rivers. These natural dams create temporary lakes, which can threaten areas downstream with catastrophic floods were the dams to fail. This was the case in several locations in Kashmir in 2005 and Sichuan in 2008.

Upland valleys are often favoured sites for reservoirs. Should an earthquake create a landslide on slopes above a reservoir, the displacement of water and the waves generated could weaken and overtop the dam.

An earthquake in northern Italy in 1963 did precisely that. It led to the collapse of a hillside above the Vaiont reservoir, generating a 100 m wave which swept over the dam and down the valley of the Piave river, drowning nearly 3000 people.

Tsunamis

Underwater earthquakes can cause the sea bed to rise vertically. This displaces the water above, producing powerful waves at the surface which spread out at high velocity from the epicentre. Because of their low height (< 1 m) and very long wavelength (up to 200 km) tsunami waves can pass underneath a ship out at sea unnoticed. However, wave height increases greatly as they approach the shore and enter shallow water. Before the wave breaks, water in front of the wave is pulled back out to sea, a process known as drawdown. Finally the tsunami wave rushes in as a wall of water that can exceed 25 m in height. The tsunami generated by the earthquake off the coast of Aceh province in Sumatra in December 2004 is estimated to have delivered about 1000 tonnes of water per metre of shoreline.

The local height of a tsunami is also affected by the shape of the sea bed and the coastline. Depending on the relief of the coastal zone, tsunamis can spread variable distances inland (Figure 14.21).

Figure 14.21 Tsunami devastation in Minamisanriku, Japan

Underwater landslides caused by earthquakes can also displace water and create tsunami waves. When a large volume of rock is shaken and slides downslope, water is dragged in behind it from all sides and collides in the centre. This can generate a tsunami wave which radiates outwards. While the resulting wave may not have enough power to cross oceans, the local effects can be devastating. For example, in 1998, 2200 villagers living in coastal communities in Papua New Guinea were killed by a local tsunami generated from an underwater landslide triggered by an earthquake. Because such events are local, warning times are short, making them particularly hazardous. Much research is currently being undertaken to map the sea bed along vulnerable coastlines, such as southern California, to identify potential underwater landslide hazards.

Stretch and challenge

1 Research the topic of mega-tsunamis caused by giant landslides. The island of La Palma in the Canary Islands is one location that has received much attention. Produce a presentation titled 'Threats from mega-tsunamis – separating fact from fiction'. Set the scene with the geological background and the possible locations of such events.
2 Present the arguments surrounding the potential effects. Several academic websites deal with this topic. Blogs, for example that of Professor David Petley, a world expert on landslides and their effects, are also useful. The American Geophysical Union maintains an interesting and up-to-date blog at http://blogs.agu.org.

✔ Review questions

1 In what tectonic settings are earthquakes most likely?
2 How are earthquakes generated? Use the concept of 'threshold' in your answer.
3 How do the Richter and Moment Magnitude scales assess an earthquake's energy?
4 What physical factors increase the effects of ground shaking?
5 Explain why liquefaction has such a severe impact on buildings.
6 Describe the hazards generated by landslides caused by earthquakes.
7 How can a tsunami be triggered by an earthquake? Use the concept of 'causality' in your answer.

14.4 What are the implications of living in tectonically active locations?

The global distribution of geophysical events such as volcanoes and earthquakes is far from random. There is a strong correlation between the location of the plate boundaries and volcanoes and earthquakes. Even so, intra-plate earthquakes and volcanic eruptions are not uncommon and can have serious impacts. The potential

hazards of these events have not, however, prevented millions of people living in locations that might be affected.

Earthquakes and volcanic eruptions cannot be prevented, so societies try to develop resilience to cope with their impacts. However, levels of preparedness vary and are positively related to economic development.

Key idea
→ There are a range of impacts people experience as a result of volcanic eruptions

Active, dormant and extinct volcanoes

On the basis of their eruption history, volcanoes are described as active, dormant or extinct. While this classification is appealing in its simplicity, in reality it is very difficult to define volcanoes according to their activity status.

Many volcanoes that have erupted in the past show no signs of activity today. In human terms they appear to be dormant, yet when viewed from the perspective of geological time they are relatively active. There is also the question of how we define an active volcano. For example, is a volcano active when there is evidence of magma movement underground and accompanying small earthquakes, but no eruption?

Some scientists consider a volcano to be active only if it has erupted in historic times. The difficulty with this approach is that the availability of documentary records of eruptions varies from region to region. The Mediterranean can claim historic records to go back over 3000 years, but in the Cascade Range in the northwest USA, historic time covers the past 300 years only. Other scientists regard a volcano as active if it is currently erupting or there is evidence of unusual earthquake patterns or gas emissions.

One widely accepted definition is that an active volcano is one that has erupted since the last glacial period or within the past 10,000 years. The Smithsonian Institute uses this definition in its Global Volcanism Programme and has compiled a list of 1500 active volcanoes. It defines a dormant volcano as one that has not erupted during the past 10,000 years but is expected to erupt some time in the future, and an extinct volcano as one that is not expected to erupt again.

Even so, Mount St Helens was considered dormant until its 1980 eruption, while the Yellowstone caldera has not erupted for 70,000 years but is certainly not considered extinct. Due to its frequent earthquakes, very active geothermal features such as geysers, and rapid rates of ground inflation, most observers consider Yellowstone an active volcano.

The impacts people experience as a result of volcanic activity

🔍 Case study: Living with volcanoes – Japan

Based on the definition of the Global Volcanism Programme, Japan has 110 active volcanoes. Included within this list are several submarine volcanoes just off shore from the Japanese mainland. Seventy per cent of Japan's land mass is mountainous, formed by volcanic activity over millions of years. Much of Japan's folklore is associated with volcanoes. Mount Fuji, along with several other volcanic peaks, is an important part of Japanese culture, tradition, myth and legend.

Japan's tectonic setting

Japan is located in one of the most tectonically active zones in the world (Figure 14.22). It is where four tectonic plates meet and widespread subduction gives rise to intense volcanic activity.

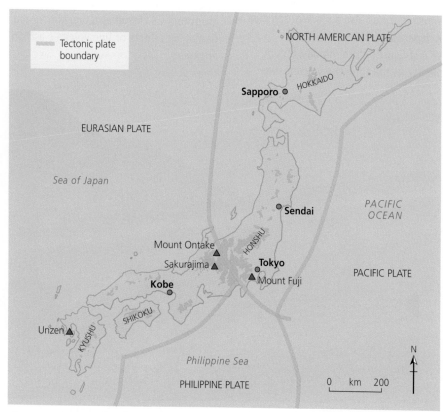

Figure 14.22 Japan's tectonic setting

Table 14.8 Scale of volcanic eruptions in Japan over the past 2000 years

VEI scale	Number of eruptions
1	56
2	227
3	551
4	113
5	41
6	10
7	0
8	0
9	0

Historical evidence records nearly 1200 volcanic eruptions in Japan in the past 2000 years. The average frequency of eruptions of different scales in Japan (Table 14.8) suggests that the more explosive the eruption, the less variable its frequency. Smaller eruptions are more variable (Table 14.9).

Mount Ontake, Japan

Mount Ontake is a classic strato-volcano situated 200 km west of Tokyo on Japan's largest island, Honshu. It rises to just over 3000 m and its summit is often snow-covered. Mount Ontake had been dormant for many centuries until a sequence of eruptions between October 1979 and April 1980. Further eruptions (some of them small

→

Table 14.9 Frequency of different scale eruptions in Japan

VEI scale	Average frequency (years)	Variation from average
2	4	Great
3	18	Great
4	50	Little
5	200	Little

phreatic eruptions) followed in 1991 and 2007. The area was popular with climbers and trekkers and was a noted tourist destination, with various facilities including a lodge close to the summit. It was also a place of spiritual pilgrimage for many Japanese people.

On 27 September 2014, just before midday, Mount Ontake erupted violently. At the time the volcano was not under any alerts or warnings, although some increase in earthquake activity had been observed. The eruption killed 63 people and large areas surrounding the volcano were affected by ash fall, pyroclastic flows, volcanic bombs and lahars.

Most casualties were climbers and hikers on the slopes of the volcano. More than 200 survivors made it down the mountain and search and rescue teams were deployed to recover those seriously injured (Figure 14.23). The Japanese prime minister, Shinzo Abe, ordered the military to assist with emergency rescue operations. For a while, air space in the vicinity of the eruption was closed as a precaution against the possible damaging effects of fine volcanic ash on aircraft jet engines.

Figure 14.23 Search and rescue teams on Mount Ontake, September 2014

Case study: Living with volcanoes – Indonesia

Like Japan, Indonesia is also an archipelago straddling tectonic plate boundaries. Indonesia has a large number of active volcanoes. There are records of around 80 that have erupted in historic times, though this figure may be an underestimate. More than three-quarters of Indonesia's inhabitants live within 100 km of a volcano that has erupted in the past 10,000 years. Indonesia has suffered more eruptions causing fatalities (114) and damage to infrastructure (195) than any other country in the world.

Similar to Japan, much of the landscape of Indonesia is dominated by volcanic peaks and Indonesia's history, culture and the personal experience of its people are closely tied to volcanoes.

Indonesia's tectonic setting

Indonesia's tectonic setting is complex, with widespread subduction along the entire 3000 km length of the Indonesian archipelago (Figure 14.24). Much of Indonesia is an island arc formed by subduction of the Indo-Australian plate beneath the Eurasian plate. Meanwhile, to the east of the chain of islands, the continental shelf of northern Australia is in collision with the Eurasian plate. These plate movements produce very high levels of seismicity and volcanicity.

A feature of volcanism in Indonesia is the large number of volcanoes that are highly explosive. This was exemplified by the cataclysmic eruption of Krakatau →

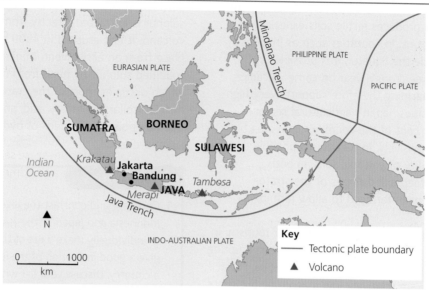

Figure 14.24 Indonesia's tectonic setting

in 1883, which also generated a devastating tsunami. Around 80 per cent of dated volcanic eruptions in Indonesia have occurred since 1900.

Merapi, Indonesia

Merapi is a strato-volcano which reaches nearly 3000 m in altitude. It is highly active, with eruption events in 1994, 1997, 2001 and 2006. On each occasion, a dome developed in the crater, which eventually collapsed, producing pyroclastic flows. The force of these eruptions was 3–4 VEI.

The VEI 4 eruption between 26 October and 12 November 2010 affected people living around the volcano. Pyroclastic flows extended 4 km to the north,

11.5 km to the west, 7 km to the east and 15 km to the south. Lava bombs were thrown 4 km from the summit in all directions. The eruption happened at the start of the rainy season, triggering lahars fed by ash from the eruption and from earlier eruptions. Ash falls forced the closure of Yogyakarta airport for over a fortnight and emissions of sulphur dioxide caused acid rain to fall over a large part of the region.

Overall, 367 people were killed by the Merapi eruption, 277 were injured, while approximately 410,000 in the vicinity of the volcano became refugees. Large numbers of livestock were killed and ash fall and pyroclastic flows amounting to 130 million m³ of material destroyed crops and infrastructure such as buildings, power lines and bridges.

Why do people choose to live in tectonically active locations?

The millions of people around the world living in areas susceptible to earthquakes and/or volcanoes do so for rational reasons. These locations offer people a place to live, to have employment and to bring up families. Those with limited economic resources can find that they have little choice but to occupy somewhere prone to earthquakes or volcanic eruptions.

Both Indonesia and Japan only exist because of volcanic and earthquake activity. The plate subduction occurring at both countries has produced vast amounts of molten material which has accumulated to form the archipelagos which make up the land mass of the countries. This has given a home to nearly 0.4 billion people (Table 14.10).

Table 14.10 Indonesia and Japan: population profiles

Country	Area (million km²)	Total population (million)	Density (persons per sq km)	Rate of population change (per year)
Indonesia	1.8	256	140	0.92%
Japan	0.365	127	349	−0.16%

Indonesia's population density should be seen in the context of a land area that includes many uninhabited islands. Thus the density of the county's inhabited area is higher than the average figure indicates. On the island of Java, for example, population density reaches just over 1000 people/km². In both Indonesia and Japan millions of people have little choice but to live alongside active or dormant volcanoes and experience many earth tremors.

The slopes of volcanoes often attract settlement. Weathered lava produces fertile soils especially in wet tropical regions. In countries such as Indonesia which sits astride the Equator, it is possible to grow two or three crops a year on some volcanic soils. Intensive farming therefore supports very high population densities. Japan's agricultural sector uses just 13 per cent of the country's land area but does so very intensively. In part this is possible due to the high fertility of volcanic soils. In both countries, the steep-sided slopes of the strato-volcanoes have been terraced to allow farming to be practised.

Volcanoes also provide opportunities for tourism. In Japan, the active volcano Sakurajima has been erupting more or less continuously since the mid-1950s. Most of the eruptions are relatively low magnitude but there are occasionally more explosive events. Its lava flows become tourist attractions.

Japan has had geothermal power stations since the 1960s but these were small scale. However, recently, geothermal resources have been receiving increasing attention. In part this is a factor in a long-term strategy to promote renewables. Interest and research into geothermal energy was also given a major boost in the aftermath of the 2011 earthquake. The resulting tsunami overwhelmed a nuclear power plant causing catastrophic leakages of radiation. One issue Japan faces in exploiting its geothermal energy resources is that nearly 80 per cent of them are located in national parks or protected hot spring locations.

Indonesia also has vast geothermal energy potential and is beginning to exploit this. Currently, only about 3 per cent of Indonesia's total electricity generation comes from geothermal but there are plans to increase this. If all goes to plan, the 350 megawatt Sarulla power plant will come online in 2022 and has the potential to be the world's largest such plant. As in Japan, much of Indonesia's geothermal potential is located in conservation zones such as national parks.

Minerals associated with volcanic eruptions, such as sulphur, are used in a variety of industrial processes such as the production of chemicals. Japan used to mine sulphur at Matsuo but production ceased in 1972. In east Java, Indonesia, sulphur is extracted directly from the crater of Ijen volcano. It last erupted in 1999 but mining takes place from an active vent. Miners cut lumps of solidified sulphur by hand from near the lake which occupies most of the crater. The water has a pH of 0.5, similar to battery acid, and the area is frequently filled by poisonous clouds of hydrogen sulphides and sulphur dioxide.

Activities

1 Compare and contrast the levels of development of Indonesia and Japan.

Individually make a list of the variables you think give a good indication of the level of development in a country. Discuss your list with another student and decide on the set of variables you are going to use. Some suggestions are:
- HDI
- GNI (ppp)
- literacy levels, especially female
- persons per doctor
- energy consumption per person
- employment structure.

2 Produce a development profile for each country. This could include tables, graphs and diagrams as appropriate.

3 Assess the economic, social, political and environmental impacts of the eruptions at Ontake and Merapi. You will need to access online material such as newspaper articles, academic articles and scientific records such as the USGS site.

Key idea
→ There are a range of impacts people experience as a result of earthquake activity

Earthquake hazards

Although earthquakes can occur anywhere, there is a much higher probability of a seismic event close to plate boundaries. In general, the more energy the earthquake releases, the greater the hazard. However, human factors also play an important role in earthquake impact.

The impacts people experience as a result of earthquake activity

Case study: Living with earthquakes – Japan

Japan experiences 400 earthquakes every day. The vast majority of these are tiny and register only on specialised recording instruments. Throughout history, Japan has also been hit by many high-magnitude quakes. These quakes have had huge impacts on its people and society. Since 2000, 23 earthquakes of 7.0 Mw (Moment Magnitude scale) have occurred in Japan causing nearly 16,000 fatalities. As with volcanoes, Japanese culture, folklore and traditions are full of references to earthquakes and their effects.

Japan's tectonic setting

The same forces that generate Japan's volcanoes also release vast amounts of seismic energy. However, unlike volcanic eruptions, earthquakes bring with them the risk of tsunamis. Movements along plate boundaries lying just offshore of the Japanese mainland create an ever-present risk of tsunamis. The Pacific coastal zone is especially vulnerable, being densely populated and containing three huge conurbations: Tokyo (36 million), Osaka (19 million) and Nagoya (9 million).

The Great East Japan Earthquake: Tōhoku

On 11 March 2011, one of the most significant seismic events in Japanese history took place. A 9.0 Mw earthquake occurred along the boundary between the Pacific and North American plates. Its epicentre was 70 km offshore of northeast Honshu Island. Some geophysical facts are given below:

- 9.0 Mw.
- Largest recorded earthquake to have struck Japan – lasted six minutes.
- One of the top five in the world since accurate recording began in 1900.
- Undersea megathrust earthquake.
- Epicentre – 70 km east of Oshika peninsula; focus – approximately 30 km.
- Honshu island moved 2.4 m east.
- The Earth shifted on its axis by 10–25 cm.
- A 400 km stretch of coastline dropped vertically by 0.6 m.
- Sea bed rose by about 7 m and moved westwards by 40–50 m.
- Very large tsunamis were triggered, reaching heights of up to 40.5 m in places when hitting the coast.
- Tsunami travelled up to 10 km inland in places (Sendai).
- Tsunami waves reached Antarctica, breaking off 125 km² of ice as giant icebergs.
- Tsunami waves of up to 2.4 m affected the coastline of California and Oregon, causing damage of US$10 million.

Figure 14.25 Tsunami impacts Japan, 2011

- Tsunami wave of 1.5 m overwhelmed Midway Atoll, drowning 110,000 nesting birds in a wildlife refuge.
- There were many aftershocks, some of which were 7.0–8.0 Mw in strength.

Social impacts

The 2011 tsunami (Figure 14.25) claimed nearly 16,000 lives and injured another 6000 people. Two-thirds of the victims were over the age of 60, a quarter were over 70; 90 per cent of deaths were due to drowning. Of the remainder, most died as a result of burns and being crushed by collapsing buildings.

Mass mortality and destruction of crematoria, morgues and the power infrastructure created problems of how to dispose of the dead bodies. Traditional funeral traditions were waived – the authorities felt they had no choice but to bury bodies in mass graves as quickly as possible to reduce the chances of diseases being spread.

People's whereabouts was a major issue following the tsunami. Children were separated from their families – Save the Children reported that 100,000 children were affected. The problem was made worse as most children were at school when the earthquake struck. Some 2000 young people were either orphaned or lost one of their parents. Some of the smaller coastal schools were devastated by the tsunami: one elementary school lost 74 of 108 students and 10 out of 13 staff.

The destruction of infrastructure, for example housing, schools and health centres, was on such a scale that reconstruction was still taking place five years after the disaster. Some communities have even been relocated from their original settlement site.

Economic impacts

The cost of the earthquake, including reconstruction, has been estimated at nearly £181 billion. Japan's National

Figure 14.26 The Fukushima Daiichi power plant, February 2012

Police Agency issued an official figure of 45,700 buildings destroyed, with a further 143,300 damaged; 230,000 vehicles were either destroyed or damaged; 15 ports were directly affected, with four destroyed in the northeast of Japan, including Sendai; 10 per cent of Japan's fishing ports were damaged, although most had reopened a year after the earthquake.

Perhaps the most serious impact was the disruption of power supplies – 4.4 million households and thousands of businesses lost electricity. The major cause of this disruption was the immediate shutdown of 11 nuclear reactors. At the Fukushima Daiichi nuclear power plant (Figure 14.26), all six of its reactors were so severely damaged by the tsunami that the plant was decommissioned. The plant's cooling systems were disabled by sea water flooding, which led to a meltdown of reactor cores and the release of radioactivity. A 30 km evacuation zone was established around the plant, and soils in the surrounding countryside were contaminated by radiation. Even today radioactivity continues to seep into the Pacific Ocean, affecting the fishing industry.

Japan experienced several weeks of power cuts (lasting 3–4 hours at a time), which added to the economic cost of the earthquake. Two oil refineries were set on fire during the earthquake, one fire taking ten days to put out.

Transport infrastructure was badly hit. Many road bridges were damaged or destroyed and in the northeast train services were badly disrupted. Twenty-three train stations were swept away, while others suffered severe damage, such as collapsed roofs.

Some 25 million tonnes of debris was created by the earthquake, requiring a costly clean-up operation. Over large areas, farmland flooded by sea water has been contaminated by salt and made uncultivable.

Japan's stock market fell as the implications for businesses such as Sony, Toyota and Panasonic were realised. Their production was hit by the shortage of electricity. Overseas operations were also affected, as supplies of parts exported from Japan were interrupted.

Political impacts

The Japanese government injected billions of yen into the economy, especially the financial sector, to bring some stability. This increased government debt at a time when its reduction was a prime political aim.

A large popular movement against nuclear power developed after the Tōhoku earthquake. Concerns over safety standards and regulation of the nuclear industry became a political issue. The government has yet to make a clear decision about the role of nuclear power in Japan's long-term energy mix. Several executives of companies involved in the Fukushima power plant have resigned. It emerged that warnings about the inadequacy of the defences against tsunami hazards had been made several years before the disaster.

Political fallout from the Fukushima nuclear accident spread around the world. For example, in western Europe the anti-nuclear lobby in Germany used the incident to support their arguments against nuclear power.

 Activities

1 On a map showing the Pacific Ocean and its surrounding countries, plot the location of the Tōhoku earthquake's epicentre 70 km east of the city of Sendai.
2 Plot the location of each of the places in Table 14.11.
3 Insert the tsunami travel time next to each location.
4 Read pages 511–12 and then plot isolines representing tsunami travel times across the Pacific at five-hour intervals.
5 Draw graphical bars at each of the locations representing the height of the tsunami wave.
6 Describe the pattern of (i) tsunami travel time and (ii) tsunami wave height across the Pacific.
7 Why did the communities around the Pacific (other than Japan) face economic losses rather than loss of life as a result of the tsunami?

Table 14.11 Data concerning the tsunami generated as a result of the Tōhoku earthquake

Location	Distance from earthquake epicentre (km)	Travel time for tsunami (hours + minutes)	Maximum tsunami wave height recorded (metres)
Scott Base, Antarctica	13,000	21–45	0.05
NE Tasmania	9,000	15–00	0.47
Vancouver Island, Canada	7,200	10–15	0.79
Antofagasta, Chile	16,500	20–20	0.96
Valparaiso, Chile	16,900	22–15	1.54
Quepos, Costa Rica	12,850	17–40	0.50
Santa Cruz, Galapagos Islands	13,200	17–45	2.26
Fiji	7,300	9–45	0.21
Tahiti	9,400	11–30	0.42
N. Sulawesi, Indonesia	4,460	5–50	0.10
Nauru	5,000	5–55	0.30
North Cape, New Zealand	8,700	12–05	0.40
Manus Island, PNG	4,500	6–00	1.03
Callao, Peru	15,150	20–15	1.44
Lagaspi, Philippines	3,350	4–35	0.30
Kuril Islands, Kamchatka, Russia	1,750	3–45	2.50
Kodiak island, Alaska, USA	5,060	8–00	0.35
Los Angeles, California, USA	8,480	11–00	0.50
San Diego, California, USA	8,630	11–20	0.63
Hilo, Hawaii, USA	6,300	7–55	1.33
Westport, Washington, USA	7,250	9–55	0.46
Saipan, Mariana Islands	2,600	3–17	1.20

Case study: Living with earthquakes – Nepal

Nepal is an earthquake-prone country. On average it receives two 7– 8 Mw earthquakes every 40 years and one 8+ Mw every 80 years. However, earthquake records are much less complete than in countries such as Japan or Indonesia and many of the small seismic events have gone unrecorded. As in Japan, earthquakes and their risks are embedded within Nepalese culture, tradition and folklore.

Nepal's tectonic setting

Nepal is situated in the middle of the collision zone where the Indo-Australian plate meets the Eurasian plate. As these plates grind against each other, pressures build up and the energy is released as earthquakes.

The geological structure of valleys in Nepal increases seismic risk. Pre-historic lakes filled many of the valleys and their legacy is hundreds of metres of relatively soft sediment. In the Kathmandu valley, the depth of sediment reaches 600 m. As seismic waves pass through this material, they are amplified, causing structures to swing violently. Depending on the water content of the soil at the time of the quake, liquefaction can be a major hazard.

The Gorkha Earthquake, Nepal

On 25 April 2015, at 11.56 local time, a major earthquake struck Nepal (Figure 14.27). Some geophysical facts are given below:

- 7.8 Mw.
- A slip along the main frontal thrust fault in the collision zone.
- Epicentre – 90 km northeast of Kathmandu.
- Focus – 15 km.
- Fifty-one aftershocks equal to or above 5 Mw and five aftershocks above 6 Mw, including a 7.3 Mw shock in May 2015.
- Tremors felt across neighbouring Indian states and into China.
- Many landslides caused.
- Snow and ice avalanches caused, including on Everest.

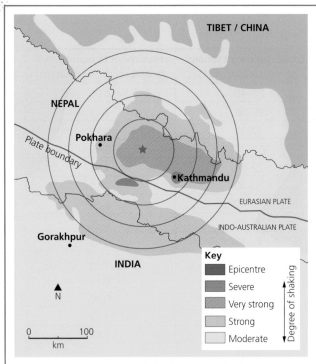

Figure 14.27 The pattern of shaking during the Gorkha earthquake, Nepal

Figure 14.28 Durbar Square, Kathmandu, 26 April 2015

Social impacts

Some 8800 fatalities resulted from the Gorkha earthquake and just over 22,000 people were injured in Nepal. There were 160 deaths in other countries, mainly in India. Avalanches on Everest killed 20 climbers.

More than 2.8 million people were displaced from the Kathmandu Valley. Many were migrants from remote areas of Nepal, who then moved away from the Kathmandu region after the quake. Some 473, 000 houses were either destroyed or badly damaged and an estimated 1 million people required food assistance. One particularly disturbing social impact concerned an increase in trafficking women and girls from the poorest families, who were homeless, to South Asian brothels.

The social impact of the earthquake was felt unevenly by different ethnic groups. The lower-caste Tibeto-Burman were hardest hit as they tended to live on the higher slopes in the Himalayas. These locations were severely affected by landslides and were also difficult to access for relief aid. Single women were also disadvantaged and struggled to obtain emergency food and medical aid – most aid was controlled by men, and women were often discriminated against.

Several culturally important locations were damaged by the quake, including the UNESCO World Heritage site of Bhaktapur (Figure 14.28). The quake also destroyed the nine-storey Dharahara Tower, one of the oldest Buddhist monuments in the Himalayas.

Economic impacts

Nepal is one of the poorest countries in the world and the 2015 earthquake had a damaging impact on its economy, though lack of reliable data makes precise estimates of its economic impact difficult. Nonetheless, the US Geological Survey put the cost to the national economy at US$10 billion. Tourism, which accounts for about 10 per cent of the economy, was badly hit, with Everest closed for the 2015 climbing season.

The Asian Development Bank provided US$200 million of aid for reconstruction. Nepal's government, businesses and individuals lacked capital reserves to cope with the scale of damage caused by the Gorkha quake. As a result of borrowing, levels of debt increased.

The timing of the earthquake also disrupted the planting season ahead of the arrival of the monsoon. There were therefore longer-term impacts of the earthquake in terms of food security.

Political impacts

Nepal was politically unstable before the earthquake struck. Even so, following a period of turmoil that included assassinations and coups, the country had shown signs of moving into a more settled period. Unfortunately, the earthquake and its economic aftermath have undermined progress towards peace and greater democracy.

Nepal received emergency aid for longer-term reconstruction from a range of countries, including EDCs and LIDCs as well as ACs. Indonesia, for example, sent assistance from its military aircraft and personnel to operate them. International aid also created tension with India, which was accused of trying to use its humanitarian aid as a way of self-promotion within the region.

1 State the differences between active, dormant and extinct volcanoes.
2 Why are Japan and Indonesia at high risk from volcanic eruptions? Use the concept of 'threshold' in your answer.
3 Draw a labelled cross-section to illustrate the tectonic setting of either Japan or Indonesia.
4 Construct a table contrasting the impacts of the Ontake and Merapi eruptions.
5 Outline the physical factors that make earthquakes in Nepal such a serious hazard.
6 Compare the potential for predicting volcanic eruptions with earthquakes.
7 How does the concept of 'resilience' help explain the longer-term impacts of earthquakes on countries at different points along the development spectrum?

14.5 What measures are available to help people cope with living in tectonically active locations?

As the world's population heads towards 10 billion by the mid-twenty-first century, and as towns and cities grow, greater numbers of people and property will be at risk from the forces of nature. Risk, however, is not evenly distributed spatially and this is particularly the case for geophysical hazards such as earthquakes and volcanic eruptions.

It is also the case that opportunities for reducing the risks from seismic and volcanic activity have never been greater. However, a range of factors need to be considered in order to understand how people can cope with living in tectonically active locations. They include societal inequality, mitigation and adaptation strategies, risk and resilience.

> **Key idea**
> → The exposure of people to risks and their ability to cope with tectonic hazards change over time

Exposure and vulnerability to tectonic hazards

Geophysical events such as volcanic eruptions and earthquakes become hazards when they pose a risk to people (Figure 14.29).

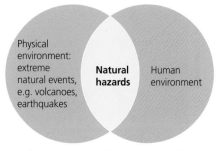

Figure 14.29 The occurrence of tectonic hazards

The disaster risk equation gives an indication of the hazard vulnerability for a location:

$$\text{Risk } (R) = \frac{\text{Frequency or magnitude of hazard } (H) \times \text{Level of vulnerability } (V)}{\text{Capacity of population to cope and adapt } (C)}$$

$$R = \frac{H \times V}{C}$$

Unlike hazard or risk, a **disaster** is an actual event, which usually involves loss of life and a great deal of damage to the human environment.

Physical exposure to earthquake and volcanic hazards depends on factors such as:

● frequency of earthquakes and volcanic activity
● magnitude of earthquakes and volcanic activity
● types of hazards generated by earthquakes and volcanoes in a particular location
● number of people living in an earthquake-prone and/or an eruption-prone area.

Important considerations when looking at the magnitude (energy released) of a tectonic event and its impact are how often the event occurs and the interval between such events (Figure 14.30).

In general, the greater the magnitude of an eruption or earthquake, the less frequently it occurs. The recurrence interval, which is the average time between two events of equal magnitude, indicates that high-magnitude events recur over longer time periods. A few high-magnitude but rare events release large amounts of energy and have the greatest impact on people and society.

Vulnerability is concerned with the ability of a person or community to withstand exposure to, and risks from, a hazard such as an earthquake or volcanic eruption. People are most vulnerable when relatively small physical changes have major socio-economic implications. But earthquake and volcanic events often cause significant physical changes and this means that even the least vulnerable people are susceptible to severe negative impacts.

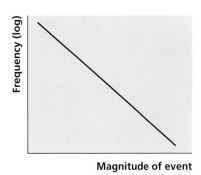

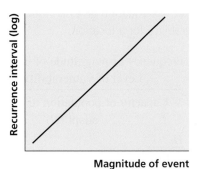

Figure 14.30 The relationship between magnitude, frequency and recurrence interval of tectonic hazards

Resilience is an indication of the rate of recovery from a hazardous event that has put an individual and/or community under stress. It is about how well an individual and society are able to function at an acceptable level when a destabilising force has affected them. Strategies for coping with changes due to volcano and earthquake hazards vary both from one place to another and between different people in the same place. Linked with this is the effectiveness of measures such as building design or warning systems designed to protect people and property against hazard effects such as ground shaking and tsunami.

How and why have risks from tectonic hazards changed over time?

It would appear that the number of natural disasters has increased through time (Figure 14.31). There has been an increase in the frequency and magnitude of some hazards such as floods and severe weather. This increase is explained by the interaction of physical and human factors. For instance, human activities such as deforestation can have a direct causative influence on hazardous events such as flooding.

Compared with flood and severe weather hazards, increases in the frequencies of earthquake and volcanic hazards events are less pronounced. This is because human activities play no part in causing earthquakes and volcanic eruptions. However, human factors do have a significant bearing on the impacts of seismic and volcanic hazard events.

Trends in tectonic hazards

The timescales over which tectonic forces operate are so long that graphs of past earthquake and volcanic events need to be interpreted with care (Figure 14.32).

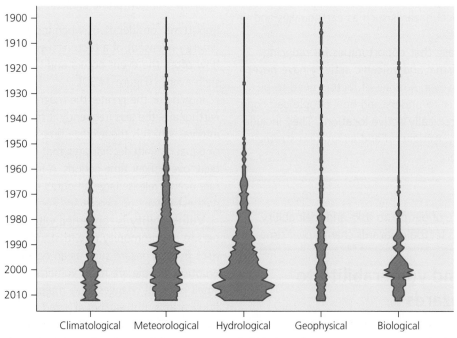

Figure 14.31 Number of natural disasters reported 1900–2012 (Source: EM-DAT, www.emdat.be)

(a)

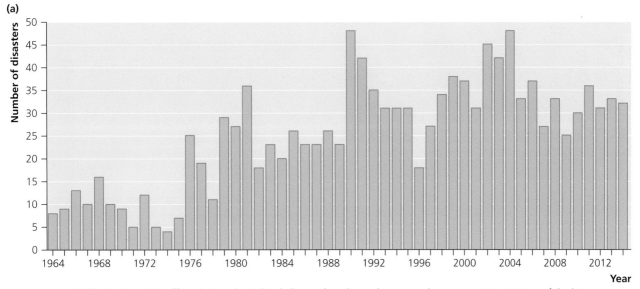

A disaster is >10 killed and/or >100 affected. Geophysical includes earthquakes, volcanoes and mass movements. Many of the latter category were triggered either by an earthquake or volcano.

(b)

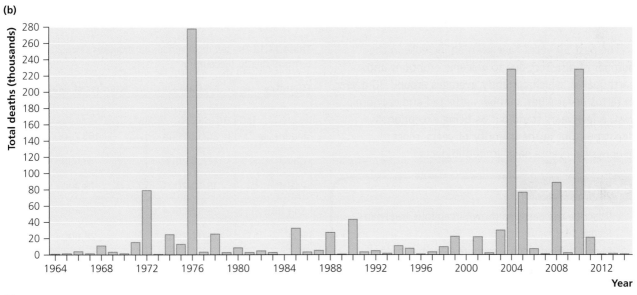

(c)

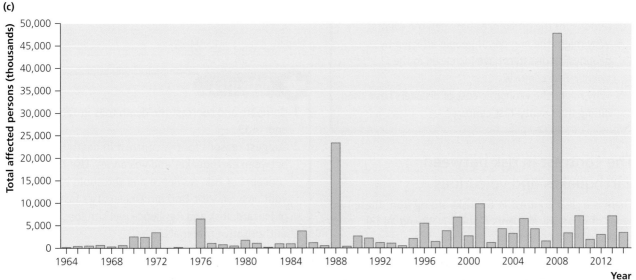

(Continued)

Figure 14.32 (*Continued*)

(d)

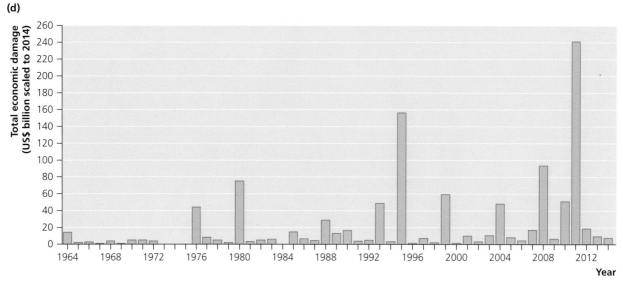

Figure 14.32 Geophysical disasters: (a) number, (b) number of people killed and (c) affected and (d) economic cost, 1964–2014 (Source: EM-DAT, www.emdat.be)

The past 50 years have seen an increase in the number of disasters due to volcanic eruptions and earthquakes, which currently averages around 30 a year. In terms of fatalities, numbers affected and economic cost, a few years stand out as being exceptionally disastrous. But in the majority of years, most geophysical events have limited impacts.

Activities

1 Describe the patterns shown in all four graphs in Figure 14.32.
2 Match the individual events responsible for the extreme impacts in 1976, 2004, 2008 and 2010.
3 For these four years, compare and contrast the numbers killed with the numbers affected. Suggest reasons for the differences between the events.
4 Why might the number of people affected by geophysical disasters have increased in the twenty-first century?
5 Why might the economic cost of disasters have risen during the twenty-first century?

The contrast in risk between earthquakes and volcanoes

There is a clear contrast in the risks arising from these two types of geophysical events (Tables 14.12 and 14.13).

Table 14.12 The ten most serious earthquakes since 2000 in terms of deaths by country

Country	Date	Number of deaths
Haiti	12/01/2010	222,500
Indonesia	26/12/2004	165,700
China	12/05/2008	87,500
Pakistan	18/10/2005	73,300
Sri Lanka	26/12/2004	35,400
Iran	26/12/2003	26,800
India	26/01/2001	20,000
Japan	11/03/2011	19,800
India	26/12/2004	16,400
Nepal	25/04/2015	8,600

Activities

1 Compare and contrast the death tolls in Tables 14.12 and 14.13.
2 Suggest reasons for the contrast in death tolls between earthquakes and volcanoes. Use the concepts of risk, resilience and threshold in your answer. Visit Durham University's Institute of Hazard, Risk and Resilience website for authoritative materials.

Table 14.13 The ten most serious eruptions since 2000 in terms of deaths by country

Country	Date	Number of deaths
Indonesia	24/10/2010	322
DRC	17/01/2002	200
Japan	27/09/2014	63
Indonesia	01/02/2014	32
Colombia	20/11/2008	16
Indonesia	14/02/2014	7
Yemen	30/09/2007	6
Ecuador	14/07/2006	5
Ethiopia	12/08/2007	5
Indonesia	04/12/2011	3

Table 14.14 Physical and human factors influencing the response to a disaster

Physical factors	Human factors
Speed of onset of event	Level of monitoring
Magnitude of event	Degree of preparation
How long the event lasts	Quality of relief Quantity of relief

The relationship between disaster and response

When a tectonic disaster strikes, its impact often follows a sequence of stages.

The shape of the disaster-response curve (Figure 14.33) changes according to different hazards. Also, the rate at which the quality of life deteriorates following a disaster will often depend on a combination of physical and human factors (Table 14.14).

> ### ➡ Activities
>
> 1 On A3 size copies of the disaster-response curve in Figure 14.33, draw amended copies to show the likely shape of the curve for the following scenarios:
>   a) An earthquake in an AC.
>   b) An earthquake in an LIDC.
>   c) A volcanic eruption in an EDC.
>   d) A volcanic eruption in an LIDC.
> 2 Add detailed annotations to each curve to highlight the contrasting experiences of countries at different points on the development spectrum.

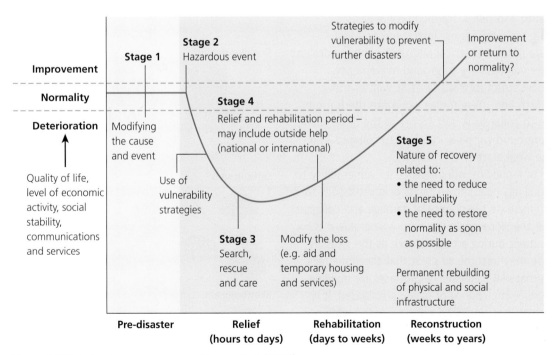

Figure 14.33 A disaster-response curve (Source: Park (1991))

Table 14.15 Some strategies for managing tectonic hazards

Modify the event	Modify people's vulnerability	Modify people's loss
Not possible for the vast majority of volcanic eruptions. However, the following have been tried with some success: • lava-diversion channels • spraying lava to cool it so it solidifies • slowing lava flows by dropping concrete blocks. Earthquakes: • Nothing can be done to modify an earthquake event.	Education – recognise signs of possible eruption; what to do when an eruption occurs, e.g. evacuation routes; drills to practise what to do when a tectonic event strikes, e.g. in an earthquake, get to open space away from buildings or shelter under a table in a doorway. Community preparedness – e.g. building of tsunami shelters and walls; strengthening of public buildings, e.g. hospitals, fire stations, schools Prediction and warning – increasing use of technology to monitor particularly active locations, e.g. individual volcanoes. Hazard-resistant building design, e.g. cross-bracing of buildings to support them during an earthquake; steep sloping rooves to prevent ash building up. Hazard mapping, e.g. predicted lahar routes; ground likely to liquefy in an earthquake. Land-use zoning to avoid building in locations identified by hazard mapping.	Emergency aid, e.g. bottled water, medical supplies, tents, food packs. Disaster-response teams and equipment, e.g. helicopters and heavy lifting machinery. Search and rescue strategies. Insurance for buildings and businesses. Resources for rebuilding public services, e.g. schools and hospitals, and help for individuals to rebuild homes and businesses.

> **Key idea**
> → There are various strategies to manage hazards from volcanic activity

Strategies for coping with hazards can be summarised as 'modify the event', 'modify the vulnerability' and 'modify the losses' (Table 14.15).

In connection with modifying vulnerability to earthquakes, much attention is now focused on aseismic design (Figure 14.34), which is architecture and engineering designed to withstand ground shaking and displacement.

Multi-storey buildings can sway, especially towards their tops. If they are built close together, one building can crash into another. A stepped profile gives considerable stability against lateral forces. Most buildings are not symmetrical and their complex designs can increase vulnerability to ground movements. Asymmetrical structures will twist as well as move back and forth and these forces can make them more susceptible to collapse. A soft storey at the base, such as an underground car park, can lead to collapse as it may be the weak part of the building.

The site of a building will influence its vulnerability to shaking. Buildings near known faults or on soft soils such as alluvium increase the chance of damage and collapse. The steeper the slope angle the more vulnerable it is to mass movement during an earthquake as the stresses on the slope may become so great that the slope fails. Excavating material to create flat surfaces for building can help reduce the chance of building collapsed. It is also important that slopes do not have high building densities as if one building falls it is likely to crash into its neighbour, setting off a chain reaction of buildings falling on each other down the slope.

Various techniques are well known and widely used to increase a building's resilience to ground shaking or displacement. Steel framed and cross-braced construction, including the basement, help hold a building together and absorb a lot of the energy when the building deforms during an earthquake. Deep foundations on soft soils help prevent the effects of liquefaction for example. The fitting of energy-absorbing pads in the foundations of buildings help absorb much of the horizontal energy from an earthquake being transmitted

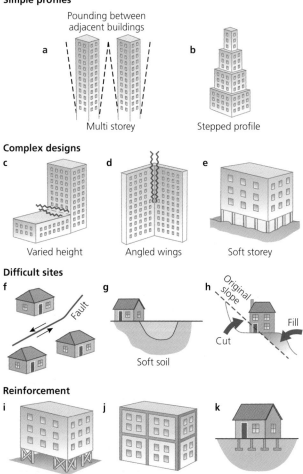

Figure 14.34 The effects of earthquakes on buildings and some aseismic designs

to the building. Rubber, steel or for private houses in EDCs and LIDCs such as Indonesia and Nepal, tyres filled with stones and sand can be very effective.

A key factor in modifying vulnerability through aseismic design is how rigorously building codes and laws are applied. In places officials have not enforced the rules but in most locations, many buildings were built before the rules became law. Retro-fitting measures are very expensive and may not always be possible even in ACs. In terms of modifying loss, it is vital that key buildings such as hospitals, power stations and water treatment plants are protected by aseismic design.

Countries vary in their ability to manage volcanic hazards

🔍 Case study: Living with volcanoes – Indonesia

Given its highly active tectonic setting, Indonesia has had to cope with frequent volcanic eruptions over many centuries. (See 'Indonesia's tectonic setting' on pages 486–87.)

Modifying vulnerability

Indonesia's Centre for Volcanology and Geological Hazard Mitigation (CVGHM) was established in 1920. Its key division is the Volcano Observation Section, which over the years has constructed permanent observatories on several active volcanoes. The first seismograph was set up on Mount Merapi in 1924, but as with monitoring elsewhere, the capability of the technology at that time was limited. Funding for the programme was also restricted owing to Indonesia's relatively weak economy. Indonesia was a Dutch colony until the end of the Second World War. Its emergence into independence was accompanied by political tension and violence among competing groups. Managing volcanoes therefore was given a low priority.

Gradually, CVGHM has been able to upgrade the monitoring instruments and widen their geographical coverage. Today more than 60 volcanoes are monitored. The threat from lahars is especially high due to the quantities of ash erupted by Indonesian volcanoes and the country's humid tropical climate with high levels of rainfall (> 1800 mm per annum). The rainy season, between November and March, is when lahars pose their biggest threat. In addition to the CVGHM, Indonesia has established a National Agency for Disaster Management. In association with university-based researchers, lahar sensors and closed-circuit television have been installed to monitor locations at greatest risk (Figure 14.35).

The CVGHM works closely with local governments, advising them on mitigation strategies such as community preparedness. Additionally, on Mount Merapi, for example, permanent settlement is forbidden on the highest slopes around the crater. Villagers on the southern slopes (at greatest risk) have been encouraged to relocate to safe zones, in some cases receiving financial assistance and a small plot of land as incentives to move.

An enduring problem is that people have a strong bond with their village site and their family fields. They often return, rebuild and accept the risks.

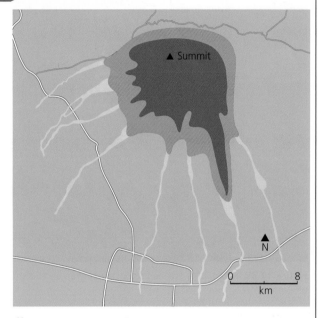

Key

- Hazard zone III (mostly affected by pyroclastic flow, lava, heavy ash fall and direct blast)
- Hazard zone II (affected by pyroclastic flow and ash fall)
- Hazard zone I (mostly lahar)

Figure 14.35 Hazard risk map, Mount Merapi, Indonesia

Modifying loss

Although Indonesia is a lower-middle-income country, the resources it can deploy once an eruption is underway are limited. Many of the farming communities affected also have limited resources. That said, Indonesia has some well-trained and well-equipped emergency services, including the military, experienced in search and rescue. Temporary shelters in safe zones are available in case of evacuation and sometimes these are upgraded into permanent residential areas.

The authorities have also cleared river channels of volcanic material and other debris in order to reduce the risk of flooding and to contain lahars. Some of this material has a commercial value (e.g. road building) and people living along the river valleys have used it for rebuilding houses and community facilities.

Italy's tectonic setting is complicated by several fracture zones in the Mediterranean where the African and Eurasian plates meet. There have been several well-documented eruptions in the past and Mount Etna on Sicily is one of the most active volcanoes in the world.

Figure 14.36 Italian army engineers constructing an earth dam on Etna, 2016

Modifying event

Italy has had some local-scale success with slowing and diverting lava flows from Etna. Earth barriers (Figure 14.36), large concrete blocks dropped into the lava flow and channels dug to divert flowing lava away from settlement have been used.

Modifying vulnerability

Monitoring plays a significant role in mitigating eruption hazards, providing accurate information to the Volcano Risk Service (CFCRV), which is the basis for decision-making such as evacuation orders. The work of the CFCRV includes:

- long-term analysis of the eruption patterns of individual volcanoes
- international comparison with similar eruptions elsewhere
- constant monitoring, e.g. seismometers, tilt meters, analysis of gas emissions, aerial and satellite surveys, e.g. infra-red to detect heat; daily and weekly bulletins are issued
- a well-established alert-level sequence understood by emergency services, all levels of government and local communities (active community preparedness, for example among the villages on the slopes of volcanoes such as Etna and Vesuvius)
- closure of airports and air space when threatened by ash in the atmosphere
- research into most effective building design to resist ash falls. Much work is carried out to identify the types of houses most at risk and to offer affordable solutions, e.g. adding a pitched roof to an existing flat roof, which will then shed ash fall.

Modifying loss

With frequent eruptions from Etna, Italy is well practised in dealing with volcanic hazards. Actions taken in modifying such events also help mitigate vulnerability and loss. Even so, houses, farms, livestock, orange and lemon groves and tourist facilities have been destroyed in recent decades. The Italian government has the resources to compensate individuals and businesses and restore infrastructure. Casualties have been minimal thanks to timely evacuation based on accurate data and well-trained public services such as the fire service, the police and the military.

Conclusion

There are benefits from living in proximity to volcanoes, although some groups, such as poor farmers, may feel they have little choice. The reliability and availability of technological monitoring means that volcanic hazards are manageable. Increasingly, the role of local communities in decision-making, together with good communications with regional and national organisations, mean that vulnerability to volcanic eruptions can be minimised.

> **Key idea**
> → There are various strategies to manage hazards from earthquake activity

Countries vary in their ability to manage earthquake hazards

In 2001, the World Agency of Planetary Monitoring and Earthquake Risk Reduction (WAPMERR) was established, bringing together researchers from around the world. The aim was to reduce the risk from earthquakes by carrying out research into earthquake prediction, tsunami risk and the construction of seismic-resistant buildings in cities most at risk.

In 2005 research carried out by WAPMERR suggested that a large-magnitude earthquake was overdue in the Himalayas. This prediction was based on records and the crustal pressures caused by plate movement since the last major earthquake. WAPMERR advised that there was an 89 per cent probability of a high-magnitude (> 8.0 Mw) event taking place in the western part of the region some time before 2100. Such an event would impact at least 3000 settlements, including large cities such as Kathmandu, and could result in as many as 200,000 fatalities.

Case study: Living with earthquakes – Nepal

There is no possibility of preventing or modifying an earthquake event. However, some of the effects of ground shaking, such as landslides, can be managed. The Nepalese are aware of the effects of deforestation in reducing slope stability and afforestation programmes operate in some parts of the country. But the demand for fuel, building materials and fodder for livestock means that Nepalese forests continue to be under pressure.

Modifying vulnerability

Although Nepal is an LIDC, it nonetheless has some resources to manage earthquake hazards. The National Society for Earthquake Technology (founded in 1993) and the Disaster Preparedness Network (founded in 1996) are Nepalese non-governmental organisations which aim to bring together science and engineering to devise strategies to reduce risks:

- Mapping of high-risk 'shake zones'
- Building codes introduced
- Encouraging households to have a 'go-bag' containing items useful in an earthquake emergency, e.g. dried foods, bottles of water, water purification tablets, whistle
- Introducing education programmes
- Organising an annual Earthquake Awareness day.

Modifying loss

The impacts of earthquakes can be severe, both for a country and for individuals. When an event of high magnitude occurs, the intervention of international disaster relief agencies is needed to mitigate its effects. Some 330 humanitarian agencies were involved in the aftermath of the 2015 earthquake in Nepal. They included international organisations such as the United Nations and the EU, individual countries such as Indonesia, Japan and the UK (Figure 14.37), and NGOs such as the Red Cross, Christian Aid and Shelter.

Figure 14.37 Japanese rescue team leaving for Nepal, April 2015

Conclusion

The Gorkha earthquake, which struck Nepal in April 2015, allows analysis of the effectiveness of disaster-planning and mitigation strategies. Given the magnitude of the quake and the casualty figures, several experts have concluded that the impacts could potentially have been much greater. For example, the risk maps for the city of Bharatpur suggested that as many as 30 per cent of its inhabitants could have been killed. In fact, there were no deaths in the city and the vast majority of the city's infrastructure and buildings escaped damage. Throughout Nepal people benefited from educational programmes and earthquake building codes.

Despite a feeling in Nepal that the disaster could have been worse, there was, nonetheless, considerable loss of life, injury and damage to buildings. Away from Kathmandu and the central valley, Nepal's mountainous topography hindered attempts to get emergency aid to remote villages. At a personal level, thousands of individual households have been devastated by the loss of family members (see pages 436, 457 and 491–92).

Case study: Living with earthquakes – Japan

Japan's economic wealth, highly developed education system and stable political environment enable it to cope with the constant threat of high-magnitude earthquake events.

Modifying vulnerability

Japan has developed a high level of preparedness to deal with earthquake hazards. Among the mitigation strategies employed are the following:

- Research and monitoring: the Japan Meteorological Agency (JMA) lies at the heart of Japan's mitigation strategies for natural disasters, including earthquakes and tsunamis and extreme weather events such as typhoons. It is responsible for providing information and warnings of impending earthquakes and tsunamis. Detailed disaster planning involves a wide range of organisations, e.g. governments, medical services, fire, military, transport, power and telecommunications companies.
- Buildings with aseismic design: e.g. steel frames and braces capable of moving without collapsing; rubber shock absorbers in foundations; very deep foundations into solid rock; a 'soft storey' at the bottom of tall buildings such as a car park, which collapses, allowing upper floors to sink down on to it; counter-weights on roofs, which move during an earthquake; suspension bridges capable of movement rather than rigid cantilever design; flexible joints in underground utility pipes, e.g. gas and water.
- Fire proofing older wooden buildings, which are common in old districts of Japanese cities.
- Land-use zoning that provides for open spaces where people can assemble after an earthquake.

- Controlling building in locations susceptible to excessive ground shaking or liquefaction.
- Tsunami warning systems off the coast.
- Refuge sites on permanent stand-by equipped with tents, bottled water, blankets.
- Community preparedness – ongoing education and training for all ages.

Modifying loss

Being one of the most advanced countries in the world, Japan has vast resources to manage losses caused by earthquakes. Well-rehearsed recovery and reconstruction plans, at national, regional and local levels, can be actioned immediately following an earthquake. The aim is to rebuild physically, economically and socially as quickly as possible.

Conclusion

Recent earthquakes have caused Japan to reappraise how it manages the impacts of earthquakes and related hazards. After the Kobe quake in 1995, much work was undertaken restrengthening structures such as bridges and roads that were previously thought to be safe from shaking. Today, following the Tōhoku earthquake, tsunami hazard management is being thoroughly reappraised.

Japan's strategies to mitigate the exposure, vulnerability and loss are constantly being updated. The country, industries and businesses and families tend to have the resources that give them a high degree of resilience. By contrast, EDCs and LIDCs have fewer resources and so their resilience is much less. As a result, the impacts of earthquake disasters in the short term are more serious, and longer lasting.

Review questions

1. What is the risk equation?
2. What factors influence a society's degree of 'vulnerability' to hazards such as volcanic eruptions and earthquakes?
3. Explain why mortality caused by tectonic hazards has fallen in recent decades.
4. Suggest reasons why the economic impacts of tectonic hazards have tended to increase through time.
5. State three measures that can reduce people's vulnerability to volcanic hazards.
6. Outline three reasons why so many people live on the slopes of volcanoes.
7. Describe and explain four measures that can reduce people's vulnerability to earthquakes.
8. What is meant by 'resilience' in the context of tectonic hazards?
9. Assess how 'inequality' influences the impacts of volcanoes and earthquakes in different countries.

Practice questions

A Level

Section A

1 Study **Table 1**.

Table 1 The ten most deadly eruptions of the twentieth century

Volcano	VEI	Year	Casualties
Mount Pelée, Martinique	4	1902	c. 33,000
Nevado del Ruiz, Colombia	3	1985	c. 23,000
Santa Maria, Guatemala	6+?	1902	c. 6,000
Kelud, Indonesia	4	1919	c. 5,000+
El Chichón, Mexico	5	1982	c. 3,500
Mount Lamington, Papua New Guinea	4	1951	c. 3,000
La Soufrière, St Vincent + the Grenadines	4?	1902	c. 1,680
Mount Agung, Indonesia	5	1963	c. 1,600
Pinatubo, Philippines	6	1991	c. 850
Mount St Helens, USA	5	1980	c. 60

Identify **three** limitations with the data evidence in Table 1. [3 marks]

2 Explain the role of earthquakes in the development of rift valleys. [6 marks]

Section B Synoptic questions

3 Assess how place profiles can be influenced by tectonic hazards. [12 marks]

Section C

4 'The causes of tectonic hazards owe more to human than physical factors.' Discuss. [33 marks]

AS Level

Section A

1 a) Explain how palaeomagnetism helps understand the processes occurring at mid-oceanic ridges. [4 marks]

b) Examine how ground shaking and ground displacement pose hazards for human activities. [6 marks]

c) Study **Table 2**, which shows statistics from selected volcanic eruptions.

Table 2

Volcano	VEI	Year	Casualties
Mount Pelée, Martinique	4	1902	c33,000
Nevado del Ruiz, Colombia	3	1985	c23,000
Santa Maria, Guatemala	6+	1902	c6,000
Kelud, Indonesia	4	1919	c5,000+
El Chichówn, Mexico	5	1982	c3,500
Mount Lamington, Papua New Guinea	4	1951	c3,000
La Soufrière, St Vincent and the Grenadines	4	1902	c1,680
Mount Agung, Indonesia	5	1963	c1,600
Pinatubo, Philippines	6	1991	c850
Mount St Helens, USA	5	1980	c60

i) Using the data above, calculate the interquartile range for the number of casualties. You must show your working. [4 marks]

ii) Using evidence from Table 2, analyse the contrasts in the casualty levels. [6 marks]

d) To what extent are tsunamis the most serious hazard arising from tectonic activity? [12 marks]

Section B Synoptic questions

2 a) With reference to Table 14.8 on page 485, suggest how living in a tectonically active region influences place profiles. [8 marks]

b) Examine how igneous rocks influence geomorphic processes in landscape systems. [8 marks]

Section C

3 How significant is the level of development of an area in determining the impacts of tectonic hazards? [20 marks]

Part 4

Investigative geography

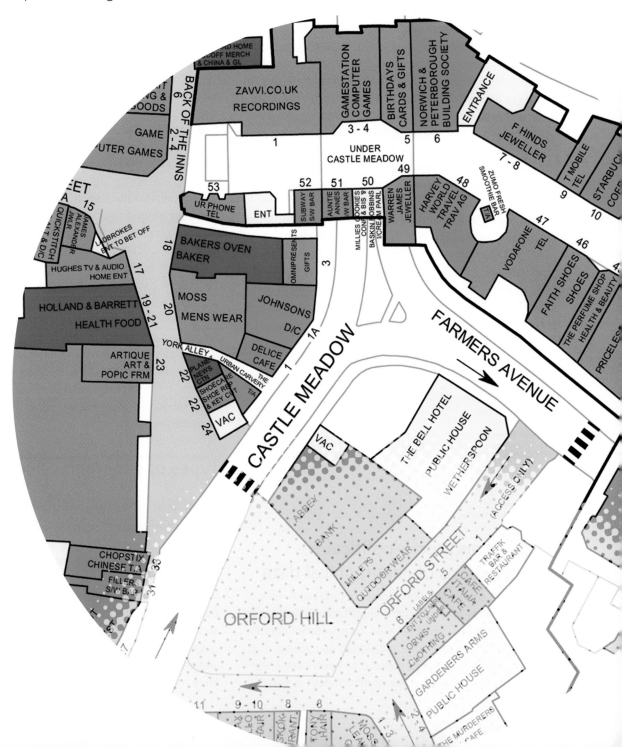

Chapter 15

Geographical skills

15.1 Geographical information

Geographical data

The defining feature of geographical data is their spatial or locational dimension. In other words, the data are linked to a specific location, whether a point or an area, on the Earth's surface. Geographical data are most often used to analyse differences between places and areas, relationships between variables located in space and movements (of people, goods and information) between places.

Geographical data may be quantitative or qualitative. Quantitative data (Table 15.1) are derived from objective observation and measurement and are the bedrock of scientific method which includes hypothesis testing and statistical analysis. Qualitative data are more subjective in nature. Most often derived from in-depth interviews, they are used to investigate individual or group opinions, feelings and experiences. Qualitative data also include field notes, field sketches, narrative descriptions, photographs and audio-visual information.

Table 15.1 Examples of quantitative geographical data used in independent investigation

Study	Data	Location
Beach profiles	Slope angles	Sand and/or shingle beaches
Scree slope analysis	Scree particle size	Location of particles on slope
Retail decline in the central business district	Discount, charity and vacant shop units	Mapping within the CBD

Ethical and socio-political implications of geographical data

Ethics are concerned with moral principles and rules of conduct. They are of primary importance in qualitative investigations involving detailed, face-to-face interviews with individuals or groups. Individuals and groups must give their informed consent to interviews. They should be aware of the type of information required;

how this information will be manipulated; where it will be published; the protection of their anonymity and privacy; and what will happen to the data after completion of the study. Careful consideration should be given to collecting data on personal details such as names, addresses, postcodes and phone numbers which might identify a respondent. Even indirect identifiers, such as occupation and place of work, could be viewed as invasions of privacy.

Some aspects of physical geography fieldwork also have an ethical dimension. Consideration of the environmental impact of data collection should precede actual fieldwork. Vegetation studies could result in trampling and damage to plants, especially in environmentally sensitive areas such as dunes and peat moorlands. Surveys on screes inevitably displace scree particles downslope and degrade these features.

Some of the socio-political implications to consider are ethical, racial, sexist, religious, discriminatory and political factors. If the topic is too sensitive it may be impossible to collect data through fieldwork or to persuade participants to join focus groups. Examples might include differences in obesity rates, diets and exercise regimens between residents in different parts of a city; spatial patterns of crime in urban areas related to socio-economic status, ethnicity, unemployment, etc.

Types of geographical information

Quantitative and qualitative

Quantitative data are numerical data obtained through direct measurement, observation or interview. They can be measured on **ratio**, **interval** (e.g. temperature), **ordinal** (e.g. ranking of service centres) and nominal (e.g. counts of land use types) scales. Quantitative data can be presented as statistical maps, charts and tables, and analysed by statistical methods.

Qualitative data are descriptive and non-numerical and are often derived from verbal or pictorial records. They may, for example, include narrative geographical descriptions of type, quality, age, condition, etc. of housing areas, shopping centres, recreational facilities; field sketch maps which show the relationships between geology, landforms and land use; and summaries of in-depth conversations and informal interviews to investigate perceptions, attitudes and motives.

Primary and secondary data

Primary data are original data collected directly either through fieldwork or desk research. By definition, all geographical fieldwork data are primary. In contrast, secondary data have been ordered, analysed, processed and often published either in textbooks, magazines and online. However, the distinction is sometimes blurred. For instance, demographic, social and economic data published on government websites such as the Office for National Statistics can be accessed, manipulated and used as primary data in research investigations.

Maps

Non-quantitative maps

Topographic maps are an essential source of geographical information. Ordnance Survey (OS) maps at 1:50,000 scale (or similar) have been available in the UK since the early nineteenth century. Large-scale maps, equivalent to 1:10,000, were published by the OS from the mid-nineteenth century, with even larger scales – 1:2,500 and 1:1,150 – covering urban areas in the twentieth century. These maps provide detailed information on relief, drainage, settlements, transport networks and land use (Figure 15.1). Increasingly,

(a) 1:50,000 Landranger Ordnance Survey map

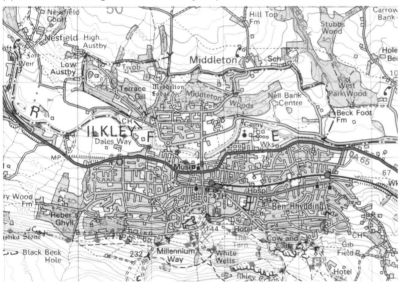

1:50,000 maps are useful in investigations at a regional scale such as:

● locating the origins of shoppers interviewed in a service centre
● delimiting small drainage basins
● understanding relief and drainage
● describing and explaining rural settlement patterns
● describing and explaining the shape and situation of settlements
● describing the distribution of woodland.

Electronic versions of the 1:50,000 series published by Anquet allow 3D transformations of the maps and also give complete satellite photo coverage.

(b) 1:25,000 Pathfinder Ordnance Survey map

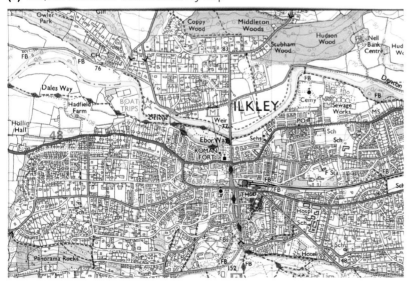

1:25,000 maps are useful in investigations at a sub-regional scale such as:

● describing and explaining the overall internal structure of towns and cities
● describing house types and the layout of suburban areas
● plotting rural land use (1:25,000 maps show field boundaries)
● detailed description of relief (contours are at 5 m intervals).

Electronic versions of the 1:25,000 series published by Anquet allow 3D transformations of the maps and also give complete satellite photo coverage.

Figure 15.1 OS maps

(c) 1:10,000 Ordnance Survey map

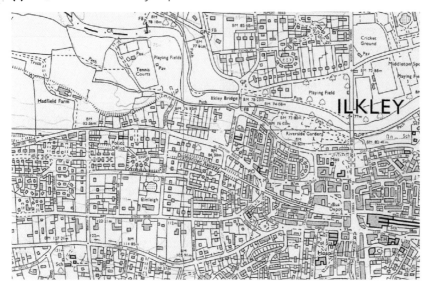

1:10,000 maps are useful in investigations at a local scale such as:
- recording the age of buildings
- recording land-use types
- recording types of housing
- delimiting flood hazard areas from spot height data
- settlement site characteristics based on spot heights.

(d) 1:2,500 Ordnance Survey map

1:2,500 maps are useful in investigations at the smallest scale to:
- identify individual houses by street name, number and house type (e.g. for spatial sampling exercises)
- record the exact location of shops and other commercial functions in a village, town or city centre
- provide detailed information of physical sites from bench marks (maps do not include contours)
- give detailed information on the functions of buildings.

Figure 15.1 (*Continued*)

geographers use maps in digital form generated from data uploaded to computers. A **raster map** is a data layer consisting of a gridded array of cells. It has a certain number of rows and columns, with a data point (or null value indicator) in each cell. Raster maps, often used as base maps in GIS, are digitised forms of conventional OS maps, street maps and other route maps, and topographic maps.

1:50,000 and 1-inch geology maps, the starting point for many investigations in physical geography, are available for most of the UK. Published by the British Geological Society (BGS), there are separate solid and drift geology sheets. Newer editions combine both solid and drift geology. These maps are also available in digital form from the BGS iGeology app.

Goad maps are large-scale plans showing commercial activities in town centres in the UK and other parts of Europe (Figure 15.2). The earliest Goad plans are available from the later nineteenth century and are an invaluable source of data on retail change and other commercial changes in town centres.

Many investigations in human geography rely on data located by postcode. Maps showing the boundaries of postcodes are available for postcode areas, districts and sectors but not for the smallest postcode units. However, the approximate location of postcode units (centroids) can be obtained from maps at www.zoopla.co.uk by entering a full postcode.

Figure 15.2 Goad map of part of Norwich

Quantitative maps

Since 2001, the census geography of England and Wales at a local scale has been based on spatial units known as output areas and super output areas. Population and socio-economic data are aggregated for these areas, whose minimum population varies from 100 residents (output areas) to 25,000 (super output areas – upper level). Digital maps are available online which show the boundaries of these census areas.

Quantitative maps

Quantitative or statistical maps are essentially stores of numerical geographical data. Most quantitative maps are a compromise between accuracy and visual impact

and their apparent objectivity can be misleading. Many quantitative maps are digital. Widely used in GIS, they store vast amounts of numerical data linked to locations on the Earth's surface.

There are five principal types of quantitative map: dot, choropleth, proportional symbol, isoline and flow maps.

Dot maps

Dot maps (Table 15.2) represent a given quantity of a variable with a dot of constant size (Figure 15.3). They have one great advantage over other quantitative maps: they show the distribution of variables within areal units such as counties, wards and parishes. As a

Table 15.2 Steps in the construction of dot maps

Step	Detail
Source a base map	A base map showing the areal units for which data are aggregated, e.g. census super output areas, wards.
Dot value	The number of items (e.g. number of people, crop area) represented by each dot. Dot values should be small enough to ensure that every areal unit has a few dots. In areas of highest density the dots should just begin to merge.
Dot size	Dot size will depend on the dot value and on the scale of the map. If drawn by hand, dots must be consistent in size.
Dot placement	Dot placement within areal units should be guided by prior knowledge of the distribution being represented. For instance, dot placement to show population distribution would be guided by the pattern of settlement.

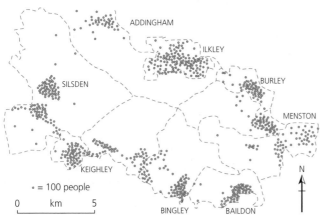

Figure 15.3 Dot map showing population distribution in mid-Airedale and mid-Wharfedale, 2001 (based on super output areas – middle level)

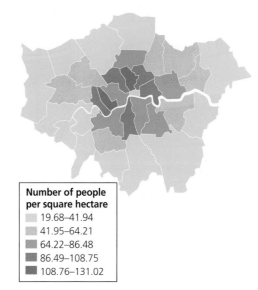

Number of people per square hectare
- 19.68–41.94
- 41.95–64.21
- 64.22–86.48
- 86.49–108.75
- 108.76–131.02

Figure 15.4 Choropleth map, London: population distribution and density, 2001

result, they provide an excellent visual impression of geographical distributions. A further advantage is that unlike choropleth maps, these distributions are not interrupted by the boundaries of areal units.

Choropleth maps

Choropleth maps show spatial variations in standardised values (i.e. percentages, densities, averages, ratios) between areas by differences in colour or shading (Figure 15.4). They are the most widely used maps in human geography, mainly because geographical data

are published in bundles for areal units from countries to census super output areas. Customised digital choropleth maps are available from the ONS website (neighbourhood statistics). They can be constructed to show a wide range of census statistical data for regional, county, local authority areas as well as for the hierarchy of census areas (e.g. super output middle and lower areas, output areas).

Table 15.3 Steps in the construction of choropleth maps

Step	Detail
Source a base map	A base map showing the areal units for which data are aggregated, e.g. census super output areas, wards.
Calculate standardised values	Record standardised values for each areal unit (e.g. percentages and densities) on the base map.
Classify the standardised values into groups	The number of classes or groups will be a compromise between excessive generalisation (too few classes) and too much detail (too many classes). Five or six classes are usually sufficient.
Define group or class intervals	Three methods are available: (1) fixed intervals, where a data set has meaningful thresholds (e.g. sex ratios of 100, zero population change); (2) intervals determined by natural breaks in the data (identified by plotting the data as a dispersion chart); (3) fixed intervals based on mathematical relationships (e.g. arithmetic, geometric, standard deviation).
Select a colouring or shading scheme	Monochrome maps use a gradation of tones from light (low values) to dark (high values). Where colours are used, softer colours from the middle part of the spectrum (yellows, greens) represent lower values. Higher values are represented by colours from the extreme ends of the spectrum (red, violet).

Table 15.4 Steps in the construction of proportional symbol maps

Steps	Detail
Source a base map	Source a base map which shows the statistical areas for which data are available.
Choose a symbol	A circle, square, triangle or bar.
Choose a scale	The largest symbols should not overlap with the symbols in adjacent areal units by more than 50%. Determine the size of the largest symbol relative to the scale of the map (e.g. 10 mm radius for a circle). Calculate the scaling factor for size of symbols based on the square root of the largest value in the data set. Thus, if a circle of 10 mm radius is used to represent a population of 625, the scaling factor is calculated thus: square root of 625 = 25 × 10/25 = 10 (mm).
Placement of symbols	Symbols should be placed centrally within each areal unit.

Choropleth maps (Table 15.3) have a number of disadvantages. First, they tell us nothing about the distribution of values within areal units. Second, areal units are highly variable in size and shape, with large units creating excessive generalisation. And third, the boundaries of areal units give the appearance of sudden discontinuities to distributions which are not reflected in reality.

Proportional symbol maps

Proportional symbol maps (Table 15.4) are based on the principle that the areas of symbols (circles, squares, triangles, bars) are proportional to the values they represent (Figure 15.5). In contrast to choropleth maps, proportional symbol maps show absolute (rather than standardised) values such as population counts, employment and retail floor space.

One disadvantage of proportional symbol maps is the difficulty of estimating values from the areas of symbols. This is because doubling the sides of a square or the radius of a circle increases the area fourfold. Another disadvantage is that the placement of symbols

and their interpretation becomes problematic where symbols overlap. Finally, proportional symbol maps convey no information on the internal distribution of values within areal units.

Isoline or isopleth maps

Isolines (or isopleths) are lines on maps that join places of equal value. Isolines are most often used in physical geography to represent continuous surfaces such as relief (contours), rainfall (isohyets) and temperature (isotherms) (Figure 15.6). In human geography their use includes journey times (isochrones), transport costs (isodapanes), rental values and pedestrian flows.

The construction of isoline maps is more subjective and generalised than other quantitative maps (Table 15.5). This is due to the transformation of a point pattern into a continuous surface. The problem is most apparent where there are few point values. Isolines are drawn between point values by **interpolation**: a technique which assumes (often incorrectly) a consistent gradient of change between points.

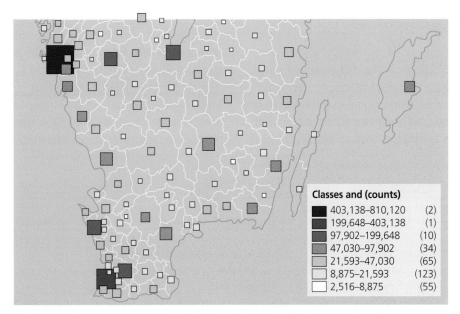

Figure 15.5 Proportional symbol map: population counts by commune in southern Sweden

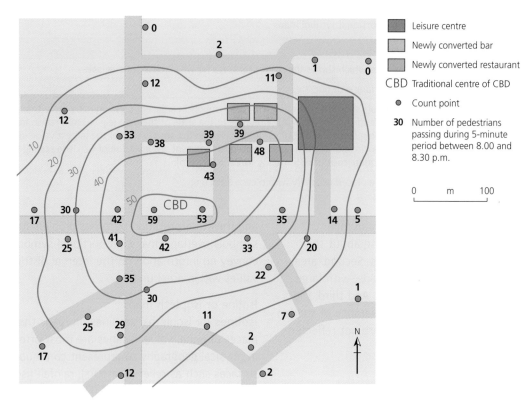

Figure 15.6 Isopleth map of pedestrian densities in a town centre

Table 15.5 Steps in the construction of isoline maps

Step	Detail
Source a base map and locate the data points	Find the location of the data points on the base map, or locate the points from GPS or grid reference data.
Plot the values	Plot the values for each data point on the map.
Decide on the number of isolines and their values	Five or six isolines are usually sufficient. Larger numbers often make construction impracticable. Generally, the fewer the point values on the map, the fewer the isolines. It is conventional to use regular intervals between isolines (e.g. 4 mb intervals between isobars on a pressure chart).
Fit the isolines	Isolines are fitted by interpolation which assumes a constant (linear) rate of change between two points. For example, a 600 mm isohyet between point values of 575 and 625 mm would pass exactly halfway between them.
Number the isolines and add layer colouring	Isolines should be numbered on the map. In order to increase clarity, areas between the isolines may be layer coloured.

Flow maps

Flow maps (Table 15.6) show the movement of people, freight and information between places. Movements are represented as lines proportional in thickness to the volume of flow. Flows paths can be either non-routed (most often shown as straight lines connecting places)

or routed, where flows follow actual pathways along streets and other transport networks.

A major drawback of flow maps is that they are often highly generalised, with flow categories covering a wide range of values, making it impossible to retrieve detailed statistical information. Non-routed

Table 15.6 Steps in the construction of flow maps

Steps	Details
Source a base map at the appropriate scale	At a local scale a base map showing the point locations where flow counts were made is needed. On a larger scale, a map showing the boundaries of areal units for which data are available is required.
Decide on the number of flow categories	In general, five or six flow categories are sufficient. Class intervals may be arithmetic or geometric but must cover the range of values. There should be no empty categories.
Select a suitable scale	Line widths will depend on the scale of the map. On non-routed maps it is important to avoid flow lines which overlap and restrict the retrieval of data.
Draw the flows	Draw the flows to scale and shade them with a single colour. In order to emphasise the flows where maps are computer-drawn, the base map should be muted.

Figure 15.7 The River Skirfare, North Yorkshire: channel changes 1992–2009

flow maps are particularly generalised. For pedestrian and traffic flows at a local scale, the corridors of movement are important and routed flow maps are more appropriate. Even so, abrupt changes often occur at check points, and rarely reflect actual changes on the ground.

Images

Satellite images and photographs contain valuable information on weather systems, land use, the shape and internal layout of settlements, river channels, coastlines and so on. Comparison of such images over time can reveal significant changes to the physical and human landscapes (e.g. coastal erosion, land use, the growth of settlements) (Figure 15.7). Remote sensing images also provide digital data on land and sea surface temperatures, vegetation growth and photosynthesis, soil moisture and rainfall intensity. These and many other images are available from government agencies and commercial sites such as NASA, NOAA, ESA and Google Earth.

Ground photographs of landscapes and townscapes provide an important historical record as well as a source of non-quantitative information for geographical narrative. Photography using mobile devices with locational tags is also an effective method of data collection. Examples include photos of shopping streets (shop types, pedestrian flows), vegetation cover and species in metre square quadrats (Figure 15.8); plant succession along transects in salt marsh and dune environments; and evidence of mass movement and rockfalls on cliff coasts.

Diagrams and graphical representations

Diagrams

Diagrams and graphical representations provide visual summaries of geographical trends, patterns, flows and cause-effect relationships. Diagrams are often used to simplify complex systems such as energy flows in food webs (Figure 15.9), changes to the physical environment during plant succession in sand dune ecosystems (Figure 15.10) and the causes of globalisation (Figure 15.11). Qualitative data can also be recorded by sketches drawn at first hand in the field (Figure 15.12).

Figure 15.8 Quadrat sampling of vegetation cover and plant species

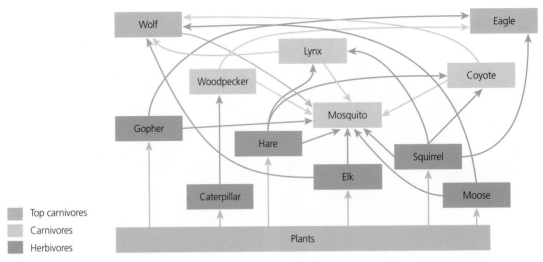

Figure 15.9 Food web

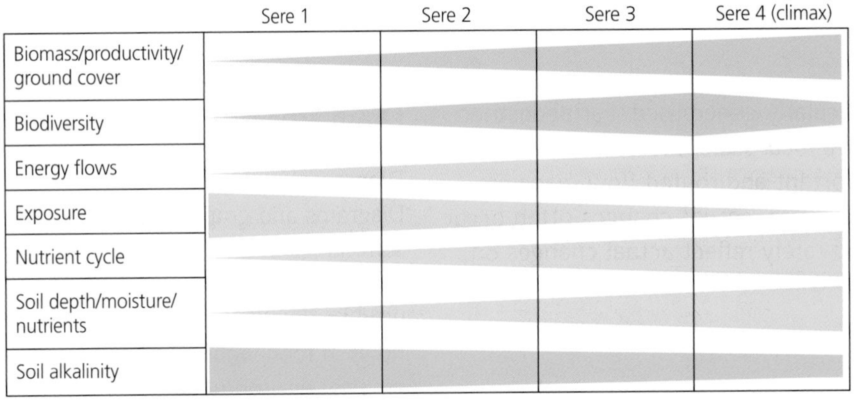

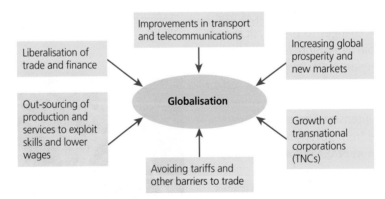

Figure 15.10 Sand dune vegetation

Figure 15.11 Globalisation

Histograms

Histograms are a type of bar chart used to represent frequency distributions. Those which are symmetrical around the mean or median value describe **normal frequency** distributions (Figure 15.13). However, many data sets in geography are asymmetrical or **skewed**. A positive skew is where the 'tail' of the distribution extends to the right (e.g. slope angles, sediment sizes) (Figure 15.14). Negative skews, which are less common, have a 'tail' extending to the left. The steepness of the curve in histograms is determined by the choice of class interval, which can lead to artificial representations/conclusions. Thus a more meaningful way to represent data is to replace the bars with a smooth **frequency distribution curve** (Figure 15.13).

Frequency distributions are sometimes plotted as **cumulative frequency curves**. They show the proportion of a data set above or below particular thresholds. Figure 15.15 shows a cumulative frequency

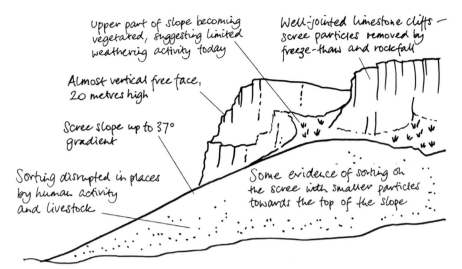

Figure 15.12 A field sketch of a scree slope at Norber Scar

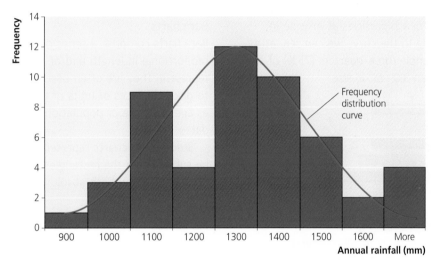

Figure 15.13 Histogram of annual rainfall at Olympia, Washington, USA, 1955–2005

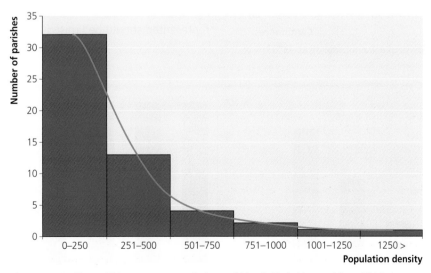

Figure 15.14 Skewed histogram – populations of North Yorkshire parishes, 2006

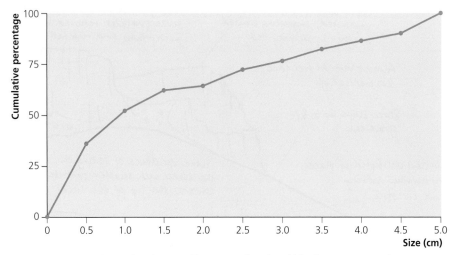

Figure 15.15 Cumulative distribution of limestone beach pebbles by size at Arnside

curve for limestone beach particles at Arnside in Cumbria. By referring to the curve we can see, at a glance, that for example three-quarters (75%) of the particles are less than 3.0 cm³ in size.

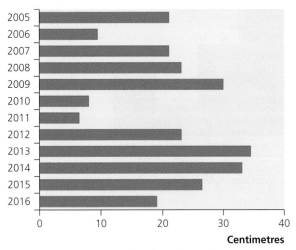

Figure 15.16 Annual growth of a walnut sapling at 53°53 N

Bar charts

Bar charts comprise a series of rectangles that are proportional in length and/or area to the values they represent. They are often used with data sets that relate to discrete places or units of time (Figure 15.16) such as age structure or population totals at different census dates.

Stacked bar charts represent two or more data sets by sub-dividing the rectangles in bar charts. They can be used to show both absolute and proportional values (Figure 15.17).

Pie charts

Pie charts are an alternative to stacked bar charts. They are circular graphs divided into segments each representing a sub-group in the population (Figure 15.18). The first segment in a pie chart always starts at 12 o'clock. The number of sub-groups represented should be limited to a maximum of seven

(a)

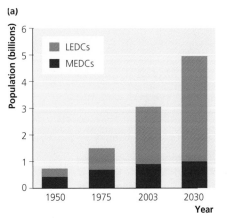

(b)

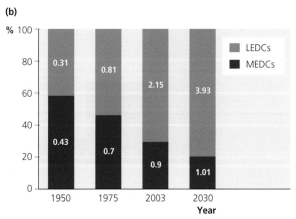

Figure 15.17 (a) Global urban growth, 1950–2030; (b) Proportion of the world's urban population (billions) in MEDCs and LEDCs, 1950–2030

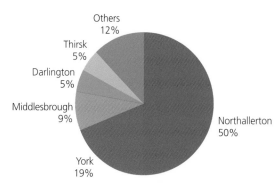

Figure 15.18 Pie chart showing destination chosen for clothes shopping by Northallerton residents

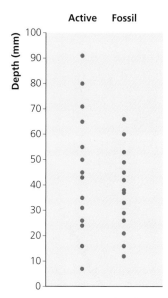

Figure 15.20 Depth of weathering pits (mm), active and fossil, on Silurian grit erratics

or eight, otherwise it becomes difficult to differentiate segmental shading.

Line charts

Line charts show continuous changes in variables in time and space, such as annual number of births (Figure 15.19) or variations in rental values with distance from a city centre.

Dispersion charts

Dispersion charts are single-axis charts which show the distribution of values in data sets. Figure 15.20 shows the depth of weathering pits at 'active' and 'fossil' sites on Silurian grit erratics at Norber in North Yorkshire. The dispersion of values suggests that where weathering is active, pits are deeper than those where weathering is no longer taking place.

Triangular charts

Triangular charts are used to plot three percentage values whose sum is 100 per cent. Examples include the proportion of a work force in primary, secondary and tertiary occupations; the proportion of sand, silt and clay particles in soils; and the proportion of children, adults and old people in a population. Triangular charts provide a visual comparison of differences between places (Figure 15.21), as well as changes at a place through time.

Scatter charts

Scatter charts are used to plot two variables, where x is the **independent variable** which causes change in y, the **dependent variable**. For example, depth of weathering can be plotted against gravestone age on a scatter chart (Figure 15.22). In this case, time (i.e. the age of gravestones) is the independent variable, which influences the depth of weathering, the dependent variable. The closer the distribution of points on a scatter chart corresponds to a straight line, the stronger the relationship between x and y. In Figure 15.22 the values trend from bottom left to top right, showing that weathering increases with time. This is known as a positive relationship. Where the points on a scatter chart trend in the opposite direction, the relationship is negative or inverse.

Factual text and discursive/creative material

Geographical information is available as a descriptive and explanatory factual narrative in textbooks, articles and other published sources. Fieldwork data may be recorded as factual text which describes and explains landscape features and human activity and behaviour in a geographical context.

Discursive and creative writing is based on the interpretation of factual information. It focuses on

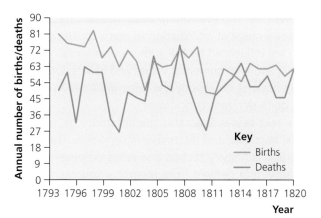

Figure 15.19 Annual number of births and deaths at St Georges-de-Reneins, 1793–1820

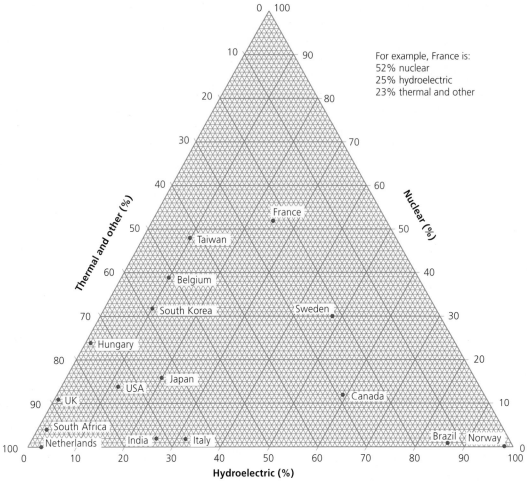

For example, France is:
52% nuclear
25% hydroelectric
23% thermal and other

Figure 15.21 Triangular graph showing the percentage of total electricity production by generating source for selected countries

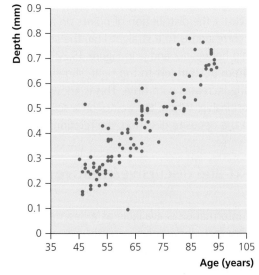

Figure 15.22 Weathering of marble gravestones at Clacton

issues such as the causes of inequality in human well-being or climate change which promote discussion and value judgements. When considering geographical information it is important to recognise the difference between factual text which is largely objective, and discursive material which includes subjective interpretation and evaluation. Discursive writing assesses a range of information to reach conclusions which are influenced by the author's own social, political and environmental values and beliefs. Other writers drawing on exactly the same information might reach quite different conclusions.

The text below illustrates the difference between factual narrative and discursive information. The first paragraph is heavily factual and relies on poverty and mortality data derived from reputable sources such as the World Bank and the UN. The second paragraph, however, goes further than merely providing facts. In considering the factors that contribute to poverty and child mortality, it makes an assessment of their relative importance. The emphasis on economic factors reflects the opinion of the author. Other writers, adopting a more overtly political stance, might reasonably conclude that poor governance or past exploitation by colonial powers are more important.

'The global distribution of poverty is heavily concentrated in the economically developing world, especially in sub-Saharan Africa (SSA) and South Asia. Of the world's poorest countries, in 2015 29 were in SSA and South Asia. Half the population of SSA and nearly 40 per cent of South Asians survive on less than one US$ a day. And child mortality rates in the least developed countries such as Somalia and Sierra Leone are up to 30 times higher than those in western Europe and North America.

'A wide range of physical, historical, political, social, economic and demographic factors contribute to poverty in the developing world. However, the most important factors are undoubtedly economic. Lack of foreign direct investment inhibits economic development, productive employment and domestic spending. It also restricts international trade and the incentive to develop essential infrastructures such as harbours, airports, railways and roads. By comparison,

the effect of poor governance, inadequate educational opportunities, the history of past colonialism, climate and disease, though important, are far less significant.'

Digital data

Information that is used and stored by computers is digital data. In geography, digital data include geographical co-ordinates such as grid references and latitude and longitude, linking them to precise locations on the Earth's surface. A wide range of digital data are available, including maps, satellite images, aerial photos and ground-based photos. Paper maps can be scanned and converted to digital form. Data recorded through fieldwork can be transferred directly via handheld devices such as smartphones and tablets to computers for processing and analysis.

The Office for National Statistics (ONS) for the UK is a major source of digital data for geographical investigation. It publishes a vast array of neighbourhood statistics, including census data and other government surveys. These data are available at various scales from output areas containing as few as 100 residents to super output areas with more than 25,000 residents (Figure 15.23). The census geography allows users to access raw data in numerical or in cartographic form as customised digital maps at appropriate scales. (Figure 15.24).

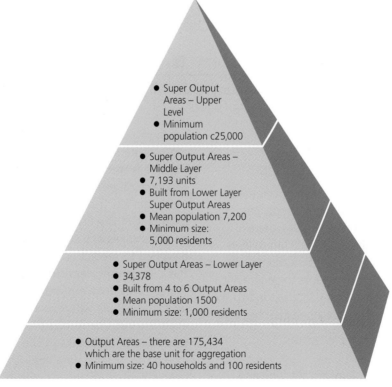

- Super Output Areas – Upper Level
- Minimum population c25,000

- Super Output Areas – Middle Layer
- 7,193 units
- Built from Lower Layer Super Output Areas
- Mean population 7,200
- Minimum size: 5,000 residents

- Super Output Areas – Lower Layer
- 34,378
- Built from 4 to 6 Output Areas
- Mean population 1500
- Minimum size: 1,000 residents

- Output Areas – there are 175,434 which are the base unit for aggregation
- Minimum size: 40 households and 100 residents

Figure 15.23 Census hierarchy of the 2011 census for neighbourhood statistics

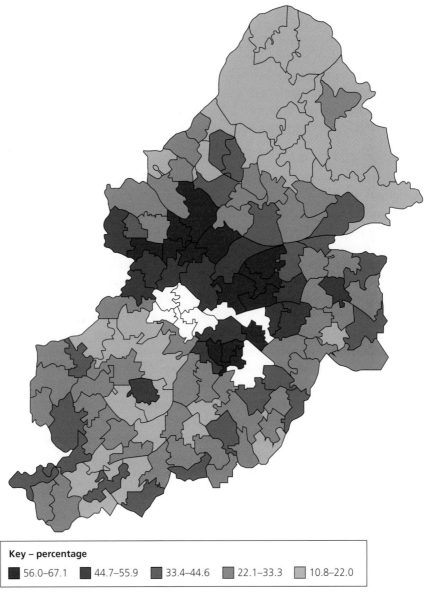

Key – percentage

■ 56.0–67.1　■ 44.7–55.9　■ 33.4–44.6　■ 22.1–33.3　■ 10.8–22.0

Figure 15.24 Distribution of poverty in Birmingham. Percentage of households below 60% of the median income (after housing costs), April 2007–March 2008

Big data and crowd-sourced data

The explosion of data in the early twenty-first century, through use of online searches, social media, satellite technology and telecommunications, has been unprecedented. For instance, in 2014 there were 1 billion Facebook visits a day and 140 million tweets. These data, amassed in real time, come with locational tags making them amenable to geographical analysis. The volume of new digital data has led to the development of two innovative technologies: big data and crowd-sourcing.

Big data

Big data refers to the technology of collecting and analysing massive new databases. However, rather than a substitute for traditional data collection and analysis, big data is seen as a complementary approach to geographical problem solving.

Google Flu Trend (GFT) is one of the best-known examples of the use of big data and illustrates both its advantages and weaknesses. GFT is an algorithm designed to predict annual patterns of influenza in the USA. It uses data from millions of online Google searches for flu symptoms, and is predicated on the idea that people will report flu symptoms they are experiencing. The results provide a reasonable match to actual flu outbreaks in the USA (Figure 15.25).

Even so, the big data algorithm significantly overestimated influenza cases in 2012–13 and in previous years.

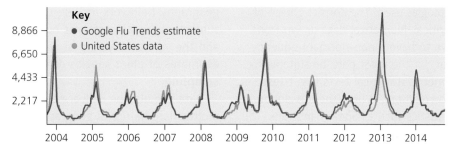

Figure 15.25 USA actual and estimated flu activity 2004–14 (Source: US Centers for Disease Control)

Other examples of big data include the analysis of daily movement patterns in cities based on social media. Tweets provide routine information on location and creation time, as well as the content of messages. Analysis of the content of tweets and the use of key words enables researchers to monitor the movements of people between home, work and recreational spaces.

Crowd-sourced data
Crowd-sourced data rely on people (on the internet or using mobile devices) voluntarily performing certain tasks which generate geographical information. Because large numbers of people participate, more accurate data are created. Crucial to the generation of crowd-sourced data is access to GPS technology, either through mobile phones or handheld GPS devices. For instance, residents of Boston, USA, have used smartphones to transmit to the highway authority's server the location of potholes in the city. In Australian cities residents have generated crowd-sourced data on noise level hotspots. The OpenStreetView website (www.openstreetview.org) is a massive global mapping project based on individually generated, crowd-sourced data made possible by the widespread availability of GPS technology.

Information from crowd-sourcing can mitigate natural hazards and prevent disasters. Between July 2007 and May 2009 four major wildfires threatened the Californian city of Santa Barbara. In the past, citizens relied on government and official agencies to provide information, manage the response to hazards and organise evacuation. Given the ease with which volunteers can create and disseminate information on the location, extent and threat of wildfire, the role of agencies is no longer crucial. Digital cameras, GPS, digital maps and other resources are available to the public and real-time information collected as photos, texts, tweets, satellite images, GPS fixes and videos can be uploaded to local newspaper and community websites and shared immediately with those at risk.

Informed and critical questioning

Data sources
An investigation is only as good as the data on which it is based. Given the massive volume of data available online, identifying accurate and reliable data sources is not straightforward. Many websites contain data which are not only inaccurate, but which for commercial or political reasons may be biased. Equally, newspaper reporting may be unreliable. This is invariably the case in the initial reporting of natural disasters. Statistics, often unverified, may be little more than guesswork; sources are unstated; and with news journalists working within strict time constraints, research may be incomplete and the situation liable to change.

The most accurate and secure sources of online data are those published by central and local governments and multilateral agencies. For example, data sourced from the UK government's Office for National Statistics, the US Census Bureau, the websites of the various UN agencies, the World Bank, WHO and many others can be used with confidence.

Historical data are often incomplete and subject to error. Pre-census population data in England and Wales rely heavily on parish registers. Yet parish registers are not a complete record of births, deaths and marriages; baptisms are not the same as births; and the rise of non-conformism, urbanisation and industrialisation means that from the mid-eighteenth century parish registers become increasingly unreliable. Regardless, whatever your data sources, they must be referenced so that the validity of your data can, if necessary, be checked.

Analytical methodologies
The characteristics of data – whether they are quantitative or qualitative – and the nature of the topic under investigation, will determine the

analytical methodologies used. The correct choice of statistical test is critical. Numerical data drawn from a population known to have a non-normal frequency distribution will not, for the most part, be suitable for analysis with parametric statistical tests (e.g. t-test, product moment correlation). Where there is doubt, non-parametric statistics such as Spearman's rank, Mann–Whitney and chi-squared should be used. The choice of test is also driven by the size of the database, whether the hypotheses are ones of association or difference, and the type of data (i.e. ratio, interval, ordinal, nominal).

Statistical analysis of qualitative data is possible where these data are presented in numerical form (e.g. people's perception of places). However, in many circumstances statistical analysis of qualitative data is often considered inappropriate – a blunt tool that misses the nuances of human behaviour. Instead, analysis focuses on more subjective methodologies involving textual description, discussion, evaluation and synthesis.

Identifying sources of error and the misuse of data

At the outset of a geographical investigation it is important to review data critically and identify possible sources of error. Historical data sources such as the early population censuses in the UK often underestimate populations, while studies of migration before 1841 may be hampered by inconsistent references to places of birth (countries, counties and towns are variably used). The changing boundaries of census enumeration areas also create problems of continuity in studies of demographic, economic and social change over time.

Sometimes the data required are unavailable and geographers are forced to use surrogate measures. For instance, urban land values are notoriously difficult to obtain, and property prices, rents or council tax values are used instead. Although related to land values, all three measure things that are slightly different.

In physical geography, investigations often encounter problems which lead to error. For instance, the measurement of widths and depths of grikes on limestone pavements is compromised where limestone particles (the result of freeze-thaw weathering) accumulate in grikes. Also, grike widths often vary with depth, being much wider closest to the surface where chemical weathering is most intense. Other typical sources of error in physical geography fieldwork include the large element of subjectivity in assessments of sediment shape, and the vertical layering of vegetation confusing estimates of plant species diversity in **quadrat sampling**.

Data can also be deliberately misused. Given the multivariate nature of geography and errors in data collection, outcomes rarely conform to textbook models and theories. There is thus the temptation to manipulate data by ignoring extreme values (viewed as exceptional) or reformulating class boundaries in tests like chi-squared until desired outcomes are achieved.

Communicating and evaluating findings

Findings should be placed in a wider context, explaining their relationship to existing theory, models and ideas. Geographical investigation should conclude with a summary of the main findings. These refer back to the initial aims and objectives of the study.

Evaluating the investigation involves critical analysis of the extent to which its aims and objectives have been achieved. Given the complexity of geographical investigation, with multiple variables, and problems of collecting accurate and representative data, it is likely that some hypotheses will remain unverified and some questions unanswered (see Chapter 16). Explaining such outcomes requires understanding of the weaknesses of the study. Were the study's aims and objectives flawed from the outset? To what extent did weaknesses in methodology and data collection undermine the investigation? With hindsight, could these weaknesses have been anticipated and addressed at an early stage? Sometimes investigations yield results which appear to contradict theory and current orthodoxy. This learning by induction might suggest alternative hypotheses and questions for future investigation.

Extended, discursive writing is an important skill for communicating information about geographical matters. As well as conveying clear and accurate description and explanation, discursive writing requires the higher level skill of evaluation. Arguments are compared and assessed on their strength, relevance and consistency, and ultimately a decision, based on value judgements, is made. In geography there are numerous examples of debatable issues such as the contribution of human activity to flood hazards, government response to coastal erosion, the siting of a third runway at Heathrow and the impact of HS2 on the economic regeneration of northern England.

15.2 Geo-located data

Smartphones and tablets

Smartphones and tablets, with in-built location and GPS functions, cameras and notes apps are valuable data-collection tools in fieldwork. Geo-data include locations given by co-ordinates (e.g. use of compass on the iPhone), field notes, field measurements and photographs. An increasing number of apps, many of which can be downloaded free, are available to allow data collection, storage and sharing of fieldwork information. Google Earth, for example, provides maps, photographs and satellite images with geotags. Large-scale maps, tailored for physical and human geography fieldwork, are available with the Fieldtrip GB app (www.fieldtripgb.blogs.edina.ac.uk). These maps can be used online, or importantly where mobile reception is poor, downloaded on to handheld devices. Locations on the map can be annotated, tagged, illustrated with photographs and tracks, saved in the Cloud and shared with other users.

Skitch (previously known as Evernote) is another free app which allows users to clip text from the web, store links, write notes, draw field sketches and annotate and geotag photos. The iGeology app of the British Geological Society provides solid and superficial geology maps covering England and Wales, which eliminates the need to carry paper maps in the field. Smartphones and tablets also mean that it is unnecessary to carry notes, paper questionnaires and some fieldwork equipment (tapes, clinometers) into the field. Meanwhile, digital data can be uploaded immediately, saving time and effort.

The Polldaddy app allows the creation of questionnaire surveys on desktops (up to ten questions) which can then be synced to smartphones and tablets. Questionnaires can then be used offline on handheld devices in the field. More specific apps include Decibel 10, which measures noise levels; Escri, which allows distance and area measurements, as well as providing locational co-ordinates on large-scale maps; and My Angle which measures slope angles.

Geographical Information Systems (GIS)

GIS systems capture, store, display, analyse and manage geo-spatial data. Data available as maps can be interrogated to search for patterns, distributions and investigate causal relationships.

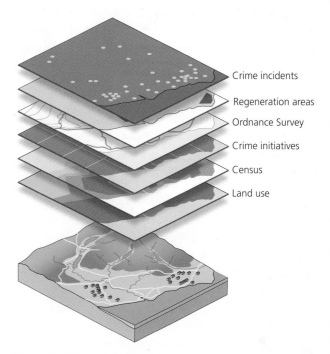

Crime incidents
Regeneration areas
Ordnance Survey
Crime initiatives
Census
Land use

Figure 15.26 A Geographical Information System (Source: adapted from ESRI)

It is estimated that 80 per cent of all data are linked to location, and handheld devices provide locational data (latitude and longitude co-ordinates, postcodes, grid references, etc.) that are sufficiently accurate to transfer to GIS systems.

GIS systems comprise three elements: (1) digital maps, (2) data located on maps, and (3) software application. At the simplest level GIS comprise maps which can be searched and where spatial data can be overlaid in layers (Figure 15.26). In this way the relationships between geographical phenomena can be displayed on a single map.

Many websites provide free geo-spatial data that can be uploaded to GIS. Examples include property prices at Zoopla; maps to support GIS at Bing Maps and Get-a-Map; and satellite images from Google Earth. Natural England's Magic site provides overlays for a wide range of OS maps covering geology, landscapes, wildlife habitats, animal species and conservation areas. At a more sophisticated level GIS can be used to optimise the location of services such as ambulance and fire stations, doctors' surgeries and supermarkets in relation to accessibility (journey times) and demand (population distribution and density), flood incidence and land use.

Crime mapping linked to GIS is used as an important tool in crime prevention by police forces throughout the UK. Crime has an inherent geographical property: it happens at a specific location and is not randomly distributed. As a result,

Figure 15.27 Marylebone High Street is one of London's crime hotspots

crime 'hotspots' are a feature of many large cities (Figure 15.27). The distribution of types of crime can be mapped and related to geo-spatial data such as socio-economic status of neighbourhoods, population densities, pedestrian flows and land

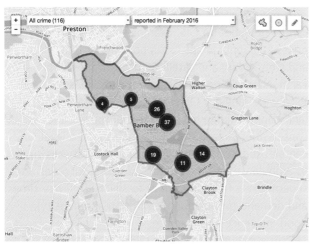

Figure 15.28 Crime distribution in Walton-le-Dale and Bamber Bridge, Lancashire, May 2015 (Source: OpenStreetMap contributors, www.openstreetmap.org/copyright)

use. In this way GIS can help the police to understand better the causes of crime, identifying trends, patterns and hotspots, and improve crime enforcement. Much of the information on crime is available to the public (www.police.uk) at neighbourhood scales (location of crimes at street level), with apps allowing scrutiny of crime over the past year and month (Figure 15.28).

15.3 Qualitative skills

Quantitative methodology, based on scientific enquiry, hypothesis testing and statistical analysis has dominated geography for the past 50 years. Qualitative methodology is an alternative approach: one which is most suitable to people-centred investigations in human geography.

Qualitative approaches rely on observation, interviews and textual analysis. Observation of people and places includes making field notes, field sketches, ground photographs and audio-visual recordings (Figure 15.29, Table 15.7). Detailed observation and narrative description is followed by classification, discussion, evaluation and the formulation of concepts.

Unlike questionnaire surveys, based on large samples, closed questions and tick boxes, qualitative methodology uses informal interviews with individuals and discussion with focus/community groups. Moreover, the selection of individuals and groups for interview does not involve objective statistical sampling. Instead, qualitative methodology focuses on in-depth research into motives, behaviours, attitudes and perceptions. It is most valuable in researching specific, locally based issues, which make no claim to universal generalisation. Qualitative methods can also be used to investigate, for example, the local impact of a new superstore in a market town or the closure of a branch library. Such an approach offers a better understanding of the depth of feeling and response of people to local issues.

Central to qualitative methodology is coding (Figure 15.30). Coding is a step-wise method of organising, labelling, processing and synthesising a mass of qualitative information (e.g. opinions, feelings, experiences) from field observation, interviews and texts. The outcome is the development of concepts and theories. Unlike quantitative approaches, coding is more empirical and relies less on preconceived models and theories. Coding also has weaknesses: it often involves great subjectivity on the part of the researcher; it can

Table 15.7 Qualitative investigation into the site of a proposed housing development at Menston, West Yorkshire (Figure 15.29)

Data collection	
Observation	Field notes, field sketches, ground photos.
Secondary data	Local geology (iGeology app), satellite images (Google Earth app), site measurements (Esri app), digital maps (Fieldtrip GB app), local newspaper reports, planning documents.
Interviews	Open-ended, in-depth interviews with note taking and/or audio data to record the views, values, perceptions, etc. of: local residents; campaign group leaders; developers; planners; elected local councillors.
Coding	Transcribing the mass of free-form data from text, field notes, public documents, interviews, recordings, etc. including the local impact on schools, doctors' surgery, public transport, traffic movement, drainage, etc.
Classifying and identifying themes	Responses classified according to different groups (interested parties); social, economic, environmental impact; positive and negative views of the development. The narrowing of the data to focus on the main emergent themes. Where possible, the presentation of some grouped data as charts.
Theoretical concepts	By the process of induction (or empiricism) defining the main ideas and concepts that emerge from the study.

Figure 15.29 Qualitative study of proposed greenfield housing development in Menston, West Yorkshire

The location of the proposed development of 120 houses in Menston. This greenfield site is approximately rectangular; covers an area around 2 ha; and measures roughly 160 × 120 m. The site has a northerly aspect and with slopes between 5 and 15 degrees. The geology comprises sandstone and shale overlain by glacial till. There is an extensive catchment above the site and heavy rainfall produces significant run-off with water pooling at the base of the slope. Springs also occur intermittently at the boundary of the sandstone and shale. Land use is permanent and largely unimproved pasture. The environmental quality of the site is high, with mature trees to the south and west, varying slope angles and extensive views across Wharfedale to the north. Late-nineteenth-century terraces adjoin the site to the north. There is a road and a local authority housing estate on the site's eastern edge.

be very time-consuming and laborious; and it may require numerous revisions of codes.

15.4 Quantitative skills

Quantitative geography is the enquiry route that follows scientific principles. It has been the dominant methodology in geography for the past 50 years. Its main characteristics are: deductive theory and hypothesis testing; objective collection of numerical data often in the form of samples; and the use of statistical methods to describe and analyse data. In this section we focus on the statistical skills used in quantitative geography.

If indicated
Theoretical concepts
emerge from saturated categories and themes

Level 3 Coding
Axial/Thematic Coding
Previous coding is studied to develop highly refined themes

Level 2 Coding
Focused Coding, Category Development
Level 2 Coding re-examines Level 1 Codes and further focuses the data

Level 1 Coding
Initial Coding, Open Coding
Large quantities of raw qualitative data are focused and labelled during Level 1 Coding

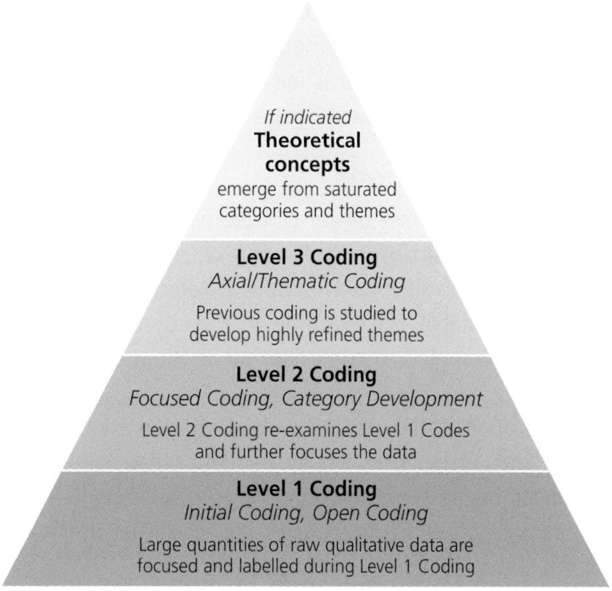

Figure 15.30 Coding

Measurement and sampling

Fieldwork measurements and error

Most aspects of fieldwork in physical geography involve data collection through measurement. Measurements can rely on standard apparatus and instruments such as ranging poles, tapes, quadrats, metre rulers, clinometers, current meters, thermometers and anemometers. Increasingly measurements can be completed using smartphones and tablets with apps that record distances, areas, routes, slopes, noise levels and so on.

Fieldwork measurements, however diligently executed, are often subject to error. For example, measuring the discharge of a small river and stream is extremely challenging. First, there is the problem of defining the boundaries of the channel in the field; second, measuring accurately the cross-section in an irregular channel; and third, recording flow velocity which varies both across the channel and with depth. As a result, estimates of discharge at the same site often vary considerably.

In human geography questionnaire surveys can measure respondents' strength of feeling towards controversial issues with rating scales (either with numbers such as 1 to 5, or with descriptors from 'very strongly in favour' to 'very strongly opposed'). In this kind of measurement there are numerous sources of error. They include differences in the interpretation of rating scales by respondents; respondents who are unable or unwilling to think through their answers carefully; and bias on the part of the interviewers who may subconsciously lead respondents to provide the kind of answers they want.

It is virtually impossible to avoid some degree of error in fieldwork measurement. However, having accepted that some error is inevitable, every effort should be made to reduce it to a minimum.

Sampling methods

A sample is a sub-set of items selected from the statistical **population**. In this context, the term population does not mean people. It refers to all of the data available. For example, in a study of a shingle beach, the population comprises all of the pebbles on the beach.

Most quantitative geographical enquiries are based on samples rather than the statistical population. However, because an investigation is only as good as the sample on which it is based, it is essential that samples are chosen objectively and represent the population accurately. A successful sample is one which enables researchers to draw valid inferences about the characteristics of a population.

The alternative to sampling is to investigate the entire statistical population. While this approach would eliminate problems associated with unrepresentative samples, it is in most cases impracticable. Most statistical populations are simply too large to study *in toto*. Moreover, a high degree of accuracy, comparable to surveying the entire population, can be achieved from a well-designed sample at a fraction of the cost.

A good sample has the following qualities:

- It is unbiased, i.e. estimates of population characteristics such as the mean and standard deviation are neither consistently larger nor consistently smaller than the true values.
- It is precise, i.e. it provides an accurate estimate of population characteristics.
- It is large enough to provide conclusive results in terms of statistical significance.
- It can be collected easily and with the minimum of resources.

Sampling methods in geographical investigations may be divided into spatial and non-spatial (Figure 15.31).

Non-spatial sampling

There are three types of non-spatial sampling: random, systematic and stratified.

The term 'random' has very specific meaning in statistical sampling: it means that samples are selected using random numbers, generated either from a calculator or a random numbers table. The basic assumption of **random sampling** is that every item or individual in a population has an equal chance of inclusion in the sample. This type of sampling is most appropriate when there is a listing of the population (e.g. a list of businesses on an industrial estate, or households in the electoral register), allowing the selection of samples by random numbers. Even without a listing, random sampling is sometimes possible. For example, high street shoppers could be chosen for interview at random intervals determined by random numbers (n), say between 1 and 9. If 100 random numbers in this range were generated prior to fieldwork, then the sampling process could be implemented quite efficiently.

Systematic sampling is an alternative to random sampling. With systematic sampling items or individuals

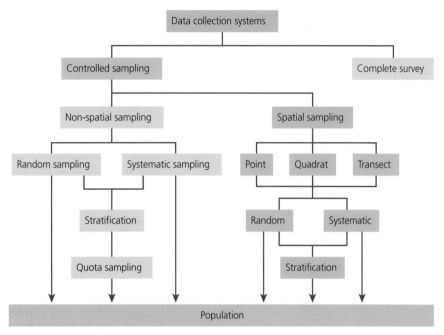

Figure 15.31 Sampling methods

are sampled from a population at fixed intervals. The fixed sample interval is usually chosen randomly. Systematic sampling is both quicker and simpler than random sampling. For these reasons it is often preferred in fieldwork investigations, especially in questionnaire surveys.

Random and systematic sampling methods are often combined with stratification. Many statistical populations are far from homogeneous, comprising various sub-groups. If some of these sub-groups are relatively small they may be underrepresented by simple random and systematic sampling. Common

Designing questionnaires

1 Initial decisions

At the outset you must decide what information is needed, who the respondents (target population) are, and the method of survey (street interviews, postal questionnaires) to be used.

2 Content and types of questions

Because respondents have limited time, and you depend on their goodwill to complete the questionnaires, only those questions that are essential should be included. Questions should be as clear and simple as possible and most should focus on facts (which are easier to collect) rather than attitudes and opinions. Questions that have political and religious connotations or are in any way economically, socially or ethnically offensive must be avoided.

As far as possible questions should be designed to elicit standardised responses so they can later be compared, tabulated and classified. For example,

responses to the question 'how often do you shop in this town centre?' can be standardised as 'several times a week', 'once a week' and 'less than once a week'. Multiple-choice questions presented in a tick-box format should be used wherever possible.

Questions dealing with attitudes and opinions generate more information if some form of rating scale is used. For example, a question such as 'do you think that traffic congestion is a problem in this town?' will produce only binary 'yes' or 'no' answers. If, however, the question is modified to read 'to what extent is traffic congestion a problem in this town' then we can use a rating scale such as 'not a problem', 'a minor problem', 'a significant problem', 'a severe problem'. A numerical scale 1 to 6 could be used in place of a nominal scale, where 1 = 'no problem' and 6 = 'a severe problem.' Purely open-ended questions are likely to produce answers that are difficult to analyse and quantify.

→

3 Layout

Questions should have a logical sequence. In the exemplar shopping questionnaire, the questions on shopping behaviour progress from goods and services purchased and frequency of shopping trips in the centre where the interviews take place, to shopping trips to other centres (Figure 15.32). Questions concerning personal characteristics are generally placed towards the end of a questionnaire. This is because respondents may regard this information as confidential and decline to answer. However, this is less likely if they have already answered the earlier questions; and even if they refuse to answer you will still have some useful information from their earlier responses.

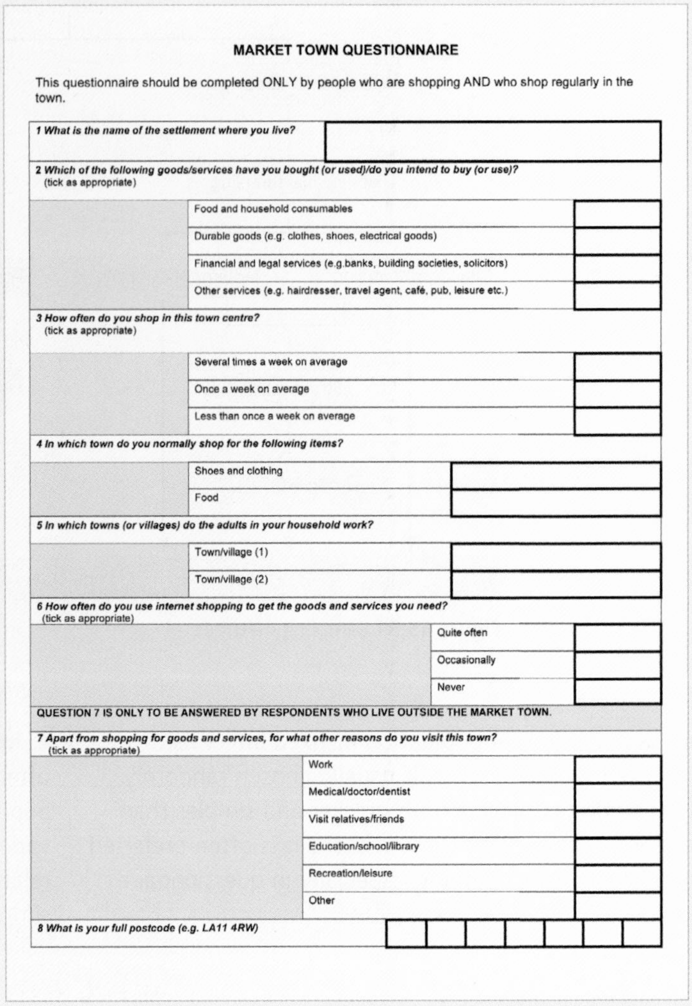

MARKET TOWN QUESTIONNAIRE

This questionnaire should be completed ONLY by people who are shopping AND who shop regularly in the town.

1 What is the name of the settlement where you live?

2 Which of the following goods/services have you bought (or used)/do you intend to buy (or use)?
(tick as appropriate)

- Food and household consumables
- Durable goods (e.g. clothes, shoes, electrical goods)
- Financial and legal services (e.g.banks, building societies, solicitors)
- Other services (e.g. hairdresser, travel agent, café, pub, leisure etc.)

3 How often do you shop in this town centre?
(tick as appropriate)

- Several times a week on average
- Once a week on average
- Less than once a week on average

4 In which town do you normally shop for the following items?

- Shoes and clothing
- Food

5 In which towns (or villages) do the adults in your household work?

- Town/village (1)
- Town/village (2)

6 How often do you use internet shopping to get the goods and services you need?
(tick as appropriate)

- Quite often
- Occasionally
- Never

QUESTION 7 IS ONLY TO BE ANSWERED BY RESPONDENTS WHO LIVE OUTSIDE THE MARKET TOWN.

7 Apart from shopping for goods and services, for what other reasons do you visit this town?
(tick as appropriate)

- Work
- Medical/doctor/dentist
- Visit relatives/friends
- Education/school/library
- Recreation/leisure
- Other

8 What is your full postcode (e.g. LA11 4RW)

Figure 15.32 Market town questionnaire

criteria used to stratify populations include age, gender, socio-economic status, income, ethnicity, employment and so on.

Two sampling strategies are designed to represent sub-groups in statistical population: **stratified random and systematic sampling**; and **quota sampling**. The difference between them relates to the selection of individuals or items for the sample. Stratified random and systematic methods select sample items or individuals objectively; whereas in quota sampling a researcher deliberately chooses items or individuals to fill the quota for each sub-group. As a result, quota sampling is less scientific than stratified random and systematic sampling. It is, however, more practical, and given constraints of time and money may often be the only feasible means of collecting stratified data. An example of stratified sampling is shown opposite.

Spatial sampling

Spatial sampling is used where location is an essential feature of the items or individuals studied. Samples are selected from points, areas and lines using random, systematic and stratified sampling strategies.

Spatial sampling procedures using points, areas and lines can be illustrated with reference to the 1:25,000 land use map in Figure 15.33. Let us assume that the purpose of sampling is to estimate the proportion of arable, pasture, forest and urban land use in the area.

Point sampling

We can sample land use at random points on the map by generating random six-figure grid references. This is done by using the random number function on a calculator and combining, say, 100 three-figure eastings between 410 and 480, with 100 three-figure northings between 450 and 490. Two problems arise with this procedure. First, generating random grid references and

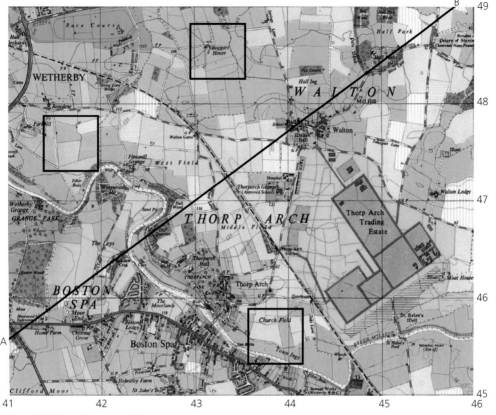

Figure 15.33 Land-use map: Boston Spa

Stratified and quota sampling

This investigation aims to determine the geographical pattern of retail spending within Preston's CBD in northwest England. Preston's CBD serves a catchment of approximately 250,000. The retail structure of the city's CBD is divided into six areas: Fishergate, Friargate, the Fishergate Mall, the St George's Mall, the St John's Mall and the covered market.

Methodology

Street interviews of shoppers provide data on where they purchase goods and services and how much they spend in each of the six shopping areas. Simple random and systematic sampling is rejected because it is unlikely to produce a representative sample. This is because some groups, such as women and older adults, are more likely to agree to be interviewed than others. A stratified or quota sampling method based on three criteria is used instead.

- A small pilot survey in the city centre shows that 65% of weekday shoppers are women and 35% are men.
- The same pilot survey gives the following estimated age profile for shoppers: 18–39 years – 40%; 40–65 years – 30%; over 65 years – 25%.

- Preston's 2001 census data give the following breakdown for housing tenure: owner occupied – 74%; local authority rented – 17%; private rented – 9%.

These criteria are selected because it is assumed that where people shop and how much they spend will be influenced by gender, age and socio-economic status (i.e. housing tenure). In addition, it is feasible to incorporate these criteria into a workable quota sampling system.

Generating quotas

The target sample is 100 shoppers divided, on the basis of gender, age and housing tenure, into the groups shown in Table 15.8.

The table provides the quotas for interviews. For example, our sample will include 21 women aged between 18 and 39 who are owner occupiers, 5 in the same age group who rent from the local authority and so on. Clearly, stratified and quota sampling are more elaborate strategies than simple random or systematic methods. Any decision to use stratified and quota sampling has to balance the desire for accuracy against the extra costs of collecting the data.

➔

Table 15.8 Sample groups for survey of retail spending

Gender	No.	Age groups (years) (no.)			Housing tenure	No.
		18–39	40–65	65+		
Women	65	21	15	12	Owner occupied	48
		5	3	2	Local authority rented	10
		3	2	2	Private rented	7
Men	35	12	3	1	Owner occupied	16
		7	2	1	Local authority rented	10
		7	1	1	Private rented	9

locating them on a map is time consuming, especially if large samples are needed. And second, unless a large enough sample is taken, it is possible that the distribution of random points on the map could give uneven coverage.

Both problems can be addressed by using systematic, rather than random, point sampling. Systematic point sampling relies on a grid of co-ordinates placed randomly over the map. In this example, the co-ordinates of the national grid system can be used. Land use is then sampled at the points of intersection on the grid.

Randomly drawn lines of transect (such as line A–B in Figure 15.33) are a third strategy for generating point samples. Sample points can either be at regular intervals along a transect, or at locations determined by random numbers. In order to collect a large enough sample several random lines of transect are needed. This raises the problem associated with random point sampling: namely that the transects may give uneven areal coverage of the map.

Area sampling

Area sampling based on quadrats is an alternative to point sampling. A quadrat is either a square of standard size on a map; or a metal frame (typically 1 metre × 1 metre), sub-divided into smaller squares, and used in the field. Quadrats are located in a study area either randomly or systematically. Within quadrats sampled items may be counted, recorded as absent or present, or their areal coverage estimated.

The three quadrats in the land-use map in Figure 15.33 have been located randomly. They provide a spatial framework for data collection. Data could be collected in several ways. We could, for example, record the absence or presence of arable land, pastureland, forest and urban land in each quadrat. More generally we might simply record for each quadrat the dominant land use (i.e. the land use covering the largest area). A more detailed approach would be to estimate the proportion of different land uses in each quadrat.

Outcomes from quadrat sampling are strongly influenced by quadrat size. In Figure 15.33 land use tends to vary by field, and quadrats need to be large enough to cover three or four fields. Quadrat sampling is often used in vegetation field surveys. In this context quadrats should be able to accommodate the largest areas occupied by individual plant species.

Quadrat size is also influenced by the geographical distribution of the population. Where populations are highly clustered, small quadrats located randomly or systematically may fail to pick up enough samples. Figure 15.34 shows the sparse and clustered distribution of yellow mountain saxifrage between 300 and 700 m above sea level. In this example small quadrats are less likely to generate sufficient samples to show the influence of altitude on geographic distribution, compared with larger ones.

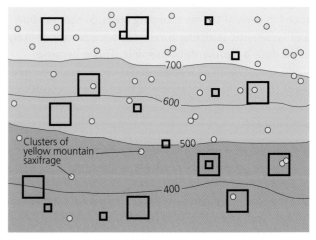

Figure 15.34 Quadrat sampling

Line sampling

Many geographical investigations use samples derived from lines or transects. Transects are particularly effective where they cross contrasting areas of relief, geology, land use, soils and so on. Although transects usually follow one or more straight lines, in urban studies they may, for practical reasons, often follow the street pattern.

Sample data are collected at random and systematic intervals, as points and areas (quadrats), along a line of transect. Figure 15.35 shows a strategy for collecting a sample of screes on a talus slope. First a base line is established at the slope foot. Then transects are located randomly and at right angles to the base line to the top of the slope. Finally, quadrats are located at systematic intervals (lower, middle, upper) on the slope.

Belt transects are a type of line sample. They are often used when populations are thinly spread across an area and where a line transect would fail to generate a large enough sample. A belt is identified on either side of a random line of transect and sample items are measured and recorded within this zone. The example in Figure 15.36 shows how belt sampling can produce an objective sample of boulders from a field of erratic perched blocks (Figure 15.37). Once selected, weathering pits on the boulders can be measured to determine rates of surface lowering by weathering processes.

Stratified spatial sampling

It might be hypothesised that rural land use in the area covered by the Boston Spa land-use map is influenced by the underlying geology (Figure 15.38). If this is so, any attempt to estimate the proportions

Figure 15.35 Stratified spatial sampling using a transect and quadrats to investigate particle size sorting on a talus slope

of different land uses by simple random or systematic sampling methods could result in error. The solution is random stratified sampling, where land use is sampled on the three different rock types in proportion to the area of the map they occupy. Let us assume that 50 per cent of the area is limestone, 22 per cent is marl and 28 per cent is sandstone. A stratified random sample of 100 would comprise 50, 22 and 28 points on the limestone, marl and sandstone respectively.

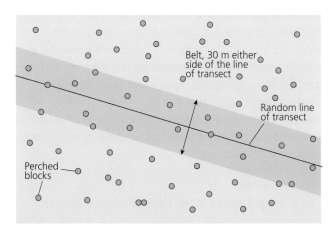

Figure 15.36 Belt sampling, Norber Scar

Figure 15.37 Perched blocks

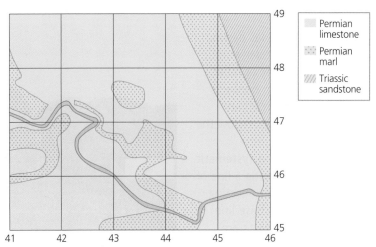

Figure 15.38 Geology sketch of the area covered by the Boston Spa land-use map (Figure 15.33)

Figure 15.38 shows how it is possible to combine line (northings and eastings), quadrat (grid squares) and stratified (rock types) sampling in fieldwork investigation.

Descriptive statistics

Although tables and charts are useful descriptions of data, we often need to summarise a data set with a single numerical value. We can do this by using measures of **central tendency** and **dispersion**.

Measures of central tendency

Measures of central tendency represent data sets by a middle value around which the other values cluster. The three standard measures of central tendency are: the arithmetic mean (or average); the median; and the principal mode.

Arithmetic mean

The arithmetic mean is the sum of values in a data set divided by its total number of values. In statistical notation the mean is shown thus:

$$\bar{x} = \sum(x) \div n$$

where $\bar{x}$ = arithmetic mean; x is a value in a data set; n is the number of values in a data set; and $\sum(x)$ is the instruction to sum all values of x.

The mean is the most widely used measure of central tendency and the only one based on mathematical principles. As we shall see, it figures prominently in many other statistical calculations. However, the mean is not always the most appropriate measure of central tendency. Because the mean weighs each value

according to its magnitude, different distributions can give similar mean values. Table 15.9 shows that Oban (Scotland) and Cotonou (Benin) have similar mean monthly rainfall values. Despite this, the monthly rainfall values are very different: at Cotonou the monthly range of values is more than three times greater than at Oban.

As a rule, the mean provides an accurate summary where data follow a normal frequency distribution with a narrow range of values. If a distribution is **skewed** by a number of exceptional high or low values, the mean is unlikely to represent a data set accurately.

Median

The median is the middle value in a data set ranked in order of size. Thus in a data set comprising 21 values the median is the 11th ranked value. Where a data set has an even number of values the median is the average of the middle two values (in Table 15.9 this would be the sixth and seventh ranked values). Unlike the mean, the median gives equal weight to each value. As a result, for data sets which are skewed it is a more representative measure than the mean. But the median has weaknesses. Even more than the mean, wildly different data sets can give similar median values. Moreover, the median has no true mathematical properties, and cannot therefore be used in any further statistical calculations.

Principal mode

The **principal mode** is the class in a histogram (or frequency table) with the most values. It is useful when classes are described nominally rather than numerically.

Table 15.9 Mean monthly rainfall at Oban, Scotland, and Cotonou, Benin

Month	Oban	Cotonou
J	146	33
F	109	33
M	83	117
A	90	125
M	72	254
J	87	366
J	120	89
A	116	38
S	141	66
O	169	135
N	146	38
D	172	13
Mean	120.9	108.9
Median	118	63.5

Table 15.10 Sex ratios for parishes in Castle Morpeth District (Northumberland) in 2001 (males/1000 females)

1658	
1408	
1247	
1194	
1162	
1134	
1093	1093 **Upper quartile**
1056	
1051	
990	
984	
982	
982	**Inter-quartile range** = 1093 − 933 = 160
980	
961	
947	
944	
944	
936	
933	933 **Lower quartile**
920	
914	
911	
889	
882	
798	

For example, Powers' roundness chart classifies beach sediments into six categories: very angular, angular, sub-angular, sub-rounded, rounded and well-rounded.

The principal mode is arguably the least useful measure of central tendency. Where numerical classes are used, it has a range of values (e.g. 4–6 cm) rather than a single value, while bimodal distributions cause particular ambiguity. Finally, the value of the principal mode depends on the arbitrary choice of class intervals.

Measures of dispersion

While measures of central tendency summarise data sets with a single value, they tell us nothing about the dispersion of values. Measures of dispersion, such as the range, inter-quartile range and the **standard deviation** do this.

Range

The range is the difference between the highest and lowest values in a data set. In Table 15.9 Oban's range of mean monthly rainfall is 100 mm exactly whereas Cotonou's is 353 mm. The main limitation of the range is that it uses just two values in a data set.

Inter-quartile range

The inter-quartile range is more representative than the range, being based on half the values in a data set. It is used alongside the median as a statement of dispersion. To find the inter-quartile range we first arrange the data in rank order of size. The median is defined by dividing the data set in half. Each half is then split into two equal parts known as **quartiles** (Table 15.10). The boundary for the upper 25 per cent of values is known as the upper quartile: the boundary which marks the lowest 25 per cent of values is the lower quartile. The difference between the upper quartile and lower quartile is the inter-quartile range.

Standard deviation

The standard deviation, also known as the root mean square deviation, is the most useful measure of dispersion. Its calculation incorporates all of the values in a data set. The standard deviation is also used in many other statistical measures such as the Student's t-test and correlation. The formula for the standard deviation is:

$$\sigma = \sqrt{\sum(x - \bar{x})^2 \div n}$$

σ = standard deviation
x = each value in the data set
$\bar{x}$ = mean of data set
n = number of values in the data set

Table 15.11 Standard deviation, December rainfall at Durham, 1989–2015

Year	December rainfall (mm), x	Actual rainfall – average, $x - \bar{x}$	Actual rainfall – average squared, $(x - \bar{x})^2$
1989	64	−0.07	0.0049
1990	93	28.93	836.94
1991	45	−19.07	363.66
1992	45	−19.07	363.66
1993	78	13.93	194.04
1994	67	2.93	8.58
1995	88	23.93	572.65
1996	63	−1.07	1.14
1997	83	18.93	358.34
1998	27	−37.07	1374.18
1999	64	−0.07	0.0049
2000	72	7.93	62.88
2001	54	−10.07	101.40
2002	92	27.93	780.08
2003	65	0.93	0.86
2004	20	−44.07	1942.16
2005	38	−26.07	679.64
2006	77	12.93	167.18
2007	53	−11.07	122.54
2008	57	−7.07	49.98
2009	81	16.93	286.62
2010	41	−23.07	532.22
2011	52	−12.07	145.68
2012	99	34.93	1220.10
2013	65	0.93	0.86
2014	27	−37.07	1374.18
2015	120	55.93	3128.16
	Mean = 64.07		**Σ = 14,667.73**

$$\sqrt{\Sigma(x - \bar{x})^2 \div n} = 543.25$$

$$\sigma = 23.31$$

Calculation of the standard deviation is shown in Table 15.11 using data on December rainfall totals at Durham between 1989 and 2015. However, the standard deviation can be more easily obtained using spreadsheets in either Microsoft Excel or Apple Numbers or from handheld calculators.

The standard deviation has a precise relationship with data sets that have a normal frequency distribution. For normal distributions 68.27 per cent of the area below the frequency curve (Figure 15.39) lies between plus or minus one standard deviation of the mean. Two standard deviations account for 95.45 per cent of the area below the curve.

To convert a value into a unit of standard deviation (known as the standard deviate) we subtract the mean and then divide by the standard deviation.

$$\text{Standard deviate} = (\text{value} - \text{mean}) \div \text{standard deviation}$$

The standard deviate is often represented by z and is referred to as a z value. For example, to estimate the likelihood that December rainfall at Durham will exceed 95 mm, we first calculate the standard deviate and then find the probability in normal deviate tables (Figure 15.40).

$$\text{Standard deviate} = (95 - 64.07) \div 23.31 = 1.33$$

According to Figure 15.40 the probability of rainfall being between the mean and the z score 1.33 (i.e. 95 mm) is 0.4082. Therefore the probability of rainfall exceeding 95 mm is equal to 0.500 − 0.4082 or 0.0918 (i.e. 9.18%).

Coefficient of variation

The value of the standard deviation is strongly influenced by the magnitude of the mean. This is a problem when comparing dispersion in two data sets with different means. We overcome this problem by using the coefficient of variation. This standardised measure expresses the standard deviation as a percentage of the mean.

$$\text{Coefficient of variation} = [\sigma \div \bar{x}] \times 100$$

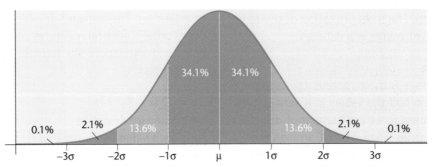

Figure 15.39 Normal frequency distribution

Table of the standard normal (z) distribution

z	0.00	0.01	0.02	0.03	0.04	0.05	0.06	0.07	0.08	0.09
0.0	0.0000	0.0040	0.0080	0.0120	0.0160	0.0199	0.0239	0.0279	0.0319	0.0359
0.1	0.0398	0.0438	0.0478	0.0517	0.0557	0.0596	0.0636	0.0675	0.0714	0.0753
0.2	0.0793	0.0832	0.0871	0.0910	0.0948	0.0987	0.1026	0.1064	0.1103	0.1141
0.3	0.1179	0.1217	0.1255	0.1293	0.1331	0.1368	0.1406	0.1443	0.1480	0.1517
0.4	0.1554	0.1591	0.1628	0.1664	0.1700	0.1736	0.1772	0.1808	0.1844	0.1879
0.5	0.1915	0.1950	0.1985	0.2019	0.2054	0.2088	0.2123	0.2157	0.2190	0.2224
0.6	0.2257	0.2291	0.2324	0.2357	0.2389	0.2422	0.2454	0.2486	0.2517	0.2549
0.7	0.2580	0.2611	0.2642	0.2673	0.2704	0.2734	0.2764	0.2794	0.2823	0.2852
0.8	0.2881	0.2910	0.2939	0.2967	0.2995	0.3023	0.3051	0.3078	0.3106	0.3133
0.9	0.3159	0.3186	0.3212	0.3238	0.3264	0.3289	0.3315	0.3340	0.3365	0.3389
1.0	0.3413	0.3438	0.3461	0.3485	0.3508	0.3531	0.3554	0.3577	0.3599	0.3621
1.1	0.3643	0.3665	0.3686	0.3708	0.3729	0.3749	0.3770	0.3790	0.3810	0.3830
1.2	0.3849	0.3869	0.3888	0.3907	0.3925	0.3944	0.3962	0.3980	0.3997	0.4015
1.3	0.4032	0.4049	0.4066	0.4082	0.4099	0.4115	0.4131	0.4147	0.4162	0.4177
1.4	0.4192	0.4207	0.4222	0.4236	0.4251	0.4265	0.4279	0.4292	0.4306	0.4319
1.5	0.4332	0.4345	0.4357	0.4370	0.4382	0.4394	0.4406	0.4418	0.4429	0.4441

Figure 15.40 Normal deviate table

For example, in Table 15.12, December rainfall at Aberporth in west Wales has a higher standard deviation than Durham, partly because its mean December rainfall is 60 per cent greater. However, the coefficient of variation takes account of differences in mean values by presenting the standard deviation as a ratio of the mean. Thus the coefficient of variation shows that the variability of December rainfall is actually greater at Aberporth than at Durham.

Table 15.12 Variability of December rainfall at Durham and Aberporth, 1989–2015

	Mean December rainfall (mm)	Mean December standard deviation (mm)	Coefficient of variation (%)
Durham	64.1	23.3	36.3
Aberporth	103.6	47.6	45.9

Inferential statistics

Inferential statistics get their name because they are used to infer population values from samples. This leads us to the concept of **statistical significance**: the probability that the outcomes of investigations based on sample data are due to chance. Normally any outcome which yields a significance level of 0.05 (or 95 per cent) and above is accepted as being statistically significant (Table 15.13).

Tests of statistical significance may be either one-tailed or two-tailed. When a hypothesis states the direction of a difference or a relationship, we use a one-tailed test. For example, a one-tailed test would be used to investigate the hypothesis that particles at the foot of a scree slope are larger than particles at the top. This hypothesis predicts the direction of the difference. Alternatively, a

Table 15.13 Statistical significance

Significance level	% significance	Probability that outcome due to chance	Statistical significance
0.001	99.9	1 in 1000	Significant
0.01	99	1 in 100	
0.025	97.5	1 in 40	
0.05	95	1 in 20	
0.1	90	1 in 10	Not significant
0.2	80	1 in 8	
0.3	70	1 in 7	

two-tailed test would simply investigate the hypothesis that there is a difference in size between particles at the top and bottom of the slope. Such a hypothesis is more general and does not state the direction of the difference. Probabilities for one-tailed tests are more demanding: exactly half the value of probabilities for two-tailed tests.

In this section we shall look at the following inferential statistics: standard error of the mean; students' t-test; Mann–Whitney U test; chi-squared test; Spearman's rank correlation; and lines of best fit. All of these functions are available for data entered in spreadsheets in Excel and Numbers.

Standard error of the mean
The standard error is used to estimate the value of the population mean from sample data sets. It is defined as the standard deviation of the sampling distribution of the mean.

$$\sigma_M = \frac{\sigma}{\sqrt{n}}$$

where σ_M is the standard deviation of the sample and n is the sample size.

The logic of the standard error is that if you take a large number of samples from a population, calculate the mean for each sample, and then plot them as a frequency curve, they will approximate a normal distribution. The standard deviation of this sampling distribution is known as the **standard error of the mean**.

The standard error enables us to estimate the limits of the population mean from sample data. This is because its relationship to the sampling distribution is the same as standard deviation to the normal frequency distribution. In other words, approximately 68 per cent of values lie within one standard error of the mean, and 95 per cent are within two standard errors and so on.

An example of the calculation of the standard error of the mean is shown in Table 15.14. Assume that we want to estimate the average size of scree particles on a talus slope from a random sample of 24 particles. The sample mean is 10.77 cm and the sample standard deviation is 5.11 cm.

Table 15.14 Particle size (cm – median axis): scree slope at Austwick

12	5	5.5	8	11
30	7	10	9	12
16	9.5	12	11	9.5
9	6	15	8	10
13	13	11	6	
Sample mean = 10.77		Sample standard deviation = 5.11		

$$\text{Standard error of sample mean } (\sigma_M) = \frac{5.11}{\sqrt{24}} = 1.04 \text{ cm}$$

Substituting values in the standard error formula gives us a standard error of the mean of 1.043. We know that at the 95 per cent confidence level the population mean lies between plus and minus two standard errors. Thus with 95 per cent probability the population mean will be $10.77 \pm 2 \times 1.043$, i.e. it will lie somewhere between 8.60 cm and 12.84 cm.

It is worth noting that the precision of the standard error varies with the square of the sample size. Thus doubling the accuracy of the standard error in our example, from 1.04 cm to 0.502 cm, would require a four-fold increase in sample size. This relationship has an important bearing on decisions concerning sample size and the desired level of precision in estimating the population mean.

Standard error of the percentage

When analysing percentages or proportions it is necessary to modify the formula for estimating the standard error of the mean. The formula for calculating the standard error of the percentage is:

$$\sqrt{(p\% \times q\%) \div n}$$

where p is the percentage of a given land use type, q is the remaining percentage (i.e. all other uses), and n is

the size of the sample. The calculation of the standard error of the percentage is shown opposite.

Testing for differences between data sets

Many hypotheses in geographical investigation focus on the differences between samples. A number of tests are available to determine whether these differences are statistically significant or due to chance. Three widely used tests of difference are the t-test, the Mann–Whitney U test and the chi-squared test.

t-test

The test compares the arithmetic means of two samples to determine the likelihood that any difference could have occurred by chance. Unlike most other inferential tests the t is a parametric test. This means that it should only be applied where samples are derived from populations that have a normal frequency distribution. The calculation of the t statistic is shown opposite.

Mann–Whitney U test

The Mann–Whitney U test measures the difference between two data sets. However, unlike the t-test, it is a non-parametric test. This means that it is distribution-free and makes no assumptions about the normality of the population. In a geographical context, where so many statistical populations are skewed, this flexibility is especially valuable.

The U test has other advantages. It can be applied to: small data sets; data measured on an ordinal (or rank order) scale; and to data sets containing unequal numbers of values. But like the t-test, the U test can only be applied to two data sets. The calculation of the U statistic is shown on pages 538–539.

To determine the significance of the U statistic (58.5) we refer to U tables. These show that the critical value of U at the 0.025 level (for a two-tailed test) when n_x and n_y are both 15, is 64. Because our U value is less than the critical value we conclude that it is statistically significant at the 0.025 level. In other words, we can be 97.5 per cent confident that our sample difference reflects a real difference in the value of farmland on small and large farms around Haslingden in 1769.

Chi-squared test

Chi-squared is used to determine whether an observed frequency distribution differs significantly from the frequencies that might be expected if the distribution were random. Like the U test, chi-squared is a non-parametric test.

There are two versions of the chi-squared test (pages 539–40): a one-sample version and a test for two or more sample distributions. Both tests are only appropriate where the following conditions apply:

- Data are in frequencies (i.e. counts for individual cells). The test is invalid for values given in percentages or proportions.
- There should not be many categories for which expected frequencies are small. For example, when there are more than two categories no more than one-fifth of expected frequencies should have values of less than five, and none should be less than one.

Two-sample chi-squared

The requirements for the two-sample version of the chi-squared test are the same as the one-sample version. However, it is important that there are few categories with expected values of less than 5. This requirement can be met by merging some categories and accepting a higher level of generalisation.

Calculating the standard error of the percentage

A systematic point sample of a 1:25,000 land-use map in the North Downs in Kent gave the following results:

Land-use type	% land use (p)	Number of sample points (n)	Standard error %
Grassland	13.8	67	4.21
Arable land	33.5	161	3.72
Woodland	24.4	118	3.95
Market gardening	8.0	39	4.34
Others	20.3	98	4.06

Standard error of the percentage for grassland:

$$\sqrt{(13.8 \times 86.2) \div 67} = 4.21\%$$

Thus with 68% confidence, the actual percentage of grassland in the area is $13.8 \pm 4.21\%$.

Calculating the t statistic

Hypotheses

1 Birth rates are higher in rural than in urban parishes (Table 15.15) in Sweden in 1875 (one-tailed test).
2 There were significant differences in birth rates between rural and urban parishes in Sweden in 1975 (two-tailed test).
3 Rural parishes:

$\bar{x}_r$ (mean) = 31.73

s_r (standard deviation) = 3.47

n_r (sample size) = 11

4 Urban parishes:

$\bar{x}_u$ (mean) = 26.82

s_u (standard deviation) = 3.97

n_u (sample size) = 11

5 t-test formula:

$$t = \bar{x}_r - \bar{x}_u \div \sqrt{(s_r^2 \div n_r) + (s_u^2 \div n_u)}$$

$$t = 31.73 - 26.82 \div (3.42^2 \div 11) + (3.97^2 \div 11)$$

$$= 3.09$$

Table 15.15 Crude birth rates (per 1000) in central Sweden, c1875

Rural parishes		Urban parishes	
Härjerad	29	Vänersborg	30
Fyrunga	32	Borås	29
Ving	29	Allingsås	26
Flo	31	Amal	27
Karaby	29	Ulricehamn	23
Längjum	33	Mariestad	23
Frammestad	28	Lidköping	24
Vara	40	Skara	24
Längjum-med-Vessby	34	Skövde	29
Lundby	30	Falköping	36
Bitterna	34	Hjo	24

Reference to statistical tables of the t distribution (Figure 15.41) determines the statistical significance of t (i.e. the likelihood that the sample difference in birth rates reflects a real difference in the population). The value of t (3.09) is checked against an estimate of sample size known as degrees of freedom. More specifically, degrees of freedom are the number of observations that are free to vary to produce a known outcome. In this instance the mean birth rate for rural areas has 10 degrees of freedom because the values for ten parishes can vary, but the eleventh is fixed. Thus we have in total $n_r - 1 + n_u - 1 = 20$ degrees of freedom for rural and urban samples.

The critical value of t at the 0.01 level for 20 degrees for a two-tailed test in Figure 15.41 is 2.845. As our t value exceeds the critical value, the difference is significant at the 0.01 (i.e. 1 in 100) level. In other words, there is only a 1 in 100 probability that the difference in birth rates between rural and urban parishes is due to chance. (Note: the t value is also significant at the 0.01 level for the one-tailed test.)

	Significance level (one-tailed)				
	0.05	0.025	0.01	0.005	0.0005
Degrees of freedom	Significance level (two-tailed)				
	0.10	0.05	0.02	0.01	0.001
1	6.314	12.71	31.82	63.66	636.62
2	2.920	4.303	6.965	9.925	31.599
3	2.353	3.182	4.541	5.841	12.924
4	2.132	2.776	3.747	4.604	8.610
5	2.015	2.571	3.365	4.032	6.869
6	1.943	2.447	3.143	3.707	5.959
7	1.895	2.365	2.998	3.499	5.408
8	1.860	2.306	2.896	3.355	5.041
9	1.833	2.262	2.821	3.250	4.781
10	1.812	2.228	2.764	3.169	4.587
11	1.796	2.201	2.718	3.106	4.437
12	1.782	2.179	2.681	3.055	4.318
13	1.771	2.160	2.650	3.012	4.221
14	1.761	2.145	2.624	2.977	4.140
15	1.753	2.131	2.602	2.947	4.073
16	1.746	2.120	2.583	2.921	4.015
17	1.740	2.110	2.567	2.898	3.965
18	1.734	2.101	2.552	2.878	3.922
19	1.729	2.093	2.539	2.861	3.883
20	1.725	2.086	2.528	2.845	3.850
21	1.721	2.080	2.518	2.831	3.819
22	1.717	2.074	2.508	2.819	3.792
23	1.714	2.069	2.500	2.807	3.768
24	1.711	2.064	2.492	2.797	3.745
25	1.708	2.060	2.485	2.787	3.725
26	1.706	2.056	2.479	2.779	3.707
27	1.703	2.052	2.473	2.771	3.690
28	1.701	2.048	2.467	2.763	3.674
29	1.699	2.045	2.462	2.756	3.659
30	1.697	2.042	2.457	2.750	3.646
40	1.684	2.021	2.423	2.704	3.551
60	1.671	2.000	2.390	2.660	3.460
80	1.664	1.990	2.374	2.639	3.416
100	1.660	1.984	2.364	2.626	3.390

Reject H_0 if calculated value of t is **greater than** critical value at chosen significance level.

Figure 15.41 Statistical table of t values

Calculating the U statistic

Hypothesis

There are significant differences in the value of land per acre between larger and smaller farms in Haslingden (Table 15.16) in 1769 (two-tailed test).

1 Arrange the values in the two data sets in rank order of size for both samples together.
2 Where values are tied the mean ranking is used. For example, if a data set comprises five values – 18, 23, 25, 25, 27 – the rank order would be 1, 2, 3.5, 3.5, 5. If the fifth value were also 25, then the ranking would be 1, 2, 4, 4, 4.
3 Sum the rank values for each of the data sets separately and then calculate the U statistic from the following equations:

Equation 1:

$$U = n_x \times n_y + \left(\frac{n_x(n_x + 1)}{2}\right) - \Sigma r_x$$

Equation 2:

$$U = n_x \times n_y + \left(\frac{n_y(n_y + 1)}{2}\right) - \Sigma r_y$$

where n_x and n_y are the number of values in data sets x and y, and Σr_x and Σr_y are the sums of the rank values in data sets x and y.

The smaller of the two values for U is used in statistical tables to estimate significance (Figure 15.42). Unlike other inferential tests, U is significant if it is less than the critical value listed in the tables. You can check the accuracy of your calculations because the sum of the two U values should equal the product of n_x and n_y.

The value of U value from Equation 1 is:

$$15 \times 15 + \left(\frac{15(15 + 1)}{2}\right) - 286.5 = 58.5$$

➜

Table 15.16 Farm sizes and land values in Haslingden, Lancashire in 1769 (£ per acre)

Sample X: farms >25 acres		Sample Y: farms <25 acres	
	Rank		Rank
1.46	9.5	1.14	22
0.77	27	1.92	3
0.25	29.5	1.21	18
0.84	26	2.05	2
1.31	15	1.44	12
0.74	28	1.81	4
0.25	29.5	1.04	24
1.03	25	1.15	21
1.34	13	1.66	7
1.11	23	1.69	5.5
1.19	19	1.16	20
1.69	5.5	2.12	1
1.46	9.5	1.25	17
1.27	16	1.56	8
1.45	11	1.38	14
	$\Sigma r_x = 286.5$		$\Sigma r_y = 178.5$

The value of U value from Equation 2 is:

$$15 \times 15 + \left(\frac{15(15+1)}{2}\right) - 178.5 = 166.5$$

	n_y 2	3	4	5	6	7	8	9	10	11	12	13	14	15
n_x 2	0	0	0	0	0	0	0	0	0	0	1	1	1	1
3	0	0	0	0	1	1	2	2	3	3	4	4	5	5
4	0	0	0	1	2	3	4	4	5	6	7	8	9	10
5	0	0	1	2	3	5	6	7	8	9	11	12	13	14
6	0	1	2	3	5	6	8	10	11	13	14	16	17	19
7	0	1	3	5	6	8	10	12	14	16	18	20	22	24
8	0	2	4	6	8	10	13	15	17	19	22	24	26	29
9	0	2	4	7	10	12	15	17	20	23	26	28	31	34
10	0	3	5	8	11	14	17	20	23	26	29	30	36	39
11	0	3	6	9	13	16	19	23	26	30	33	37	40	44
12	1	4	7	11	14	18	22	26	29	33	37	41	45	49
13	1	4	8	12	16	20	24	28	30	37	41	45	50	54
14	1	5	9	13	17	22	26	31	36	40	45	50	55	59
15	1	5	10	14	19	24	29	34	39	44	49	54	59	64

Reject H_0 if calculated value is **less than or equal to** critical value at chosen significance level.

Figure 15.42 Statistical table of U values

To determine the significance of the U value (58.5) we refer to U tables (Figure 15.42). These show that the critical value of U at the 0.025 level (for a two-tailed test) when n_x and n_y are both 15, is 64. Because our U value is less than the critical value we conclude that it is statistically significant at the 0.025 level. In other words, we can be 97.5% confident that our sample difference reflects a real difference in the value of farmland on small and large farms in the Haslingden area in 1769.

Calculating a one-sample chi-squared

Hypothesis

The distribution of bracken shows a preference for steeper slopes (one-tailed).

A study of the distribution of bracken on Muggleswick Common, an area of moorland in County Durham, suggested a relationship between slope angle and frequency of occurrence (Table 15.17).

The question is whether the distribution of bracken simply reflects frequency of slope angles on the moor, or whether the plant shows a preference for particular slopes. If we assume the distribution is random we can generate an expected distribution based on the frequency of slopes (Table 15.18). Thus, 44% of bracken samples (i.e. 14.52) will occur in the slope class 0–6.9°, 12.54 in class 7–13.9°, and 5.94 in the class of more than 13.9°. Our expected distribution looks very different from the observed distribution.

The formula for calculating the chi-squared statistic is:

$$\Sigma((O - E)^2 \div E)$$

The final step is to find the significance of chi-squared from statistical tables (Figure 15.43). Degrees of freedom are obtained by multiplying the number of columns (k) minus one, by the number of rows (r) minus one. In this example, with three columns and two rows there are (3 – 1)(2 – 1) or 2 degrees of freedom. The chi-squared tables show that at the 95% confidence level with two degrees of freedom the critical value of chi-squared is 5.99. Our chi-squared value exceeds the critical value, which means that it is statistically significant (in fact it is significant at the 0.001 level). Thus we can assume that bracken is not randomly distributed on the moorland and appears to show a preference for steeper (and therefore better drained) slopes.

→

Table 15.17 Slope angle and bracken distribution on Muggleswick Common

Slope classes (°)	% area occupied by each slope category	Number of quadrats dominated by bracken
0–6.9	44	4
7–13.9	38	18
>13.9	18	11

Table 15.18 Observed and expected distribution of bracken on Muggleswick Common

	0–6.9°	7–13.9°	>13.9°
Observed (O)	4	18	11
Expected (E)	14.52	12.54	5.94
($O - E$)	−10.52	5.46	5.06
($O - E$)^2	110.7	29.8	25.6
($O - E$)$^2 \div E$	7.62	2.38	4.31
$\Sigma((O - E)^2 \div E) = 7.62 + 2.38 + 4.31 = 14.31$			

df	0.10	0.05	0.01	0.005	0.001
1	2.71	3.84	6.63	7.88	10.83
2	4.61	5.99	9.21	10.60	13.82
3	6.25	7.81	11.34	12.84	16.27
4	7.78	9.49	13.23	14.86	18.47
5	9.24	11.07	15.09	16.75	20.51
6	10.64	12.53	16.81	13.55	22.46
7	12.02	14.07	18.48	20.28	24.32
8	13.36	15.51	20.09	21.95	26.12
9	14.68	16.92	21.67	23.59	27.83
10	15.99	18.31	23.21	25.19	29.59
11	17.29	19.68	24.72	26.76	31.26
12	18.55	21.03	26.22	28.30	32.91
13	19.81	22.36	27.69	29.82	34.53
14	21.06	23.68	29.14	31.32	36.12
15	22.31	25.00	30.58	32.80	37.70

Figure 15.43 Statistical table of chi-squared values

Calculating the two-sample chi-squared

Historically, most towns and cities grow outwards from their core. As a result, until the mid-twentieth century, the fabric of the urban environment gets newer with distance from the centre. This idea is illustrated in Table 15.19. It shows how in Preston in 1910, the distribution of cotton mills built before and after 1845 varied with distance from the town centre.

Hypothesis
The ages of mills decrease with distance from the town centre.

1 Sum the row values, column values and the total number of values in the data set.
2 Calculate the expected frequencies for each cell by multiplying its row value by its column value and dividing by the total number of values (Table 15.20). For example, the expected value for cell A1 in Table 15.19 is: $(14 \times 42) \div 83 = 7.08$; B1 = $(47 \times 42) \div 83 = 23.78$.
3 Substitute the expected values for each cell in the chi-squared formula and sum the results:

$$\Sigma((O - E)^2 \div E)$$

Chi-squared = $(12 - 7.08)^2 \div 7.08 + (26 - 23.78)^2 \div 23.78 + (4 - 11.13)^2 \div 11.13 + (2 - 6.92)^2 \div 6.92 + (21 - 23.22)^2 \div 23.22 + (18 - 10.87)^2 \div 10.87 = 16.59$.

The significance of the chi-squared value is checked in statistical tables (Figure 15.43) against $(k - 1)(r - 1)$ degrees of freedom — in this example 2 degrees, where k is the number of columns and r is the number of rows. The critical value for chi-squared at the 0.001 level with 2 degrees of freedom is 13.82. Since our value exceeds the critical value, we accept it as statistically significant. Thus we can accept the hypothesis that the age of mills declines with distance from the centre of Preston in 1910.

Table 15.19 Age of cotton mills in Preston in 1910 and distance from the town centre

		A	B	C	
		0–1 km	1.1–2 km	2.1–3 km	Σr
1	Pre-1845	12	26	4	42
2	Post-1845	2	21	18	41
	Σk	14	47	22	83

Table 15.20 Calculating expected frequencies

$(14 \times 42) \div 83 = 7.08$	$(47 \times 42) \div 83 = 23.78$	$(22 \times 42) \div 83 = 11.13$
$(14 \times 41) \div 83 = 6.92$	$(47 \times 41) \div 83 = 23.22$	$(22 \times 41) \div 83 = 10.87$

Analysing relationships between data sets: correlation

Correlation measures the statistical association between two variables, *x* and *y*. The statistic is used to determine how variations in *x* influence *y*. It is assumed that variable *x* is the **independent variable** responsible for changes in the **dependent variable**, *y* (Table 15.21).

Although the relationships in Table 15.21 are clear, identifying independent and dependent variables is not always easy. For example, does the volume of pedestrian flow in a high street explain the types of shops found there, or is it the shops themselves that influence flows; do wave types influence beach gradients or vice versa?

There are several measures of statistical correlation. We shall concentrate on two: Spearman's rank correlation and Pearson's product moment correlation. Both tests require two variables arranged in paired values and assume that one variable (*x*) influences change in the other variable (*y*).

Correlation coefficients measure the strength of a relationship between two variables. They vary on a scale from +1 to −1, where +1 is a perfect positive correlation (Figures 15.44–15.46), −1 a perfect negative or inverse correlation. A correlation coefficient close to zero suggests little or no relationship.

Table 15.21 Examples of independent and dependent variable relationships

Independent variable (*x*)	Dependent variable (*y*)
Altitude	Temperature
Drainage basin area	Stream discharge
Gradient	Velocity of stream flow
Distance	Number of commuters
Types of beach sediment	Beach gradient
Slope angle	Frequency of terracettes

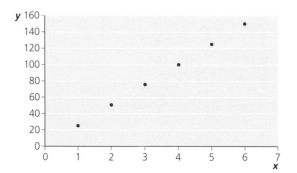

Figure 15.44 Scatter chart (a): perfect positive correlation: +1

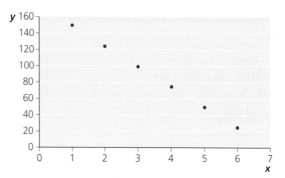

Figure 15.45 Scatter chart (b): perfect inverse correlation: −1

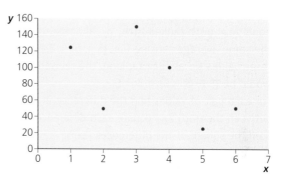

Figure 15.46 Scatter chart (c): weak correlation: 0–0.2

Spearman's rank correlation

Spearman's rank correlation is a non-parametric test. It thus belongs to the same group of statistical tests as chi-squared and the U test, and has the advantage of being distribution free. The calculation is shown on page 542.

Pearson's product moment correlation

Pearson's product moment correlation is an alternative to Spearman's rank correlation. Its outcome is a correlation coefficient with the same properties as the Spearman's rank correlation coefficient. Because Pearson's correlation is a parametric test it should only be used on sample data from a statistical population with a normal distribution. Calculation of Pearson's correlation is available in Excel and shown on page 543.

Coefficient of determination

The coefficient of determination is the product moment correlation coefficient squared, expressed as a percentage. It measures the statistical variation in *y* 'explained' by *x*. The Spearman rank correlation coefficient for wind speed and wave height (Table 15.22) is 0.61. This gives a coefficient of determination of $0.61^2 \times 100\% = 37.2\%$. We can therefore state that 37.2% of the variation in wave height is 'explained' by wind speed. Other factors that influence wave height include fetch (direction from which the wind blows), water depth and the configuration of the sea bed.

Calculating the Spearman rank correlation

The Spearman rank correlation, like the U test, is calculated from ordinal rather than ratio or interval data. This means that the original data are arranged in rank order of magnitude and the calculation is based on the rank values.

Hypothesis
Wave height is influenced by wave speed.

Calculation of the Spearman rank correlation (r_s) involves the following steps:

1. Rank each pair of values for *x* and *y* (Table 15.22) from 1 (largest) to 14 (smallest). If two values are equal (e.g. both ranked fifth) we allocate to them the same average ranking (i.e. $(5 + 6) \div 2 = 5.5$). If three values tie as fifth ranking, the revised ranking would be $(5 + 6 + 7) \div 3 = 6$.
2. For each pair of values find the difference in rank between them (*d*) and square each difference (d^2).
3. Sum the square of the differences (Σd^2).
4. Complete the calculation of the Spearman correlation using the formula:

$$r_s = 1 - (6\Sigma d^2) \div n^3 - n$$

In this example the Spearman rank correlation between wind speed (*x*) and wave height (*y*) $= 1 - (6 \times 170) \div 2730$:

$$r_s = 1 - 0.37 = +0.63$$

The significance of the correlation coefficient is obtained from Table 15.47. In this example there are 14 pairs of values and 12 degrees of freedom ($n - 2$). The critical value of r_s at the 0.05 level with 12 degrees of freedom is 0.591. Because our correlation coefficient (0.63) exceeds the critical value, it is statistically significant at the 0.05 level. However, being less than the critical values at 0.02 and 0.01, it is not significant at these higher levels.

	Significance level (one-tailed)		
	0.025	0.01	0.005
Degrees of freedom	Significance level (two-tailed)		
	0.05	0.02	0.01
5	1.000	1.000	
6	0.886	0.943	1.000
7	0.786	0.893	0.929
8	0.738	0.833	0.881
9	0.683	0.783	0.833
10	0.648	0.746	0.794
12	0.591	0.712	0.777
14	0.544	0.645	0.715
16	0.506	0.601	0.665
18	0.475	0.564	0.625
20	0.45	0.534	0.591
22	0.428	0.508	0.562
24	0.409	0.485	0.537
26	0.392	0.465	0.515
28	0.377	0.448	0.496
30	0.364	0.432	0.478

Figure 15.47 Statistical table for Spearman rank correlation coefficient

Table 15.22 Wave height and wind speed around the coastline of the British Isles (recorded at 14 moored buoys)

Wind speed *x* (m/sec)	Wave height *y* (m)	Rank *x*	Rank *y*	Rank *x* – rank *y*	(Rank *x* – rank *y*)²
10.3	3.2	7	5	+2	4
12.4	2.8	4	7	−3	9
11.3	3.6	5.5	4	+1.5	2.25
16.0	2.5	2	8	−6	36
11.3	4.4	5.5	1	+4.5	20.25
5.7	2.3	14	9.5	+4.5	20.25
9.3	1.9	11	11	0	0
10.0	1.2	9	13	−4	16
7.7	1.6	12	12	0	0
6.2	0.3	13	14	−1	1
9.8	4.1	10	3	+7	49
13.4	3.0	3	6	−3	9
17.0	4.2	1	2	−1	1
10.1	2.3	8	9.5	−1.5	2.25

Procedure for calculating Pearson's product moment correlation in Excel

1 Open an Excel document and insert pairs of data for x and y variables.
2 Select 'autosum' on the standard toolbar and click 'more functions'.
3 Select the category 'statistical', scroll down and select 'PEARSON', and click 'OK'.
4 Place the cursor in the box 'array 1' and select all the data for variable x in your spreadsheet.
5 Place the cursor in the box 'array 2' and repeat for the dependent variable y in your spreadsheet.
6 The correlation coefficient will appear under formula result in the functions argument box.

The Pearson product moment correlation coefficient for wind speed and wave height (Table 15.22) is 0.56. This compares with 0.61 returned by the Spearman rank correlation. However, Pearson's correlation is more mathematically demanding than Spearman. Thus, despite being lower, Pearson's product moment correlation is still statistically significant at the 0.05 level (Figure 15.48).

Degrees of freedom	Significance level (one-tailed)			
	0.05	0.025	0.01	0.005
	Significance level (two-tailed)			
	0.10	0.05	0.02	0.01
1	0.9877	0.9969	0.9995	0.9999
2	0.900	0.950	0.980	0.990
3	0.805	0.878	0.934	0.959
4	0.729	0.811	0.882	0.917
5	0.669	0.754	0.833	0.874
6	0.622	0.707	0.789	0.834
7	0.582	0.666	0.750	0.798
8	0.549	0.632	0.716	0.765
9	0.521	0.602	0.685	0.735
10	0.497	0.576	0.658	0.708
11	0.476	0.553	0.634	0.684
12	0.458	0.532	0.612	0.661
13	0.441	0.514	0.592	0.641
14	0.426	0.497	0.574	0.628
15	0.412	0.482	0.558	0.606
16	0.400	0.468	0.542	0.590
17	0.389	0.456	0.528	0.575
18	0.378	0.444	0.516	0.561
19	0.369	0.433	0.503	0.549
20	0.360	0.423	0.492	0.537

Figure 15.48 Statistical table of Pearson's correlation coefficient

Lines of best-fit: simple linear regression

Simple linear regression, involving two variables, x and y, is a technique for fitting a straight line to points on a scatter chart (see page 544). The regression line is known as 'least squares' because it minimises the sum of the squares of the deviations from the line, and is statistically the 'best-fit'. Regression lines are expressed in the form of an equation:

$$y = a \pm bx$$

where regression coefficient a is the value of y when x is zero, and b is the gradient of the trend line. An equation with a plus sign describes a positive correlation; a minus sign describes an inverse correlation.

Regression has two important uses. First, it allows us to predict a value of y from a known value of x. This is especially useful when we have plenty of data on variable x, but relatively little on variable y. For

example, from a simple regression model we might predict rainfall from altitude, or river flow from rainfall (Figure 15.49). Second, a regression equation provides us with a precise model of the relationship between two variables, and allows us to make comparisons with the same variables in other geographical locations.

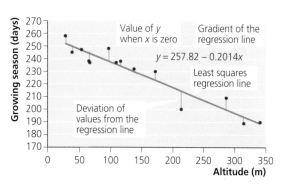

Figure 15.49 Least square linear regression for growing season and altitude in northern England

The simplest way to fit a simple regression line to points on a scatter chart is to use an Excel spreadsheet.

1 Open an Excel document and insert pairs of data for x and y variables.
2 Select the chart wizard.
3 Select scatter chart.
4 Complete the scatter chart with appropriate title and labels for x and y axes and select 'as new sheet' and then press 'finish'.

5 Select 'chart', then 'add trendline', then 'linear regression'. A trend line will appear on your scatter chart.
6 Under chart options select 'equation' and 'r squared'. The equation for the linear regression line will appear on your chart, as well as the coefficient of determination.

📑 Exam-style question

1 a) Study the land-use map extract in Figure 15.33 on page 529 which shows part of North Yorkshire near the market town of Wetherby.

   i) Suggest a geographical question or issue that could be investigated in the area shown in Figure 15.33. Justify your choice using map evidence. [4 marks]

   ii) State one type of primary data you would collect in your investigation. [2 marks]
   iii) Explain one suitable method you would use to collect your primary data. [6 marks]

   b) With reference to a fieldwork investigation you have carried out, assess the extent to which the investigation achieved its aims. [12 marks]

Chapter 16

Independent investigation

The independent investigation consists of a written report with a recommended length of 3000–4000 words which assesses the process of geographical enquiry. It requires the individual student to:

- define research questions or hypotheses which relate to part of the specification
- include data from fieldwork investigations (collected individually or in groups), and secondary data
- contextualise, analyse and summarise data
- draw conclusions and communicate them by means of extended writing and the presentation of data.

The report must also have a clear structure, be written in continuous prose and include primary and secondary data. Where the report includes digital material this must be referenced to or evidenced through web links and/or screenshots.

16.1 Stages of geographical enquiry

This section outlines the stages of geographical enquiry. Depending on the topic, the enquiry may follow either the strict, *sequential* route of scientific investigation or the 'more fluid', *recursive* strategy of qualitative investigation (Figure 16.1). In the

standard scientific model hypotheses are formulated in advance from existing theory and knowledge. In this **deductive** model, quantitative data are then collected and analysed systematically. This is an approach that relies heavily on statistical analysis. However, it is not suitable for all investigations. Topics which are more personal and focus on unique cases may be based on informal, in-depth interviews, observation and geographical narrative. With this **inductive** approach, hypotheses and generalisations may be the outcome rather than the starting point of study.

The sequential model of investigation

Stage 1: Identifying a suitable geographical question or hypothesis

Geographical scale of the investigation

A successful outcome depends partly on the choice of an appropriate geographical scale for the investigation. For instance, an investigation into sediment movement by longshore drift is more likely to produce significant results if it is conducted over 2–3 kilometres of coastline rather than just a few hundred metres. At the smaller scale, the scarcity

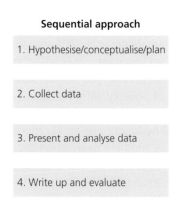

Sequential approach

1. Hypothesise/conceptualise/plan

2. Collect data

3. Present and analyse data

4. Write up and evaluate

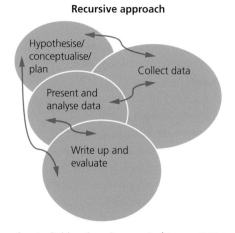

Recursive approach

Hypothesise/conceptualise/plan

Collect data

Present and analyse data

Write up and evaluate

Figure 16.1 Sequential and recursive approaches to fieldwork and research (Source: D Holmes, RGS www.rgs.org/OurWork/Schools/Fieldwork+and+local+learning/Planning+your+fieldtrip/Fieldwork+ideas/Qualitative+fieldwork.htm)

of tracer sediments (e.g. derived from a localised, point-specific source) may fail to demonstrate the direction and trend of longshore movement.

Feasibility of the research, the hypothesis or question

At the outset it is essential to establish that an investigation is feasible. Some hypotheses may be impossible to test due to practical problems of measurement (e.g. rates of erosion or mass movement on coastal cliffs), inaccessibility (e.g. research locations on private property) or lack of a secondary database. For example, an attempt to investigate migration in early nineteenth-century England would almost certainly prove impossible because the earliest censuses (1801 and 1811) did not include a place of birth question.

Problems can also arise as a result of boundary changes to the small spatial units used to aggregate census population data. In 2001 the old enumeration districts for small areas were replaced by new super output areas. Thus the lack of comparable spatial data would make it extremely difficult to tackle an investigation of population change in a suburb or market town for, say, the period 1971 to 2011.

Defining the hypothesis or question

A **hypothesis** is a statement whose accuracy can be tested objectively using **scientific methodology**. In general, hypotheses fall into two main types: those that focus on spatial or areal differences and those that focus on relationships between variables. Most geographical hypotheses usually have either a clearly defined spatial dimension or an equally clear emphasis on people-environment relationships. A well-defined hypothesis should show the direction of a difference or a relationship. For instance, the hypothesis that shingle beaches are steeper than sand beaches is more clearly defined than the alternative that the gradients of shingle beaches differ from the gradients of sand beaches.

The geographical connection must be obvious in hypotheses used in the independent investigation. In a study of shopping patterns in a market centre we might consider the frequency of shopping and the type of goods and services bought. As a result, the following hypothesis might be formulated: most shoppers purchase goods and services at least once a week. However, this hypothesis lacks any obvious connection with either location or people-environment relationships – it is not therefore intrinsically geographical. Simply by adding a reference to the origin of shoppers, we can modify the hypothesis to give it an unequivocal geographical character. Thus the revised

hypothesis is: the frequency with which shoppers purchase goods and services is inversely related to journey times between shoppers' place of residence and the market centre.

Questions/hypotheses based on wider theories, ideas and concepts

Theory is a set of statements or principles which have been tested and proved valid. In an investigative study it may be possible to formulate questions and hypotheses based on a theory. For example, central place theory provides a conceptual understanding of settlement hierarchies and how individual settlements function as service centres; and how distance and travel costs influence consumers' use of these services. From this theory we might derive a hypothesis such as: the number and range of services provided by a central place depends on the number of people it serves.

Sometimes a topic for investigation, while lacking formal theory, is supported by a detailed understanding of processes. This understanding allows us to formulate research questions and hypotheses. For example, prior knowledge tells us that scree slope development is influenced by gravity, particle size and friction. This enables us to formulate a hypothesis such as: rock particles on scree slopes are sorted by size.

Stage 2: Developing a strategy for conducting an independent investigation

Identifying the data needed to examine/test the question or hypothesis

The need to examine a question or test a hypothesis relies on data that must be:

- available at an appropriate scale
- available in sufficient quantity
- easy to collect.

The type of data will determine how their collection is undertaken: whether by desk research, or fieldwork involving measurement, observation and interview.

Let us assume that you decide to investigate the geography of poverty within a town or city in the UK. At an early stage the feasibility of the investigation is determined by identifying an appropriate database. In this example, multiple deprivation data for 2010 and 2015 are available for output areas at the Office for National Statistics (ONS) website. If the investigation aims to explain as well as describe spatial patterns of deprivation, it would be useful to provide information on education, ethnicity or unemployment at the same scale. These and other relevant data are available from the 2011 census of population on the ONS website.

An investigation into the sorting of shingle on beaches around a headland where variations in wave energy exist would need to ascertain: the location and number of beaches with contrasting aspects; whether the lithology of beach material was similar for all beaches; and the accessibility of beaches at low tide.

Strategies and methods are needed for successful data collection

Careful thought needs to be given to the timing and location of data collection. Questionnaire surveys should be timed to maximise the number of potential respondents. Thus shopping centre investigations should consider the time of day and whether data collection should take place on a weekday or at the weekend. A similar questionnaire survey in a commuter village might be timed for a weekend when most people are at home. Beach studies need to take account of tide times, and are best conducted at low tide or on an ebb tide. Thought should also be given to the location of street interviews. Interviews of shoppers conducted exclusively at bus stops or in car parks are unlikely to produce representative data.

Access and the problem of private land ownership may create obstacles to successful investigation. A heath or moorland which is otherwise suitable for vegetation studies may be inaccessible due to a lack of public footpaths or unco-operative private landowners.

Because most statistical populations in geography are huge, investigations are usually based on sample data. Decisions have to be made on the quantity of data to collect (i.e. sample size) and objective sampling methods that ensure representativeness. In the context of questionnaire surveys the aims of the investigation will affect decisions on whether the appropriate strategy is street interviews, door-to-door interviews or postal questionnaires. The choice of scale also becomes a strategic decision. An enquiry into the spatial pattern of consumer spending in a central shopping area (CBD) requires a town or city large enough to have at least five or six clearly defined shopping streets/areas. Thus, an urban centre of 300,000 inhabitants is likely to be more suitable than one of 30,000.

Limitations of time and resources

Constraints of time and resources impose limits on geographical investigations. Thus most investigations must strike a balance between the amount of data collected and the time and resources expended on data collection.

Some considerations connected to time and resources are essentially practical. For instance, how much time will be devoted to data collection and how much potential fieldwork time will be spent travelling (and walking) to and from the fieldwork site? Will data be collected by a group or by an individual? If in a group, how many students will be needed for the data collection process? This last consideration is important in questionnaire surveys, where rejection rates are often high and where each interview may take several minutes.

Minimising the potential risks of data collection through fieldwork

Geography fieldwork, whether in urban or rural areas, always involves an element of risk. While we cannot eliminate risk entirely, we can take sensible precautions to minimise it. Thus, before embarking on fieldwork you should do three things. First, identify the potential hazards. Second, assess the level of risk presented by each hazard. And third, devise a strategy or plan for dealing with each hazard. Usually risk assessment is formalised by completing a risk assessment sheet which is then submitted to your teacher/safety officer for approval.

Although the nature of risk varies according to the environment, there are a number of general safety guidelines which should be observed.

- Always try to work in groups. Ideally, groups should comprise at least three people. In the event of an accident one member of the group can assist the injured person while the other can get help (remember that in remote environments mobile phone networks may not be available).
- Carry a mobile phone. Leave your phone switched on and make sure that your teacher or a responsible adult has your phone number.
- Wear (or carry with you) suitable outdoor clothing including waterproofs and several layers of clothing. For coastal studies wellies and hard hats are essential, and for work on slopes and rough terrain, it is advisable to wear boots that provide grip and protection for your ankles.
- If you are working in a remote area you should carry a torch, a survival bag, a whistle and emergency rations.
- Carry a small first-aid kit, a map of the area (at a suitable scale) and a compass at all times.
- Leave precise details of your itinerary with your teacher or a responsible adult, including your time of departure and return.

- Wear safety helmets when working in environments such as cliffed coastlines, screes and deep upland valleys.

Stage 3: Collecting and recording data

Primary and secondary data

The independent investigation must be based on **primary data**. Primary data are new data, which have not previously been collected or processed. All data collected first-hand through fieldwork are by definition primary data. Documents can also be a source of primary data. Trade directories provide information on retailing and other businesses (e.g. types, number of outlets). Census data for individual households are available from 1841 to 1911. Parish registers are primary historical documents which provide a wealth of detail on baptisms, marriages and deaths before the introduction of civil registration in 1837.

Secondary data are information that are available in published documents such as textbooks, articles, maps, charts and diagrams. Today, much of this data is available online. Unlike primary data, secondary data have been processed, ordered and analysed before publication. In the context of an investigation, you must acknowledge and reference the sources of secondary data you have used.

The collection and recording of data

Primary data collection relies on observation, measurement and interview. Observation and the systematic recording is the simplest data collection technique. Field sketches, complemented with labels and annotations, rely on direct observation. Data on traffic flows, land uses, shop types, environmental assessment and plant species counts are also sourced through observation.

Fieldwork in physical geography often involves data collection through measurement. Slope surveys, using basic instruments such as clinometers, ranging poles and tapes, generate numerical data on valley and beach profiles; anemometers, thermometers and flow meters are used to measure wind speed, temperature and water flow velocity respectively; soil pH is measured using a BDH soil testing kit; and callipers measure the long, short and median axes of rock particles. These and other measurements are recorded as quantitative data to answer research questions.

Data which relate to people's attitudes, behaviour and personal characteristics can only be accessed through questionnaires or interviews. There are two approaches to questionnaire-type surveys. First, those completed by a respondent at home or in the office where the researcher delivers the questionnaires either by hand or by post. And second, those where the researcher is present to ask the questions and record the responses. These interviews are most often conducted in public spaces (e.g. a shopping street), though occasionally doorstep interviews are undertaken.

Ensuring the accuracy and reliability of primary data collection

Accurate and reliable data are essential to a successful investigation: it is impossible to have confidence in the outcome of an investigation if doubts exist about the quality of data. In most investigations the challenge is to acquire objective **sample** data which accurately represent the statistical population as a whole. Three objective sampling techniques are available: **random** sampling; **systematic** sampling; and **stratified** sampling (see Chapter 15, pages 526–29).

Central to many geographical investigations are the locational characteristics of statistical populations. In these circumstances locational data are included by spatial sampling. Spatial samples are selected from either points, areas (quadrats) or transects (lines) using random, systematic and stratified methods (see Chapter 15, pages 528–32).

Stage 4: Data presentation

Organising and presenting data for analysis

At the initial stage of analysis data are processed and presented as tables, charts and maps. The research questions or hypotheses under scrutiny will determine the organisation and grouping of data. For example, an enquiry into household movements originating within a suburb in a large city might begin by mapping the location of new residences, and then presenting the data first as a frequency table showing distances moved, and then as a histogram. This data presentation stage is an essential prelude to more rigorous analysis based on statistical methods. It also provides guidance on whether there are any significant patterns or trends and if they merit further analysis.

Mapping methods

Statistical maps are the most effective way to present and store spatial data. There are five types of statistical map: **dot, choropleth**, **isopleth**, **proportional symbol** and **flow** maps (see Chapter 15, pages 509–13). Choice of mapping type is influenced by the following criteria:

- accuracy and detail
- visual impact and clarity
- effectiveness in depicting spatial patterns
- ease with which statistical information can be retrieved.

All statistical maps involve compromise, a trade-off between accuracy and detail, and visual impact and clarity.

Graphical methods

Data presented as charts, rather than data sets and tables, have two main advantages. First, they provide a convenient summary of geographical data; and second, through their visual impact, they assist assimilation of data and understanding of patterns and trends. Charting data is an important stage in geographical investigation, preparing data for hypothesis testing. The most widely used charts in geographical investigation are **histograms**, bar charts, pie charts, line charts and scatter charts (see Chapter 15, pages 514–18).

The choice of graphical techniques must first be appropriate to the data; and second, it should aim to achieve a balance between preserving as much detail as possible and allowing generalisation to reveal patterns and trends.

Stage 5: Analysing and interpreting data in geographical investigations

Describing data presented as tables, charts and maps

As a preliminary to statistical analysis, the information presented in charts and maps is first described. This involves (a) a description of the main patterns and trends, (b) exemplification of the patterns and trends citing data from the relevant tables, charts or maps, and (c) references to exceptions or anomalies which deviate from the main patterns and trends. Descriptions should not be overgeneralised, but at the same time should give a coherent and clear summary of patterns and trends.

Descriptive statistical techniques

Descriptive statistical techniques include measures of central tendency and dispersion. They are used to describe or summarise data sets numerically.

It is often useful to represent a data set with a single value. To do this we choose a middle value around which the other values cluster. This is known as a measure of **central tendency**. Three standard measures of central tendency are used: the arithmetic mean or average; the **median**; and the **principal mode** (see Chapter 15, pages 532–33).

Unfortunately, measures of central tendency tell us nothing about the dispersion of values in a data set. It is important to have a summary measure of dispersion because very different data sets can yield similar values for the mean, median and mode. The main measures of dispersion are the range, inter-quartile range and **standard deviation** (see Chapter 15, pages 533–35).

Inferential statistical techniques

Whatever the results of an investigation, there is always the possibility that outcomes could be due to chance. This is because most geographical enquiries rely on samples rather than the **statistical population**. Using inferential statistical tests we can establish the probability that the results are due to chance. If this probability is small we are able to accept our results with a high level of confidence. The OCR specification names three inferential statistics for testing hypotheses of difference: the Mann–Whitney U test, the chi-squared test and the t-test; and two tests for relationships: the Spearman rank correlation and best-fit trend lines (see Chapter 15, pages 536–44).

Interpreting statistical outcomes

We interpret the results of geographical investigation by assessing their **statistical significance**. Essentially this means establishing the probability that the results are not due to chance. Statistical significance is normally set at the 95 per cent (or 0.05) threshold. At this level, there is only a 5 per cent probability (i.e. 1 in 20) that the result has occurred purely by chance. With this level of confidence we can accept a hypothesis. Alternatively, a hypothesis which failed to reach the 95 per cent level would be rejected.

Explaining outcomes and anomalous results

If the results of an investigation are consistent with the hypothesis being tested, then the hypothesis, and the theory or logic which underpins it, are verified. Sometimes a general pattern or trend is clear, but some values deviate from the expected. Such anomalies could be explained by the multivariate nature of many geographical features (i.e. their explanation involves more than one causal factor) or problems of data collection such as sampling accuracy.

Stage 6: Summarising the findings and evaluating the methodology of an investigation

Providing a conclusion and linkage to the original questions/hypotheses posed

In the first stage of an investigation we define its aims and objectives; in the final stage we provide a conclusion which sets out the main findings and considers the extent to which the stated aims and objectives have been achieved. Often our conclusions are equivocal. This is not an admission of failure. Most often it arises because geographical problems are complex, and data collection cannot be controlled to ensure complete accuracy.

The extent to which a study supports general theories, ideas and concepts

Research questions and hypotheses derive from a general body of theory, ideas or concepts. It is useful to place an investigation within its wider context and comment on the extent to which its findings are consistent with previous studies. For instance, studies of step-pool sequences on boulder-bed streams in California have shown a regularity of spacing, and a strong relationship between the ratio of channel width to distance between steps, and channel gradient. An independent investigation of step-pools might, in its conclusion, compare its findings with other similar studies, noting and explaining the similarities and differences.

The limitations of methods used and data collected, and possible improvements

It is almost certain that an investigation at A-level will encounter some limitations of methodology. The most obvious concern the choice and execution of the sampling strategy, and, as a result, the amount and quality of data collected. In questionnaire surveys obtaining representative samples is extremely difficult. Rejection rates for street interviews are high and students may resort to interviewing anyone willing to respond. Any deviation from the objective sampling strategy will create data of dubious accuracy, which may undermine the validity of the study. Even where objective statistical sampling is adhered to, without stratification, interview data are often biased. It is well-known that some groups of people are more likely to respond to questionnaire surveys than others.

Insufficient sample data impose further limitations on many investigations. An inadequate database makes statistical analysis difficult, which in turn provides little confidence in the reliability of results. This problem can be tackled at the planning stage by identifying the statistical test used for analysis and the minimum size of sample needed to obtain statistically significant results. And where there is doubt it is always best to collect a larger rather than a smaller sample.

The recursive model of investigation

The recursive model to geographical enquiry has many similarities with the sequential approach (Figure 16.1, page 545). At the outset, aims, objectives and hypotheses must be formulated. Data sources will then be identified and sample data collected. Data presentation will use conventional methods such as charts, tables and maps, and enquiries will conclude with a summary of findings and an evaluation of outcomes.

The recursive approach is, however, better suited to investigations that involve human behaviours which are complex, subtle and difficult to measure and quantify than the more formal and objective sequential approach. Often the focus of enquiry is people's opinions, experiences, perceptions and feelings. They invariably involve in-depth, informal conversations and interviews with individuals or groups, which are more time consuming than questionnaire surveys. Using this approach, it is possible to explore in detail people's motives, views and actions. Observation and recording also plays an important role, through field notes, sketches, audio-visual material and narrative descriptions.

An investigation into differences in voter participation in local elections between two urban wards would lend itself to a recursive, non-quantitative approach. At the outset there are no hypotheses or presumptions. Data collection would comprise conversations and interviews with individuals or groups, recorded with field notes or by Dictaphone, possibly at community venues. In this way, the researcher tries to understand how each individual or group perceives the world and how this influences their behaviour. Although the results may provide a basis for hypotheses, the overall objective is not generalisation but to investigate and understand a specific issue in a particular place and the responses and behaviours of individuals.

Glossary

Ablation – the loss of ice and snow, especially from a glacier, through melting, evaporation and sublimation.

Absolute humidity – the mass of water vapour in a given volume of air.

Abstraction – the extraction of water from rivers and boreholes for public water supply, agriculture etc.

Abyssal plain – the deepest part of the ocean covering vast areas of ocean floor. Submarine mountain chains and trenches interrupt the relatively flat plain.

Accumulation – the addition or gain of snow and ice to a glacier over time.

Acidification – decreasing pH, e.g. of the oceans due to an increase in the uptake of CO_2 from the atmosphere.

Active layer – the near surface layer in a periglacial environment which seasonally freezes and thaws.

Adiabatic expansion – the expansion of a parcel of air due to a decrease in pressure. Expansion causes cooling.

Advanced countries (ACs) – countries that share a number of important economic characteristics, including well-developed financial markets, high degrees of financial intermediation and diversified economic structures with rapidly growing service sectors.

Advection – the horizontal movement of an air mass which often results in either heating or cooling.

Aeolian – erosional, transportational and depositional processes by the wind.

Aggradation – the disposition of sediment in a river channel which results in a rise in the bed elevation.

Agribusiness – a large scale farming practice run on business lines.

Agro-ecosystems – an ecosystem that is managed to produce food.

Alases – flat-floored, steep-side depressions in periglacial environments.

Albedo – proportion of sunlight reflected from a surface.

Algal bloom – a rapid increase in algae in a water system.

Alkaloid – a naturally occurring chemical compound. The active ingredient in many medicinal drugs, e.g. morphine.

Alluvial fan – a cone of sediment deposited by a river where it leaves a steep upland course and enters a lowland area.

Anthropocene – the current geological period where humankind is the main driver of environmental change.

Appropriate technology – technology which is 'appropriate' in its context of use. It is small scale, can be managed locally and often uses skills available in the local community.

Aquifer – a water-bearing band of porous or permeable rock.

Archipelago – a closely grouped cluster of islands.

Arête – a narrow, 'knife-edged' ridge between two corries.

Artesian aquifer – a confined aquifer containing groundwater that when tapped will rise to the surface under its own pressure.

Artesian pressure – the hydrostatic pressure exerted on groundwater in a confined aquifer occupying a synclinal structure.

Aspect – the direction a slope faces.

Asthenosphere – the layer in the Earth's mantle below the lithosphere. The high temperatures cause the rocks to soften and become viscous meaning that they can easily deform.

Asylum seeker – a person who seeks entry to another country by claiming to be a refugee.

Atmosphere – the thin envelope of gases (mainly nitrogen and oxygen) that surrounds the Earth.

Attrition – the erosion of sediment transported by rivers, glaciers, waves and wind.

Backwash – flow of water down a beach after a wave has broken.

Bajada – a series of alluvial fans that merge to form a continuous apron of sediment along a mountain front in dryland regions.

Balance of payments – difference between a country's inflows and outflows of money, including all transactions with the rest of the world for goods and services, flows of FDI and migrant remittances, over a period of time.

Barchan – a discrete, crescent-shaped sand dune, with horns pointing downwind.

Barrage – a dam built across an estuary or bay so that the rise and/or fall in the tides can be used to generate energy by the turning of turbines.

Base flow – water input to streams and rivers from natural reservoirs such as aquifers, peat bogs, soils etc.

Beach recharge – the addition of sediment to a beach by humans to increase its size and volume.

Benioff zone – boundary between a subducting ocean plate and the over-riding continental plate at a destructive boundary.

Bio-prospecting – the discovery and commercial exploitation of biological resources, e.g. genetic material in organisms.

Biodegrade – the breakdown of a substance through biochemical reactions or the actions of organisms such as bacteria into different compounds.

Biodiversity – the number of different plant, animal, fungi etc. species in a given area.

Biodiversity hotspot – a region that is rich in biodiversity and where species and habitats are threatened with destruction.

Biological weathering – the breakdown of rocks through the chemical and physical action of living organisms, e.g. burrowing, tree roots etc.

Bioluminescence – the ability of some organisms to generate light from their bodies.

Biopiracy – the exploitation of medicinal drugs from wild environments by pharmaceutical companies with little or no benefit to indigenous people.

Biosphere – the space at the Earth's surface and within the atmosphere occupied by living organisms.

Black smoker – hydrothermal vent on the ocean floor at constructive or destructive plate boundaries. The water carries high amounts of metal sulphides.

Block disintegration – physical weathering of massively jointed rocks which results in boulder-sized rock debris.

Blockfield – a large expanse of boulders strewn across a level surface, often in mountain environments.

Blue water navy – a navy capable of operating away from its home bases and across the deep oceans.

Butte – an isolated, steep-sided, flat-topped rock outcrop.

Caldera – large-scale volcanic crater formed as a result of an explosive eruption which emptied the magma chamber causing the volcano sides to subside.

Canopy – the uppermost layer of treetops and branches in a forest or woodland ecosystem.

Cap and trade – an internal scheme to control carbon emissions. A market-based solution to climate change where polluters either cut their emissions or incur extra costs by buying tradable carbon credits.

Capital intensive – high levels of investment in manufacturing in plant and machinery per employee.

Capital – accumulated wealth of any kind used in producing more wealth; this may include tangibles such as machinery, buildings, land, and intangibles such as money, investment finance, company stocks.

Capitalist – the socio-economic system in which production of goods and services takes place to generate profit. A key driving force in the system is people's desire for gain and self-interest.

Caprock – a resistant layer of rock often on a plateau-like surface, which protects older weaker rocks from denudation.

Carbon capture and storage – the removal of CO_2 from emissions by thermal power stations and its storage in disused oil and gas wells underground.

Carbon credits – allowances that permit given levels of CO_2 emissions by businesses. Excess emissions must by covered by trading carbon credits.

Carbon fertilisation – rising CO_2 levels in the atmosphere which increase photosynthesis and stimulate plant growth.

Carbon offset – market-based approach to limiting carbon emissions. Businesses receive annual carbon quotas (credits). These can be sold/bought on international carbon markets.

Carbon sink – a long-term store of carbon in ocean sediments, carbonate rocks, forests etc.

Carbon source – inputs of carbon to the atmosphere by respiration, combustion and decomposition.

Carbonaceous rock – rocks mainly comprising the fossilised remains of plants, e.g. coal, lignite.

Carbonate rock – rocks comprising carbonate minerals (e.g. $CaCO_3$) such as limestone and chalk.

Carrying capacity – the maximum number of a species that can be supported in a given area in a sustainable way.

Catchment – the area drained by a river and its tributaries i.e. a drainage basin.

Central tendency – the middle value in a data set, e.g. mean, median, mode.

Chelation – a type of chemical weathering caused by acids derived from rainwater and dead organic material.

Chemical weathering – the in situ breakdown of rocks by chemical processes such as oxidation, solution and hydrolysis.

Choropleth map – a numerical map with areas shaded in proportion to the values they represent.

Cirrus cloud – a high altitude, feather-like cloud made of ice crystals.

Civil society – NGOs and other organisations that are independent of governments, working voluntarily, either individually or collectively, in support of citizens and communities throughout the world.

Closed system – a system with inputs and outputs of energy, but without any movement of materials across system boundaries.

Coastal squeeze – the erosion of mudflats, sand flats and beaches, trapped between rising sea level and hard coastal defence structures (e.g. sea walls).

Coefficient of determination – the square of the correlation coefficient showing the % of change in y explained by x.

Cold seeps – low temperature mineral-rich springs flowing from the ocean floor.

Commodity – a good or material (not a service) interchangeable through trade or commerce such as a raw material.

Communicable disease – an infectious disease that spreads from host to host.

Comparative advantage – the principle that countries or regions benefit from specialising in an economic activity in which they are relatively more efficient or skilled.

Composite cone volcano – see strato-volcano.

Comprehensive redevelopment – the planning and rebuilding of a substantial part of an urban area involving the demolition of nearly all of the previous buildings and infrastructure.

Compressing flow – the movement of glacial ice down a gentle gradient, during which it thickens.

Concordant – a coastline with bands of different geologies lying parallel to the shore.

Condensation – the phase change of water vapour (gas) to water (liquid).

Contagious diffusion – the process by which a disease spreads through direct contact with a carrier.

Contagious disease – a disease spread by contact or indirect contact between people, e.g. Ebola.

Containerisation – the shipping of goods (by road, rail and sea) in standard-sized metal boxes. It allows efficient, mechanised handling of large volumes of goods and lowers transport costs.

Continental drift – the theory that the continents are mobile and have moved across the Earth's surface through geological time.

Continental rise – gently sloping ocean floor between the continental slope and the abyssal plain.

Continental shelf – the gently sloping offshore extension of a continent extending into the ocean as far as the continental slope.

Continental slope – where the continental shelf becomes steeper as it descends to the deep ocean.

Conurbation – large urban area in population and areal terms, made up of the merging together of previously separate towns and cities.

Convection – the motion of a gas or liquid which when warmed rises until eventually it cools and sinks in a continuous circulation.

Coral bleaching – the loss of algae (which give coral its colours) from coral due to an increase in water temperature.

Core region – an area where economic activity is concentrated and living standards are relatively high.

Corporate social responsibility – commitment and initiative of a corporation to assess and take responsibility for its social and environmental impact. This includes its ethical behaviour towards the quality of life of its work force, their families and local communities, and its contribution to economic development and the natural environment.

Corrasion – the scouring and erosion of rock surfaces by sediments transported by rivers, glaciers, waves and wind (also known as abrasion).

Crevasse splays – low-lying areas of deposited sediment between levées.

Cryoturbation – frost churning of layers of regolith in periglacial environments.

Cryptobiotic crust – communities of cyanobacteria, fungi, lichens and mosses which form a fragile surface layer in drylands.

Cumulative frequency curve – a frequency curve that shows the proportion of data above (or below) a specific threshold (e.g. zero).

Cumuliform cloud – a cloud formed by convection with a rounded-top, lumpy appearance and flat base.

Cusps – a pointed and regular arc pattern of sediment on a beach.

Cyber conflict – use of the internet, cellular technologies and space-based communications for malevolent and/or destructive purposes in order to change or modify political, military and economic interactions between entities such as states, corporations and individuals.

De-industrialisation – the absolute or relative decline in the importance of manufacturing in the economy of a country or region.

Debris flow – mass movements of coarse boulders, mud, timber etc. in channels following periods of torrential rainfall.

Decarbonation – the reduction or removal of carbon emissions from energy sources.

Deductive approach – a scientific method of problem solving based on testing a theory, model or idea.

Deflation – erosion of clay and silt-sized particles by wind action in drylands.

Degradation (of land) – a process leading to a significant reduction in the production capacity of land.

Denudation – the wearing away of the Earth's surface by weathering and erosion.

Dependent variable – a variable affected by change in a related (independent) variable, e.g. rainfall affected by altitude.

Deposition – the laying down of sediment transported by rivers, waves, glaciers and wind, as energy levels decline.

Deprivation – when a person's well-being falls below a generally regarded minimum. A range of factors are usually included to measure this such as employment, housing, health and education.

Desert pavement – an almost continuous surface of rocky particles in drylands, resulting from the removal of finer particles by wind action.

Desertification – the reduction in agricultural capacity due to overexploitation of resources and natural processes such as drought. Only in extreme cases does it result in desert-like conditions.

Development gap – difference in prosperity and well-being between rich and poor countries. This could be measured, for example, by GDP per capita and HDI.

Dew – deposits of moisture on the ground and vegetation due to condensation caused by radiative cooling, most often at night.

Dew point – the critical temperature at which condensation occurs.

Diagenesis – the process by which snow becomes ice due to compression.

Diaspora – the spread of an ethnic or national group from their homeland, e.g. Jews from Israel or Kurds from Kurdistan.

Diffusion – the process by which a disease spreads outwards from its geographical source.

Disaster – natural hazards (e.g. earthquakes, floods) that result in major loss of life, injury and economic damage.

Discordant – a coastline with bands of different geologies lying perpendicular to the shore.

Dispersion – the distribution of values in a data set around a central value.

Disposable income – the proportion of a person's income that is left after essentials such as housing, food, clothing, heating and taxes have been paid.

Distributaries – small branching stream channels that flow away from a main stream or river.

Dot map – a numerical map where locational values are represented by dots of constant size.

Downwelling – the sinking of dense, salty (or cold) water in the oceans.

Drift – the collective term for all glacial deposits, including till and outwash.

Drumlins – streamlined mounds of glacial drift.

Dyke – a vertical or near vertical minor intrusion of magma through surrounding older rocks.

Dynamic equilibrium – a system displaying unrepeated average states through time.

Earthquake focus – point in the crust where rocks fracture, releasing energy.

Economic migrant – a person who moves from another country, region or place, involving a permanent or semi-permanent change of residence, to improve their standard of living or job opportunities.

Economic restructuring – the change in proportions of people working in various economic sectors, e.g. the change in ACs from secondary to tertiary employment.

Economies of scale – internal economies of scale are savings in unit costs that arise from large-scale production, derived from within a plant; external economies of scale are savings made by a firm that arise from outside the firm itself, such as benefits of proximity to other firms or infrastructure.

Ecosystem – a community of living organisms and their relationships with each other and the environment.

Ecosystem service – the processes by which the environment produces resources used by humans such as oxygen, water, food and materials.

Edge city – substantial urban development on the fringe of an existing conurbation. They are often formally planned and are relatively self-sufficient.

Effusive eruption – a gentle free-flowing basic eruption of lava, e.g. basalt.

Emerging and developing countries – countries that do not share all the economic development characteristics required to be advanced, and also are not eligible for the Poverty Reduction and Growth Trust, identified by the IMF.

Emigration – out-migration of people from a country, which involves permanent change of residence.

Endemic disease – a disease that exits permanently in a geographical area or human group.

Energy security – the uninterrupted availability of energy sources at an affordable price.

Enhanced greenhouse effect – increasing levels of CO_2 and other greenhouse gases in the atmosphere amplifying the natural greenhouse effect.

Enhanced weathering – artificial crushing of rocks to increase surface areas and absorption of atmospheric CO_2 by silicates to form carbonate minerals.

Entitlements – the purchasing and bargaining power that gives people access to food and other basic needs.

Epicentre – point at the surface directly above an earthquake focus.

Epidemic – a disease outbreak that spreads quickly through the population of a geographical area.

Equilibrium – a long-term balance between inputs and outputs in a system.

Equinox – the dates when the Sun is overhead at the Equator (21 March and 21 September) and when the length of day and night are the same everywhere on the planet.

Erg – a sand desert.

Erosion – the wearing away and/or removal of rock and other material by a moving force.

Escarpment – a tilt block forming an extensive upland area, with a short, steep (scarp) slope and a long, gentle (dip) slope.

Esker – a long, sinuous ridge composed of stratified sand and gravel.

Eustatic – worldwide change in sea level.

Eustatic sea level rise – a worldwide rise in sea level caused by melting of glaciers and ice sheets and increases in ocean temperatures.

Eutrophication – the process whereby nutrient enrichment in water (streams, rivers, lakes) leads to a fall in oxygen levels and the subsequent death of species which are O_2 dependent.

Evaporation – the process by which liquid water is converted into a gaseous state.

Evapotranspiration – combined loss of water at the surface through evaporation and transpiration by plants.

Exfoliation – the peeling away of outer rock layers of boulders and rock outcrops by weathering in drylands.

Expansion diffusion – the process by which a disease spreads outwards into a new area while carriers in the source area remain infected.

Explosive eruption – a violent eruption owing to the build-up of pressure within a volcano, due to viscous magma (e.g. andesite) preventing the escape of gases (especially steam).

Extending flow – the movement of glacial ice down a steep gradient, during which it thins.

Farmers' markets – a market where local farmers sell their produce directly to the consumer.

Fault scarp – a cliff or escarpment formed directly by rocks being displaced either side of a fault.

Ferrous minerals – minerals with a significant iron content, e.g. iron ores.

Fetch – the distance of open water in one direction from a coastline, over which the wind can blow.

Flash flood – a sudden and violent flood event caused by intense rainfall and rapid run-off.

Flocculation – a process by which salt causes the aggregation of minute clay particles into larger masses that are too heavy to remain suspended in water.

Flocs – a mass formed in a fluid by the aggregation of suspended particles.

Flood basalt – a large area of basaltic lava erupted over a long (thousands of years) from multiple vents, e.g. the Deccan Plateau.

Flow map – a numerical map that represents the movement of people, goods or information by lines where widths are proportional to flows.

Flow resources – renewable natural resources that can be used and are replenished at the same time, e.g. solar, wind and tidal energy.

Flux – the rate of energy transfer per unit area.

Fog – cloud at ground level caused by radiative cooling and advection.

Food chain – a series of organisms through which food energy moves before it is completely expended.

Food miles – the distance food items travel from farmer to consumer.

Food security – when there is access to sufficient food for individuals to lead a healthy life.

Food supply chain – the process whereby food moves from producer to consumer. It involves production, processing, distribution, storage, consumption and disposal.

Food web – a series of food chains linked together across different trophic levels within an ecosystem.

Forced labour – when people are coerced to work through use of violence or intimidation, or by more subtle means of detention such as retention of identity papers.

Foreign Direct Investment (FDI) – inward investment by a foreign company (usually a large TNC) in a country.

Freeze-thaw – a mechanical weathering process caused by water, confined in rock joints, expanding as it freezes, and as a result breaking rocks into smaller particles.

Frequency distribution curve – a type of line graph showing the distribution of values in a data set by classes or categories.

Friedmann's core-periphery model – four-stage model of spatial economic development where development is initially concentrated in the economic core and eventually diffuses to the periphery.

Frost heave – the downslope displacement of soil particles that results from cycles of freeze-thaw.

Gelifluction – the slow, downslope mass flow of saturated regolith resting on a layer of permafrost.

Genetically modified crops – genetic engineering applied to food crops to increase production, quality, and resistance to disease or drought.

Gentrification – process by which former low-income inner city housing districts in ACs are invaded by higher-income groups and refurbished.

Geographic Information Systems (GIS) – integrated computer tools for gathering, storing, processing and analysing geographical data that can be plotted on maps.

Geomorphic – relating to the formation and shaping of landforms and landscapes by natural processes.

Geopolitics – term used in the early twentieth century in the 'Heartland' work of Sir Halford Mackinder, referring to ways in which geographical factors were central in shaping international politics.

Ghetto – concentration of people with similar socio-economic, cultural or ethnic background within a well-defined small part of an urban area.

Gini coefficient – a statistical measure of the degree of similarity between two sets of percentage data.

Glacial – a prolonged cold climatic phase lasting for tens of thousands of years and causing continental glaciation in middle and high latitudes.

Glacier mass balance – the difference between the amount of snow and ice accumulation and the amount of ablation occurring in a glacier over one year.

Glacio-fluvial – relating to meltwater from a glacier.

Global commons – the Earth's shared natural resources, e.g. the oceans and the atmosphere.

Global dimming – release of aerosols to the atmosphere by burning fossil fuels. The aerosols reflect insolation and lower global temperatures.

Global governance – intervention by the global community, attempting to regulate issues, such as human rights, sovereignty and territorial integrity.

Global governance: Option C – Human rights

Global governance: Option D – Power and borders

Global shift – the locational movement of manufacturing production in particular from ACs to EDCs and LIDCs from the 1970s onwards.

Global supply chains – flows of materials, products, information, services and finance in a network of suppliers, manufacturers, distributors and customers around the world.

Global value chains – formed when the different stages of production are located across different countries. Companies attempt to optimise their operations by locating various stages of production across different locations.

Globalisation – the growing integration and interdependence of people's lives in a complex process with economic, social (cultural), political and environmental components.

Globesity – the widespread increase in excess average weight due to poor diet and lack of exercise.

Glycosides – a compound formed from sugar and another compound. Many medicinal drugs are glycosides derived from plants.

Graben – the downfaulted section of a rift valley.

Granular disintegration – the breakdown of rocks by weathering into coarse, granular particles.

Green belt – zone of predominantly rural land use on the periphery of an urban area where strict controls on development apply.

Greenfield site – land not previously built on and on which new developments are proposed or constructed.

Greenhouse gas – gases in the atmosphere such as CO_2, CH_4 and water vapour which absorb long-wave terrestrial radiation.

Groundwater – water stored underground in permeable and porous rocks known as aquifers.

Groundwater flow – the horizontal movement of water within aquifers.

Guyot – flat-topped seamount with its summit well below the ocean surface. Many appear to have volcanic origin, having formed at mid-ocean ridges.

Gyre – a very large circulation of surface water between 20° and 30° N and S in the oceans.

Halocline – the depth in the ocean where there is a rapid change in salinity of sea water.

Hard engineering – the controlled disruption of natural processes by using man-made structures, e.g. sea wall, levées.

Heat balance – the difference between inputs of solar energy to the Earth–atmosphere system and energy outputs from terrestrial radiation and gases in the atmosphere. Currently inputs exceed outputs and the global climate responds by warming.

Heat island – elevated temperatures found in large urban areas caused by emissions of heat, low albedo of urban surfaces, air pollution etc.

Hierarchical diffusion – the process by which a disease spreads through a structured order of places, e.g. from major cities to towns and villages.

Histogram – a type of bar chart where the frequency of values is represented in classes or categories.

Host – an animal that sustains a reservoir for pathogens such as bacteria and viruses, e.g. bats and rabies.

Hot spot – an area of intense volcanic activity where a mantle plume reaches the Earth's surface causing eruptions. Located away from plate boundaries, e.g. Hawaii.

Household – one person living alone, or a group of people (not necessarily related) living at the same address.

Housing tenure – the system under which housing is occupied, e.g. owner-occupiers or tenants renting from a landlord.

Human rights – basic rights and freedoms inherent to all human beings, to which all people are entitled without discrimination.

Human rights norms – established customary behaviour based on moral principles and ways of living inculcated into the culture of a country or area over a long period of time.

Humanitarian intervention – action taken (often by a third-party country or multilateral agency like the UN) in a sovereign state to protect people at risk from war, famine, flood, genocide etc.

Hydration – the breakdown of rocks by cycles of wetting (expansion) and drying (contraction).

Hydrocarbons – the main chemical compounds making up fossil fuels.

Hydrothermal vent – very hot water springs, rich in dissolved minerals, flowing from the ocean floor at mid-ocean ridges or hot spots.

Hydrostatic pressure – the pressure exerted by a confined fluid, such as water under or in a glacier.

Hypothesis – a statement whose validity can be tested using scientific methodology.

Ice shelf – a floating sheet of ice permanently attached to a land mass.

Ice wedge cast – a downward tapering body of sediment differentiated from surrounding regolith by texture and/or colour. Casts are evidence of former ice wedges which developed during past periglacial conditions.

Ice-contact drift – sediment deposited under or against ice.

Icon – something that has meaning as being representative of a place, culture or religion. In the urban landscape a building can become an icon, e.g. Buckingham Palace.

Immigration – in-migration of people into a country, which involves a permanent change of residence.

Import substitution – promotion and development of industries within a country aimed at reducing manufacturing imports.

Independence – a situation in which the people of a country exercise self-government and sovereignty over their state territory having gained political freedom from outside control.

Independent variable – a variable causing change in another (dependent) variable, e.g. rainfall influencing river flow.

Indigenous – something that originates naturally from a particular location.

Inductive approach – a scientific method that aims to generate new theories from data.

Infant mortality rate – annual number of deaths of infants under one year old per 1000 live births.

Infectious disease – a disease spread by parasites, bacteria, viruses, fungi etc.

Infiltration – the vertical movement of rainwater through the soil.

Infiltration capacity – the maximum rate at which water, under the pull of gravity, soaks into the soil.

Informal sector – those parts of the economy outside official recognition and record. People do not need formal qualifications to be employed in it, neither is there regulation of it.

Input – an addition of energy and/or materials to a system.

Inselberg – an isolated, steep-sided mountain, often surrounded by extensive plains.

Insolation weathering – the in situ breakdown of rocks in tropical and sub-tropical deserts caused by extreme diurnal changes in surface temperature.

Inter-glacial – a period of climatic warming (lasting c.10,000 years) between glacials.

Inter-regional trade – the flow of international trade among major world regions such as Europe, North America and Asia.

Inter-tropical convergence zone – area of permanent low pressure around the Equator which forms the rising limb of the Hadley cell. It is associated with instability, convection and thunderstorms.

Interception loss – rainwater stored temporarily on the leaves, stems and branches of vegetation which is evaporated and does not reach the ground surface.

Interdependence – interrelationships between ACs, EDCs and LIDCs through trade, FDI, foreign aid and migration.

International border – geographical boundary of a sovereign state, defined and recognised by international law, and identified on the political map of the world.

International community – all countries whose identity and sovereignty are recognised under the auspices of the UN, plus other international organisations that choose to participate in global discussions and decision-making and which act collectively to resolve humanitarian issues.

International law – body of law that governs international relations between states or nations. This provides the framework for the obligations of states to be maintained.

International migrant stock – the number of people born in a country other than that in which they live. This also includes refugees.

International treaties – international agreements concluded between states, in written form and governed by international law.

Interpolation – the insertion of isolines (e.g. contours) to a point pattern which assumes that the gradient of change between fixed points is constant.

Interval scale – a numerical scale in which zero is arbitrary (e.g. temperature).

Intervening obstacles – physical, economic, social and political factors which may disrupt or terminate a migration at any point between origin and destination.

Intervention – actions of a state, group of states or international organisations in a foreign territory to end gross violations of human rights. This includes military force, economic sanctions and the assistance of NGOs.

Intra-regional trade – the flow of international trade within one or other of the major world regions such as Europe or Asia.

IPCC – Intergovernmental Panel on Climate Change is a body made up of hundreds of scientists from around the world who research and regularly report on climate change.

Island arc – chain of volcanic islands formed along a subduction zone.

Isopleth map – a numerical map where values are represented by continuous lines on a surface, e.g. contours.

Isostatic – relating to vertical movements of the Earth's crust.

Isostatic changes – changes in the absolute level of the land. They are localised and result from either tectonic activity or the addition or removal of weight from the land.

557

Jökulhaulps – extreme glacial meltwater outbursts caused by geothermal or volcanic activity beneath glaciers.

'Just in time' (JIT) – a system companies use to increase efficiency and decrease storage costs and waste. Products are delivered immediately (just) as they are required.

Key settlement – rural settlement where services (e.g. schools, doctors' surgeries, shops) are concentrated to meet thresholds that will ensure their economic viability.

Kinetic energy – the capacity to do work as a result of motion.

Knowledge economy – wealth creating activities that gather, store and analyse knowledge, e.g. high-tech manufacturing, finance, telecommunications, business services, design, education and health.

Krill – small, shrimp-like crustacea that feed on plankton and are then themselves a major source of food for other organisms

Lag deposit – extensive, coarse, rocky particles in deserts, too large to be transported by wind action.

Lag time – the difference in time between maximum rainfall and peak river discharge.

Land degradation – the deterioration of land suitability for agriculture by soil erosion, desertification and salinisation.

Land grabbing – a process whereby rich countries acquire land in poorer countries.

Landscape – the visible features (landforms) of an area of the Earth's surface.

Lateral moraine – a ridge of till running along the edge of a glacial valley.

Leaching – soluble materials draining away in soil.

Levées – ridges of coarse deposits found alongside stream channels and elevated above the floodplain.

Life cycle – the progress of a person through various stages based on age and family unit, from infancy to old age.

Linear dune – narrow, ridge-like dunes, often tens of kilometres in length and aligned in the direction of the prevailing wind.

Liquefaction – the process by which sediments and soils lose their mechanical strength from a sudden loss of cohesion. The material is temporarily transformed into a fluid as the result of being violently shaken during an earthquake.

Lithification – the transformation of sediments into rock.

Lithology – the chemical and physical characteristics of rock types.

Lithosphere – layer in the Earth's mantle above the asthenosphere together with the crust which is divided into a series of tectonic plates. The lithosphere is rigid and is moved by the flows of semi-molten rock in the asthenosphere.

Loam – soil with a balanced textural composition of sand, silt and clay.

Loess – a wind-blown deposit of fine particles and dust that weathers to form fertile but easily eroded soils.

Longshore drift – the movement of sediment by waves and currents along a coastline.

Low-income developing countries – countries that are eligible for financial support from the IMF through the Poverty Reduction and Growth Trust.

Malnutrition – shortages of proteins and essential vitamins caused by an unbalanced diet.

Marine snow – the remains of organisms living in the ocean's upper layer which falls to deeper water. It represents an important transfer of energy from the photic zone to the ocean depths.

Mass movement – the downslope transportation of material under gravity.

Maternal mortality rate – annual number of deaths of women while pregnant, or within 42 days of termination of pregnancy, from any cause related to or aggravated by the pregnancy or its management. Measured per 100,000 live births.

Mean – the average value in a data set.

Median – the middle value in a data set where values are ranked in order of magnitude.

Merchandise – commodities or products (not services) available for sale.

Mesa – an extensive, steep-sided, flat-topped upland.

Methane hydrate – a structure of ice that contains methane between ice crystals and is found in ocean floor sediments.

Microclimate – local climates whose main characteristics are determined by topography and land use.

Mid-ocean ridge – the boundary between two diverging oceanic plates. It consists of two parallel chains of submarine mountains separated by a graben, and offset in places by transform faults.

Migrant remittances – money transferred from one country to another, sent home by migrants to their family, friends and community.

Militarisation – a significant increase in military activity in a place.

Millennium Development Goals – targets set by the UN in 2000 to improve people's lives in areas such as child mortality, gender equality, poverty and hunger.

Modified Mercalli scale – a subjective measure of earthquake intensity using factors such as what was felt by people and the type and scale of damage to buildings and infrastructure.

Moho – short for Mohorovičić discontinuity, the boundary between the crust and mantle. It lies at c 35 km beneath the continents and 10–15 km beneath the ocean floor.

Moment Magnitude Scale (Mw) – a measure of earthquake strength using the amount of physical movement caused by a quake.

Monoculture – the cultivation of a single crop.

Multinational corporation – a firm with the power to co-ordinate and control operations in several countries.

Multiplier effect – the process by which a new or expanding economic activity in an area creates additional employment as its employees have money to spend on goods and services. As the wealth of an area increases it stimulates more economic activity.

Myrdal's model of circular and cumulative causation – initial advantages to economic development in a core region, such as natural resources or accessibility, trigger a series of virtuous growth cycles through positive feedback, leading to further economic growth and the acquired advantages of agglomeration.

Nation building – processes by which a state government promotes nationality, for example through its education system or the media.

Nation-state – a nation which has its own independent state; the boundaries of the state coincide with the area inhabited by the nation.

Nation – large group of people with strong bonds of identity, united by shared descent, history, traditions, culture and language.

Natural capital – natural resources with a value to humans.

Natural income – annual yield from natural resources such as timber, fish, plants and mineral ores.

Nearshore zone – the area of coastal environment between mean high tide and mean low tide.

Negative feedback – an automatic response to change in a system that restores equilibrium.

Net migration – difference between the number of people moving permanently into an area and out of that area.

Net primary productivity (NPP) – the rate at which plants accumulate energy in the form of organic matter taking into account the energy used in respiration.

New International Division of Labour – reorganisation of production at the global scale, as a result of deindustrialisation in advanced countries and the global spread of MNCs. This has produced an overall pattern of higher-paid managerial jobs in ACs and lower-paid labouring jobs in LIDCs.

Nivation hollow – a depression formed by freeze-thaw and meltwater transport of weathered rock particles beneath a permanent snow patch.

Non-communicable disease – diseases such as CVD and cancer which cannot be spread between people.

Non-ferrous minerals – minerals that do not contain iron, e.g. manganese and copper.

Non-infectious disease – a non-communicable disease due to age, genetic defects, e.g. cancer.

Non-point source pollution – release of pollutants from numerous, dispersed sources, e.g. gases from vehicles or ships or chemical run-off from agriculture.

Non-renewable resources – resources that are finite on human timescales. Once used they cannot be replaced, e.g. fossil fuels and minerals.

Normal frequency curve – a symmetrical, bell-shaped frequency curve where most values cluster around the mean.

Norms – moral principles, customs and ways of living that are universally accepted as standard behaviour.

Nutrition spectrum – a scale of diet and nutrition that marks critical thresholds in human health, e.g. starvation and obesity.

Ocean conveyor belt – see thermohaline circulation.

Ocean currents – the large scale horizontal flow of ocean water (at the surface and at depth) driven by planetary winds and contrasts in water temperature and salinity.

Ocean trench – narrow deep depression on the ocean floor adjacent to a subduction zone.

Ognip – the remains of a collapsed pingo, forming a depression surrounded by circular earth ramparts.

Open system – a type of system whose boundaries are open to both inputs and outputs of energy and matter.

Ordinal scale – a numerical scale where values are represented by their rank order in a data set.

Orthogonals – imaginary lines, perpendicular to wave fronts, representing the transfer of energy as a wave moves towards the coast.

Output – the transfer of energy and/or materials out of a system.

Outsourcing – cost saving strategy where a company that has comparative advantage provides goods or services for another company even though they could be produced in-house.

Outwash – material deposited by glacial meltwater.

Outwash plain – a flat expanse of glacio-fluvial sediment located beyond an existing or former glacier or ice sheet front.

Overcultivation – cultivation which, given environmental resources, is not sustainable in the long term and is evidenced by declining yields, soil exhaustion and soil erosion.

Overgrazing – excessive grazing of land by livestock which destroys or degrades pasture and is not sustainable.

Overland flow – rainfall that runs off the ground surface either because the soil is saturated or the intensity of rainfall exceeds the soil's infiltration capacity.

Overnutrition – prolonged excessive food intake, which increases body weight.

Oxidation – a chemical process that weathers certain types of rock and involves the absorption of oxygen from either the atmosphere or water by rock minerals.

Palaeomagnetism – traces of changes in the Earth's magnetic field in the alignment of magnetic minerals in sedimentary and igneous rocks.

Pandemic – an epidemic which spreads worldwide, e.g. Spanish flu

Peak discharge – the maximum flow (in cumecs) of a river in response to a rainfall event.

Pediment – an extensive, gently sloping rock platform extending from the foot of a plateau-like feature in dryland environments.

Pelagic – the environment of the open ocean but not including the ocean floor.

Percolation – the movement of surface and soil water into underlying permeable rocks.

Permaculture – sustainable, high yielding and intensive agriculture based on natural processes.

Permafrost – permanently frozen soil and regolith.

Permeability – the ability of a rock to absorb water through rock joints, pores etc.

Permeable rock – a type of rock that is penetrated by water, either through mineral pores (air spaces) or along joints, faults and fissures.

Photic zone – shallow layer of water in the ocean that light penetrates, on average to about 50 m.

Photoperiod – length of day, i.e. from sunrise to sunset.

Photosynthesis – process by which green plants convert water and CO_2 into starch and sugar in the presence of sunlight.

Phreatic – an eruption of steam from a volcano caused when groundwater meets magma.

Phytoplankton – tiny photosynthesising marine organisms in the surface waters of the oceans.

Pillow lava – rounded mounds of lava erupted along mid-oceanic ridges, which cool rapidly on contact with sea water.

Pinch points – a point where congestion is likely to occur.

Pingo – a conical ice-cored hill in periglacial environments.

Plant succession – the sequence of changes within a plant community as it develops through time.

Playa – a basin-like area of inland drainage in a dryland region. After heavy rain it may contain a shallow, temporary lake.

Player – individual or organisation with an interest and or influence in actions, decisions or operations. Also known as a stakeholder.

Pluton – mass of igneous rock injected into overlying rock deep in the crust.

Pluvial – a climatic period associated with higher rainfall and more humid conditions in the recent geological past.

Point source pollution – release of pollutants from a single clearly identifiable source, e.g. a factory chimney or sewage pipe.

Pollution – the process by which human activity contaminates the environment, with adverse effects on the quality of air, water etc. and the health of people and other organisms.

Polyculture – the cultivation of several different crops, often simultaneously in the same field.

Porosity – the ability of a rock, like chalk, to absorb water through tiny air spaces (or pores) between mineral particles.

Porous rock – rocks which contain pores or air spaces between mineral particles, where water is stored.

Positive feedback – an automatic response to change in a system which generates further change.

Post-industrial – a society and economy no longer dominated by the secondary sector but one where the great majority of people are involved in tertiary activities.

Potential energy – the capacity to do work that a body possesses by virtue of its position and that is potentially transformable into another form of energy.

Potential evapotranspiration – the amount of moisture that could in theory be evaporated and transpired from a surface, assuming that moisture is freely available all year round.

Potentiometric surface – an imaginary surface that defines the theoretical level to which water would rise in a confined aquifer.

Precipitation – moisture (rain, snow, hail) falling from clouds towards the ground.

Pressure melting point – the temperature at which ice melts when under pressure.

Price shocks – an unexpected and unpredictable change in prices. It can be positive or negative.

Primary data – information that is unprocessed and often collected by research.

Primary producer – green plants (both terrestrial and marine) that convert sunlight, CO_2 and mineral nutrients in photosynthesis, into organic matter (chemical energy).

Primary sector – economic activities that produce food, fuel and raw materials, e.g. agriculture, forestry, fishing, mining, quarrying and water supplies.

Principal mode – the class of values that occurs most frequently in a data set.

Producer organisms – organisms that capture energy from the Sun in the process of photosynthesis and store this energy as organic matter.

Proglacial lakes – a body of water impounded in front of a glacier.

Proportional symbol map – a numerical map where the size or magnitude of values is shown by symbols (circles, squares) proportional in area to the values they represent.

Protozoa – single-celled micro-organisms that get their energy from the surrounding environment.

Pull factors – positive attributes of a place or destination which attract migrants.

Push factors – negative attributes of a migrant's place of origin which force a migrant to leave.

Pyramidal peak – an angular, sharply pointed mountain peak which results from corrie erosion.

Quadrat sampling – a type of spatial sampling employing squares as sampling units. In fieldwork quadrats are often metre-square frames.

Quartile – a quarter part of a data set ranked in order of size or magnitude.

Quaternary sector – economic activities that provide services to other economic activities, e.g. finance, research and development, advertising and consultancy.

Quota sampling – a type of sampling method that pre-selects a numerical target of samples based on criteria such as age, residence, economic status etc. of respondents.

Rain shadow – an area of below average precipitation situated in the lee of an upland.

Random sampling – a type of sampling that uses random numbers and where every individual in a population has an equal chance of inclusion in the sample.

Raster map – a type of map used in GIS made up of points, lines and polygons.

Ratio scale – a numerical scale that has a meaningful zero value.

Rebranding – developments aimed at changing negative perceptions of a place, making it more attractive to investment.

Recessional moraines – a series of ridges running transversely across a glacial trough.

Recharge – net input of water into an aquifer causing a rise in the water table.

Refugee – a person who has moved outside the country of his/her nationality or usual domicile because of genuine fear of persecution or death.

Reg – a stony, rocky desert.

Regeneration – the investment of capital and ideas into an area to revitalise and renew its socio-economic and environmental status.

Regolith – a loose layer of rocky material overlying bedrock.

Reimaging – developments associated with rebranding and usually involving cultural, artistic or sporting elements.

Relative humidity – the mass of water vapour in a given volume of air as a ratio of the mass needed to saturate it.

Relocation diffusion – the process by which a disease spreads into a new area and evacuates its area of origin.

Renewable resource – a resource capable of regeneration on human timescales, such as most forests and fish stocks.

Residence time – the length of time that a molecule of water or carbon dioxide etc. remains in natural storage (e.g. in the atmosphere or oceans).

Resilience – the degree to which an area can recover from the impacts of a hazard. The level of vulnerability affects a community's resilience.

Respiration – the process in living organisms where the intake of oxygen oxidises organic substances to produce energy and release carbon dioxide.

Responsibility to Protect – individual states have the primary responsibility to protect their populations from genocide, war crimes, crimes against humanity and ethnic cleansing.

Richter scale – system measuring the magnitude of an earthquake by using the energy released when overstrained rocks suddenly fracture. Uses a logarithmic scale, with just over 9.0 being the largest quake measured to date.

Rift valley – a valley formed by downfaulting between parallel faults. Examples are found on land (East African Rift Valley) and along mid-oceanic ridges.

Rip currents – strong and relatively narrow currents of water that flow seaward against breaking waves.

River discharge – the volume of water flowing in a river channel, measured in cubic metres per second (cumecs).

Rock flour – fine material derived from abrasion by a glacier.

Rock glacier – coarse rock particles forming linear accumulations and moving slowly downslope due to the formation and melting of interstitial ice.

Rock groynes – low walls made of resistant pieces of rock and built out from the coast into the sea.

Run-off – the movement of water across the land surface.

Salinisation – the accumulation of salts in soil.

Salinity – level of concentration of dissolved salts in water expressed as parts per thousand.

Salt weathering – the breakdown of permeable rocks by the internal growth of salt crystals precipitated out of solution.

Saltpans – flat areas of ground covered with salt deposited by the evaporation of saline water.

Sand sea – a vast area of sand dunes, completely covering the solid rock surfaces at depth.

Saturated overland flow – the movement of water across the ground – often as a thin film – when the soil is saturated.

Scientific methodology – a systematic approach to investigation which includes hypothesis formulation, controlled data collection and statistical analysis.

Sea-floor spreading – lateral movement of new oceanic crust away from a mid-ocean ridge (constructive plate boundary). A key process in the theory of continental drift.

Seamount – a volcanic peak rising steeply from the ocean floor. Some are isolated while others occur in chains extending away from a mid-ocean ridge or hot spot.

Secession – transfer of part of a state's area and population to another state.

Secondary data – data which have been processed/and or published, e.g. books, articles, blogs etc.

Secondary sector – economic activities involving manufacturing industries, e.g. processing raw materials, making semi-finished and finished goods.

Sediment budget – the balance of the sediment volume entering and exiting a particular section of coast.

Sediment cell – a stretch of coastline and its associated nearshore area within which the movement of coarse sediment, sand and shingle is largely self-contained.

Self-determination – right of a group with a distinctive territorial identity to freely determine its political status and freely pursue its economic, social and cultural development.

Separatism – claims for, or practice of, separation of a group of people from a larger state on the basis of their ethnicity or unified national culture, traditions, religion and language.

Services – tradeable activities providing expertise which relies on work-force skills, experience and knowledge rather than processing or production of physical materials. Producer services include legal and financial services for other firms. Consumer services include education and health care for local communities.

Settling velocity – the speed required for suspended particles of a given size, transported by rivers, the wind and tidal currents, to be deposited.

Shield volcano – a volcano with a broad base and gently sloping sides. It forms from the effusive eruption of fluid basalt lava.

Shifting cultivation – a traditional method of cultivation in tropical forests which involves rotation of land rather than rotation of crops.

Sill – a minor, approximately horizontal, intrusion of magma into surrounding older rocks.

Single market – an economic union of countries trading with each other without any internal borders or tariffs.

Sink – anything that absorbs more of a particular substance than it releases, e.g. the oceans act as a sink for CO_2.

Skewness – the extent to which a frequency curve deviates from a normal, symmetrical pattern.

Social inequality – the unequal distribution of factors such as income, education or health across a population.

Soft engineering – the deployment of natural, sustainable processes to protect the environment from erosion,

flooding etc., rather than hard engineering structures, e.g. beaches and salt marshes rather than sea walls.

Solar output – the amount of radiant energy emitted by the Sun and intercepted by the Earth. Most solar output is short-wave.

Solifluction – the slow flow of fine, water-saturated regolith from higher to lower ground.

Solution – the chemical weathering process by which rock minerals are dissolved.

Sovereignty – the absolute authority that independent states exercise in the government of the land and peoples in their territory.

Spatial inequality – the unequal distribution of factors such as income, education or health across geographic space at any scale.

Stability – the atmospheric condition where an air parcel displaced from its original level returns to the same level (i.e. because it is heavier/denser than the surrounding air).

Stadial – a sudden and brief period of glacial conditions lasting several hundred years.

Stakeholder – *see* player.

Standard deviation – the most mathematical measurement of dispersion in a data set (i.e. the root mean square deviation).

Standard error of the mean – the standard deviation of a sampling distribution.

Star dune – a pyramid-shaped sand dune which develops where there is no dominant wind direction.

State – area of land, of an independent country, with well-defined boundaries, within which there is a politically organised body of people under a single government.

State apparatus – set of state institutions and organisations through which state power is exercised; these include legal mechanisms, administrative organisations, police and armed forces, and health, education and welfare services.

Statistical population – the entire pool from which a statistical sample is drawn (e.g. all the pebbles on a shingle beach).

Statistical significance – the likelihood that the outcome from a sample survey is not due to chance. It is normally set at 95% probability.

Stemflow – the flow of water along the branches and stems of trees and other plants to the ground.

Stone garlands – elongated accumulations of stones on a slope.

Storm surge – elevated sea surface near the coast most often caused by extreme storms (e.g. hurricanes) and tsunamis.

Storm wave – a wave generated locally by high wind energy.

Strata – the layers or beds found in sedimentary rock.

Stratified sampling – a sample drawn from a heterogeneous statistical population divided into groups.

Stratiform clouds – clouds lying in a level sheet.

Strato-volcano – a steep-sided volcano made up of layers of lava and ash emitted during explosive eruptions. Also known as a composite cone.

Striations – scratches or grooves on rock surfaces formed by glacial abrasion.

Structure – the physical characteristics of rocks, including their jointing, bedding, faulting, angle of dip etc.

Sub-aerial process – a collective term for weathering and mass movement processes.

Subduction – the tectonic process found at convergent plate margins where an oceanic plate descends into the Earth's mantle and is destroyed.

Sublimation – the phase change of water from ice to vapour.

Super-volcano – a volcano that erupts more than 1000 km^3 of material in a single eruption event.

Superpower – states with a dominant position in the international system capable of exerting their power (economic, cultural, political) globally.

Surface wash – rainwater that runs-off the ground surface but is not confined within a channel.

Swash – the movement of water up a beach after a wave has broken.

Swell wave – a relatively smooth ocean wave that travels some distance from the area of its generation.

Symbiosis – the close association between two or more organisms of different species, often but not necessarily benefiting each member.

Syncline – a downfolded, basin-like geological structure.

System – a group of objects and the relationships between them.

Systematic sampling – a sample drawn at regular intervals from a population (e.g. every 10 m, every nth person).

Tailings – waste produced by mining when the rock containing the mineral has been washed.

Talik – unfrozen ground.

Talus – a steep, concave debris slope at the foot of a cliff or free-face, comprising angular rock particles (also known as a scree slope).

Tectonic plate – a large slab of the Earth's lithosphere and crust.

Temperature inversion – an increase in temperature with height, i.e. the reverse of the normal temperature profile.

Terminal moraine – a ridge of till extending across a glacial trough.

Terms of trade – value of a country's exports relative to that of its imports. This is measured as: average price of exports ÷ average price of imports × 100. If export prices rise relative to import prices, there is improvement in a country's terms of trade

Territorial integrity – principle that the defined territory of a state, over which it has exclusive and legitimate control, is inviolable. This is enshrined in the Charter of the UN and an important part of international law.

Territory – extent of land under the jurisdiction of a sovereign state.

Tertiary sector – economic activities providing services, e.g. education and health, legal, financial, insurance, government.

Theory – a set of statements or principles that has been tested and proved valid.

Thermal energy – the capacity to do work as a result of heat.

Thermal expansion of water – increase in volume of water due to its rise in temperature.

Thermocline – the depth in the ocean where there is a rapid change in water temperature.

Thermohaline circulation – the global pattern of ocean currents (surface and sub-surface) driven primarily by differences in temperature and salinity.

Threshold – the minimum number of people or spending required to support a good or service.

Threshold (coastal) – the relatively shallow part of a ria or fjord at its seaward end.

Throughfall – rainfall, initially intercepted by vegetation, which drips to the ground.

Throughflow – water flowing horizontally through the soil to stream and river channels.

Tidal range – the vertical difference in height between consecutive high and low waters over a tidal cycle.

Till – unsorted material deposited directly by glacial ice.

Time-space compression – a set of processes leading to a 'shrinking world' caused by reductions in the relative distance between places, e.g. travel time.

Tipping point – the critical point or threshold which if passed leads to irreversible change, e.g. the moment when summer sea ice in the Arctic disappears, which would alter the Atlantic thermohaline circulation.

Trade liberalisation – the process of making international trade free from barriers such as taxes on imports. The WTO has been instrumental in promoting free trade between member countries and has therefore played an important role in economic globalisation and the expansion of world trade.

Trade winds – prevailing winds that blow southeast and northeast towards the Equator from the sub-tropics.

Tragedy of the commons – a metaphor illustrating how individuals can overexploit a resource in common ownership (e.g. the atmosphere, fish in the deep ocean), leading to its depletion or degradation.

Transform faults – large-scale faults in the crust at right angles to a mid-ocean ridge, which range from a few tens of kilometres to several hundred. Earthquakes occur along their lengths as they slip.

Transnational corporation – very large company with factories and offices in more than one country, which markets products and services worldwide.

Transpiration – the evaporation of moisture from pores on the leaf surfaces of plants.

Transportation – The movement of material by the kinetic energy of a medium such as water, wind or ice.

Treaty – formally concluded and ratified agreement between states.

Tree line – The latitudinal and altitudinal limit of tree growth.

Trench – a narrow, deep depression on the ocean floor, adjacent to a subduction zone.

Trophic level – the level at which energy in the form of food is transferred from one organism to another as part of a food chain.

Troposphere – the lowest layer of the atmosphere, which contains virtually all of the planet's weather.

Tundra – treeless region in the sub-Arctic and in high mountains which has a short growing season and severe winter temperatures.

Turbid – cloudy or muddy conditions owing to sediments held in suspension.

Turbidity – the cloudiness of a fluid caused by large numbers of individual particles in suspension.

UN Global Compact – initiative that invites companies to align their strategies and operations according to universal principles on human rights, labour, environment and anti-corruption, and to take actions that advance societal goals.

Undernutrition – too little food intake to maintain body weight.

Upper mantle – the layer of the Earth's interior extending 75–200 km below the surface (also known as the asthenosphere). The upper mantle is able to flow under pressure.

Urban heat island – a town or city that is significantly warmer than its surrounding rural areas due to human activities.

Vector – a carrier, e.g. mosquitoes, that transmits an infectious disease, e.g. malaria.

Venture capital – investment in small or medium sized enterprises involving relatively high levels of risk but with the potential for significant gains. The enterprises are those not yet secure enough to raise their own finance.

Virtual water – the volume of freshwater needed to make a product, measured at the place where the product was manufactured.

Viscosity – an indication of how well a substance flows. Acid lavas have high viscosity as they are sticky and do not flow far from a vent.

Volcanic Explosivity Index (VEI) – combines magnitude and intensity of an eruption on a logarithmic scale, 0 = least explosive, 8 = most explosive.

Vulnerability – measure of the level of risk an area faces from the impacts of a hazard. Levels of vulnerability are influenced by physical, social, economic and political factors.

Water balance – the relationship between precipitation, streamflow, evapotranspiration, and soil moisture and groundwater storage in a drainage basin over a year.

Water cycle budget – the annual volume of movement of water by precipitation, evapotranspiration, run-off etc. between stores such as oceans, permeable rocks, ice sheets, vegetation, soil etc.

Water table – the upper surface of the zone of saturation in permeable rocks and the soil.

Wave period – the time interval between successive wave crests arriving at a given point.

Wave refraction – the reorientation of wave fronts as they enter shallow water so that they approach parallel (or roughly parallel) to the shoreline.

Weathering – the *in situ* breakdown of rocks exposed at, or near, the land surface by physical, chemical and biological processes.

Westphalian model – The Peace of Westphalia (1648) marked the formal recognition of states as sovereign and independent political entities; it established the principle of sovereign equality of states, forming the basis of international law that governs the global political system today.

Xerophyte – a plant with adaptations that enable it to survive in an environment with little water.

Yardang – a streamlined, parallel rock ridge in tropical deserts, aligned with the direction of the prevailing wind.

Zero tillage – a type of arable cultivation designed to minimise moisture loss from the soil. Its main feature is the absence of ploughing.

Zeugen – tabular-shaped rock outcrops in hot deserts, attributed to wind erosion.

Index

Note: page numbers for a particular topic include pages on which diagrams, photos and tables are to be found.